Raja

SURINDERPAL SINGH BADESHA

CARPENTRY

CARPENTRY

FOURTH EDITION

FLOYD VOGT

THOMSON

DELMAR LEARNING

Australia ■ Canada ■ Mexico ■ Singapore ■ Spain ■ United Kingdom ■ United States

Carpentry, Fourth Edition
by
Floyd Vogt

Vice President, Technology and Trades ABU:
Alar Elken

Director of Learning Solutions:
Sandy Clark

Acquisitions Editor:
Alison Weintraub

Product Manager:
Jennifer A. Thompson

Marketing Director:
David Garza

Channel Manager:
Erin Coffin

Marketing Coordinator:
Mark Pierro

Production Director:
Mary Ellen Black

Production Manager:
Andrew Crouth

Senior Project Editor:
Christopher Chien

Art/Design Specialist:
Mary Beth Vought

Illustrations:
Terrel V. Broiles

Technology Project Manager:
Kevin Smith

Technology Project Specialist:
Linda Verde

Editorial Assistant:
Maria Conto

Library of Congress Cataloging-in-Publication Data:
Vogt, Floyd.
  Carpentry/Floyd Vogt; Doug Holman, contributor.—4th ed.
    p. cm.
  Rev. ed. of: Carpentry / Gaspar Lewis, Floyd Vogt. c2001
    Includes bibliographical references and index.
  ISBN 1-4018-7069-4
  1. Carpentry. I. Holman, Doug. II. Lewis, Gaspar J. Carpentry. III. Title.
TH5604.L44 2006
694—dc22
                                        2005048589

ISBN: 1-4018-7069-4

NOTICE TO THE READER

Contents

SECTION 4

INTERIOR FINISH   644

Preface

Welcome to *Carpentry, 4th Edition;* a modern approach to residential construction. Designed for students enrolled in carpentry courses at secondary and 2-year, 4-year post-secondary schools, *Carpentry, 4th Edition* walks the student step-by-step through the various principles and practices associated with constructing a residential building.

APPROACH

A unique blend of traditional and up-to-date construction practices, the author focuses topics on need-to-know information in four comprehensive sections: **Section 1—Tools and Materials, Section 2—Rough Carpentry, Section 3—Exterior Finish, Section 4— Interior Finish.** Beginning with the layout of the building and finishing with trim carpentry, each section features step-by-step procedures of key carpentry jobs, critical safety information, tips of the trade, and insight into the construction industry.

Section I

This section describes wood as a building material in its many forms—from boards to engineered sheets and beams. Fasteners and tools critical to securing and shaping building products are also covered, while a unit on prints and codes helps students to understand the written language of construction.

Section II

Beginning with the basics, this section explains how to locate a building on a site using surveying equipment; explores concrete applications for foundations, slabs, and stairs; and steps the students through the framing of floors, walls, roofs, and stairs. In addition, a section on energy conservation considerations for insulation and ventilation helps students keep pace with industry standards.

Section III

Exterior carpentry jobs, including applying various roof and siding finishes, installing windows and doors, and constructing decks with rails and fences, is covered in this section.

Section IV

The final trim carpentry jobs involved in finishing a residential building are described in this section. The section begins with the application of drywall and various finishes on floors, walls, and ceilings. Next, fitting trim to windows, doors, and walls is explained. Also, constructing and finishing a staircase with balustrade is covered. Lastly, to finish the job, the text explains how to install manufactured cabinets and construct countertops, drawers, and doors.

Key Features

The *Introduction* provides essential information on working in the construction field, including *General Safety Guidelines*, critical *Soft Skills*, and notable *Organizations* to help keep students informed of industry expectations.

Success Stories open each of the four sections featuring stories of success from individuals across the nation who currently work in the field.

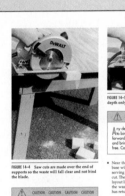

Safety Reminders and **Cautions** integrated throughout the units continuously reinforce the importance of proactively practicing safety in order to prevent injury or fatalities on the job.

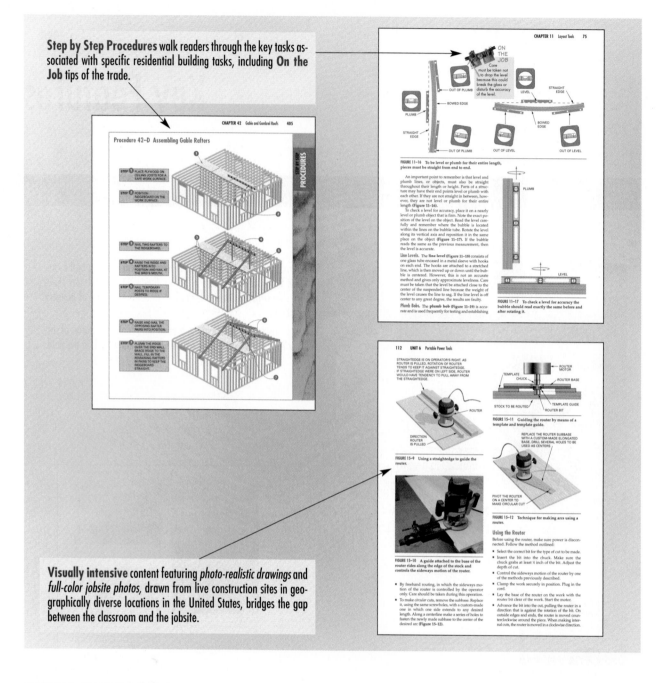

Step by Step Procedures walk readers through the key tasks associated with specific residential building tasks, including **On the Job** tips of the trade.

Visually intensive content featuring *photo-realistic drawings* and *full-color jobsite photos*, drawn from live construction sites in geographically diverse locations in the United States, bridges the gap between the classroom and the jobsite.

NEW TO THIS EDITION

- An *all new full-color design* for the fourth edition brings the jobsite to life—integrating photo-realistic drawings, drawn to scale, and on-the-job photos drawn from construction projects across the nation.

- *Safety* information has been greatly expanded, including a section on *General Safety Guidelines* in the *Introduction, Safety Reminders* to open units alerting students to potentially dangerous situations on the job, and *Cautions* to help prevent accidents when working with various tools of the trade.

- A section in the *Introduction* covers the critical **Soft Skills** required of successful and proficient carpenters. Coverage of important *organizations,* including OSHA and Skills-USA, that help students keep up to date on industry expectations is also included.

- **Success Stories** open each of the four sections, and highlight successful accounts of individuals working on carpentry jobs across the nation, providing insight into the industry.

- A revised section on **Construction Prints** features a brand new set of residential drawings to step students through the process of reading and interpreting prints accurately on the job.

- *New procedures* and *examples* in *Roof and Stair Framing* bring students up to date with the latest methods in construction while clarifying important concepts.
- The section on *Steel Framing* has been revised to include coverage of structural framing. These additions reflect the changing nature of construction materials and processes.
- Up-to-date coverage of *Insulation and Ventilation, Roofing,* and *Siding* emphasizes geographic differences and encourages careful consideration of these various methods prior to installation.
- *Revised Review Questions* and additional *Key Terms* have been incorporated at the end of unit sections to evaluate student knowledge of important concepts presented in the text and to highlight important terminology.

SUPPLEMENTAL PACKAGE

Included with the fourth edition of *Carpentry* is an extensive supplemental package to aid student learning, and help instructors prepare lessons and evaluations:

An *Instructor's Resource Guide* provides many helpful tools to facilitate effective classroom presentations of material—comprehensive *Lesson Plans,* integrated PowerPoint references and Teaching Tips, a review of important math, blueprint reading to help evaluate student knowledge at the start of the course, and *Answers to Review Questions* and *Workbook Questions* are included for reference. The *Instructor's Guide* is also available in electronic format on the *E.resource.*

A *Workbook* for students provides a wide range of practice problems to reinforce concepts learned in each chapter, as well as to prepare students for exams. Question types include multiple choice, completion, and identification as well as critical math problems and soft skill activities. Instructors may assign these questions as homework, to ensure full comprehension of the material.

E.resource—a complete CD-ROM for the instructor, this electronic resource contains several components for classroom prep:

- *Note to the Instructor* includes valuable information on effectively teaching construction students—everything from learning styles to safety on the worksite is covered in this introduction to the course.
- An important *Math Review,* including brief lessons and practice problems for students, allows instructors to review important math concepts and ensure that their students are up to speed prior to introduction of the material.
- The *Instructor's Guide* in Word format allows instructors to customize the lesson plans, or add their own notes to further enhance classroom presentation. Answers to the Review Questions in the book and workbook are also provided.
- *PowerPoint® presentations* include an outline of each of the chapters in the text, and features drawings, photos, and procedures from the book.
- A *Testbank* available in ExamView format provides 1,500 questions, covering the 30 units in the text, for evaluating student comprehension of important concepts.
- An *Image Library* contains a multitude of photos, drawings, and procedures included in the book, allowing instructors to supplement presentations in class. A set of *Auto-CAD* drawings of the house plan presented in the book is also included.
- *Additional Resources* provides a listing of important industry organizations, as well as websites for further information on the subject. This feature is also available to students on the *Online Companion.*

An *Online Companion* is also available to students, and links them to important information on acquiring a career in the carpentry field—including advice on writing resumes and interviewing for a job, related industry links to keep current on the latest in the field, various construction job descriptions, and more!

Acknowledgments

The publisher and author wish to sincerely thank those who contributed to the *Carpentry* book and helped to enhance the text for the fourth edition. Our gratitude is extended to those reviewers who contributed to the revision and previous editions of the *Carpentry* book—your insights and recommendations were invaluable:

Louis Bermudes
Instructor
San Jose City College
San Jose, CA

Richard Cappelmann
Technical College of the Low Country
Beaufort, South Carolina

Carl Gamarino
Industrial Technical Education Center
Brockport, Pennsylvania

Kirk Garrison
Dept. Co-Chair
Portland Community College
Portland, OR

Rick Glanville
Camosun College
Victoria, British Columbia, Canada

Robert Gresko
Instructor, Construction Technology
Pennsylvania College of Technology
Williamsport, PA

Douglas A. Holman
East Tennessee Carpenters Joint Apprenticeship
    and Training Program
Knoxville, Tennessee

Larry Kness
Southeast Community College
Milford, Nebraska

Jim Loosle
Kearns High School
Taylorsville, Utah

John E. Mackay
Training Coordinator
New Jersey Carpenters Technical
    Training Centers
Kenilworth, New Jersey

Dave Rainforth
Southeast Community College
Milford, Nebraska

Terry Schaefer
Western Wisconsin Technical College
LaCrosse, Wisconsin

Kathy Swan
Carpenter's Training Trust of
    Western Washington
Renton, Washington

David Vancise
Master Carpentry Instructor
Indian River Community College
Ft. Pierce, FL

Robert Wilcke
Carpentry Instructor
Western Iowa Tech Community College
Sioux City, IA

Special thanks also to Margaret Magnarelli for her diligence in acquiring and writing the "Success Stories" to provide additional information to the students on the industry, to Dave Hultenius for his residential drawings, and to Terrel Broiles for his tireless effort to bring the drawings contained within these pages to life.

Thanks to Ed Snider, David Sitton, Gary Gafford, Richard Sergent, Murray Nevel, and Eric Van Nattan of Beazer Homes, and Ron Ferschke, formally of Leveit & Sons, for sharing their time and knowledge of construction. With their help, this edition discusses regional variations in construction methods.

Thanks to Richard Harrington, of State University of New York at Delhi, for his support and inspiring desire to brainstorm; Steve Munson, of Munson's Building Supplies, for lending access to the latest tools; and Robert Braun for his help in the overwhelming clerical duties. A special thanks to all the numerous trade people who were willing to share their trade techniques and pause in their tasks for pictures. Last and most, thanks to my wife, Pamela for her ubiquitous support.

The supplemental package could also not be completed without the assistance of various talented instructors—to Kerry Reinhackel for the creation of the PPT and Lesson Plans, and to David Vancise, master carpentry instructor at Indian River Community College in Ft. Pierce FL, for his completion of the Testbank, our many thanks.

And to the Thomson Delmar Learning team, whose dedication to the project produced quality learning materials for aspiring carpenters everywhere—Alison Weintraub, Acquisitions Editor; Jennifer Thompson, Product Manager; Andrew Crouth, Production Manager; Mary Beth Vought, Art and Design Specialist; and Chris Chien, Project Editor—thanks to you all!

About the Author

Floyd Vogt is a sixth generation carpenter/builder. He was raised in a family with a small business devoted to all phases of home construction and began working in the family business at age 15.

After completing a B.A. in chemistry from the State University of New York at Oneonta, Mr. Vogt returned to the field as a self-employed remodeler. In 1985, he began teaching at State University of New York at Delhi of Technology in Delhi, New York, www.delhi.edu. He is currently an Associate Professor of carpentry at Delhi, where he has taught many courses, including Light Framing, Advanced Framing, Math, Energy Efficient Construction, Finish Carpentry, Finish Masonry, and Estimating. Mr. Vogt is a carpentry regional coordinator for Skills-USA and serves as a post-secondary Skills-USA student advisor.

Introduction

The history of carpentry goes back to 8000 B.C., when primitive people used stone axes to shape wood to build shelters. Stone Age Europeans built rectangular timber houses more than 100 feet long—proving the existence of carpentry even at this early date. The Egyptians used copper woodworking tools as early as 4000 B.C. By 2000 B.C. they had developed bronze tools and were proficient in the drilling, dovetailing, mitering, and mortising of wood.

In the Roman Empire, two-wheeled chariots, called *carpentum* in Latin, were made of wood. A person who built such chariots was called a *carpentarius*, from which the English word *carpenter* is derived. Roman carpenters handled iron adzes, saws, rasps, awls, gouges, and planes.

During the Middle Ages, most carpenters were found in larger towns where work was plentiful. They would also travel with their tools to outlying villages or wherever there was a major construction project in progress. By this time, they had many efficient, steel-edged hand tools. During this period skillful carpentry was required for the building of timber churches and castles.

In the twelfth century, carpenters banded together to form **guilds.** The members of the guild were divided into **masters, journeymen,** and **apprentices.** The master was a carpenter with much experience and knowledge who trained apprentices. The apprentice lived with the master and was given food, clothing, and shelter and worked without pay. After a period of five to nine years, the apprentice became a journeyman who worked and traveled for wages. Eventually, a journeyman could become a master. Guilds were the forerunners of the modern labor unions and associations.

Starting in the fifteenth century, carpenters used great skill in constructing the splendid buildings of the Renaissance period and afterward. With the introduction of the balloon frame in the early nineteenth century, more modern construction began to replace the slower mortise-and-tenon frame. In 1873, electric power was used for the first time to drive machine tools. The first electric hand drill was developed in 1917, and in 1925, electric portable saws were being used.

At present, many power tools are available to the carpenter to speed up the work. Although the volume of the carpenter's work has been reduced by the use of manufactured parts, some of the same skills carpenters used in years past are still needed for the intricate interior finish work in buildings.

Carpenters construct and repair structures and their parts using wood, plywood, and other building materials. They lay out, cut, fit, and fasten the materials to erect the framework and apply the finish. They build houses, factories, banks, schools, hospitals, churches, bridges, dams, and other structures. In addition to new construction, a large part of the industry is engaged in remodeling and repair of existing buildings.

The majority of workers in the construction industry are carpenters **(Figure I–1).** They are the first trade workers on the job, laying out excavation and building lines. They take part in every phase of the construction, working below the ground, at ground level, or at great heights. They are the last to leave the job when they put the key in the lock.

SPECIALIZATION

In large cities, where there is a great volume of construction, carpenters tend to specialize in one area of the trade. They may be specialists in rough carpentry, which are called rough carpenters or framers **(Figure I–2).** Rough carpentry does not mean that the workmanship is crude. They typically measure to within 1/16 inch. Just as much care is taken in the rough work as in any other work. Rough carpentry will be covered eventually by the finish work or dismantled, as in the case of concrete form construction.

Finish carpenters specialize in applying exterior and interior finish, sometimes called trim **(Figure I–3).** Many materials require the finish carpenter to

FIGURE I–1 Carpenters make up the majority of workers in the construction industry.

measure to tolerances of ½ or ¼ of an inch. Other specialties are constructing concrete forms **(Figure I–4)**, laying finish flooring, building stairs, applying gypsum board **(Figure I–5)**, roofing, insulating, and installing suspended ceilings.

In smaller communities, where the volume of construction is lighter, carpenters often perform tasks in all areas of the trade from the rough to the finish. Such communities may have workers that perform all aspects of construction including plumbing, electrical work, and excavating. The general carpenter needs a more complete knowledge of the trade than the specialist does.

REQUIREMENTS

Every job has specific requirements of the workforce. These include the skills necessary to perform each task and an attitude or mindset the workers should have while on the job. Construction skills vary to fit the type of work being performed. They tend to be clearly defined and it is easy to determine

FIGURE I–2 Some carpenters specialize in framing.

if a worker possesses these skills. The construction attitude is less tangible, but no less important. While work may continue if the workers have only the required skills, minor or severe slowdowns are inevitable if workers do not have the proper attitude.

Skill

Carpenters need to know how to use and maintain hand and power tools. They need to know the kinds, grades, and characteristics of the materials with which they work—how each can be cut, shaped, and most satisfactorily joined. Carpenters must be familiar with the many different fasteners available and choose the most appropriate for each task.

Carpenters should know how to lay out and frame floors, walls, stairs, and roofs. They must know how to install windows and doors, and how to apply numerous kinds of exterior and interior finish. They must use good judgment to decide on proper procedures to do the job at hand in the most efficient and safest manner.

Carpenters need to be in good physical condition because much of the work is done by hand and sometimes requires great exertion. They must be able to lift large sheets of plywood, heavy wood timbers, and bundles of roof shingles; they also have to climb ladders and scaffolds.

Carpenters need reading and math skills. Much of construction begins as an idea put on paper. A

FIGURE I–4 Erecting concrete formwork may be a specialty.

FIGURE I-5   Some carpenters choose to do only gypsum board application.

carpenter must be able to interpret these ideas from the written form to create the desired structure. This is done by reading prints and using a ruler. The quantity of material needed must be estimated using math and geometry. Accurate measurements and calculations speed the construction process and reduce wasted materials.

Communication skills are also very important. Carpenters must communicate with many people during the construction process. Work to be done as determined by the owner or architect must be accurately understood; otherwise costly delays and expenses may result. Efficiency of work relies heavily on workers' understanding of what others are doing and what work must be done next. Communication is vital for a jobsite to be safe.

Attitude

The proper construction attitude is not as clearly defined as job skills. Regional variations and requirements will affect the expected jobsite attitude. For example in regions where heat is a concern, workers develop a steady rhythm to their work that survives the heat. In regions where winters are harsh, workers develop a faster style of work when the weather is warm knowing they may have downtime in the winter. Other attitudes, such as having a good work ethic, are universal to all jobsites.

A good ethic is not easily defined in one sentence. It involves the person as a whole, the way he or she approaches life and their work. A person with a good work ethic has respect and lets it show. Respect for the jobsite, fellow workers, tools and materials, and themselves reveals care and concern for the construc-

tion process. Workers demonstrating this form of respect are safe and more pleasant to work alongside.

Workers with a good work ethic show up fifteen minutes early, not a few minutes late. When finished with a task, they look for something else to do that promotes the job, even if it's using a broom. They perform their tasks as well as expected, up to the standard required for that application. They finish the task, not leaving something undone for someone else to fix. They are interested in learning, looking around at the work of others for ways to improve their own skills. They cooperate with other workers and tradespersons to make the jobsite a pleasant work place. They are also honest with the material and their time, never cheating on the accepted method. They feel that nothing less than a first-class job is acceptable.

Jobsite humor makes work easier to do and more fun. Some tasks, by their very nature, are boring and unpleasant. Humor can make difficult jobs seem to get done faster. Unfortunately, humor has a bad side, because humor for one person can be pain for another. Jobsite humor can be tasteless. It can single out a person, making them feel alienated. This type of humor can have severe effects on jobsite safety. If someone feels like a victim, they defend themselves, and do not concentrate on what they are doing. During these times unintended things can happen and someone may get hurt. Keep jobsite humor suitable for everyone present.

There is no replacement for teamwork on the jobsite. Someone once said, "Two people working together can outperform three people working alone." This is easily seen when one person holds material to be fastened or cut for another. But the difference between the team and individuals is more dramatic when each member of the team anticipates the next move of the other. For example, while holding a board that is being fastened, that person looks to see what else can be done. For example, is more material needed? Does the horse or ladder have to be moved into a better position? Will there be another tool needed? How can the tool best be handed to the fellow worker? It can be easy to miss this type of teamwork when it is done without words and without being asked.

Another way a team works better is in the mental energy used on the job. Two pairs of eyes working on one task can ward off errors and mistakes. Two minds can find the better, faster, safer method. Teamwork is a major reason why people stay in construction for a lifetime.

TRAINING

Vocational training in carpentry is offered by many high schools for those who become seriously interested at an early age. For those who have completed high school, carpentry training programs are offered at many postsecondary vocational schools and col-

APPRENTICESHIP AND TRAINING SYSTEM OF THE CARPENTRY TRADE

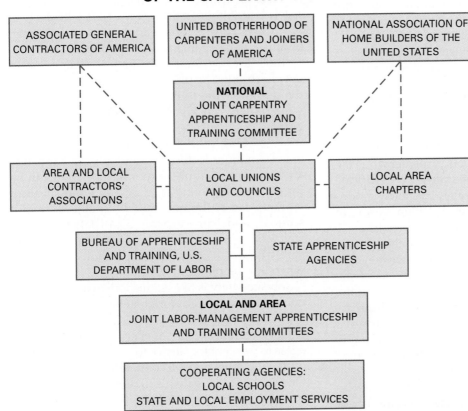

FIGURE I–6 Industry, labor, and government work together in carpentry apprenticeship programs.

leges. Because there is so much to learn, it can take years to learn carpentry.

Some schools participate in programs in which students go to school part-time and work as on-the-job apprentices part-time. Upon graduation, they may continue with the same employer as full-time apprentices.

Apprenticeship training programs (usually four years in length) are offered by the United Brotherhood of Carpenters and Joiners of America, a carpenters' union, and by contractors' associations, such as the National Association of Home Builders of the United States (NAHB) and the Associated General Contractors of America (AGC) in cooperation with the Bureau of Apprenticeship and Training of the U.S. Department of Labor **(Figure I–6)**. In Canada, apprenticeships are administered by the provincial governments with a final exam called the interprovincial examination written for national accreditation. Usually these organizations give apprenticeship credit for completion of previous vocational school training in carpentry, resulting in a shortened apprenticeship and a higher starting wage.

The apprentice **(Figure I–7)** must be at least seventeen years of age, in some areas eighteen, and must learn the trade while working on the job with experienced journeymen. Under the apprenticeship agreement, basic standards provide, among other things,

FIGURE I–7 Many opportunities lie ahead for the apprentice carpenter.

that there can be only a certain number of apprentices hired by a particular contractor in relation to the number of journeymen. This is to ensure that apprentices receive proper supervision and training on the job.

Basic standards also provide that there is a progressively increasing schedule of wages. Starting pay for the apprentice carpenter is usually about 50 percent of the journeyman carpenter's rate. Due to periodic increases about every six months, the apprentice should receive 95 percent of the journeyman's wage during the last six months of the apprenticeship.

Under the terms of agreement, the apprentice is required to attend classes for a certain number of hours each year. A minimum of 144 hours per year is normally considered necessary. These classes are usually held at local schools, twice a week for about thirty-six weeks during the school year, for the length of the apprenticeship. The apprentice becomes accepted as a journeyman carpenter when the training is completed, and is awarded a Certificate of Completion of Apprenticeship. The newly graduated apprentice is now expected, within reason, to do jobs required of the journeyman.

Although it is in the best interests of the apprentice to be indentured, that is, to have a written contract with an organization, with conditions of the apprenticeship agreed on, it is possible to learn the trade as a helper until enough skills have been acquired to demand the recognition and rewards of a journeyman carpenter. However, self-discipline is required to gain knowledge of both the practical and theoretical aspects of the trade. There may be time on the job to explain to a helper or apprentice how to do a certain task, but there usually is not time to explain concepts.

Many opportunities exist for the journeyman carpenter. Advancement depends on dependability, skill, productivity, and ingenuity, among other characteristics. Carpentry foremen, construction superintendents, and general contractors usually rise from the ranks of the journeyman carpenters. Many who start as apprentice carpenters eventually operate their own construction firms. A survey revealed that 90 percent of the top officials (presidents, vice presidents, owners, and partners) of construction companies who replied began their careers as apprentices. Many of the project managers, superintendents, and craft supervisors employed by these companies also began as apprentices.

National student organizations like Skills-USA, formally VICA (Vocational Industrial Clubs of America, Inc.), offer students training in leadership, teamwork, citizenship, and character development. The organization helps build and reinforce work attitudes, self-confidence, and communications skills for the future workforce. It emphasizes total quality at work, high ethical standards, superior work skills, and pride in the dignity of work. SkillsUSA also promotes involvement in community service activities. Yearlong student activities culminate with skill competitions where local winners move up to regional, state, and then national and international levels.

SAFETY AT THE WORKSITE

Much of the work performed in construction carries risk with it. Although the risk will vary with the type of work, all construction workers, at sometime, may be at risk of being maimed or killed on the job.

Tools are, by their very nature, dangerous to use. Each must be operated in a fashion suited to the design of the tool. Some tools can cause cuts while others can kill. Safety cannot be taken for granted and must be built into the methods and process being used for any particular task. Unusual situations leading to accidents can happen at anytime. For example, a roofing carpenter can be thrown to the ground by a gust of wind, cleanup workers can be injured by falling objects from a scaffold, and trim carpenters can fall through a stairway opening that is under construction. Safety is like air, it must be continuous and everywhere for people to survive.

Job-site safety is like team sportsmanship. Everyone on the team must work together and play by the same rules or someone is at risk of being injured. Safety is like silence; just as everyone in a room must agree to be silent for silence to exist, so too everyone on the job must agree to be safe for safety to exist. One person can create a situation where dozens of workers will be at risk. All safety programs are successful only because workers join the team.

Safe Work Practices

Safe methods and practices learned correctly today will develop into habits that will become second nature for a lifetime. Always approach new tasks thoughtfully and carefully. If a new tool is being used, become familiar with its requirements. Adapt to the tool because it cannot adapt to you. Always read and follow the directions associated with the tool or task. Read the manufacturer's recommendations for use and installation. Manufacturers clearly define the risks and recommended uses. Failure to follow their instructions is foolish and short-sighted. Always use the safety devices designed into a tool. They are there to protect the operator, not make it harder to use the tool. Always remember where you are and the possible dangers of the work site. Horseplay and practical jokes are distracting to everyone around. When horseplay happens, the work environment is less safe. Workers have suffered permanent back injuries while lifting an object too fast because someone touched them inappropriately. Respect is the key word to safety. Always listen and communicate fully. Think of possible ways for misunderstanding and clear them up. The only stupid question is the one not asked.

Safe Work Conditions

Safe work conditions are not hard to create, yet they can be easily neglected. Always maintain clear areas for working and walking. Store materials in an orderly fashion and dispose of waste materials as they are produced. If you stumble and struggle for a place to stand, then safety risks have increased. Always work to keep the area as dry as possible. Sweep water puddles away or make small trenches to divert the water. Keep electrical cords and tools dry. Always maintain tools and equipment according to manufacturer's recommendations. Equipment needs may include lubrication, cleaning, drying, or inspection. Always understand the risks associated with building materials. All workers are to have access to information on the hazards of materials they are working with through the **Material Safety Data Sheet (MSDS).** Read them. Always look for ways to keep fires from starting. Any possible heat source should be carefully studied and isolated from starting fires. Always be aware of the effect of air temperature on safety. Cold or frozen areas can be slippery to work on. Hot areas can have soft material that is easily marred. Always wear personal protective devices. Eyes, ears, and airways should be protected when injury is possible. It is easier to understand why eyes are important to protect. Most of our perception of the world comes through our eyes. But ear, lung, and sinus damage often occurs slowly. Most times they do not show any effects for years. Personal protection is a personal responsibility.

Construction efficiency and jobsite safety require good listening skills. Listening to what is asked and performing that task efficiently is one form of good listening. It is just as important to listen to the sounds of the job. Most sounds are normal, some are not. A saw cutting wood, a mixer mixing mortar, an air compressor providing power, and the bangs, taps, and thuds of construction are examples of normal noises. As workers become acquainted and accustomed to these sounds, it is important to be able to tell when these sounds change. A change in sound can be an early warning signal that something is wrong. It could be that something is about to break or someone is in trouble and needs assistance.

OSHA

OSHA (Occupational Safety and Health Administration) was created in 1970 to improve workplace safety. OSHA's goal is to save lives, prevent injuries, and protect the health of America's workers. They have been given the authority to work with state agencies to enforce acceptable work site safety standards. OSHA provides research information, education, and training in the field of occupational safety and health. Their efforts and safety in general will only succeed if every worker agrees to be a team player and does their best to maintain safe work habits and site conditions.

SUMMARY

Carpentry is a trade in which there is a great deal of self-satisfaction, pride, and dignity associated with the work. It is an ancient trade and the largest of all trades in the building industry.

Skilled carpenters who have labored to the best of their ability can take pride in their workmanship, whether the job was a rough concrete form or the finest finish in an elaborate staircase **(Figure I–8).** At the end of each working day, carpenters can stand back and actually see the results of their labor. As the years roll by, the buildings that carpenters' hands had a part in creating still can be viewed with pride in the community.

FIGURE I–8 After completing a complicated piece of work, such as this intricate staircase, carpenters can take pride in and view their accomplishment. *(Courtesy of L. J. Smith, Inc.)*

SECTION ONE
TOOLS AND MATERIALS

Success Story

Victor Sanchez
Title: Superintendent
Company: Tract Homes, JTS Communities, Sacramento, California

EDUCATION
Back in 1999, Vic was a high school student struggling to find himself. When graduation rolled around, he found himself confronted with two options: the military or a job training program. He chose the latter and ended up at the Sacramento Job Corps program, where he focused on carpentry and discovered he was a natural at the construction industry.

HISTORY
Two and a half years later, he had graduated from the program and began working for JTS Communities, a Sacramento-based homebuilder, first as a carpenter. He worked his way up to assistant superintendent and then was promoted to his current position.

ON THE JOB
The workday begins at 6:30 a.m., when Vic goes to the site of a home he's working on. As a superintendent, it's his job to keep the project on schedule and below budget. He also orchestrates the construction team, and is constantly reviewing their work to be sure it's top quality. He often walks around the interior and exterior of the house, carefully making

sure it fits what was in the prints, from baseboard to framing. "I refer back to the floor plan, and if anything seems wrong, I will work with any of the thirty different trades involved to correct it," says Vic. Knowing carpentry is part of what makes him good at his job—he can spot things that other people might not. "There are certain tricks carpenters use to cover where they've mishandled a job—caulking or putty, for example—and I'm able to call them on it!"

BEST ASPECTS
Getting respect from his peers, and even those much older than him, has given Vic a lot of confidence. He's also glad he's found a career that he can excel at.

CHALLENGES
The biggest challenge has been time management. "The job really has to be done by its deadline," he says. And in cases where they remain on schedule, the project tends to be smooth throughout. "But if you get off track, you get very overwhelmed very quickly."

IMPORTANCE OF EDUCATION
Getting training is "immensely" important says Vic; he found vocational school to be a great way to figure out he was moving in the right career direction. "You get the hands-on experience, instead of wasting several months of an employer's time only to learn you don't actually

want to do that." For those looking to get into a superintendent career, Sanchez says a college engineering degree is becoming more essential: "You'll come in to the industry where I am at now."

FUTURE OPPORTUNITIES
"I'd like to be a contractor," says Vic—it's a goal he's planning to realize in the next decade. But right now, he's enjoying working within a large company, where he has the opportunity to build up a network of high-quality subcontractors.

WORDS OF ADVICE
"Knowledge is power," says Vic. "Get an education."

UNIT 1

Wood and Lumber

CHAPTER **1** **Wood**

CHAPTER **2** **Lumber**

The construction material most often associated with a carpenter is wood. Wood has properties that make it the first choice in many applications in home construction. Wood is easy to tool and work with, pleasing to look at and smell, and has strength that will last a long time.

Lumber is manufactured from the renewable resources of the forest. Trees are harvested and sawn into lumber in many shapes and sizes having a variety of characteristics. It is necessary to understand the nature of wood to get the best results from the use of it. With this knowledge, the carpenter is able to protect lumber from decay, select it for appropriate use, work it with proper tools, and join and fasten it to the best advantage.

Wood comes from many tree species having many different characteristics. A good carpenter knows the different applications for specific wood species.

OBJECTIVES

After completing this unit, the student should be able to:

- name the parts of a tree trunk and state their function.

- describe methods of cutting the log into lumber.

- define hardwood and softwood, give examples of some common kinds, and list their characteristics.

- explain moisture content at various stages of seasoning, tell how wood shrinks, and describe some common lumber defects.

- state the grades and sizes of lumber and compute board measure.

1   Wood

The carpenter works with wood more than any other material and must understand its characteristics in order to use it intelligently. Wood is a remarkable substance. It can be more easily cut, shaped, or bent into almost any form than just about any other structural material. It is an efficient insulating material. It takes almost 6 inches of brick, 14 inches of concrete, or more than 1,700 inches of aluminum to equal the insulating value of only 1 inch of wood.

There are many kinds of wood that vary in strength, workability, elasticity, color, grain, texture, and smell. It is important to keep these qualities in mind when selecting wood. For instance, baseball bats, diving boards, and tool handles are made from hickory and ash because of their greater ability to bend without breaking (elasticity). Oak and maple are used for floors because of their beauty, hardness, and durability. Redwood, cedar, cypress, and teak are used in exterior situations because of their resistance to decay **(Figure 1–1).** Cherry, mahogany, and walnut are typically chosen for their beauty.

With proper care, wood will last indefinitely. It is a material with beauty and warmth that has thousands of uses. Wood is one of our greatest natural resources. With wise conservation practices, wood will always be in abundant supply. It is fortunate that we have perpetually producing forests that supply this major building material to construct homes and other structures that last for hundreds of years.

When those structures have served their purpose and are torn down, the wood used in their construction can be salvaged and used again (recycled) in new building, remodeling, or repair. Wood is biodegradable and when it is considered not feasible for reuse, it is readily absorbed back into the earth with no environmental harm.

STRUCTURE AND GROWTH

Wood is a material cut from a complex living organism, called trees. Trees are made up of many different kinds of cells and growth areas that are visible in the cross-sectional view **(Figure 1–2).** Wood is made up of many hollow cells held together by a natural substance called **lignin.** The size, shape, and arrangement of these cells determine the strength, weight, and other properties of wood. Tree growth takes place in the **cambium layer,** which is just inside the protective shield of the tree called the *bark.* The tree's roots absorb water that passes upward through the **sapwood** to the leaves, where it is combined with carbon dioxide from the air. Sunlight causes these materials to change into food, which is then carried down and distributed toward the center of the trunk through the **medullary rays.**

As the tree grows outward from the **pith** (center), the inner cells become inactive and turn into **heartwood.** This older section of the tree is the central part of the tree and usually is darker in color and more

FIGURE 1–1   Redwood is often used for exterior trim and siding. *(Courtesy of California Redwood Association)*

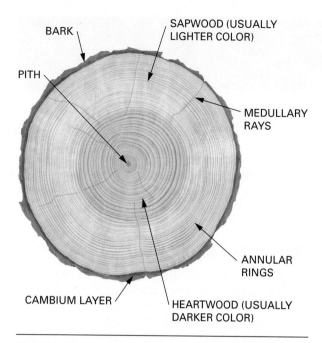

BARK

PITH

SAPWOOD (USUALLY LIGHTER COLOR)

MEDULLARY RAYS

ANNULAR RINGS

CAMBIUM LAYER

HEARTWOOD (USUALLY DARKER COLOR)

FIGURE 1–2   A cross section of a tree showing its structure. *(Courtesy of Western Wood Products Association)*

durable than sapwood. The heartwood of cedar, cypress, and redwood, for instance, is extremely resistant to decay and is used extensively for outdoor furniture, patios, and exterior siding. Used for the same purposes, sapwood decays more quickly.

Each spring and summer, a tree adds new layers to its trunk. Wood grows rapidly in the spring; it is rather porous and light in color. In summer, tree growth is slower; the wood is denser and darker, forming distinct rings. Because these rings are formed each year, they are called **annular rings.** By counting the dark rings, the age of a tree can be determined. By studying the width of the rings, periods of abundant rainfall and sunshine or periods of slow growth can be discerned. Some trees, like the Douglas fir, grow rapidly to great heights and have very wide and pronounced annular rings. Mahogany, which grows in a tropical climate where the weather is more constant, has annular rings that do not contrast as much and sometimes are hardly visible.

HARDWOODS AND SOFTWOODS

Woods are classified as either **hardwood** or **softwood**. There are different methods of classifying these woods. The most common method of classifying wood is by its source. Hardwood comes from **deciduous** trees that shed their leaves each year. Softwood is cut from **coniferous,** or cone-bearing, trees, commonly known as evergreens **(Figure 1–3)**. In this method of classifying wood, some of the softwoods may actually be harder than the hardwoods. For instance, fir, a softwood, is harder and stronger than basswood, a hardwood. There are other methods of classifying hardwoods and softwoods, but this method is the one most widely used.

FIGURE 1–3   Softwood is from cone-bearing trees, hardwood from broad-leaf trees.

Softwoods									
Kind	**Color**	**Grain**	**Hardness**	**Strength**	**Work-ability**	**Elasticity**	**Decay Resistance**	**Uses**	**Other**
Red Cedar	Dark Reddish Brown	Close Medium	Soft	Low	Easy	Poor	Very High	Exterior	Cedar Odor
Cypress	Orange Tan	Close Medium	Soft to Medium	Medium	Medium	Medium	Very High	Exterior	
Fir	Yellow to Orange Brown	Close Coarse	Medium to Hard	High	Hard	Medium	Medium	Framing Millwork Plywood	
Ponderosa Pine	White with Brown Grain	Close Coarse	Medium	Medium	Medium	Poor	Low	Millwork Trim	Pine Odor
Sugar Pine	Creamy White	Close Fine	Soft	Low	Easy	Poor	Low	Pattern-making Millwork	Large Clear Pieces
Western White Pin	Brownish White	Close Medium	Soft to Medium	Low	Medium	Poor	Low	Millwork Trim	
Southern Yellow Pine	Yellow Brown	Close Coarse	Soft to Hard	High	Hard	Medium	Medium	Framing Plywood	Much Pitch
Redwood	Reddish Brown	Close Medium	Soft	Low	Easy	Poor	Very High	Exterior	Light Sapwood
Spruce	Cream to Tan	Close Medium	Medium	Medium	Medium	Poor	Low	Siding Subflooring	Spruce Odor

FIGURE 1–4 Common species of softwood and their characteristics.

Some common hardwoods are ash, birch, cherry, hickory, maple, mahogany, oak, and walnut. Some common softwoods are pine, fir, hemlock, spruce, cedar, cypress, and redwood.

Wood can also be divided into two groups according to cell structure. **Open-grained** wood has large cells that show tiny openings or pores in the surface. To obtain a smooth finish, these pores must be filled with a specially prepared paste wood filler. Examples of open-grained wood are oak, mahogany, and walnut. Some **close-grained** hardwoods are birch, cherry, maple, and poplar. All softwoods are close grained. (See **Figure 1–4** for common types of softwoods and their characteristics and **Figure 1–5** for common hardwoods.)

Identification of Wood

Identifying different kinds of wood can be very difficult because some closely resemble each other. For instance, ash and white oak are hard to distinguish from each other, as are some pine, hemlock, and spruce. Not only are they the same color, but the grain pattern and weight are about the same. Only the most experienced workers are able to tell the difference.

It is possible to get some clues to identifying wood by studying the literature, but the best way to learn the different kinds of wood is by working with them. Each time you handle a piece of wood, examine it. Look at the color and the grain; feel if it is heavy or light, if it is soft or hard; and smell it for a characteristic odor. Aromatic cedar, for instance, can always be identified by its pleasing moth-repelling odor, if by no other means. After studying the characteristics of the wood, ask or otherwise find out the kind of wood you are holding, and remember it. In this manner, after a period of time, those kinds of wood that are used regularly on the job can be identified easily. Identification of kinds of wood that are seldom worked with can be accomplished in the same manner, but, of course, the process will take a little longer.

Hardwoods

Kind	Color	Grain	Hardness	Strength	Work-ability	Elasticity	Decay Resistance	Uses	Other
Ash	Light Tan	Open Coarse	Hard	High	Hard	Very High	Low	Tool Handles Oars Baseball Bats	
Basswood	Creamy White	Close Fine	Soft	Low	Easy	Low	Low	Drawing Bds Veneer Core	Imparts No Taste or Odor
Beech	Light Brown	Close Medium	Hard	High	Medium	Medium	Low	Food Containers Furniture	
Birch	Light Brown	Close Fine	Hard	High	Medium	Medium	Low	Furniture Veneers	
Cherry	Lt. Reddish Brown	Close Fine	Medium	High	Medium	High	Medium	Furniture	
Hickory	Light Tan	Open Medium	Hard	High	Hard	Very High	Low	Tool Handles	
Lauan	Lt. Reddish Brown	Open Medium	Soft	Low	Easy	Low	Low	Veneers Paneling	
Mahogany	Russet Brown	Open Fine	Medium	Medium	Excellent	Medium	High	Quality Furniture	
Maple	Light Tan	Close Medium	Hard	High	Hard	Medium	Low	Furniture Flooring	
Oak	Light Brown	Open Coarse	Hard	High	Hard	Very High	Medium	Flooring Boats	
Poplar	Greenish Yellow	Close Fine	Medium Soft	Medium Low	Easy	Low	Low	Furniture Veneer Core	
Teak	Honey	Open Medium	Medium	High	Excellent	High	Very High	Furniture Boat Trim	Heavy Oily
Walnut	Dark Brown	Open Fine	Medium	High	Excellent	High	High	High Quality Furniture	

FIGURE 1–5 Common species of hardwood and their characteristics.

2 Lumber

MANUFACTURE OF LUMBER

When logs arrive at the sawmill, the bark is removed first. Then a huge bandsaw slices the log into large planks, which are passed through a series of saws. The saws slice, edge, and trim them into various dimensions, and the pieces become **lumber.**

Once trimmed of all uneven edges, the lumber is stacked according to size and grade and taken outdoors where **stickering** takes place. Stickering is the process of restacking the lumber on small crosssticks that allow air to circulate between the pieces. This air-seasoning process may take six months to two years due to the large amount of water found in lumber.

Following air-drying, the lumber is then dried in huge ovens. Once dry, the rough lumber is surfaced to standard sizes and shipped **(Figure 2–1).**

The long, narrow surface of a piece of lumber is called its edge, the long, wide surface is termed its side, and its extremities are called ends. The distance across the edge is called its thickness, across its side is called its width, and from end to end is called its length **(Figure 2–2).** The best-appearing side is called the face side, and the best-appearing edge is called the face edge.

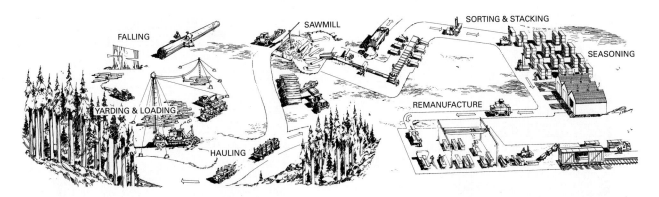

FIGURE 2–1 Manufacturing path of lumber from forest to finished product. *(Courtesy of California Redwood Association)*

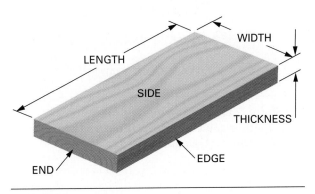

FIGURE 2–2 Lumber and board surfaces are distinguished by specific names.

In certain cases, the surfaces of lumber may acquire different names. For instance, the distance from top to bottom of a beam is called its depth, and the distance across its top or bottom may be called its width or its thickness. The length of posts or columns, when installed, may be called their height.

Plain-Sawed Lumber

A common way of cutting lumber is called the **plain-sawed** method, in which the log is cut tangent to the annular rings. This method produces a distinctive grain pattern on the wide surface **(Figure 2–3)**. This method of sawing is the least expensive and produces greater widths. However, plain-sawed lumber shrinks more during drying and warps easily. Plain-sawed lumber is sometimes called slash-sawed lumber.

Quarter-Sawed Lumber

Another method of cutting the log, called quarter-sawing, produces pieces in which the annular rings are at or almost at right angles to the wide surface. **Quarter-sawed** lumber has less tendency to warp

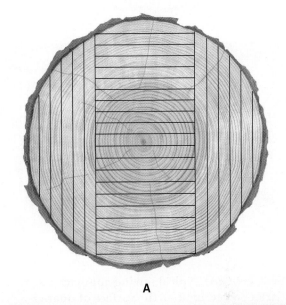

A

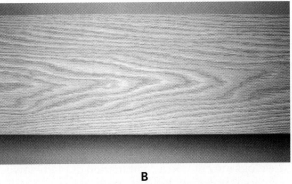

B

FIGURE 2–3 (A) Typical sawing approach for plain-sawed lumber. (B) Surface of plain-sawed lumber.

and shrinks less and more evenly when dried. This type of lumber is durable because the wear is on the edge of the annular rings. Quarter-sawed lumber is frequently used for flooring.

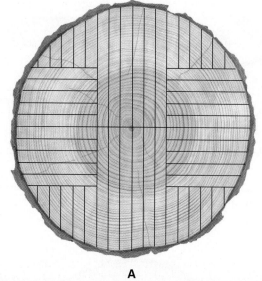

A

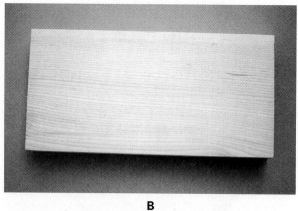

B

FIGURE 2–4 **(A) Typical sawing approach for quarter-sawed lumber. (B) Surface of quarter-sawed lumber.**

A distinctive and desirable grain pattern is produced in some wood, such as oak, because the lumber is sawed along the length of the medullary rays. Quarter-sawed lumber is sometimes called vertical-grain or edge-grain **(Figure 2–4)**.

Combination Sawing

Most logs are cut into a combination of plain-sawed and quarter-sawed lumber. With computers and laser-guided equipment, the **sawyer** determines how to cut the log with as little waste as possible in the shortest amount of time to get the desired amount and kinds of lumber **(Figure 2–5)**.

MOISTURE CONTENT AND SHRINKAGE

When a tree is first cut down, it contains a great amount of water. Lumber, when first cut from the

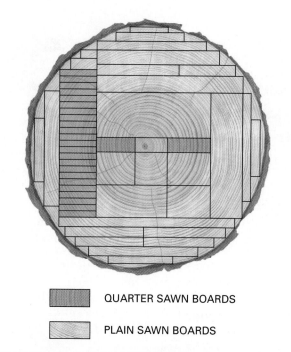

QUARTER SAWN BOARDS

PLAIN SAWN BOARDS

FIGURE 2–5   **Typical sawing approach for combination-sawed lumber.** *(Courtesy of Western Wood Products Association.)*

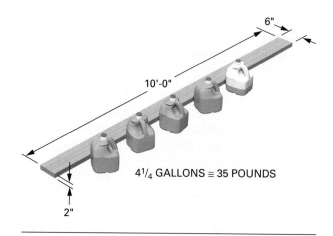

6"

10'-0"

4¼ GALLONS ≅ 35 POUNDS

2"

FIGURE 2–6   **Green lumber contains a large amount of water.**

log, is called **green lumber** and is very heavy because most of its weight is water. A piece 2 inches thick, 6 inches wide, and 10 feet long may contain as much as 4¼ gallons of water weighing about 35 pounds **(Figure 2–6)**.

Green lumber should not be used in construction because it shrinks as it dries to the same moisture content as the surrounding air. This shrinking is considerable and unequal because of the large amount of water that leaves it. When it shrinks, it usually warps, depending on the way it was cut from the log. The use of green lumber in construction results in cracked ceilings and walls, squeaking floors, sticking doors, and many other problems caused by shrinking and warping of the lumber as it dries. Therefore, lumber must be dried to a suitable degree before it can be surfaced and used.

Green lumber is also subject to decay. Decay is caused by *fungi,* low forms of plant life that feed on wood. This decay is commonly known as dry rot because it usually is not discovered until the lumber has dried. Decay will not occur unless wood moisture content is in excess of 19 percent. Wood construction maintained at moisture content of less than 20 percent will not decay. It is important that lumber with an excess amount of moisture be exposed to an environment that will allow the moisture to evaporate. Seasoned lumber must be protected to prevent the entrance of moisture that allows the growth of fungi and decay of the wood. (The subject of fungi and wood decay is discussed in detail in Chapter 33.)

Moisture Content

The **moisture content** (MC) of lumber is expressed as a percentage, and indicates how much of the weight of a wood sample is actually water. It is derived by determining the difference in the weight of the sample before and after it has been oven-dried and dividing that number by the dry weight. For example, if a wood sample weighs 16 ounces prior to drying and 13 ounces after drying, we assume that there were $16 - 13 = 3$ ounces of water in the wood. To determine the moisture content of the sample before drying, we divide the evaporated water weight by the dry weight and multiply by one hundred: $(3 \div 13) \times 100 = 23.0769$ percent. Thus, the moisture content of the sample before drying was roughly 23 percent. Lumber used for framing and exterior finish should have an MC that does not exceed 19 percent, preferably 15 percent. For interior finish, an MC of 10 to 12 percent is recommended.

Green lumber has water in the hollow part of the wood cells as well as in the cell walls. When wood starts to dry, the water in the cell cavities, called *free water,* is first removed. When all of the free water is gone, the wood has reached the **fiber-saturation point;** approximately 30 percent MC. No noticeable shrinkage of wood takes place up to this point.

As wood continues to dry, the water in the walls of the cells is removed and the wood starts to shrink. It shrinks considerably from its size at the fiber-saturation point to its size at the desired MC of less than 19 percent. The actual shrinkage of the lumber will vary with regional climate conditions and the MC of the lumber when delivered **(Figure 2–7).** Lumber at this stage is called *dry* or **seasoned** and now must be protected from getting wet.

Lumber Size	Actual Width	Width @ 19% MC	Width @ 11% MC	Width @ 8% MC
2 × 4	3½″	3½″	3 7/16″	3⅜″
2 × 6	5½″	5½″	5⅜″	5 5/16″
2 × 8	7½″	7¼″	7⅛″	7 1/16″
2 × 10	9¼″	9¼″	9 1/16″	9″

Source: U.S. Span Book for Major Wood Species [available from Canadian Wood Council, (800)463-5091; http://www.cwc.ca/publications/US_Span_book].

FIGURE 2–7 Lumber dimensions change with the moisture content of the wood.

It is important to understand not only that wood shrinks as it dries, but also how it shrinks. Little shrinkage occurs along the length of lumber; therefore, shrinkage in that direction is not considered. Most shrinkage occurs along the length of each annular ring, with the longer rings shrinking more than the shorter ones. When viewing plain-sawed lumber in cross-section, it can be seen that the piece warps as it shrinks because of the unequal length of the annular rings. A cross-section of quarter-sawed lumber shows annular rings of equal length. Therefore, although the piece shrinks, it shrinks evenly with little warp **(Figure 2–8).** Wood warps as it dries according to the way it was cut from the tree. Cross-sectional views of the annular rings are different along the length of a piece of lumber; therefore, various kinds of warp result when lumber dries and shrinks.

When the moisture content of lumber reaches that of the surrounding air (about 10 to 12 percent MC), it is at **equilibrium moisture content.** At this point, lumber shrinks or swells only slightly with changes in the moisture content of the air. Realizing that lumber undergoes certain changes when moisture is absorbed or lost, the experienced carpenter uses techniques to deal with this characteristic of wood **(Figure 2–9).**

Drying Lumber

Lumber is either **air-dried, kiln-dried,** or a combination of both. In air-drying the lumber is stacked in piles with spacers, which are called stickers, placed between each layer to permit air to circulate through

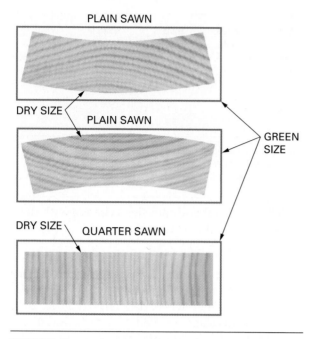

PLAIN SAWN

PLAIN SAWN

DRY SIZE

GREEN SIZE

DRY SIZE   QUARTER SAWN

FIGURE 2–8   As lumber dries, the annular rings become shorter sometimes causing wood to deform.

the pile **(Figure 2–10).** Kiln-dried lumber is stacked in the same manner but is dried in buildings, called kilns, which are like huge ovens **(Figure 2–11).**

Kilns provide carefully controlled temperatures, humidity, and air circulation to remove moisture. First, the humidity level is raised and the temperature is increased; the humidity is then gradually decreased. Kiln-drying has the advantage of drying lumber in a shorter period of time, but is more expensive than air-drying.

The recommended moisture content for lumber to be used for exterior finish at the time of installation is 12 percent, except in very dry climates, where 9 percent is recommended. Lumber with low moisture content (8 to 10 percent) is necessary for interior trim and cabinet work.

The moisture content of lumber is determined by the use of a **moisture meter (Figure 2–12).** Points on the ends of the wires of the meter are driven into the wood, and the moisture content is read off the meter.

Experienced workers know when lumber is green (because it is much heavier than dry lumber), and

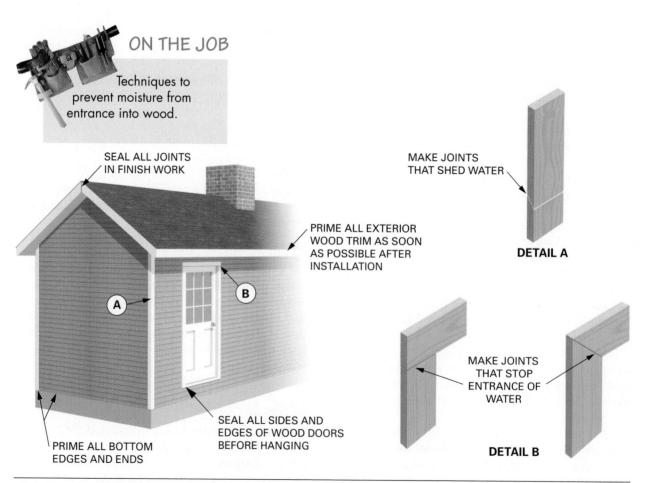

ON THE JOB

Techniques to prevent moisture from entrance into wood.

SEAL ALL JOINTS IN FINISH WORK

PRIME ALL EXTERIOR WOOD TRIM AS SOON AS POSSIBLE AFTER INSTALLATION

MAKE JOINTS THAT SHED WATER

DETAIL A

PRIME ALL BOTTOM EDGES AND ENDS

SEAL ALL SIDES AND EDGES OF WOOD DOORS BEFORE HANGING

MAKE JOINTS THAT STOP ENTRANCE OF WATER

DETAIL B

FIGURE 2–9   Techniques to prevent water from getting in behind the wood surface.

FIGURE 2–10 Lumber is stickered and stacked to allow air drying. *(Courtesy of Northwest Hardwoods, a Weyerhaeuser Business)*

FIGURE 2–11 Lumber is placed in a kiln to reduce moisture content. *(Courtesy of American Wood Dryers)*

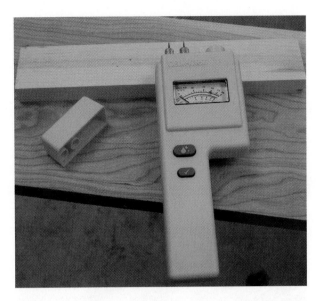

FIGURE 2–12 A moisture meter is used to determine moisture content in wood.

they can estimate fairly accurately the moisture content of lumber simply by lifting it.

Lumber is brought to the planer mill, where it is straightened, smoothed, and uniformly sized. This process can be done when the lumber is dry or green. Most construction lumber is surfaced on four sides (S4S) to standard thicknesses and widths. Some may be surfaced on only two sides (S2S) to required thicknesses.

Lumber Storage

Lumber should be delivered to the job site so materials are accessible in the proper sequence; that is, those that are to be used first are on the top and those to be used last are on the bottom.

Lumber stored at the job site should be adequately protected from moisture and other hazards. A common practice that must be avoided is placing unprotected lumber directly on the ground. Use short lengths of lumber running at right angles to the length of the pile and spaced close enough to keep the pile from sagging and coming into contact with the ground. The base on which the lumber is to be placed should be fairly level to keep the pile from falling over.

Protect the lumber with a tarp or other type of cover. Leave enough room at the bottom and top of the pile for circulation of air. A cover that reaches to the ground will act like a greenhouse, trapping ground moisture within the stack.

Keep the piles in good order. Lumber spread out in a disorderly fashion can cause accidents as well as subject the lumber to stresses that may cause warping.

LUMBER DEFECTS

A defect in lumber is any fault that detracts from its appearance, function, or strength. One type of defect is called a **warp.** Warps are caused by, among other things, drying lumber too fast, careless handling and storage, or surfacing the lumber before it is thoroughly dry. Warps are classified as **crooks, bows, cups,** and **twists (Figure 2–13).**

Splits in the end of lumber running lengthwise and across the annular rings are called **checks (Figure 2–14).** Checks are caused by faster drying of the end than of the rest of the stock. Checks can be prevented to a degree by sealing the ends of lumber with paint,

A

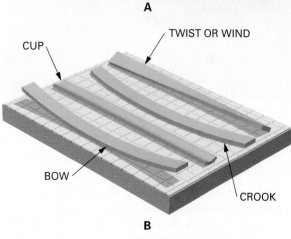

B

FIGURE 2–13   Board deformations have names that depend on the type of warp.

wax, or other material during the drying period. Cracks that run parallel to and between the annular rings are called **shakes** and may be caused by weather or other damage to the tree.

The *pith* is the spongy center of the tree. It contains the youngest portion of the lumber, called **juvenile wood.** Juvenile wood is the portion of wood that contains the first seven to fifteen growth rings. The wood cells in this region are not well aligned and are therefore unstable when they dry. They shrink in different directions, causing internal stresses. If a board has a high percentage of juvenile wood, it will warp and twist in remarkable ways. **Knots** are cross-sections of branches in the trunk of the tree. Knots are not necessarily defects unless they are loose or weaken the piece. **Pitch pockets** are small cavities that hold pitch, which sometimes oozes out. A **wane** is bark on the edge of lumber or the surface from which the bark has fallen. *Pecky* wood has small grooves or channels running with the grain. This is common in cypress. Pecky cypress is often used as an interior wall paneling when that effect is desired. Some other defects are *stains, decay,* and *wormholes.*

LUMBER GRADES AND SIZES

Lumber grades and sizes are established by wood products associations of which many wood mills are members. Member wood mills are closely supervised by the associations to ensure that standards are maintained. The **grade** stamp of the association is assurance that lumber grade standards have been met.

Member mills use the association grade stamp to indicate strict quality control. A typical grade stamp is shown in **Figure 2–15** and shows the association trademark, the mill number, the lumber grade, the species of wood, and whether the wood was green or dry when it was planed.

FIGURE 2–14   Examples of common wood defects.

— Twist

— Cup

— Heartwood Stain

— Pith & check

— Wane

— Rot

— Check

A) ASSOCIATION'S TRADEMARK (WESTERN WOOD PRODUCTS ASSN)

B) MILL NUMBER

C) GRADE OF LUMBER (IN THIS CASE IT IS STANDARD)

D) KIND OF WOOD (DOUGLAS FIR)

E) MOISTURE CONTENT (S-DRY STANDS FOR 19 PERCENT M.C.)

FIGURE 2–15   Typical softwood lumber grade stamp. *(Courtesy of Western Wood Products Association)*

Softwood Grades

The largest softwood association of lumber manufacturers is the Western Wood Products Association (WWPA), which grades lumber in three categories: **boards** (under 2 inches thick), **dimension** (2 to 4 inches thick), and **timbers** (5 inches and thicker). The board group is divided into boards, sheathing, and form lumber. The dimension group is divided into light framing, studs, structural light framing, and structural joists and planks. Timbers are divided into beams and stringers. The three main categories are further classified according to strength and appearance as shown in **Figure 2–16.**

Hardwood Grades

Hardwood grades are established by the National Hardwood Lumber Association. **Firsts and seconds** *(FAS)* is the best grade of hardwood and must yield about 85 percent clear cutting. Each piece must be at least 6 inches wide and 8 feet long. The next best grade is called *select.* For this, the minimum width is 4 inches and the minimum length is 6 feet. **No. 1 common** allows even narrower widths and shorter lengths, with about 65 percent clear cutting.

Note that the price and quality of lumber ar often linked. Some large home centers sell lumber at lower prices because it is of a lower quality. Understanding the information on lumber grade stamps will help the builder wisely purchase material.

Lumber Sizes

Rough lumber that comes directly from the sawmill is close in size to what it is called, **nominal size.** There are slight variations to nominal size because of the heavy machinery used to cut the log into lumber. When rough lumber is planed, it is reduced in thickness and width to standard and uniform sizes. Its nominal size does not change even though the actual size does. Therefore, when *dressed* (surfaced), although a piece may be called a 2 × 4, its actual size is 1½ inches (38 mm) by 3½ inches (89 mm). The same applies to all surfaced lumber; the nominal size (what it is called) and the actual size are not the same. **Figure 2–17** shows the standard nominal and dressed sizes of softwood lumber, based on WWPA rules. Hardwood lumber is usually purchased in the rough and straightened, smoothed, and sized as needed.

BOARD MEASURE

Softwood lumber is usually purchased by specifying the number of pieces—thickness(") × width(") × length(') (i.e., 35–2" × 6" × 16')—in addition to the grade. This is referred to as material list form. Often, when no particular lengths are required, the thickness, width, and total number of linear feet (length in feet) are ordered. The length of the pieces then may vary and are called *random* lengths. Another method of purchasing softwood lumber is by specifying the thickness, width, and total number of *board feet.* Lumber purchased in this manner may also contain random lengths.

Hardwood lumber is purchased by specifying the grade, thickness, and total number of board feet. Large quantities of both softwood and hardwood lumber are priced and sold by the board foot.

A **board foot** is a measure of lumber volume. It is defined as the volume of wood equivalent to a piece of wood that measures 1-inch thick by 1-foot wide by 1-foot long. It allows for a comparison of different size pieces. For example, the volume of wood in a 2 × 6 that is 1 foot long is also one board foot **(Figure 2–18).**

To calculate board feet, use the following formula: number of pieces × thickness in inches × width in inches × length in feet ÷ 12 = number of board feet. This formula uses the dimensions used to identify the board, for example 2 × 6–12 feet. Always use the nominal dimensions, not the actual dimensions; for example, 16 pieces of 2 × 4–8 feet is 85⅓ board feet (bd ft).

$$\frac{\overset{4}{\cancel{16}} \times 2 \times 4 \times 8}{\underset{3}{\cancel{12}}} = \frac{256}{3} = 85\tfrac{1}{3} \text{ bd ft}$$

Note that this formula seems to ignore the basic rules of arithmetic by multiplying feet and inches together. This is because it calculates board feet, not cubic feet or cubic inches.

Grade Selector Charts

Boards

HIGHEST QUALITY APPEARANCE GRADES	**SELECTS**	B & BETTER (IWP—SUPREME)* C SELECT (IWP—CHOICE) D SELECT (IWP—QUALITY)	
	FINISH	SUPERIOR PRIME E	
	PANELING	CLEAR (ANY SELECT OR FINISH GRADE) NO. 2 COMMON SELECTED FOR KNOTTY PANELING NO. 3 COMMON SELECTED FOR KNOTTY PANELING	
	SIDING (Bevel, Bungalow)	SUPERIOR PRIME	

Specification Checklist

☐ Grades listed in order of quantity.
☐ Include all species suited to project.
☐ Specify lowest grade that will satisfy job requirement.
☐ Specify surface texture desired.
☐ Specify moisture content suited to project.
☐ Specify Ⓦ grade stamp. For finish and exposed pieces, specify stamp on back or ends.

BOARDS SHEATHING & FORM LUMBER	NO. 1 COMMON (IWP—COLONIAL) NO. 2 COMMON (IWP—STERLING) NO. 3 COMMON (IWP—STANDARD) NO. 4 COMMON (IWP—UTILITY) NO. 5 COMMON (IWP—INDUSTRIAL)
	ALTERNATE BOARD GRADES SELECT MERCHANTABLE CONSTRUCTION STANDARD UTILITY ECONOMY

Western Red Cedar

FINISH PANELING AND CEILING	CLEAR HEART A B
BEVEL SIDING	CLEAR—V.G. HEART A—BEVEL SIDING B—BEVEL SIDING C—BEVEL SIDING

*Idaho White Pine (IWP) carries its own comparable grade designations.

Dimension/All Species   2″ to 4″ thick (also applies to finger-jointed stock)

STRUCTURAL LIGHT FRAMING 2″ to 4″ Thick 2″ to 4″ Wide	SELECT STRUCTURAL NO. 1 NO. 2 NO. 3	These grades are designed to fit those engineering applications where higher bending strength ratios are needed in light framing sizes. Typical uses would be for trusses, concrete pier wall forms, etc.
LIGHT FRAMING 2″ to 4″ Thick 2″ to 4″ Wide	CONSTRUCTION STANDARD UTILITY	This category for use where high strength values are NOT required; such as studs, plates, sills, cripples, blocking, etc.
STUDS 2″ to 4″ Thick 2″ and Wider	STUD	An optional all-purpose grade. Characteristics affecting strength and stiffness values are limited so that the "Stud" grade is suitable for vertical framing members, including load bearing walls.
STRUCTURAL JOISTS & PLANKS 2″ to 4″ Thick 5″ and Wider	SELECT STRUCTURAL NO. 1 NO. 2 NO. 3	These grades are designed especially to fit in engineering applications for lumber five inches and wider, such as joists, rafters and general framing uses.

Timbers   5″ and thicker

BEAMS & STRINGERS 5″ and thicker Width more than 2″ greater than thickness	DENSE SELECT STRUCTURAL* DENSE NO. 1* DENSE NO. 2* SELECT STRUCTURAL NO. 1 NO. 2** NO. 3**	**POSTS & TIMBERS** 5″ × 5″ and larger Width not more than 2″ greater than thickness	DENSE SELECT STRUCTURAL * DENSE NO. 1* DENSE NO. 2* SELECT STRUCTURAL NO. 1 NO. 2** NO. 3**

*Douglas Fir or Douglas Fir–Larch only.

FIGURE 2–16   Softwood lumber grades. (*Courtesy of Western Wood Products Association*)

Standard Lumber Sizes/Nominal, Dressed, Based on WWPA Rules

ProductDescription	Thickness In.	Nominal Size Width In.	Nominal Size Surfaced Dry	Dressed Dimensions Thicknesses and Widths In. Surfaced Unseasoned	Dressed Dimensions Lengths Ft.	
DIMENSION	S4S . . . . Other surface combinations are available. See "Abbreviations" below.	2 3 4	2 3 4 5 6 8 10 12 Over 12	1½ 2½ 3½ 4½ 5½ 7½ 9½ 11½ ¾ off normal	1⁹⁄₁₆ 2⁹⁄₁₆ 3⁹⁄₁₆ 4⅝ 5⅝ 7½ 9½ 11½ Off ½	6' and longer, generally shipped in multiples of 2'
SCAFFOLD PLANK	Rough Full Sawn or S4S . . . . (Usually shipped unseasoned)	1¼ & Thicker	8 and Wider	If Dressed refer to "DIMENSION" sizes		6' and longer, generally shipped in multiples of 2'
TIMBERS	Rough or S4S . . . . (Shipped unseasoned)	5 and Larger		½ off nominal (S4S) See 3.20 of WWPA Grading Rules for Rough		6' and longer, generally shipped in multiples of 2'

		Nominal Size Thickness In.	Nominal Size Width In.	Dressed Dimensions Thickness In.	Dressed Dimensions Width In.	Lengths Ft.
DECKING	2" Single T&G	1	5 6 8 10 12	1½	4 5 6¾ 8¾ 10¾	6' and longer, generally shipped in multiples of 2'
	3" and 4" Double T&G	3 4	6	2½ 3½	5¼	
FLOORING	(D & M), (S2S & CM)	⅜ ½ ⅝ 1 1¼ 1½	2 3 4 5 6	⁵⁄₁₆ ⁷⁄₁₆ ⁹⁄₁₆ ¾ 1 1¼	1⅛ 2⅛ 3⅛ 4⅛ 5⅛	4' and longer, generally shipped in multiples of 2'
CEILING AND PARTITION	(S2S & CM)	⅜ ½ ⅝ ¾	3 4 5 6	⁵⁄₁₆ ⁷⁄₁₆ ⁹⁄₁₆ ¹¹⁄₁₆	2⅛ 3⅛ 4⅛ 5⅛	4' and longer, generally shipped in multiples of 2'
FACTORY AND SHOP LUMBER	S2S	1 (4/4) 1¼ (5/4) 1½ (5/4) 1¾ (7/4) 2 (8/4) 2½ (10/4) 3 (12/4) 4 (16/4)	5" and wider (except 4" and wider in 4/4 No. 1 Shop and 4/4 No. 2 Shop, and 2" and wider in 5/4 & Thicker No. 3 Shop)	¾ (4/4) 1⁵⁄₃₂ (4/4) 1¹³⁄₃₂ (6/4) 1¹⁹⁄₃₂ (7/4) 1¹³⁄₁₆ (8/4) 2⅜ (10/4) 2¾ (12/4) 3¾ (16/4)	Usually sold random width	6' and longer, generally shipped in multiples of 2'

Abbreviations

Abbreviated descriptions appearing in the size table are explained below.
S1S—Surfaced one side.
S2S—Surfaced two sides.

S4S—Surfaced four sides.
S1S1E—Surfaced one side, one edge.
S1S2E—Surfaced one side, two edges.
CM—Center matched.

D & M—Dressed and matched.
T & G—Tongue and grooved.
Rough Full Sawn—Unsurfaced green lumber cut to full specified size.

Product Classification

	thickness in.	width in.		thickness in.	width in.
board lumber	1"	2" or more	beams & stringers	5" and thicker	more than 2" greater than thickness
light framing	2" to 4"	2" to 4"	posts & timbers	5" × 5" and larger	not more than 2" greater than thickness
studs	2" to 4"	2" to 6" 10' and shorter	decking	2" to 4"	4" to 12" wide
structural light framing	2" to 4"	2" to 4"	siding	thickness expressed by dimension of butt edge	
structural joists & planks	2" to 4"	5" and wider	mouldings	size at thickest and widest points	

Nailing Diagram

BOARD ON BOARD — Over 8" wide, use 2 nails 3-4" apart

TONGUE AND GROOVE — Over 8" wide, use 2 nails 3-4" apart

BOARD AND BATTEN — Over 8" Wide use 2 nails 3-4" apart in center

CHANNEL RUSTIC — Over 8" wide, use 2 nails 3-4" apart at exposed edge

Lengths of lumber generally are 6 feet and longer in multiples of 2'

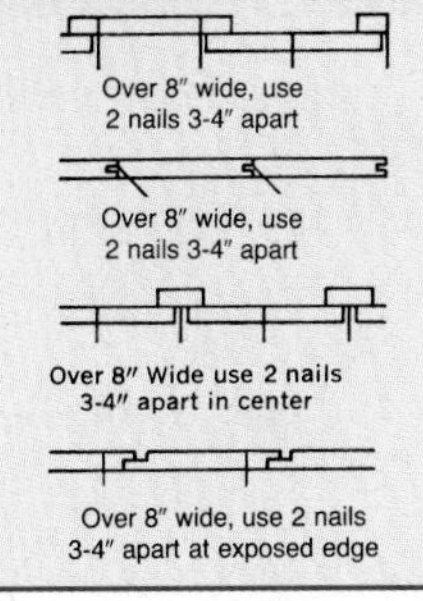

FIGURE 2–17 Softwood lumber sizes. (*Courtesy of Western Wood Products Association*)

BOARD FEET = NUMBER OF PIECES X THICKNESS" X WIDTH" X LENGTH' ÷ 12

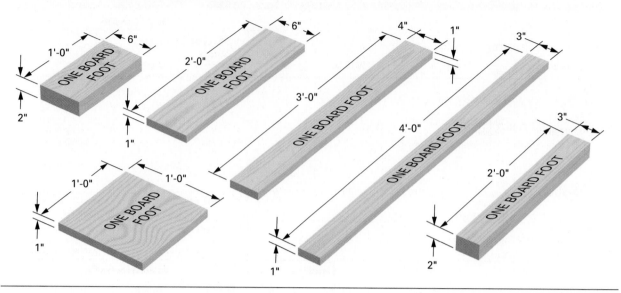

FIGURE 2–18 One board foot of lumber is a volume of wood; it can have many shapes.

Key Terms

air-dried lumber	equilibrium moisture content	lignin	quarter-sawed
annular rings		lumber	sapwood
board foot	fiber-saturation point	lumber grades	sawyer
boards	firsts and seconds	medullary rays	seasoned
bows	grade	moisture content	shakes
cambium layer	green lumber	moisture meter	softwood
close-grained	hardwood	nominal size	stickering
coniferous	heartwood	No. 1 common	timbers
crooks	juvenile wood	open-grained	twists
cups	kiln-dried	pitch pocket	wane
deciduous	knots	pith	warp
dimension lumber		plain-sawed	

Review Questions

Select the most appropriate answer.

1. The center of a tree in cross-section is called the
   - **a.** heartwood.
   - **b.** lignin.
   - **c.** pith.
   - **d.** sapwood.

2. New wood cells of a tree are formed in the
   - **a.** heartwood.
   - **b.** bark.
   - **c.** medullary rays.
   - **d.** cambium layer.

3. Tree growth is faster in the
   - **a.** spring.
   - **b.** summer.
   - **c.** fall.
   - **d.** winter.

4. Quarter-sawed lumber is sometimes called
   - **a.** tangent-grained.
   - **b.** slash-sawed.
   - **c.** vertical-grained.
   - **d.** plain-sawed.

5. Lumber is called "green" when
   - **a.** it is stained by fungi.
   - **b.** the tree is still standing.
   - **c.** it is first cut from the log.
   - **d.** it has reached equilibrium moisture content.

6. Wood will not decay unless its moisture content is in excess of
   - **a.** 15 percent.
   - **b.** 19 percent.
   - **c.** 25 percent.
   - **d.** 30 percent.

7. When all of the free water in the cell cavities of wood is removed and before water is removed from cell walls, lumber is at the
   - **a.** fiber-saturation point.
   - **b.** 30 percent moisture content.
   - **c.** equilibrium moisture content.
   - **d.** shrinkage commencement point.

8. Air-dried lumber cannot be dried to less than
   - **a.** 8 to 10 percent moisture content.
   - **b.** equilibrium moisture content.
   - **c.** 25 to 30 percent moisture content.
   - **d.** its fiber-saturation point.

9. A commonly used and abundant softwood is
   - **a.** ash.
   - **b.** fir.
   - **c.** basswood.
   - **d.** birch.

10. The wood species most resistant to decay is
    - **a.** pine.
    - **b.** spruce.
    - **c.** cypress.
    - **d.** hemlock.

UNIT 2

Engineered Panels

Growing concern over efficient use of forest resources led to the development of reconstituted wood. The resulting wood products are referred to as engineered lumber. One form of engineered lumber is manufactured from peeled logs and reconstituted wood into large sheets referred to as engineered panels, commonly called plywood. The tradesperson must know the different kinds, sizes and recommended applications to use these materials to the best advantage.

Plywood is a general term used to cover a variety of materials. A carpenter must understand the uses for the various materials to make optimum use of the engineered panel.

OBJECTIVES

After completing this unit, the student should be able to describe the composition, kinds, sizes, grades, and several uses of:

- plywood
- oriented strand board
- composite panels
- particleboard
- hardboard
- medium-density fiberboard
- softboard

3   Rated Plywood and Panels

The term **engineered panels** refers to man-made products in the form of large reconstituted wood sheets, sometimes called **panels.** In some cases, the tree has been taken apart and its contents have been redistributed into sheet or panel form. The panels are widely used in the construction industry. They are also used in the aircraft, automobile, and boat-building industries, as well as in the making of road signs, furniture, and cabinets.

With the use of engineered panels, construction progresses at a faster rate because a greater area is covered in a shorter period of time **(Figure 3–1).** These panels, in certain cases, present a more attractive appearance and give more protection to a surface than does solid lumber. It is important to know the kinds and uses of various engineered panels in order to use them to the best advantage.

APA-RATED PANELS

Many mills belong to associations that inspect, test, and allow mills to stamp the product to certify that it conforms to government and industrial standards. The grade stamp assures the consumer that the product has met the rigid quality and performance requirements of the association.

The trademarks of the largest association of this type, the Engineered Wood Association (formerly called the American Plywood Association and still know by the acronym APA), appear only on products manufactured by APA member mills **(Figure 3–2).** This association is concerned not only with quality supervision and testing of **plywood** (cross-laminated wood veneer) but also of **composites** (veneer faces bonded to reconstituted wood cores) and nonveneered panels, commonly called **oriented strand board (OSB) (Figure 3–3).**

Plywood

One of the most extensively used engineered panels is plywood. Plywood is a sandwich of wood. Most plywood panels are made up of sheets of veneer (thin pieces) called **plies.** These plies, arranged in layers, are bonded under pressure with glue to form a very strong panel. The plies are glued together so that the grain of each layer is at right angles to the next one. This cross-graining results in a sheet that is as strong as or stronger than the wood it is made from. Plywood usually contains an odd number of layers so that the face grain on both sides of the sheet runs in the direction of the long dimension of the panel **(Figure 3–4).** Softwood plywood is commonly made with three, five, or seven layers. Because of its construction, plywood is more stable with changes of humidity

FIGURE 3–1   Sheet material covers a greater area in a shorter period of time than does solid lumber.

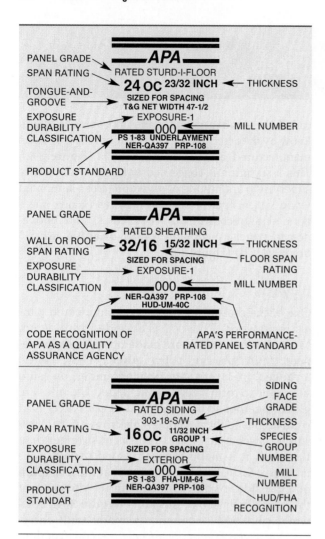

PANEL GRADE → *APA*
SPAN RATING → RATED STURD-I-FLOOR
TONGUE-AND- → **24 OC** 23/32 INCH ← THICKNESS
GROOVE → SIZED FOR SPACING
T&G NET WIDTH 47-1/2
EXPOSURE → EXPOSURE-1
DURABILITY → 000 ← MILL NUMBER
CLASSIFICATION → PS 1-83 UNDERLAYMENT
NER-QA397 PRP-108
PRODUCT STANDARD

PANEL GRADE → *APA*
RATED SHEATHING
WALL OR ROOF → **32/16** 15/32 INCH ← THICKNESS
SPAN RATING → SIZED FOR SPACING ← FLOOR SPAN RATING
EXPOSURE → EXPOSURE-1
DURABILITY → 000 ← MILL NUMBER
CLASSIFICATION → NER-QA397 PRP-108
HUD-UM-40C
CODE RECOGNITION OF   APA'S PERFORMANCE-
APA AS A QUALITY   RATED PANEL STANDARD
ASSURANCE AGENCY

PANEL GRADE → *APA* SIDING FACE GRADE
RATED SIDING
303-18-S/W
SPAN RATING → **16 OC** 11/32 INCH ← THICKNESS
GROUP 1 ← SPECIES GROUP NUMBER
EXPOSURE → SIZED FOR SPACING
DURABILITY → EXTERIOR
CLASSIFICATION → 000 ← MILL NUMBER
PRODUCT → PS 1-83 FHA-UM-64
STANDAR → NER-QA397 PRP-108 ← HUD/FHA RECOGNITION

FIGURE 3–2 The grade stamp is assurance of a high-quality, performance-rated panel. *(Courtesy of APA–The Engineered Wood Association)*

FIGURE 3–3 APA performance-rated panels. *(Courtesy of APA–The Engineered Wood Association)*

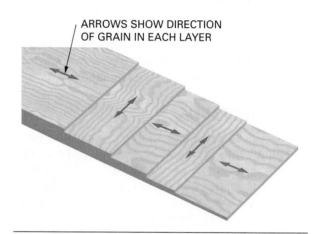

ARROWS SHOW DIRECTION OF GRAIN IN EACH LAYER

FIGURE 3–4 In the construction of plywood, the grains of veneer plies are placed at right angles to each other.

FIGURE 3–5 The veneer is peeled from the log like paper unwinding from a roll. *(Courtesy of APA–The Engineered Wood Association)*

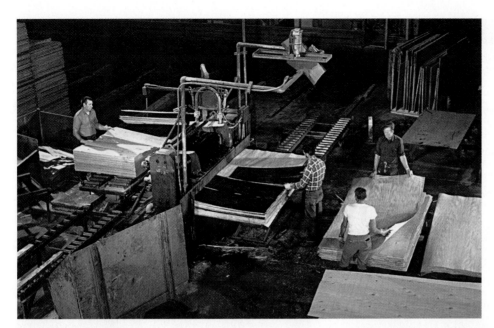

FIGURE 3–6 Gluing and assembling plywood veneers into panels. *(Courtesy of APA– The Engineered Wood Association)*

and is more resistant to shrinking and swelling than wood boards.

Manufacture of Veneer Core Plywood. Specially selected "peeler logs" are mounted on a huge lathe in which the log is rotated against a sharp knife. As the log turns, a thin layer is peeled off like paper unwinding from a roll **(Figure 3–5)**. The entire log is used. The small remaining spindles are utilized for making other wood products.

The long ribbon of veneer is then cut into desired widths, sorted, and dried to a moisture content of 5 percent. After drying, the veneers are fed through glue spreaders that coat them with a uniform thickness. The veneers are then assembled to make panels **(Figure 3–6)**. Large presses bond the assembly under controlled heat and pressure. From the presses, the panels are either left unsanded, touch-sanded or smooth-sanded, then cut to size, inspected, and stamped.

Veneer Grades. In declining order, the letters *A, B, C plugged, C,* and *D* are used to indicate the appearance quality of panel veneers. Two letters are found in the grade stamp of veneered panels. One letter indicates the quality of one face, while the other letter indicates the quality of the opposite face. The exact description of these letter grades is shown in **Figure 3–7**. Panels with B-grade or better veneer faces are always sanded smooth. Some panels, such as APA-Rated Sheathing, are unsanded because their intended use does not require sanding. Other panels used for such purposes as subflooring and underlayment require only a touch-sanding to make the panel thickness more uniform.

Veneer Grades	
A	Smooth, paintable. Not more than 18 neatly made repairs, boat, sled, or router type, and parallel to grain, permitted. Wood or synthetic repairs permitted. May be used for natural finish in less demanding applications.
B	Solid surface. Shims, sled or router repairs, and tight knots to 1 inch across grain permitted. Wood or synthetic repairs permitted. Some minor splits permitted.
C Plugged	Improved C veneer with splits limited to ⅛-inch width and knotholes or other open defects limited to ¼ × ½ inch. Wood or synthetic repairs permitted. Admits some broken grain.
C	Tight knots to 1½ inch. Knotholes to 1 inch across grain and some to 1½ inch if total width of knots and knotholes is within specified limits. Synthetic or wood repairs, discoloration and sanding defects that do not impair strength permitted. Limited splits allowed. Stiching permitted.
D	Knots and knotholes to 2½ inch width across grain and ½ inch larger within specified limits. Limited splits are permitted. Stiching permitted. Limited to Exposure 1 or interior panels.

FIGURE 3–7 Veneer letter grades define veneer appearance. *(Courtesy of APA–The Engineered Wood Association)*

Strength Grades. Softwood veneers are made of many different kinds of wood. These woods are classified in groups according to their strength **(Figure 3–8)**. Group 1 is the strongest. Douglas fir and southern pine are in Group 1 and are used to make most of the softwood plywood. The group number is also shown in the grade stamp.

Classification of Species

Group 1	Group 2	Group 3	Group 4	Group 5
Apitong	Cedar	Alder	Aspen	Basswood
Beech	Port	Red	Bigtooth	Poplar,
American	Orford	Birch	Quaking	balsam
Birch	Cypress	Paper	Cativo	
Sweet	Douglas	Cedar	Cedar	
Yellow	fir 2[a]	Alaska	Incense	
Douglas	Fir	Fir	Western	
fir 1[a]	Balsam	Subalpine	red	
Kapur	California	Hemlock,	Cottonwood	
Keruing	red	Eastern	Eastern	
Larch,	Grand	Maple	Black	
Western	Noble	Bigleaf	(Western	
Maple, sugar	Pacific	Pine	poplar)	
Pine	silver	Jack	Pine	
Caribbean	White	Lodgepole	Eastern	
Ocote	Hemlock,	Ponderosa	white	
Pine	Western	Spruce	Sugar	
South	Lauan	Redwood		
Loblolly	Almon	Spruce		
Longleaf	Bagtikan	Engelmann		
Shortleaf	Mayapis	White		
Slash	Red			
Tanoak	Tangile			
	White			
	Maple, black			
	Mengkulang			
	Meranti,			
	red[b]			
	Mersawa			
	Pine			
	Pond			
	Red			
	Virginia			
	Western			
	white			
	Spruce			
	Black			
	Red			
	Sitka			
	Sweetgum			
	Tamarack			
	Yellow-			
	poplar			

(a) Douglas fir from trees grown in the states of Washington, Oregon, California, Idaho, Montana, Wyoming, and the Canadian Provinces of Alberta and British Columbia shall be classed as Douglas Fir No. 1. Douglas Fir from trees grown in the states of Nevada, Utah, Colorado, Arizona, and New Mexico shall be classed as Douglas Fir No. 2.

(b) Red meranti shall be limited to species having a specific gravity of 0.41 or more based on green volume and oven dry weight.

FIGURE 3–8   Plywood is classified into five groups according to strength and stiffness, with Group 1 being the strongest. (Courtesy of APA– The Engineered Wood Association)

Oriented Strand Board

Oriented strand board (OSB) is a nonveneered performance-rated structural panel composed of small oriented (lined up) strand-like wood pieces arranged in three to five layers with each layer at right angles to the other **(Figure 3–9)**. The cross-lamination of the layers achieves the same advantages of strength and stability as in plywood.

Manufacture of OSB. Logs from specially selected species are debarked and sliced into strands that are

FIGURE 3–9   Oriented strand board is often used for sheathing. (Courtesy of Louisiana Pacific Corporation)

between $\frac{25}{1000}$ and $\frac{30}{1000}$ of an inch thick, ¾ and 1 inch wide, and between 2½ and 4½ inches long. The strands are dried, loaded into a blender, coated with liquid resins, formed into a mat consisting of three or more layers of systematically oriented wood fibers, and fed into a press where, under high temperatures and pressure, they form a dense panel. As a safety measure, the panels have one side textured to help prevent slippage during installation.

Nonveneered panels have previously been sold with such names as waferboard, structural particleboard, and others. At present almost all panels manufactured with oriented strands or wafers are called oriented strand board. Various manufacturers have their own particular brand names for OSB, such as Oxboard, Aspenite, and many others.

Composite Panels

Composite panels are manufactured by bonding veneers of wood to both sides of reconstituted wood panels. More efficient use of wood is thus allowed with this product while retaining the wood grain appearance on both faces of the panel.

Composite panels rated by the American Plywood Association are called *COM-PLY* and are manufactured in three or five layers. A three-layer panel has a reconstituted wood core with wood veneers on both sides. A five-layer panel has a wood veneer in the center as well as on both sides.

PERFORMANCE RATINGS

A performance-rated panel meets the requirements of the panel's end use. The three end uses for which panels are rated are single-layer flooring, exterior siding, and sheathing for roofs, floors, and walls. Names given to designate end uses are *APA-Rated Sheathing, Structural I, APA-Rated Sturd-I-Floor,* and *APA-Rated Siding* **(Figure 3–10)**. Panels are tested to meet standards in areas of resistance to moisture, strength, and stability.

Exposure Durability

APA performance-rated panels are manufactured in three exposure durability classifications: *Exterior, Exposure 1,* and *Exposure 2.* Panels marked Exterior are designed for permanent exposure to the weather or moisture. Exposure 1 panels are intended for use where long delays in construction may cause the panels to be exposed to the weather before being protected. Panels marked Exposure 2 are designed for use when only moderate delays in

APA RATED SHEATHING Typical Trademark		Specially designed for subflooring and wall and roof sheathing. Also good for a broad range of other construction and industrial applications. Can be manufactured as OSB, plywood, or a composite panel. BOND CLASSIFICATIONS: Exterior, Exposure 1, Exposure 2. COMMON THICKNESSES: ⁵⁄₁₆, ⅜, ⁷⁄₁₆, ¹⁵⁄₃₂, ½, ¹⁹⁄₃₂, ⅝, ²³⁄₃₂, ¾. (a)
APA STRUCTURAL I RATED SHEATHING(c) Typical Trademark		Unsanded grade for use where shear and cross-panel strength properties are of maximum importance, such as panelized roofs and diaphragms. Can be manufactured as OSB, plywood, or a composite panel. BOND CLASSIFICATIONS: Exterior, Exposure 1. COMMON THICKNESSES: ⁵⁄₁₆, ⅜, ⁷⁄₁₆, ¹⁵⁄₃₂, ½, ¹⁹⁄₃₂, ⅝, ²³⁄₃₂, ¾.
APA RATED STURD-I-FLOOR Typical Trademark		Specially designed as combination subfloor-underlayment. Provides smooth surface for application of carpet and pad and possesses high concentrated and impact load resistance. Can be manufactured as OSB, plywood, or a composite panel. Available square edge or tongue-and-groove. BOND CLASSIFICATIONS: Exterior, Exposure 1, Exposure 2. COMMON THICKNESSES: ¹⁹⁄₃₂, ⅝, ²³⁄₃₂, ¾, 1, 1⅛.
APA RATED SIDING Typical Trademark		For exterior siding, fencing, etc. Can be manufactured as plywood, as a composite panel or as an overlaid OSB. Both panel and lap siding available. Special surface treatment such as V-groove, channel groove, deep groove (such as APA Texture 1-11), brushed, rough sawn and overlaid (MDO) with smooth- or texture-embossed face. Span Rating (stud spacing for siding qualified for APA Sturd-I-Wall applications) and face grade classification (for veneer-faced siding) indicated in trademark. BOND CLASSIFICATION: Exterior, COMMON THICKNESSES: ¹¹⁄₃₂, ⅜, ⁷⁄₁₆, ¹⁵⁄₃₂, ½, ¹⁹⁄₃₂, ⅝.

(a) Specific grades, thicknesses and bond classifications may be in limited supply in some areas. Check with your supplier before specifying.

(b) Specify Performance Rated Panels by thickness and Span Rating. Span Ratings are based on panel strength and stiffness. Since these properties are a function of panel composition and configuration as well as thickness, the same Span Rating may appear on panels of different thickness. Conversely, panels of the same thickness may be marked with different Span Ratings.

(c) All plies in Structural I plywood panels are special improved grades and panels marked PS 1 are limited to Group 1 species. Other panels marked Structural I Rated qualify through special performance testing.

FIGURE 3–10 Guide to APA Performance-Rated Panels. (*Courtesy of APA– The Engineered Wood Association*)

providing protection from the weather are expected. The exposure durability of a panel may be found in the grade stamp.

Span Ratings

The span rating in the grade stamp on APA-Rated Sheathing appears as two numbers separated by a slash, such as $^{32}/_{16}$ or $^{48}/_{24}$. The left number denotes the maximum recommended spacing of supports when the panel is used for roof or wall sheathing. The right number indicates the maximum recommended spacing of supports when the panel is used for subflooring. In both cases, the long dimension of the panel must be placed across three or more sup-

ports. A panel marked $^{32}/_{16}$, for example, may be used for roof sheathing over rafters not more than 32 inches on center, or for subflooring over joists not more than 16 inches on center.

The span ratings on APA-Rated Sturd-I-Floor and APA-Rated Siding appear as a single number. APA-Rated Sturd-I-Floor panels are designed specifically for combined subflooring-underlayment applications and are manufactured with span ratings of 16, 20, 24, and 48 inches.

APA-Rated Siding is produced with span ratings of 16 and 24 inches. The rating applies to vertical installation of the panel. All siding panels may be applied horizontally direct to studs 16 to 24 inches on center provided horizontal joints are blocked.

4   Nonstructural Panels

PLYWOOD

All the rated products discussed in the previous chapter may be used for nonstructural applications. In addition, other plywood products, grade-stamped by the APA–The Engineered Wood Association, are available for nonstructural use. They include sanded and touch-sanded plywood panels **(Figure 4–1)** and specialty plywood panels **(Figure 4–2).**

Hardwood Plywood

Plywood is available with hardwood face veneers, of which the most popular are birch, oak, and lauan. Beautifully grained hardwoods are sometimes matched in a number of ways to produce interesting face designs. Hardwood plywood is used in the interior of buildings for such things as wall paneling, built-in cabinets, and fixtures.

Particleboard

Particleboard is a reconstituted wood panel made of wood flakes, chips, sawdust, and planer shavings **(Figure 4–3).** These wood particles are mixed with an adhesive, formed into a mat, and pressed into sheet form. The kind, size, and arrangement of the wood particles determine the quality of the board.

The highest quality particleboard is made of large wood flakes in the center. The flakes become gradually smaller toward the surfaces where finer particles are found. This type of construction results in an extremely hard board with a very smooth surface. Softer and lower quality boards contain the same

size particles throughout. These boards usually have a rougher surface texture. In addition to the size, kind, and arrangement of the particles, the quality of the board is determined by the method of manufacture.

Particleboard Grades. The quality of particleboard is indicated by its density (hardness), which ranges from 28 to 55 pounds per cubic foot. Nonstructural particleboard is used in the construction industry for the construction of kitchen cabinets and countertops, and for the core of veneer doors and similar panels.

Fiberboards

Fiberboards are manufactured as *high-density, medium-density,* and *low-density* boards.

Hardboards

High-density fiberboards are called **hardboards** and are commonly known by the trademark *Masonite* regardless of the manufacturer. The hardboard industry makes almost complete use of the great natural resource of wood by utilizing the wood chips and board trimmings, which were once considered waste.

Wood chips are reduced to fibers and water is added to make a soupy pulp. The pulp flows onto a traveling mesh screen where water is drawn off to form a mat. The mat is then pressed under heat to weld the wood fibers back together by utilizing lignin, the natural adhesive in wood.

APA A-A Typical Trademark `A-A•G-1•EXPOSURE 1-APA•000•PS 1-83`		Use where appearance of both sides is important for interior applications such as built-ins, cabinets, furniture, partitions; and for exterior applications such as fences, signs, boats, shipping containers, tanks, ducts, etc. Smooth surfaces suitable for painting. EXPOSURE DURABILITY CLASSIFICATIONS: Interior, Exposure 1, Exterior. COMMON THICKNESSES: ¼, ¹¹⁄₃₂, ⅜, ¹⁵⁄₃₂, ½, ¹⁹⁄₃₂, ⅝, ²³⁄₃₂, ¾.
APA A-B Typical Trademark `A-B•G-1•EXPOSURE 1-APA•000•PS 1-83`		For use where appearance of one side is less important but where two solid surfaces are necessary. EXPOSURE DURABILITY CLASSIFICATIONS: Interior, Exposure 1, Exterior. COMMON THICKNESSES: ¼, ¹¹⁄₃₂, ⅜, ¹⁵⁄₃₂, ½, ¹⁹⁄₃₂, ⅝, ²³⁄₃₂, ¾.
APA A-C Typical Trademark	─── APA ─── **A-C** GROUP 1 EXTERIOR ─── 000 ─── PS 1-83	For use where appearance of only one side is important in exterior applications, such as soffits, fences, farm buildings, etc.[c] EXPOSURE DURABILITY CLASSIFICATION: Exterior. COMMON THICKNESSES: ¼, ¹¹⁄₃₂, ⅜, ¹⁵⁄₃₂, ½, ¹⁹⁄₃₂, ⅝, ²³⁄₃₂, ¾.
APA A-D Typical Trademark	─── APA ─── **A-D** GROUP 1 EXPOSURE 1 ─── 000 ─── PS 1-83	For use where appearance of only one side is important in interior applications, such as paneling, built-ins, shelving, partitions, flow racks, etc.[c] EXPOSURE DURABILITY CLASSIFICATIONS: Interior, Exposure 1. COMMON THICKNESSES: ¼, ¹¹⁄₃₂, ⅜, ¹⁵⁄₃₂, ⅝, ²³⁄₃₂, ¾.
APA B-B Typical Trademark `B-B•G-2•EXPOSURE 1-APA•000•PS 1-83`		Utility panels with two solid sides. EXPOSURE DURABILITY CLASSIFICATIONS: Interior, Exposure 1, Exterior. COMMON THICKNESSES: ¼, ¹¹⁄₃₂, ⅜, ¹⁵⁄₃₂, ½, ¹⁹⁄₃₂, ⅝, ²³⁄₃₂, ¾.
APA B-C Typical Trademark	─── APA ─── **B-C** GROUP 1 EXTERIOR ─── 000 ─── PS 1-83	Utility panel for farm service and work buildings, boxcar and truck linings, containers, tanks, agricultural equipment, as a base for exterior coatings and other exterior uses or applications subject to high or continuous moisture.[c] EXPOSURE DURABILITY CLASSIFICATION: Exterior. COMMON THICKNESSES: ¼, ¹¹⁄₃₂, ⅜, ¹⁵⁄₃₂, ½, ¹⁹⁄₃₂, ⅝, ²³⁄₃₂, ¾.
APA B-D Typical Trademark	─── APA ─── **B-D** GROUP 2 EXPOSURE 1 ─── 000 ─── PS 1-83	Utility panel for backing, sides of built-ins, industry shelving, slip sheets, separator boards, bins and other interior or protected applications.[c] EXPOSURE DURABILITY CLASSIFICATIONS: Interior, Exposure 1. COMMON THICKNESSES: ¼, ¹¹⁄₃₂, ⅜, ¹⁵⁄₃₂, ½, ¹⁹⁄₃₂, ⅝, ²³⁄₃₂, ¾.
APA **Underlayment** Typical Trademark	─── APA ─── UNDERLAYMENT GROUP 1 EXPOSURE 1 ─── 000 ─── PS 1-83	For application over structural subfloor. Provides smooth surface for application of carpet and pad and possesses high concentrated and impact load resistance. For areas to be covered with resilient flooring, specify panels with "sanded face."[b] EXPOSURE DURABILITY CLASSIFICATIONS: Interior, Exposure 1. COMMON THICKNESSES[a]: ¼, ¹¹⁄₃₂, ⅜, ¹⁵⁄₃₂, ½, ¹⁹⁄₃₂, ⅝, ²³⁄₃₂, ¾.

FIGURE 4–1 Guide to APA sanded and touched-sanded plywood panels. *(Continued)*
(Courtesy of APA–The Engineered Wood Association)

| APA
C-C Plugged[d]
Typical Trademark | ——**APA**——
C-C PLUGGED
GROUP 2
EXTERIOR
000
PS 1-83 | For use as an underlayment over structural subfloor, refrigerated or controlled-atmosphere storage rooms, pallet fruit bins, tanks, boxcar and truck floors and linings, open soffits, and other similar applications where continuous or severe moisture may be present. Provides smooth surface for application of carpet and pad and possesses high concentrated and impact load resistance. For areas to be covered with resilient flooring, specify panels with "sanded face."[b] EXPOSURE DURABILITY CLASSIFICATION: Exterior. COMMON THICKNESSES:[a] $^{11}\!/_{32}$, $\%$, $^{15}\!/_{32}$, $\frac{1}{2}$, $^{19}\!/_{32}$, $\%$, $^{23}\!/_{32}$, $\frac{3}{4}$. |
| APA
C-D Plugged
Typical Trademark | ——**APA**——
C-D PLUGGED
GROUP 2
EXPOSURE1
000
PS 1-83 | For open soffits, built-ins, cable reels, separator boards, and other interior or protected applications. Not substitute for Underlayment or APA Rated Sturd-I-Floor as it lacks their puncture resistance. EXPOSURE DURABILITY CLASSIFICATIONS: Interior, Exposure 1. COMMON THICKNESSES: $\%$, $^{15}\!/_{32}$, $\frac{1}{2}$, $^{19}\!/_{32}$, $\%$, $^{23}\!/_{32}$, $\frac{3}{4}$. |

(a) Panels 1/2 inch and thicker are Span Rated and do not contain species group number in trademark.

(b) Also available in Underlayment A-C or Underlayment B-C grades, marked either "touch sanded" or "sanded face."

(c) For nonstructural floor underlayment, or other applications requiring improved inner ply construction, specify panels market either "plugged inner plies" (May also be designated plugged crossbands under face or plugged crossbands or core); or "meets underlayment requirements."

(d) Also may be designated APA Underlayment C-C Plugged.

FIGURE 4–1 *(Continued)*

Some panels are **tempered** (coated with oil and baked to increase hardness, strength, and water resistance). Carbide-tipped saws trim the panels to standard sizes.

Sizes of Hardboard.
The most popular thicknesses of hardboard range from $\frac{1}{8}$ to $\%$ inch. The most popular sheet size is 4 feet by 8 feet, although sheets can be ordered in practically any size.

Classes and Kinds of Hardboard.
Hardboard is available in three different classes: tempered, standard, and service tempered **(Figure 4–4).** It may be obtained smooth-one-side (S1S) or smooth-two-sides (S2S). Hardboard is available in many forms, such as perforated, grooved, and striated.

Uses of Hardboard.
Hardboard may be used inside or outside. It is used for exterior siding and interior wall paneling. It is also used extensively for cabinet backs and drawer bottoms. It can be used wherever a dense, hard panel is required. Because of the composition of hardboard, it is important to seal all sides and edges in exterior applications.

Because it is a wood-based product, hardboard can be sawed, routed, shaped, and drilled with standard woodworking tools. It can be securely fastened with glue, screws, staples, or nails.

Medium-Density Fiberboard

Medium-density fiberboard (MDF) is manufactured in a manner similar to that used to make hardboard except that the fibers are not pressed as tightly together. The refined fiber produces a fine-textured, homogeneous board with an exceptionally smooth surface. Densities range from 28 to 65 pounds. It is available in thicknesses ranging from $\%_{6}$ to $1\frac{1}{2}$ inches and comes in widths of 4 and 5 feet. Lengths run from 6 to 18 feet.

MDF can be used for case goods, drawer parts, kitchen cabinets, cabinet doors, signs, and some interior wall finish.

Softboard

Low-density fiberboard is called **softboard.** Softboard is very light and contains many tiny air spaces because the particles are not compressed tightly.

The most common thicknesses range from $\frac{1}{2}$ to 1 inch. The most common sheet size is 4 feet by 8 feet, although many sizes are available.

Uses of Softboard.
Because of their lightness, softboard panels are used primarily for insulating or sound control purposes. They are used extensively as decorative panels in suspended ceilings and as ceiling tiles **(Figure 4–5).** Softboard can be used for exterior wall sheathing. This type may be coated or

APA **DECORATIVE** Typical Trademark	──APA── DECORATIVE GROUP 2 INTERIOR ── 000 ── PS 1-83	Rough-sawn, brushed, grooved, or striated faces. For paneling, interior accent walls, built-ins, counter facing, exhibit displays. Can also be made by some manufacturers in Exterior for exterior siding, gable ends, fences and other exterior applications. Use recommendations for Exterior panels vary with the particular product. Check with manufacturer. EXPOSURE DURABILITY CLASSIFICATIONS: Interior Exposure 1, Exterior. COMMON THICKNESSES: 5⁄16, 3⁄8, 1⁄2, 5⁄8.
APA **HIGH DENSITY** **OVERLAY (HDO)**[(a)] Typical Trademark HDO•A-A•G-1•EXT-APA•000•PS 1-83		Has a hard semi-opaque resin-fiber overlay on both faces. Abrasion resistant. For concrete forms, cabinets, countertops, signs, tanks. Also available with skid-resistance screen-grid surface. EXPOSURE DURABILITY CLASSIFICATION: Exterior, COMMON THICKNESSES: 3⁄8, 1⁄2, 5⁄8, 3⁄4.
APA **MEDIUM DENSITY** **OVERLAY (MDO)**[(a)] Typical Trademark	──APA── M. D. OVERLAY GROUP 1 EXTERIOR ── 000 ── PS 1-83	Smooth, opaque, resin-fiber overlay on one or both faces. Ideal base for paint, both indoors and outdoors. For exterior siding, paneling, shelving, exhibit displays, cabinets, signs. EXPOSURE DURABILITY CLASSIFICATION: Exterior. COMMON THICKNESSES: 11⁄32, 3⁄8, 15⁄32, 1⁄2, 19⁄32, 5⁄8, 23⁄32, 3⁄4.
APA **MARINE** TYPICAL TRADEMARK MARINE•A-A•EXT-APA•PS 1-83		Ideal for boat hulls. Made only with Douglas fir or western larch. Subject to special limitations on core gaps and face repairs. Also available with HDO faces. EXPOSURE DURABILITY CLASSIFICATION: Exterior. COMMON THICKNESSES: 1⁄4, 3⁄8, 1⁄2, 5⁄8, 3⁄4.
APA B-B **PLYFORM** **CLASS 1**[(a)] Typical Trademark	──APA── PLYFORM B - B CLASS1 EXTERIOR ── 000 ── PS 1-83	Concrete form grades with high reuse factor. Sanded both faces and mill-oiled unless otherwise specified. Special restrictions on species. Also available in HDO for very smooth concrete finish, and with special overlays. EXPOSURE DURABILITY CLASSIFICATION: Exterior. COMMON THICKNESSES: 19⁄32, 5⁄8, 23⁄32, 3⁄4.
APA **PLYRON** Typical trademark PLYRON•EXPOSURE 1-APA•000		Hardboard face on both sides. Faces tempered, untempered, smooth, or screened. For countertops, shelving, cabinet doors, flooring. EXPOSURE DURABILITY CLASSIFICATIONS: Interior, Exposure 1, Exterior. COMMON THICKNESSES: 1⁄2, 5⁄8, 3⁄4.

(a) Can also be manufactured in Structural 1 (all plies limited to Group 1 species).

FIGURE 4-2 Guide to APA specialty plywood panels. *(Courtesy of APA– The Engineered Wood Association)*

FIGURE 4-3 Particleboard is made from wood flakes, shavings, resins, and waxes. *(Courtesy of Duraflake Division, Williamette Industries, Inc.)*

Class	Surface	Nominal Thickness
1	S1S	⅛
	and	¼
Tempered	S2S	
2	S1S	⅛
	and	¼
Standard	S2S	⅜
3	S1S	⅛
	and	¼
Service tempered	S2S	⅜

FIGURE 4–4 **Kinds and thicknesses of hardboard.**

FIGURE 4–5 **Softboards are used extensively for decorative ceiling panels.** *(Courtesy of Armstrong World Industries, Inc.)*

impregnated with asphalt to protect it from moisture during construction.

Softboard panels can easily be cut with a knife, hand saw, or power saw. They cannot be handplaned with any satisfactory results. Wide-headed nails, staples, or adhesive are used to fasten softboards in place, depending on the type and use.

OTHER

The preceding chapters have been limited to engineered wood panels and boards. Many other are used in the construction industry besides those already mentioned. It is recommended that the student study *Sweet's Architectural File* to become better acquainted with the thousands of building material products on the market. This reference is well known by architects, contractors and builders, and is revised and published annually. Sweet's may be found online at http://www.sweets.com. Two publications, *Products for General Building and Renovations* and *Products for Home Building and Remodeling,* may be available at the school or city library.

Key Terms

fiberboard	**panels**	**plywood**	**tempered**
hardboard	**particleboard**	**softboard**	

Review Questions

Select the most appropriate answer.

1. Plywood usually contains
   a. three layers.
   b. four layers.
   c. an even number of layers.
   d. an odd number of layers.

2. Most softwood plywood is made of
   a. cedar.   c. western pine.
   b. fir.   d. spruce.

3. The best-appearing face veneer of a softwood plywood panel is indicated by the letter
   a. A.   c. E.
   b. B.   d. Z.

4. Which is a good selection of plywood for exterior wall sheathing?
   a. APA Structural Rated Sheathing, Exposure 1
   b. APA A-C, Exterior
   c. APA-Rated Sheathing, Exposure 2
   d. CD, Plugged, Exterior

5. The fraction 32/16 on a APA grade stamp refers to
   a. panel thickness.
   b. maximum rafter spacing.
   c. maximum wall stud spacing.
   d. maximum wall stud and floor joist spacing.

6. Particleboard not rated as structural may be used for
   a. countertops.   c. wall sheathing.
   b. subflooring.   d. roof sheathing.

7. Hardboard may be used in
   a. interior applications only.
   b. exterior and interior applications.
   c. applications protected from moisture.
   d. cabinet and furniture work only.

8. Much of the softboard is used in the construction industry for
   a. underlayment for wall-to-wall rugs.
   b. roof covering.
   c. decorative ceiling panels.
   d. interior wall finish.

9. For use as an underlayment over structural subflooring where continuous or severe moisture may be present, select
   a. A-C, Group 1, Exterior plywood.
   b. B-D, Group 2, Exposure 1 plywood.
   c. Underlayment, Group 1, Exposure 1 plywood.
   d. C-C Plugged, Group 2, Exterior plywood.

10. The recommended selection of plywood where the appearance of one side is important for interior applications such as built-ins and cabinets is
   a. A-A, Group 1, Exposure 1 plywood.
   b. M.D. Overlay, Group 1, Exterior plywood.
   c. Decorative, Group 2, Interior plywood.
   d. A-D, Group 2, Exposure 2 plywood.

UNIT 3

Engineered Lumber Products

An increased demand for lumber and a diminishing supply of large old-growth trees has forced the wood industry to come up with solutions to ensure the survival of our natural resources. Structural lumber previously sawn from large logs is now produced from reconstituted wood in many shapes and sizes. These produces are collectively referred to as engineered lumber.

SAFETY REMINDER

Engineered wood is stronger than wood sawed from logs. Yet it must be handled, cut and fastened properly to achieve the safe and desirable characteristics.

OBJECTIVES

After completing this unit, the student should be able to describe the manufacture, composition, uses, and sizes of:

- laminated veneer lumber
- parallel strand lumber and laminated strand lumber
- wood I-beams
- glue-laminated beams

5 Laminated Veneer Lumber

Old-growth trees are large, tall, tight-grain trees that take more than two hundred years to mature. The lumber produced from these trees is of the highest quality. Large-sized lumber can be efficiently cut from these trees. Due to centuries of logging, the number of old-growth trees is decreasing. Second- and third-growth trees as well as trees planted during reforestation efforts are more abundant. These are smaller and produce fewer large-sized pieces. Lumber from these trees sometimes has undesirable wood characteristics, such as a tendency to warp.

An inevitable result of the decreasing supply of large old-growth trees and the abundance of smaller trees is the development and use of **engineered lumber products.**

Engineered lumber products, or ELPs, are reconstituted wood products and assemblies designed to replace traditional structural lumber. Engineered lumber products consume less wood and can be made from smaller trees than can traditional lumber. Traditional lumber processes typically convert 40 percent of a log to structural solid lumber. Engineered lumber processes convert up to 75 percent of a log into structural lumber. In addition, the manufacturing processes of engineered lumber consume less energy than those of solid lumber. Also, some ELPs make use of abundant, fast-growing species not currently harvested for solid lumber.

The final engineered lumber product has greater strength and consequently can span greater distances. It is predicted that engineered lumber will be used more than solid lumber in the near future. It is important that present and future builders be thoroughly informed about engineered lumber products.

This unit describes how engineered lumber products are made, where they are used, and what sizes are available. Construction details for ELPs are shown in later units on floor, wall, and roof framing.

LAMINATED VENEER LUMBER

Laminated veneer lumber, commonly called LVL, is one of several types of engineered lumber products **(Figure 5–1).** It was first used to make airplane propellers during World War II. The world's first commercially produced LVL for building construction was patented as MICRO-LAM laminated veneer lumber in 1970. LVL is now widely used in wood frame construction.

FIGURE 5–1   There are several types of engineered lumber. *(Courtesy of Trus Joist MacMillan)*

Manufacture of LVL

Like plywood, LVL is a wood veneer product. The grain in each layer of veneer in LVL runs in the same direction, parallel to its length **(Figure 5–2).** This is unlike plywood, in which each layer of veneer is laid with the grain at right angles to each other.

Laminated veneer lumber is made from sheets of veneer peeled from logs, similar to the first step in the manufacture of plywood (see Figure 3–5). Douglas fir or southern pine is used because of its strength and stiffness. The veneer is peeled in widths of 27 or 54 inches and from 1/10 to 3/16 inch in thickness. It is then

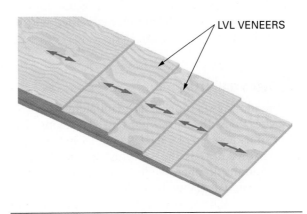

FIGURE 5–2   In LVL, the grain in each layer of veneer runs in the same direction.

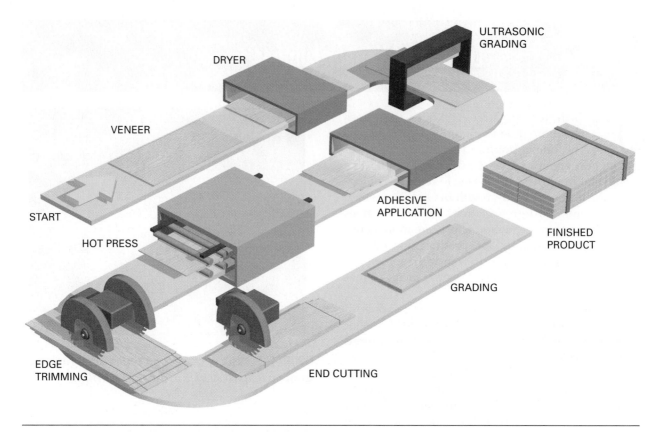

FIGURE 5–3 LVL manufacturing process, simplified. *(Courtesy of Trus Joist MacMillan)*

dried, cut into sheets, ultrasonically graded for strength, and sorted **(Figure 5–3).**

The veneer sheets are laid in a staggered pattern so that the ends overlap. They are then permanently bonded together with an exterior type adhesive in a continuous press under precisely controlled heat and pressure. Unlike plywood, the LVL veneers are **densified;** that is, the thickness is compressed and made more compact. Fifteen to twenty layers of veneer make up a typical 1¾-inch thick beam **(Figure 5–4).** The edges of the bonded veneers are then edge trimmed to specified widths and end-cut to specified lengths.

LVL Sizes and Uses

Laminated veneer lumber is manufactured up to 3½ inches thick, 18 inches wide, and 80 feet long. The usual thickness are 1½ inches and 1¾ inches. The 1½-inch thickness is the same as nominal 2-inch-thick framing lumber, while doubling the 1¾-inch thickness equals the width of nominal 4-inch framing. LVL widths are usually 9¼, 11¼, 11⅞, 14, 16, and 18 inches. LVL beams may be fastened together to make a thicker and stronger beam **(Figure 5–5).**

FIGURE 5–4 Close-up view of LVL.

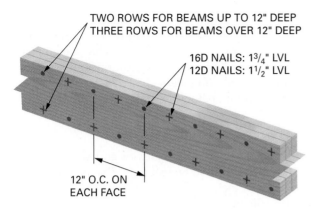

TWO ROWS FOR BEAMS UP TO 12" DEEP
THREE ROWS FOR BEAMS OVER 12" DEEP

16D NAILS: 1³/₄" LVL
12D NAILS: 1¹/₂" LVL

12" O.C. ON
EACH FACE

NOTES:
THESE FASTENING RECOMMENDATIONS APPLY:
• TO TWO-PLY AND THREE-PLY BEAMS ONLY.
• WHEN BEAM IS LOADED UNIFORMLY ON THE
  TOP EDGE OF EACH PLY.

FIGURE 5–5 **Recommended nailing pattern for fastening LVL beams together.** *(Courtesy of Louisiana Pacific Corporation)*

Laminated veneer lumber is intended for use as high-strength, load-carrying beams to support the weight of construction over window and door openings, and in floor and roof systems **(Figure 5–6).** Although LVLs are heavier than dimension lumber, they are still easy to handle and suitable for use in concrete forming in manufactured housing, and in

FIGURE 5–6 **LVL is designed to be used for load-carrying beams.**

many other specialties where a lightweight, strong beam is required. It can be cut with regular tools and requires no special fasteners.

6 Parallel Strand Lumber and Laminated Strand Lumber

PARALLEL STRAND LUMBER

Parallel strand lumber (PSL) commonly known by its brand name, Parallam **(Figure 6–1),** was developed, like all engineered lumber products, to meet the need of the building industry. PSL provides large-dimension lumber (beams, planks, and posts). PSL also utilizes small-diameter, second-growth trees, thus protecting the diminishing supply of old-growth trees.

Manufacture of PSL

Parallam PSL is manufactured by peeling veneer of Douglas fir and southern pine from logs in much the same manner as for plywood and LVL. The veneer is then dried and clipped into *strands* (narrow strips) up to 8 feet in length and ⅛ or ¹⁄₁₀ inch in thickness **(Figure 6–2).** Small defects are removed and the strands are then coated with a waterproof adhesive. The oriented strands are fed into a rotary belt press

FIGURE 6–1   Parallel strand lumber is commonly called Parallam and is used as beams and posts to carry heavy loads. *(Courtesy of Trus Joist MacMillan)*

and bonded using a patented microwave pressing process. The result is a continuous timber, up to 11 inches thick by 17 inches wide, which can then be factory-ripped into widths and thicknesses to fit builders' needs **(Figure 6–3).** The four surfaces are sanded smooth before the product is shipped.

PSL Sizes and Uses

PSL comes in many thicknesses and widths and is manufactured up to 66 feet long. PSL is available in square and rectangular shapes for use as posts and beams. Beams are sold in a convenient 1¾-inch thickness for installation of single and multiple laminations. Solid 3½-inch thicknesses are compatible with 2 × 4 wall framing. A list of beam and post sizes is shown in **Figure 6–4.** Also available is Parallam 269, which measures 2¹¹⁄₁₆ inches thick.

Parallel strand lumber can be used wherever there is a need for a large beam or post. The differences between PSL and solid lumber are many. Solid lumber beams may have defects, like knots, checks, and shakes, which weaken them, whereas PSL is consistent in strength throughout its length. PSL is readily available in longer lengths and its surfaces

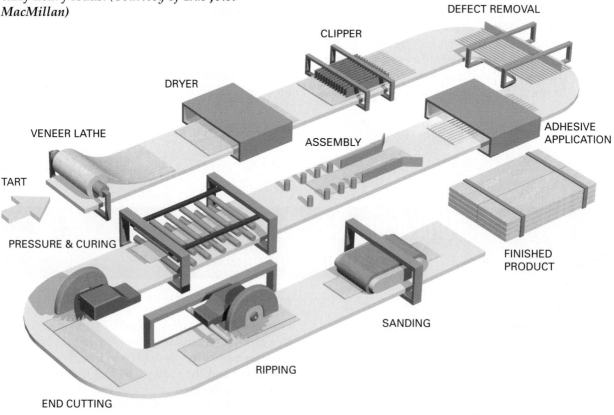

FIGURE 6–2   The manufacture of parallel strand lumber, simplified. *(Courtesy of Trus Joist MacMillan)*

FIGURE 6–3   Close-up view of PSL.

PSL Column & Post Sizes	
Thickness	**Width**
3½″	3½″
3½″	5¼″
3½″	7″
5¼″	5¼″
5¼″	7″
7″	7″

PSL Beam Sizes	
Thickness	**Depths**
1¾″	7¼″
3½″	9¼″
5¼″	11¼″
7″	11½″
	11⅞″
	12″
	12½″
	14″
	16″
	18″

FIGURE 6–4   Available sizes of PSL used for posts and beams.

are sanded smooth, eliminating the need to cover them by boxing the beams.

LAMINATED STRAND LUMBER

A registered brand name of **laminated strand lumber** (LSL) is *TimberStrand* **(Figure 6–5).** While LVL and PSL are made from Douglas fir and southern pine, LSL can be made from very small logs of practically any species of wood; its strands are much shorter than those of parallel strand lumber. At present, LSL is be-ing manufactured from surplus, overmature aspen trees that usually are not large, strong, or straight enough to produce ordinary wood products.

PSL

LSL

FIGURE 6–5   Laminated strand lumber is used for non-load-bearing situations such as rim joists.

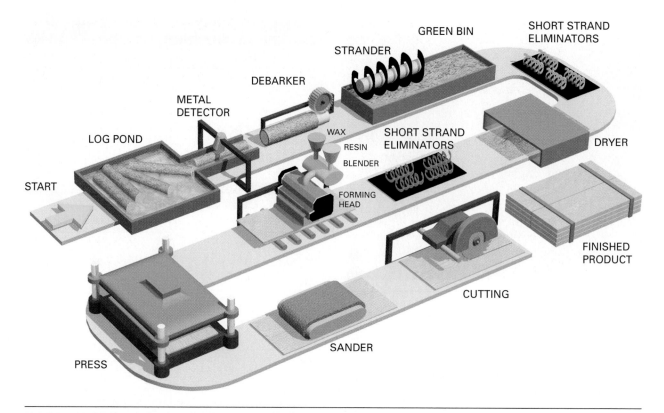

FIGURE 6–6   Laminated strand lumber can be manufactured from practically any species of wood. *(Courtesy of Trus Joist MacMillan)*

Manufacture of LSL

The TimberStrand LSL manufacturing process begins by cleaning and debarking 8-foot aspen logs **Figure 6–6).** The wood is then cut into strands up to 12 inches long, dried, and treated with a resin. The treated strands are aligned parallel to each other to take advantage of the wood's natural strength. The strands are pressed into solid *billets* (large blocks) up to 5½ inches thick, 8 feet wide, and 35 feet long **(Figure 6–7).** Scraps from the process fuel the furnace that provides heat and steam for the plant.

LSL Sizes and Uses

The long billet is resawn and sanded to sizes as required by customers. It is used for a wide range of **millwork,** such as doors, windows, and virtually any product that requires high-grade lumber. It is also used for truck decks, manufactured housing, and some structural lumber, such as window and door headers.

LSL is made from wood that is less strong and stiff than the wood used to make PSL. For this reason LSL is not designed to be a structural framing member that will carry heavy loads over long spans. It is often used as a **rim joist** attached to engineered floor joists.

FIGURE 6–7   Close-up view of LSL.

7 Wood I-Joists

WOOD I-JOISTS

The wood I-beam joist was invented in 1969, as a substitute for solid lumber **joists (Figure 7–1)**. Wood I-joists are engineered wood assemblies that utilize an efficient "I" shape, common in steel beams, which gives them tremendous strength in relation to their size and weight. Consequently, they are able to carry heavy loads over long distances while using considerably less wood than solid lumber of a size necessary to carry the same load over the same span.

The flanges of the joist may be made of laminated veneer lumber or specially selected **finger-jointed** solid wood lumber **(Figure 7–2)**. The web of the beam may be made of plywood, laminated veneer lumber, or oriented strand board.

The Manufacture of Wood I-Joists

Regardless of who makes them, the manufacturing process of wood I-joists consists of gluing top and bottom flanges to a connecting center web. First the web material is ripped to a specified width and then the edges and ends are shaped for joining to flanges and adjacent web sections. The ends of the flange material are finger-jointed for gluing end to end. One side of the flange material is grooved to receive the beam's web. Flanges and webs are then assembled with waterproof glue by pressure-fitting the web into the flanges. The wood I-joist is end-trimmed and the adhesive cured in an oven or at room temperature **(Figure 7–3)**. As with most engineered wood products, wood I-joists are produced to an approximate equilibrium moisture content.

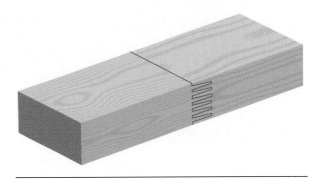

FIGURE 7–2 **Finger-joints are used to join the ends of short pieces of lumber to make a longer piece.**

FIGURE 7–1 **Wood I-joists are available in many sizes. (*Courtesy of Louisiana Pacific Corporation*)**

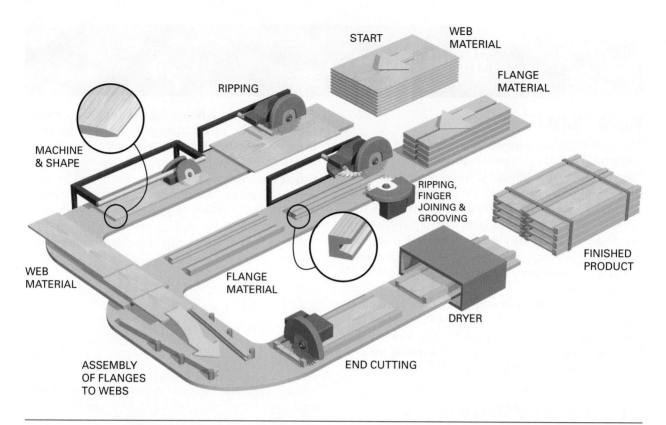

FIGURE 7–3   Manufacture of wood I-joists, simplified. *(Courtesy of Trus Joist MacMillan)*

Wood I-Joist Sizes and Uses

Wood I-joists may have webs of various thicknesses, and flanges may vary in thickness and width, depending on intended end use and the manufacturer. Joist depths are available from 9¼ to 24 inches **(Figure 7–4)**. Joists with larger webs and flanges are designed to carry heavier loads. Wood I-joists are available up to 80 feet long.

Wood I-joists are intended for use in residential and commercial construction as floor joists and roof rafters **(Figure 7–5)**.

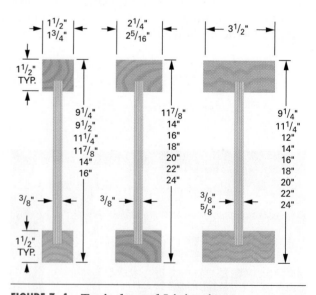

FIGURE 7–4   Typical wood I-joist sizes.

A

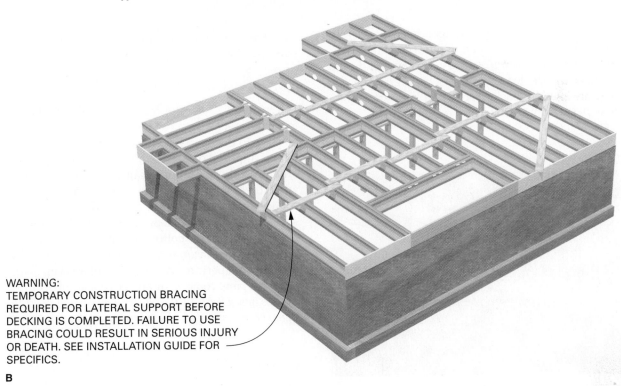

WARNING:
TEMPORARY CONSTRUCTION BRACING
REQUIRED FOR LATERAL SUPPORT BEFORE
DECKING IS COMPLETED. FAILURE TO USE
BRACING COULD RESULT IN SERIOUS INJURY
OR DEATH. SEE INSTALLATION GUIDE FOR
SPECIFICS.

B

FIGURE 7–5   Wood I-joists are used as (A) roof rafters. (B) for floor joists and window and door headers. *(Courtesy of Louisiana Pacific Corporation)*

8   Glue-Laminated Lumber

GLUE-LAMINATED LUMBER

Glue-laminated lumber, commonly called *glulam,* is constructed of solid lumber glued together, side against side, to make beams and joists of large dimensions that are stronger than natural wood of the same size **(Figure 8–1).** Even if it were possible, it would not be practical to make solid wood beams as large as most glulams. Glulams are used for structural purposes, but architectural appearance grade glulams are decorative as well and, in most cases, their surfaces are left exposed to show the natural wood grain.

Manufacture of Glulams

Glulam beams are made by gluing stacks of *lam stock into large beams,* **headers,** and columns **(Figure 8–2).** Lams are the individual pieces in the glued-up stack. The lams, which are at approximate equilibrium moisture content, are glued together with exterior adhesives. Thus glulams are rated in terms of weather resistance.

Lam Layup. The lams are arranged in a certain way for maximum strength and minimum shrinkage. Different grade lams are placed where they will do the most good under load conditions. High-grade *tension* lams are used in tension faces, and high-grade *compression* lams are used in compression faces.

FIGURE 8–1   Glue-laminated lumber is commonly called glulam. (Courtesy of APA–The Engineered Wood Association)

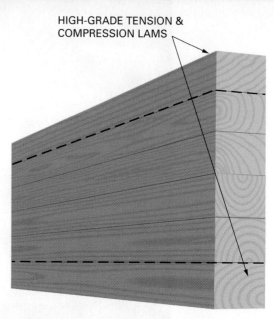

HIGH-GRADE TENSION & COMPRESSION LAMS

THE SEQUENCE OF LAM GRADES, FROM BOTTOM TO TOP OF A GLULAM, IS REFERRED TO AS A LAM LAYUP AND IS A VITALLY IMPORTANT FACTOR IN GLULAM PERFORMANCE.

FIGURE 8–2   Glulams are made by gluing stacks of solid lumber.

Tension is a force applied to a member that tends to increase its length, while compression is a force tending to decrease its length **(Figure 8–3).** When a load is imposed on a glulam beam that is supported on both ends, the topmost lams are in compression while those at the bottom are in tension.

More economical lams are used in the lower-stressed middle sections of the glulam. The sequence of lam grades, from bottom to top of a glulam, is referred to as *lam layup,* and it is a vitally important factor in glulam performance. Because of this lam layup, glulam beams come with one edge stamped "TOP." *Always remember to install glulam beams with the "TOP" stamp facing toward the sky* **(Figure 8–4).**

Glulam Grades

The American Plywood Association–Engineered Wood Systems trademark, APA-EWS **(Figure 8–5),** appears on all beams manufactured by American Wood Systems (AWS) member mills. The AWS is a

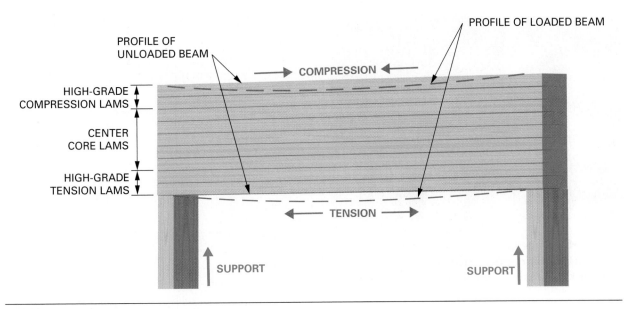

FIGURE 8-3 The load on a beam places lams in tension and compression. *(Courtesy of Bohemia, Inc.)*

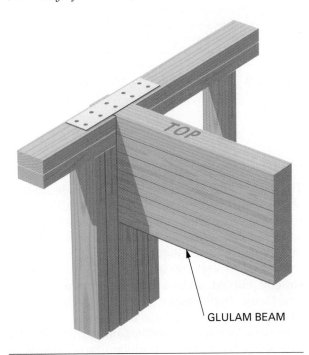

FIGURE 8-4 Glulam beams must be installed with the edge stamped "TOP" pointed toward the sky.

1) INDICATES STRUCTURAL USE:
   B - SIMPLE SPAN BENDING MEMBER.
   C - COMPRESSION MEMBER.
   T - TENSION MEMBER.
   CB- CONTINUOUS OR CANTILEVERED SPAN BENDING MEMBER.
2) MILL NUMBER.
3) IDENTIFICATION OF ANSI STANDARD A190.1, STRUCTURAL GLUED LAMINATED TIMBER.
4) CODE RECOGNITION OF *AMERICAN WOOD SYSTEMS* AS A QUALITY ASSURANCE AGENCY FOR GLUED STRUCTURAL MEMBERS.
5) APPLICABLE LAMINATING SPECIFICATION.
6) APPLICABLE COMBINATION NUMBER.
7) SPECIES OF LUMBER USED.
8) DESIGNATES APPEARANCE GRADE (INDUSTRIAL ARCHITECTURAL, PREMIUM).

FIGURE 8-5 The grade stamp assures that the beam has met all the necessary requirements. *(Courtesy of APA–The Engineered Wood Association)*

related corporation of the APA. The trademark guarantees that the glulams meet all the requirements of the American National Standards Institute (ANSI).

Glulams are manufactured in three grades for appearance. The *industrial appearance grade* is used in warehouses, garages, and other structures in which appearance is not of primary importance or when the beams are not exposed. *Architectural appearance grade* is for projects where appearance is important, and the *premium appearance grade* is used when appearance is critical **(Figure 8–6).** There is no difference in strength among the different appearance grades. All glulams with the same design values are rated for the same loadings, regardless of appearance grade.

FIGURE 8–6 Some glulam beams are also manufactured for appearance. *(Courtesy of Willamette Industries, Inc.)*

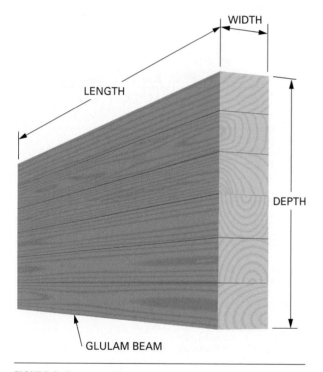

FIGURE 8–7 The dimensions of glulam beams are indicated by width, depth, and length.

Depth in Inches	Width						
	2½″	3⅛″	3½″	5⅛″	5½″	6¾″	8¾″
6		*	*				
7½	*	*	*				
9	*	*	*	*	*		
10½	*	*	*	*	*		
12	*	*	*	*	*	*	*
13½	*	*	*	*	*	*	*
15	*	*	*	*	*	*	*
16½	*	*	*	*	*	*	*
18	*	*	*	*	*	*	*
19½	*	*	*	*	*	*	*
21	*	*	*	*	*	*	*
22½	*	*	*	*	*	*	*
24				*	*	*	*
25½				*	*	*	*
27						*	*
28½						*	*

Sizes generally available from 10 to 40 feet long in 2′-0″ increments

FIGURE 8–8 Available glulam beam sizes.

Glulam Sizes and Uses

The dimensions of glulam beams are indicated by width, depth, and length **(Figure 8–7)**. Widths range from 2½ to 8¾ inches, depths from 6 to 28½ inches, and lengths are generally available from 10 to 40 feet in 2-foot increments **(Figure 8–8)**.

Various wood species are used to produce straight, curved, arched, and special shapes for all structures—from elegant homes and churches to large malls, warehouses, and civic centers. In all, the American Institute of Timber Construction (AITC) recognizes more than 100 glulam beam combinations, most of which are for specialized applications.

Key Terms

engineered lumber products

finger-jointed

glue-laminated lumber

headers

joist

laminated strand lumber

laminated veneer lumber

millwork

parallel strand lumber

rim joist

Review Questions

Select the most appropriate answer.

1. Engineered lumber is produced today because of
   a. the decreasing supply of old-growth trees.
   b. the increasing supply of younger, smaller trees.
   c. a need to make efficient use of natural resources.
   d. all of the above.

2. Unlike plywood, the veneers of LVL are
   a. deciduous.
   b. densified.
   c. diversified.
   d. double faced.

3. Laminated veneer lumber is often used for
   a. high-strength load-carrying beams.
   b. posts and columns.
   c. low-strength rim joists.
   d. floor joists.

4. Parallel strand lumber is made from
   a. Alaskan cedar and California redwood.
   b. Douglas fir and southern pine.
   c. Idaho pine and eastern hemlock.
   d. Englemann spruce and western pine.

5. Parallel strand lumber is often used for
   a. high-strength load-carrying beams.
   b. posts and columns.
   c. low-strength rim joists.
   d. floor joists.

6. Laminated strand lumber is often used for
   a. high-strength load-carrying beams.
   b. posts and columns.
   c. low-strength rim joists.
   d. floor joists.

7. The web of wood I-beams may be made of
   a. hardboard.
   b. particleboard.
   c. solid lumber.
   d. strand board.

8. The flanges of wood I-beams are generally made from
   a. glue-laminated lumber.
   b. laminated veneer lumber.
   c. parallel strand lumber.
   d. laminated strand lumber.

9. The lams used in glue-laminated lumber are at approximately
   a. the fiber-saturation point.
   b. 19 percent moisture content.
   c. equilibrium moisture content.
   d. 30 percent moisture content.

10. In glulam beams, specially selected tension lams are placed
    a. at the top of the beam.
    b. at the bottom of the beam.
    c. in the center of the beam.
    d. throughout the beam.

UNIT 4

Fasteners

The simplicity of fasteners can be misleading to students of construction. It is easy to believe only that nails are driven, screws are turned, and sticky stuff is used to glue. Although this tends to be true, joining material together so it will last a long time is more challenging. Many times, a fastener is used for just one type of material. Some fasteners should never be used with certain other materials. The fastener selected often separates a quality job from a shoddy one.

Fasteners have been evolving for centuries. Today they come in many styles, shapes, and sizes requiring different fastening techniques. It is important for the carpenter to know what fasteners are available, which securing technique should be employed, and how to wisely select the most appropriate fastener for various materials under different conditions.

There are many kinds and styles of fasteners made of various materials. Selecting the proper fastener is important for strength and durability.

OBJECTIVES

After completing this unit, the student should be able to name and describe the following commonly used fasteners and select them for appropriate use:

- nails
- screws and lag screws
- bolts
- solid and hollow wall anchors
- adhesives.

9 Nails, Screws, and Bolts

Nails, screws, and bolts are the most widely used of all fasteners. They come in many styles and sizes. The carpenter must know what is available and wisely select those most appropriate for fastening various materials under different conditions.

NAILS

There are hundreds of kinds of nails manufactured for just about any kind of fastening job. They differ according to purpose, shape, material, coating, and in other ways. Nails are made of aluminum, brass, copper, steel, and other metals. Some nails are hardened so that they can be driven into masonry without bending. Only the most commonly used nails are described in this chapter **(Figure 9–1)**.

Uncoated steel nails are called **bright** nails. Various coatings may be applied to reduce corrosion, increase holding power, and enhance appearance. To prevent rusting, steel nails are coated with *zinc.* These nails are called **galvanized** nails. They may be coated by being dipped in molten zinc (*hot-dipped galvanized* nails), or they may be *electroplated* with corrosion-resistant metal (*plated* nails). Hot-dipped nails have a heavier coating than plated nails and are more resistant to rusting. Many manufacturers specify that their products be fastened with hot-dipped nails because of the heavier coating.

When fastening metal that is going to be exposed to the weather, use nails of the same material. For ex-

ample, when fastening aluminum, copper, or galvanized iron, use nails made of the same metal. Otherwise, a reaction with moisture and the two different metals, called **electrolysis,** will cause one of the metals to disintegrate over time.

When fastening some woods, such as cedar, redwood, and oak, that will be exposed to the weather, use stainless steel nails. Otherwise, a reaction between the acid in the wood and bright nails causes dark stains to appear around the fasteners.

Nail Sizes

The sizes of some nails are designated by the **penny** system. The origin of this system of nail measurement is not clear. Although many people think it should be discarded, it is still used in the United States. Some believe it originated years ago when nails cost a certain number of pennies for a hundred of a specific length. Of course, the larger nails cost more per hundred than smaller ones, so nails that cost 8 pennies per hundred were larger than those that cost 4 pennies per hundred. The symbol for penny is *d;* perhaps it is the abbreviation for *denarius,* an ancient Roman coin.

In the penny system the shortest nail is 2d and is 1 inch long. The longest nail is 60d and is 6 inches long **(Figure 9–2).** A sixpenny nail is written as 6d and is 2 inches long. Eventually, a carpenter can determine the penny size of nails just by looking at them.

The thickness or diameter of the nail is called its gauge. In the penny system, gauge depends on the kind and length of the nail. The gauge increases with the length of the nail. Long nails, 20d and over, are called *spikes.* The length of nails not included in the penny system is designated in inches and fractions of an inch, and the gauge may be specified by a number.

Kinds of Nails

Most nails, cut from long rolls of metal wire, are called **wire nails.** *Cut nails,* used only occasionally, are wedge-shaped pieces stamped from thin sheets of metal. The most widely used wire nails are the common, box, and finish nails **(Figure 9–3).**

Common Nails. Common nails are made of wire, are of heavy gauge, and have a medium-sized head. They have a pointed end and a smooth shank. A barbed section just under the head increases the holding power of common nails.

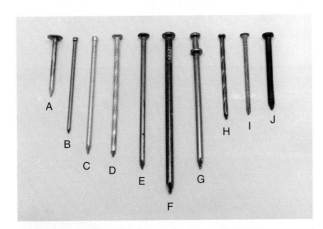

FIGURE 9–1 Kinds of commonly used nails: (A) roofing, (B) finish, (C) galvanized, (D) galvanized spiral, (E) box, (F) common, (G) duplex, (H) spiral, (I) coated box, and (J) masonry.

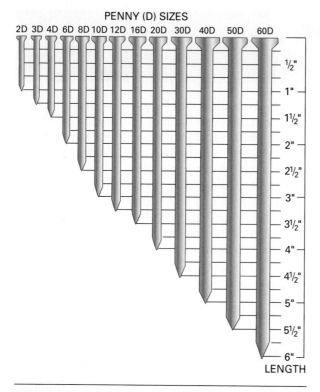

FIGURE 9–2   Most nails are sized according to the penny system.

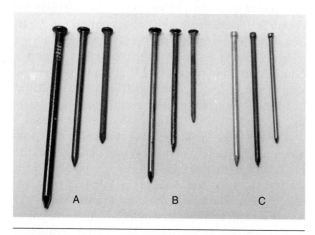

FIGURE 9–3   The most widely used nails are the (A) common, (B) box, and (C) finish nails.

Box Nails.
Box nails are similar to common nails, except they are thinner. Because of their small gauge, they can be used close to edges and ends with less danger of splitting the wood. Many box nails are coated with resin cement to increase their holding power.

Finish Nails.
Finish nails are of light gauge with a very small head. They are used mostly to fasten interior trim. The small head is sunk into the wood with a nail set and covered with a filler. The small head of the finish nail does not detract from the appearance of a job as much as would a nail with a larger head.

Casing Nails.
Casing nails are similar to finish nails. Many carpenters prefer them to fasten exterior finish. The head is cone shaped and slightly larger than that of the finish nail, but smaller than that of the common nail. The shank is the same gauge as that of the common nail.

Duplex Nails.
On temporary structures, such as wood scaffolding and concrete forms, the **duplex nail** is often used. The lower head ensures that the piece is fastened tightly. The projecting upper head makes it easy to pry the nail out when the structure is dismantled **(Figure 9–4)**.

Brads.
Brads are small finishing nails **(Figure 9–5)**. They are sized according to length in inches and gauge. Usual lengths are from ½ inch to 1½ inches, and gauges run from #14 to #20. The higher the gauge number, the thinner the brad. Brads are used for fastening thin material, such as small molding.

Roofing Nails.
Roofing nails are short nails of fairly heavy gauge with wide, round heads. They are used for such purposes as fastening roofing material and softboard wall sheathing. The large head holds thin or soft material more securely.

Some roofing nails are coated to prevent rusting. Others are made from noncorrosive metals such as aluminum and copper. The shank is usually barbed to increase holding power. Usual sizes run from ¾ inch to 2 inches. The gauge is not specified when ordering.

Masonry Nails.
Masonry nails may be cut nails or wire nails **(Figure 9–6)**. These nails are made from

FIGURE 9–4   Duplex nails are used for temporary fastening.

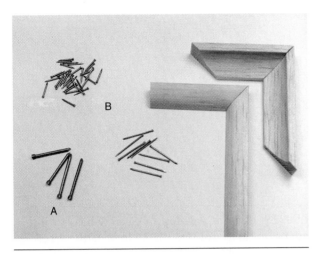

FIGURE 9–5 (A) Finishing nails, and (B) brad nails.

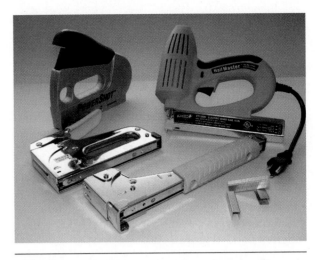

FIGURE 9–7 Staples can be driven by a variety of guns.

hardened steel to prevent them from bending when being driven into concrete or other masonry. The cut nail has a blunt point that tends to prevent splitting when it is driven into hardwood. Some masonry and flooring nails have round shanks of various designs for better holding power.

Staples. Staples are U-shaped fasteners used to secure a variety of materials. They come in a number of sizes and are designated by their length. Sold in boxes of 1,000 or more, they are driven by several types of tools, including the squeeze stapler, hammer tacker, and electric staplers **(Figure 9–7).** They are used to fasten thin material.

Heavy-duty staples are similar in design only larger in size and gauge. These types of staples are often used to install some roofing materials or during cabinet construction **(Figure 9–8).**

FIGURE 9–8 Using a power stapler to fasten wood.

SCREWS

Wood screws are sued when greater holding power is needed and when the work being fastened must at times be removed. For example, door hinges must be applied with screws because nails would pull loose after a while, and the hinges may, at times, need to be removed. Screws cost more than nails and require more time to drive. When ordering screws, specify the length, gauge, type of head, coating, kind of metal, and screwdriver slot.

Kinds of Screws

A wood screw is identified by the shape of the screwhead and screwdriver slot. For instance, a

FIGURE 9–6 Masonry nails are made of hardened steel and may chip apart when driven.

screw may be called a *flat head Phillips* or a *round-head* screw. Three of the most common shapes of screwheads are the *flat head, round head,* and *oval head.*

The pointed end of a screw is called the *gimlet point.* The threaded section is called the *thread.* The smooth section between the head and thread is called the *shank.* Screw lengths are measured from the point to that part of the head that sets flush with the wood when fastened **(Figure 9–9).**

Screw Slots. A screwhead that is made with a straight, single slot is called a **common screw.** A **Phillips head** screw has a crossed slot. There are many other types of screwdriver slots, each with a different name **(Figure 9–10).**

Sheet Metal Screws. Wood screws are threaded, in most cases, only partway to the head. The threads of sheet metal screws extend for the full length of the screw and are much deeper. Sheet metal screws are used for fastening thin metal.

Another type of screw, used with power screwdrivers, is the **self-tapping screw,** which is used extensively to fasten metal framing. This screw has a cutting edge on its point to eliminate the need to predrill a hole **(Figure 9–11).** It is important that the drilling process be completed before the threading

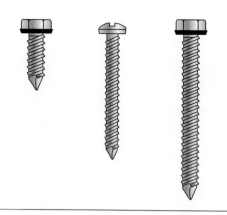

FIGURE 9–11 The self-tapping screw has cutting edges on the point that drill a hole as the screw is driven.

process begins. Drill points are available in various lengths and must be equal to the thickness of the metal being fastened.

Drywall screws are designed to fasten drywall to wood or metal framing members. The head is shaped like a bugle with threads extending nearly to the head. Drywall screws come in coarse or fine threads and sharp or drill points. Deck screws are similarly shaped but have an epoxy coating to resist corrosion **(Figure 9–12).**

Many other screws are available that are designed for special purposes. Like nails, screws come in a variety of metals and coatings. Steel screws with no coating are called bright screws.

Screw Sizes

Screws are made in many different sizes. Usual lengths range from ¼ inch to 4 inches. Gauges run from 0 to 24 **(Figure 9–13).** Unlike some nails, the higher the

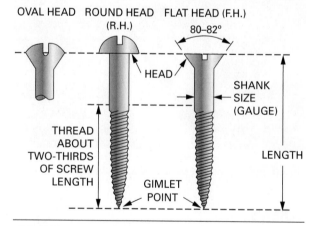

OVAL HEAD   ROUND HEAD   FLAT HEAD (F.H.)
(R.H.)   80–82°

HEAD
SHANK SIZE (GAUGE)
THREAD ABOUT TWO-THIRDS OF SCREW LENGTH
LENGTH
GIMLET POINT

FIGURE 9–9 Common styles of slotted screws and screw terms.

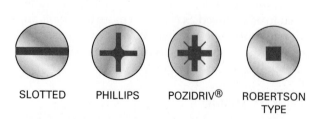

SLOTTED   PHILLIPS   POZIDRIV®   ROBERTSON TYPE

FIGURE 9–10 Common styles of screwhead slots.

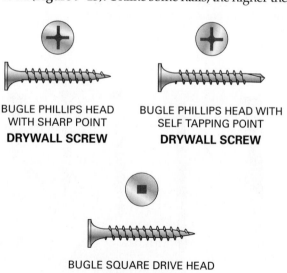

BUGLE PHILLIPS HEAD WITH SHARP POINT
DRYWALL SCREW

BUGLE PHILLIPS HEAD WITH SELF TAPPING POINT
DRYWALL SCREW

BUGLE SQUARE DRIVE HEAD
DECK SCREW

FIGURE 9–12 Typical drywall and deck screws.

gauge number, the greater the diameter of the screw. Screw lengths are not available in every gauge. The lower gauge numbers are for shorter, thinner screws. Higher gauge number are for longer screws.

Lag Screws

Lag screws (Figure 9–14) are similar to wood screws except that they are larger and have a square or hex head designed to be turned with a wrench instead of a screwdriver. This fastener is used when great holding power is needed to join heavy parts and where a bolt cannot be used.

Lag screws are sized by diameter and length. Diameters range from ¼ inch to 1 inch, with lengths from 1 inch to 12 inches and up. Shank and pilot holes are drilled to receive lag screws in the same manner as for wood screws. (See Chapter 13.) Place a flat washer under the head to prevent the head from digging into the wood as the lag screw is tightened down. Apply a little wax to the threads to allow the screw to turn more easily and to prevent the head from twisting off.

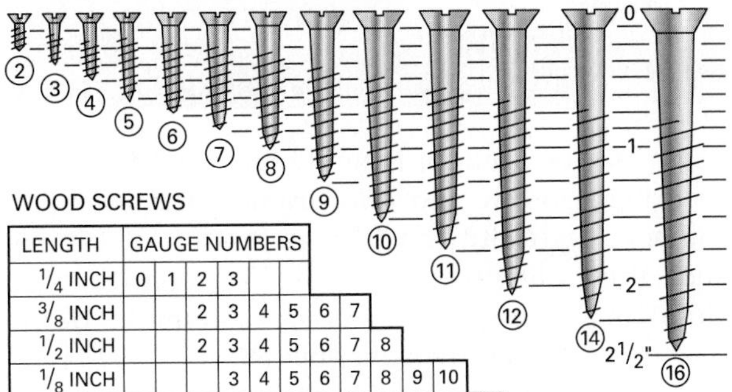

BOLTS

Most bolts are made of steel. To retard rusting, galvanized or stainless steel bolts are used. As with nails and screws, they are available in different kinds of metals and coatings. Many kinds are used for special purposes, but only a few are generally used. Commonly used bolts are the carriage, machine, and stove bolts **(Figure 9–15).**

Carriage Bolts

The **carriage bolt** has a square section under its oval head. The square section is embedded in wood and prevents the bolt from turning as the nut is tightened.

Machine Bolts

The **machine bolt** has a square or hex head. This is held with a wrench to keep the bolt from turning as the nut is tightened.

Stove Bolts

Stove bolts have either round or flat heads with a screwdriver slot. They are usually threaded all the way up to the head. *Machine screws* are very similar to stove bolts.

WOOD SCREWS

LENGTH	GAUGE NUMBERS													
¼ INCH	0	1	2	3										
⅜ INCH			2	3	4	5	6	7						
½ INCH			2	3	4	5	6	7	8					
⅝ INCH				3	4	5	6	7	8	9	10			
¾ INCH					4	5	6	7	8	9	10	11		
⅞ INCH							6	7	8	9	10	11	12	
1 INCH							6	7	8	9	10	11	12	14
1¼ INCH								7	8	9	10	11	12	14	16
1½ INCH							6	7	8	9	10	11	12	14	16	18
1¾ INCH									8	9	10	11	12	14	16	18	20
2 INCH									8	9	10	11	12	14	16	18	20
2¼ INCH										9	10	11	12	14	16	18	20
2½ INCH													12	14	16	18	20
2¾ INCH														14	16	18	20
3 INCH															16	18	20
3½ INCH																18	20	24
4 INCH																18	20	24

WHEN YOU BUY SCREWS, SPECIFY (1) LENGTH, (2) GAUGE NUMBER, (3) TYPE OF HEAD – FLAT, ROUND, OR OVAL, (4) MATERIAL – STEEL, BRASS, BRONZE, ETC., (5) FINISH – BRIGHT, STEEL, CADMIUM, NICKEL, OR CHROMIUM PLATED.

FIGURE 9–13 Wood screw sizes.

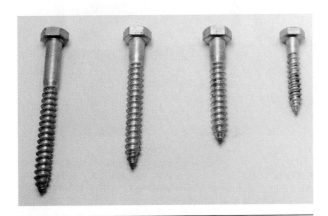

FIGURE 9–14 Lag screws are large screws with a square or hex head.

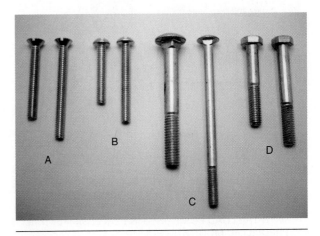

FIGURE 9–15 Commonly used bolts include (A) flat-head stove, (B) round-head stove, (C) carriage, and (D) machine.

Bolt Sizes

Bolt sizes are specified by diameter and length. Carriage and machine bolts range from ¾ inch to 20 inches in length and from ³⁄₁₆ to ¾ inch in diameter. Stove bolts are small in comparison to other bolts. They commonly come in lengths from ⅜ inch to 6 inches and from ⅛ to ⅜ inch in diameter.

Drill holes for bolts the same diameter as the bolt. Use flat washers under the head (except for carriage bolts) and under the nut to prevent the nut from cutting into the wood and to distribute the pressure over a wider area. Be careful not to overtighten carriage bolts. The head need only be drawn snug, not pulled below the surface.

---

10 Anchors and Adhesives

ANCHORS

Special kinds of fasteners used to attach parts to solid masonry and hollow walls and ceilings are called **anchors.** There are hundreds of types available. Those most commonly used are described in this chapter.

Solid Wall Anchors

Solid wall anchors may be classified as heavy, medium, or light duty. Heavy-duty anchors are used to install such things as machinery, hand rails, dock bumpers, and storage racks. Medium-duty anchors may be used for hanging pipe and ductwork, securing window and door frames, and installing cabinets. Light-duty anchors are used for fastening such things as junction boxes, bathroom fixtures, closet organizers, small appliances, smoke detectors, and other lightweight objects.

Heavy-Duty Anchors. The **wedge anchor (Figure 10–1)** is used when high resistance to pullout is required. The anchor and hole diameter are the same, simplifying installation. The hole depth is not critical as long as the minimum is drilled. Proper installation requires cleaning out the hole.

INSERT – CLEAN HOLE, THEN DRIVE THE ANCHOR FAR ENOUGH INTO THE HOLE SO THAT AT LEAST SIX THREADS ARE BELOW THE TOP OF THE SURFACE OF THE FIXTURE.

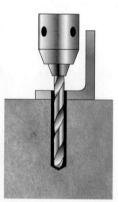
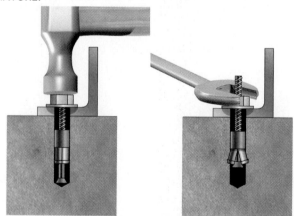

DRILL – SIMPLY DRILL A HOLE THE SAME DIAMETER AS THE ANCHOR. DO NOT WORRY ABOUT DRILLING TOO DEEP BECAUSE THE ANCHOR WORKS IN A "BOTTOMLESS HOLE." YOU CAN DRILL INTO THE CONCRETE WITH THE LOAD POSITIONED IN PLACE; SIMPLY DRILL THROUGH THE PREDRILLED MOUNTING HOLES.

ANCHOR – MERELY TIGHTEN THE NUT. RESISTANCE WILL INCREASE RAPIDLY AFTER THE THIRD OR FOURTH COMPLETE TURN.

FIGURE 10–1   The wedge anchor has high resistance to pullout. (*Courtesy of U.S. Anchor*)

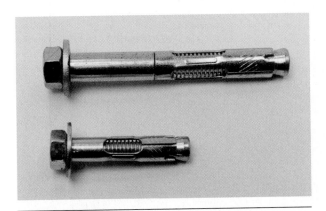

FIGURE 10–2 Sleeve anchors eliminate the problem of exact hole depth requirements.

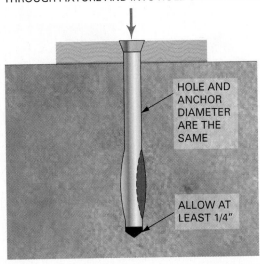

DRILL HOLE AND DRIVE ANCHOR WITH HAMMER THROUGH FIXTURE AND INTO HOLE UNTIL FLUSH.

HOLE AND ANCHOR DIAMETER ARE THE SAME

ALLOW AT LEAST 1/4"

FIGURE 10–4 The split fast is a one-piece, all-steel anchor for hard masonry.

The **sleeve anchor (Figure 10–2)** and its hole size are the same and the hole depth need not be exact. After inserting the anchor in the hole, it is expanded by tightening the nut. This anchor can be used in material such as brick that may have voids or pockets.

The **drop-in anchor (Figure 10–3)** consists of an expansion shield and a cone-shaped, internal expander plug. The hole must be drilled at least equal to the length of the anchor. A setting tool, supplied with the anchors, must be used to drive and expand the anchor. This anchor takes a machine screw or bolt.

Medium-Duty Anchors. Split fast anchors (Figure 10–4) are one-piece steel with two sheared expanded halves at the base. When driven, these halves are compressed and exert immense outward force on the inner walls of the hole as they try

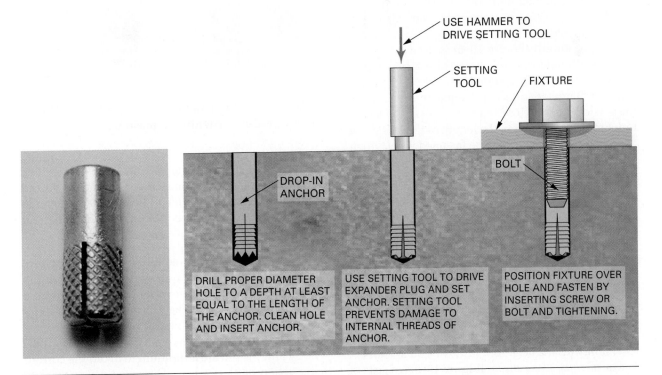

USE HAMMER TO DRIVE SETTING TOOL

SETTING TOOL

FIXTURE

DROP-IN ANCHOR

BOLT

DRILL PROPER DIAMETER HOLE TO A DEPTH AT LEAST EQUAL TO THE LENGTH OF THE ANCHOR. CLEAN HOLE AND INSERT ANCHOR.

USE SETTING TOOL TO DRIVE EXPANDER PLUG AND SET ANCHOR. SETTING TOOL PREVENTS DAMAGE TO INTERNAL THREADS OF ANCHOR.

POSITION FIXTURE OVER HOLE AND FASTEN BY INSERTING SCREW OR BOLT AND TIGHTENING.

FIGURE 10–3 The drop-in anchor is expanded with a setting tool. *(Right: courtesy of U.S. Anchor)*

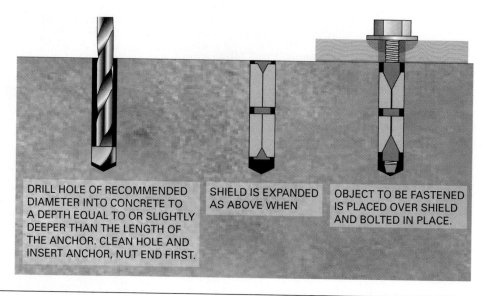

DRILL HOLE OF RECOMMENDED DIAMETER INTO CONCRETE TO A DEPTH EQUAL TO OR SLIGHTLY DEEPER THAN THE LENGTH OF THE ANCHOR. CLEAN HOLE AND INSERT ANCHOR, NUT END FIRST.

SHIELD IS EXPANDED AS ABOVE WHEN

OBJECT TO BE FASTENED IS PLACED OVER SHIELD AND BOLTED IN PLACE.

FIGURE 10–5 Two opposing wedges of the double expansion anchor pull toward each other, expanding the full length of the anchor body. *(Right: courtesy of U.S. Anchor)*

to regain their original shape. They come in both flat- and round-head styles.

Single and *double expansion anchors* **(Figure 10–5)** are used with machine screws or bolts. Drill a hole of recommended diameter to a depth equal to the length of the anchor. Place the anchor into the hole, flush with or slightly below the surface. Position the object to be fastened and bolt into place. Once fastened, the object may be unbolted, removed, and refastened, if desired.

The **lag shield (Figure 10–6)** is used with a lag screw. The shield is a split sleeve of soft metal, usually a zinc alloy. It is inserted into a hole of recommended diameter and a depth equal to the length of the shield plus ½ inch or more. The lag screw length is determined by adding the length of the shield, the thickness of the material to be fastened, plus ¼ inch. The tip of the lag screw must protrude from the bottom of the anchor to ensure proper expansion. As the fastener is threaded in, the shield expands tightly and securely in the drilled hole.

The *concrete screw* **(Figure 10–7)** utilizes specially fashioned high and low threads that cut into a properly sized hole in concrete. Screws come in ³⁄₁₆ and ¼ inch diameters and up to 6 inches in length. The hole diameter is important to the performance of the screw. It is recommended that a minimum of 1 inch and a maximum of 1¾ inch embedment be used to determine the fastener length. The concrete screw system eliminates the need for plastic or lead anchors.

The *machine screw anchor* **(Figure 10–8)** consists of two parts. A lead sleeve slides over a threaded, cone-shaped piece. Using a special setting punch that

FIGURE 10–6 Lag shields are designed for light- to medium-duty fastening in masonry.

comes with the anchors, the lead sleeve is driven over the cone-shaped piece to expand the sleeve and hold it securely in the hole. A hole of recommended diameter is drilled to a depth equal to the length of the anchor.

Light-Duty Anchors. Three kinds of **drive anchors** are commonly used for quick and easy fastening in solid masonry. They differ only in the material from which they are made. The *hammer drive anchor* **(Figure 10–9)** has a body of zinc alloy containing a steel expander pin. In the *aluminum drive anchor*, both the body and the pin are aluminum to avoid the corroding action of electrolysis. The *nylon nail anchor* utilizes a nylon body and a threaded steel expander pin. All are installed in a similar manner.

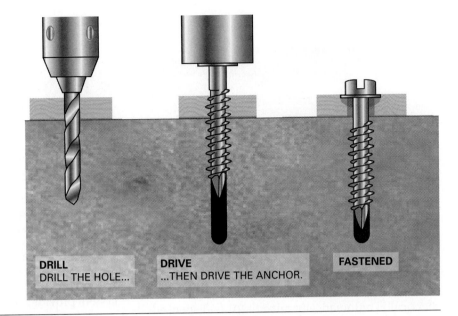

DRILL
DRILL THE HOLE...

DRIVE
...THEN DRIVE THE ANCHOR.

FASTENED

FIGURE 10–7   Concrete screws eliminate the need for an anchor when fastening into concrete.

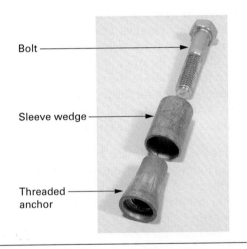

Bolt

Sleeve wedge

Threaded anchor

FIGURE 10–8   Machine screw anchors have a threaded bottom held by a sleeve wedged over the top.

Lead and *plastic anchors*, also called **inserts (Figure 10–10)**, are commonly used for fastening lightweight fixtures to masonry walls. These anchors have an unthreaded hole into which a screw is driven. The anchor is placed into a hole of recommended diameter and ¼ inch or more deeper than the length of the anchor. As the screw is turned, the threads of the screw cut into the soft material of the insert. This causes the insert to expand and tighten in the drilled hole. Ribs on the sides of the anchors prevent them from turning as the screw is driven.

Chemical Anchoring Systems. Threaded studs, bolts, and rebar (concrete reinforcing rod) may be anchored in solid masonry with a chemical bond using an *epoxy resin compound.* Two types of systems commonly used

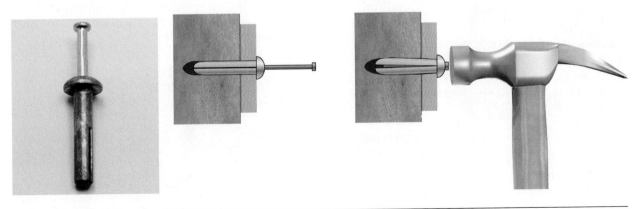

FIGURE 10–9   Hammer drive anchors come assembled for quick and easy fastening.

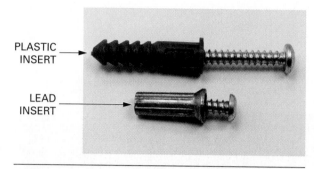

FIGURE 10–10   Lead and plastic anchors or inserts are used for light-duty fastening.

are the *epoxy injection* **(Figure 10–11)** and *chemical capsule* **(Figure 10–12).**

In the injection system, a dual cartridge is inserted into a tool similar to a caulking gun. The chemical is automatically mixed as it is dispensed. Small cartridges are available to accurately dispense epoxy from an ordinary caulking gun.

Chemical capsules contain the exact amount of all chemicals needed for one installation. Each capsule is marked with the appropriate hole size to be used. Drill holes according to diameter and depth indicated on each capsule. It is important to thoroughly clean and clear the hole of all concrete dust before inserting the capsule.

With all solid masonry anchors, follow the specifications in regard to hole diameter and depth, minimum embedment, maximum fixture thickness, and allowable load on anchor.

EPOXY INJECTION

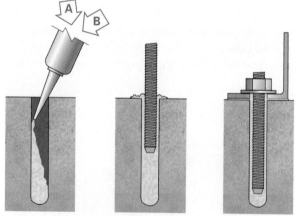

MIX THE TWO-COMPONENT ADHESIVE SYSTEM AND PLACE IN HOLE. PUSH THE ANCHOR ROD INTO THE HOLE AND ROTATE SLIGHTLY TO COAT WITH ADHESIVE. ALLOW TO CURE.

FIGURE 10–11   The epoxy injection system is designed for high-strength anchoring. (*Courtesy of U.S. Anchor*)

CHEMICAL CAPSULE ANCHOR

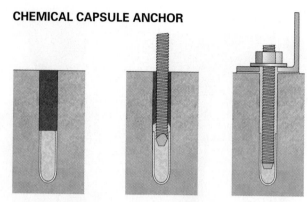

ATTACH THE ANCHOR ROD TO A ROTARY HAMMER ADAPTER. THE COMBINATION OF THE ROTATION AND HAMMERING ACTION MIXES THE CAPSULE CONTENTS TOGETHER. ALLOW TO CURE.

FIGURE 10–12   The chemical capsule anchoring system provides for easy, premeasured application.

Hollow Wall Fasteners

Toggle Bolts.   Toggle bolts (Figure 10–13) may have a wing or a tumble toggle. The wing toggle is fitted with springs, which cause it to open. The tumble toggle falls into its open position when passed through a drilled hole in the wall. The hole must be drilled large enough for the toggle of the bolt to slip through. A disadvantage of using toggle bolts is that, if removed, the toggle falls off inside the wall.

Plastic Toggles. The **plastic toggle (Figure 10–14)** consists of four legs attached to a body that has a hole through the center and fins on its side to prevent turning during installation. The legs collapse to allow insertion into the hole. As sheet metal screws are turned through the body, they draw in and expand the legs against the inner surface of the wall.

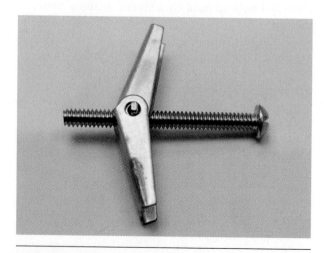

FIGURE 10–13   Toggle bolts are used for fastening in hollow walls.

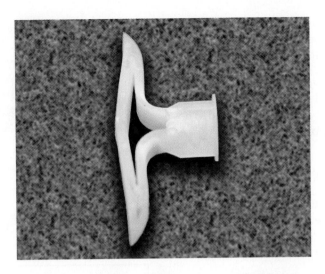

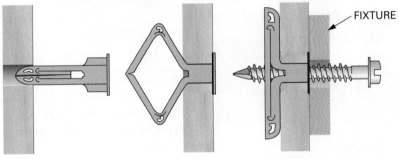

SQUEEZE TOGGLE
WINGS FLAT AND
PUSH INTO HOLE
DRILLED IN WALL.

TAP ANCHOR IN AND
FLUSH WITH WALL.

PLACE FIXTURE OVER
HOLE, INSERT SHEET
METAL SCREW, AND
TIGHTEN.

FIXTURE

FIGURE 10–14 The plastic toggle hollow wall anchor. *(Botttom: courtesy of U.S. Anchor)*

Expansion Anchors. Hollow wall **expansion anchors (Figure 10–15)** consist of an expandable sleeve, a machine screw, and a fiber washer. The collar on the outer end of the sleeve has two sharp prongs that grip into the surface of the wall material. This prevents the sleeve from turning when the screw is tightened to expand the anchor. After expanding the sleeve, the screw is removed, inserted through the part to be attached, and then screwed back into the anchor. Some types require that a hole be drilled, while other types have pointed ends that may be driven through the wall material.

Installed fixtures may be removed and refastened or replaced by removing the anchor screw without disturbing the anchor expansion. Anchors are manufactured for various wall board thicknesses. Make sure to use the right size anchor for the wall thickness in which the anchor is being installed.

Conical Screws. The deep threads of the **conical screw** anchor **(Figure 10–16)** resist stripping out when screwed into gypsum board, strand board, and

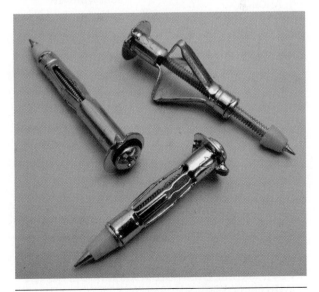

FIGURE 10–15 Hollow wall expansion anchors.

similar material. After the plug is seated flush with the wall, the fixture is placed over the hole and fastened by driving a screw through the center of the plug.

Nylon Plugs. The **nylon plug (Figure 10–17)** is used for a number of hollow wall and some solid wall applications. A hole of proper diameter is drilled. The plug is inserted, and the screw is driven to draw or expand the plug.

Connectors

Widely used in the construction industry are devices called **connectors.** Connectors are metal pieces formed into various shapes to join wood to wood, or wood to concrete or other masonry. Connectors are

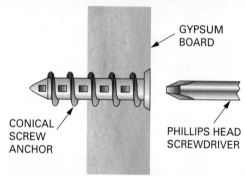

GYPSUM
BOARD

CONICAL
SCREW
ANCHOR

PHILLIPS HEAD
SCREWDRIVER

DRIVE ANCHOR IN WALL BY TURNING WITH
SCREWDRIVER UNTIL HEAD IS FLUSH WITH SURFACE.

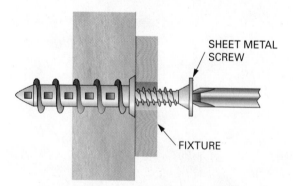

SHEET METAL
SCREW

FIXTURE

PLACE FIXTURE OVER HOLE IN ANCHOR AND FASTEN
WITH PROPER SIZE SHEET METAL SCREW.

FIGURE 10–16 The conical screw anchor is a self-drilling, hollow wall anchor for lightweight fastenings. *(Bottom: courtesy of U.S. Anchor)*

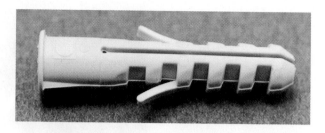

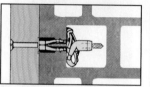

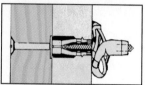

HOLLOW BRICK

PLASTER BOARD

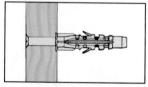

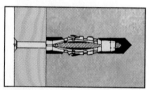

HIGH VALUES IN PLASTER

AERATED CONCRETE

FIGURE 10–17 The nylon plug is used for many types of hollow wall fastening. *(Courtesy of U.S. Anchor)*

available in hundreds of shapes and styles and have specific names depending on their function. Only a few are discussed here.

Wood-to-Wood. *Framing anchors* and *seismic* and *hurricane ties* **(Figure 10–18)** are used to join parts of a wood frame. *Post* and *column caps* and *bases* are used at the top and bottom of those members **(Figure 10–19).** *Joist hangers* and *beam hangers* are available in many sizes and styles **(Figure 10–20).** It is important to use the proper style, size, and quantity of nails in each hanger.

Wood-to-Concrete. Some wood-to-concrete connectors are *sill anchors, anchor bolts,* and *hold-downs* **(Figure 10–21).** A *girder hanger* and a *beam seat* **(Figure 10–22)** make beam-to-foundation wall connections. *Post bases* come in various styles. They are used to anchor posts to concrete floors or footings.

Many other specialized connectors are used in frame construction. Some are described in the framing sections of this book.

ADHESIVES

The carpenter seldom uses any glue in the frame or exterior finish. Glue is used on some joints and other parts of the interior finish work. A number of **mastics** (heavy, paste-like adhesives) are used throughout the construction process.

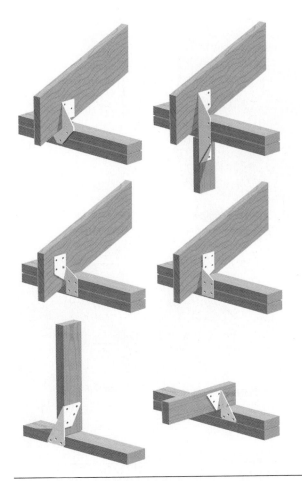

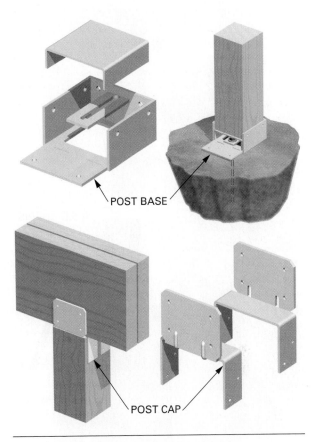

POST BASE

POST CAP

FIGURE 10–19 Caps and bases help fasten tops and bottoms of posts and columns. *(Courtesy of Simpson Strong-Tie Co.)*

FIGURE 10–18 Framing ties and anchors are manufactured in many unique shapes. *(Courtesy of Simpson Strong-Tie Co.)*

Glue

White and Yellow Glue. Most of the glue used by the carpenter is the so-called white glue or yellow glue. The white glue is *polyvinyl acetate;* the yellow glue is *aliphatic resin.* Neither type is resistant to moisture. Both are fast setting, so joints should be made quickly after applying the glue. They are available under a number of trade names and are excellent for joining wood parts not subjected to moisture.

Urethane Glue. **Urethane glue** is a fine all-purpose glue available for bonding wood, stone, metal, ceramics, and plastics. Its strong, waterproof bond cures with exposure to moisture in material and air. It can be used for interior or exterior work and does not become brittle over time. It tends to expand while curing, filling gaps and spaces in material.

Because urethane glue sticks to just about anything, care should be taken when working with it. It cannot be dissolved by common solvents and clean up can be difficult. It often requires days to scrape and rub it from skin.

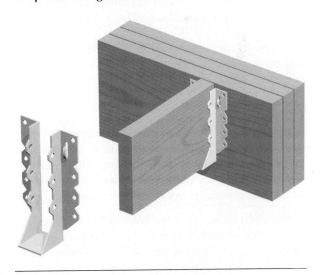

FIGURE 10–20 Hangers are used to support joists and beams. *(Courtesy of Simpson Strong-Tie Co.)*

Contact Cement. **Contact cement** is so named because pieces coated with it bond on contact and need not be clamped under pressure. It is extremely important that pieces be positioned accurately before contact is made. Contact cement is widely used to

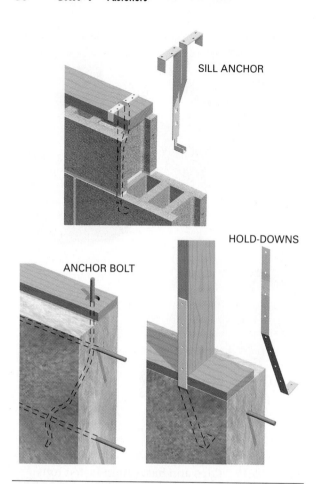

SILL ANCHOR

HOLD-DOWNS

ANCHOR BOLT

FIGURE 10–21   Sill anchors, anchor bolts, and hold-downs connect frame members to concrete. *(Courtesy of Simpson Strong-Tie Co.)*

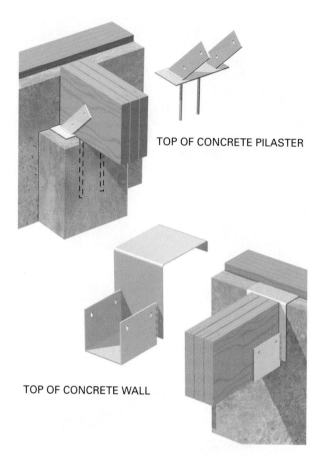

TOP OF CONCRETE PILASTER

TOP OF CONCRETE WALL

FIGURE 10–22   Girder and beam seats provide support from concrete walls. *(Courtesy of Simpson Strong-Tie Co.)*

apply plastic laminates for kitchen countertops. It is also used to bond other thin or flexible material that otherwise might require elaborate clamping devices.

Mastics

Several types of mastics are used throughout the construction trades. They come in cans or cartridges used in hand or air guns and in large quantities that are troweled into place. With these adhesives, the bond is made stronger and fewer fasteners are needed.

Construction Adhesive. One type of mastic is called *construction adhesive.* It is used in a glued floor system, described in a following unit on floor framing. It can be used in cold weather, even on wet or frozen wood. It is also used on stairs to increase stiffness and eliminate squeaks.

Panel Adhesive. *Panel adhesive* **(Figure 10–23)** is used to apply such things as wall paneling, foam insulation, gypsum board, and hardboard to wood, metal, and masonry. It is usually dispensed with a caulking gun. It is important to use the adhesives matched for the material being installed.

FIGURE 10–23   Applying panel adhesive to stud with a caulking gun.

Troweled Mastics. Other types of mastics may be applied by hand for such purposes as installing vinyl base, vinyl floor tile, or ceramic wall tile. A notched trowel is usually used to spread the adhesive. The depth and spacing of the notches along the edges of the trowel determine the amount of adhesive left on the surface.

It is important to use a trowel with the correct notch depth and spacing. Failure to follow recommendations will result in serious consequences. Too much adhesive causes the excess to squeeze out onto the finished surface. This leaves no alternative but to remove the applied pieces, clean up, and start over. Too little adhesive may result in loose pieces.

Key Terms

anchors	contact cement	machine bolt	sleeve anchor
box nails	drive anchor	masonry nails	solid wall anchors
brads	drop-in anchor	mastics	split fast anchor
bright	duplex nail	nylon plug	staple
carriage bolt	electrolysis	penny	stone bolt
casing	expansion anchors	Phillips head	toggle bolts
common nails	finish nails	plastic toggle	urethane glue
common screw	galvanized	roofing nails	wedge anchor
conical screw	lag screws	self-tapping screw	wire nails
connectors	lag shield		

Review Questions

Select the most appropriate answer.

1. The length of an eight penny nail is
   a. 1½ inches.
   b. 2 inches.
   c. 2 ½ inches.
   d. 3 inches.

2. Fasteners coated with zinc to retard rusting are called
   a. coated.
   b. dipped.
   c. electroplated.
   d. galvanized.

3. Brads are
   a. types of screws.
   b. small box nails.
   c. small finishing nails.
   d. kinds of stove bolts.

4. When a moisture-resistant exterior glue is required use
   a. white glue.
   b. urethane glue.
   c. yellow glue.
   d. rubber cement.

5. Many carpenters prefer to use casing nails to fasten
   a. interior finish.
   b. exterior finish.
   c. door casings.
   d. roof shingles.

6. The blunt point on the end of a cut nail helps
   a. drive the nail straight.
   b. prevent splitting of the wood.
   c. hold the fastened material more securely.
   d. start the nail in the material.

7. On temporary structures, such as wood scaffolding, and concrete forms, use
   a. common nails.
   b. duplex nails.
   c. galvanized nails.
   d. spikes.

8. The common name for a fastener with tapered threads is
   a. bolt.
   b. screw.
   c. lag.
   d. all of the above.

9. An example of a solid wall anchor is a
   a. wedge anchor.
   b. conical screw.
   c. toggle bolt.
   d. all of the above.

10. The term *connector* refers to a
   a. metal device used to fasten wood to masonry.
   b. wire used to make nails and screws.
   c. worker using a nail or screw gun.
   d. all of the above.

UNIT 5
Hand Tools

One of the many benefits to working in the field of construction is the variety and diversity of tools available. Tools are the means by which construction happens.

Knowing how to choose the proper tool and how to keep it in good working condition is essential. A tradesperson should never underestimate the importance of tools and never neglect their proper use and care. Tools should be kept clean and in good condition. If they get wet on the job, dry them as soon as possible, and coat them with light oil to prevent them from rusting.

Carpenters are expected to have their own hand tools and to keep them in good working condition. Tools vary in quality, which is related to cost. Generally, expensive tools have better quality than inexpensive tools. For example, inferior tools cannot be brought to a sharp, keep edge and will dull rapidly. They will bend or break under normal use. Quality tools are worth the expense. The condition of a tool reveals the attitude of the owner toward his or her profession.

SAFETY REMINDER

Carpentry as a trade was created using hand tools, some having a long history. Each tool has a specific purpose and associated risk of use.

The use of tools requires the operator to be knowledgeable about how to safely manipulate the tools. This applies to hand tools as well as power tools. Safety is an attitude—an attitude of acceptance of a tool and all of its operational requirements. Safety is a blend of ability, skill, and knowledge—a blend that should always be present when working with tools.

OBJECTIVES

After completing this unit, the student should be able to:

- identify and describe the hand tools that are commonly used by the carpenter.
- use each of the hand tools in a safe and appropriate manner.
- sharpen and maintain hand tools in suitable working condition.

11 Layout Tools

LAYOUT TOOLS

Much of the work a carpenter does must first be laid out, measured, and marked. Layout tools are used to measure distances, mark lines and angles, test for depths, and align various material into the proper positions.

Measuring Tools

The ability to take measurements quickly and accurately must be mastered early in the carpenter's training. Practice reading the rule or tape to gain skill in fast and precise measuring.

Most industrialized countries use the metric system of measure. Linear metric measure centers on the meter, which is slightly larger than a yard. Smaller parts of a meter are denoted by the prefix *deci-* (¹⁄₁₀), which is used instead of *feet*. *Centi-* (¹⁄₁₀₀) and *milli-* (¹⁄₁₀₀₀) are used instead of inches and fractions. The prefix *kilo-* represents 1,000 times larger and is used instead of *miles*. For example, in metric measure a 2 × 4 is 39 mm (millimeter) × 89 mm. The metric system is easier to use than the English system because all measurements are in decimal form and there are no fractions.

Pocket Tapes. Most measuring done by tradespeople is done with **pocket tapes (Figure 11–1)**. These are painted steel ribbons wound around a spool with a spring inside. The spring returns the tape after it is extended. They are made as small as 3 feet but typical professional models are 16, 25, and 33 feet long.

They are divided into feet, inches, and sixteenths of an inch. They have clearly marked increments of 12 and 16 inches, the spacing for standard framing members, to speed up the layout. Markings are usually black for each 12 inches and red for every 16 inches. Some tapes also have small black dots at increments of 19.2 inches **(Figure 11–2)**. This spacing is typically only used for layout of some engineered floor members.

FIGURE 11–1 Pocket tape. (*Courtesy of Stanley Tools*)

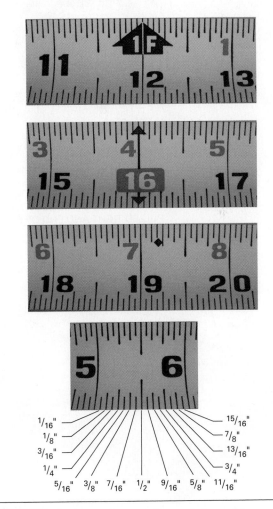

FIGURE 11–2 Tapes have color coded markings at 12-, 16- and 19.2-inch intervals. Each inch is typically broken into sixteenth of an inch increments.

Each inch on a tape is divided into fractions of an inch. Each fraction line has a name that must be memorized (see Figure 11–2). Most measuring done by a carpenter is to the nearest 16th, while a cabinetmaker will work to a 64th. A carpenter should be able to read a ruler quickly and accurately.

Steel tapes in 50- and 100-foot lengths are commonly used to lay out longer measurements. They are not spring loaded, so they must be rewound by hand. The end of the tape has a steel ring with a folding hook attached. The hook may be unfolded to go over the edge of an object. It may also be left in the folded position and the ring placed over a nail when extending the tape. Remember to place the nail so that the *outside* of the ring, which is the actual end of the tape, reaches to the desired mark **(Figure 11–3).** Rewind the

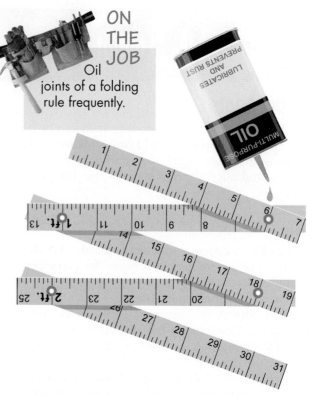

ON THE JOB
Oil joints of a folding rule frequently.

FIGURE 11–4   All tools require some form of preventive maintenance.

tape when not using it. If the tape is kinked, it will snap. Keep it out of water. If it gets wet, dry it thoroughly while rewinding.

Some tradespeople prefer the 6- or 8-foot *folding rule.* The folding rule sometimes has a metal extension on one end for making inside measurements. You should also oil the joints occasionally to prevent breaking the rule when opening and closing it **(Figure 11–4).**

Squares

The carpenter has the use of a number of different kinds of squares to measure and lay out for square and other angle cuts.

Combination Squares. The **combination square (Figure 11–5)** consists of a movable blade, 1 inch wide and 12 inches long, that slides along the body of the square. It is used to lay out or test 90- and 45-degree angles. Hold the body of the square against the edge of the stock and mark along the blade **(Figure 11–6).** It can function as a depth gauge to lay out or test the depth of **rabbets, grooves,** and **dadoes.** It can also be used with a pencil as a marking gauge to draw lines parallel to the edge of a board. Drawing lines in this manner is called *gauging* lines. Lines may also be gauged by holding the pencil and riding the finger along the edge of the board. Finger-gauging takes practice, but

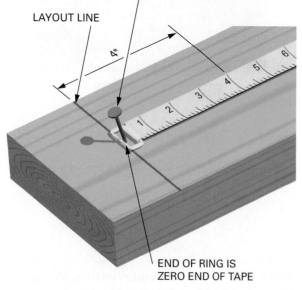

DRIVE NAIL ON ANGLE SO TAPE WILL SLIDE TO SURFACE WHEN PULLED TIGHT AND END OF RING WILL BE ON LAYOUT LINE

LAYOUT LINE

4"

END OF RING IS ZERO END OF TAPE

FIGURE 11–3   Steel tape. *(Top: courtesy of Stanley Tools)*

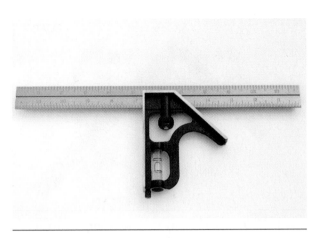

FIGURE 11–5 **The body and blade of a combination square are adjustable.**

once mastered saves a lot of time. Be sure to check the edge of the wood for slivers first.

Speed Squares. Some carpenters prefer to use a triangular-shaped square known by the brand name **Speed Square (Figure 11–7).** Speed Squares are made of one-piece plastic or aluminum alloy and are available in two sizes. They can be used to lay out 90- and 45-degree angles and as guides for portable power saws. A degree scale allows angles to be laid out; other scales may be used to lay out rafters.

Framing Squares. The **framing square,** often called the *steel, or rafter square* **(Figure 11–8),** is an L-shaped tool made of thin steel or aluminum. The longer of the two legs is called the *blade* or *body* and is 2 inches

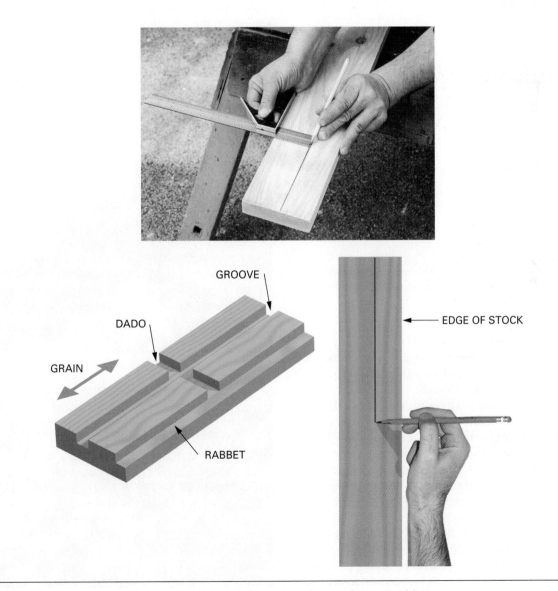

GROOVE

DADO

GRAIN

RABBET

EDGE OF STOCK

FIGURE 11–6 **The combination square is useful for squaring and as a marking gauge. A pencil held in one hand is a quick way to draw a parallel line. Check the wood first to reduce the potential for splinters.**

FIGURE 11–7 Speed Squares are used for layout of rafters and other angles.

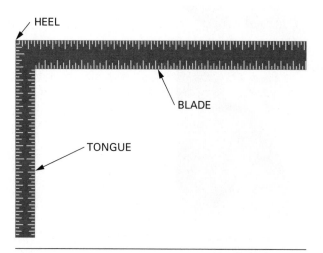

FIGURE 11–8 Framing or rafter square. (*Courtesy of Stanley Tools*)

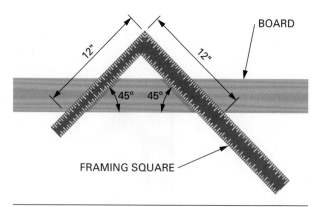

FIGURE 11–9 Laying out 45-degree angles with a framing square.

table is printed on the body. On the same side of the square, on the tongue, can be found the *octagon scale*, which is used to lay out eight-sided timbers from square ones **(Figure 11–10).**

On the back side of the square, the *Essex board foot table* is used to calculate the number of board feet in lumber. The *brace table* is used to figure the length of

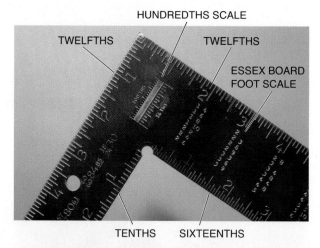

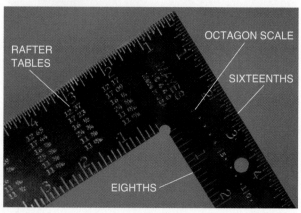

FIGURE 11–10 Both sides of a rafter square have tables and scales to assist the carpenter.

(50 mm) wide and 24 inches (600 mm) long. The shorter leg is called the *tongue* and it is 1½ inches (38 mm) wide and 16 inches (400 mm) long. The outside corner is called the *heel.*

The framing square is a centuries-old tool. Entire books have been written about it. Based on the use of the right triangle, many layout techniques have been devised and used. These techniques and necessary scales, tables, and graduations stamped on the square were designed to assist the carpenter in the many calculations needed. Today a pocket calculator has virtually replaced these aids. The only exception is the rafter table, which is still useful today and will be discussed in more detail in Unit 15.

The framing square is useful in laying out roof rafters, bridging, and stairs. It is also used to lay out 90- and 45-degree angles **(Figure 11–9).**

The side that has the manufacturer's name stamped on it is referred to as the *face side.* The rafter

PROCEDURE 11–A Checking a Straightedge

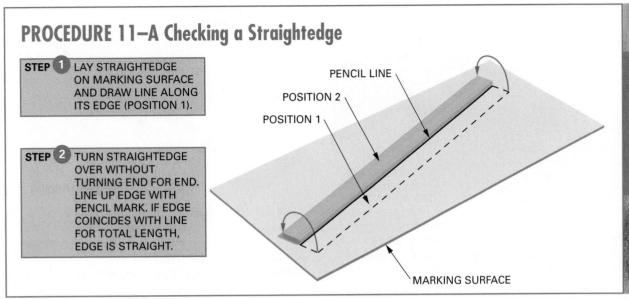

STEP 1 LAY STRAIGHTEDGE ON MARKING SURFACE AND DRAW LINE ALONG ITS EDGE (POSITION 1).

STEP 2 TURN STRAIGHTEDGE OVER WITHOUT TURNING END FOR END. LINE UP EDGE WITH PENCIL MARK. IF EDGE COINCIDES WITH LINE FOR TOTAL LENGTH, EDGE IS STRAIGHT.

PENCIL LINE

POSITION 2

POSITION 1

MARKING SURFACE

diagonal braces. The *hundredths scale,* consisting of an inch divided into one hundred parts, is used to find ¹⁄₁₀₀ths of an inch. This scale may be used to convert fractions to decimals and vice versa.

On the face side, the edges are divided into inches that are graduated into ⅛ths on the inside and ¹⁄₁₆ths on the outside. The edges on the back are divided into inches and ½ths on the outside, while one inside edge is graduated into ¹⁄₁₆ths and the other into ¹⁄₁₀ths.

Sliding T-Bevels. The *sliding T-bevel,* sometimes called a *bevel square* or just a **bevel (Figure 11–11)**, consists of a body and a sliding blade that can be turned to any angle and locked in position. It is used to lay out or test angles other than those laid out with squares. The body of the tool is held against the edge of the stock, and the angle is laid out by marking along the blade.

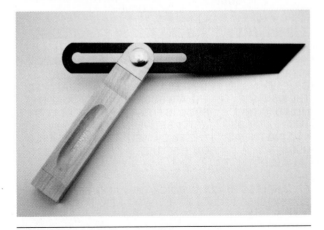

FIGURE 11–11   Sliding T-bevel. *(Courtesy of Stanley Tools)*

Straightedges

A **straightedge** can be made of metal or wood. It can have any thickness, width, or length, as long as the size is convenient for its intended use and it has at least one edge that is absolutely straight from one end to the other. To determine if it is straight, sight along the edge. Another way is to lay the piece on its side and mark along its edge from one end to the other. Turn the piece over. Hold each end on the line just marked, and mark another line. If both lines coincide, the edge is straight **(Procedure 11–A).**

Straightedges are useful for many purposes. The framing square, the blade of the combination square, or the back of a saw could be used as a straightedge for drawing short, straight lines.

To determine a straight distance over large distances, a line (or string) and gauge blocks can be used. This method uses a taut string held away by offset blocks from the surface of the material being straightened. Another block of the same thickness as the offset blocks is then used periodically to test the material's distance from the line **(Figure 11–12)**. This method is easy to do and can be very accurate.

Trammel Points. A pair of tools called **trammel points** may be used to draw circles or parts of circles, called arcs **(Figure 11–13),** which may be too large for a compass. They can be clamped to a strip of wood any distance apart according to the desired radius of the circle to be laid out. One trammel point can be set on the center while the other, which may have a pencil attached, is swung to lay out the circle or arc.

In place of trammel points, the same kinds of layouts can be made by using a thin strip of wood with a brad or small finish nail through it for a center

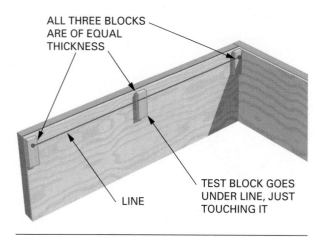

FIGURE 11–12   Use of a line and gauge blocks is an effective method to determine a straight line.

FIGURE 11–13   Trammel points are used to lay out arcs of large diameter.

point. Measure from the end of the strip a distance equal to the desired radius. Drive the brad through the strip until the point comes through. Set the point of the brad on the center, and hold a pencil against the end while swinging the strip to form the circle or arc **(Figure 11–14)**. To keep the pencil from slipping, a small V may be cut on the end of the strip or a hole may be drilled near the end to insert the pencil. Make sure measurements are taken from the bottom of the V or the center of the hole.

Levels

In construction, the term **level** is used to indicate that which is *horizontal*, and the term **plumb** is used to mean the same as *vertical*. The term *level* also refers to a tool that is used to achieve both level and plumb.

Carpenter's Levels.

The **carpenter's level (Figure 11–15)** is used to test both level and plumb surfaces. Accurate use of the level depends on accurate read-

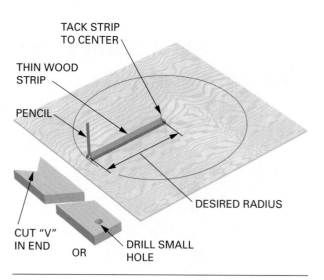

FIGURE 11–14   A thin strip of wood can be used to lay out circles or arcs.

ing. The air bubble in the slightly crowned glass tube of the level must be exactly centered between the lines marked on the tube. The tubes of a level are oriented in two directions for testing level and plumb. The number of tubes in a level depends on the level length and manufacturer.

Levels are made of wood or metal, usually aluminum. They come in various lengths from 12 to 78 inches. It is wise to use the longest level practical to improve accuracy.

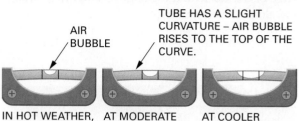

IN HOT WEATHER, OR IF LAYING IN THE SUN, THE LIQUID IN THE TUBE EXPANDS MAKING THE AIR BUBBLE SMALLER.

AT MODERATE TEMPERATURES THE AIR BUBBLE SHOULD FIT EXACTLY BETWEEN THE TWO LINES ON THE TUBE.

AT COOLER TEMPERATURES THE LIQUID IN THE TUBE CONTRACTS MAKING THE AIR BUBBLE LARGER.

REGARDLESS OF CONDITIONS, THE AIR BUBBLE MUST BE CENTERED BETWEEN THE TWO LINES ON THE TUBE.

FIGURE 11–15   The bubble size of a carpenter's level can be affected by temperature.

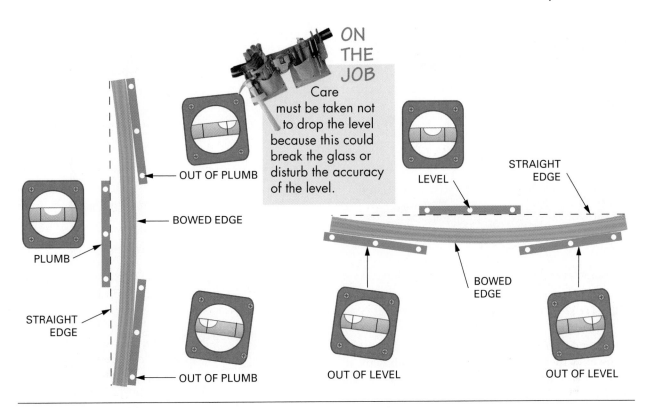

ON THE JOB Care must be taken not to drop the level because this could break the glass or disturb the accuracy of the level.

FIGURE 11–16 To be level or plumb for their entire length, pieces must be straight from end to end.

An important point to remember is that level and plumb lines, or objects, must also be straight throughout their length or height. Parts of a structure may have their end points level or plumb with each other. If they are not straight in between, however, they are not level or plumb for their entire length **(Figure 11–16).**

To check a level for accuracy, place it on a nearly level or plumb object that is firm. Note the exact position of the level on the object. Read the level carefully and remember where the bubble is located within the lines on the bubble tube. Rotate the level along its vertical axis and reposition it in the same place on the object **(Figure 11–17).** If the bubble reads the same as the previous measurement, then the level is accurate.

Line Levels. The **line level (Figure 11–18)** consists of one glass tube encased in a metal sleeve with hooks on each end. The hooks are attached to a stretched line, which is then moved up or down until the bubble is centered. However, this is not an accurate method and gives only approximate levelness. Care must be taken that the level be attached close to the center of the suspended line because the weight of the level causes the line to sag. If the line level is off center to any great degree, the results are faulty.

Plumb Bobs. The **plumb bob (Figure 11–19)** is accurate and is used frequently for testing and establishing

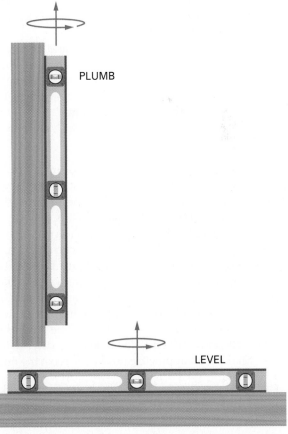

FIGURE 11–17 To check a level for accuracy the bubble should read exactly the same before and after rotating it.

FIGURE 11–18 Line level.

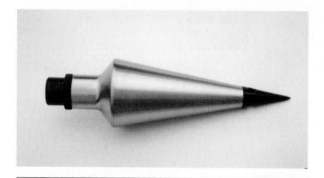

FIGURE 11–19 Plumb bob.

plumb lines. Suspended from a line, the plumb bob hangs absolutely vertical. However, it is difficult to use outside when the wind is blowing because it will move with the wind. Plumb bobs come in several different weights. Heavy plumb bobs stop swinging more quickly than lighter ones. Some have hollow centers that are filled with heavy metal to increase the weight without enlarging the size.

The plumb bob is useful for quick and accurate plumbing of vertical members of a structure **(Figure 11–20).** It can be suspended from a great height to establish a point that is plumb with another. Its only limitation is the length of the line.

Chalk Lines

Long straight lines are laid out by using a **chalk line.** A line coated with chalk dust is stretched tightly between two points and snapped against the surface **(Figure 11–21).** The chalk dust is dislodged from the line and remains on the surface.

A *chalk box* or *chalk line reel* is filled with chalk dust that comes in a number of colors **(Figure 11–22).** The most popular colors are blue, yellow, red, and white. The dust saturates the line, which is on a reel inside the box. The line is ready to be snapped when it is pulled out of the box. After several snaps, the line will need more chalk and must be reeled in to be recoated with chalk. Shaking or tapping the box helps recoat the line.

Chalk Line Techniques. When unwinding and chalking the line, keep it off the surface until snapped. Otherwise many lines will be made on the surface, and this could be confusing. Make sure lines are stretched tight before snapping in order to snap a straight and true line. Sight long lines by eye for straightness to make sure there is no sag in the line.

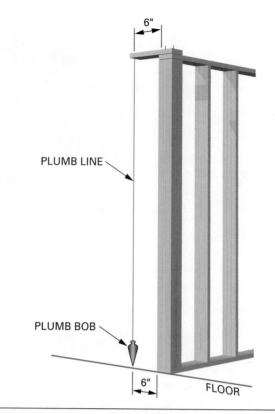

FIGURE 11–20 The post is plumb when the distance between it and the plumb line is the same.

FIGURE 11–21 Snapping a chalk line.

FIGURE 11–22 Chalk line reel.

If there is a sag, take it out by supporting the line near the center. Press the center of the line to the material and snap the line on both sides of the center. Keep the line from getting wet. If it does get wet, leave it outside the box until it dries.

Wing Dividers

Wing dividers can be used as a compass to lay out circles and arcs and as dividers to space off equal distances. However, this tool is used mainly for **scribing** and is often called a *scriber*. Scribing is the technique of laying out stock to fit against an irregular surface **(Figure 11–23)**. For easier and more accurate scribing, heat and bend the end of the solid metal leg outward **(Figure 11–24)**. Pencils are usually used in place of the interchangeable steel marking leg. Use pencils with hard lead that keep their points longer.

Butt Markers

Butt markers (Figure 11–25) are available in three sizes. They are often used to mark hinge gains. The marker is laid on the door edge at the hinge location and tapped with a hammer to outline the cutout for the hinge.

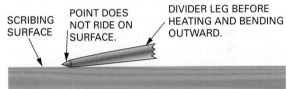

ON THE JOB

Bend the leg of the dividers as shown for easier and more accurate scribing.

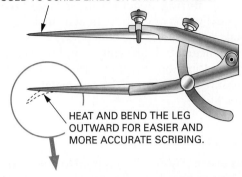

METAL LEG IS INTERCHANGEABLE WITH PENCIL AND IS USED TO SCRIBE LINES ON DARK SURFACES.

HEAT AND BEND THE LEG OUTWARD FOR EASIER AND MORE ACCURATE SCRIBING.

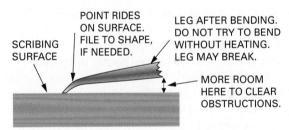

SCRIBING SURFACE

POINT DOES NOT RIDE ON SURFACE.

DIVIDER LEG BEFORE HEATING AND BENDING OUTWARD.

SCRIBING MAY NOT BE ACCURATE BECAUSE POINT IS NOT RIDING ON SURFACE.

POINT RIDES ON SURFACE. FILE TO SHAPE, IF NEEDED.

LEG AFTER BENDING. DO NOT TRY TO BEND WITHOUT HEATING. LEG MAY BREAK.

SCRIBING SURFACE

MORE ROOM HERE TO CLEAR OBSTRUCTIONS.

SCRIBING IS MORE ACCURATE WHEN POINT RIDES ON SURFACE.

FIGURE 11–24 Adjusting one of the metal legs of a scriber makes it a more accurate tool.

FIGURE 11–23 Scribing is laying out a piece to fit against an irregular surface.

FIGURE 11–25 Butt hinge markers make mortising the hinge into a door easier.

12 Boring and Cutting Tools

BORING TOOLS

The carpenter is often required to cut holes in wood and metal. **Boring** denotes cutting large holes in wood. **Drilling** is often thought of as making holes in metal or small holes in wood. Boring tools include those that actually do the cutting and those used to turn the cutting tool. The hole size and the bit size are measured according to their diameters. (See also Chapter 14.)

Bit Braces

The **bit brace (Figure 12–1)** is used to hold and turn auger bits to bore holes in wood. Its size is determined by its *sweep* (the diameter of the circle made by its handle). Sizes range from 8 to 12 inches. Most bit braces come with a ratchet that can be used when there is not enough room to make a complete turn of the handle.

Bits

Auger Bits. **Auger bits (Figure 12–2)** are available with coarse or fine *feed screws*. Bits with coarse feed screws are used for fast boring in rough work. Fine feed bits are used for slower boring in finish work. As the bit is turned, the feed screw pulls the bit through the wood so little or no pressure on the bit is necessary. The *spurs* score the outer circle of the hole in advance of the *cutting lips*. The cutting lips lift the chip up and through the twist of the bit.

A full set of auger bits ranges in sizes from ¼ to 1 inch, graduated in ¹⁄₁₆-inch increments. The bit size is designated by the number of ¹⁄₁₆-inch increments in its diameter. For instance, a #12 bit has 12 sixteenths. Therefore, it will bore a ¾-inch diameter hole.

FIGURE 12–1   Bit brace.

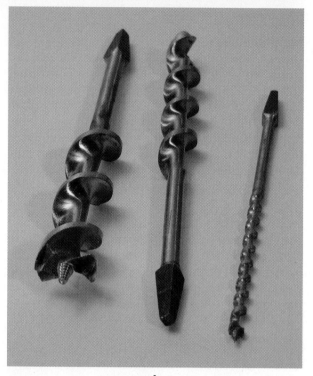

A

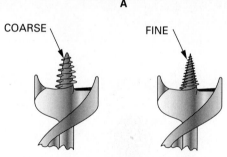

COARSE   FINE

COARSE AND FINE AUGER BIT FEED SCREWS

B

FIGURE 12–2   (A) Auger bit and (B) coarse and fine feed screws.

Expansive Bits. To bore holes over 1 inch in diameter, the carpenter may use an **expansive bit (Figure 12–3).** With two interchangeable and adjustable cutters, large-diameter holes may be bored. This tool is handy for boring holes in doors for locksets when power tools are not available. Usually these locksets require a hole of 2⅛ inches in diameter.

Boring Techniques. To avoid splintering the back side of a piece when boring all the way through, stop when the point comes through. Finish by boring

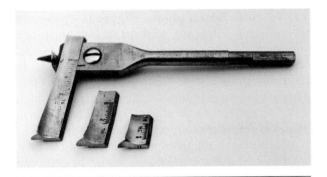

FIGURE 12–3   Expansive bit.

from the opposite side. This is especially important when both sides are exposed to view as in the case of a door. Care must be taken not to strike any nails or other objects that might cause blunting and shortening of the spurs. If the spurs become too short, the auger bit is ruined.

Drills

Twist Drills. **Twist drills** range in size from ¹⁄₁₆ to ¼ inch in increments of ¹⁄₆₄ inch **(Figure 12–4).** High-speed twist bits are made of hardened steel that allows the bit to drill holes in various materials including mild steel. The general rule is to drill in wood at high speed and lower pressure, but to drill in mild steel at slow speeds and high pressure. Cutting oil can be used to cool the bit, keeping it sharper longer.

Countersinks. The **countersink (Figure 12–5)** may be turned in a power drill. It forms a recess for a flat-head screw to set flush with the surface of the material in which it is driven.

FIGURE 12–5   Countersink boring bit.

Combination Drills. *Combination drills* and *countersinks* are used to drill shank and pilot holes for screws and countersink in one operation **(Figure 12–6).**

CUTTING TOOLS

The carpenter uses many kinds of cutting tools and must know which to select for each job, as well as how to use, sharpen, and maintain them.

Edge-Cutting Tools

Wood Chisels. The **wood chisel (Figure 12–7),** is used to cut recesses in wood for such things as door hinges and locksets and to make joints.

Chisels are sized according to the width of the blade and are available in widths of ⅛ inch to 2 inches. Most carpenters can do their work with a

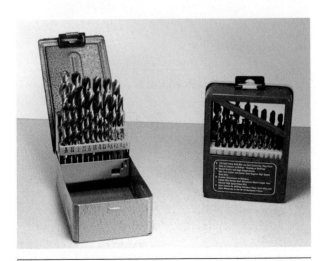

FIGURE 12–4   Twist bit sets have bits that may range in size from ¹⁄₁₆ to ½ in increments of ¹⁄₆₄.

FIGURE 12–6   Combination drill and countersink.

FIGURE 12–7 Wood chisel.

set consisting of chisels that are ¼, ½, ¾, 1, and 1½ inches in size.

Firmer chisels have long, thick blades and are used on heavy framing. *Butt chisels* are short, with a thinner blade. They are preferred for finish work.

⚠ **CAUTION CAUTION CAUTION CAUTION**

Improper use of chisels has caused many accidents. When using chisels, keep both hands behind the cutting edge at all times **(Figure 12–8)**. When not in use, the cutting edge should be shielded. Never put or carry chisels or other sharp or pointed tools in pockets. ■

FIGURE 12–8 Keep both hands in back of the chisel's cutting edge.

Bench Planes. Bench planes (Figure 12–9) come in several sizes. They are used for smoothing rough surfaces and bringing work down to the desired size. Large planes are used, for instance, on door edges to produce a straight surface over a long distance. Long planes will bridge hollows in a surface and cut high spots better than short plane **(Figure 12–10)**. Small planes are more easily used for shorter work.

Bench planes are given names according to their length. The longest is called the *jointer.* In declining order are the *fore, jack,* and *smooth* planes. It is not necessary to have all the planes. The jack plane is 14 inches long and of all the bench planes is considered the best for all-around work.

Block Planes. Block planes are small planes designed to be held in one hand. They are often used to smooth the edges of short pieces and for trimming end grain to make fine joints **(Figure 12–11)**.

Block planes are designed differently than bench planes. On bench planes, the cutting edge bevel is on the bottom side. On block planes, it is on the top. In

FIGURE 12–9 The jack plane is a general-purpose bench plane.

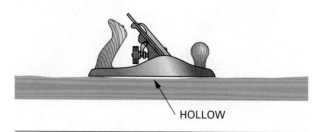

HOLLOW

FIGURE 12–10 Longer planes bridge hollows to allow for planing of long, straight edges.

FIGURE 12-11 A block plane is small and often has a low blade angle.

addition, the **bench plane iron** has a plane iron cap attached to it, while the block plane iron has none **(Figure 12–12)**.

Unlike bench planes, block planes are available with their blades set at a high angle or at a low angle. Most carpenters prefer the low-angle block plane because this type of plane cuts end grain more effectively. They also have a smoother cutting action and fit into the hand more comfortably.

Using Planes. When planing, have the stock securely held against a stop. Always plane with the grain. When starting, push forward while applying pressure downward on the **toe** (front). When the **heel** (back) clears the end, apply pressure downward on both ends while pushing forward. When the opposite end is approached, relax pressure on the toe and continue pressure on the heel until the cut is complete **(Figure 12–13)**. This method prevents tilting the plane over the ends of the stock and helps ensure a straight, smooth edge.

Sharpening Chisels and Plane Irons. To produce a keen edge, the tool must be **whetted** (sharpened) using an oilstone or waterstone. Hold the tool on a well-oiled stone so that the bevel rests flat on it. Move the tool back and forth across the stone for a few strokes. Then, make a few strokes with the flat side of the chisel or plane iron held absolutely flat on the stone. Continue whetting in this manner until as keen an edge as possible is obtained. To obtain a keener edge, repeat the procedure on a finer stone or on a piece of leather. The edge is sharp when, after having whetted

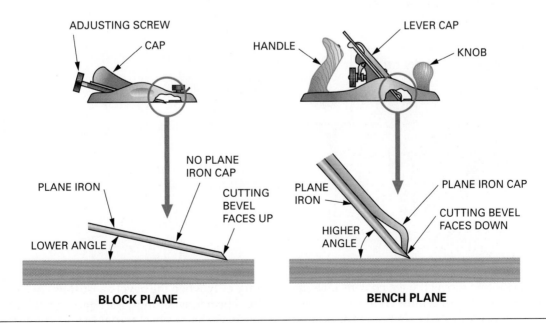

FIGURE 12-12 Difference between a block plane and a bench plane.

2 PRESSURE
POINTS

3 PRESSURE
POINTS

2 PRESSURE
POINTS

FIGURE 12–13   Correct method for using a plane.

PROCEDURE 12–A Sharpening a Plane Iron

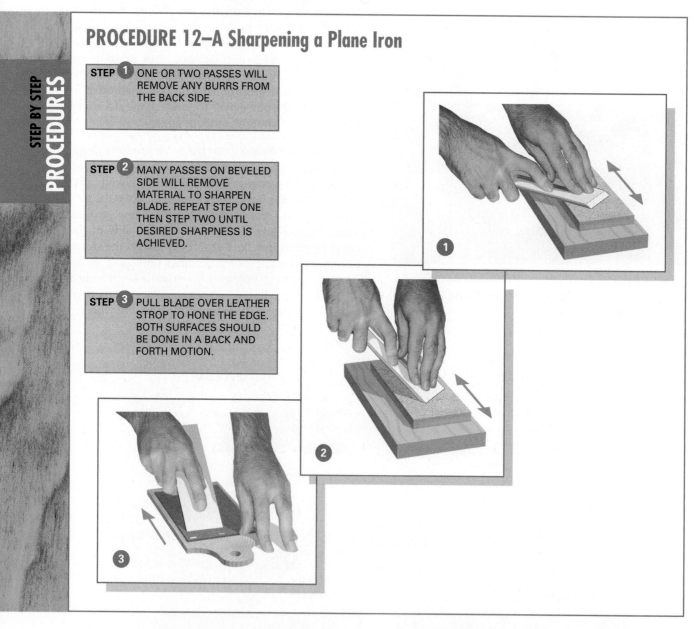

STEP 1 ONE OR TWO PASSES WILL REMOVE ANY BURRS FROM THE BACK SIDE.

STEP 2 MANY PASSES ON BEVELED SIDE WILL REMOVE MATERIAL TO SHARPEN BLADE. REPEAT STEP ONE THEN STEP TWO UNTIL DESIRED SHARPNESS IS ACHIEVED.

STEP 3 PULL BLADE OVER LEATHER STROP TO HONE THE EDGE. BOTH SURFACES SHOULD BE DONE IN A BACK AND FORTH MOTION.

STEP BY STEP PROCEDURES

the bevel and before tuning it over, no wire edge can be felt on the flat side **(Procedure 12–A).**

Chisels and plane irons do not have to be ground each time they need sharpening. Grinding is neces-

sary only when the bevel has lost its concave shape by repeated whettings, the edge is badly nicked, or the bevel has become too short and blunt. The edge of a blade may be whetted many times before it needs grinding. After many whettings the bevel of the wood chisel or plane iron may need to be shaped by a grinding wheel. The bevel should have a concave surface, which is called a *hollow grind.* To obtain a hollow grind, a grinding attachment may be used.

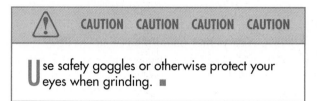

⚠ CAUTION CAUTION CAUTION CAUTION

Use safety goggles or otherwise protect your eyes when grinding. ■

If a grinding attachment is not available, hold the chisel or plane iron by hand on the tool rest at the proper angle. A general rule is that the width of the bevel is approximately twice the thickness of the blade **(Figure 12–14).** Move the blade up on the grinding wheel for a longer bevel and down for a shorter bevel.

Let the index finger of the hand holding the chisel ride against the outside edge of the tool rest as the chisel is moved back and forth across the revolving wheel. Dip the blade in water frequently to prevent overheating. Do not move the position of your index finger, making sure that the tool can be replaced on the wheel at exactly the same angle to obtain a smooth hollow to the bevel. Grind the chisel or plane iron until an edge is formed **(Figure 12–15).** A burr or *wire edge* will be formed on the edge on the flat side. This can be felt by lightly rubbing your thumb along the flat side toward the edge.

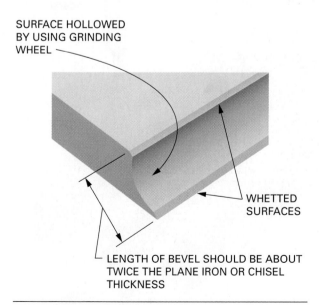

SURFACE HOLLOWED
BY USING GRINDING
WHEEL

WHETTED
SURFACES

LENGTH OF BEVEL SHOULD BE ABOUT
TWICE THE PLANE IRON OR CHISEL
THICKNESS

FIGURE 12–14 Details of the cutting edge of chisels and planes.

FIGURE 12–15 Grinding a wood chisel.

Snips. **Straight tin snips** are generally used to cut straight lines on thin metal, such as roof flashing and metal roof edging. Three styles of **aviation snips** are available for straight metal cutting and for left and right curved cuts **(Figure 12–16).** The color of

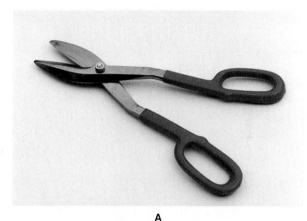

A

B

FIGURE 12–16 (A) Metal shears and (B) left-, straight-, and right-cutting aviation snips.

the handle denotes the differences in the design of the snips. Yellow handles are for straight cuts, green are for cutting curves to the right, and red are for cutting to the left.

Hatchets. For wood shingling of side walls and roofs, among other purposes, the **shingling hatchet** is used. In addition to the shingling hatchet, many carpenters carry a slightly heavier hatchet for such uses as pointing stakes or otherwise tapering rough stock. A special drywall hatchet is also used for the installation of gypsum board **(Figure 12–17)**.

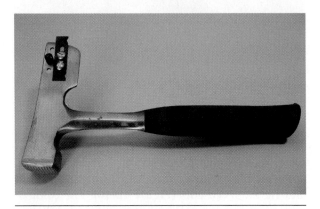

FIGURE 12–17   Shingling hatchet.

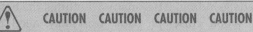

> ⚠ CAUTION   CAUTION   CAUTION   CAUTION
>
> When using hatchets for driving fasteners, make sure there are no workers in the path of the backswing. ■

Knives. A carpenter usually has a jackknife of good quality. The jackknife is used mostly for sharpening pencils and for laying out recessed cuts for some types of finish hardware, such as door hinges. The jackknife is used for laying out this type of work because a finer line can be obtained with it than with a pencil. In addition to marking, it also scores the layout line that is helpful when chiseling the recess **(Procedure 12–B)**.

The **utility knife (Figure 12–18)** is frequently used instead of a jackknife for such things as cutting gypsum board and softboards. Replacement blades are carried inside the handle.

Scrapers. The **hand scraper (Figure 12–19)** is very useful for removing old paint, dried glue, pencil, crayon, and other marks from wood surfaces. The

---

STEP BY STEP PROCEDURES

PROCEDURE 12–B Chiseling Square Edges

STEP ① SCORE LAYOUT LINE WITH KNIFE.

STEP ② LIFT CHIP WITH CHISEL.

STEP ③ CHIP BREAKS OFF AT SCORED LINE.

STEP ④ A SHOULDER IS PROVIDED TO REST THE CHISEL AGAINST WHEN DEEPENING THE CUT.

ON THE JOB
Score layout lines with a knife for more accurate chiseling of recesses.

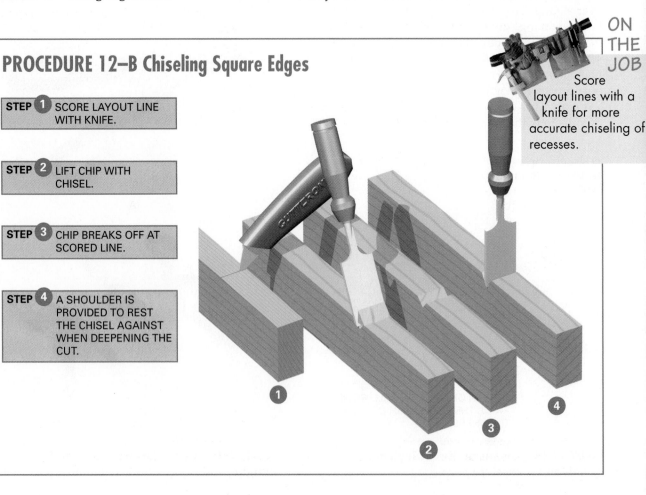

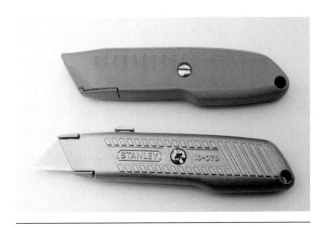

FIGURE 12–18 Utility knives.

FIGURE 12–19 Hand scraper.

scraper blades are reversible, removable, and replaceable. They dull quickly, but can be easily sharpened by filing on the bevel and against the cutting edge.

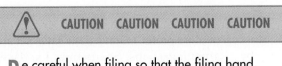

⚠ CAUTION CAUTION CAUTION CAUTION

B e careful when filing so that the filing hand does not come in contact with the cutting edge. Also, care should be taken to file evenly the entire cutting edge and not to hollow out the center of the blade or the outside corners. ■

Tooth-Cutting Tools

The carpenter uses several kinds of saws to cut wood, metal, and other material. Each one is designed for a particular purpose.

Handsaws

Handsaws **(Figure 12–20)** used to cut across the grain of lumber are called **crosscut saws.** To cut with the grain, **ripsaws** are sometimes used. The difference in the cutting action is in the shape of the teeth. The crosscut saw has teeth shaped like knives. These teeth cut through the wood fibers to give a smoother action and surface when cutting across the grain. The ripsaw has teeth shaped like rows of tiny chisels that cut the wood ahead of them **(Figure 12–21).** Another design to handsaw teeth, called a **shark tooth saw,** makes the teeth longer and able to cut in both directions of blade travel.

To keep the saw from binding, the teeth are *set*, that is, alternately bent, to make the saw cut or *kerf* wide enough to give clearance for the blade.

A 7-point or 8-point (number of tooth points to the inch) saw is designed to cut across the grain of framing and other rough lumber. A 10-point or 11-point saw is designed for fine crosscuts on finish work. The number of points is usually stamped on the saw blade at the heel.

Using Handsaws. Stock is handsawed with the face side up because the back side is splintered, along the cut, by the action of the saw going through the stock. This is not important on rough work. However, on finish work, it is essential to identify the face side of a piece and to make all layout lines and saw cuts with the face side up.

The saw cut is made on the waste side of the layout line by cutting away part of the line and leaving the rest. This takes some practice, especially when it is important to make thin layout lines rather than broad, heavy ones. Press the blade of the saw against the thumb when starting a cut. Make sure the thumb is above the teeth; steady it with the

FIGURE 12–20 Handsaws are still useful on the jobsite. Some handsaws are made with deeper teeth, which are designed to cut in both directions.

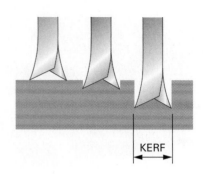
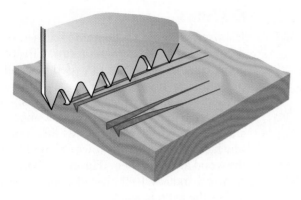

HOW A CROSSCUT SAW CUTS

CROSS SECTION OF CROSSCUT TEETH

KERF

CROSSCUT SAW CUTS

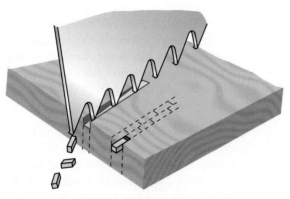
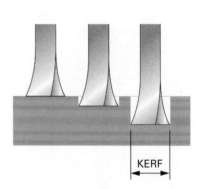

HOW A RIP SAW CUTS

CROSS SECTION OF RIP TEETH

KERF

RIP SAW CUTS

FIGURE 12–21   Cutting action of ripsaws and crosscut saws. *(Courtesy of Disston)*

index finger, with the rest of the hand on the work. Move the thumb until the saw is aligned as desired and start the cut on the upstroke **(Figure 12–22).** Move the hand away when the cut is deep enough. When handsawing, hold crosscut saws at about a 45-degree angle.

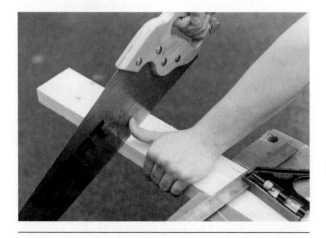

FIGURE 12–22   **Starting a cut with a handsaw.**

> ⚠️ **CAUTION   CAUTION   CAUTION   CAUTION**
>
> Do not use a ripsaw for cutting across the grain. It can jump at the start of the cut, possibly causing injury. ■

Most carpenters prefer to have their saws set and filed by sharpening shops. Sharpening handsaws requires special tools, much skill, and experience to do a professional job.

Special-Purpose Saws

Compass and Keyhole Saws. The **compass saw** is used to make circular cuts in wood. The **keyhole saw** is similar to the compass saw except its blade is narrower for making curved cuts of smaller diameter **(Figure 12–23).** To start the saw cut, a hole needs to be bored (except in soft material, when the point of the saw blade can be pushed through).

FIGURE 12–23   Compass saw.

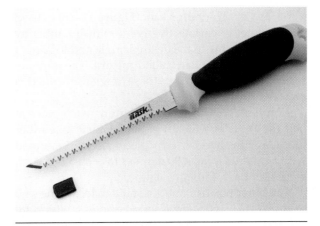

FIGURE 12–26   Wallboard saw.

Coping Saws. The **coping saw (Figure 12–24)** is used primarily to cut molding to make coped joints. A *coped joint* is made by cutting and fitting the end of a molding against the face of a similar piece. (Coping is explained in detail in Unit 27.) The coping saw is also used to make any small, irregular curved cuts in wood or other soft material.

The coping saw blade has fine teeth that may be installed either with the teeth pointing toward or away from the handle. Which is best depends on the operator and the situation. The blade cuts only in the direction the teeth point.

Hacksaws. Hacksaws **(Figure 12–25)** are used to saw metal. Hacksaw blades are available with 18, 24,

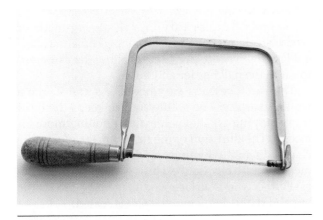

FIGURE 12–24   Coping saw.

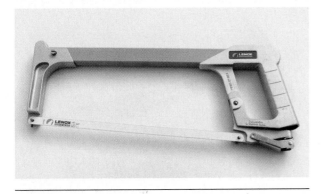

FIGURE 12–25   Hacksaw.

and 32 points to the inch. Coarse-toothed blades are used for fast cutting in thick metal. Fine-toothed blades are used for smooth cutting of thin metal. At least three teeth of the blade should be in contact with thin metal or cutting will be difficult. Make sure that blades are installed with the teeth pointing away from the handle.

Wallboard Saws. The **wallboard saw (Figure 12–26)** is similar to the compass saw but is designed especially for gypsum board. The point is sharpened to make self-starting cuts for electric outlets, pipes, and other projections. Another type with a handsaw handle is also used frequently.

Pullsaws. The Japanese-style **pullsaws** are gaining popularity. They have a unique design in that they cut on the up (pull) stroke instead of the down stroke **(Figure 12–27).** Some models cut in both directions. The pulling action of the pull saw is actually easier to use. It takes less effort and gives more control. They cut fast and smooth with thin kerfs. Many styles are available for cutting rough and fine work.

FIGURE 12–27   Pullsaws cut when they are pulled, which is the reverse of handsaws. (*Courtesy of Shark Corporation*)

Miter Boxes. The **miter box (Figure 12–28)** is used to cut angles of various degrees on finish lumber by swinging the saw to the desired angle and locking it in place. These cuts are called **miters.** The joint between the pieces cut at these angles is called a **mitered joint.**

A mitered joint is made by cutting each piece at half the angle at which it is to be joined to another piece. For instance, if two pieces are to be joined with a mitered joint at 90 degrees to each other, each piece is cut at a 45-degree angle.

The miter box has built-in stops to locate the saw to cut 90-, 67½-, 60-, and 45-degree angles, which are commonly used miter angles. The back saw in a miter box should only be used in the miter box and for no other purpose.

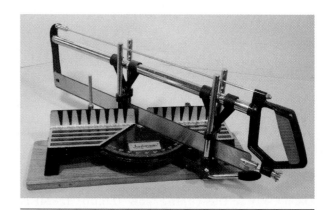

FIGURE 12–28   A handsaw miter box.

---

13 Fastening and Dismantling Tools

Discussed in this chapter are those tools used to drive nails and turn screws and other fasteners. Tools used to clamp, hold, pry, and dismantle workpieces are also included.

FASTENING TOOLS

The carpenter must decide which fastening tool to select and be able to use it competently and safely for the job at hand.

Hammers

The carpenter's **claw hammer** is available in a number of styles and weights. The claws may be straight or curved. Head weights range from 7 to 32 ounces. Most popular for general work is the 16-ounce, curved claw hammer **(Figure 13–1).** For rough work, a 20- or 22-ounce **framing hammer (Figure 13–2),** is often used. This has a longer handle and may have a straight or curved claw. In some areas, a 28- or 32-ounce framing hammer is preferred for extra driving power.

Hammer handles vary in styles being made of wood, steel, or fiberglass. They also come in different lengths. The longer handles allow the carpenter to drive the nail harder and faster.

The hammerheads are smooth or serrated into a waffled surface. The waffle surface keeps the head from slipping off the nail, making nailing more effective. The direction of the driven nail can even be changed slightly by twisting the wrist. These hammers should be used exclusively for framing because the wood surface is damaged by the waffle imprint that is left when the nail is seated.

FIGURE 13–1   Curved claw hammer.

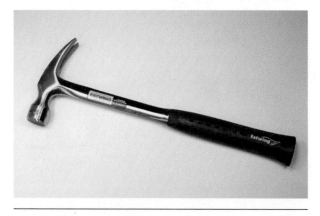

FIGURE 13–2   Straight claw framing hammer.

Nail Sets. **Nail sets (Figure 13–3)** are used to set nail heads below the surface. The most common sizes are $\frac{1}{32}$, $\frac{2}{32}$, and $\frac{3}{32}$ inch. The $\frac{1}{4}$-inch nail set is used to drive the large-headed nails typically used for exterior finish work. The size refers to the diameter of the tip. The surface of the tip is concave to prevent it from slipping off the nail head. If the tip becomes flattened, the nail set is more difficult to keep on the nail being driven.

Nailing Techniques. Hold the hammer firmly, close to the end of the handle, and hit the nail squarely. If the hammer frequently glances off the nail head, try cleaning the hammer face **(Figure 13–5).** As a general rule, use nails that are three times longer than the thickness of the material being fastened. To swing a hammer, the entire arm and shoulder are used. It is important to use the wrist too. During the

FIGURE 13–3 Nail set.

FIGURE 13–4 Wear eye protection when driving nails.

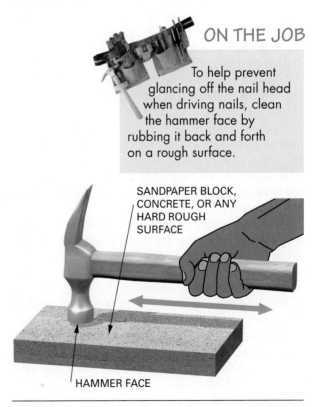

ON THE JOB

To help prevent glancing off the nail head when driving nails, clean the hammer face by rubbing it back and forth on a rough surface.

SANDPAPER BLOCK, CONCRETE, OR ANY HARD ROUGH SURFACE

HAMMER FACE

FIGURE 13–5 Roughing up the hammerhead face helps keep the hammerhead from glancing off the nail.

FIGURE 13–6 Toenailing is the technique of driving nails at an angle.

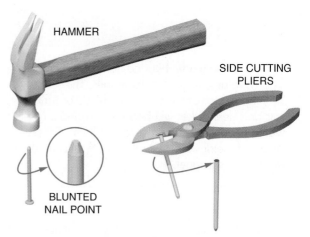

HAMMER

SIDE CUTTING PLIERS

BLUNTED NAIL POINT

SOMETIMES BLUNTING THE POINT OF A NAIL WILL PREVENT SPLITTING THE STOCK. TAP THE POINT LIGHTLY SO AS NOT TO BEND THE NAIL.

IF A TWIST DRILL IS NOT HANDY, CUT THE HEAD OFF A NAIL OF THE SAME GAUGE AS THE NAILS BEING USED, AND USE IT TO DRILL HOLES.

FIGURE 13–7 Methods to avoid splitting wood.

latter part of the swing, as the hammer nears the nail, the wrist is rotated quickly, giving more speed to the hammerhead. This increased speed generates more nail-driving force, all with less arm effort.

Toenailing is the technique of driving nails at an angle to fasten the end of one piece to another **(Figure 13–6).** It is used when nails cannot be driven into the end, called **face nailing.** Toenailing generally uses smaller nails than face nailing and offers greater withdrawal resistance of the pieces joined. Start the nail about ¾ to 1 inch from the end and at an angle of about 30 degrees from the surface.

Drive finish nails almost all the way. Then set the nail below the surface with a nail set to avoid making hammer marks on the surface. Set finish nails at least ⅛ inch deep so the filler will not fall out.

In hardwood, or close to edges or ends, drill a hole slightly smaller than the nail shank to prevent the wood from splitting or the nail from bending. If a twist drill of the desired size is not available, cut the head off a finish nail of the same gauge and use it for making the hole.

Blunting or cutting off the point of the nail also helps prevent splitting the wood **(Figure 13–7).** The point spreads the wood fibers as the nail is driven, while the blunt end pushes the fibers ahead of it and reduces the possibility of splitting.

Holding the nail tightly with the thumb and as many fingers as possible while driving the nail in hardwood helps prevent bending the nail. Of course, hold the nail in this manner only as long as possible. Be careful not to glance the hammer off the nail and hit the fingers.

When nailing along the length of a piece, stagger the nails from edge to edge, rather than in a straight line. This avoids splitting and provides greater strength **(Figure 13–8).** Drive nails at an angle into end grain for greater holding power. This is called dovetail nailing. When fastening pieces side to side, nails are driven at an angle for greater strength **(Figure 13–9).** In addition, this may keep the nail points from protruding if using 12d or 3¼-inch nails to fasten 2-inch nominal stock.

When fastening two pieces of stock together, the alignment is usually very important. To make this

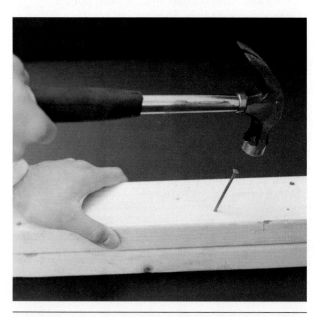

FIGURE 13–8 Stagger and angle nails for greater strength and to avoid splitting the stock.

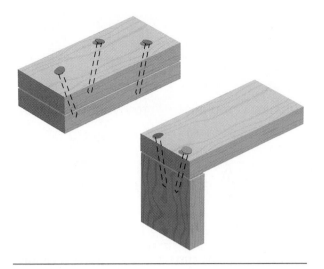

FIGURE 13–9 **Driving nails at an angle increases holding power.**

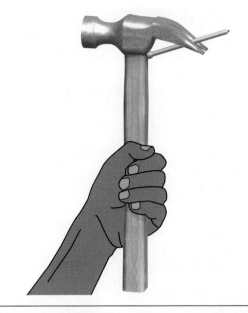

FIGURE 13–10 **Method of starting a nail with one hand.**

process easier, drive the first nail only to where it protrudes from the first layer slightly **(Procedure 13–A).** Then align the pieces as necessary. Push or tap the first layer with small protruding point into the second layer. This will act as a tack holding the proper alignment until the nail is driven.

When it is necessary to start a nail higher than you can hold it, use the nail starter located in the head of many framing hammers. If your hammer does not have one, press the nail tightly between the claws of the hammer, with the head of the nail against the handle **(Figure 13–10).** Turn the claws of the hammer toward the surface. Reach up and swing the hammer to start and hold the nail in the stock. Pull the hammer claws away from the nail,

turn the hammer around, and drive the nail (all the way in).

Screwdrivers

Screwdrivers are manufactured to fit all types of screw slots. The carpenter generally uses only the *slotted* screwdriver **(Figure 13–11),** which has a straight tip to drive common screws, and the *Phillips* screwdriver. Other screwdrivers include the Robertson screwdriver, which has a squared tip (see Figure 9–10).

PROCEDURE 13–A Aligning Boards with a Protruding Nail

STEP 1 START NAIL UNTIL IT PROTRUDES SLIGHTLY THROUGH THE FIRST LAYER.

STEP 2 ALIGN PIECES AND TAP THEM TOGETHER. THE NAIL SERVES AS A TACK. FINISH DRIVING THE NAIL.

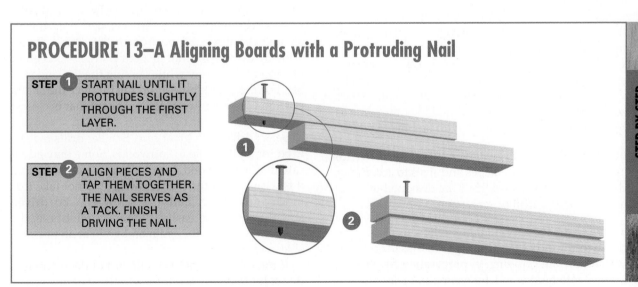

STEP BY STEP PROCEDURES

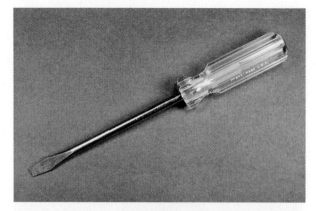

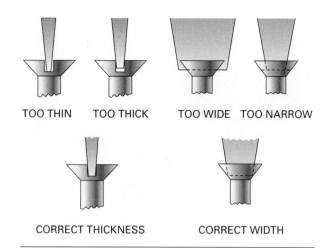

FIGURE 13–12   The correct size screwdriver for the screw being driven is best.

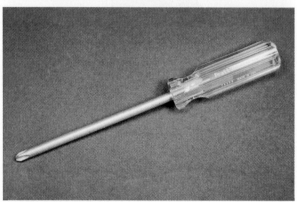

FIGURE 13–11   Slotted and Phillips screwdrivers.

FIGURE 13–13   Screwgun drive bits for various screw head styles.

Screwdriver Sizes.

Slotted screwdrivers are sized by the length of the blade and by the type. Lengths generally run from 3 to 12 inches. Phillips screwdrivers are sized by their length and point size. Commonly used sizes are lengths that run the same as common screwdrivers and points that come in numbers 0, 1, 2, 3, and 4. The higher number indicates a point with a larger diameter.

Screwdrivers should fit snugly, without play, into the slot of the screw being driven. The screwdriver tip should not be wider than the screwhead, nor should it be too narrow **(Figure 13–12)**. The correct size screwdriver helps ensure that the screw will be driven without slipping out of the slot. When seated, the screwdriver slot should look the same as before the screw was driven, with no burred edges.

Screwdriver Bits.

Screwdriver bits (Figure 13–13) are available in many shapes and sizes to accommodate a variety of screws. They are designed to drive a screw using a drill or screw gun.

Screwdriving Techniques.

If possible, select screws so that two-thirds of their length penetrates the piece in which they are gripping. In preparation for driving a screw into hard wood, for example, a *pilot hole* and a *shank hole* must be drilled. The pilot hole allows the screw to be driven into place without splitting the material. The shank hole, very importantly, allows the two pieces to come together tight when the screw is driven.

In addition, if the screw has a flat head, the shank hole may be countersunk so the head will be flush or set below the surface when driven **(Procedure 13–B)**.

To select the pilot hole drill bit size, hold the bit against the threaded portion of the screw. Determine by eye the bit that just covers the solid center portion of the screw while leaving the threads visible. For the shank hole, hold the bit over the shank portion of the screw, selecting the bit that is closest in size.

Select drills with great care. Smaller drills may be used for a pilot hole in softwoods. Some may advocate that in softwood no pilot hole is necessary, but it is wise to drill them anyhow. It does not take that much more time, and the screw can be driven straight and more easily. Without a pilot hole, the screw may follow the grain and go in at an undesirable angle.

If the pilot hole is too small or not deep enough, difficulty may be encountered in driving the screw. This causes slipping and damage to the screw slots.

PROCEDURE 13-B Making a Pilot Hole

STEP 1 DETERMINE SIZE OF AND DRILL PILOT HOLE. USE STOP BLOCK, IF NEEDED, TO PREVENT GOING THROUGH. DEPTH OF PILOT HOLE MUST BE SAME OR DEEPER THAN SCREW LENGTH.

STEP 2 DRILL SHANK HOLE SAME DIAMETER AS SCREW SHANK. USE STOP BLOCK OF APPROPRIATE LENGTH TO PREVENT DRILLING SHANK HOLE TOO DEEP.

STEP 3 COUNTERSINK DEEP ENOUGH SO SCREWHEAD WILL BE SLIGHTLY BELOW SURFACE WHEN DRIVEN.

STEP 4 DRIVE SCREW WITH SCREWDRIVER OF PROPER SIZE UNTIL SCREW IS WELL SEATED.

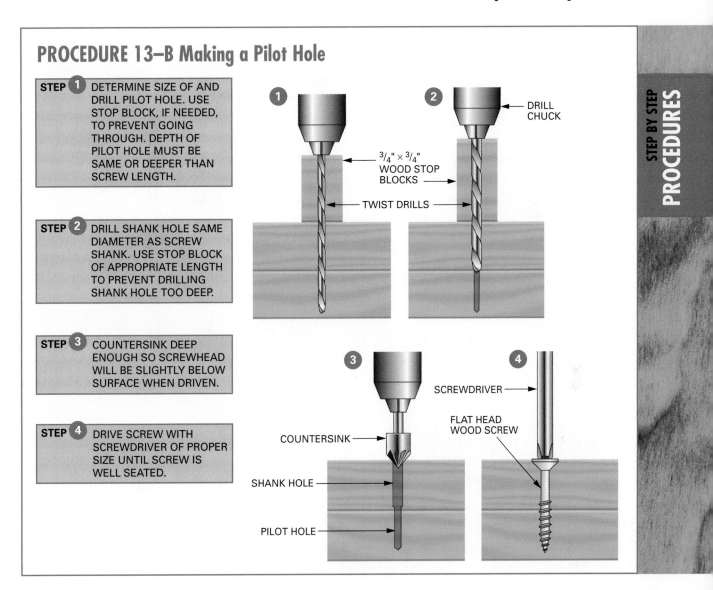

Also, if too much pressure is applied when driving the screw, the head may be twisted off. This is particularly true when driving screws of soft metal, such as aluminum or brass. It might be wise to first drive a steel screw, remove it, and then drive the screw of softer metal. Rub some wax (paraffin) on the threads of the screw to make driving easier. Remember that if the pilot hole is too large, the screw will not grip.

Use a stop when drilling the shank hole to make sure it will not be drilled too deep. This will prevent it from going through the material when drilling the pilot hole. A simple stop can be made by drilling a hole lengthwise through a piece of nominal 1 × 1 stock, cutting it to the desired length, and inserting it on the twist drill against the chuck.

If the material to be fastened is thick, the screw may be set below the surface by **counterboring** to gain additional penetration without resorting to a longer screw. To set the screwhead below the surface, bore the counterbored hole first. Drilling pilot and shank holes first leaves no stock to guide the center point of the bit used to make the counterbored hole. Use a stop block to ensure desired depth. The diameter of the hole should be equal to or slightly larger than the diameter of the screwhead. Next, drill the pilot hole; then, the shank hole **(Procedure 13–C)**.

DISMANTLING TOOLS

Dismantling tools are used to take down staging and scaffolding, concrete forms, and other temporary structures. In addition, they are used for tearing out sections of a building when remodeling. Carpenters must be skilled in the use of dismantling tools, and in the work, so that the dismantled members are not damaged any more than necessary.

PROCEDURE 13–C Counterboring a Screw Hole

STEP 1 DRILL COUNTERBORED HOLE OF DESIRED DIAMETER.

STEP 2 DRILL SHANK AND PILOT HOLES AS DESCRIBED PREVIOUSLY. DRIVE SCREW UNTIL WELL SEATED.

STEP 3 COUNTERBORED HOLES MAY BE PLUGGED WITH VARIOUS SHAPED PLUGS AS SHOWN BELOW.

FLAT PLUG OR BUNG (GRAIN IS HORIZONTAL)

TAPERED PLUG

RABBETED PLUG

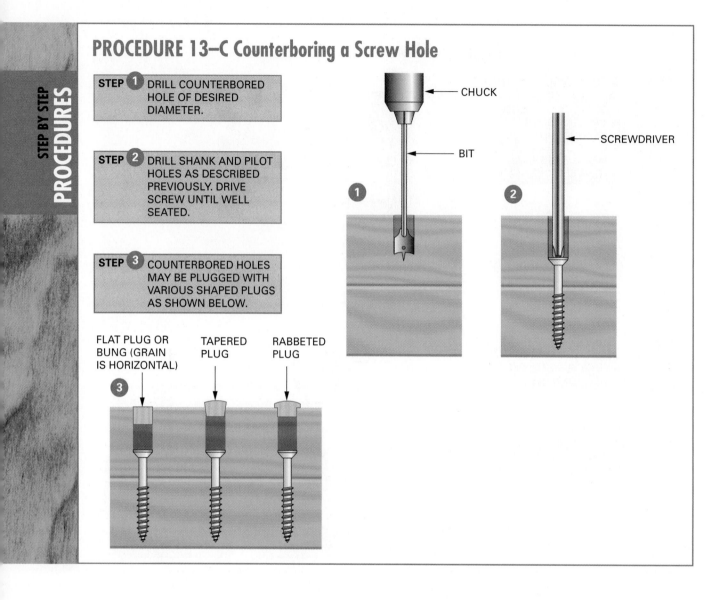

Hammers

In addition to fastening, hammers are often used for pulling nails to dismantle parts. To increase leverage and make nail pulling easier, place a small block of wood under the hammer head **(Figure 13–14)**.

Bars and Pullers

The **wrecking bar (Figure 13–15)** is used to withdraw spikes and to pry when dismantling parts of a structure **(Figure 13–16)**. They are available in lengths from 12 to 36 inches, with the 30-inch size often preferred for construction work.

Carpenters need a small **flat bar,** similar to those shown in **Figure 13–17,** to pry small work and pull small nails. To extract nails that have been driven home, a **nail claw,** commonly called a *cat's paw,* is used **(Figure 13–18)**.

FIGURE 13–14   Pull a nail more easily by placing a block of wood under the hammer.

FIGURE 13–15   Wrecking bars.

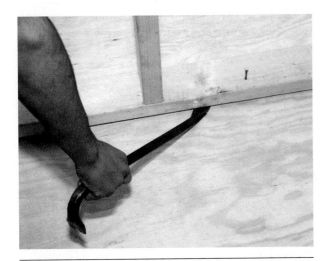

FIGURE 13–16   Using a wrecking bar to pry stock loose.

FIGURE 13–17   Flat bars.

FIGURE 13–18   Nail claw.

FIGURE 13–19   Adjustable wrench.

Holding Tools

To turn nuts, lag screws, bolts, and other objects, an **adjustable wrench** is often used (**Figure 13–19**). The wrench is sized by its overall length. The 10-inch adjustable wrench is the one most widely used.

For extracting, turning, and holding objects, a pair of pliers is often used. Many kinds are manufactured, but the **combination pliers (Figure 13–20)** is designed for general use.

Clamps come in a variety of styles and sizes (**Figure 13–21**). They are useful for holding objects together while they are being fastened, glued, and used as temporary guides. *Spring clamps* are quick and easy to set. They are spring loaded for ease of closing their jaws. Simply squeezing the handles opens the jaws; releasing sets them. *C-clamps* are named for their shape. The size is designated by the throat opening. *Quick clamps* are named for the speed at which they can be adjusted and set. One side of the jaws is stationary and the other slides on the bar. After the material is placed in the jaws, the handles are squeezed

to tighten and set the clamp. The small trigger is pulled to release the clamp. The *wood screw,* also called *parallel clamps,* is made of wood blocks and large screws. These clamps are used primarily for holding wood pieces while they are glued. It takes some practice to set this clamp quickly. The center screw is turned one way to tighten and the other screw is turned the opposite way.

FIGURE 13–20   (A) Combination pliers and (B) adjustable pliers.

FIGURE 13–21   (A) Spring clamp, (B) C-clamp, (C) quick clamp, and (D) wood screw.

Key Terms

adjustable wrench	coping saw	level	shark tooth
auger bits	counterboring	line level	shingling hatchet
aviation snips	countersink	miter box	speed square
bench plane iron	crosscut saws	mitered joint	steel tapes
bench planes	dadoes	miters	straightedge
bevel	drilling	nail claw	straight tin snips
bit brace	expansive bit	nail set	toe
block planes	face nailing	plumb	toenailing
boring	flat bar	plumb bob	trammel points
butt markers	framing hammer	pocket tapes	twist drills
carpenter's level	framing square	pullsaws	utility knife
chalk line	grooves	rabbets	wallboard saw
clamps	hacksaws	ripsaws	whetted
claw hammer	hand scraper	screwdriver bits	wing dividers
combination pliers	heel	screwdrivers	wood chisel
combination square	keyhole saw	scribing	wrecking bar
compass saw			

Review Questions

Select the most appropriate answer.

1. A safe worker attitude that promotes a safe jobsite comes from
   a. ability.
   b. skill.
   c. knowledge.
   d. all of the above.

2. When stretching a steel tape to lay out a measurement, place the ring so the
   a. 1-inch mark is on the starting line.
   b. end of the steel tape is on the starting line.
   c. inside of the ring is on the starting line.
   d. outside of the ring is on the starting line.

3. In construction, the term *plumb* means perfectly
   a. horizontal.
   b. level.
   c. straight.
   d. vertical.

4. A large, L-shaped squaring tool that has tables stamped on it is called a
   a. framing square.
   b. Speed Square.
   c. bevel square.
   d. combination square.

5. The layout tool that may be adjusted to serve as a marking gauge is the
   a. framing square.
   b. Speed Square.
   c. bevel square.
   d. combination square.

6. To adjust a carpenter's level into a level position when the bubble is found to be touching the right line, the
   a. right side should be raised.
   b. left side should be raised.
   c. left side should be lowered.
   d. entire level should be raised.

7. When snapping a long chalk line care should be taken to
   a. dampen the string.
   b. keep the string from sagging.
   c. hold the string loosely.
   d. let the string touch the surface as it unwinds.

8. The tool used to mark material to conform to an irregular surface is called a
   a. pen.
   b. chisel.
   c. scriber.
   d. chalk line.

9. The name of the one-handed plane with a low blade angle is the
   a. block plane.
   b. bench plane.
   c. chisel.
   d. plane iron.

10. The color of the handle on aviation snips indicates
    a. which hand to use.
    b. direction in which curves can easily be cut.
    c. the manufacturer.
    d. what material may be easily cut.

UNIT 6

Portable Power Tools

The sound of construction has changed over the years. The rhythmic whoosh of a handsaw has virtually been replaced by the whir and ring of a circular saw. Power tools have been created to increase the productivity of most job-site tasks.

The number and style of power tools available today for the carpenter is vast and the list continues to grow. Power tools enable the carpenter to do more work in a shorter time period with less effort.

SAFETY REMINDER

With increased speed and production comes an increase in personal risk. This danger can come from a spectrum of human shortcomings that range from a lack of knowledge and skill to overconfidence and carelessness. Safe operation of power tools requires knowledge and discipline.

Learn the safe operating techniques from the manufacturer's recommended instructions before operating any tool. Once you understand these procedures, follow them every time the tool is used. Don't take chances, life is too short as it is.

Being aware of the dangers of operating power tools is the first step in avoiding accidents. This begins with eye and ear protection.

Portable power tools are everywhere on the construction site today. All tools must be carefully used and properly maintained to keep workers using the tool and those nearby safe.

OBJECTIVES

After completing this unit, the student should be able to:

- state general safety rules for operating portable power tools.
- identify, describe, and safely use the following portable power tools: circular saws, jigsaws, reciprocating (also called saber) saws, drills, hammer-drills, screwdrivers, planes, routers, sanders, staplers, nailers, and powder-actuated drivers

Safety is as important as breathing. Have a complete understanding of a tool before attempting to operate it. Read the manufacturer's operating instructions.

Maintain a proper attitude of safety at all times by following these guidelines:

- Wear eye and ear protection. Eyes and ears do not grow back.
- Do not become distracted by others or distracting to others when tools are being operated.
- Do not wear loose-fitting clothes or jewelry that might become caught in the tool.
- Make sure the material being tooled is securely held and supported.
- Remember, using sharp tools and cutters is actually safer than using dull ones.
- Stay alert and develop an attitude of care and respect for yourself and others.

Use the proper power source:

- Electricity used to power a tool can be fatal to humans. Use ground fault interrupter circuits (GFICs) at all times. These will trip before any electricity can leak out of a tool and cause a shock.
- Do not use frayed or badly worn power cords.
- Use properly sized extension cords that are rated for the power requirements of the tool.
- Avoid using a cord longer in length than necessary. Voltage to the tool drops as the cord gets longer.
- Keep extension cords out of the path of all construction traffic.
- Unplug the tool whenever touching the cutting surface of the tool.

Safety is a team sport. It requires all workers to take part. ■

14 Saws, Drills, and Drivers

SAWS

The carpenter uses several kinds of portable saws for crosscutting and **ripping,** for making circular cuts, and for cutting openings in floors, walls, and ceilings.

Electric Circular Saws

Commonly called the **skilsaw,** the portable electric circular saw **(Figure 14–1)** is used often by the carpenter. The circular saw blade is driven by an electric motor. The saw has a base that rests on the work to be cut. A handle with a trigger switch is provided for the operator to control the tool. The saw is adjustable for depth of cut. A retractable safety guard is provided over the blade, extending under the base.

The base may be tilted for making *bevel* cuts. Bevel or miter cuts are those where the edge, end, or face of a board is cut at an angle. Edge bevels run along the length of the board or with the grain. End bevels run along the width of the board or across the

FIGURE 14–1 Using a portable electric circular saw to cut compound angles.

grain cutting. Face bevels are angle cuts made on the face **(Figure 14–2)**. Compound bevels are cuts with two angles, usually a combination of face and end bevels.

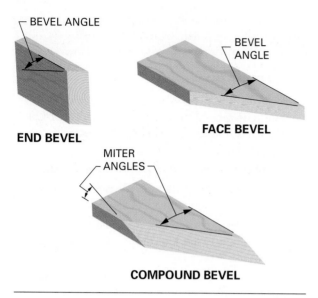

END BEVEL

FACE BEVEL

COMPOUND BEVEL

FIGURE 14–2   Edge, flat, and compound bevel cuts.

Saws are manufactured in many styles and sizes. The size is determined by the diameter of the blade, which ranges from 4½ to 16 inches. The most common size of circular saw uses a 7¼-inch-diameter blade. The handle and switch may be located on the top or in back. The blade may be driven directly by the motor or through a worm gear **(Figure 14–3)**.

The forward end of the base is notched in two places to serve as guides for following layout lines. One notch is used to follow layout lines when the base is tilted to 45 degrees and the other when the base is not tilted.

⚠ **CAUTION   CAUTION   CAUTION   CAUTION**

Make sure the saw blade is installed with the teeth pointing in the correct direction. The teeth of the saw blade projecting below the base should point toward the base as the blade rotates. ■

To loosen the bolt that holds the blade in place, first unplug the saw and lock the arbor. This is done by pushing the arbor locking slide of the saw, usually found between the blade shield and the handle. While pushing the slide, rotate the blade by hand until the slide locks the arbor. With the proper wrench, turn the bolt in the same direction as the rotation of the blade. To tighten, turn the bolt in a direction opposite to the rotation of the blade.

On most models an adjustable attachment that fits into the base is used for ripping narrow pieces

FIGURE 14–3   **Direct drive and worm gear drive portable electric circular saws.** (*Top: Courtesy of Porter Cable; bottom: Courtesy of S-B Power Tools*)

parallel with the edge. The saw may also be guided by tacking or clamping a straightedge to the material and running the edge of the saw base against it. Allowance must be made for the distance from the saw blade to the edge of the saw base when positioning the straightedge.

Circular Saw Blades. Circular saw blades are available in a number of styles. The shape and number of teeth around the circumference of the blade determine their cutting action. Carbide-tipped blades are used more than high-speed steel blades. They stay sharper longer. More complete information on saw blades can be found in Unit 7.

Using the Portable Circular Saw. Safe and efficient cutting follows an established method:

■ Make sure the work is securely held and that the waste will fall clear and not bind the saw blade **(Figure 14–4)**.

■ Adjust the depth of cut so that the blade just cuts through the work. Never expose the blade any more than is necessary **(Figure 14–5)**. This will reduce the sawdust spray into the operator's face when cutting thinner material.

FIGURE 14-4 Saw cuts are made over the end of supports so the waste will fall clear and not bind the blade.

Make sure the guard operates properly. Be aware that the guard may possibly stick in the open position. Never wedge the guard back in an open position. ■

- Mark the stock. Put on safety goggles. Rest the forward end of the base on the work. With the blade clear of the material, start the saw.
- When it has reached full speed, advance the saw into the work. Make sure to observe the line to be followed. With the saw cut in the waste side of the material, cut as close to the line as possible for a short distance.
- Stop the saw advancement into the material and check the alignment of the edge of the base to the line being cut. They should be parallel.
- Follow the line closely.

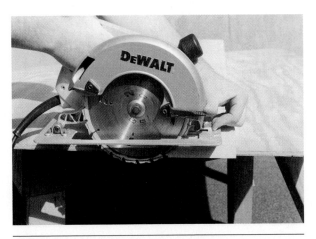

FIGURE 14-5 The blade of the saw is adjusted for depth only enough to cut through the work.

Any deviation from the line may cause the saw to bind and kick back. Do not force the saw forward. If the saw does bind, stop the motor and bring the saw back to where it will run free. Continue following the line closely. ■

- Near the end of the cut, the forward end of the base will go off the work. Guide the saw by observing the line at the saw blade and finish the cut. The saw may also be guided by watching the layout line at the saw cut for the whole length. Let the waste drop clear and ensure that the guard has returned. Release the switch.

Keep the saw clear of your body until the saw blade has completely stopped. ■

- When starting cuts across stock at an angle, it may be necessary to retract the guard by hand. A handle is provided for this purpose **(Figure 14-6)**. Release the handle after the cut has been started and continue as above.
- Compound miter cuts may be made by cutting across the stock at an angle with the base tilted.

Portable circular saws cut on the upstroke. The saw blade rotates upward through the material. As the teeth of the saw blade come through the top surface, splintering of the stock occurs at the layout line. The severity of the splintering depends on

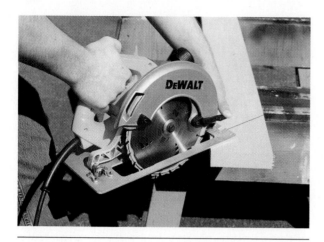

FIGURE 14–6   Retracting the guard of the portable circular saw by hand.

FIGURE 14–7   Making a plunge cut with a portable circular saw. First retract the guard, place the front edge of the saw base on the material and then pivot, running the saw slowly into the material.

the kind of blade used, kind and thickness of the material being cut, and other factors. More splintering occurs when cutting across grain than with the grain.

On *finish work,* that is, work that will ultimately be exposed to view, any splintering along the cut is unacceptable. One way to prevent this is to mark and cut from the back side. If it is not possible to cut from the back side, or if both sides may be exposed to view, mark the layout lines on the face side. Then score along the layout lines with a sharp knife. Make the cuts just outside the scored lines. Another way to avoid splintering is to place a strip of masking tape over the cut before marking it. The tape helps hold the wood fibers in place while being cut.

Making Plunge Cuts.   Many times it is necessary to make internal cuts in the material such as for sinks in countertops or openings in floors and walls. To make these cuts with a portable electric circular saw, the saw must be plunged into the material:

- Accurately lay out the cut to be made. Adjust the saw for depth of cut.

- Wearing eye protection, hold the guard open and tilt the saw up with the front edge of the base resting on the work. Have the saw blade over, and in line with, the cut to be made **(Figure 14–7).**

- Make sure the teeth of the blade are clear of the work and start the saw. Lower the blade slowly into the work, following the line carefully, until the entire base rests squarely on the material.

- Advance the saw into the corner. Release the switch and wait until the saw stops before removing it from the cut.

| ⚠ | CAUTION   CAUTION   CAUTION   CAUTION |

Make sure the tool comes to a complete stop before withdrawing it from the material being cut. ■

- Reverse the direction and cut into the corner. Again, wait until the saw stops before removing it from the cut.

| ⚠ | CAUTION   CAUTION   CAUTION   CAUTION |

Never move the saw backwards while cutting. The direction of the turning blade will make the saw want to rise up out of the cut, jumping backward across everything in its path. ■

- Proceed in like manner to cut the other sides. Finish the cut into the corners with a handsaw or jigsaw.

Jigsaws

The **jigsaw** is widely used to make curved cuts **(Figure 14–8).** The teeth of the blade point upward when installed, so the saw cuts on the upstroke. To produce a splinter-free cut on the face side, it is best to cut on the side opposite the face of the work, if possible, or to score with a knife along the layout line.

FIGURE 14–8 The jigsaw can make either straight or orbital cutting actions.

There are many styles and varieties of jigsaws. The length of the stroke along with the amperage of the motor determine its size and quality. Strokes range from ½ to 1 inch. The longest stroke is the best for faster and easier cutting. Some saws can be switched from straight up-and-down strokes to several orbital (circular) motions to provide more effective cutting action for various materials **(Figure 14–9)**. The base of the saw may be tilted to make bevel cuts.

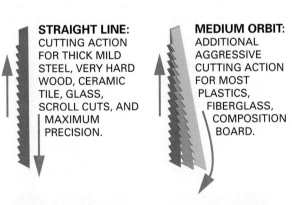

STRAIGHT LINE: CUTTING ACTION FOR THICK MILD STEEL, VERY HARD WOOD, CERAMIC TILE, GLASS, SCROLL CUTS, AND MAXIMUM PRECISION.

MEDIUM ORBIT: ADDITIONAL AGGRESSIVE CUTTING ACTION FOR MOST PLASTICS, FIBERGLASS, COMPOSITION BOARD.

SMALLEST ORBIT: FOR FASTER CUTTING ON WOOD AND HARDBOARD AND MORE AGGRESSIVE CUTTING IN MILD STEEL AND ALUMINUM.

LARGEST ORBIT: FOR ROUGH HIGH-SPEED CUTS IN WOOD, PLASTER-BOARD, COUNTERTOPS; THE MOST AGGRESSIVE CUTTING ACTION.

FIGURE 14–9 Some jigsaws have an additional orbital cutting action during each stroke of the blade.

APPLICATION	PART NUMBER	TEETH PER INCH	OVERALL LENGTH (INCHES)	WIDTH (INCHES)
Metal cutting (High speed steel)				
Non-ferrous metal cutting ¼" to ¾" thick	94612	8	3⅝	5⁄16
Metal cutting over ⅛" thick	94613	12	2¾	5⁄16
Metal cutting up to ⅛" thick	94614	21	2¾	5⁄16
Metal cutting up to 1⁄16" thick	94615	36	2¾	5⁄16
Wood cutting (Taper ground, high speed steel)				
Fast, smooth curve cutting in wood up to 2" thick	94618	6	3⅝	¼
Fast, smooth cutting in wood up to 2" thick	94619	6	3⅝	5⁄16
Very smooth curve cutting in wood up to 1" thick	94622	10	3⅝	¼
Very smooth cutting in wood up to 1" thick	94623	10	3⅝	5⁄16
Wood cutting (Alternate set—ground, high carbon steel)				
Fast, medium curve cutting in wood up to 2" thick	94627	6	3⅝	¼
Fast, medium cutting in wood up to 2" thick	94628	6	3⅝	5⁄16
Wood cutting (Alternate set—high carbon steel)				
Fast, rough cutting in wood up to 2" thick	94631	6	3⅝	5⁄16
Medium cutting in wood up to 2" thick	94634	8	3⅝	5⁄16
General purpose (Alternate set—high speed steel)				
General purpose cutting in wood, plastic, metal, etc.	94637	12	3⅝	5⁄16
Wood scroll cutting (Taper ground, high carbon steel)				
Smooth, intricate scroll cuts in wood up to 1¾" thick	94640	20	2¾	3⁄16

FIGURE 14–10 Jigsaw blade selection guide.

Blade Selection. Many blades are available for fine or coarse cutting in wood or metal **(Figure 14–10)**. Wood-cutting blades have teeth that are from 6 to 12 points to the inch. Blades with coarse teeth (fewer points to the inch) cut faster, but rougher. Blades with more teeth to the inch may cut slower, but produce a smoother cut surface. They do not splinter the work as much. Proper blade selection and use will produce the best cuts.

Using the Jigsaw. Follow a safe and established procedure:

- Outline the cut to be made. Secure the work either by hand, tacking, clamping, or by some other method.
- Using eye protection, hold the base of the saw firmly on the work. With the blade clear, squeeze the trigger.

■ Push the saw into the work, following the line closely. Make the saw cut into the waste side, and cut as close to the line as possible without completely removing it.

■ Keep the saw moving forward, holding the base down firmly on the work. This will allow the saw to cut faster and more efficiently by keeping saw vibration to a minimum. Turn the saw as necessary in order to follow the line to be cut. Feeding the saw into the work as fast as it will cut, but not forcing it, finish the cut. Keep the saw clear of your body until it has stopped.

Making Plunge Cuts. Plunge cuts may be made with the jigsaw in a manner similar to that used with the circular saw:

■ Tilt the saw up on the forward end of its base with the blade in line and clear of the work **(Figure 14–11).**

■ Start the motor, holding the base steady. Very gradually and slowly, lower the saw until the blade penetrates the work and the base rests firmly on it.

⚠ CAUTION CAUTION CAUTION CAUTION
Hold the saw firmly to prevent it from jumping when the blade makes contact with the material and to make a successful plunge cut. Thicker material may require a pilot hole to be drilled prior to blade insertion. ■

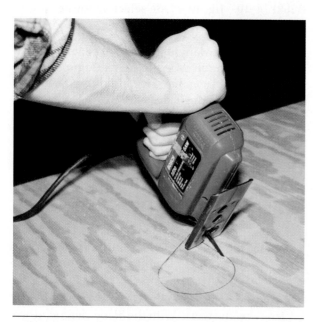

FIGURE 14–11   **Making an internal cut by plunging the jigsaw.**

■ Cut along the line into the corner. Back up for about an inch, turn the corner, and cut along the other side and into the corner.

■ Continue in this manner until all the sides of the opening are cut.

■ Turn the saw around and cut in the opposite direction to cut out the corners.

Reciprocating Saws

The **reciprocating saw (Figure 14–12),** sometimes called a *sawzall* or saber saw, is used primarily for *roughing in* work. This work consists of cutting holes and openings for such things as pipes, heating and cooling ducts, and roof vents. It can be likened to a powered compass saw.

Most models have a variable speed of from 0 to 2,400 strokes per minute. Like jigsaws, some models may be switched to several *orbital* cutting strokes from a straight back and forth to an orbital cutting action. Ordinarily, the orbital cutting mode is used for fast cutting in wood. The reciprocating stroke should be used for cutting metal.

Reciprocating Saw Blades.   Common blade lengths run from 4 to 12 inches. They are available for cutting practically any type of material. Blades are available to cut wood, metal, plaster, fiberglass, ceramics, and other material. They are made of a hardened steel that allows the blades to occasionally cut nails with ease.

Using the Reciprocating Saw.   The reciprocating saw is used in a manner similar to the jigsaw. The difference is that the reciprocating saw is heavier and more powerful. It can be used more efficiently to cut through rough, thick material, such as walls when remodeling. With a long, flexible blade, it can be used to cut flush with a floor or along the side of a stud.

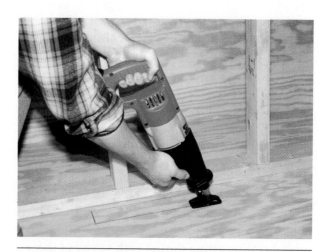

FIGURE 14–12   **Using the reciprocating saw to cut an opening in the subfloor.**

To use the saw, lines must be laid out and followed. The base or shoe of the saw is held firmly against the work whenever possible. Like the jigsaw, this reduces saw vibration and allows the saw to cut faster and more efficiently. To make cutouts, first drill a hole in the material. Then insert the blade, start the motor, and follow the layout lines. The blades can be reversed for cutting in confined areas.

DRILLS AND DRIVERS

Portable power drills, manufactured in a great number of styles and sizes, are widely used to drill holes and drive fasteners in all kinds of construction materials.

Drills

The drills used in the construction industry are classified as light duty or heavy duty. Light-duty drills usually have a *pistol-grip* handle. Heavy-duty drills have a *spade-shaped* or *D-shaped* handle **(Figure 14–13).**

Drill Sizes. The size of a drill is determined by the capacity of the *chuck,* its maximum opening. The chuck is that part of the drill that holds the cutting tool. The most popular sizes for light-duty models are ¼ and ⅜ inch. Heavy-duty drills have a ½-inch chuck or larger.

Drill Speed and Rotation. Most drills have *variable speed* and *reversible* controls. Speed of rotation can be controlled from 0 to maximum rpm (revolutions per minute) by varying the pressure on the trigger switch. Slow speeds are desirable for drilling larger holes or holes in metal. Faster speeds are used for drilling smaller holes in softer material. A reversing switch changes direction of the rotation for removing screws or withdrawing bits and drills from clogged holes.

Bits and Twist Drills. Twist drills are used in electric drills to make small holes in wood or metal **(Figure 14–14).** For larger holes in wood, plastics, and composition materials, a variety of wood-cutting bits are used.

For boring holes in rough work, the *spade bit* is commonly used. For a hole with a cleaner edge in finish work, the *Power Bore* bit may be used **(Figure 14–15).**

FIGURE 14–13 **Portable power drills are available in a number of styles** (*Top: Courtesy of Porter Cable*)

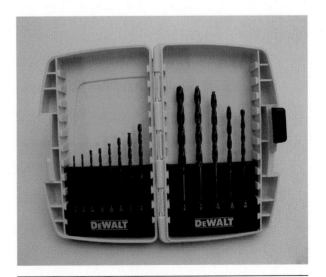

FIGURE 14–14 **High-speed twist drills are used to drill holes in any material.**

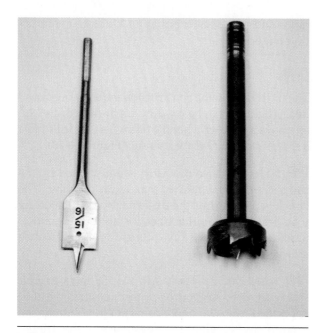

FIGURE 14–15 **Spade and Power Bore bits are used to drill larger holes in wood and similar material.**

These bits make fine clean holes in wood with diameters from ¼" to 2⅛".

Other Drill Accessories. Occasionally, carpenters may use **hole saws (Figure 14–16).** These saws cut holes through material from ⅝ inch to 6 inches in diameter. One disadvantage of the hole saw is that a hole cannot be made partially through the material. Only the circumference is cut and the waste is not expelled.

Masonry drill bits have carbide tips for drilling holes in concrete, brick, tile, and other masonry. They are frequently used in portable power drills. They are more efficiently used in **hammer-drills.**

FIGURE 14–16 **Hole saws actually saw holes in material leaving a circular center plug.**

Using Portable Electric Drills. Select the proper size and type of bit and insert it. Tighten the chuck with the chuck key or by holding the chuck of a *keyless* chuck. Holes in metal must be center-punched because the drill will wander off center.

> ⚠ CAUTION   CAUTION   CAUTION   CAUTION
>
> Hold small pieces securely by clamping or other means. When drilling through metal, especially, the drill has a tendency to hang up when it penetrates the underside. If the piece is not held securely, the hangup will cause the piece to rotate with the drill. It could then hit anything in its path and possibly cause serious injury to a person before power to the drill can be shut off. ∎

Place the bit on the center of the hole to be drilled. Hold the drill at the desired angle and start the motor. Apply pressure as required, but do not force the bit. Drill into the stock, being careful not to wobble the drill. Failure to hold the drill steady may result in breakage of small twist drills.

> ⚠ CAUTION   CAUTION   CAUTION   CAUTION
>
> While drilling a hole, withdraw the turning bit periodically to clear the shavings. This will keep the bit cooler and it will last longer. More importantly it will help prevent the bit from jamming in the hole. Jamming causes the bit to stop suddenly. Personal injury may occur if the drill is powerful and the jammed bit is large. Therefore, be ready to release the trigger at any time. ∎

Hammer-drills

Hammer-drills (Figure 14–17) are similar to other drills. However, they can be changed to a hammering action as they drill, quickly making holes in concrete or other masonry. Some models deliver as many as 50,000 hammer blows per minute. Most popular are the ⅜- and ½-inch sizes.

A depth stop is usually attached to the side of the hammer-drill. It can be converted to a conventional drill by a quick-change mechanism. Most models have a variable speed of from 0 up to 2,600 rpm.

The hammer-drill has the same type chuck and is used in the same manner as conventional portable power drills.

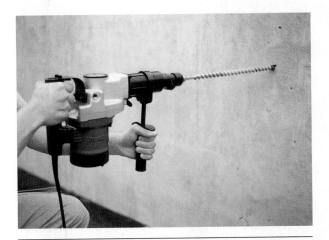

FIGURE 14–17 The hammer-drill is used to make holes in masonry.

FIGURE 14–18 The drywall driver is used to fasten wallboard with screws. (*Top: Courtesy of Porter Cable*)

Screwguns

Screwguns or *drywall drivers* **(Figure 14–18)** are used extensively for fastening gypsum board to walls and ceilings with screws. They are similar in appearance to the light-duty drills, except for the chuck. They have a pistol-type grip for one-hand operation and controls for varying the speed and reversing the rotation.

The chuck is made to receive special screwdriver bits of various shapes and sizes. A screwgun has an adjustable nosepiece, which surrounds the bit. When the forward end of the nosepiece touches the surface, the clutch is separated and the bit stops turning with a vibrating noise. Adjusting the nosepiece makes variations in the screw depth.

Accessories are available for driving screws with different heads. Hex head nutsetters are available in magnetic or nonmagnetic styles, in sizes ranging from ³⁄₁₆ to ³⁄₈ inch.

CORDLESS TOOLS

Cordless power tools are widely used due to their convenience, strength, and durability **(Figure 14–19)**. The tools' power source is a removable, rechargeable battery usually attached to the handle of the tool. The batteries range in voltage from 4 to 24 volts and, in general, the higher the voltage, the stronger the tool.

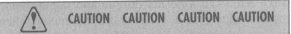

⚠ **CAUTION CAUTION CAUTION CAUTION**

It is easy to think that because these tools are battery powered, they are safer to use than higher voltage tools with cords. While they are safe to use, the operator should never forget the proper techniques and requirements for using the tools. Always wear personal protection equipment. ∎

Downtime is practically eliminated with spare batteries and improvements in the chargers, which can quick charge in as little as 15 minutes **(Figure 14–20)**. Improved chargers can charge batteries of various voltages. They stop charging when the correct voltage level is reached, before the battery begins to overheat. This prolongs the life of the battery.

Cordless tools tend to make the job site safer by eliminating extension cords. Yet the extension cord sometimes serves as a safety line for tools that fall. Take care to set the tool down in a safe place after each use particularly when working on scaffolds.

Cordless drills come with variable-speed reversing motors and a positive clutch, which can be adjusted to

FIGURE 14–19 **Cordless models are available for many tools.** *(Courtesy of Porter Cable)*

drive screws to a desired torque and depth. Some models allow for a hammer-drilling mode when drilling masonry. Cordless circular saws are powerful and able to easily cut dimension lumber as well as trim material **(Figure 14–21).** Sizes include 4½- and

6½-inch blades. Cordless jigsaws and reciprocating saws offer the same features as corded models. They also cut and handle in much the same manner giving the same performance.

FIGURE 14–20 **Cordless tool batteries can be recharged in 15 minutes.**

FIGURE 14–21 **Cordless circular saws easily cut dimension lumber.**

15 Planes, Routers, and Sanders

PORTABLE POWER PLANES

Portable power planes make planing jobs much easier for the carpenter. Planing the ends and edges of hardwood doors and stair treads, for example, takes considerable effort with hand planes, even with razor-sharp cutting edges.

Jointer Planes

The **jointer plane** is used primarily to smooth and straighten long edges, such as fitting doors in openings **(Figure 15–1)**. It is manufactured in lengths up to 18 inches. The electric motor powers a cutter head that may measure up to 3¾ inches wide. The planing depth, or the amount that can be taken off with one pass, can be set for 0 up to ⅛ inch.

An adjustable fence allows planing of squares, beveled edges to 45 degrees, or **chamfers (Figure 15–2).** A rabbeting guide is used to cut rabbets up to ⅞ inch deep.

Operating Power Planes

The operation and feel of the power plane is similar to that of the bench plane. The major differences are the vibration and the ease of cutting with the power plane.

> ⚠ **CAUTION   CAUTION   CAUTION   CAUTION**
>
> Extreme care must be taken when operating power planes. There is no retractable guard, and the high-speed cutterhead is exposed on the bottom of the plane. Keep the tool clear of your body until it has completely stopped. Keep extension cords clear of the tool. ■

■ Set the side guide to the desired angle, and adjust the depth of cut.

■ Hold the toe (front) firmly on the work, with the plane cutterhead clear of the work.

■ Start the motor. With steady, even pressure make the cut through the work for the entire length. Guide the angle of the cut by holding the guide against the side of the stock. Apply pressure to

FIGURE 15–1   A portable electric jointer plane.

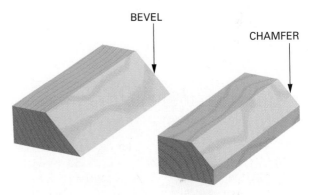

BEVEL

CHAMFER

A BEVEL IS A CUT AT AN ANGLE THROUGH THE TOTAL THICKNESS

A CHAMFER IS AN ANGLED CUT PART WAY THROUGH THE THICKNESS

FIGURE 15–2   A bevel and chamfer.

the toe of the plane at the beginning of the cut. Apply pressure to the heel (back) at the end of the cut to prevent tipping the plane over the ends of the work.

■ To plane a **taper,** that is, to take more stock off one end than the other, make a number of passes. Each pass should be shorter than the preceding one. Lift the plane clear of the stock at the end of the pass. Make the last pass completely from one end to the other **(Figure 15–3).**

PORTABLE ELECTRIC ROUTERS

One of the most versatile portable tools used in the construction industry is the **router (Figure 15–4)**. It is available in many models, ranging from ¼ hp to

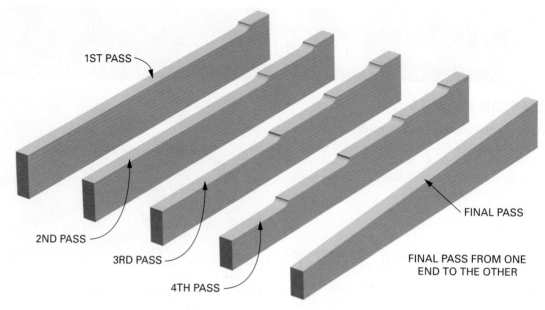

1ST PASS

2ND PASS

3RD PASS

4TH PASS

FINAL PASS

FINAL PASS FROM ONE END TO THE OTHER

DEPTH OF CUT EXAGGERATED FOR ILLUSTRATIVE PURPOSES

FIGURE 15–3   **Technique for planing a taper.**

FIGURE 15–4   **Using a portable electric router.**

over 3 hp with speeds of from 18,000 to 30,000 rpm. These tools have high-speed motors that enable the operator to make clean, smooth-cut edges.

The motor powers a ¼- or ½-inch chuck in which cutting bits of various sizes and shapes are held **(Figure 15–5).** An adjustable base is provided to control the depth of cut. A trigger or toggle switch controls the motor.

The router is used to make many different cuts, including grooves, dadoes, rabbets, and **dovetails**

(Figure 15–6). It is also used to shape edges and make cutouts, such as for sinks in countertops. It is extensively used with accessories, called **templates,** to cut recesses for hinges in door edges and door frames. When operating the router, it is important to be mindful of the bit at all times. Watch what you are cutting and keep the router moving. Stalling the movement of the router will cause the bit to burn or melt the material.

Laminate Trimmers

A light-duty specialized type of router is called a **laminate trimmer**. It is used almost exclusively for trimming the edges of *plastic laminates* **(Figure 15–7).** Plastic laminate is a thin, hard material used primarily as a decorative covering for kitchen and bathroom cabinets and countertops. (The installation of this material is described in a following chapter.)

Guiding the Router

Controlling the sideways motion of the router is accomplished by the following methods:

- By using a router bit with a pilot (guide) **(Figure 15–8).** The pilot may be solid and rotate with the bit or may have ball-bearing pilots. These guide the router along the uncut portion of the material being routed.

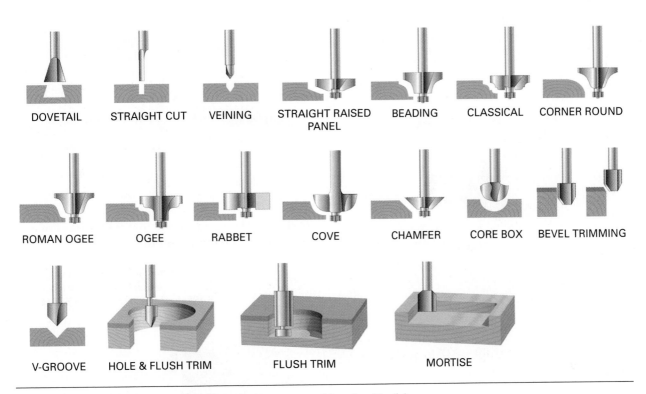

DOVETAIL **STRAIGHT CUT** **VEINING** **STRAIGHT RAISED PANEL** **BEADING** **CLASSICAL** **CORNER ROUND**

ROMAN OGEE **OGEE** **RABBET** **COVE** **CHAMFER** **CORE BOX** **BEVEL TRIMMING**

V-GROOVE **HOLE & FLUSH TRIM** **FLUSH TRIM** **MORTISE**

FIGURE 15–5 Router bit selection guide. *(Courtesy of Stanley Tools)*

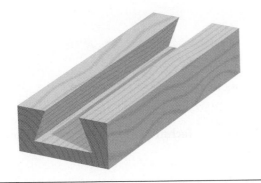

FIGURE 15–6 A dovetail cut is easily made with a router.

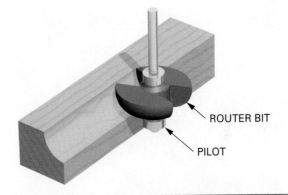

ROUTER BIT

PILOT

FIGURE 15–8 Router bits may have a pilot bearing to guide the cut.

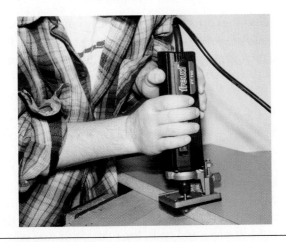

FIGURE 15–7 The laminate trimmer is used to trim the edges of plastic laminates.

- By guiding the edge of the router base against a straightedge **(Figure 15–9).** Be sure to keep the router tight to the straightedge, and do not rotate the router during the cut, because its base may not be centered.

- By using an adjustable guide attached to the base of the router **(Figure 15–10).** The guide rides along the edge of the stock. Make sure the edge is in good condition.

- By using a template (pattern) with template guides attached to the base of the router **(Figure 15–11).** This is the method widely used for cutting recesses for door hinges. (This process is explained in detail in a following unit on installing doors.)

STRAIGHTEDGE IS ON OPERATOR'S RIGHT. AS ROUTER IS PULLED, ROTATION OF ROUTER TENDS TO KEEP IT AGAINST STRAIGHTEDGE. IF STRAIGHTEDGE WERE ON LEFT SIDE, ROUTER WOULD HAVE TENDENCY TO PULL AWAY FROM THE STRAIGHTEDGE.

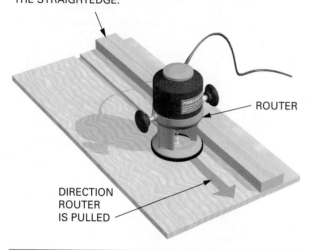

ROUTER

DIRECTION ROUTER IS PULLED

FIGURE 15–9   Using a straightedge to guide the router.

FIGURE 15–10   A guide attached to the base of the router rides along the edge of the stock and controls the sideways motion of the router.

- By freehand routing, in which the sideways motion of the router is controlled by the operator only. Care should be taken during this operation.

- To make *circular* cuts, remove the subbase. Replace it, using the same screwholes, with a custom-made one in which one side extends to any desired length. Along a centerline make a series of holes to fasten the newly made subbase to the center of the desired arc **(Figure 15–12).**

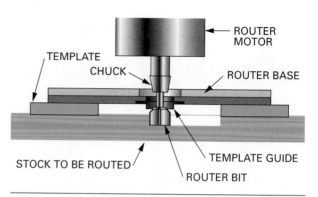

TEMPLATE

CHUCK

ROUTER MOTOR

ROUTER BASE

STOCK TO BE ROUTED

TEMPLATE GUIDE

ROUTER BIT

FIGURE 15–11   Guiding the router by means of a template and template guide.

REPLACE THE ROUTER SUBBASE WITH A CUSTOM-MADE ELONGATED BASE. DRILL SEVERAL HOLES TO BE USED AS CENTERS

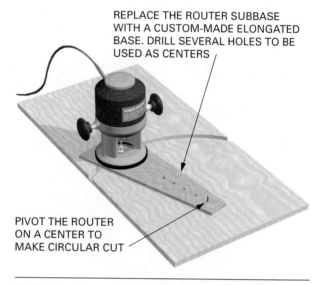

PIVOT THE ROUTER ON A CENTER TO MAKE CIRCULAR CUT

FIGURE 15–12   Technique for making arcs using a router.

Using the Router

Before using the router, make sure power is disconnected. Follow the method outlined:

- Select the correct bit for the type of cut to be made.

- Insert the bit into the chuck. Make sure the chuck grabs at least ½ inch of the bit. Adjust the depth of cut.

- Control the sideways motion of the router by one of the methods previously described.

- Clamp the work securely in position. Plug in the cord.

- Lay the base of the router on the work with the router bit clear of the work. Start the motor.

- Advance the bit into the cut, pulling the router in a direction that is against the rotation of the bit. On outside edges and ends, the router is moved counterclockwise around the piece. When making internal cuts, the router is moved in a clockwise direction.

CAUTION   CAUTION   CAUTION   CAUTION

Finish the cut, keeping the router clear of your body until it has stopped. Be aware that the router bit is unguarded. ■

PORTABLE ELECTRIC SANDERS

Interior trim, cabinets, and other finish should be sanded before any finishing coats of paint, stain, polyurethane, or other material are applied. It is shoddy workmanship to coat finish work without sanding. In too many cases, expediency seems to take precedence over quality. Trim needs to be sanded because the grain probably has been *raised*. This happens because the stock has been exposed to moisture in the air between the time it was milled and the time of installation. Also, rotary planing of lumber leaves small ripples in the surface. Although hardly visible before a finish is applied, they become very noticeable later.

Some finishing coats require more sanding than others. If a penetrating stain is to be applied, extreme care must be taken to provide a surface that is evenly sanded with no cross-grained scratches. If paints or transparent coatings are to be applied, surfaces need not be sanded as thoroughly. Portable sanders make sanding jobs less tedious.

Portable Electric Belt Sanders

The **belt sander** is used frequently for sanding cabinetwork and interior finish **(Figure 15–13)**. The size of the belt determines the size of the sander. Belt widths range from 2½ to 4 inches. Belt lengths vary from 18 to 24 inches. The 3-inch by 21-inch belt sander is a popular, lightweight model. Some sanders have a bag to collect the sanding dust. Remember to wear eye protection.

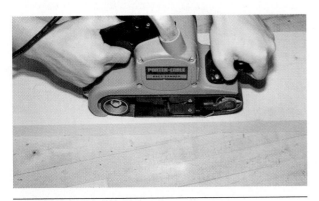

FIGURE 15–13 Using a portable electric belt sander.

FIGURE 15–14 Sanding belts should be installed in the proper direction, as indicated by the arrow on the inside of the belt.

Installing Sanding Belts. Sanding belts are usually installed by retracting the forward roller of the belt sander. An arrow is stamped on the inside of some sanding belts to indicate the direction in which the belt should run. The sanding belt should run *with*, not against, the lap of the joint **(Figure 15–14)**. Sanding belts joined with butt joints may be installed in either direction. Install the belt over the rollers. Then release the forward roller to its operating position.

To keep the sanding belt centered as it is rotating, the forward roller can be tilted. Stand the sander on its back end. Hold it securely and start it. Turn the adjusting screw one way or the other to track the belt and center it on the roller **(Figure 15–15)**.

Using the Belt Sander. More work has probably been ruined by improper use of the portable belt sander than any other tool. It is wise to practice on scrap stock until enough experience in its use is gained to ensure an acceptable sanded surface. Care must be taken to sand squarely on the sander's pad. Allowing the sander to tilt sideways or to ride on either roller results in a gouged surface.

CAUTION   CAUTION   CAUTION   CAUTION

Make sure the switch of the belt sander is off before plugging the cord into a power outlet. Some trigger switches can be locked in the "ON" position. If the tool is plugged in when the switch is locked in this position, the sander will travel at high speed across the surface. This could cause damage to the work and/or injury to anyone in its path. ■

FIGURE 15–15   The belt should be centered on its rollers by using the tracking screw.

- Secure the work to be sanded. Make sure the belt is centered on the rollers and is tracking properly.

- Hold the tool with both hands so that the edge of the sanding belt can be clearly seen.

- Start the machine. Place the pad of the sander flat on the work. Pull the sander back and lift it just clear of the work at the end of the stroke.

> ⚠ **CAUTION   CAUTION   CAUTION   CAUTION**
>
> Be careful to keep the electrical cord clear of the tool. Because of the constant movement of the sander, the cord may easily get tangled in the sander if the operator is not alert. ■

- Place the sander back on the material and bring the sander forward. Continue sanding using a skimming motion that lifts the sander just clear of the work at the end of every stroke. Sanding in this manner prevents overheating of the sander and helps to remove material evenly. It also allows debris to be cleared from the work, and the operator can see what has been done.

- Do not sand in one spot too long. Be careful not to tilt the sander in any direction. Always sand with the pad flat on the work. Do not exert excessive pressure. The weight of the sander is enough. Always sand with the grain to produce a smooth finish.

- Make sure the sander has stopped before setting it down. It is a good idea to lay it on its side to prevent accidental traveling.

Finishing Sanders

The finishing sander or palm sander **(Figure 15–16)** is used for the final sanding of interior work. These tools are manufactured in many styles and sizes. They are available in cordless models.

Finishing sanders either have an orbital motion, an oscillating (straight back and forth) motion, or a combination of motions controlled by a switch. The orbital motion has faster action, but leaves scratches across the grain. The *random orbital* sander reduces this problem with a design that randomly moves the center of the rotating paper at high speed. This allows the paper to sand in all directions at once. The straight line motion is slower, but produces no cross-grain scratches on the surface.

Most sanders take ¼ or ½ sheet of sandpaper. It is usually attached to the pad by some type of friction or spring device. Some sanding sheets come precut with an adhesive backing for easy attachment to the sander pad.

To use the finishing sander, proceed as follows:

- Select the desired grit sandpaper. Attach it to the pad, making sure it is tight. A loose sheet will tear easily.

- Start the motor and sand the surface evenly, *slowly* pushing and pulling the sander with the grain. Let the action of the sander do the work. Do not use excessive pressure because this may overload the machine and burn out the motor. Always hold the sander flat on its pad.

FIGURE 15–16   A portable electric finishing sander. *(Courtesy of S-B Power Tools)*

Abrasives. The quality of the abrasives on the sand paper is determined by the length of time it is able to retain its sharp cutting edges. *Flint* and *garnet* are natural minerals used as abrasives. Although sandpaper made with flint is less expensive, it does not last as long as garnet. Synthetic (man-made) abrasives include *aluminum oxide* and *silicon carbide.* Sandpaper coated with aluminum oxide is probably the most widely used for wood.

Grits. Sandpaper **grit** refers to the size of the abrasive particles. Sandpaper with large abrasive particles is considered coarse. Small abrasive particles are used to make fine sandpaper.

Sandpaper grits are designated by a grit number. The grit numbers range from No. 12 (coarse) to No. 400 (fine) **(Figure 15–17).** Commonly used grits are 60 or 80 for rough sanding, and 120 or 180 for finish sanding.

Sand with a coarser grit until a surface is uniformly sanded. Do not switch to a finer grit too soon. Do not use worn or clogged abrasives. Their use causes the surface to become glazed or burned.

Description	Grit No.
Very Fine	400
	360
	320
	280
Fine	240
	220
	180
	150
Medium	120
	100
Coarse	80
	60
	50
	40
Very coarse	36
	30
	24
	20
	16
	12

FIGURE 15–17   Grits of coated abrasives.

16   Fastening Tools

Portable fastening tools included in this chapter are those called **pneumatic** (powered by compressed air) and **powder-actuated** (drive fasteners with explosive powder cartridges). They are widely used throughout the construction industry for practically every fastening job from foundation to roof.

PNEUMATIC STAPLERS AND NAILERS

Pneumatic **staplers** and **nailers** are commonly called *guns* **(Figure 16–1).** They are used widely for quick fastening of framing, subfloors, wall and roof sheathing, roof shingles, exterior finish, and interior trim. A

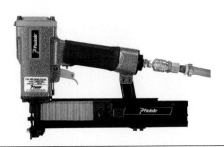

FIGURE 16–1   Pneumatic nailers and staplers are widely used to fasten building materials. *(Courtesy of Paslode)*

number of manufacturers make a variety of models in several sizes for special fastening jobs. For instance, a *framing gun,* although used for many fastening jobs, is not used to apply interior trim.

Older models required frequent oiling, either by hand or by an oiler installed in the air line. With improvements in design, some newer models require no oiling at all.

Remember to wear eye protection. It is also recommended to wear ear protection. Prolonged exposure to loud noises will damage the ear.

Nailing Guns

The heavy-duty framing gun **(Figure 16–2)** drives smooth-shank, headed nails up to 4 inches, ring-

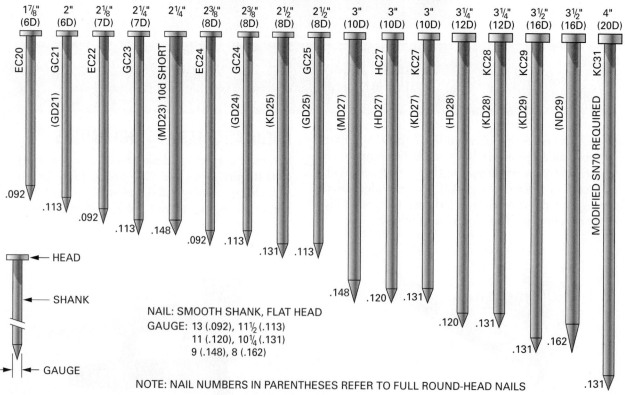

NAIL: SMOOTH SHANK, FLAT HEAD
GAUGE: 13 (.092), 11½ (.113)
11 (.120), 10¼ (.131)
9 (.148), 8 (.162)

NOTE: NAIL NUMBERS IN PARENTHESES REFER TO FULL ROUND-HEAD NAILS

FIGURE 16–2 **Heavy-duty framing guns are used for floor, wall, and roof framing.** *(Courtesy of Senco Products, Inc.)*

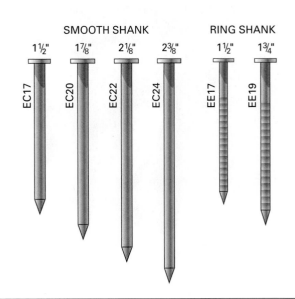

FIGURE 16–3   A light-duty nailer is used to fasten light framing, subfloors, and sheathing. *(Courtesy of Senco Products, Inc.)*

shank nails 2⅜ inches, and screw shank nails 3 inches to fasten all types of framing. A light-duty version **(Figure 16–3)** drives smooth-shank nails up to 2⅜ inches and ring-shank nails up to 1¾ inches to fasten light framing, subfloor, sheathing, and similar components of a building. Nails come glued in strips for easy insertion into the magazine of the gun **(Figure 16–4).** Check local building codes and specifications for the correct nail styles and sizes. For example, clip-headed nails cannot be used to fasten plywood that is used as a shear (high-stress) panel.

The *finish nailer* **(Figure 16–5)** drives finish nails from 1 to 2 inches long. It can be used for the application of practically all kinds of exterior and interior finish work. It sets or flush drives nails as desired. A nail set is not required, and the possibility of marring the wood is avoided.

The *brad nailer* **(Figure 16–6)** drives both slight-headed and medium-headed brads ranging in length from ⅝ inch to 1⅜ inches. It is used to fasten small moldings and trim, cabinet door frames and panels, and other miscellaneous finish carpentry.

The *coil roofing nailer* **(Figure 16–7)** is designed for fastening asphalt and fiberglass roof shingles. It drives five different sizes of wide, round-headed roofing nails from ⅞ to 1¾ inches. The nails come in coils of 120 **(Figure 16–8),** which are easily loaded in a nail canister.

Staplers

Like nailing guns, *staplers* are manufactured in a number of models and sizes and are classified as light-, medium-, and heavy-duty staplers.

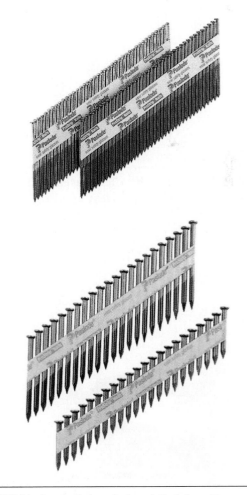

FIGURE 16–4   Both headed and finish nails used in nailing guns come glued together in strips. *(Courtesy of Paslode)*

FINISH NAILS

1"	1¼"	1½"	1¾"	2"
(2D)	(3D)	(4D)	(5D)	(6D)

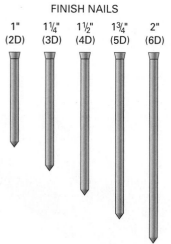

FIGURE 16–5 The finish nailer is used to fasten all kinds of interior trim. (*Courtesy of Senco Products, Inc.*)

SLIGHT-HEADED BRADS   MEDIUM-HEADED BRADS

⅝"	¾"	1"		1"	1¼"	1½"	1⅝"

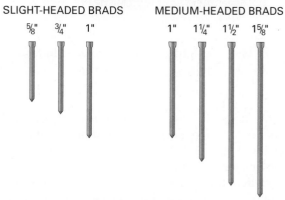

FIGURE 16–6 A light-duty brad nailer is used to fasten thin molding and trim. (*Courtesy of Senco Products, Inc.*)

COIL ROOFING NAILS

⅞"	1"	1¼"	1½"	1¾"

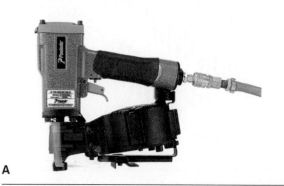

A

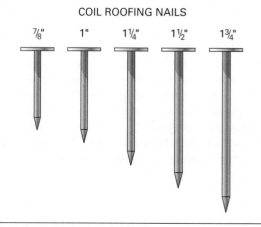

B

FIGURE 16–7 A coil roofing nailer is used to fasten asphalt roof shingles. (*a) Courtesy of Paslode (b) Courtesy of Senco Products, Inc.*)

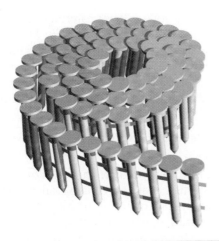

FIGURE 16–8 **Roofing nails come in coils for use in the roofing nailer.**

A popular tool is the *roofing stapler* **(Figure 16–9)**, which is used to fasten roofing shingles. It comes in several models and drives 1-inch wide-crown staples in lengths from ¾ inch to 1½ inches. It can also be used for fastening other materials, such as lath wire, insulation, furniture, and cabinets. The staples, like nails, come glued together in strips **(Figure 16–10)** for quick and easy reloading. Most stapling guns can hold up to 150 staples.

No single model stapler drives all widths and lengths of staples. Models are made to drive narrow-crown, intermediate-crown, and wide-crown staples. A popular model drives ⅜-inch intermediate-crown staples from ¾ inch to 2 inches in length.

Cordless Guns

Conventional pneumatic staplers and nailers are powered by compressed air. The air is supplied by an air compressor **(Figure 16–11)** through long lengths of air hose stretched over the construction site. The development of **cordless nailing** and **stapling guns** **(Figure 16–12)** eliminates the need for air compressors and hoses. The cordless gun utilizes a disposable fuel cell. A battery and spark plug power an internal combustion engine that forces a piston down to drive the fastener. Another advantage is the time saved in setting up the compressor, draining it at the end of the day, coiling the hoses, and storing the equipment.

The *cordless framing nailer* drives nails from 2 to 3¼ inches in length. Each fuel cell will deliver energy to drive about 1,200 nails. The battery will last long enough to drive about 4,000 nails before recharging is required.

The *cordless finish nailer* drives finish nails from ¾ to 2½ inches in length. It will drive about 2,500 nails before a new fuel cell is needed and about 8,000 nails before the battery has to be recharged.

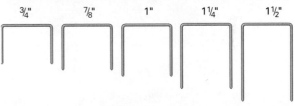

FIGURE 16–9 **The wide-crown stapler, being used in this photograph to fasten roof shingles, can be used to fasten a variety of materials.** *(Courtesy of Senco Products, Inc.)*

FIGURE 16–10 **Staples, like nails, come glued together in strips for use in stapling guns.** *(Courtesy of Senco Products, Inc.)*

FIGURE 16–11 **Compressors supply air pressure to operate nailers and other pneumatic tools.** *(Courtesy of Porter Cable)*

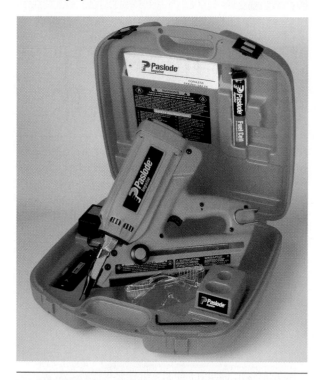

FIGURE 16–12 **Each cordless gun comes in its own case with battery, battery charger, safety glasses, instructions, and storage for fuel cells.** *(Courtesy of Paslode)*

The *cordless stapler* drives intermediate-crown staples from ¾ inch to 2 inches in length. It will drive about 2,500 staples with each fuel cell and about 8,000 staples with each charge of the battery.

Using Nailers and Staplers

Because of the many designs and sizes of staplers and nailers, you should study the manufacturer's directions and follow them carefully. Use the right nailer or stapler for the job at hand. Make sure all safety devices are working properly and always wear eye protection. A work contact element allows the tool to operate only when this device is firmly depressed on a work surface and the trigger is pulled, promoting safe tool operation.

- Load the magazine with the desired size staples or nails.
- Connect the air supply to the tool. For those guns that require it, make sure there is an oiler in the air supply line, adequate oil to keep the gun lubricated during operation, and an air filter to keep moisture from damaging the gun. Use the recommended air pressure. Larger nails require more air pressure than smaller ones.

 CAUTION CAUTION CAUTION CAUTION

Exceeding the recommended air pressure may cause damage to the gun or burst air hoses, possibly causing injury to workers. ■

- Press the trigger and tap the nose of the gun to the work. When the trigger is depressed, a fastener is driven each time the nose of the gun is tapped to the work. The fastener may also be safely driven by first pressing the nose of the gun to the surface and then pulling the trigger.
- Upon completion of fastening, disconnect the air supply.

⚠ CAUTION CAUTION CAUTION CAUTION

Never leave an unattended gun with the air supply connected. Always keep the gun pointed toward the work. Never point it at other workers or fire a fastener except into the work. Serious injury can result from horseplay with the tool. ■

POWDER-ACTUATED DRIVERS

Powder-actuated drivers (Figure 16–13) are used to drive specially designed pins into masonry or steel. They are used in a manner similar to firing a gun. Powder charges of various strengths drive the pin when detonated.

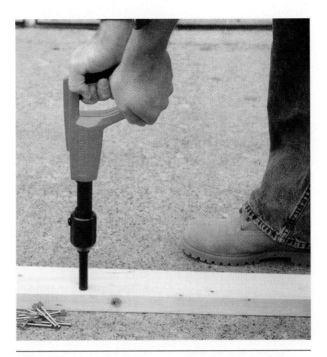

FIGURE 16–13 Powder-actuated drivers are used for fastening into masonry or steel.

Drivepins

Drivepins are available in a variety of sizes. Three styles are commonly used. The *headed* type is used for fastening material. The *threaded* type is used to bolt an object after the pin is driven. The *eyelet* type is used when attachments are to be made with wire.

Powder Charges

Powder charges are color coded according to strength. Learn the color codes for immediate recognition of the strength of the charge. A stronger charge is needed for deeper penetration or when driving into hard material. The strength of the charge must be selected with great care.

Because of the danger in operating these guns, many states require the operator to be certified. Certificates may be obtained from the manufacturer's representative after a brief training course.

Using Powder-Actuated Drivers

- Study the manufacturer's directions for safe and proper use of the gun. Use eye and ear protection.

> ⚠ CAUTION CAUTION CAUTION CAUTION
>
> Treat this tool as if it were a gun. Never use it if the end that ejects the fastener is facing toward other workers. ■

- Make sure the drivepin will not penetrate completely through the material into which it is driven. This has been the cause of fatal accidents.
- To prevent ricochet hazard, make sure the recommended shield is in place on the nose of the gun. A number of different shields are available for special fastening jobs.
- Select the proper fastener for the job. Consult the manufacturer's drivepin selection chart to determine the correct fastener size and style.
- Select a powder charge of necessary strength. Always use the weakest charge that will do the job and gradually increase strength as needed. Load the driver with the pin first and the cartridge second.
- Keep the tool pointed at the work. Wear safety goggles. Press hard against the work surface, and pull the trigger. The resulting explosion drives the pin. Eject the spent cartridge.

> ⚠ CAUTION CAUTION CAUTION CAUTION
>
> If the gun does not fire, hold it against the work surface for at least 30 seconds. Then remove the cartridge according to the manufacturer's directions. Do not attempt to pry out the cartridge with a knife or screwdriver; most cartridges are rim fired and could explode.

Key Terms

belt sander	grit	jointer plane	pneumatic
chamfers	hammer-drills	laminate trimmer	powder-actuated
cordless nailing	hole saws	masonry drill bits	powder-actuated drivers
dovetails	jigsaw	nailers	

reciprocating saw	screwguns	staplers	taper
ripping	skilsaw	stapling guns	templates
router			

Review Questions

Select the most appropriate answer.

1. To use a power tool properly, the operator should always
   a. wear eye protection.
   b. wear ear protection.
   c. understand the manufacturer's recommended instructions.
   d. all of the above.

2. The guard of the portable electric saw should never be
   a. lubricated.
   b. adjusted.
   c. retracted by hand.
   d. wedged open.

3. When selecting an extension cord for a power tool
   a. use a longer cord to keep the cord from heating up.
   b. keep the cord evenly spread out around the work area.
   c. use one with a GFCI or plugged into a GFCI outlet.
   d. all of the above.

4. When using a power tool for cutting, the operator should wear
   a. safety contact lenses and steel-toed work boots.
   b. ear and eye protection.
   c. stereo headphones.
   d. all of the above.

5. Sharp tools
   a. put less stress on the operator than dull tools.
   b. cut slower than dull tools.
   c. are more dangerous to use than dull tools.
   d. all of the above.

6. The jigsaw is used primarily for making
   a. curved cuts.
   b. compound miters.
   c. cuts in drywall.
   d. long straight cuts.

7. The saw primarily used for rough-in work is the
   a. reciprocating saw.
   b. sawzall.
   c. saber saw.
   d. all of the above.

8. The tool that is best suited for drilling metal as well as wood is the
   a. auger bit.
   b. high-speed twist drill.
   c. expansive bit.
   d. speed bit.

9. To produce a neat and clean hole in wood
   a. use a fast spinning, sharp bit.
   b. use a slower travel speed.
   c. finish the hole by drilling from the back side.
   d. all of the above.

10. When using the router to shape four outside edges and ends of a piece of stock, the router is guided in a
    a. direction with the grain.
    b. clockwise direction.
    c. counterclockwise direction.
    d. all of the above.

11. The tool for which some codes require the operator to be certified because of the potential danger is the
    a. powder-actuated driver.
    b. hammer drill.
    c. cordless pneumatic nailer.
    d. screwgun.

12. The saw arbor nuts that hold circular saw blades in position are loosened by rotating the nut
    a. clockwise.
    b. with the rotation of the blade.
    c. counterclockwise.
    d. against the rotation of the blade.

UNIT 7

Stationary Power Tools

Many kinds of stationary power woodworking tools are used in wood mills and cabinet shops for specialized work. On the building site, usually only table and miter saws are available for use by carpenters. These are not heavy-duty machines. They must be light enough to be transported from one job to another. These saws are ordinarily furnished by the contractor.

SAFETY REMINDER

Stationary power tools are strong and durable, designed for heavy duty tasks. While using power tools the worker must keep a constant awareness of the risks of using the tool.

OBJECTIVES

After completing this unit, the student should be able to:

- describe different types of circular saw blades and select the proper blade for the job at hand.
- describe, adjust, and operate the radial arm saw and table saw safely to crosscut lumber to length, rip to width, and make miters, compound miters, dadoes, grooves, and rabbets.
- operate the table saw in a safe manner to taper rip and to make cove cuts.
- describe, adjust, and operate the power miter saw to crosscut to length, making square and miter cuts safely and accurately.

⚠ CAUTION   CAUTION   CAUTION   CAUTION   CAUTION   CAUTION   CAUTION   CAUTION   CAUTION   CAUTION

- Be trained and competent in the use of stationary power tools before attempting to operate them without supervision.
- Make sure power is disconnected when changing blades or cutters.
- Make sure saw blades are sharp and suitable for the operation. Ensure that safety guards are in place and that all guides are in proper alignment and secured.
- Wear eye and ear protection. Wear appropriate, properly fitted clothing.
- Keep the work area clear of scrap that might present a tripping hazard.

- Keep stock clear of saw blades before starting a machine.
- Do not allow your attention to be distracted by others or be distracting to others while operating power tools.
- Turn off the power and make sure the machine has stopped before leaving the area.

Safety precautions that apply to specific operations are given when those operations are described in this unit. ■

17 Circular Saw Blades

All of the stationary power tools described in this unit use circular saw blades. To ensure safe and efficient saw operation, it is important to know which type to select for a particular purpose **(Figure 17–1).**

CIRCULAR SAW BLADES

The more teeth a saw blade has, the smoother the cut. However, a fine-toothed blade does not cut as fast. This means that the stock must be fed more slowly. A coarse-toothed saw blade leaves a rough surface, but cuts more rapidly. Thus, the stock can be fed faster by the operator.

If the feed is too slow, the blade may overheat. This will cause it to lose its shape and wobble at high speed. This is a dangerous condition that must be avoided. An overheated saw will probably start to bind in the cut, possibly causing kickback and serious operator injury.

The same results occur when trying to cut material with a dull blade. Always use a sharp blade. Use fine-toothed blades for cutting thin, dry material, and coarse-toothed blades for cutting heavy, rough lumber.

Types of Circular Saw Blades

Many different types of circular saw blades are used for various cutting applications. Blades are made to cut wood and wood products, plastic laminates, ma-

sonry, and steel. Each material has a different requirement for cutting and the blades are designed accordingly (Figure 17–1).

Wood cutting blades are diverse in design styles. They are loosely classified into two groups, **high-speed steel blades** and **tungsten carbide-tipped blades.** High-speed steel blades are the original generation of circular blades. Teeth may be sharpened with a file and tend not to stay sharp as long as carbide-tipped blades. The carbide-tipped blades are superior and well suited for cutting material that contains adhesives and other foreign material that rapidly dull high-speed steel blades. Tungsten carbide is a hard metal that can only be sharpened with diamond-impregnated grinding wheels. Most saw blades being used today are carbide-tipped. Resurfacing of the tips should be done by a professional blade sharpener.

In both groups of saw blades, the number and shape of the teeth vary to give different cutting actions according to the kind, size, and condition of the material to be cut. Masonry and steel cutting blades are made in general types and also abrasive and steel types. Abrasive blades are made of composite materials that are consumed as the blade cuts. The blade diameter gets smaller with each cut and it is easy to tell when the blade should be replaced. Steel blades designed to cut steel and masonry are carbide tipped or diamond coated. They tend to last longer and are more expensive than abrasive blades.

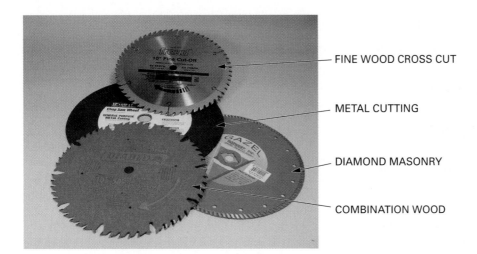

FIGURE 17–1 Circular saw blades are made to cut different materials.

- FINE WOOD CROSS CUT
- METAL CUTTING
- DIAMOND MASONRY
- COMBINATION WOOD

High-Speed Circular Saw Blades

High-speed steel blades are classified as *rip, crosscut,* or *combination* blades. They may be given other names, such as plywood, panel, or flooring blades, but they are still in one of the three classifications.

Ripsaws. The **ripsaw blade (Figure 17–2)** usually has fewer teeth than crosscut or combination blades. As in the hand ripsaw, every tooth of the circular ripsaw blade is filed or ground at right angles to the face of the blade. This produces teeth with a cutting edge all the way across the tip of the tooth. These

teeth act like a series of small chisels that cut and clear the stock ahead of them.

Use a ripsaw for cutting solid lumber with the grain when a smooth edge is not necessary. Also, use a ripsaw when cutting unseasoned or green lumber and lumber of heavy dimension with the grain.

Crosscut Saws. The teeth of a **crosscut circular saw blade** are shaped like those of the crosscut handsaw **(Figure 17–3).** The sides of the teeth are alternately filed or ground on a bevel. This produces teeth that come up to points instead of edges.

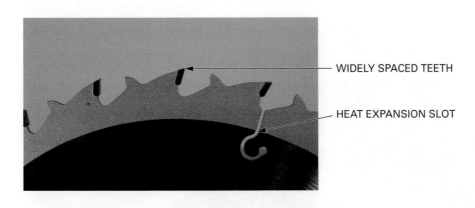

FIGURE 17–2 The teeth of a ripsaw have square-edged cutting tips with wide spaces between them.

- WIDELY SPACED TEETH
- HEAT EXPANSION SLOT

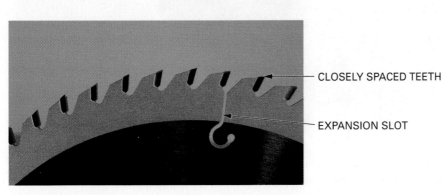

FIGURE 17–3 The teeth of a crosscut blade have beveled sides and pointed tips.

- CLOSELY SPACED TEETH
- EXPANSION SLOT

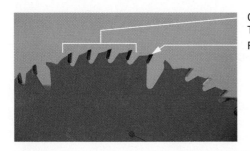

CROSS CUT
TEETH
RIP TOOTH

FIGURE 17–4 The teeth of a combination blade include both crosscut and rip teeth.

Crosscut teeth slice through the wood fibers smoothly. A crosscut blade is an ideal blade for cutting across the grain of solid lumber. It also cuts plywood satisfactorily with little splintering of the cut edge.

Combination Saws. The **combination blade** is used when a variety of ripping and crosscutting is to be done. It eliminates the need to change blades for different operations.

There are several types of combination blades. One type has groups of teeth around its circumference **(Figure 17–4)**. The leading tooth in each group is a rip tooth. The ones following are crosscut teeth.

Carbide-Tipped Blades

Carbide Blade Component Parts. The component parts of a **carbide blade** have functions that make the blade

perform well **(Figure 17–5)**. The teeth consist of bits of carbide and are silver soldered to the metal blade. They should be the only part to actually touch the material. Carbide is made in different hardnesses for various applications. Drilling stone and masonry often requires hammering, thus softer carbide is used. The hardest carbide is used for cutting metal. The harder the carbide material, the more brittle the blade, which should be handled gently. A carbide blade can break like glass if it is dropped.

The *gullet* is a gap created to allow for blade expansion when it heats up from normal use. The anti-vibration slots are cut in the blade with a laser to help keep the blade true while it is being run under heavy load. The anti-kickback teeth are designed to keep the tooth that follows it from biting too deeply into the material. Otherwise the blade might kick back at the operator. They usually are located before any large space in the blade. The hook angle is the angle at which the tooth is pitched forward. Larger hook angles cut more aggressively, making a faster but less smooth cut.

Carbide Tooth Style. The teeth of a carbide blade are ground to different shapes giving the blade different cutting abilities **(Figure 17–6)**.

Square Grind. The *square grind* is similar to the rip teeth in a steel blade. It is used primarily to cut solid wood with the grain. It can also be used on composition boards when the quality of the cut surface is not important.

FIGURE 17–5 Parts of a typical carbide combination blade.

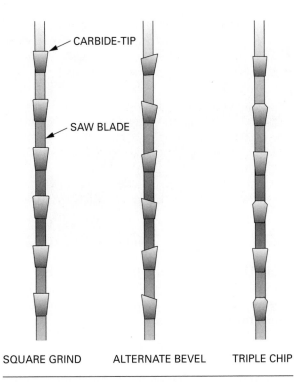

SQUARE GRIND   ALTERNATE BEVEL   TRIPLE CHIP

FIGURE 17–6   **Three main tooth styles of carbide-tipped saw blades.**

Alternate Top Bevel. The *alternate top bevel* grind is used with excellent results for crosscutting solid lumber, plywood, hardboard, particleboard, fiberboard, and other wood composite products. It can also be used for ripping operations. However, the feed is slower than that of square or combination grinds.

Triple Chip. The *triple chip grind* is designed for cutting brittle material without splintering or chipping the surface. It is particularly useful for cutting an extremely smooth edge on plastic laminated material, such as countertops. It can also be used like a planer blade to produce a smooth cut surface on straight, dry lumber of small dimension.

Combination. Carbide-tipped saws also come with a combination of teeth. The leading tooth in each set is square ground. The following teeth are ground at alternate bevels. These saws are ideally suited for ripping and crosscutting solid lumber and all kinds of engineered wood products. Because of their versatility, these saw blades are probably the most widely used by carpenters.

Carbide-tipped teeth are not set (bent slightly outward). The carbide tips are slightly thicker than the saw blade itself. Therefore, they provide clearance for the blade in the saw cut. In addition, the sides of the carbide tips are slightly beveled back to provide clearance for the tip.

Dado Blades

Dadoes, grooves, and *rabbets* can be cut using a single saw blade **(Figure 17–7),** but this would require making several passes of the material with the single blade. A *dado set,* which consists of more than one blade, is commonly used to make these cuts faster because only one pass needs to be made through the stock. One type, called the stack dado head, consists of two outside circular saw blades with several *chippers* of different thicknesses placed between **(Figure 17–8).**

Most dado sets make cuts from ⅛ to ¹³⁄₁₆ inch wide. This is done by using one or more of the blades and chippers together. Wide cuts are obtained by adding shims between chippers or by making more passes through the material. When installing this type of dado set, make sure the tips of the chippers

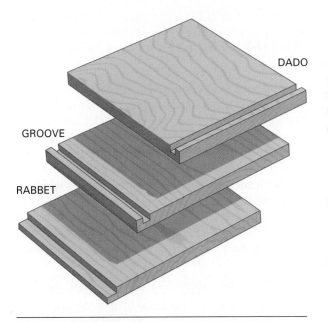

FIGURE 17–7   **Notches in wood have different names depending on where they are located.**

FIGURE 17–8   **A stacked dado head set.**

FIGURE 17–9 A wobbler head dado blade.

FIGURE 17–10 Always unplug the saw before touching blade or cutters. The saw arbor nut is loosened by turning it in the same direction in which the blade rotates.

are opposite the gullets and not against the side of the blade. Chipper tips are *swaged* (made wider than the body of the chipper) to ensure a clean cut across the width of the dado or groove.

Another type of dado blade is the one-unit *adjustable dado head* (**Figure 17–9**), commonly called a *wobbler head*. This type can be adjusted for width of dado by rotating the sections of the head. The head can be adjusted by loosening the arbor nut and need not be removed from the saw arbor. This type of dado head does not make perfectly square-bottomed slots.

Removing and Replacing Circular Saw Blades

Carbide circular saw blades will at some time need to be sharpened or replaced. Several clues will help the operator realize that the blade is dull:

- The feeding pressure of the material into the saw will seem to be increasing. The ability to notice this takes some experience with the saw.

- Burn marks are left on the material after the cut. Note that some material, like cherry, burns easily even with a sharp blade. Keep the feed rate constant to reduce burning.

- Pitch and sawdust will build up on the blade. This buildup will cause blade drag, resulting in more heating up than normal.

- Broken or chipped teeth are the most obvious sign of a dull blade. Remember the teeth are brittle. Protect them from being struck by or dropped on hard objects.

Saw arbor shafts may have a right- or left-hand thread for the nut, depending on which side of the blade it is located. No matter what direction the arbor shaft is threaded, the arbor nut is loosened in the same direction in which the saw blade rotates (**Figure 17–10**). The arbor nut is always tightened against the rotation of the saw blade. This design prevents the arbor nut from loosening during operation.

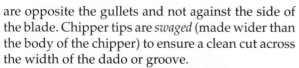

18 Radial Arm and Miter Saws

The radial arm saw and the miter saw are similar in purpose. Materials are placed against a fence and crosscutting operations are similar. While many cutting operations can be performed with the radial arm saw, it is used primarily for crosscutting. The miter saw is specifically designed to crosscut.

RADIAL ARM SAWS

The major function of a **radial arm saw** (Figure 18–1) is crosscutting. The stock remains stationary while the saw moves across it. The operator does not have to push or pull the stock through the blade; instead, the saw is moved. The end of long lengths are easily

Ripping

Although it is possible to rip lumber using the radial arm saw, it should be avoided if a table saw can be used. The setup for a radial arm saw takes some time and the ripping setup must be followed very carefully to protect the operator and the material being cut.

For ripping operations, the arm is locked in the straight crosscutting position at right angles to the fence. The motor unit is pulled out and rotated horizontally. The ripping width may be measured with a rule or tape from the fence to the blade. It may also be set by reading the rip scale built in on the arm of the saw.

Lock the saw carriage in the desired position by tightening the *rip lock* against the side of the arm. Lower the motor until the saw teeth are just below the table surface.

Adjust the blade shield guard so that the in-feed end almost touches the material to be cut. This is important because the guard holds the stock to the table and prevents the saw from picking up the stock during the rip and binding the motor. It also keeps most of the sawdust from flying at the operator.

Lower the kickback assembly so that the kickback fingers are about ⅛ inch lower than the material to be ripped. Using the fence as a guide, feed the stock evenly into the saw blade **(Figure 18–6).**

FIGURE 18–6 **Ripping operation using a radial arm saw.**

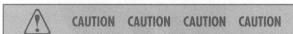

> ⚠ CAUTION   CAUTION   CAUTION   CAUTION
>
> Do not feed stock from the kickback end of the guard. *Feed the stock against the rotation of the blade.* Feeding stock in the wrong direction may cause it to be pulled from the operator's hands and through the saw with great force. This could cause serious or fatal injury to anyone in its path. ∎

Do not force the stock into the saw. Feed it with a continuous motion. When ripping narrow stock, use a *push stick* on the end of the stock to push it between the saw blade and the fence.

Bevel Ripping. **Bevel ripping** is done in the same manner as straight ripping, except that the blade is tilted to the desired angle. A V-groove may be made in this position by cutting partway through the stock. Then turn it around, end for end, and run it through again. The depth, width, and angle of the groove may be varied with different adjustments.

POWER MITER SAW

The term **power miter saw,** also called a *power miter box,* refers the a tool that has a circular saw blade mounted above the base. It comes in variety of styles and sizes. Operation of the saw is to push it down into the material using a chopping action. These tools are used in most phases of construction. A retractable blade guard provides operator protection and should be in place when the saw is being used.

The simplest of these designs is often referred to as a *chop box* **(Figure 18–7).** It comes in sizes with 10-, 12-, 14-, and 16-inch-diameter blades. The saw cannot be adjusted to cut angles; it can cut only in the up-and-down direction. To cut angles, the material itself must be clamped in place at the desired angle. Chop boxes are designed to cut metal, typically metal studs and pipe. An abrasive blade or a special carbide blade is used. Carbide blades cut cooler and do not spray sparks as do abrasive blades.

The power miter saw designed for wood has more possible adjustments **(Figure 18–8).** The saw blade may be adjusted 45 degrees to the right or left. Some models may even tilt to the side: left, right, or both directions. It is possible with these saws to cut compound miters. Positive stops are located in the saw angle and the tilt angle adjustments to help the operator cut common angles. Other angles are cut by reading the degree scale and locking in the desired position. Typical sizes include 8½-, 10-, and 12-inch-diameter blades.

In the more sophisticated models, the saw slides out on rails to increase the width of the cut. These models are referred to as *sliding compound miter saws* **(Figure 18–9).** The 12-inch models can cut a 4½-inch by 12-inch block at 90 degrees, a 4½-inch by 8½-inch block at 45 degrees, and a 3½-inch by 8½-inch block

FIGURE 18–8   Power miter box. *(Courtesy of Delta)*

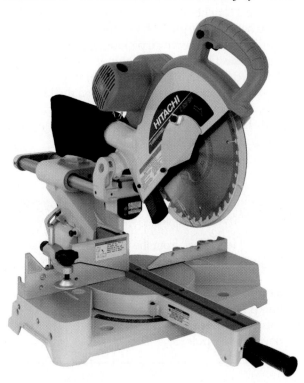

FIGURE 18–7   A chop saw is designed to cut with a chopping action. *(Courtesy of DeWalt)*

FIGURE 18–9   Sliding compound miter saw. *(Courtesy of Hitachi)*

to a compound 45-degree miter. They can also be adjusted to cut compound angles that exceed 45 degrees. Some will cut up to 60 degrees in either the left or right direction. Some tilt for a bevel cut in both directions.

The sliding compound miter saw allows for cutting large crown moldings with the stock laying flat on the base of the saw. The blade is adjusted to the compound angles using the preset stops on the saw. Check the manufacturer's instructions for the saw to determine crown molding cutting procedures. Some models have a laser attachment that lays a red line on

the material before the cut is made showing where the blade will enter the material **(Figure 18–10)**.

The operation of all of these types of miter saws is primarily the same. Adjustments are made to the blade travel and then made secure with locking clamps. Material should be held firmly against the base and fence of the saw. Some models have a material hold-down clamp to ensure the piece does not move during the cut **(Figure 18–11)**. The saw is

FIGURE 18–10 Some miter saws have lasers to assist the operator. *(Courtesy of Hitachi)*

FIGURE 18–11 Hold-down clamps are used to keep material from shifting during the cut. *(Courtesy of Delta)*

started and the blade eased into the material. With sliding models the saw may be pulled toward the operator then pushed down and back to cut the material. Rate of feed depends on the material being

cut. Listen to the saw to determine the best feed rate. The saw should keep nearly its full rpm running sound when cutting; that is, it should not slow down much.

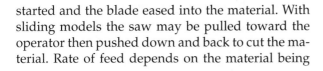

19 Table Saws

The **table saw** is one of the most frequently used woodworking power tools. In many cases, it is brought to the job site when the interior finish work begins. It is a useful tool because so many kinds of work can be performed with it. Common table saw operations, with different jigs to aid the process, are discussed in this chapter.

TABLE SAWS

The size of the table saw **(Figure 19–1)** is determined by the diameter of the saw blade. It may measure up to 16 inches. A commonly used table saw on the construction site is the 10-inch model.

The blade is adjusted for depth of cut and tilted up to 45 degrees by means of handwheels. A **rip fence** guides the work during ripping operations. A **miter gauge** is used to guide the work when cutting square ends and miters. The miter gauge slides in

FIGURE 19–1 A table saw.

grooves in the table surface. It may be turned and locked in any position up to 45 degrees.

A guard should always be placed on the blade to protect the operator. Exceptions to this include some table saw operations like dados, rabbets, and cuts where the blade does not penetrate the entire stock thickness. A general rule is if the guard can be used for any operation on the table saw, it should be used.

Ripping Operations

For ripping operations, the table saw is easier to use and safer than the radial arm saw. To rip stock to width, adjust the rip fence to the desired width. To check the accuracy of the scale under the *rip fence*, measure from the rip fence to the point of a saw tooth set closest to the fence. Lock the fence in place. Adjust the height of the blade to about ¼ inch above the stock to be cut. With the stock clear of the blade, turn on the power.

> **CAUTION   CAUTION   CAUTION   CAUTION**
>
> - Stand to either side of the blade's cutting line. Avoid standing directly behind the saw blade. Make sure no one else is in line with the saw blade in case of kickback.
> - Be prepared to turn the saw blade off at any moment should the blade bind.
> - A splitter and anti-kickback devices should be used when the cutting operation allows it. ∎

Hold the stock against the fence with the left hand. Push the stock forward with the right hand, holding the end of the stock **(Figure 19–2)**. As the

end approaches the saw blade, let it slip through the left hand. Remove the left hand from the work. Push the end all the way through the saw blade with the right hand, if the stock is of sufficient width (at least 5 inches wide). If the stock is not wide enough, use a *push stick* **(Figure 19–3)**.

> **CAUTION   CAUTION   CAUTION   CAUTION**
>
> - Make sure the stock is pushed all the way through the saw blade. Leaving the cut stock between the fence and a running saw blade may cause a kickback, injuring anyone in its path.
> - Use a push stick especially when ripping narrow pieces.
> - Keep small cutoff pieces away from the running blade. Do not use your fingers to remove cutoff pieces; use the push stick or wait until the saw has stopped to do so.
> - Always use the rip fence for ripping operations. Never make freehand cuts.
> - Never reach over a running saw blade. ∎

Bevel Ripping

Ripping on a bevel is done in the same manner as straight ripping except that the blade is tilted **(Figure 19–4)**. The blade is adjustable from 0 to 45 degrees. Take care not to let the blade touch the rip fence. Also keep the stock firmly in contact with the table. If the stock is allowed to lift off from the table during the cut, the width of the stock will vary.

Taper Ripping

Tapered pieces (one end narrower than the other) can be made with the table saw by using a **taper ripping jig.** The jig consists of a wide board with the length and amount of taper cut out of one edge. The other edge is held against the rip fence. The stock to be tapered is held in the cutout of the jig. The taper is cut by holding the stock in the jig as both are passed through the blade **(Figure 19–5)**.

By using taper ripping jigs, tapered pieces can be cut according to the design of the jig. A handle on the jig makes it safer to use. Also, if the jig is the same thickness as the stock to be cut, then the cutout section of the jig can be covered with a thin strip of

FIGURE 19–2   Using the table saw to rip lumber (guard has been removed for clarity).

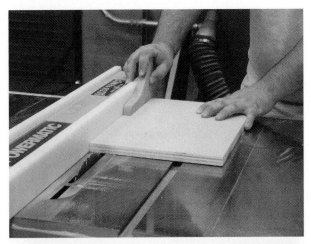

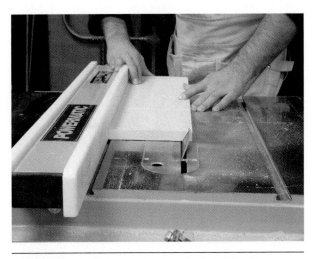

FIGURE 19–4 When bevel ripping, the blade is tilted (guard has been removed for clarity).

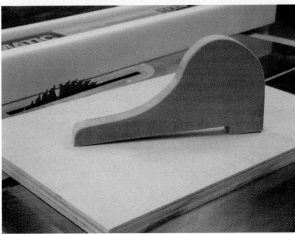

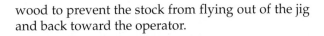

FIGURE 19–3 Use a push stick to rip narrow pieces (guard has been removed for clarity).

FIGURE 19–5 Using a taper ripping jig to cut identical wedges (guard has been removed for clarity).

wood to prevent the stock from flying out of the jig and back toward the operator.

Rabbeting and Grooving

Making *rabbets* is usually done using a dado head which is comprised of multiple blades and cutters. This tool makes fast, accurate grooves with one pass **(Figure 19–6).** Make sure firm even pressure is applied to the table and fence. This will provide consistent and accurate depth cuts.

Rabbets and narrow *grooves* can also be made with a single blade when a dado head is unavailable or when only a few are needed. The operation takes two settings of the rip fence and two passes through the saw blade are required **(Figure 19–7).** For narrow grooves, one or more passes are required. Move the rip fence slightly with each pass until the desired width of groove is obtained.

Crosscutting Operations

For most crosscutting operations, the *miter gauge* is used. To cut stock to length with both ends squared, first check the miter gauge for accuracy. Hold a framing square against it and the side of the saw blade. Usually the miter gauge is operated in the left-hand groove. The right-hand groove is used only when it is more convenient.

Square one end of the stock by holding the work firmly against the miter gauge with one hand while pushing the miter gauge forward with the other hand. Measure the desired distance from the squared end. Mark on the front edge of the stock. Repeat the procedure, cutting to the layout line **(Figure 19–8).**

FIGURE 19–6 Rabbeting an edge using a single saw blade by making two passes.

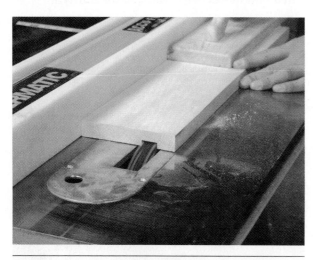

FIGURE 19–7 Cutting a groove with a dado head installed on the table saw.

FIGURE 19–8 Using the miter gauge as a guide to crosscut (guard has been removed for clarity).

 CAUTION   CAUTION   CAUTION   CAUTION

Do not use the miter gauge and the rip fence together on opposite sides of the blade unless a stop block is used. Pieces cut off between the blade and the fence can easily bind and be hurled across the room or at the operator. ∎

Cutting Identical Lengths

When a number of identical lengths need to be cut, first square one end of each piece. Clamp a stop to an *auxiliary wood fence* installed on the miter gauge. Place the square end of the stock against the stop block. Then make the cut. Slide the remaining stock across the table until its end comes in contact with the stop block. Then

make another cut. Continue in this manner until the desired number of pieces is cut **(Figure 19–9)**.

Another method for cutting identical lengths uses a block clamped to the rip fence in a location such that once the cut is made, there is clearance between the cut piece and the rip fence. The fence is adjusted so that the desired distance is between the face of the stop block and saw blade.

Square one end of the stock. Slide the squared end against the stop block. Then make a cut. Continue making cuts in this manner until the desired number of pieces is obtained **(Figure 19–10).**

Mitering

Flat miters are cut in the same manner as square ends, except the miter gauge is turned to the desired

FIGURE 19–9 Cutting identical lengths using a stop block on an auxiliary fence of the miter gauge (guard has been removed for clarity).

FIGURE 19–10 Using the rip fence and block as a stop to cut identical lengths (guard has been removed for clarity).

angle. *End miters* are made by adjusting the miter gauge to a square position and making the cut with the blade tilted to the desired angle.

Compound miters are cut with the miter gauge turned and the blade tilted to the desired angles.

A *mitering jig* can be used with the table saw. **Procedure 19–A** shows the construction and use of the jig. Use of such a jig eliminates turning the miter gauge each time for left- and right-hand miters.

Dadoing

Dadoing is done in a similar manner as crosscutting except a dado set is used **(Figure 19–11)**. The dado

set is only used to cut *partway* through the stock thickness.

TABLE SAW AIDS

A very useful aid is an *auxiliary fence.* A straight piece of ¾-inch plywood about 12 inches wide and as long as the ripping fence is screwed or bolted to the metal fence. When cuts must be made close to the fence, the use of an auxiliary wood fence prevents the saw blade from cutting into the metal fence. Also, the additional height provided by the fence gives a broader surface to steady wide work when its edge

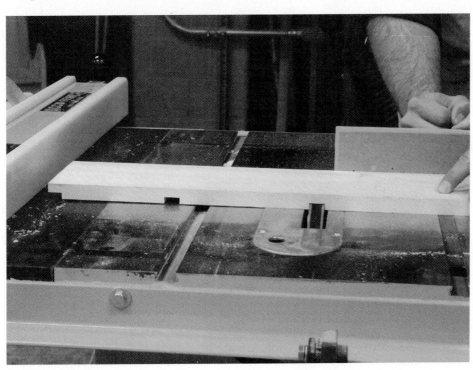

FIGURE 19–11 Cutting dadoes using a dado head.

PROCEDURE 19–A Making a Table Saw Jig to Cut 45-Degree Miters

STEP 1 FASTEN WOOD STRIPS, SIZED, SPACED, AND LOCATED TO FIT IN MITER GAUGE GROOVES, TO UNDERSIDE OF PLYWOOD. RUB WAX ON PLYWOOD AND WOOD STRIPS.

STEP 2 WITH SAW BLADE BELOW TABLE SURFACE PLACE JIG SO WOOD STRIPS FIT IN GROOVES AND SLIDE BACK AND FORTH. JIG SHOULD MOVE EASILY, BUT WITH NO PLAY. MAKE A SAW CUT ABOUT HALF-WAY ACROSS THE WIDTH.

STEP 3 HOLD FRAMING SQUARE AT 12 AND 12 ON IN-FEED EDGE WITH HEEL CENTERED ON THE SAW CUT. MARK LAYOUT LINES ON OUTSIDE EDGES OF SQUARE.

STEP 4 ATTACH GUIDE STRIPS SO EDGES ARE TO LAYOUT LINES.

NOTE:
USE SAW GUARD WHEN MAKING CUTS–GUARD IS NOT SHOWN FOR CLARITY.

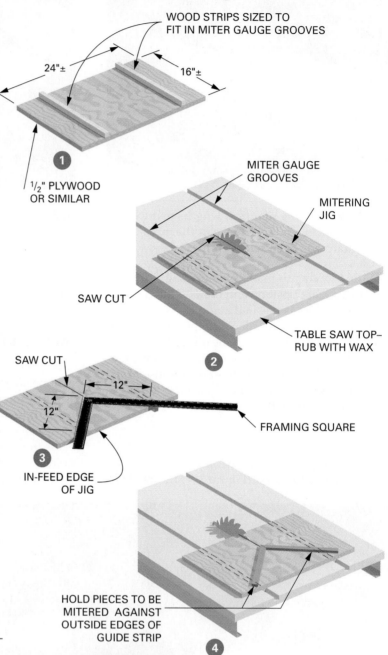

WOOD STRIPS SIZED TO FIT IN MITER GAUGE GROOVES

24"±   16"±

1

½" PLYWOOD OR SIMILAR

MITER GAUGE GROOVES

MITERING JIG

SAW CUT

TABLE SAW TOP–RUB WITH WAX

2

SAW CUT

12"

12"

FRAMING SQUARE

3

IN-FEED EDGE OF JIG

HOLD PIECES TO BE MITERED AGAINST OUTSIDE EDGES OF GUIDE STRIP

4

is being cut. The auxiliary wood fence also provides a surface on which to clamp **feather boards.** An auxiliary fence is also useful when attached to the miter gauge.

Feather boards are useful aids to hold work against the fence as well as down on the table surface during ripping operations **(Figure 19–12).** Feather boards may be made of 1 × 6 nominal lumber, with one end cut at a 45-degree angle. Saw cuts are made in this end about ¼ inch apart. This gives the end some spring and allows pressure to be applied to the piece being ripped while also allowing the ripped piece to move.

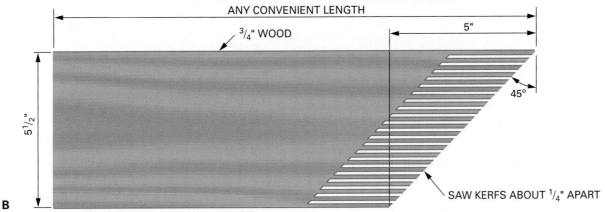

FIGURE 19–12 (A) Feather boards are useful aids to hold work during table saw operations (guard has been removed for clarity). Note that the locations of the feather boards do not cause the saw blade to bind. (B) Typical feather board design.

Key Term

bevel ripping

carbide blade

combination blade

crosscut circular saw
   blade

feather boards

high-speed steel
   blade

miter gauge

power miter saw

radial arm saw

rip fence

ripsaw blade

table saw

taper ripping

tungsten carbide-
   tipped blade

Review Questions

Select the most appropriate answer.

1. To use a power tool properly, the operator should always
   a. wear eye protection.
   b. wear ear protection.
   c. understand the manufacturer's recommended instructions.
   d. all of the above.

2. The most frequently used blade in general carpentry is the
   a. combination planer blade.
   b. combination carbide-tipped blade.
   c. square-grind carbide-tipped blade.
   d. ripsaw blade.

3. The alternate top bevel grind, carbide-tipped circular saw blade is designed to
   a. rip solid lumber.
   b. crosscut solid lumber.
   c. cut in either direction of the grain.
   d. cut green lumber of heavy dimension.

4. A device used with stationary saws to make repeated cuts of the same length is called a
   a. rabbet.
   b. feather board.
   c. stop block.
   d. all of the above.

5. The saw best suited to crosscut and rip material is the
   a. power miter box.
   b. radial arm saw.
   c. table saw.
   d. all of the above.

6. When using a table saw it is a good idea to
   a. stand away from the line of the blade.
   b. know the location of the shut off switch.
   c. keep wood scraps away from the blade.
   d. all of the above.

7. The table saw guide used for cutting material with the grain is called
   a. rip fence.   c. tilting arbor.
   b. miter gauge.   d. ripping jig.

8. The tool designed to hold a piece of stock safely while being cut with a table saw is a
   a. miter gauge.
   b. push stick.
   c. feather board.
   d. all of the above.

9. A dado is a wide cut partway through the thickness of the material and
   a. across the grain.
   b. with the grain.
   c. in either direction of the grain.
   d. close to the edge.

10. The table saw tool that should not be used at the same time as a rip fence is the
    a. push stick
    b. miter gauge
    c. dado head
    d. none of the above

UNIT 8

Architectural Plans and Building Codes

The ability to interpret architectural plans and to understand building codes, zoning ordinances, building permits, and inspection procedures is necessary for promotion on the job. With this ability, the carpenter has the competence needed to handle jobs of a supervisory nature, including those of foreman, superintendent, and contractor.

SAFETY REMINDER

A set of building prints is the vehicle used to communicate to the tradesperson the desired structure. Failing to read and understand prints properly can have costly and potentially disastrous results.

OBJECTIVES

After completing this unit, the student should be able to:

- describe and explain the function of the various kinds of drawings contained in a set of architectural plans.
- demonstrate how specifications are used.
- read and use an architect's scale.
- identify various types of lines and read dimensions.
- identify and explain the meaning of symbols and abbreviations used on architectural plans.
- read and interpret plot, foundation, floor, and framing plans.
- locate and explain information found in exterior and interior elevations.
- identify and utilize information from sections and details.
- use schedules to identify and determine location of windows, doors, and interior finish.
- define and explain the purpose of building codes and zoning laws.
- explain the requirements for obtaining a building permit and the duties of a building inspector.

20   Understanding Architectural Plans

ARCHITECTURAL PLANS

Blueprinting is an early method of creating drawings and making copies. It uses a water wash to produce prints with white lines against a dark blue background. Although these are true blueprints, the process is seldom used today. The word *blueprint* remains in use, however, to mean any copy of the original drawing.

A later method of making prints uses the original black line drawings of graphite or ink on translucent paper, cloth, or polyester film. Copies are usually made with a *diazo* printer. The copying process utilizes an ultraviolet light that passes through the paper, but not the lines. The light causes a chemical reaction on the copy paper, except where the lines are drawn. The copy paper is then exposed to ammonia vapor. This vapor develops the remaining chemical into blue, black, or brown lines against a white background. The diazo printing process is faster and less expensive than the early blueprinting methods.

Today most architectural plans, also called architectural drawings or construction drawings, are done using a CAD (computer-aided drafting) program. CAD drawings are produced faster and more easily than drawings done by hand. Changes to the plan can be made faster, allowing them to be easily customized to the desires of the customer. Once the plans are finished, they are sent to a plotter, which prints as many copies as needed.

Prints are drawn using symbols and a standardized language. To adequately read and interpret these prints, the builder must be able to understand the language of architectural plans. While this is not necessarily difficult, it can be confusing.

TYPES OF VIEWS

The architect can choose from any of several types of views to describe a building with clarity, and often multiple views are prepared. Because many people and many different trades are needed to build a house, different views are required that contain different information.

Pictorial View

Pictorial drawings are usually three-dimensional (3D) *perspective* or **isometric** views **(Figure 20–1).** The lines in a perspective view diminish in size as

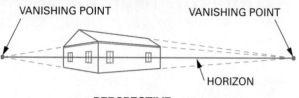

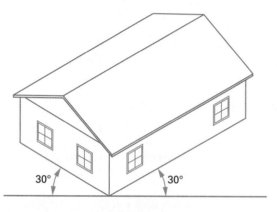

FIGURE 20–1   **Pictorial views used in architectural drawings.**

they converge toward vanishing points located on a line called the *horizon*. In an isometric drawing, the horizontal lines are drawn at 30-degree angles. All lines are drawn to actual scale. They do not diminish or converge as in perspective drawings.

A *presentation drawing* is usually a perspective. It shows the building from a desirable vantage point to display its most interesting features **(Figure 20–2).**

FIGURE 20–2   **A presentation drawing is usually a perspective view.**

Walks, streets, shrubs, trees, vehicles, and even people may be drawn. Presentation drawings are often colored for greater eye appeal. Presentation drawings provide little information for construction purposes. They are usually used as a marketing tool to show the client how the completed building will look.

Multiview Drawings

Different kinds of *multiview* drawings, also called **orthographics,** are required for the construction of a building. Multiview drawings are two dimensional and offer separate views of the building **(Figure 20–3).** The bottom view is never used in architectural drawings.

These different views are called **plan view, elevation, section view,** and **details.** When put together, they constitute a set of architectural plans. *Plan views* show the building from above, looking down. There are many kinds of plan views for different stages of construction **(Figure 20–4).** *Eleva-*

MULTIVIEWS

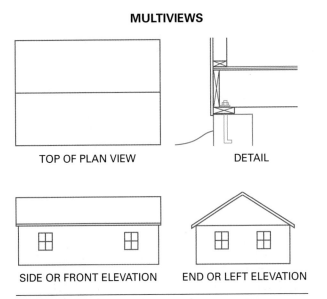

TOP OF PLAN VIEW   DETAIL

SIDE OR FRONT ELEVATION   END OR LEFT ELEVATION

FIGURE 20–3 **The two-dimensional views used in architectural drawings.**

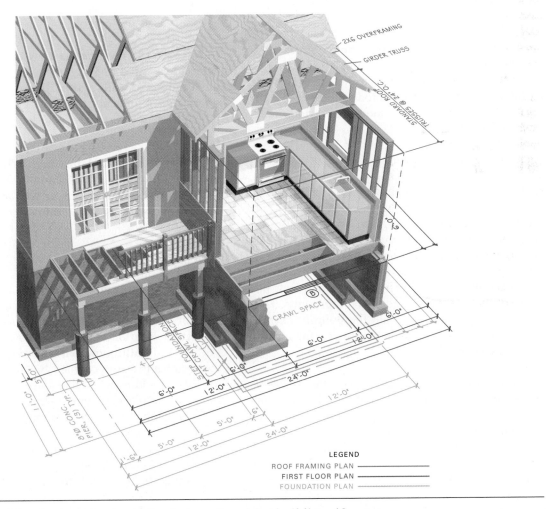

LEGEND
ROOF FRAMING PLAN ―――――――
FIRST FLOOR PLAN ―――――――
FOUNDATION PLAN ―――――――

FIGURE 20–4 **Plan views are horizontal cut views through a building.** (*Courtesy of David Hultenius, structural engineer*)

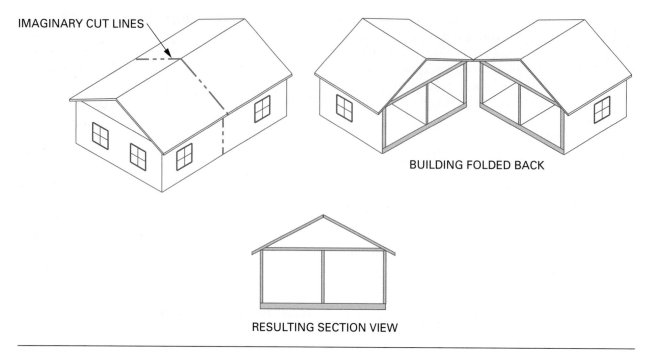

IMAGINARY CUT LINES

BUILDING FOLDED BACK

RESULTING SECTION VIEW

FIGURE 20–5 Section views are cutaway views.

tions show the building as seen from the street. They can show front, back, right side, and left side. *Section views* show a cross-section as if the building were sliced open to reveal its skeleton **(Figure 20–5)**. *Details* are blown up, zoomed-in views of various items to show a closer view.

Plan Views

Plot Plan. The *plot plan* shows information about the lot, such as property lines and directions and measurements for the location of the building, walks, and driveways **(Figure 20–6)**. It shows elevation heights and direction of the sloping ground. The drawing simulates a view looking down from a considerable height. It is made at a small scale because of the relatively large area it represents. This view helps the builder locate the building on the site and helps local officials to estimate the impact of the project on the lot.

Foundation Plan. The *foundation plan* **(Figure 20–7)**, (page 146) shows a horizontal cut through the foundation walls. It shows the shape and dimensions of the foundation walls and footings. Windows, doors, and stairs located in this level are included. First floor framing material size, spacing, and direction are sometimes included.

Floor Plan. The *floor plan* is view of the horizontal cut made about 4 to 5 feet above the floor. It shows the locations of walls and partitions, windows, and doors and fixtures appropriate for each room **(Figure 20–8)**. Dimensions are included.

Framing Plan. *Framing plans* are not always found in a set of architectural drawings. When used, they may be of the floor or roof framing. They show the support beams and girders as well as the size, direction, and spacing of the framing members **(Figure 20–9)**.

Elevations

Elevations are a group of drawings **(Figure 20–10)** that show the shape and finishes of all sides of the exterior of a building. *Interior elevations* are drawings of certain interior walls. The most common are kitchen and bathroom wall elevations. They show the design and size of cabinets built on the wall **(Figure 20–11)**. Other walls that have special features, such as a fireplace, may require an elevation drawing. Occasionally found in some sets of plans are *framing elevations*. Similar to framing plans, they show the spacing, location, and sizes of wall framing members. No further description of framing drawings is required to be able to interpret them.

Section Views

A set of architectural plans may have many *section* views. Each is designed to reveal the structure or skeleton view of a particular part of the building **(Figure 20–12).** A section reference line is found on the plans or elevations to identify the section being viewed.

Details

To make parts of the construction more clear, it is usually necessary to draw **details.** Details are small parts drawn at a very large scale, even full size if necessary. Their existence is revealed on the plan and elevation views using a symbol **(Figure 20–13).**

The location of the symbol shows the location of the vertical cut through the building. This symbol may have different shapes yet all have numbers or letters that refer to the page where the detail is shown.

Other Drawings

Drawings relating to electrical work, plumbing, heating, and ventilating may be on separate sheets in a set of architectural plans. For smaller projects, separate plans are not always needed. All necessary information can usually be found on the floor plan. The carpenter is responsible for building to accommodate wiring, pipes, and ducts. He or she must be able to interpret these plans proficiently to understand the work involved.

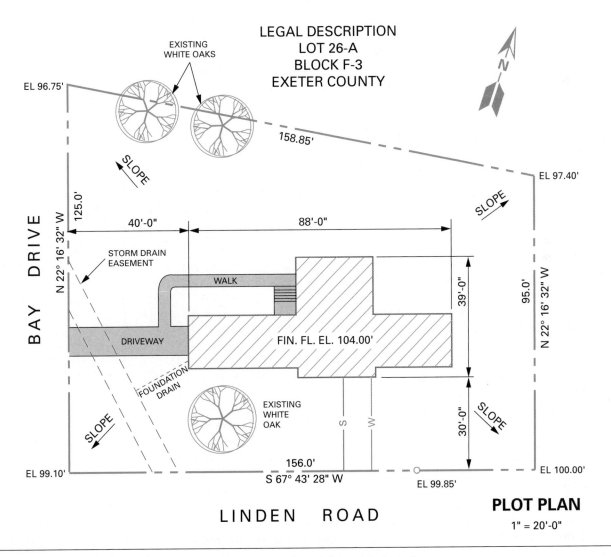

FIGURE 20–6   A plot plan is a view of construction from about 500 feet above.

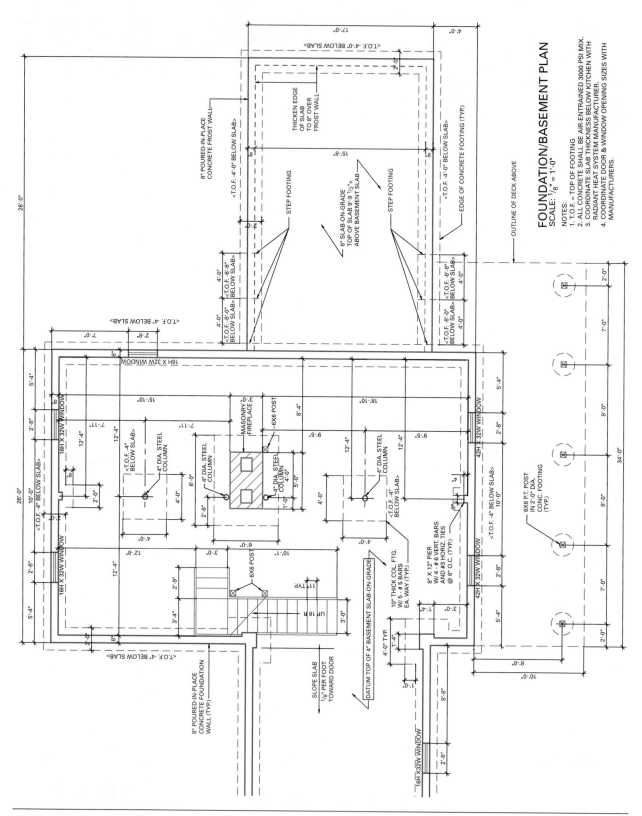

FIGURE 20–7 A foundation plan. (*Courtesy of David Hultenius, structural engineer*)

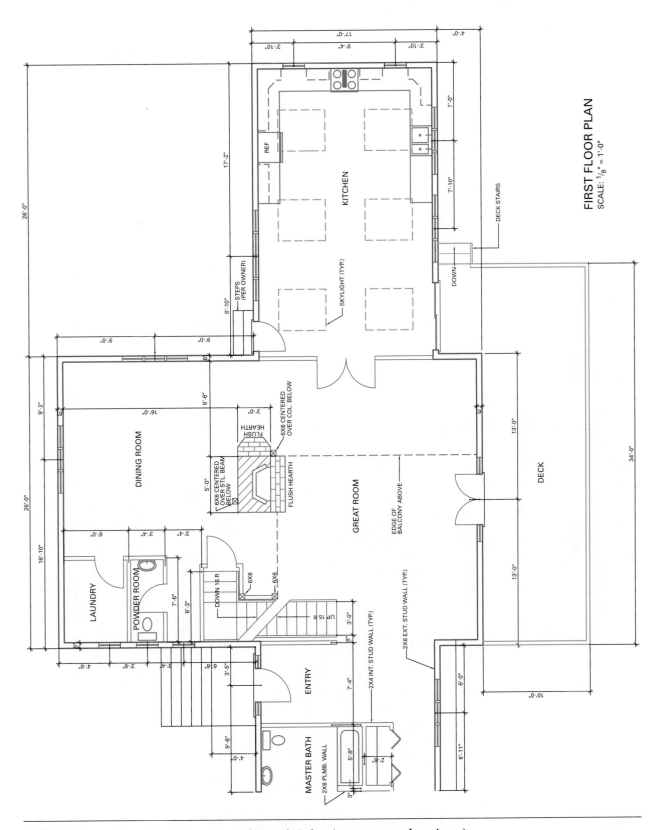

FIGURE 20–8 A floor plan. *(Courtesy of David Hultenius, structural engineer)*

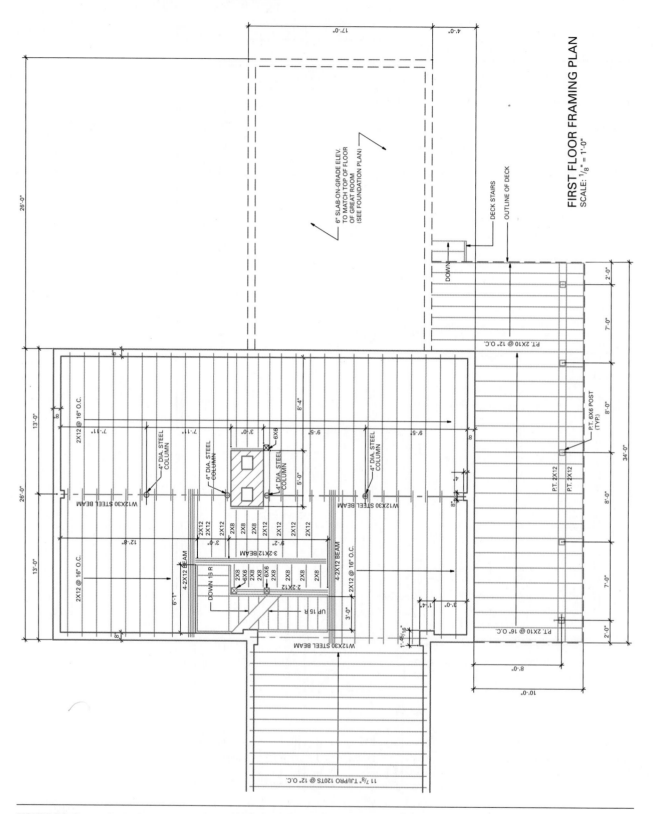

FIGURE 20–9   A first-floor framing plan. (*Courtesy of David Hultenius, structural engineer*)

FIGURE 20–10 Elevations. *(Courtesy of David Hultenius, structural engineer)*

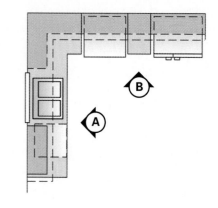

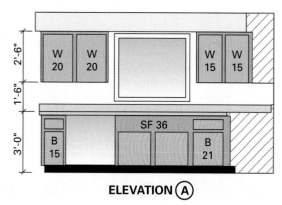

ELEVATION (A)

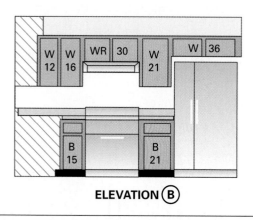

ELEVATION (B)

FIGURE 20–11 Interior wall elevations. *(Courtesy of PTEC–Clearwater Architectural Drafting Department)*

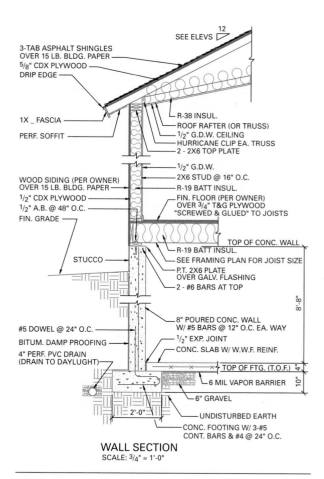

SEE ELEVS

3-TAB ASPHALT SHINGLES
OVER 15 LB. BLDG. PAPER
5/8" CDX PLYWOOD
DRIP EDGE

1X _ FASCIA
PERF. SOFFIT

R-38 INSUL.
ROOF RAFTER (OR TRUSS)
1/2" G.D.W. CEILING
HURRICANE CLIP EA. TRUSS
2 - 2X6 TOP PLATE

WOOD SIDING (PER OWNER)
OVER 15 LB. BLDG. PAPER
1/2" CDX PLYWOOD
1/2" A.B. @ 48" O.C.
FIN. GRADE

1/2" G.D.W.
2X6 STUD @ 16" O.C.
R-19 BATT INSUL.
FIN. FLOOR (PER OWNER)
OVER 3/4" T&G PLYWOOD
"SCREWED & GLUED" TO JOISTS

TOP OF CONC. WALL
R-19 BATT INSUL.
SEE FRAMING PLAN FOR JOIST SIZE
P.T. 2X6 PLATE
OVER GALV. FLASHING
2 - #6 BARS AT TOP

STUCCO

#5 DOWEL @ 24" O.C.
BITUM. DAMP PROOFING
4" PERF. PVC DRAIN
(DRAIN TO DAYLUGHT)

8" POURED CONC. WALL
W/ #5 BARS @ 12" O.C. EA. WAY
1/2" EXP. JOINT
CONC. SLAB W/ W.W.F. REINF.
TOP OF FTG. (T.O.F.)
6 MIL VAPOR BARRIER
6" GRAVEL
UNDISTURBED EARTH
CONC. FOOTING W/ 3-#5
CONT. BARS & #4 @ 24" O.C.

2'-0"

WALL SECTION
SCALE: 3/4" = 1'-0"

FIGURE 20–12 A section is a view of a vertical cut through part of the construction. *(Courtesy of David Hultenius, structural engineer)*

SCHEDULES

Besides drawings, printed instructions are included in a set of drawings. **Window schedules** and **door schedules (Figure 20–14)** give information about the location, size, and kind of windows and doors to be installed in the building. Each of the different units is given a number or letter. A corresponding number or letter is found on the floor plan to show the location of the unit. Windows may be identified by letters and doors by numbers. The letters and numbers may be framed with various geometric figures, such as circles and triangles.

A **finish schedule (Figure 20–15)** may also be included in a set of plans. This schedule gives information on the kind of finish material to be used on the floors, walls, and ceilings of the individual rooms.

Window, door, and finish schedules are used for easy understanding and to conserve space on floor plans.

SPECIFICATIONS

Specifications, commonly called *specs*, are written to give information that cannot be completely provided in the drawings or schedules. They supplement the working drawings with more complete descriptions of the methods, materials, and quality of construction. If there is a conflict, the specifications take precedence over the drawings. Any conflict should be pointed out to the architect so corrections can be made.

The amount of detail contained in the specs will vary, depending on the size of the project. On small

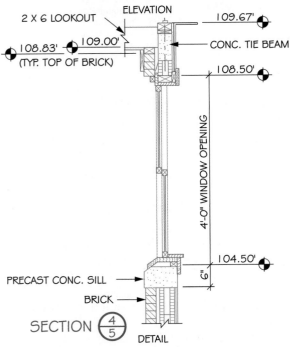

FIGURE 20–13 Detail of a window. *(Courtesy of PTEC–Clearwater Architectural Drafting Department)*

Window Schedule

Sym	Size	Model	Rough Open	Quan.
A	1′ × 5′	Job Built	Verify	2
B	8′ × 5′	W 4 N 5 CSM.	8′-0¾ × 5′-0⅞	1
C	4′ × 5′	W 2 N 5 CSM.	4′-0¾ × 5′-0⅞	2
D	4′ × 3⁶′	W 2 N 3 CSM.	4′-0¾ × 3′-6½	2
E	3⁶ × 3⁶	2 N 3 CSM.	3′-6½ × 3′-6½	2
F	6′ × 4′	G 64 Sldg.	6′-0½ × 4′-0½	1
G	5′ × 3⁶	G 536 Sldg.	5′-0½ × 3′-6½	4
H	4′ × 3⁶	G 436 Sldg.	4′-0½ × 3′-6½	1
J	4′ × 2′	A 41 Awn.	4′-0½ × 2′-0⅞	3

Door Schedule

Sym	Size	Type	Quan.
1	3′ × 6⁸	S.C. RP. Metal Insulated	1
2	3′ × 6⁸	S.C. Flush Metal Insulated	2
3	2⁸ × 6⁸	S.C. Self Closing	2
4	2⁸ × 6⁸	Hollow Core	5
5	2⁶ × 6⁸	Hollow Core	5
6	2⁶ × 6⁸	Pocket Sldg.	2

FIGURE 20–14 A typical window and door schedule.

Interior Finish Schedule

Room	Floor					Walls				Ceil.		
	Vinyl	Carpet	Tile	Hardwood	Concrete	Paint	Paper	Texture	Spray	Smooth	Brocade	Paint
Entry					●							
Foyer			●			●			●		●	●
Kitchen			●					●		●		●
Dining				●		●			●		●	●
Family		●				●			●		●	●
Living		●				●		●			●	●
Mstr. Bath			●				●			●		●
Bath #2			●						●	●	●	●
Mstr. Bed		●				●		●			●	●
Bed #2		●				●			●		●	●
Bed #3		●				●			●		●	●
Utility	●					●			●	●	●	●

FIGURE 20–15 A typical finish schedule.

jobs, they may be written by the architect. On larger jobs, a specifications writer, trained in the construction process, may be required. For complex commercial projects, a **specifications guide,** used by spec writers, has been developed by the Construction Specifications Institute (CSI). The guide has sixteen major divisions, each containing a number of subdivisions **(Figure 20–16).**

DIVISION 1—GENERAL REQUIREMENTS
- 01010 Summary of Work
- 01100 Alternatives
- 01150 Measurement & Payment
- 01200 Project Meetings
- 01300 Submittals
- 01400 Quality Control
- 01500 Temporary Facilities & Controls
- 01600 Material & Equipment
- 01700 Project Closeout

DIVISION 2—SITE WORK
- 02010 Subsurface Exploration
- 02100 Clearing
- 02110 Demolition
- 02200 Earthwork
- 02250 Soil Treatment
- 02300 Pile Foundations
- 02350 Caissons
- 02400 Shoring
- 02500 Site Drainage
- 02550 Site Utilities
- 02600 Paving & Surfacing
- 02700 Site Improvements
- 02800 Landscaping
- 02850 Railroad Work
- 02900 Marine Work
- 02950 Tunneling

DIVISION 3—CONCRETE
- 03100 Concrete Formwork
- 03150 Forms
- 03200 Concrete Reinforcement
- 03250 Concrete Accessories
- 03300 Cast-in-Place Concrete
- 03350 Specially Finished (Architectural) Concrete
- 03360 Specially Placed Concrete
- 03400 Precast Concrete
- 03500 Cementitious Decks
- 03600 Grout

DIVISION 4—MASONRY
- 04100 Mortar
- 04150 Masonry Accessories
- 04200 Unit Masonry
- 04400 Stone
- 04500 Masonry Restoration & Cleaning
- 04550 Refractories

DIVISION 5—METALS
- 05100 Structural Metal Framing
- 05200 Metal Joists
- 05300 Metal Decking
- 05400 Lightgage Metal Framing
- 05500 Metal Fabrications
- 05700 Ornamental Metal
- 05800 Expansion Control

DIVISION 6—WOOD & PLASTICS
- 06100 Rough Carpentry
- 06130 Heavy Timber Construction
- 06150 Trestles
- 06170 Prefabricated Structural Wood
- 06200 Finish Carpentry
- 06300 Wood Treatment
- 06400 Architectural Woodwork
- 06500 Prefabricated Structural Plastics
- 06600 Plastic Fabrications

DIVISION 7—THERMAL & MOISTURE PROTECTION
- 07100 Waterproofing
- 07150 Dampproofing
- 07200 Insulation
- 07300 Shingles & Roofing Tiles
- 07400 Preformed Roofing & Siding
- 07500 Membrane Roofing
- 07570 Traffic Topping
- 07600 Flashing & Sheet Metal
- 07800 Roof Accessories
- 07900 Sealants

DIVISION 8—DOOR & WINDOWS
- 08100 Metal Doors & Frames
- 08200 Wood & Plastic Doors
- 08300 Special Doors
- 08400 Entrances & Storefronts
- 08500 Metal Windows
- 08600 Wood & Plastic Windows
- 08650 Special Windows
- 08700 Hardware & Specialties
- 08800 Glazing
- 08900 Window Walls/Curtain Walls

DIVISION 9—FINISHES
- 09100 Lath & Plaster
- 09250 Gypsum Wallboard
- 09300 Tile
- 09400 Terrazzo
- 09500 Acoustical Treatment
- 09540 Ceiling Suspension Systems
- 09550 Wood Flooring
- 09650 Resilient Flooring
- 09680 Carpeting
- 09700 Special Flooring
- 09760 Floor Treatment
- 09800 Special Coatings
- 09900 Painting
- 09950 Wall Covering

DIVISION 10—SPECIALTIES
- 10100 Chalkboards & Tackboards
- 10150 Compartments & Cubicles
- 10200 Louvers & Vents
- 10240 Grilles & Screens
- 10260 Wall & Corner Guards
- 10270 Access Flooring
- 10280 Specialty Modules
- 10290 Pest Control
- 10300 Fireplaces
- 10350 Flagpoles
- 10400 Identifying Devices
- 10450 Pedestrian Control Devices
- 10500 Lockers
- 10530 Protective Covers
- 10550 Postal Specialties
- 10600 Partitions
- 10650 Scales
- 10670 Storage Shelving
- 10700 Sun Control Devices (Exterior)
- 10750 Telephone Enclosures
- 10800 Toilet & Bath Accessories
- 10900 Wardrobe Specialties

DIVISION 11—EQUIPMENT
- 11050 Built-in Maintenance Equipment
- 11100 Bank & Vault Equipment
- 11150 Commercial Equipment
- 11170 Checkroom Equipment
- 11180 Darkroom Equipment
- 11200 Ecclesiastical Equipment
- 11300 Educational Equipment
- 11400 Food Service Equipment
- 11480 Vending Equipment
- 11500 Athletic Equipment
- 11500 Industrial Equipment
- 11600 Laboratory Equipment
- 11630 Laundry Equipment
- 11650 Library Equipment
- 11700 Medical Equipment
- 11800 Mortuary Equipment
- 11830 Musical Equipment
- 11850 Parking Equipment
- 11860 Waste Handling Equipment
- 11870 Loading Dock Equipment
- 11880 Detention Equipment
- 11900 Residential Equipment
- 11970 Theater & Stage Equipment
- 11990 Registration Equipment

DIVISION 12—FURNISHINGS
- 12100 Artwork
- 12300 Cabinets & Storage
- 12500 Window Treatment
- 12550 Fabrics
- 12600 Furniture
- 12670 Rugs & Mats
- 12700 Seating
- 12800 Furnishing Accessories

DIVISION 13—SPECIAL CONSTRUCTION
- 13010 Air Supported Structures
- 13050 Integrated Assemblies
- 13100 Audiometric Room
- 13250 Clean Room
- 13350 Hyperbaric Room
- 13400 Incinerators
- 13440 Instrumentation
- 13450 Insulated Room
- 13500 Integrated Ceiling
- 13540 Nuclear Reactors
- 13550 Observatory
- 13600 Prefabricated Structures
- 13700 Special Purpose Rooms & Buildings
- 13750 Radiation Protection
- 13770 Sound & Vibration Control
- 13800 Vaults
- 13850 Swimming Pools

DIVISION 14—CONVEYING SYSTEMS
- 14100 Dumbwaiters
- 14200 Elevators
- 14300 Hoists & Cranes
- 14400 Lifts
- 14500 Material Handling Systems
- 14570 Turntables
- 14600 Moving Stairs & Walks
- 14700 Tube Systems
- 14800 Powered Scaffolding

DIVISION 15—MECHANICAL
- 15010 General Provisions
- 15050 Basic Materials & Methods
- 15180 Insulation
- 15200 Water Supply & Treatment
- 15300 Waste Water Disposal & Treatment
- 15400 Plumbing
- 15500 Fire Protection
- 15600 Power or Heat Generation
- 15650 Refrigeration
- 15700 Liquid Heat Transfer
- 15800 Air Distribution
- 15900 Controls & Instrumentation

DIVISION 16—ELECTRICAL
- 16010 General Provisions
- 16100 Basic Materials & Methods
- 16200 Power Generation
- 16300 Power Transmission
- 16400 Service & Distribution
- 16500 Lighting
- 16600 Special Systems
- 16700 Communications
- 16850 Heating & Cooling
- 16900 Controls & Instrumentation

FIGURE 20–16 CSI format for specifications.

Section 06200—Finish Carpentry

General: This section covers all finish woodwork and related items not covered elsewhere in these specifications. The contractor shall furnish all materials, labor, and equipment necessary to complete the work, including rough hardware, finish hardware, and specialty items.

Protection of Materials: All millwork (finish woodwork*) and trim is to be delivered in a clean and dry condition and shall be stored to insure proper ventilation and protection from dampness. Do not install finish woodwork until concrete, masonry, plaster, and related work is dry.

Materials: All materials are to be the best of their respective kind. Lumber shall bear the mark and grade of the association under whose rules it is produced. All millwork shall be kiln dried to a maximum moisture content of 12%.

1. Exterior trim shall be select grade white pine, S4S.

2. Interior trim and millwork shall be select grade white pine, thoroughly sanded at the time of installation.

Installation: All millwork and trim shall be installed with tight fitting joints and formed to conceal future shrinkage due to drying. Interior woodwork shall be mitered or coped at corners (cut in a special way to form neat joints*). All nails are to be set below the surface of the wood and concealed with an approved putty or filler.

*(explanations in parentheses have been added to aid the student.)

FIGURE 20–17   Sample specifications following the CSI format.

Using the specification guide, under Division 6—WOOD & PLASTICS, Section 06200, an example of the content and the manner in which specifications are written is shown in **Figure 20–17.**

For some light commercial and residential construction, many sections of the spec guide would not apply. A shortened version is then used, eliminating divisions 12,000, 13,000, and 14,000. On simpler plans, notations made on the same sheets as the drawings may take the place of specifications. The ability to read notations and specifications accurately is essential in order to conform to the architect's design.

ARCHITECTURAL PLAN LANGUAGE

Carpenters must be able to read and understand the combination of lines, dimensions, symbols, and notations on the architectural drawings. Only then can they build exactly as the architect has designed the construction. No deviation from the plans may be made without the approval of the architect.

Scales

It would be inconvenient and impractical to make full-sized drawings of a building. Therefore, they are drawn to **scale.** This means that each line in the drawing is reduced proportionally to a size that clearly shows the information and can be handled conveniently. Not all drawings in a set of plans, or even on the same page, are drawn at the same scale. The scale of a drawing is stated in the title block of the page. It can also be listed directly below the drawing.

Architect's Scale. The **architect's scale** is commonly used to scale lines when making drawings **(Figure 20–18).** It has a triangular cross-section giving space for two scales on each face. Six faces are produced, allowing room for a potential of six scales. But it doesn't stop there. The main scale is simply a ruler divided into 1-inch increments and fractions in $\frac{1}{16}$-inch increments. The other five scales actually show two scales in each. Each scale is doubled up with another scale. The desired scale used depends on which end of the scale is read.

Each of the five scales is divided into fractional increments depending on the number labeled at the end scale. For example, the $\frac{1}{4}$ scale is read from right to left. At the other end, the $\frac{1}{8}$ scale is read from left to right **(Figure 20–19).** One scale is twice as big (or half as big depending) as its counterpart. To draw a line in $\frac{1}{4}$ scale, each foot of actual size is drawn as $\frac{1}{4}$ inch. Read the scales in one direction from the starting end, being careful not to confuse it with the scale running in the opposite direction.

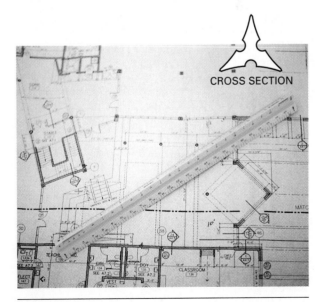

CROSS SECTION

FIGURE 20–18   The architect's scale is used to draw plans.

The scales are paired as follows:

1½" = 1'-0"	and	3" = 1'-0"
1" = 1'-0"	and	½" = 1'-0"
⅜" = 1'-0"	and	¾" = 1'-0"
⅛" = 1'-0"	and	¼" = 1'-0"
³⁄₃₂" = 1'-0"	and	³⁄₁₆" = 1'-0"

On the end of each scale, a space representing one foot at that scale is divided up. This space is used when scaling off dimensions that include fractions of a foot. What these lines represent depends on the scale. In some cases, it is divided into fractions of an inch and in others it is divided into whole inches. For example, in Figure 20–19, the 1½ scale's smallest

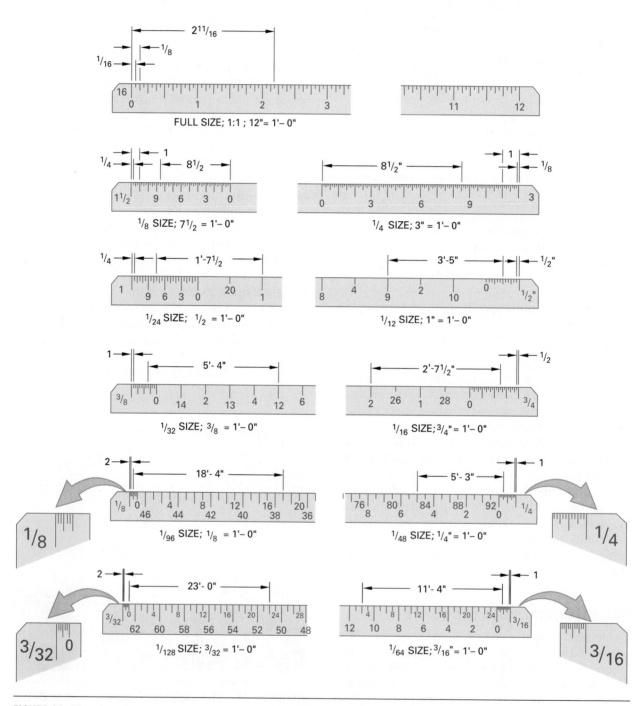

FIGURE 20–19   The eleven scales found on an architect's scale.

line represents an actual ¼ inch. In the ⅜ scale, the smallest line represents 1 actual inch.

Commonly Used Scales.

Probably the most commonly used scale found on architectural plans is ¼ inch equals 1 foot. This is indicated as ¼" = 1'-0". It is often referred to as a "quarter-inch scale." This means that every ¼ inch on the drawing will equal one foot in the building. Floor plans and exterior elevations for most residential buildings are drawn at this scale.

To show the location of a building on a lot and other details of the site, the architect may use a scale of ¹⁄₁₆" = 1'-0". This reduces the size of the drawing to fit it on the paper. To show certain details more clearly, larger scales of 1½" = 1'-0" or 3" = 1'-0" are used. Complicated details may be drawn full size or half size. Other scales are used when appropriate. Section views showing the elevation of interior walls are often drawn at ½" = 1'-0" or ¾" = 1'-0".

Drawing plans to scale is important. The building and its parts are shown in true proportion, making it easier for the builder to visualize the construction. However, the use of a scale rule to determine a dimension should be a last resort. Dimensions on plans should be determined either by reading the dimension or adding and subtracting other dimensions to determine it. The use of a scale rule to determine a dimension will result in inaccuracies.

Types of Lines

Some lines in an architectural drawing look darker than others. They are broader so they stand out clearly from other lines. This variation in width is called *line contrast*. This technique, like all architectural drafting standards, is used to make the drawing easier to read and understand **(Figure 20–20)**.

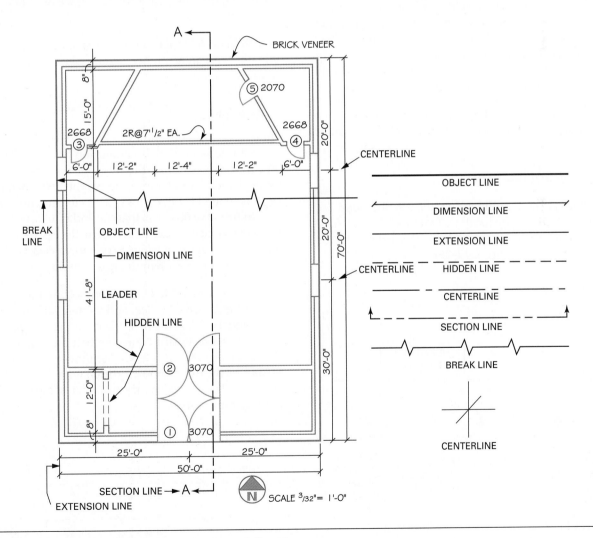

FIGURE 20–20 Types of lines on architectural drawings.

Object Line. Lines that outline the object being viewed are broad, solid lines called object lines. These lines represent the portion of the building visible in this view.

Hidden Line. To indicate an important object not visible in the view, a hidden line consisting of short, fine, uniform dashes is used. Hidden lines are used only when necessary. Otherwise the drawing becomes confusing to read.

Centerline. Centerlines are indicated by a fine, long dash, then a short dash, then a long dash, and so on. They show the centers of doors, windows, partitions, and similar parts of the construction.

Section Line. A section reference or cutting-plane line is, sometimes, a broad line consisting of a long dash followed by two short dashes. At its ends are arrows. The arrows show the direction in which the cross-section is viewed. Letters identify the cross-sectional view of that specific part of the building. More elaborate methods of labeling section reference lines are used in larger, more complicated sets of plans **(Figure 20–21).** The sectional drawings may be on the same page as the reference line or on other pages.

Break Line. A break line is used in a drawing to terminate part of an object that, in actuality, continues. It can only be used when there is no change in the drawing at the break. Its purpose is to shorten the drawing to utilize space.

Dimension Line. A dimension line is a fine, solid line used to indicate the location, length, width, or thickness of an object. It is terminated with arrowheads, dots, or slashes **(Figure 20–22).**

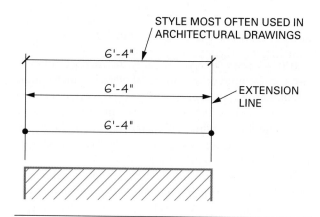

FIGURE 20–22 **Several methods of terminating dimension lines.**

Extension Line. Extension lines are fine, solid lines projecting from an object to show the extent of a dimension.

Leader Line. A leader line is a fine solid line. It terminates with an arrowhead, and points to an object from a notation.

Dimensions

Dimension lines on a blueprint are generally drawn as continuous lines. The dimension appears above and near the center of the line. All dimensions on vertical lines should appear above the line when the print is rotated ¼ turn clockwise. Extension lines are drawn from the object so that the end point of the dimension is clearly defined. When the space is too small to permit dimensions to be shown clearly, they may be drawn as shown in **Figure 20–23.**

Kinds of Dimensions. Dimensions on architectural blueprints are given in feet and inches, such as 3'-6", 4'-8", and 13'-7". A dash is always used to separate the foot measurement from the inch measurement. When the dimension is a whole number of feet with no inches, the dimension is written with zero inches, as 14'-0". The use of the dash prevents mistakes in reading dimensions.

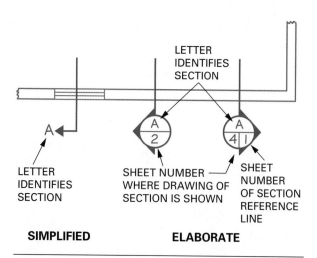

SIMPLIFIED   **ELABORATE**

FIGURE 20–21 **Several ways of labeling section reference lines.**

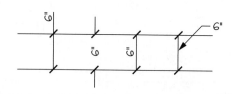

FIGURE 20–23 **Methods of dimensioning small spaces.**

Dimensions of 1 foot and under are given in inches, as 10″, 8″, and so on. Dimensions involving fractions of an inch are shown, for example, as 1′-0½″, 2′-3¾″, or 6½″. If there is a difference between a written dimension and a scaled dimension of the same distance, the written dimension should be followed.

Modular Measure

In recent years, **modular measurement** has been used extensively. A grid with a unit of 4 inches is used in designing buildings **(Figure 20–24).** The idea is to draw the plans to use material manufactured to

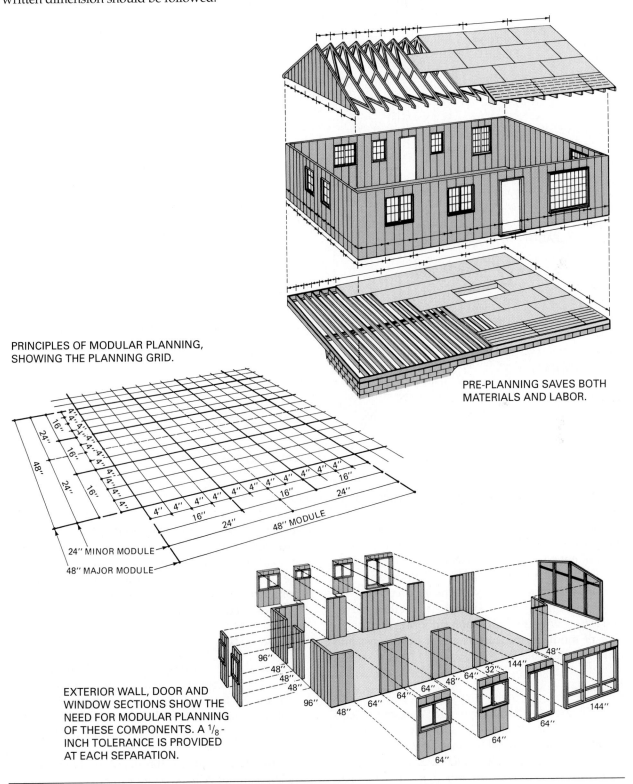

PRE-PLANNING SAVES BOTH MATERIALS AND LABOR.

PRINCIPLES OF MODULAR PLANNING, SHOWING THE PLANNING GRID.

24″ MINOR MODULE

48″ MAJOR MODULE

48″ MODULE

EXTERIOR WALL, DOOR AND WINDOW SECTIONS SHOW THE NEED FOR MODULAR PLANNING OF THESE COMPONENTS. A ⅛-INCH TOLERANCE IS PROVIDED AT EACH SEPARATION.

FIGURE 20–24 **Modular measurement uses a grid of 4 inches.**

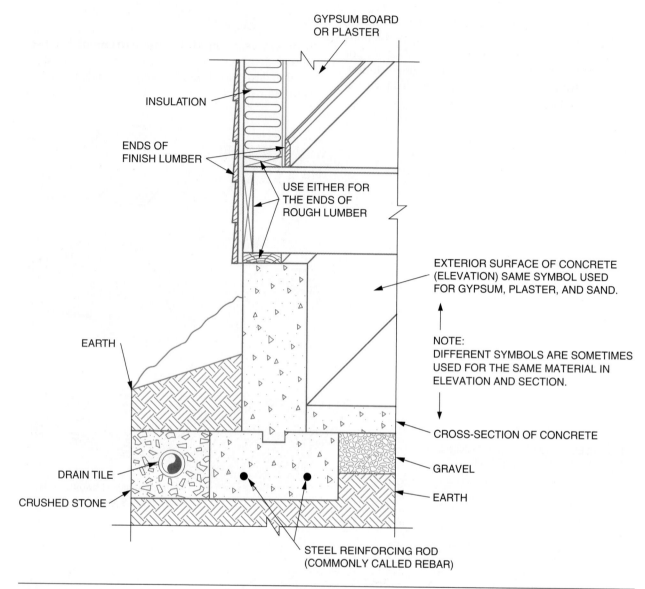

FIGURE 20–25   **Symbols for commonly used construction materials.**

fit the grid spaces. Drawing plans to a modular measure enables the builder to use manufactured component parts with less waste, such as 4 × 8 sheet materials and manufactured wall, floor, and roof sections that fit together with greater precision.

The spacing of framing members and the location and size of windows and doors adhering to the concept of modular measurement cut down cost and conserve materials.

Symbols

Symbols are used on drawings to represent objects in the building, such as doors, windows, cabinets, plumbing, and electrical fixtures. Others are used in regard to the construction, such as for walls, stairs, fireplaces, and electrical circuits. They may be used for identification purposes, such as those used for

section reference lines. The symbols for various construction materials, such as lumber, concrete, sand, and earth **(Figure 20–25),** are used when they make the drawing easier to read. (More detailed illustration, description, and use of architectural symbols are presented in following units where appropriate.)

Abbreviations

Architects find it necessary to use abbreviations on drawings to conserve space. Only capital letters, such as DR for door, are used. Abbreviations that make an actual word, such as FIN. for finish, are followed by a period. Several words may use the same abbreviation, such as W for west, width, or with. The **location** of these abbreviations is the key to their meaning. A list of commonly used abbreviations is shown in **Figure 20–26.**

Access Panel .... AP
Acoustic .... ACST
Acoustical Tile .... AT
Aggregate .... AGGR
Air Conditioning .... AIR COND
Aluminum .... AL
Anchor Bolt .... AB
Angle ....
Apartment .... APT
Approximate .... APPROX
Architectural .... ARCH
Area .... A
Area Drain .... AD
Asbestos .... ASB
Asbestos Board .... AB
Asphalt .... ASPH
Asphalt Tile .... AT
Basement .... BSMT
Bathroom .... B
Bathtub .... BT
Beam .... BM
Bearing Plate .... BRG PL
Bedroom .... BR
Blocking .... BLKG
Blueprint .... BP
Boiler .... BLR
Book Shelves .... BK SH
Brass .... BRS
Brick .... BRK
Bronze .... BRZ
Broom Closet .... BC
Building .... BLDG
Building Line .... BL
Cabinet .... CAB
Calking .... CLKG
Casing .... CSG
Cast Iron CI
Cast Stone .... CS
Catch Basin .... CB
Cellar .... CEL
Cement .... CEM
Cement Asbestos Board ..... CEM AB
Cement Floor .... CEM FL
Cement Mortar ...... CEM MORT
Center .... CTR
Center to Center .... C TO C
Center Line .... or CL
Center Matched .... CM
Ceramic .... CER
Channel .... CHAN
Cinder Block .... CIN BL
Circuit Breaker .... CIR BKR
Cleanout .... CO
Cleanout Door .... COD
Clear Glass .... CL GL
Closet .... C, CL or CLO
Cold Air .... CA
Cold Water .... CW
Collar Beam .... COL B
Concrete .... CONC
Concrete Block .... CONC B
Concrete Floor .... CONC FL
Conduit .... CND
Construction .... CONST
Contract .... CONT
Copper .... COP
Counter .... CTR
Cubic Feet .... CU FT
Cut Out .... CO
Detail .... DET
Diagram .... DIAG
Dimension .... DIM
Dining Room .... DR
Dishwasher .... DW
Ditto .... DO
Double-Acting .... DA
Double Strength Glass .... DSG
Down .... DN
Downspout .... DS
Drain .... D or DR
Drawing .... DWG

Dressed and Matched .... D & M
Dryer .... D
Electric Panel .... EP
End to End .... E to E
Excavate .... EXC
Expansion Joint .... EXP JT
Exterior .... EXT
Finish .... FIN
Finished Floor .... FIN FL
Firebrick .... FBRK
Fireplace .... FP
Fireproof .... FPRF
Fixture .... FIX
Flashing .... FL
Floor .... FL
Floor Drain .... FD
Flooring .... FLG
Fluorescent .... FLUOR
Flush .... FL
Footing .... FTG
Foundation .... FND
Frame .... FR
Full Size .... FS
Furring .... FUR
Galvanized Iron .... GI
Garage .... GAR
Gas .... G
Glass .... GL
Glass Block .... GL BL
Grille .... G
Gypsum .... GYP
Hardware .... HDW
Hollow Metal Door .... HMD
Hose Bib .... HB
Hot Air .... HA
Hot Water .... HW
Hot Water Heater .... HWH
I Beam I
Inside Diameter .... ID
Insulation .... INS
Interior .... INT
Iron I
Jamb .... JB
Kitchen .... K
Landing .... LDG
Lath .... LTH
Laundry .... LAU
Laundry Tray .... LT
Lavatory .... LAV
Leader L
Length .... L, LG or LNG
Library .... LIB
Light .... LT
Limestone .... LS
Linen Closet .... L CL
Lining .... LN
Living Room .... LR
Louver LV
Main .... MN
Marble .... MR
Masonry Opening .... MO
Material .... MATL
Maximum .... MAX
Medicine Cabinet .... MC
Minimum .... MIN
Miscellaneous .... MISC
Mixture .... MIX
Modular .... MOD
Mortar .... MOR
Moulding .... MLDG
Nosing .... NOS
Obscure Glass .... OBSC sL
On Center .... OC
Opening .... OPNG
Outlet .... OUT
Overall .... OA
Overhead .... OVHD
Pantry .... PAN
Partition .... PTN
Plaster .... PL or PLAS
Plastered Opening .... PO

Plate .... PL
Plate Glass .... PL GL
Platform .... PLAT
Plumbing .... PLBG
Plywood .... PLY
Porch .... P
Precast .... PRCST
Prefabricated .... PREFAB
Pull Switch .... PS
Quarry Tile Floor .... QTF
Radiator .... RAD
Random .... RDM
Range .... R
Recessed .... REC
Refrigerator .... REF
Register .... REG
Reinforce or Reinforcing ..... REINF
Revision .... REV
Riser .... R
Roof .... RF
Roof Drain .... RD
Room .... RM or R
Rough .... RGH
Rough Opening .... RO
Rubber Tile .... R TILE
Scale .... SC
Schedule .... SCH
Screen .... SCR
Scuttle .... S
Section .... SECT
Select .... SEL
Service .... SERV
Sewer .... SEW
Sheathing .... SHTHG
Sheet .... SH
Shelf and Rod .... SH & RD
Shelving .... SHELV
Shower .... SH
Sill Cock .... SC
Single Strength Glass .... SSG
Sink .... SK or S
Soil Pipe .... SP
Specification .... SPEC
Square Feet .... SQ FT
Stained .... STN
Stairs .... ST
Stairway .... STWY
Standard .... STD
Steel .... ST or STL
Steel Sash .... SS
Storage .... STG
Switch .... SW or S
Telephone .... TEL
Terra Cotta .... TC
Terrazzo .... TER
Thermostat .... THERMO
Threshold .... TH
Toilet .... T
Tongue and Groove .... T & G
Tread .... TR or T
Typical .... TYP
Unfinished .... UNF
Unexcavated .... UNEXC
Utility Room .... URM
Vent V
Vent Stack .... VS
Vinyl Tile .... V TILE
Warm Air .... WA
Washing Machine .... WM
Water .... W
Water Closet .... WC
Water Heater .... WH
Waterproof .... WP
Weather Stripping .... WS
Weephole .... WH
White Pine .... WP
Wide Flange .... WF
Wood .... WD
Wood Frame .... WF
Yellow Pine .... YP

FIGURE 20–26 Commonly used abbreviations found in construction drawings.

21   Floor Plans

A house starts out as an idea drawn on paper. The 3D vision of the structure is converted to many 2D views **(Figure 21–1)**. These views must be then read and interpreted by the builder, who makes the idea come alive.

FLOOR PLANS

Floor plans (Figure 21–2) contain a substantial amount of information. They are used more than any other kind of drawing. After consideration of many factors that determine the size and shape of the building, floor plans are drawn first. Others, such as the foundation plan and elevations, are derived from it. They are generally drawn at a scale of ¼″ = 1′-0″ or 1:48. A separate plan is made for each floor of buildings with more than one story.

Floor Plan Symbols

To make the plan as uncluttered as possible, numerous *symbols* are used. Recognition of commonly used symbols makes it easier to read the floor plan as well as other plans that use the same symbols. Symbols used in elevation and section drawings are different from plan symbols. They are described in following chapters.

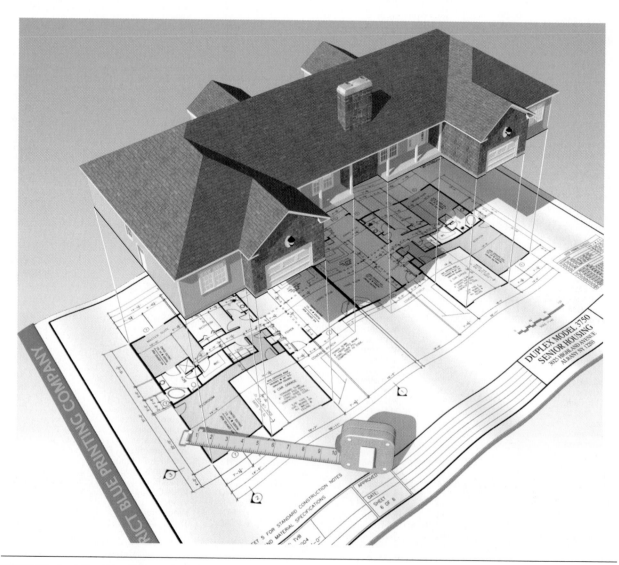

FIGURE 21–1   The 3D vision of a building is described on 2D paper.

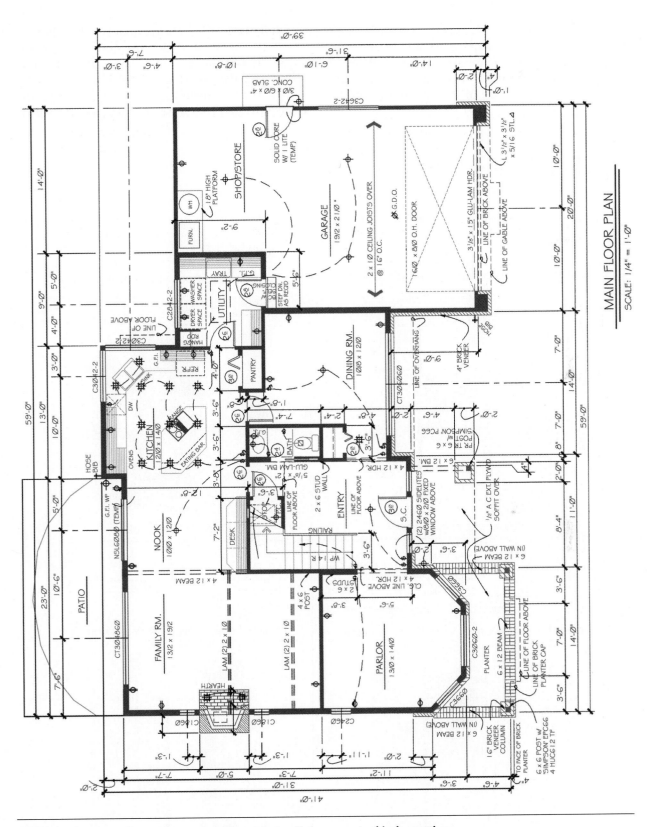

FIGURE 21–2 First-floor plans contain a substantial amount of information.

Door Symbols. Symbols for exterior doors are drawn with a line representing the outside edge of the sill. Interior door symbols show no sill line. The symbols in **Figure 21–3** identify the **swing** and show on which side of the opening to hang the door.

Similarly, exterior **sliding** door symbols show the sill line. The symbols for interior sliding doors, called **bypass** doors, show none. **Pocket doors** slide inside the wall **(Figure 21–4)**.

Bifold doors open to almost the full width of a closet opening. They are used when complete access

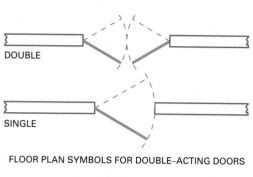

DOUBLE

SINGLE

FLOOR PLAN SYMBOLS FOR DOUBLE-ACTING DOORS

FLOOR PLAN SYMBOL FOR AN EXTERIOR DOOR

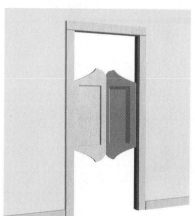

DOUBLE-ACTING DOORS PICTORIAL

EXTERIOR DOOR PICTORIAL

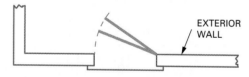

EXTERIOR WALL

FLOOR PLAN SYMBOL FOR AN EXTERIOR DUTCH DOOR

FLOOR PLAN SYMBOL FOR AN INTERIOR SWINGING DOOR

DUTCH DOOR PICTORIAL

INTERIOR SWINGING DOOR PICTORIAL

FIGURE 21–3 Floor plan symbols for exterior and interior swinging doors.

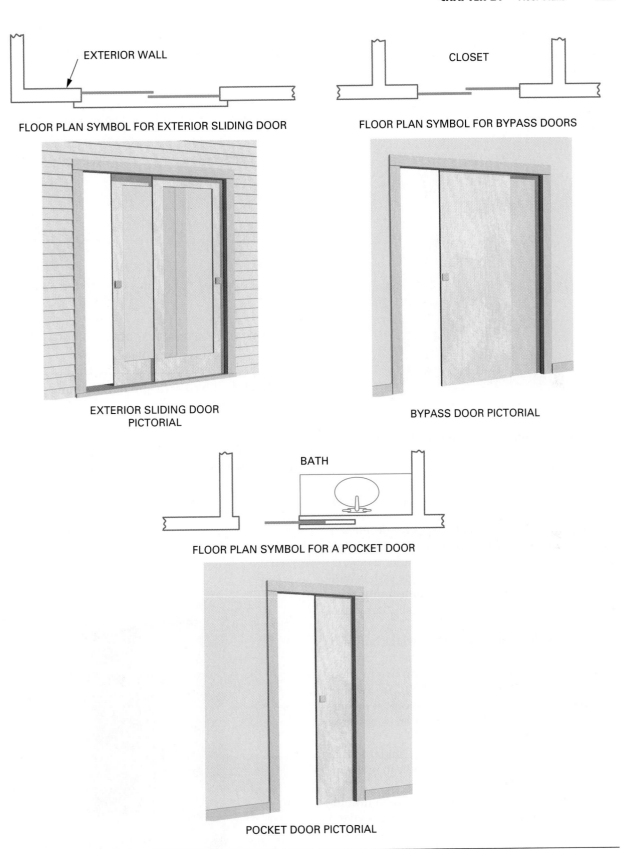

FLOOR PLAN SYMBOL FOR EXTERIOR SLIDING DOOR

EXTERIOR WALL

FLOOR PLAN SYMBOL FOR BYPASS DOORS

CLOSET

EXTERIOR SLIDING DOOR
PICTORIAL

BYPASS DOOR PICTORIAL

FLOOR PLAN SYMBOL FOR A POCKET DOOR

BATH

POCKET DOOR PICTORIAL

FIGURE 21–4 Symbols for exterior and interior sliding door.

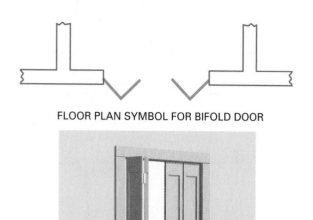

FLOOR PLAN SYMBOL FOR BIFOLD DOOR

BIFOLD DOOR PICTORIAL

FIGURE 21–5 **Bifold doors are sometimes used on closets and wardrobes.**

to the closet is desired. The sections or panels of the doors are clearly seen in the symbols **(Figure 21–5)**.

Window Symbols. The inside and outside lines of window symbols represent the edges of the window sill. In between, other lines are drawn for the panes of glass. A window with a fixed, single sash is indicated by one line. Because the **double-hung window** has two sashes that slide vertically, its symbol shows two lines **(Figure 21–6)**.

The **casement window,** which swings outward, is depicted by a symbol similar to that for a swinging door. They may be shown having two or more units in each window **(Figure 21–7)**. An **awning window** is similar to a casement except it swings outward from the top. Its open position is indicated by dashed lines **(Figure 21–8)**.

The symbol for a *sliding window* is similar to that for a sliding door except for a line indicating the inside edge of the sill **(Figure 21–9)**.

Different kinds of windows may be used in combination. **Figure 21–10** shows a window with a fixed sash, with casements on both sides. Main entrances may consist of a door with a **sidelight** on one or both sides **(Figure 21–11)**.

Close to the window and door symbols are letters and numbers that identify the units in the window and door schedules.

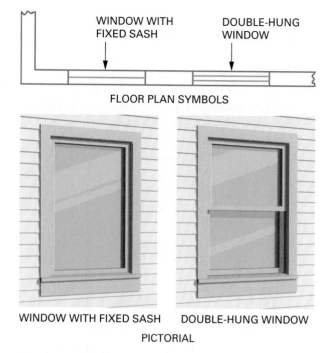

WINDOW WITH FIXED SASH   DOUBLE-HUNG WINDOW

FLOOR PLAN SYMBOLS

WINDOW WITH FIXED SASH   DOUBLE-HUNG WINDOW

PICTORIAL

FIGURE 21–6 **Fixed sash and double-hung window symbols.**

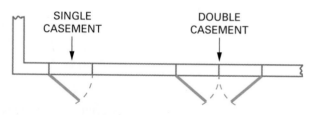

SINGLE CASEMENT   DOUBLE CASEMENT

FLOOR PLAN SYMBOLS

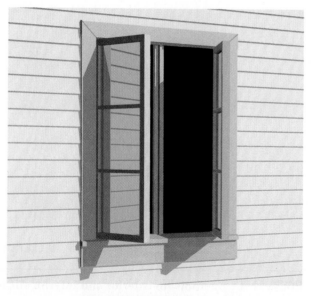

DOUBLE CASEMENT PICTORIAL

FIGURE 21–7 **Symbols for casement windows.**

FLOOR PLAN SYMBOL

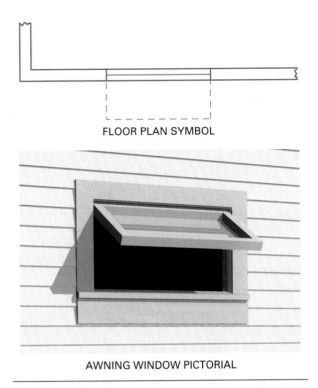

AWNING WINDOW PICTORIAL

FIGURE 21–8 The open position of an awning window is shown with dashed lines.

FLOOR PLAN SYMBOL

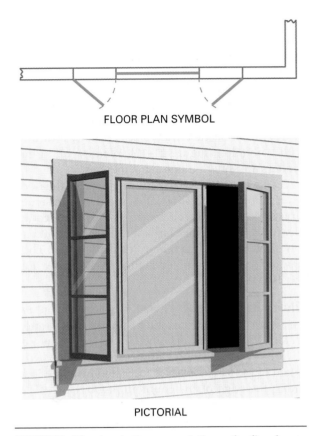

PICTORIAL

FIGURE 21–10 A window consisting of a fixed sash and casement units.

FLOOR PLAN SYMBOL

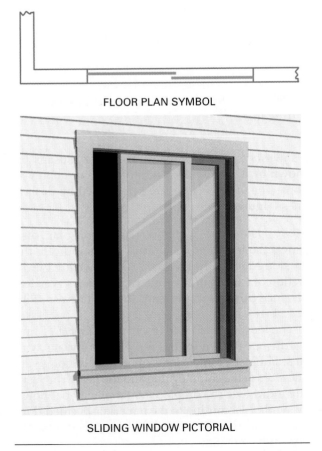

SLIDING WINDOW PICTORIAL

FIGURE 21–9 Sliding window floor plan symbol.

FLOOR PLAN SYMBOL

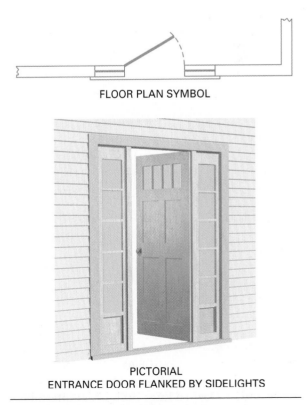

PICTORIAL
ENTRANCE DOOR FLANKED BY SIDELIGHTS

FIGURE 21–11 An entrance door flanked by sidelights.

Structural Members. Openings without doors in interior walls for passage from one area to another are indicated by dashed lines. Notations are given if the opening is to be *cased* (trimmed with molding) and if the top is to be *arched* or treated in any other manner **(Figure 21–12).**

The location of garage door *headers* may be shown by a series of dashes. Their size is usually indicated with a notation **(Figure 21–13).** Window and other headers are shown in the same manner.

Ceiling beams above the cutting plane of the floor, which support the ceiling joists, are also represented by a series of dashes. *Ceiling joists* or **trusses** are identified in the floor plan with a double-ended arrow showing the direction in which they run. Their size and spacing are noted alongside the arrow **(Figure 21–14).**

Kitchen, Bath, and Utility Room. The location of *bathroom* and *kitchen fixtures,* such as sinks, tubs, refrigerators, stoves, washers, and dryers, are shown by obvious symbols, abbreviations, and notations. The extent of the *base cabinets* is indicated by a line indicating the edge of the countertop. Objects such as

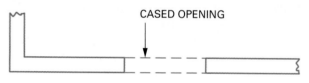

CASED OPENING

FLOOR PLAN SYMBOL

PICTORIAL

FIGURE 21–12   Dashed lines indicate an interior wall opening without a door.

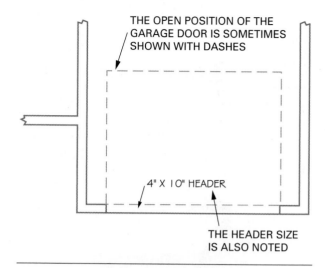

THE OPEN POSITION OF THE GARAGE DOOR IS SOMETIMES SHOWN WITH DASHES

4" X 10" HEADER

THE HEADER SIZE IS ALSO NOTED

FIGURE 21–13   The symbol for a garage door header is a dashed line.

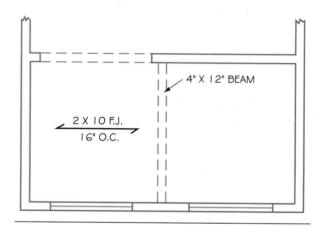

4" X 12" BEAM

2 X 10 F.J.
16" O.C.

FIGURE 21–14   Symbols for ceiling beams and ceiling joists.

dishwashers, trash compactors, and lazy susans are shown by a dashed line and notations or abbreviations. The *upper cabinets* are symbolized by dashed lines **(Figure 21–15).**

Other Floor Plan Symbols. The floor plan also shows the location of *stairways.* Lines indicate the outside edges of the **treads.** Also shown are the direction of travel and the number of **risers** (vertical distance from tread to tread) in the staircase **(Figure 21–16).**

The location and style of *chimneys, fireplaces,* and hearths are shown by the use of appropriate symbols. Fireplace dimensions are generally not given. The sizes of the chimney flue and the hearth are usually indicated. The fireplace material may be shown by symbols according to the material specified **(Figure 21–17).** The kind of material may also be identified with a notation.

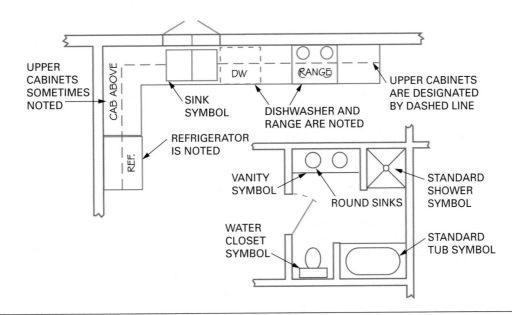

FIGURE 21-15 Floor plan symbols for kitchen and bath cabinets.

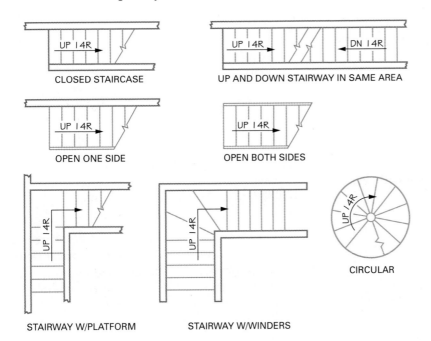

FIGURE 21-16 Stair symbols vary according to the style of the staircase.

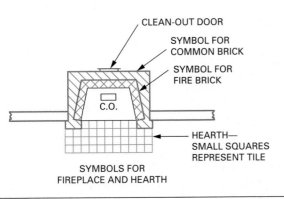

FIGURE 21-17 Symbols for fireplace and hearth.

An *attic access,* also called a **scuttle,** is usually located in a closet, hall, or garage ceiling. It is outlined with dashed lines. It may also be identified by a notation **(Figure 21-18).**

The floor plan symbol for a *floor drain* is a small circle, square, or circle within a square. The slope of the floor is shown by straight lines from the corners of the floor to the center of the drain. Floor drains are appropriately installed in utility rooms where washers, dryers, laundry tubs, and water heaters are located **(Figure 21-19).**

The location of outdoor water faucets, called **hose bibbs,** is shown by a symbol **(Figure 21-20)** projecting from exterior walls where desired. For clarity, the symbol is labeled.

Electrical outlets, switches, and lights may be shown on the floor plan of simpler structures by the use of curved, dashed lines running to *switches, outlets,* and *fixtures* (**Figure 21–21**). The symbols for these electrical components are shown in **Figure 21–22**. Complex buildings require separate electrical plans, as well as prints for plumbing and for heating and ventilation.

Dimensions

Dimensions are placed and printed so that they are as easy to read as possible. Read dimensions carefully. A mistake in reading a dimension early in the

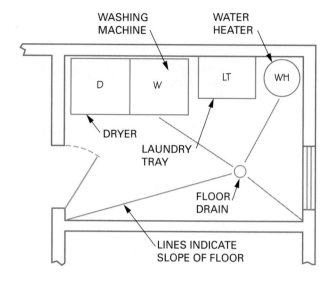

FIGURE 21–19 Some symbols found in a utility room.

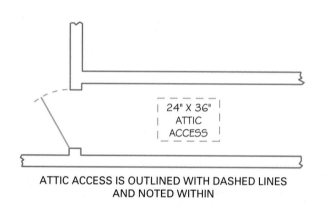

ATTIC ACCESS IS OUTLINED WITH DASHED LINES AND NOTED WITHIN

FIGURE 21–18 Attic access is outlined with dashed lines.

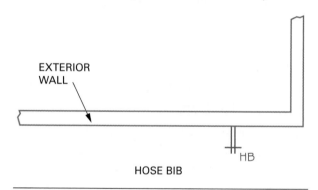

HOSE BIB

FIGURE 21–20 An exterior hose bibb.

FIGURE 21–21 Part of a typical electrical floor plan.

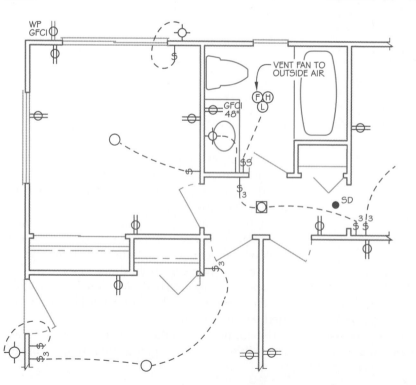

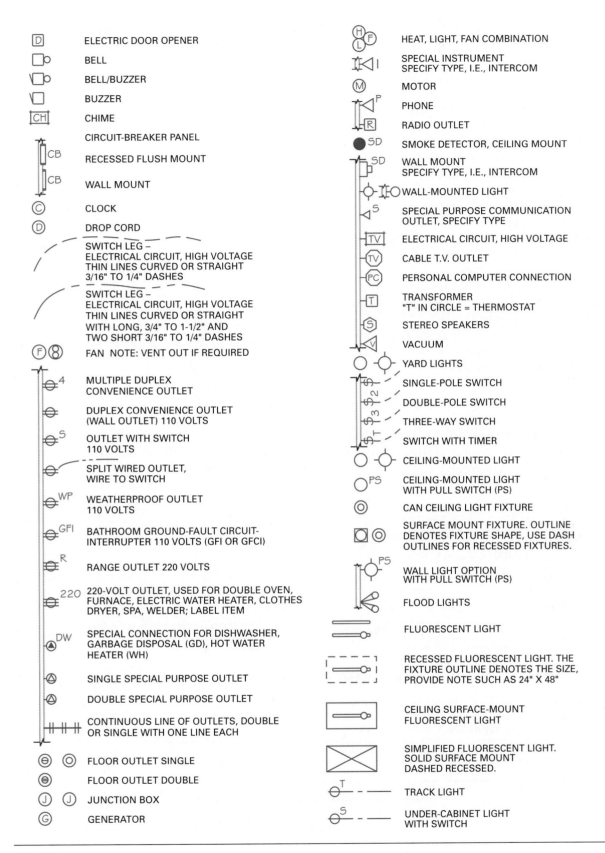

FIGURE 21–22 **Electrical symbols used on plans.**

construction process could have serious consequences later.

Exterior Dimensions. On floor plans, the *overall* dimensions of the building are found on the outer dimension lines. In a *wood frame,* the dimensions are to the outside face of the frame. *Concrete block* walls are dimensioned to their outside face. *Brick veneer* walls are dimensioned to the outside face of the wood frame, with an added dimension and notation for the veneer **(Figure 21–23).**

Multiple dimensions lines are sometimes needed form the same corner. When this happens they are stacked. The dimension lines closest to the exterior walls are used to show the location of *windows* and *doors.* In a wood frame, they are dimensioned to their centerline. In concrete block walls, the dimensions are to the edges of the openings and also show the opening width **(Figure 21–24).**

The second dimension line from the exterior wall is used to locate the centerline of *interior partitions,* which intersect the exterior wall.

Interior Dimensions. Dimensions are given from the outside of a wood frame or from the inside of concrete block walls, to the centerlines or edges of **partitions.** Interior *doors* and other openings are dimensioned to their centerline similar to exterior walls.

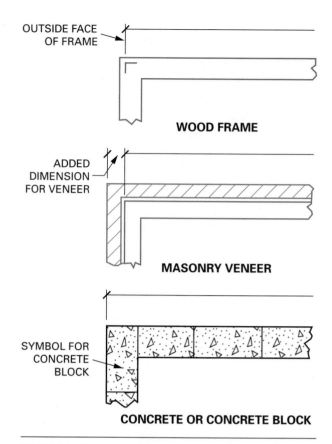

FIGURE 21–23 Overall dimensions are made to the structural portion of a building.

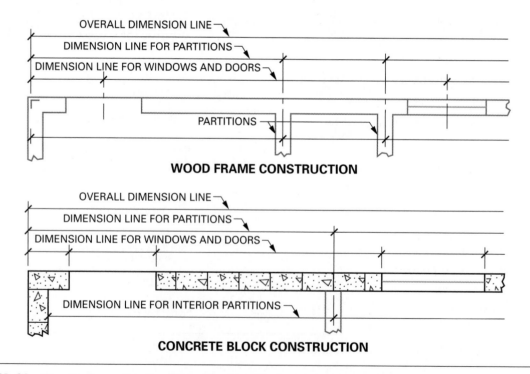

FIGURE 21–24 Standard practice for dimensioning windows, doors, partitions, and then the overall dimension.

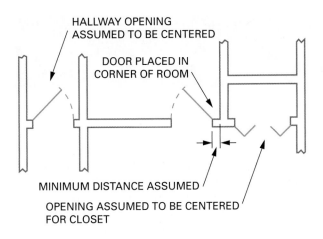

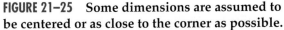

FLOOR PLAN

FIGURE 21–25 **Some dimensions are assumed to be centered or as close to the corner as possible.**

Not all interior dimensions are required. Some may be assumed. For instance, it can be clearly seen if a door is centered between two walls of a closet or hallway **(Figure 21–25).**

The minimum distance from the corner is typically determined by framing the jack and king studs starting at the corner of the room **(Figure 21–26).** This allows sufficient room for the door and **casing** to be applied. If wider custom casing is used, more room may be needed. The goal is to place the door in the corner with room to finish the corner, but not so close that the casing must be scribed to the wall.

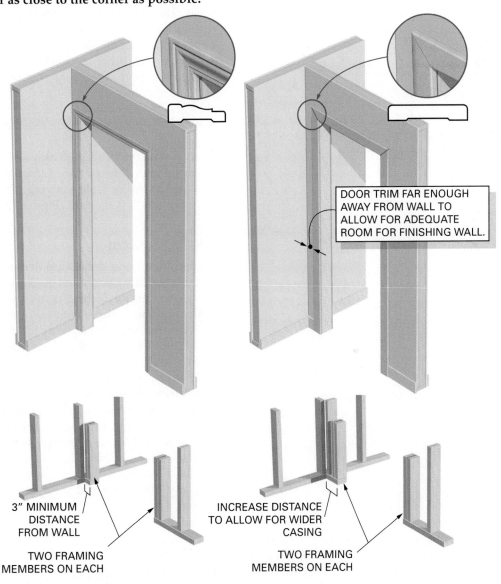

FIGURE 21–26 **Locating a door to allow room for finish.**

22 Sections and Elevations

SECTIONS

Floor plans are views of a horizontal cut. Sections show *vertical* cuts called for on the floor, framing, and foundation plans **(Figure 22–1).** Sections provide information not shown on other drawings. The number and type of section drawings in a set of prints depend on what is required for a complete understanding of the construction. They are usually drawn at a scale of ⅜″ = 1′-0″ or 1:25.

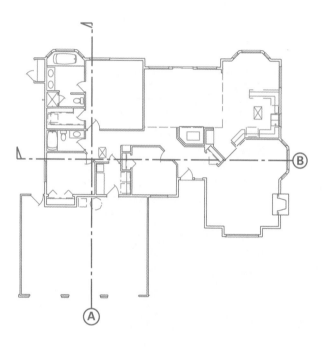

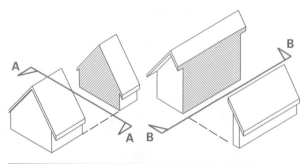

FIGURE 22–1 Sections are views of vertical cutting planes across the width or through the length of a building. Section reference lines identify the location of the section and the direction from which it is being viewed.

Kinds of Sections

Full sections cut across the width or through the length of the entire building **(Figure 22–2).** For a small residence, only one full section may be required to fully understand the construction. Commercial structures may require several full and many partial sections for complete understanding.

A **partial section** shows the vertical relationship of the parts of a small portion of the building. Partial sections through exterior walls are often used to give information about materials from foundation to roof **(Figure 22–3).** They are drawn at a larger scale. The information given in one wall section does not necessarily apply for all walls. It may not apply, in fact, for all parts of the same wall. The wall section, or any other section being viewed, applies only to that part of the construction located by the section reference lines. Because the construction changes throughout the building, many section views are needed to provide clear and accurate information.

An enlargement of part of a section is required when enough information cannot be given in the space of a smaller scale drawing. These large-scale drawings are called *details* **(Figure 22–4).**

Reading Sections

Section views and details are densely packed with information and offer much guidance to the builder. Measurements for the height, thickness, and spacing of the building components are often given. Material types and specifications are indicated. Fastening instructions for normal and special situations are also labeled.

The section view may be referenced for information during all phases of construction. The requirements for the footings and foundation, floor and walls, ceiling and roofing are all drawn. Also included are the finish materials and any special installation instructions. The location and size of all building materials such as steel, wood, masonry, and any other material used in the building are shown.

ELEVATIONS

Elevations are orthographic drawings. They are usually drawn at the same scale as the floor plan. They show each side of the building as viewed from

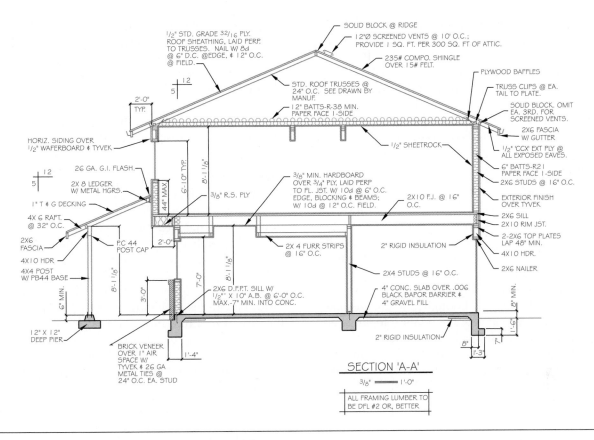

1/2" STD. GRADE 32/16 PLY. ROOF SHEATHING, LAID PERP. TO TRUSSES. NAIL W/ 8d @ 6" O.C. @EDGE, & 12" O.C. @ FIELD.

SOLID BLOCK @ RIDGE

12"Ø SCREENED VENTS @ 10' O.C.; PROVIDE 1 SQ. FT. PER 300 SQ. FT OF ATTIC.

235# COMPO. SHINGLE OVER 15# FELT.

PLYWOOD BAFFLES

STD. ROOF TRUSSES @ 24" O.C. SEE DRAWN BY MANUF.

TRUSS CLIPS @ EA. TAIL TO PLATE.

SOLID BLOCK. OMIT EA. 3RD. FOR SCREENED VENTS.

12" BATTS-R-38 MIN. PAPER FACE 1-SIDE

2X6 FASCIA W/ GUTTER

HORIZ. SIDING OVER 1/2" WAFERBOARD & TYVEK

1/2" 'CCX' EXT PLY @ ALL EXPOSED EAVES.

26 GA. G.I. FLASH.

1/2" SHEETROCK

6" BATTS-R21 PAPER FACE 1-SIDE

2X 8 LEDGER W/ METAL HGRS.

3/8" MIN. HARDBOARD OVER 3/4" PLY, LAID PERP TO FL. JST. W/ 10d @ 6" O.C. EDGE, BLOCKING & BEAMS; W/ 10d @ 12" O.C. FIELD.

2X6 STUDS @ 16" O.C.

1" T & G DECKING

3/8" R.S. PLY

2X10 F.J. @ 16" O.C.

EXTERIOR FINISH OVER TYVEK

4X 6 RAFT. @ 32" O.C.

2X6 SILL

2X10 RIM JST.

2X6 FASCIA

P.C. 44 POST CAP

2X 4 FURR STRIPS @ 16" O.C.

2-2X6 TOP PLATES LAP 48" MIN.

4X10 HDR.

2" RIGID INSULATION

4X10 HDR.

4X4 POST W/ PB44 BASE

2X4 STUDS @ 16" O.C.

2X6 NAILER

2X6 D.F.P.T. SILL W/ 1/2"× X 10" A.B. @ 6'-0" O.C. MAX.-7" MIN. INTO CONC.

4" CONC. SLAB OVER .006 BLACK BAPOR BARRIER & 4" GRAVEL FILL

12" X 12" DEEP PIER

2" RIGID INSULATION

BRICK VENEER OVER 1" AIR SPACE W/ TYVEK & 26 GA METAL TIES @ 24" O.C. EA. STUD

SECTION 'A-A'

3/8" = 1'-0"

ALL FRAMING LUMBER TO BE DFL #2 OR, BETTER

FIGURE 22–2   A typical section view of a residence.

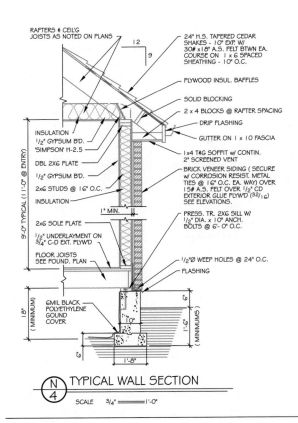

RAFTERS & CEIL'G JOISTS AS NOTED ON PLANS

24" H.S. TAPERED CEDAR SHAKES - 10" EXP. W/ 30# x18" A.S. FELT BTWN EA. COURSE ON 1 x 6 SPACED SHEATHING - 10" O.C.

PLYWOOD INSUL. BAFFLES

SOLID BLOCKING

2 x 4 BLOCKS @ RAFTER SPACING

DRIP FLASHING

INSULATION 1/2" GYPSUM B'D. 'SIMPSON' H-2.5

GUTTER ON 1 x 10 FASCIA

1x4 T&G SOFFIT w/ CONTIN. 2" SCREENED VENT

DBL 2X6 PLATE

1/2" GYPSUM B'D.

BRICK VENEER SIDING (SECURE w/ CORROSION RESIST. METAL TIES @ 16" O.C. EA. WAY) OVER 15# A.S. FELT OVER 1/2" CD EXTERIOR GLUE PLYW'D (32/16) SEE ELEVATIONS.

2x6 STUDS @ 16" O.C.

INSULATION

1" MIN.

PRESS. TR. 2X6 SILL W/ 1/2" DIA. x 10" ANCH. BOLTS @ 6'- 0" O.C.

2x6 SOLE PLATE

1/2" UNDERLAYMENT ON 3/4" C-D EXT. PLYW'D

1/2"Ø WEEP HOLES @ 24" O.C.

FLOOR JOISTS SEE FOUND. PLAN

FLASHING

6MIL BLACK POLYETHYLENE GOUND COVER

18" (MINIMUM)

6" 6" 1'-6" (MINIMUMS)

6"

1'-8"

N 4

TYPICAL WALL SECTION

SCALE   3/4" = 1'-0"

FIGURE 22–3   A partial section through an exterior wall.

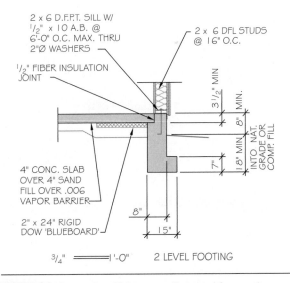

2 x 6 D.F.P.T. SILL W/ 1/2" x 10 A.B. @ 6'-0" O.C. MAX. THRU 2"Ø WASHERS

2 x 6 DFL STUDS @ 16" O.C.

1/2" FIBER INSULATION JOINT

3 1/2" MIN.

8" MIN.

4" CONC. SLAB OVER 4" SAND FILL OVER .006 VAPOR BARRIER

2" x 24" RIGID DOW 'BLUEBOARD'

18" MIN. INTO NAT. GRADE OR COMP. FILL

7"

8"

15"

3/4" = 1'-0"   2 LEVEL FOOTING

FIGURE 22–4   A detail is a small part of a section drawn at a large scale.

outside at a distance of about 100 feet. Generally four elevations, one for each side, are included in a set of drawings. They are titled Front, Rear, Left Side, and Right Side. They may also be titled according to the compass direction that they face, for instance, North, South, East, and West. From the exterior elevations, the general shape and design of the building can be determined **(Figure 22–5).**

Symbols

Elevation symbols are different than floor plan symbols for the same object. In elevation drawings, the symbols represent, as closely as possible, the actual object as it would appear to the eye. To make the drawing more clear, the symbols are usually identified with a notation.

The location of any steps, porches, dormers, skylights, and chimneys, although not dimensioned, can be seen in elevations. Foundation footings and walls below the grade level may be shown with hidden lines. The kind and size of exterior siding, railings, entrances, and special treatment around doors and windows are shown **(Figure 22–6).**

The elevations show the windows and doors in their exact location. Other openings, such as **louvers,** are shown in place. Their style and size are identified by appropriate symbols and notations **(Figure 22–7).**

The type of roofing material, the roof pitch, and the cornice style may also be determined from the exterior elevations **(Figure 22–8).**

Dimensions

In relation to other drawings, elevations have few dimensions. Some dimensions usually given are floor to floor heights, distance from grade level to finished floor, height of window openings from the finished floor, and distance from the ridge to the top of the chimney.

A number of other things may be shown on exterior elevations, depending on the complexity of the structure. Little information is given in elevations that cannot be seen in more detail in plans and sections. However, elevations serve an important purpose in making the total construction easier to visualize.

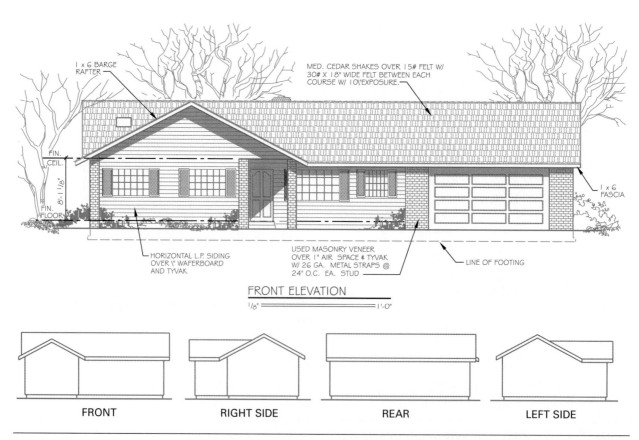

FIGURE 22–5 Elevations show the exterior of a building.

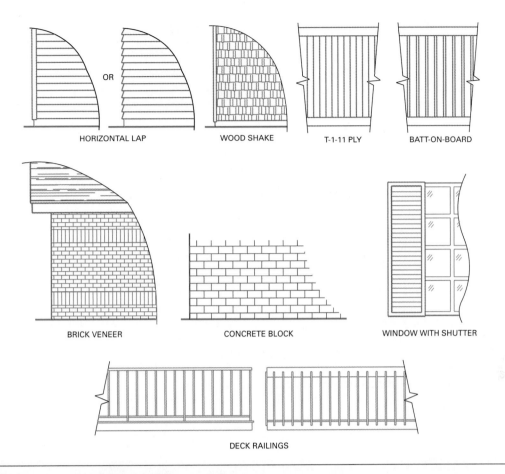

FIGURE 22–6 Symbols for siding, railings, and shutters.

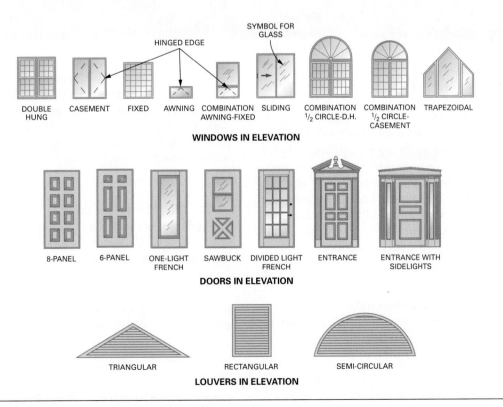

FIGURE 22–7 Symbols for windows, doors, and louvers.

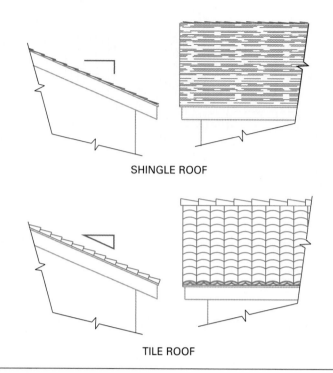

SHINGLE ROOF

TILE ROOF

FIGURE 22–8 Symbols for roofing and roof slope.

23 Plot and Foundation Plans

PLOT PLANS

A **plot plan** is a map of a section of land used to show the proposed construction **(Figure 23–1)**. Depending on the size, the scale of plot plans may vary from 1″ = 10′ to 1″ = 200′ or 1:100 to 1:250. It is a required drawing when applying for a permit to build in practically every community. It is a necessary drawing to plan construction that may be affected by various features of the land. The plan must show compliance with zoning and health regulations. Although plot plan requirements may vary with localities, certain items in the plan are standard.

Property Lines

The property line *measurements* and *bearings,* known as **metes and bounds,** show the shape and size of the parcel. They are standard in every plot plan.

Measurements. The boundary lines are measured in *feet, yards, rods, chains,* or *meters.* Typically, in the United States, the foot is the most commonly used measurement, while most of the world and Canada use meters.

3 feet equals one yard.
16½ feet or 5½ yards equal one rod.
66 feet or 22 yards or 4 rods equal one chain.

Parts of measurement units are expressed as decimals. For instance, a boundary line dimension is expressed as 100.50 feet, not 100 feet, 6 inches. The measurement is shown centered on, close to, and inside the line.

North. The North compass direction is clearly marked on every plot plan. In a clear space, an arrow of any style labeled with the letter "N" is pointed in the north direction **(Figure 23–2)**.

Bearings. In addition to the length of the boundary line, its **bearing** is shown. The bearing is a compass direction given in relation to a *quadrant* of a circle. There are 360 degrees in a circle and 90 degrees in each quadrant. Degrees are divided into *minutes* and *seconds.*

One degree equals 60 minutes (60′)
One minute equal 60 seconds (60″)

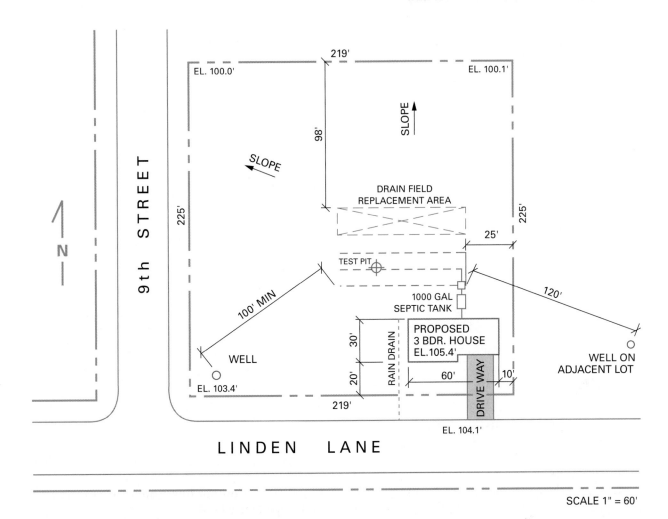

FIGURE 23–1   A typical plot plan.

FIGURE 23–2   Typical North direction symbols.

The boundary line bearing is expressed as a certain number of degrees clockwise or counterclockwise from either North or South. For instance, a bearing may be shown as N 30° W, N 45° E, S 60° E, S 30° W **(Figure 23–3).** No bearings begin with East or West as a direction. The bearing is shown centered close to and outside the boundary line opposite its length.

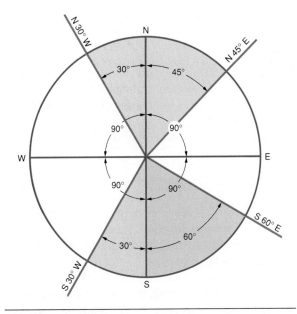

FIGURE 23–3   Method of indicating bearings for property lines.

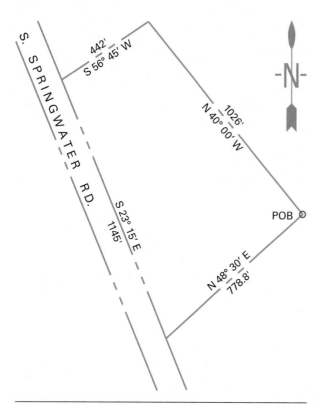

FIGURE 23–4 Measurements, bearings, and legal description of a parcel of land.

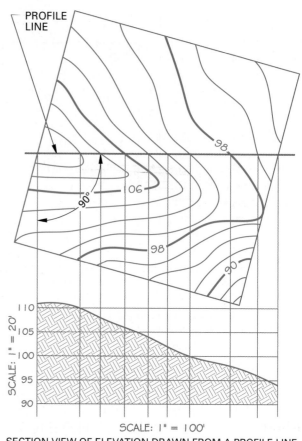

SECTION VIEW OF ELEVATION DRAWN FROM A PROFILE LINE

FIGURE 23–5 Contour lines show the elevation and slope of the land.

Point of Beginning. An object that is unlikely to be moved easily, such as a large rock, tree, or iron rod driven into the ground, is used for a **point of beginning.** To denote this point on the plot plan, one corner of the lot may be marked with the abbreviation POB. It is from this point that the lot is laid out and drawn **(Figure 23–4).**

Topography

Topography is the detailed description of the land surface. It includes any outstanding physical features and differences in *elevation* of the building site. Elevation is the height of a surface above sea level. It is expressed in decimal feet and ¹⁄₁₀₀ of a foot or meters and decimals.

Contour Lines. **Contour lines** are irregular, curved lines connecting points of the same elevation of the land. The vertical distance between contour lines is called the **contour interval.** It may vary depending on how specifically the contour of the land needs to be shown on the plot plan.

When contour lines are close together, the slope is steep. Widely spaced contour lines indicate a gradual slope. At intervals, the contour lines are broken and the elevation of the line inserted in the

space **(Figure 23–5).** On some plans, dashed contour lines indicate the existing grade and solid lines depict the new grade. Topography is not always a requirement on plot plans. This is especially true for sites where there is little or no difference in elevation of the land surface. The slope of the finished grade may be shown by arrows instead of contour lines **(Figure 23–6).**

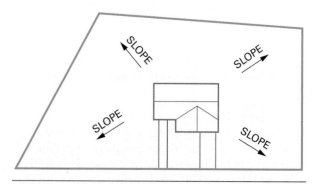

FIGURE 23–6 Arrows are sometimes used in place of contour lines to show the slope of the land.

Elevations

The height of several parts of the site and the construction are indicated on the plot plan. It is necessary to know these elevations for grading the lot and construction of the building and accessories.

Benchmark. Before construction begins, a reference point, called a **benchmark,** is established on or close to the site. It is used for conveniently determining differences in elevation of various land and building surfaces.

The benchmark is established on some permanent object, which will not be moved or destroyed, at least until the construction is complete. It may be the actual elevation in relation to sea level. It may also be given an arbitrary elevation of 100.00 feet. All points on the lot, therefore, would either be above, level with, or below the benchmark **(Figure 23–7).** The location of the benchmark is clearly shown on the plot plan with the abbreviation BM.

Finish Floor. The elevation of the finished floor or the top of the foundation levels may be shown and noted on the plot plan. This elevation helps the contractor to determine the bottom of the excavation. The bottom of excavation elevation must be calculated. To do this, subtract the building component heights from the finish floor height given. For exam-

ple, what is the bottom of excavation elevation for the building shown in **Figure 23–8?**

To solve, add up the components:

$$10'' + 8'-0'' + 1'-2\tfrac{3}{4}'' = 10'-\tfrac{3}{4}''$$

Convert ¾″ to decimal feet

$$\tfrac{3}{4} \div 12 = 0.0625$$

Add 0.0625 to 10 feet

$$0.0625 + 10 = 10.0625 \text{ (rounded off to 10.06 feet)}$$

Next, subtract

$$104.50' - 10.06' = 94.44'$$

The bottom of excavation elevation is 94.44 feet.

Converting Decimals to Fractions. Calculations, especially those dealing with elevations and roof framing, require the carpenter to convert decimals of a foot to feet, inches, and 16ths of an inch as found on the rule or tape. To convert, use the following method:

- Multiply a decimal of a foot by 12 (the number of inches in a foot) to get inches.

- Multiply any remainder decimal of an inch by 16 (the number of 16ths in an inch) to get 16ths of an inch. Round off any remainder to the nearest 16th of an inch.

- Combine whole feet, whole inches, and 16ths of an inch to make the conversion.

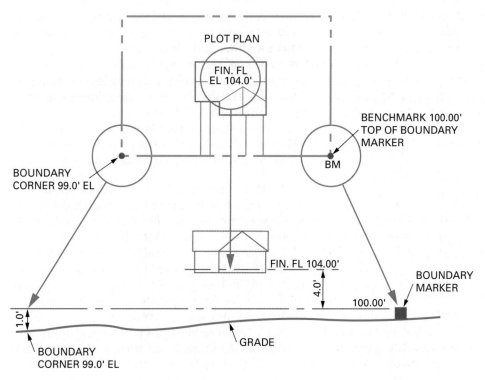

FIGURE 23–7 A benchmark is a reference point used for determining differences in elevation.

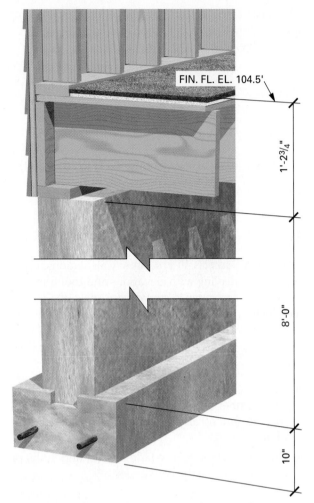

FIN. FL. EL. 104.5'

1'-2³/₄"

8'-0"

10"

ADD COMPONENTS: 10" + 8'-0" + 1'-2³/₄" = 10'-0³/₄".

CONVERT TO DECIMAL FEET: ³/₄" ÷ 12 = 0.0625 FEET.

ADD TO 10 FEET = 10.0625 OR ROUNDED TO 10.06 FEET.

SUBTRACT 104.50' – 10.06' = 94.44 FEET IS THE EXCAVATION ELEVATION.

FIGURE 23–8 The distance from the finished floor elevation to the bottom of the footing may need to be calculated.

For example, convert the finished floor elevation of 104.65 feet to feet, inches, and 16ths of an inch.

1. Multiply .65 ft. × 12 = 7.80 inches.
2. Multiply .80 inches × 16 = 12.8 16ths of an inch.
3. Round off 12.8 16ths of an inch to 13 16ths of an inch.
4. Combine feet, inches, and sixteenths = 104'-7¾₆".

It is more desirable to remember the method of conversion rather than use conversion tables. Reliance on conversion tables requires access to the ta-

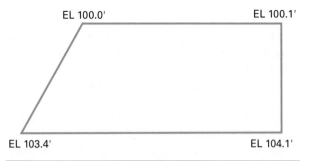

EL 100.0'   EL 100.1'

EL 103.4'   EL 104.1'

FIGURE 23–9 The elevations of property line corners are indicated on the plot line.

bles and knowledge of their use, encourages dependence on them, and results in helplessness without them.

Other Elevations. In addition to contour lines, the elevation of each corner of the property is noted on the plot plan **(Figure 23–9).** The top of one of the boundary corner markers makes an excellent benchmark.

Existing and proposed roads adjacent to the property are shown as well as any **easements.** Easements are right-of-way strips running through the property. They are granted for various purposes, such as access to other property, storm drains, or utilities. Elevations of a street at a driveway and at its centerline are usually required. One of these is sometimes used as the benchmark.

The Structure

The shape and location of the building are shown. Distances, called *setbacks*, are dimensioned from the boundary lines to the building.

The shape, width, and location of patios, walks, driveways, and parking areas may also be shown on plot plans. Details of their construction are found in another drawing.

A plot plan may also show any *retaining walls*. These walls are used to hold back earth to make more level surfaces instead of steep slopes.

Utilities. The water supply and public sewer connections are shown by noted lines from the structure to the appropriate boundary line. If a private sewer disposal system is planned, it is shown on the plot plan **(Figure 23–10).** Although there are many kinds of sewer disposal systems, those most commonly used consist of a septic tank and leach or drain field, which are usually subject to strict regulations with regard to location and construction.

The location of gas lines is shown, if applicable. Sometimes the location of the nearest utility pole is

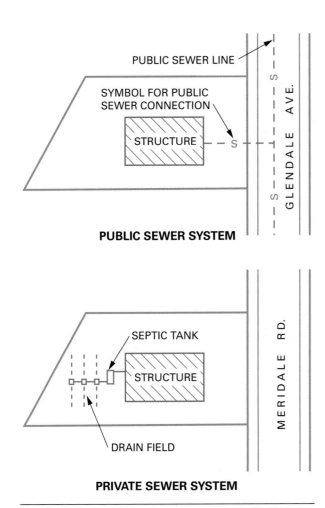

PUBLIC SEWER SYSTEM

PUBLIC SEWER LINE

SYMBOL FOR PUBLIC
SEWER CONNECTION

STRUCTURE

GLENDALE AVE.

SEPTIC TANK

STRUCTURE

DRAIN FIELD

MERIDALE RD.

PRIVATE SEWER SYSTEM

FIGURE 23–10   The style of sewage disposal system is shown on the plot plan.

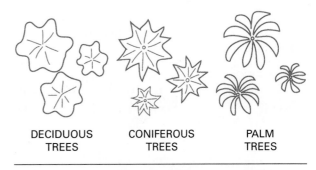

DECIDUOUS TREES   CONIFEROUS TREES   PALM TREES

FIGURE 23–11   Various symbols are used to indicate different kinds of trees.

given. It is shown using a small solid circle as a symbol and is noted. Foundation drain lines leading to a storm drain, drywell, or other drainage may be shown and labeled.

Landscaping. The location and kind of existing and proposed trees are shown on the plot plan. Existing trees are noted, whether they are to be saved or removed. Those that are to be saved are protected by barriers during the construction process. Various symbols are used to show different kinds of trees **(Figure 23–11).**

Identification. Included on the plot plan are the name and address of the property owner, the title and scale of the drawing, and a legal description of the property **(Figure 23–12).**

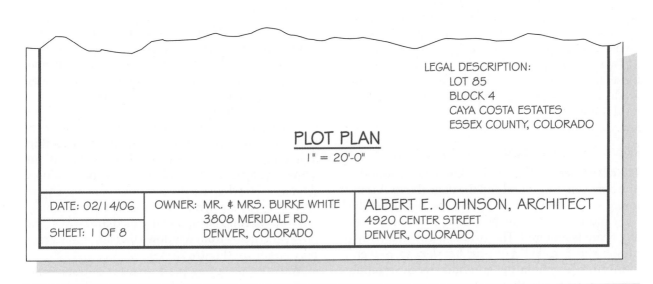

LEGAL DESCRIPTION:
LOT 85
BLOCK 4
CAYA COSTA ESTATES
ESSEX COUNTY, COLORADO

PLOT PLAN
1" = 20'-0"

DATE: 02/14/06

SHEET: 1 OF 8

OWNER: MR. & MRS. BURKE WHITE
3808 MERIDALE RD.
DENVER, COLORADO

ALBERT E. JOHNSON, ARCHITECT
4920 CENTER STREET
DENVER, COLORADO

FIGURE 23–12   Certain identification items are needed on the plot plan.

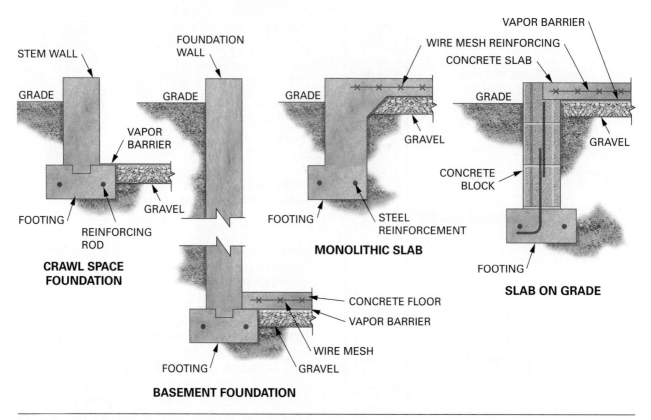

FIGURE 23–13   **Commonly used foundation styles.**

Foundation Plans

The foundation plan is drawn at the same scale as the floor plan. It is a view from above of a horizontal cut through the foundation. Great care must be taken when reading the foundation plan so no mistakes are made. A mistake in the foundation affects the whole structure, and generally requires adjustments throughout the construction process.

Two commonly used types of foundations are those having a crawl space, or basement below grade, and those with a concrete slab floor at grade level **(Figure 23–13).**

Crawl Space and Basement Foundations

The **crawl space** is the area enclosed by the foundation between the ground and the floor above. A minimum distance of 18 inches from the ground to the floor and 12 inches from the ground to the bottom of any beam is required. The ground is covered with a plastic sheet, called a **vapor barrier,** to prevent moisture rising from the ground from penetrating into the floor frame above.

A foundation enclosing a basement is similar to that of a crawl space except the walls are higher, windows and doors may be installed, and a concrete floor is provided below grade. The basement may be used for additional living area, garage, utility room, or workshop.

Reading Plans. Whether the foundation supports a floor using closely spaced floor joists or more widely spaced post-and-beam construction, the information given in the foundation plan is similar. A typical foundation plan is shown in **Figure 23–14.**

The inside and outside of the foundation wall are clearly outlined. Dashed lines on both sides of the wall show the location of the foundation footing. The type, size, and spacing of anchor bolts are shown by a notation. Wall openings for windows, doors, or crawl space access and vents are shown with appropriate symbols and noted. Small retaining walls of concrete or metal, called **areaways,** that hold earth away from windows that are below grade may be shown **(Figure 23–15).**

Walls for *stoops* or platforms for entrances are shown. A notation is made in regard to the material

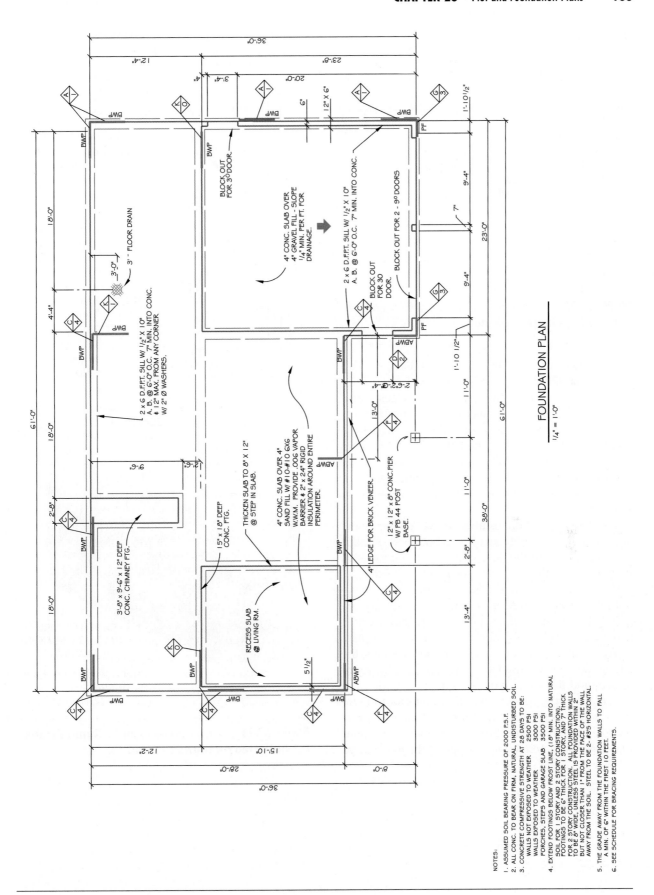

FOUNDATION PLAN
1/4" = 1'-0"

NOTES:

1. ASSUMED SOIL BEARING PRESSURE OF 2000 P.S.F.
2. ALL CONC. TO BEAR ON FIRM, NATURAL, UNDISTURBED SOIL.
3. CONCRETE COMPRESSIVE STRENGTH AT 28 DAYS TO BE:
   WALLS NOT EXPOSED TO WEATHER   2500 PSI
   WALLS EXPOSED TO WEATHER   3000 PSI
   PORCHES, STEPS AND GARAGE SLAB   3500 PSI
4. EXTEND FOOTINGS BELOW FROST LINE, (18" MIN. INTO NATURAL
   SOIL, FOR 1 STORY AND 2 STORY CONSTRUCTION).
   FOOTINGS TO BE 6" THICK FOR 1 STORY, AND 7" THICK
   FOR 2 STORY CONSTRUCTION. ALL FOUNDATION WALLS
   TO BE 8" WIDE, UNLESS STEEL IS PROVIDED WITHIN 2"
   BUT NOT CLOSER THAN 1" FROM THE FACE OF THE WALL
   AWAY FROM THE SOIL. STEEL TO BE 2- #35 HORIZONTAL.
5. THE GRADE AWAY FROM THE FOUNDATION WALLS TO FALL
   A MIN. OF 6" WITHIN THE FIRST 10 FEET.
6. SEE SCHEDULE FOR BRACING REQUIREMENTS.

FIGURE 23–14 A foundation plan for a partial basement.

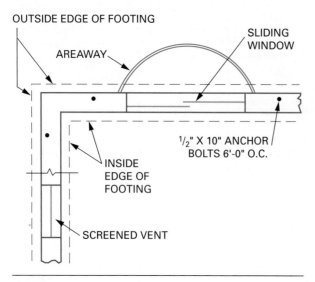

FIGURE 23–15 Partial plan of a foundation wall with various items indicated.

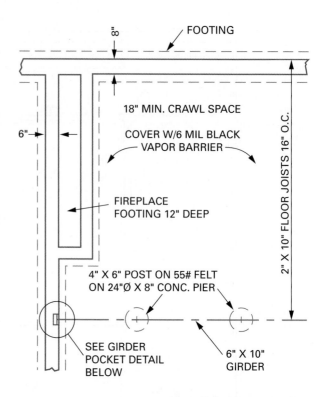

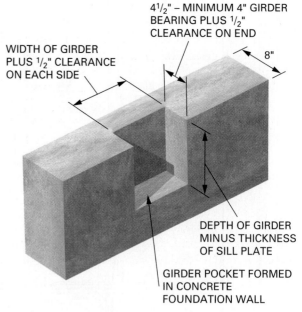

FIGURE 23–16 Girder pockets and column details are indicated on foundation plans.

with which to fill the enclosed area and cap the surface. Other footings shown by dashed lines include those for chimneys, fireplaces, and columns or posts. Columns or posts support girders shown by a series of long and short dashes directly over the center of the columns. A recess in the foundation wall, called a **beam pocket,** that is used to support the ends of the girder is shown. Notations are made to identify all of these items **(Figure 23–16).**

Floor joist direction installed above the plan view is shown by a line with arrows on both ends similar to those shown in the floor plan for ceiling joists. A notation gives the size and spacing of the joists.

The composition, thickness, and underlying material of the basement floor or crawl space surface are noted. The location of a stairway to the basement is shown with the same symbol as used in the floor plan.

Although it may be stated in the specifications, the strength of concrete used for various parts of the foundation, and the wood type and grade, in addition to size, may be specified by a notation. Plans for foundations with basements may show the location of furnaces and other items generally found in the floor plan if the basement is used as part of the living area.

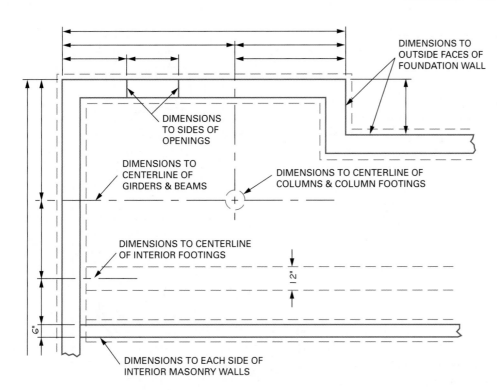

FIGURE 23–17 Typical dimensioning of crawl space and basement foundations.

Dimensions. It is important to understand how parts of the foundation are dimensioned. Foundation walls are dimensioned face to face. Interior footings, columns, posts, girders, and beams are dimensioned to their centerline **(Figure 23–17).**

SLAB-ON-GRADE FOUNDATION

The **slab-on-grade foundation** is used in many residential and commercial buildings. It takes less labor and material than foundations to support beam and joist floor framing. Often the concrete for the footing, foundation, and slab can be placed at the same time. There are several kinds, but slab-on-grade foundation plans **(Figure 23–18)** show common components:

- The shape and size of the slab are shown with solid lines, as are patios and similar areas. Changes in floor level, such as for a fireplace hearth or a sunken living area, are indicated by solid lines. A notation gives the depth of the recess.

- Exterior and interior footing locations ordinarily below grade are indicated with dashed lines. Footings outside the slab and ordinarily above grade are shown with solid lines.

- Appropriate symbols and notations are used for fireplaces, floor drains, and ductwork for heating and ventilation. Blueprints for larger buildings usually have separate drawings for electrical, plumbing, and mechanical work.

- Notations are written for the slab thickness, wire mesh reinforcing, fill material, and vapor barrier under the slab. Interior footing, mudsill, reinforcing steel, and anchor bolt size, location, and spacing are also noted.

Dimensions. Overall dimensions are to the outside of the slab. Interior piers are located to their centerline. Door openings are dimensioned to their sides.

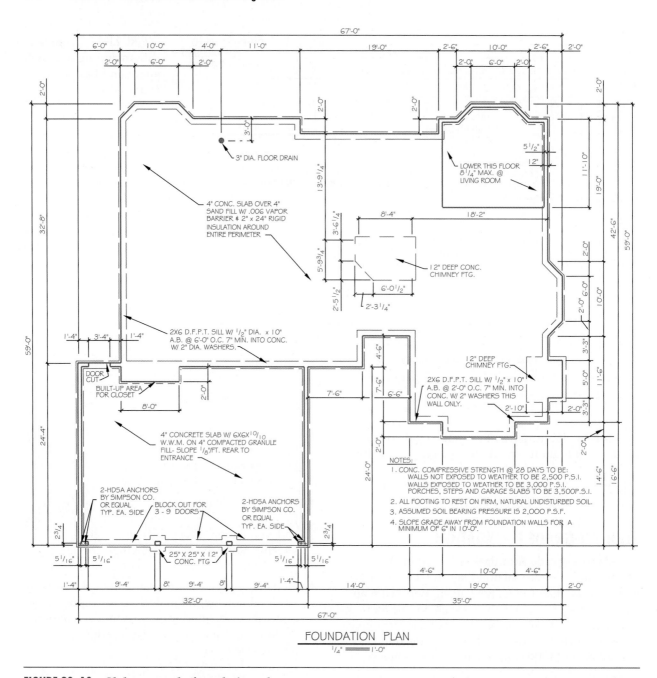

FIGURE 23–18   Slab-on-grade foundation plan.

24   Building Codes and Zoning Regulations

Cities and towns have laws governing many aspects of new construction and remodeling. These laws protect the consumer and the community. Codes and regulations provide for safe, properly designed buildings in a planned environment. Contractors and carpenters should have knowledge of local zoning regulations and building codes.

ZONING REGULATIONS

Zoning regulations, generally speaking, deal with keeping buildings of similar size and purpose in areas for which they have been planned. They can also regulate the space in each of the areas. The community is divided into areas called **zones,** shown on *zoning maps.*

Zones

The names given to different zones vary from community to community. The zones are usually abbreviated with letters or a combination of letters and numbers. A large city may have thirty or more zoning districts.

There may be several *single-family residential zones.* Some zones have less strict requirements than others. Other areas may be zoned as *multifamily residential.* They may be further subdivided into areas according to the number of apartments. Other residential zones may be set aside for *mobile home parks,* and those that allow a combination of *residences, retail stores, and offices.*

Other zones may be designated for the *central business district,* or various kinds of *commercial districts* and different *industrial* zones.

Lots

Zoning laws regulate buildings and building sites. Most cities specify a *minimum lot size* for each zone and a *maximum ground coverage* by the structure. The *maximum height* of the building for each zoning district is stipulated. A *minimum lot width* is usually specified as well as *minimum yards.*

Minimum yard refers to the distance buildings must be kept from property lines. These distances are called **setbacks.** They are usually different for front, rear, and side.

Some communities require a certain amount of landscaped area, called **green space,** to enhance the site. In some residential zones, as much as half the lot must be reserved for green space. In a central business area, only 5 to 10 percent may be required.

In most zones, off-street parking is required. For instance, in single-family residential zones, room for two parking spaces on the lot is required.

Nonconforming Buildings

Because some cities were in existence before the advent of zoning laws, many buildings and businesses may not be in their proper zone. They are called **nonconforming.** It would be unfair to require that buildings be torn down, or to stop businesses, in order to meet the requirements of zoning regulations.

Nonconforming businesses or buildings are allowed to remain. However, restrictions are placed on rebuilding. If partially destroyed, they may be allowed to rebuild, depending on the amount of destruction. If 75 percent or more is destroyed, they are not usually allowed to rebuild in the same manner or for the same purpose in the same zone.

Any hardships imposed by zoning regulations may be relieved by a **variance.** Variances are granted by a Zoning Board of Appeals within each community. A public hearing is held after a certain period of time. The general public, and, in particular, those abutting the property are notified. The petitioner must prove certain types of hardship specified in the zoning laws before the zoning variance can be granted.

BUILDING CODES

Building codes regulate the design and construction of buildings by establishing minimum safety standards. They prevent such things as roofs being ripped off by high winds, floors collapsing from inadequate support, buildings settling because of a poor foundation, and tragic deaths from fire due to lack of sufficient exits from buildings. In addition to building codes, other codes govern the mechanical, electrical, and plumbing trades.

Some communities have no building codes. Some write their own. Some have codes, but exempt residential construction. Some have adopted one of three national model building codes. Some use one of the national codes supplemented with their own. Some states have building codes that supersede national ones. There are literally hundreds of building codes.

It is important to have a general knowledge of the building code used by a particular community. Construction superintendents and contractors must have extensive knowledge of the codes.

National Building Codes

Many **national building codes** are in existence today. Many are being updated almost every year. These codes are written by various organizations whose purpose it is to standardize construction over a broad area. These organizations include the International Conference of Building Officials (ICBO), Southern Building Code Congress International, Inc. (SBCCI), and Building Officials and Code Administrators International, Inc. (BOCA). The codes produced by each organization are as follows:

> ICBO—*Uniform Building Code*
> SBCCI—*Standard Building Code*
> BOCA—*National Building Code.*

Recently another code was created that merges these three codes and includes input from the International Code Council (ICC). This code is called the *International Residential Code* (IRC). A goal of this code is to reduce the confusion of so many different codes. Many states have adopted this code or created another one using it as a model.

In Canada, the *National Building Code* sets the minimum standard. Some provinces augment this code with more stringent requirements and publish the combination as a *Provincial Building Code.* A few cities have charters, which allow them to publish their own building codes.

Use of Residential Codes

In addition to structural requirements, major topics covered by residential codes include:

- exit facilities, such as doors, halls, stairs, and windows as emergency exits, and smoke detectors.
- room dimensions, such as ceiling height and minimum area.
- light, ventilation, and sanitation, such as window size and placement, maximum limits of glass

area, fans vented to the outside, requirements for baths, kitchens, and hot and cold water.

Use of Commercial Codes

Codes for commercial work are much more complicated than those for residential work. The structure must first be defined for code purposes. To define the structure, six classifications must be used.

1. The *occupancy group* classifies the structure by how and whom it will be used. The classification

---

CITY OF ANYWHERE, USA
APPLICATION FOR BUILDING PERMIT

RADON GAS FEE _____

FOR OFFICE USE ONLY

Permit Type _____   Permit # _____

Permit Class of Work _____   Log # _____

Permit Use Code _____   Issue Date _____

Lot _____ Block _____ Sub _____ Permit Cost _____

Fire Zone: IN _____   OUT _____   Zone: _____   T.I.F. Due (Y/N or NA) _____

Utility Notification   1. FL Power   _____   B of A (Y/N) _____ Case No.
                              2. Peoples Gas _____   E.D.C. (Y/N) _____ Case No.
NOTE: Items with*   3. Water Dept. _____   C.R.A. (Y/N) _____ Case No.
must be entered in computer.   H.P.C. (Y/N) _____ Case No.

*Plat Page _____   *Sec _____   *Township _____   *Range _____   Zone _____

*Dept of Commerce Code _____   *Const. Type _____ Protected _____ Unprotected _____

*Additional Permits Required:
   Building _____   Plumbing _____   No. of W.C. _____   No. of Meters _____

   Electrical _____   Mechanical _____   Gas _____   Fire Sprk. _____   Landscape _____

   Park/Paving _____   Total Spaces _____   Handicap _____

*Flood Zone _____   *Setbacks: Front _____   Left Side _____   Right Side _____
                                          Rear _____   Other Requirements _____
Threshold Building YES _____ NO _____   _____

Special Notes/Comments to Inspector: _____
_____
_____

APPLICANT PLEASE FILL OUT THIS SECTION

JOB ADDRESS _____   Suite or Apt. No. _____

CONTRACTOR _____   Cert./Reg. No. _____   Telephone _____

PROPERTY OWNER'S Name _____   Address _____

          City _____   State _____   Zip _____   Telephone _____

Building Description: Total Sq. Ft. _____   Estimated Job Value _____
   LF-SF or Dimensions _____   Building Use _____
   Valuation of Work _____   Former Use _____
   No. of Units _____   No. of Suites _____   No. of Stories _____

Special Notes or Comments: _____
_____

PHONE 555-1234 FOR ALL INSPECTIONS   HCS-12 Rev. 6-1-88
                    (OVER)

FIGURE 24–1   A typical form used to apply for a building permit.

is designated by a letter, such as R, which includes not only single-family homes but apartments and hotels.

2. The size and location of the building.
3. The type of construction. Five general types are given numbers 1 through 5. Types 1 and 2 require that all structural parts be noncombustible. Construction in types 3, 4, or 5 can be made of either masonry, steel, or wood.
4. The floor area of the building.
5. The height of the building. Zoning regulations may also affect the height.
6. The number of people who will use the building, called the *occupant load*, determines such things as the number and location of exits.

Once the structure is defined, the code requirements may be studied.

BUILDING PERMITS

A **building permit** is needed before construction can begin. Application is made to the office of the local building official. The building permit application form **(Figure 24–1)** requires a general description of the construction, legal description and location of the property, estimated cost of construction, and information about the applicant.

Drawings of the proposed construction are submitted with the application. The type and kind of drawings required depend on the complexity of the building. For commercial work, usually five sets of plot plans and two sets of other drawings are required. The drawings are reviewed by the building inspection department. If all is in order, a permit **(Figure 24–2)** is granted upon payment of a fee. The fee is usually based on the estimated cost of the construction. Electrical, mechanical, plumbing, water, and sewer permits

ANYWHERE, USA
DEPARTMENT OF HOUSING & CONSTRUCTION SERVICES

BUILDING PERMIT

THIS PERMIT BECOMES INVALID IF NO INSPECTIONS HAVE BEEN MADE DURING ANY 3 MONTH PERIOD.

Flood Elevation - ________ Lowest Floor Minimum Required

☐ New Construction ☐ Moving ☐ Siding
☐ Grounds Improvements ☐ Fences ☐ Walls
☐ Utility Building ☐ Pool
☐ Reroofing ☐ Other ____________

Permit No. ____________ (ZONE) ____________

Job Address ____________

Lot ________ **Blk.** ________ **Sub.** ________

Date ____________ This permit covers building construction only. Additional permits are required for electric, plumbing, gas and/or mechanical installations.

BUILDING			ELECTRICAL			PLUMBING			MECHANICAL-GAS		
Type of Inspection	Date	Inspector	Type of Inspection	Date	Inspector	Type of Inspection	Date	Inspector	Type of Inspection	Date	Inspector

NOTE: Building, Electrical, Plumbing and Mech/Gas Inspections shall be dated and initialed by inspectors before walls and ceilings are covered.

BUILDING OK TO COVER		ELECTRICAL OK TO COVER		PLUMBING OK TO COVER		MECH/GAS OK TO COVER	
Date	Inspector	Date	Inspector	Date	Inspector	Date	Inspector

NOTE: This card shall remain posted at the job site until all final inspections have been dated and initialed by inspectors.

BUILDING FINAL OK		ELECTRICAL FINAL OK		PLUMBING FINAL OK		MECH/GAS FINAL OK	
Date	Inspector	Date	Inspector	Date	Inspector	Date	Inspector

THIS CARD MUST BE POSTED IN AN EASILY SEEN LOCATION.

For INSPECTIONS, call 555-1234. ■ For other information, call 555-6789.

FIGURE 24–2 A typical building permit.

are usually obtained by subcontractors. The permit card must be displayed on the site in a conspicuous place until the construction is completed.

Inspections

Building inspectors visit the job site to perform code inspections at various intervals. These inspections may include:

1. A *foundation inspection* takes place after the trenches have been excavated and forms erected and ready for the placement of concrete. No reinforcing steel or structural framework of any part of any building may be covered without an inspection and a release.

2. A *frame inspection* takes place after the roof, framing, fire blocking, and bracing are in place, and all concealed wiring, pipes, chimneys, ducts, and vents are complete.

3. The *final inspection* occurs when the building is finished. A Certificate of Occupancy or Completion is then granted.

Some communities require many more inspections. These are designed to verify that the building meets the code of that area. For example, in southern Florida a separate inspection is made of windows after installation to ensure that they are installed to withstand severe wind loads from hurricanes. In California, anchor bolts and metal shear walls require a separate inspection to ensure that they will withstand seismic loads from earthquakes.

It is the responsibility of the contractor to notify the building official when the construction is ready for a scheduled inspection. If all is in order, the inspector signs the permit card in the appropriate space and construction continues. If the inspector finds a code violation, it is brought to the attention of the contractor or architect for compliance.

These inspections ensure that construction is proceeding according to approved plans. They also make sure construction is meeting code requirements. This protects the future occupants of the building and the general public. In most cases, a good rapport exists between inspectors and builders, enabling construction to proceed smoothly and on schedule.

Key Terms

architect's scale
areaways
awning window
beam pocket
bearing
benchmark
bifold
blueprinting
building codes
building permits
bypass
casement window
casing
contour interval
contour lines
crawl space
details

door schedules
dormers
double-hung window
easement
elevation
finish schedules
floor plans
full sections
green space
hearth
hose bibbs
isometric
louvers
metes and bounds
modular measurement

national building codes
non-conforming variance
orthographics
partial section elevations
partition
plan view
plot plan
pocket doors
point of beginning
risers
scale
scuttle
section view
setbacks

sidelight
skylight
slab-on-grade foundation
sliding
specifications
specifications guide
specifications writer
swing
topography
treads
trusses
vapor barrier
window schedules
zones
zoning regulations

Review Questions

Select the most appropriate answer.

1. A drawing view looking from the top downward is called a(n)
   a. elevation.   c. plan.
   b. perspective.   d. section.

2. A drawing view showing a vertical cut through the construction is called a(n)
   a. elevation.   c. plan.
   b. perspective.   d. section.

3. The more commonly used scale for floor plans is
   a. ¼″ = 1′-0″.   c. 1½″ = 1′-0″.
   b. ¾″ = 1′-0″.   d. 3″ = 1′-0″.

4. The length of a line on a ¼″ = 1′-0″ scaled print that represents an actual distance of 14′-0″ is
   a. 3½″.   c. 14″.
   b. 4½″.   d. 56″.

5. The symbol "FL" written on a set of prints means
   a. floor.   c. flashing.
   b. flush.   d. all of the above.

6. To determine a dimension that is not written on a set of plans it is best to
   a. use an architect's scale to measure.
   b. calculate it.
   c. read the specifications.
   d. use the plot plan.

7. Centerlines are indicated by a
   a. series of short, uniform dashes.
   b. series of long then short dashes.
   c. long dash followed by two short dashes.
   d. solid, broad, dark line.

8. Which of the style of dimensioning below would most likely be found on a set of prints?

   a. 3′.   c. 3′-0″.
   b. 3 ft.   d. all of the above.

9. The setback of a building from the property lines would be found on a
   a. floor plan.   c. elevation drawing.
   b. plot plan.   d. foundation plan.

10. To find out which edge of a door is to be hinged, look on the
    a. elevations.   c. floor plan.
    b. specifications.   d. wall section.

11. The direction, size, and spacing of the first floor joists is found on the
    a. foundation plan. c. second floor plan.
    b. first floor plan. d. all of the above.

12. The finished floor height is usually found on the
    a. plot plan.   c. foundation plan.
    b. floor plan.   d. framing plan.

13. An exterior wall stud height can best be determined from the
    a. floor plan.   c. wall section.
    b. framing elevation. d. specifications.

14. The view of a set of prints most helpful in determining the material installed behind a brick veneer is a(n)
    a. section view.   c. foundation plan.
    b. elevation.   d. all of the above.

15. The laws that guide what type of building may be built in a particular area are called
    a. Zoning Regulations.
    b. National Building Codes.
    c. Residential Codes.
    d. Building Permits.

SECTION TWO
ROUGH CARPENTRY

Success Story

Blaine Powell

Title: Third Year Apprentice
Company: Imboden Construction, Hobart, Indiana

EDUCATION

"I've always wanted to be a carpenter," says Blaine, who recalls helping his dad fix things around the house as a kid. In his junior year of high school, Blaine entered Hammond Area Career center and took up building trades. "The program gave me a knowledge of how to use tools, but also some field experience."

HISTORY

Blaine began working with Imboden Construction, an Indiana-based home builder, in 2002, as a first-year apprentice. He's been with the company ever since. Every year, he's gained more responsibility and respect.

ON THE JOB

As an apprentice, Blaine reports in every day at 7 a.m. to the foreman in charge of the project he's working on. That person will give him a task that needs to be done that day, be it framing walls, cutting rafters or cutting stairs. Having been with the company for a few years, Blaine is relatively senior, and has had several opportunities to serve as foreman of his own job. In those cases, he's responsible for managing a crew and getting the project done.

BEST ASPECTS

"I love a challenge," says Blaine. "In this field there's always a new problem that needs to be attacked with ingenuity." His favorite task: framing roofs, because while the principles are always the same, the jobs aren't.

CHALLENGES

Blaine admits that he has very high standards, and that sometimes makes it difficult for him to accept the work of people below him. "I know how I want the job done," he says. "But I'm just learning how to get other people to do work on higher a level. It's about giving clear instructions and getting to know the strengths of your crew."

IMPORTANCE OF EDUCATION

It helps you develop a diverse skill set. "I once worked with a carpenter who'd only done cabinets and countertops, and we had first year apprentices who knew more about framing walls than he did," Blaine says. "Specialization in one field is good, but you have to have a broad knowledge of construction before you can specialize."

FUTURE OPPORTUNITIES

"I want to be the boss," jokes Blaine. The opportunities he's had to be the foreman have convinced him to work toward leadership. This year, he'll be taking the test to become a journeyman, which means he'll have even more opportunity to manage a crew.

WORDS OF ADVICE

Don't slack off. "In the union, once you get a bad name you're not called to jobs. But if your boss can trust you, he will keep you working."

UNIT 9

Building Layout

CHAPTER **25** **Leveling and Layout Tools**

CHAPTER **26** **Laying Out Foundation Lines**

Before construction begins, lines must be laid out showing the location and elevation of the building foundation. Accuracy in laying out these lines is essential in order to comply with local zoning ordinances. In addition, accurate layout lines provide for a foundation that is level and to specified dimensions. Accuracy in the beginning makes the work of the carpenter and other construction workers easier later. Layout for the location of a building and its component parts must be done properly. Failure to do so can be costly and time consuming.

OBJECTIVES

After completing this unit, the student should be able to:

- establish level points across a building area using a water level and using a carpenter's hand spirit level in combination with a straightedge.

- accurately set up and use the builder's level, transit-level, and laser level for leveling, determining and establishing elevations, and laying out angles.

- lay out building lines by using the Pythagorean theorem method for squaring corners and check the layout for accuracy.

- build batter boards and accurately establish layout lines for building using building layout instruments.

25 Leveling and Layout Tools

Building layout requires leveling lines as well as laying out various angles over the length and width of the structure. The carpenter must be able to set up, adjust, and use a variety of leveling and layout tools.

LEVELING TOOLS

Several tools, ranging from simple to state-of-the-art tools, are used to level the layout. More sophisticated leveling and layout tools, although preferred, are not always available.

Levels and Straightedges

If sophisticated leveling tools are not available, simple leveling tools may be used to level a building area. Such tools include a *carpenter's hand level* and a long straightedge. This leveling process begins at some point or stake that is at the desired elevation. Stakes are then placed across the building area to the desired distance from the starting point. This method can be an accurate, although time-consuming, method of leveling over a long distance. It can also be done by one person. Care should be taken to be sure each step is performed properly because slight errors, multiplied by each succeeding step, can grow into large ones.

Begin by selecting a length of lumber for the straightedge, sighting it carefully to make sure it is straight. It should be wide enough that it will not sag when placed on its edge and supported only on its ends. Place the straightedge on edge with one end on the first stake or a surface at the desired elevation. Drive a second stake at the other end slightly higher than level. Reposition the straightedge to the top of each stake. Place the level on top and carefully drive the second stake until level is achieved. Remember it is easier to drive the stake further than it is to raise it, so don't go too far. Recheck the levelness of the two stakes.

Continue across the building area to the desired distance by moving the straightedge one stake at a time. Use the last stake drive as the new starting stake **(Figure 25–1)**. Place the other end on another driven stake until the straightedge is again level.

If the ground is so hard that stakes are difficult to drive, a crow or shale bar may be used. This will also help to keep the stakes straight with relative ease. Continue moving the straightedge from stake to stake until the desired distance is leveled **(Figure 25–2)**.

If you want to level to the corners of building layouts, start by driving a stake near the center so its top is to the desired height. Level from the center stake to each corner in the manner described in **Procedure 25–A**.

Water Levels

A *water level* is a very accurate tool, dating back centuries. It is used for leveling from one point to another. Its accuracy, within a pencil point, is based on the principle that water seeks its own level **(Figure 25–3)**.

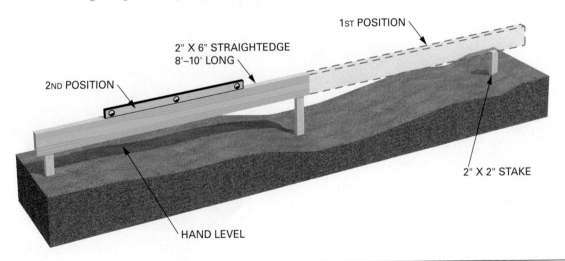

FIGURE 25–1   Leveling with a straightedge from stake to stake.

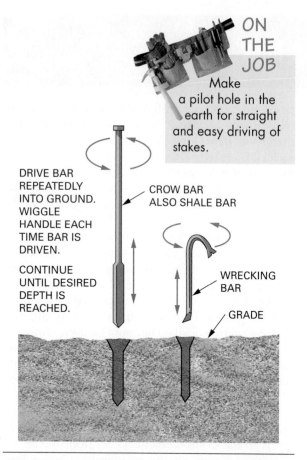

DRIVE BAR REPEATEDLY INTO GROUND. WIGGLE HANDLE EACH TIME BAR IS DRIVEN.

CONTINUE UNTIL DESIRED DEPTH IS REACHED.

CROW BAR ALSO SHALE BAR

WRECKING BAR

GRADE

FIGURE 25–2   Methods of starting a stake in hard ground.

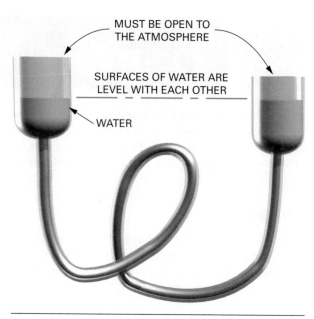

MUST BE OPEN TO THE ATMOSPHERE

SURFACES OF WATER ARE LEVEL WITH EACH OTHER

WATER

FIGURE 25–3   Water seeks its own level. Both ends of the water level must be open to the atmosphere.

One commercial model consists of 50 feet of small-diameter, clear vinyl tubing and a small tube storage container. A built-in reservoir holds the colored water that fills the tube. One end is held to the starting point. The other end is moved down until the water level is seen and marked on the surface to be leveled **(Figure 25–4)**.

PROCEDURE 25–A Leveling Corners Using a Level

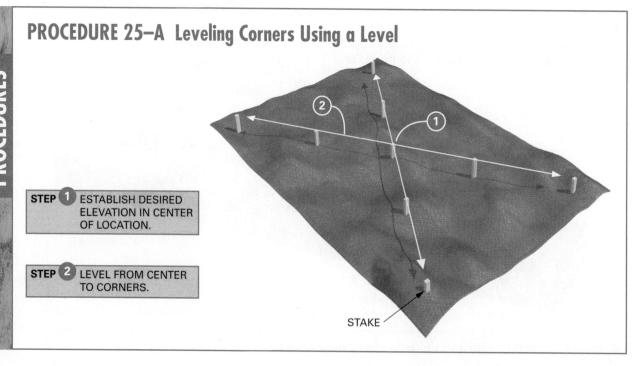

STEP 1 ESTABLISH DESIRED ELEVATION IN CENTER OF LOCATION.

STEP 2 LEVEL FROM CENTER TO CORNERS.

STAKE

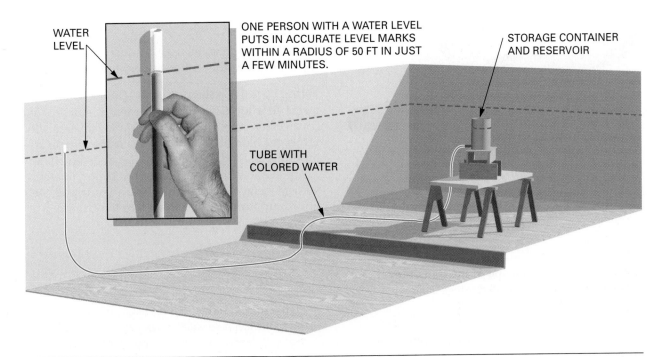

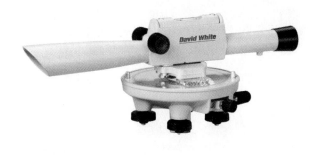

FIGURE 25–4 The water level is a simple yet effective leveling tool.

Although highly accurate, the water level is somewhat limited by the length of the plastic tube. However, extension tubings are available. Also, though slightly inconvenient, the water level may be moved from point to point.

A water level is accurate only if both ends of the tube are open to the air and there are no air bubbles in the length of the tube. Because both ends must be open, there may occasionally be some loss of liquid. However, this is replenished by the reservoir. Any air bubbles can be easily seen with the use of colored water.

In spite of these drawbacks, the water level is an extremely useful, inexpensive, simple tool for leveling from room to room, where walls obstruct views, down in a hole, or around obstructions. Another advantage is that leveling can be done by one person.

OPTICAL LEVELS

Commonly used instruments for leveling, plumbing, and angle layout are **optical levels,** which include the builder's level and transit-level.

Builder's Levels

The **builder's level (Figure 25–5)** consists of a *telescope* to which a *spirit level* is mounted. The telescope is fixed in a horizontal position. It can rotate 360 degrees for measuring horizontal angles but cannot be tilted up or down.

FIGURE 25–5 The builder's level. *(Courtesy of David White)*

Transit-Levels

The **transit-level (Figure 25–6)** is similar to the builder's level. However, its telescope can be moved up and down 45 degrees in each direction. This feature enables it to be used more effectively than the builder's level.

Automatic Levels

Automatic levels and *automatic transit-levels* **(Figure 25–7)** are similar to those previously described except that they have an internal *compensator*. This compensator uses gravity to maintain a true level line of sight. Even if the instrument is jarred, the line of sight stays true because gravity does not change.

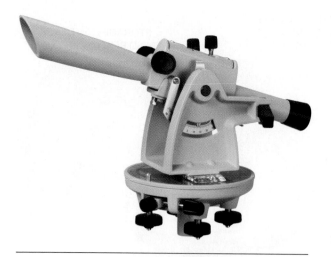

FIGURE 25–6 The telescope of the transit-level can be moved up and down 45 degrees each way. (*Courtesy of David White*)

Many models of leveling instruments are available. To become familiar with more sophisticated levels, study the manufacturers' literature. No matter what type of level is used, the basic procedures are the same.

Using Optical Levels

Before the level can be used, it must be placed on a *tripod* or some other solid support and leveled.

Setting Up and Adjusting the Level. The telescope is adjusted to a level position by means of four *leveling screws* that rest on a *base leveling plate.* In higher quality levels, the base plate is part of the instrument. In less expensive models, the base plate is part of the tripod.

Open and adjust the legs of the tripod to a convenient height. Spread the legs of the tripod well apart, and firmly place its feet into the ground.

> ⚠️ **CAUTION CAUTION CAUTION CAUTION**
>
> On a smooth surface it is essential that the points on the feet hold without slipping. Make small holes or depressions for the tripod points to fit into. Or, attach wire or light chain to the lower ends of each leg. **(Figure 25–8).** ■

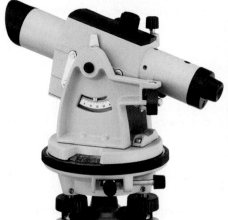

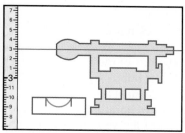

1) CONVENTIONAL INSTRUMENT CORRECTLY LEVELED. ROD READING IS 3'-3".

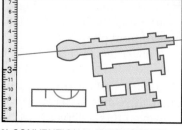

2) CONVENTIONAL INSTRUMENT SLIGHTLY OUT OF LEVEL. VIAL BUBBLE IS OFF CENTER AND INCORRECT ROD READING IS 3'-1$\frac{1}{2}$".

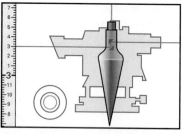

3) AUTOMATIC LEVEL-TRANSIT CORRECTLY LEVELED. ROD READING IS 3'-3".

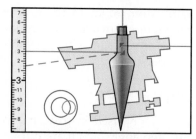

4) AUTOMATIC LEVEL-TRANSIT SLIGHTLY OUT OF LEVEL. CIRCULAR BUBBLE IS OFF CENTER, BUT THE COMPENSATOR CORRECTS FOR THE VARIATION FROM LEVEL AND MAINTAINS A CORRECT ROD READING OF 3'-3".

FIGURE 25–7 Automatic levels and automatic transit-levels level themselves when set up nearly level. (*Courtesy of David White*)

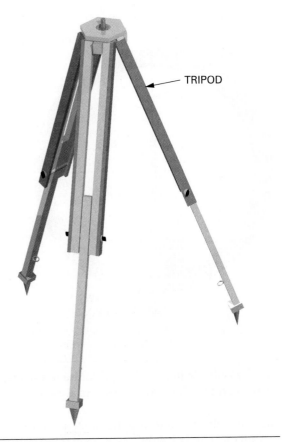

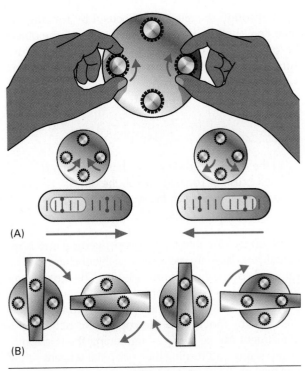

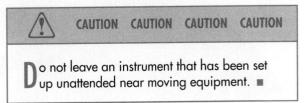

FIGURE 25–9 **(A) Level the instrument by moving thumbs toward or away from each other. (B) The instrument is level when the bubble remains centered as the telescope is revolved in a complete circle.**

FIGURE 25–8 **Make sure the feet of the tripod do not slip on smooth or hard surfaces.**

When set up, the top of the tripod should be close to level. Sight by eye and tighten the tripod wing nuts. With the top of the tripod close to level, adjustment of the instrument is made easier.

Lift the instrument from its case by the frame. Note how it is stored so it can be replaced in the case in the same position. Make sure the horizontal clamp screw is loose so the telescope revolves freely. While holding onto the frame, secure the instrument to the tripod.

Accurate leveling of the builder's level is important. Line up the telescope directly over two opposite leveling screws. Adjust these opposite screws so they are nearly snug. Back one screw off slightly to free both screws for leveling. Then turn the screws in opposite directions with forefingers and thumbs. Move the thumbs toward or away from each other, as the case may be, to center the bubble in the spirit level **(Figure 25–9)**. The bubble will always move in the same direction as your left thumb is moving.

Rotate the telescope 90 degrees over the other two opposite leveling screws and repeat the procedure. Make sure each of the screws has the same, but not too much, tension. Return to the original position, check, and make minor adjustments. Continue adjustments until the bubble remains exactly centered when the instrument is revolved in a complete circle.

⚠ CAUTION CAUTION CAUTION CAUTION

Care must be taken not to damage the instrument. Never use force on any parts of the instrument. All moving parts turn freely and easily by hand. Excessive pressure on the leveling screws may damage the threads of the base plate. Unequal tension on the screws will cause the instrument to wobble on the base plate, resulting in leveling errors. Periodically use a toothbrush dipped in light instrument oil to clean and lubricate the threads of the adjusting screws. ■

⚠ CAUTION CAUTION CAUTION CAUTION

Do not leave an instrument that has been set up unattended near moving equipment. ■

Sighting the Level. To sight an object, rotate the telescope and sight over its top, aiming it at the object.

Look through the telescope. Focus it by turning the focusing knob one way or the other, until the object becomes clear. Keep both eyes open. This eliminates squinting, does not tire the eyes, and gives the best view through the telescope.

⚠	CAUTION CAUTION CAUTION CAUTION

If the lenses need cleaning, dust them with a soft brush or rag. Do *not* rub the dirt off. Rubbing may scratch the lens coating. ■

When looking into the telescope, vertical and horizontal *cross-hairs* are seen. They enable the target to be centered properly **(Figure 25–10).** The cross-hairs themselves can be brought into focus by turning the eyepiece one way or the other. Center the cross-hairs on the object by moving the telescope left or right. A fine adjustment can be made by tightening the horizontal clamp screw and turning the horizontal tangent screw one way or the other. The horizontal cross-hair is used for reading elevations. The vertical cross-hair is used when laying out angles and aligning vertical objects.

Leveling

When the instrument is leveled, a given point on the line of sight is exactly level with any other point.

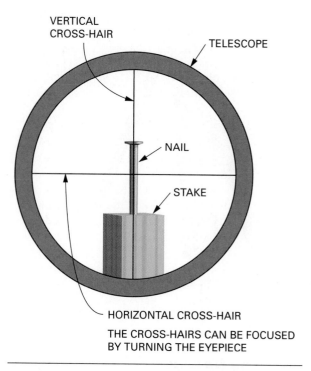

FIGURE 25–10   When looking into the telescope, vertical and horizontal cross-hairs are seen.

Any line whose points are the same distance below or above the line of sight is also level **(Figure 25–11).** To level one point with another, a helper must hold a *target* on the point to be leveled. A reading is taken.

FIGURE 25–11   Any line parallel to the established level line is also level.

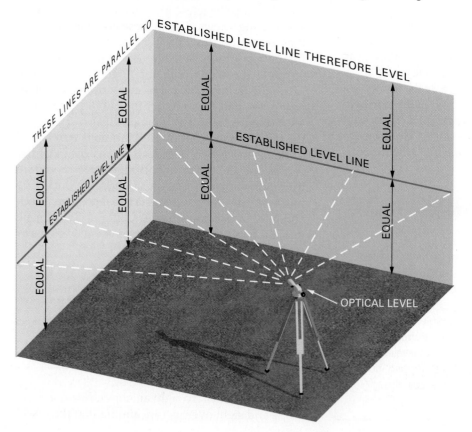

The target is then moved to selected points that are brought to the same elevation by moving those points up or down to get the same reading.

Targets. A tape is often used as a target. The end of the tape is placed on the point to be leveled. The tape is then moved up or down until the same mark is read on the tape as was read at the starting point.

Because of its flexibility, the tape may need to be backed up by a strip of wood to hold it rigid (**Figure 25–12**).

The simplest target is a plain 1 × 2 strip of wood. The end of the stick is held on the starting point of desired elevation. The line of sight is marked on the stick. The end of the stick is then placed on top of various points. They are moved up or down to bring the mark to the same height as the line of sight (**Procedure 25–B**). A stick of practically any length can be used.

Cut the stick to a length so that the mark to be sighted is a noticeable distance off from the center of its length. It is then immediately noticeable if the stick is inadvertently turned upside down (**Figure 25–13**).

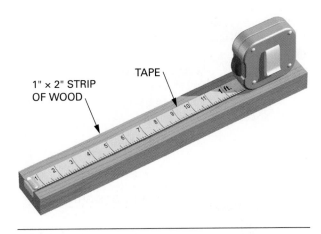

1" × 2" STRIP OF WOOD   TAPE

FIGURE 25–12   A tape can be backed up by a strip of wood to make it a stiff and steady target.

If, for some reason, it is not desirable to cut the stick, clearly mark the top and bottom ends.

Leveling Rods. For longer sightings, the *leveling rod* is used because of its clearer graduations. A variety of rods are manufactured of wood or fiberglass for

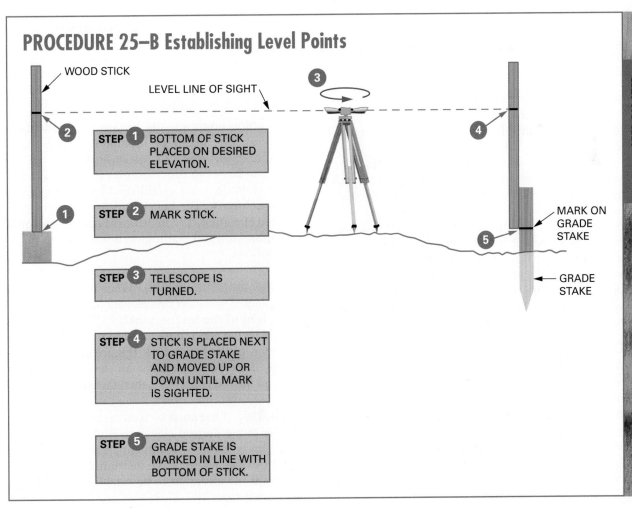

PROCEDURE 25–B Establishing Level Points

WOOD STICK

LEVEL LINE OF SIGHT

STEP 1 BOTTOM OF STICK PLACED ON DESIRED ELEVATION.

STEP 2 MARK STICK.

STEP 3 TELESCOPE IS TURNED.

STEP 4 STICK IS PLACED NEXT TO GRADE STAKE AND MOVED UP OR DOWN UNTIL MARK IS SIGHTED.

STEP 5 GRADE STAKE IS MARKED IN LINE WITH BOTTOM OF STICK.

MARK ON GRADE STAKE

GRADE STAKE

STEP BY STEP PROCEDURES

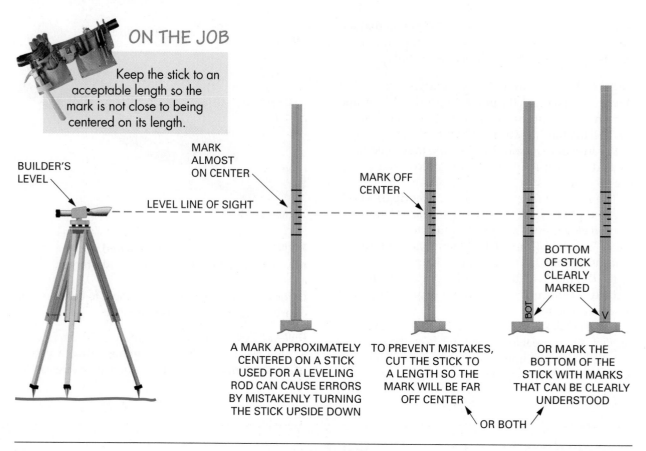

ON THE JOB

Keep the stick to an acceptable length so the mark is not close to being centered on its length.

BUILDER'S LEVEL

MARK ALMOST ON CENTER

LEVEL LINE OF SIGHT

MARK OFF CENTER

BOTTOM OF STICK CLEARLY MARKED

A MARK APPROXIMATELY CENTERED ON A STICK USED FOR A LEVELING ROD CAN CAUSE ERRORS BY MISTAKENLY TURNING THE STICK UPSIDE DOWN

TO PREVENT MISTAKES, CUT THE STICK TO A LENGTH SO THE MARK WILL BE FAR OFF CENTER

OR MARK THE BOTTOM OF THE STICK WITH MARKS THAT CAN BE CLEARLY UNDERSTOOD

OR BOTH

FIGURE 25–13   Techniques for creating an easy-to-use marking stick.

several leveling purposes. They are made with two or more sections that extend easily and lock into place. Rods vary in length—from two-section rods extending 9'-0" up to seven-section rods extending 25'-0".

The builder's rod has feet, inches, and 8ths of an inch. The graduations are ⅛ inch wide and ⅛ inch apart. The engineer's rod is very similar yet the scale is slightly different. It is in feet, tenths, and hundredths of a foot. Instead of inches, the number markings represent a tenth of a foot. The smaller graduations are 1/100 of a foot wide and 1/100 foot apart, which is slightly smaller than ⅛ inch. They are both designed for easy reading. An oval-shaped, red and white, movable target is available to fit on any rod for easy reading **(Figure 25–14).**

Communication. A responsible rod operator holds the rod vertical and faces the instrument so it can be read with ease and accuracy. Sighting distances are not usually over 100 to 150 feet, yet sometimes voice commands cannot be used. Hand signals are then given to the rod operator to move the target as desired by the instrument operator. Usually appropriate hand signals are given even when distances are not great. Shouting on the job site is unnecessary, unprofessional, and creates confusion.

Establishing Elevations

Many points on the job site, such as the depth of excavations, the height of foundation footing and walls, and the elevation of finish floors, are required to be set at specified elevations or grades. These elevations are established by starting from the **benchmark.** The benchmark is a point of designated elevation. The instrument operator records elevations and rod readings in a notebook to make calculations.

Height of the Instrument (HI). When it is necessary to set a point at some definite elevation, first determine the **height of the instrument** (HI). To find HI, place the rod on the benchmark and add the reading to the elevation of the benchmark **(Figure 25–15).** For instance, if the benchmark has an elevation of 100.00 feet and the rod reads 5'-8", then the HI is 105'-8".

Grade Rod. What must be read on the rod when its base is at the desired elevation is called the **grade rod.** This is found by subtracting the desired elevation from the height of the instrument (HI). For instance, if the elevation to be established is 102'-0", subtract it from 105'-8" (HI) to get 3'-8" (the grade

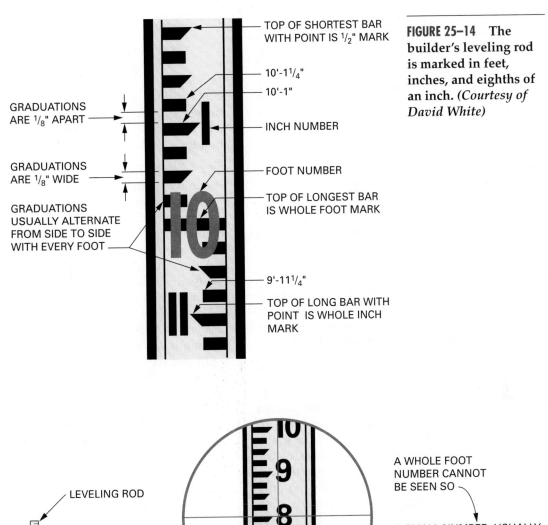

TOP OF SHORTEST BAR WITH POINT IS ¹/₂" MARK

10'-1¹/₄"

10'-1"

INCH NUMBER

GRADUATIONS ARE ¹/₈" APART

FOOT NUMBER

GRADUATIONS ARE ¹/₈" WIDE

TOP OF LONGEST BAR IS WHOLE FOOT MARK

GRADUATIONS USUALLY ALTERNATE FROM SIDE TO SIDE WITH EVERY FOOT

9'-11¹/₄"

TOP OF LONG BAR WITH POINT IS WHOLE INCH MARK

FIGURE 25–14 The builder's leveling rod is marked in feet, inches, and eighths of an inch. (*Courtesy of David White*)

LEVELING ROD

A WHOLE FOOT NUMBER CANNOT BE SEEN SO

A SMALL NUMBER, USUALLY BETWEEN THE 3 AND 4 AND THE 6 AND 7 INCH MARK, INDICATES THE NUMBER OF FEET

THE ROD READING IS 5' - 8"

SEE ENLARGEMENT

BUILDER'S LEVEL

5' - 8"

READING ON ROD

LEVEL LINE OF SIGHT

ADD THE ELEVATION OF THE BENCHMARK → 100' - 0"
TO THE ROD READING → 5' - 8"
TO GET THE HEIGHT OF THE INSTRUMENT → 105' - 8"

BENCHMARK

BM 100' - 0"

FIGURE 25–15 Determining the height of the instrument (HI).

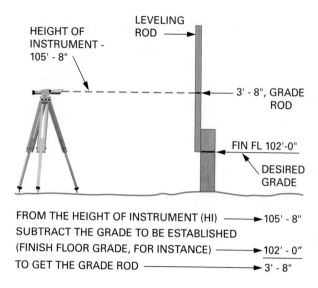

HEIGHT OF INSTRUMENT - 105' - 8"

LEVELING ROD

3' - 8", GRADE ROD

FIN FL 102'-0"

DESIRED GRADE

FROM THE HEIGHT OF INSTRUMENT (HI) ⟶ 105' - 8"
SUBTRACT THE GRADE TO BE ESTABLISHED
(FINISH FLOOR GRADE, FOR INSTANCE) ⟶ 102' - 0"
TO GET THE GRADE ROD ⟶ 3' - 8"

THE "GRADE ROD" IS WHAT THE ROD MUST READ WHEN ITS BASE IS AT THE DESIRED GRADE OR HEIGHT

FIGURE 25–16 **Calculating the grade rod and establishing a desired elevation.**

rod). The rod operator places the rod at the desired point. He or she then moves it up or down, at the direction of the instrument operator, until the grade rod of 3'-8" is read on the rod. The base of the builder's rod is then at the desired elevation **(Figure 25–16).** A mark, drawn at the base of the rod on a stake or other object, establishes the elevation.

Determining Differences in Elevation

Differences in elevation need to be determined for such tasks as grading driveways, sidewalks, and parking areas, laying out drainage ditches, plotting contour lines, and estimating cut and fill requirements. The difference in elevation of two or more points is easily determined with the use of the builder's level or transit-level.

Single Setup. To find the difference in elevation between two points, set up the instrument about midway between them. Place the rod on the first point. Take a reading, and record it. Swing the level to the other point, take a reading, and record. The difference in elevation is the difference between the recordings.

When making many readings, keeping track of the readings becomes more difficult. Using the surveying technique of tracking *backsight* and *foresight* makes this easier. Backsight is the reading from a level to a known or previously measured point. Foresight is a reading to a new location. All backsights are *plus* (+) *sights* and all foresights are *minus* (−) *sights*. These are recorded on a table for each setup of the transit or level **(Figure 25–17).**

Place the rod on point A and record the backsight reading as a plus (+) sight. Place the rod on point B, take the foresight reading, and record as a minus (−) sight. Add the plus and minus sights to get the difference in elevation.

Sometimes a reading is made above the level line of the transit. To keep this straight a rule is applied:

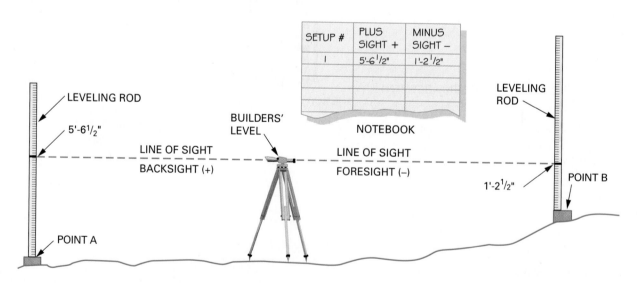

SETUP #	PLUS SIGHT +	MINUS SIGHT −
1	5'-6 1/2"	1'-2 1/2"

NOTEBOOK

LEVELING ROD

5'-6 1/2"

BUILDERS' LEVEL

LEVELING ROD

LINE OF SIGHT
BACKSIGHT (+)

LINE OF SIGHT
FORESIGHT (−)

1'-2 1/2"

POINT B

POINT A

THE DIFFERENCE BETWEEN THE PLUS SIGHT AND MINUS SIGHT IS 4'-4".
THE PLUS SIGHT IS LARGER, SO POINT B IS 4'-4" HIGHER THAN POINT A.

FIGURE 25–17 **Determining a difference of elevation between two points requiring only one setup.**

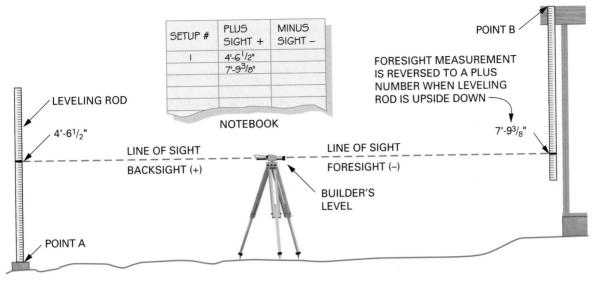

SETUP #	PLUS SIGHT +	MINUS SIGHT −
I	4'-6¹/₂"	
	7'-9³/₈"	

NOTEBOOK

A MINUS SIGHT BECOMES A PLUS SIGHT IF THE ROD IS READ UPSIDE DOWN.
POINT B IS 12'-3⁷/₈" HIGHER THAN POINT A.

FIGURE 25–18   For any readings taken with the rod upside down, the plus and minus signs of the sighting measurements are reversed.

For any backsight or foresight reading where the rod is flipped upside down, the plus or minus sign is reversed **(Figure 25–18)**.

Multiple Setup. Sometimes the difference in elevation of two points is too great or the distance is too far apart. Then it is necessary to make more than one instrument setup to determine the difference. The procedure is similar to a series of one setup operations until the final point is reached.

CAUTION   CAUTION   CAUTION   CAUTION

When carrying a tripod-mounted instrument, handle with care. Carry it in an upright position. Do not carry it over the shoulder or in a horizontal position. Be careful when going through buildings or close quarters not to bump the instrument. ■

Record all backsights as plus sights and all foresights as minus sights, unless the rod is upside down when read **(Figure 25–19)**. Find the sum of all minus sights and all plus sights. The difference between them is the difference in elevation of the beginning and ending points. If the sum of the plus sights is larger, then the end point is higher than the starting point. If the sum of the minus sights is larger, then the end point is lower than the starting point.

Measuring and Laying Out Angles

To measure or lay out angles, the instrument must be set over a particular point on the ground. A hook, centered below the instrument, is provided for suspending a plumb bob. The plumb bob is used to place the level directly over this point. In more sophisticated instruments, a built-in *optical plumb* allows the operator to sight to a point below, exactly plumb with the center of the instrument. This enables quick and accurate setups over a point **(Figure 25–20)**.

Setting Up over a Point. Suspend the plumb bob from the instrument. Secure it with a slip knot. Move the tripod and instrument so that the plumb bob appears to be over the point.

Press the legs of the tripod into the ground. Lower the plumb bob by moving the slip knot until it is about ¼ inch above the point on the ground. The final centering of the instrument can be made by loosening any two adjacent leveling screws and slowly shifting the instrument until the plumb bob is directly over the point **(Figure 25–21)**. Retighten the same two leveling screws that were previously loosened, and level the instrument. Shift the instrument on the base plate until the plumb bob is directly over

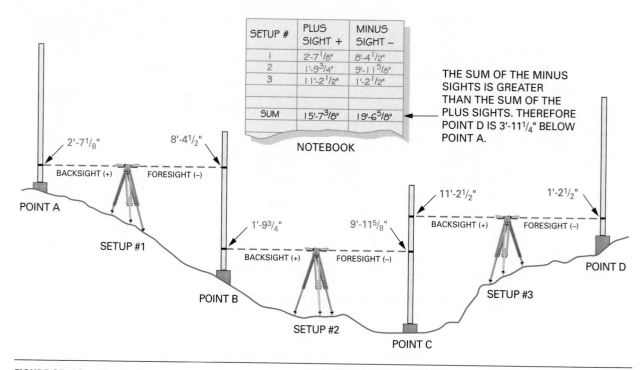

SETUP #	PLUS SIGHT +	MINUS SIGHT –
1	2'-7$\frac{1}{8}$"	8'-4$\frac{1}{2}$"
2	1'-9$\frac{3}{4}$"	9'-11$\frac{5}{8}$"
3	11'-2$\frac{1}{2}$"	1'-2$\frac{1}{2}$"
SUM	15'-7$\frac{3}{8}$"	19'-6$\frac{5}{8}$"

NOTEBOOK

THE SUM OF THE MINUS SIGHTS IS GREATER THAN THE SUM OF THE PLUS SIGHTS. THEREFORE POINT D IS 3'-11$\frac{1}{4}$" BELOW POINT A.

FIGURE 25–19   Determining a difference of elevation between two points requiring more than one setup.

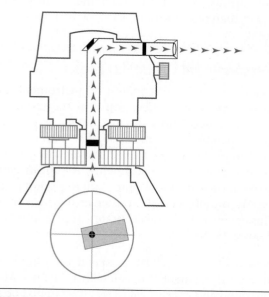

FIGURE 25–20   Some instruments have a device called an optical plumb for setting the instrument directly over a point. (*Courtesy of David White*)

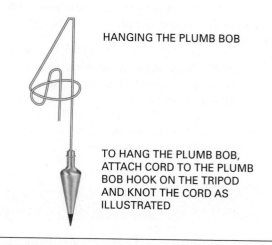

HANGING THE PLUMB BOB

TO HANG THE PLUMB BOB, ATTACH CORD TO THE PLUMB BOB HOOK ON THE TRIPOD AND KNOT THE CORD AS ILLUSTRATED

FIGURE 25–21   To locate the instrument directly over a point, a plumb bob is suspended from the level.

the point. Check the levelness of the instrument. Adjust, if necessary.

Circle Scale and Index.

A **horizontal circle scale** (outside ring) is divided into 90-degree quadrants. A pointer, or *index,* turns with the telescope. The circle scale remains stationary and indicates the number of degrees the telescope is turned. When desired, the horizontal circle may be turned by hand for setting to zero degrees, no matter which way the telescope is pointing. By starting at zero and rotating the telescope on it, any horizontal angle can be easily measured **(Figure 25–22)**.

Reading the Horizontal Vernier.

For more precise readings, the **horizontal vernier** is used to read minutes of a degree **(Figure 25–23)**. A vernier is a smaller scale used to make more precise measurements when the zero index, the point where the reading is made, falls

THE HORIZONTAL CIRCLE SCALE IS DIVIDED INTO QUADRANTS OF 90° EACH, AND REMAINS STATIONARY AS THE TELESCOPE IS TURNED.

IT IS GRADUATED IN DEGREES AND NUMBERED EVERY 10 DEGREES.

IT MAY BE ROTATED BY HAND TO ADJUST THE FIRST READING TO ZERO.

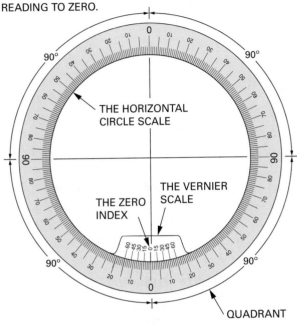

FIGURE 25–22 Reading the horizontal circle scale.

THE VERNIER ROTATES WITH THE TELESCOPE

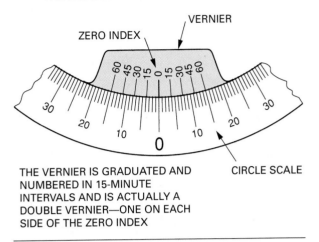

THE VERNIER IS GRADUATED AND NUMBERED IN 15-MINUTE INTERVALS AND IS ACTUALLY A DOUBLE VERNIER—ONE ON EACH SIDE OF THE ZERO INDEX

FIGURE 25–23 The vernier scale is used for reading minutes of a degree.

between the lines of the circle scale. The vernier is actually two verniers, one on each side of the vernier zero index. This makes it possible to read any angle, whether turned to the right or to the left.

The vernier scale turns with the telescope. If the eye end of the telescope is turned to the left (clockwise), the vernier scale on the left side of the zero index is used. If the eye end of the telescope is turned to the right (counterclockwise), the vernier scale on the right side of the zero index is used. Use either the left or right vernier scale according to the direction in which the eye end of the telescope is turned when measuring or laying out angles **(Figure 25–24).**

To read the vernier scale, first lock the transit at the desired position. Read the degrees where the zero index lines up with the circle scale. Read the smaller of the two numbers **(Procedure 25–C).** In this case the measurement is 75 plus degrees to the right. Therefore the right vernier is used. Now locate the vernier line that happens to line up best with the larger circle scale lines. In this case, the 45-minute line is best aligned with a degree line, thus the measurement is 75 degrees, 45 minutes.

Measuring a Horizontal Angle. After leveling the instrument over the point of an angle, called its *vertex,* loosen the horizontal clamp screw. Rotate the

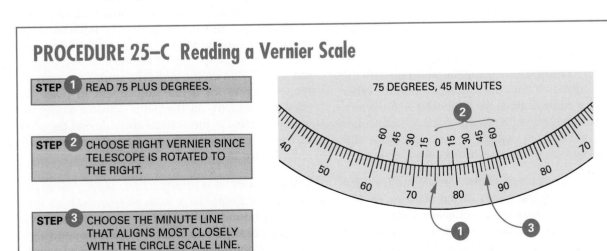

PROCEDURE 25–C Reading a Vernier Scale

STEP ❶ READ 75 PLUS DEGREES.

STEP ❷ CHOOSE RIGHT VERNIER SINCE TELESCOPE IS ROTATED TO THE RIGHT.

STEP ❸ CHOOSE THE MINUTE LINE THAT ALIGNS MOST CLOSELY WITH THE CIRCLE SCALE LINE.

75 DEGREES, 45 MINUTES

STEP BY STEP
PROCEDURES

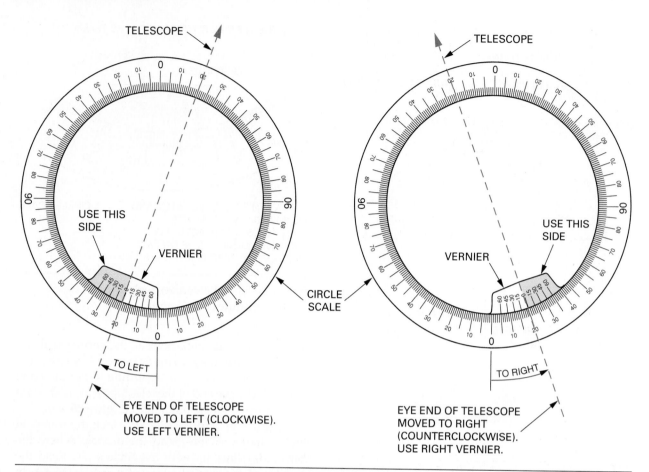

FIGURE 25–24 Method of determining which vernier scale to read.

instrument until the vertical cross-hair is nearly in line with a distant point on one side of the angle. Tighten the clamp screw. Then turn the tangent screw to line up the vertical cross-hair exactly with the point.

If the point is above or below the line of sight, and a transit-level is not available, sight a straightedge held plumb from the point. Or, sight the line of a plumb bob line suspended over the point. If using a transit-level, release the locking lever and tilt the telescope to sight the point.

By hand, turn the horizontal circle scale to zero. Loosen the clamp screw. Swing the telescope until the vertical cross-hair lines up with a point on the other side of the angle. Tighten the horizontal clamp screw. Then turn the tangent screw for a fine adjustment, if necessary. Read the degrees on the circle scale and minutes on the vernier scale **(Procedure 25–D)**.

Laying Out a Horizontal Angle. Center and level the instrument over the vertex of the angle to be laid out. Sight the telescope on a distant point on one side of the angle. Tighten the horizontal clamp, and set the circle scale to zero. Loosen the clamp. Then

turn the telescope until close to the desired number of degrees. Tighten the clamp. Use the tangent screw and make a fine adjustment until the index reads exactly the desired number of degrees. Sight the vertical cross-hair to lay out the other side of the angle **(Procedure 25–E)**.

Measuring Vertical Angles. The **vertical arc** scale is attached to the telescope. It measures vertical angles to 45 degrees above and below the horizontal. By tilting and rotating the telescope, set the horizontal cross-hair on the points of the vertical angle being measured. Tighten the vertical clamp. Then turn the tangent screw for a fine adjustment to place the cross-hair exactly on the point. Vertical angles are read by means of the vertical arc scale and the obvious vernier similar to the reading of horizontal angles **(Figure 25–25)**.

Setting Points in a Straight Line. It may be necessary to set points in a straight line, such as on a property boundary line. Set up the instrument over the one point. Rotate and tilt the telescope to sight the other point fairly close. Tighten the horizontal

PROCEDURE 25–D Using a Transit to Measure Horizontal Angles

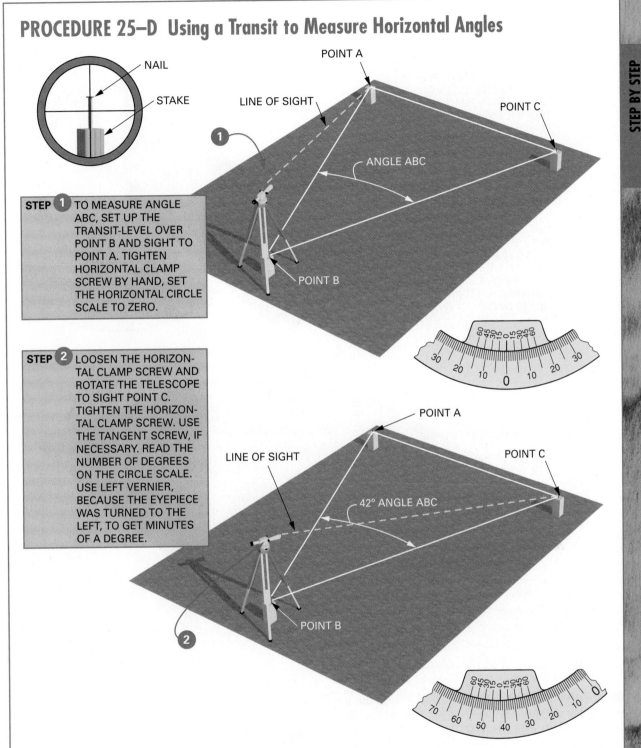

**STEP ① ** TO MEASURE ANGLE ABC, SET UP THE TRANSIT-LEVEL OVER POINT B AND SIGHT TO POINT A. TIGHTEN HORIZONTAL CLAMP SCREW BY HAND, SET THE HORIZONTAL CIRCLE SCALE TO ZERO.

**STEP ② ** LOOSEN THE HORIZONTAL CLAMP SCREW AND ROTATE THE TELESCOPE TO SIGHT POINT C. TIGHTEN THE HORIZONTAL CLAMP SCREW. USE THE TANGENT SCREW, IF NECESSARY. READ THE NUMBER OF DEGREES ON THE CIRCLE SCALE. USE LEFT VERNIER, BECAUSE THE EYEPIECE WAS TURNED TO THE LEFT, TO GET MINUTES OF A DEGREE.

clamp. Then turn the tangent screw until the vertical cross-hair is exactly on the far point. With the horizontal clamp tight, release the lock levers and depress the telescope, as required, to sight points between the corners and along the boundary line **(Procedure 25–F).**

If it is necessary to continue in a straight line beyond the far point, move and set up the instrument over the far point, point F. Sight back to the first point. Tighten the horizontal clamp. Then turn the horizontal circle scale to zero. Loosen the horizontal clamp. Turn the telescope 180 degrees, and tighten

PROCEDURE 25–E Using a Transit to Lay Out a 90-Degree Angle

STEP 1 TO LAY OUT A 90° ANGLE FROM LINE EF, SET UP THE TRANSIT-LEVEL OVER POINT E AND SIGHT TO POINT F. TIGHTEN THE HORIZONTAL CLAMP SCREW. TURN THE HORIZONTAL CIRCLE SCALE TO ZERO.

STEP 2 LOOSEN THE HORIZONTAL CLAMP SCREW AND ROTATE THE TELESCOPE UNTIL 90° IS READ ON THE HORIZONTAL CIRCLE SCALE. TIGHTEN THE HORIZONTAL CLAMP SCREW. DRIVE A STAKE WITH A NAIL CENTERED IN ITS END TO SIGHT POINT D. AN ANGLE OF 90° IS LAID OUT.

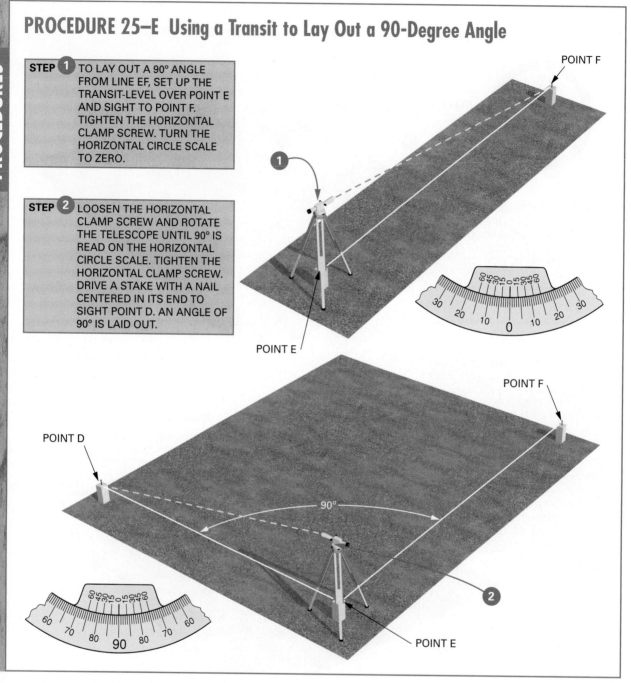

the horizontal clamp again. Depress or raise the telescope as needed to sight additional points in a straight line **(Procedure 25–G)**.

Storing the Instrument. If the instrument gets wet, dry it before returning it to its case. Keep it in its carrying case when it is not being used or when being transported in a vehicle over long distances.

LASER LEVELS

A **laser** is a device in which light energy is released in a narrow beam. The light beam is absolutely straight. Unless interrupted by an obstruction or otherwise disturbed, the light beam can be seen for a long distance.

Lasers have many applications in space, medicine, agriculture, and engineering. The **laser level** has been developed for the construction industry to provide more efficient layout work **(Figure 25–26)**.

Kinds and Uses of Laser Levels

Several manufacturers make laser levels in a number of different models. The least expensive models are the least sophisticated. A low-price unit is lev-

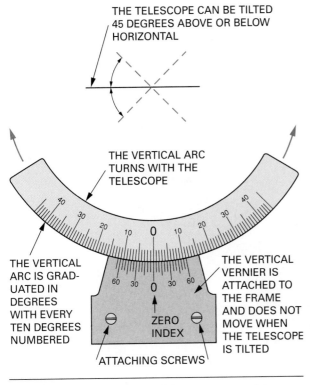

THE TELESCOPE CAN BE TILTED 45 DEGREES ABOVE OR BELOW HORIZONTAL

THE VERTICAL ARC TURNS WITH THE TELESCOPE

THE VERTICAL ARC IS GRADUATED IN DEGREES WITH EVERY TEN DEGREES NUMBERED

THE VERTICAL VERNIER IS ATTACHED TO THE FRAME AND DOES NOT MOVE WHEN THE TELESCOPE IS TILTED

ZERO INDEX

ATTACHING SCREWS

FIGURE 25–25 Vertical angles are measured by using the vertical arc and vernier scales.

eled and adjusted manually. More expensive ones are automatically adjusted to and maintained in level. Power sources include batteries, a rechargeable battery pack, or an AC/DC converter for 110 or 220 volts.

FIGURE 25–26 Laser levels have been developed for use in the construction industry. *(Courtesy of Trimble)*

PROCEDURE 25–F Using a Transit to Set Points in a Straight Line

STEP 1 TO SET POINTS IN A STRAIGHT LINE BETWEEN POINTS A AND F, FIRST SET UP THE TRANSIT-LEVEL OVER POINT A. SIGHT TO POINT F AND TIGHTEN THE HORIZONTAL CLAMP SCREW.

STEP 2 DEPRESS THE TELESCOPE TO SET POINTS AT B, C, D, AND E.

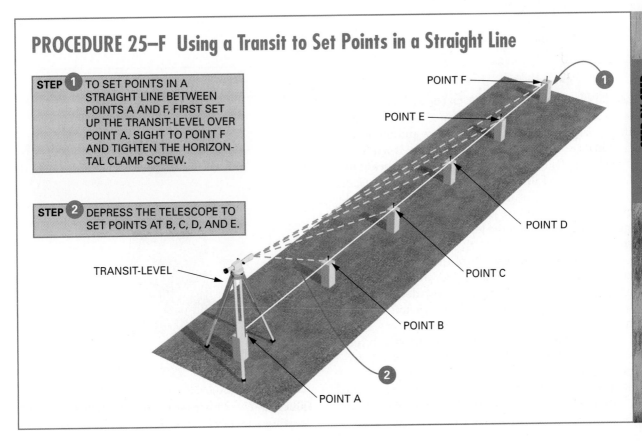

POINT F

POINT E

POINT D

POINT C

TRANSIT-LEVEL

POINT B

POINT A

PROCEDURE 25-G Using a Transit to Set Many Straight Points

STEP 1 TO CONTINUE POINTS IN A STRAIGHT LINE: SET THE TRANSIT-LEVEL OVER POINT F, THE FARTHEST POINT SIGHTED FROM POINT A.

STEP 2 SIGHT BACK TO POINT A. TIGHTEN HORIZONTAL CLAMP. TURN CIRCLE SCALE TO ZERO.

STEP 3 LOOSEN HORIZONTAL CLAMP AND TURN TELESCOPE 180 DEGREES TO SIGHT POINT L. DEPRESS TELESCOPE TO SIGHT INTERMEDIATE POINTS G, H, I, J, AND K.

POINT L

POINT F

POINT A

Establishing and Determining Elevations. A simple, easy-to-use model is mounted on a tripod or solid flat surface and leveled like manual or automatic optical instruments. The laser is turned on. It will emit a red beam, usually ⅜ inch in diameter. The beam rotates through a full 360 degrees, creating a level *plane* of light. As it rotates, it establishes equal points of elevation over the entire job site, similar to a line of sight being rotated by the telescope of an optical instrument **(Figure 25–27)**.

Depending on the quality of the instrument, the laser head may rotate at various revolutions per second (RPS), up to 40 RPS. Its quality also determines its working range. This may vary from a 75- to 1,000-foot radius.

Laser beams are difficult to see outdoors in bright sunlight. To detect the beam, a battery-powered electronic *sensor* target, also called a *receiver* or *detector,* is attached to the leveling rod or stick. Most sensors have a visual display with a selectable audio tone to indicate when it is close to or on the beam **(Figure 25–28)**.

FIGURE 25–27 The laser beam rotates 360 degrees, creating a level plane of light.

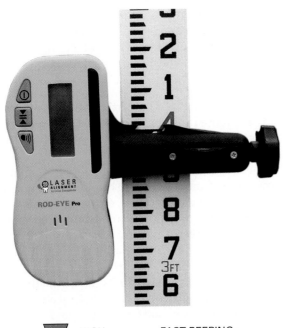

HIGH .... FAST BEEPING

ON-GRADE .... SOLID TONE

LOW .... SLOW BEEPING

FIGURE 25–28 An electronic target senses the laser beam. An audio feature provides tones to match the visual display. *(Courtesy of Leica Geosystems)*

In addition to electronic sensor targets, specially designed targets are used for interior work, such as installing ceiling grids and leveling floors.

The procedures for establishing and determining elevations with laser levels are similar to those with optical instruments. To establish elevations, the sensor is attached to the grade rod. The grade rod is moved up or down until the beam indicates that the base of the rod is at grade.

One of the great advantages of using laser levels is that, in most cases, only one person is needed to do the operations **(Figure 25–29)**. Another advantage is that certain operations are accomplished more easily in less time.

Special Horizontal Operations. For leveling *suspended ceiling grids,* the laser level is mounted to an adjustable grid mount bracket. Once the first strip of angle trim has been installed, the laser and bracket can be attached easily and clamped into place.

The unit is leveled. The height of the laser beam is then adjusted for use with a special type of magnetic or clip-on target. Magnetic targets are used on steel grids. Clip-on targets are used on aluminum or plastic. Once the laser is set up, the ceiling grid is quickly and easily leveled using the rotating beam as a ref-

FIGURE 25–29 When using laser levels, only one person is required for leveling operations. *(Courtesy of Leica Geosystems)*

erence and viewing the beam through the target **(Figure 25–30).** Sprinkler heads, ceiling outlets, and similar objects can be set in a similar manner.

Special Vertical Operations. Most laser levels are designed to work laying on their side using special brackets or feet. The unit is placed on the floor and leveled. To lay out a partition, rotate the head of the laser downward. Position the laser beam over one end of the partition. Align the beam toward the far point by turning the head toward and adjusting to the second point. Recheck the alignment and turn on to rotate the beam. The beam will be displayed as continuous straight and plumb lines on the floor, ceiling, and walls to both align and plumb the partition at the

FIGURE 25–30 Leveling suspended ceiling grids. The beam is viewed through the target. *(Courtesy of Leica Geosystems)*

FIGURE 25–31   Placed on its side, the laser level is used to lay out and align partitions and walls. (*Courtesy of Leica Geosystems*)

same time **(Figure 25–31).** Mounted on excavation equipment, lasers simplify and speed up construction work **(Figure 25–32).**

Layout Operations. Some laser levels emit a plumb line of light projecting upward from the top at a right angle (90 degrees) to the plane of the rotating

FIGURE 25–32   Attaching a laser detector to a bulldozer allows operator to level the ground quickly. (*Courtesy of Leica Geosystems*)

beam. The plumb reference beam allows one person to lay out 90-degree cross-walls or building lines. Set the unit on its side as for laying out a partition. The reference beam establishes a 90-degree corner with the rotating laser beam.

Plumbing Operations. The vertical beam provides a ready reference for plumbing posts, columns, elevator shafts, slip forming, and wherever a plumb reference beam is required.

Mount the laser unit on a tripod over a point of known offset from the work to be plumbed. Suspend a plumb bob from the center of the tripod directly over the point. Apply power to the unit. The vertical beam that is projected is ready for use as a reference. Move the top of the object until it is offset from the beam the same distance as the bottom point.

The heads on some laser units can be tilted so the rotating beam produces a plane of light at an angle to the horizontal. Such units are used for laying out slopes. Other units are manufactured for special purposes such as pipelaying and tunnel guidance. Marine laser units, with ranges up to 10 miles, are used in port and pier construction and offshore work.

Laser Safety

With a little common sense, the laser can be used safely. All laser instruments are required to have

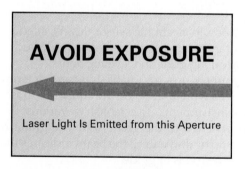

APERTURE LABEL

WARNING LABEL

FIGURE 25–33   Warning labels must be attached to every laser instrument.

warning labels attached **(Figure 25–33)**. The following are safety precautions for laser use:

- Only trained persons should set up and operate laser instruments.

- Never stare directly into the laser beam or view it with optical instruments.

- When possible, set the laser up so it is above or below eye level.

- Turn the laser off when not in use.

- Do not point the laser at others.

26 Laying Out Foundation Lines

Before any layout can be made, the builder must determine the dimensions of the building and its location on the site from the plot plan. This task often falls to a professional surveyor, but in many areas may be done by any qualified person. Care should be taken to position the building properly to avoid costly changes. It is usually the carpenter's responsibility to lay out building lines.

STAKING THE BUILDING

Proper layout begins with locating the property corners, then placing stakes at each corner of the building. Some lots are large enough that the building is only measured from one property line, while other lots are small enough to make it necessary to check all property lines with the building **(Figure 26–1)**.

Begin by finding the survey rods that mark the corners of the property. Do not guess where the property lines are. Sometimes it is a good idea to

stretch and secure lines between each corner, laying out all the property boundary lines.

Locate the front building line by measuring in from the front property line the specified front setback. Measure from both ends of the building line and drive a stake at each end. Stretch a line between these stakes to better show the front edge of the building **(Figure 26–2)**.

Along the front building line, measure in from the side property line the specified side setback. Drive a stake, Stake A, firmly into the ground. Place a nail in the top of the stake to aid in making precise measurements **(Figure 26–3)**. From this nail, measure the front dimension of the building along the front building line. Drive Stake B directly under the front building line string. Drive a nail in the top of the stake marking the exact length of the building **(Figure 26–4)**.

The third stake, Stake C, is placed to locate the back corner of the building. It must be square or at a

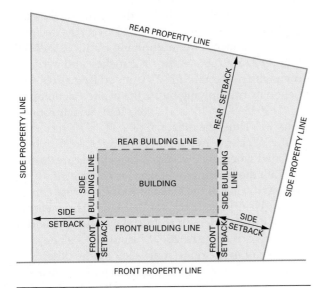

FIGURE 26–1 Locating the building on a lot from the dimensions found on the plot plan.

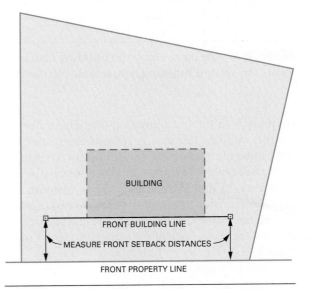

FIGURE 26–2 Locating the front building line.

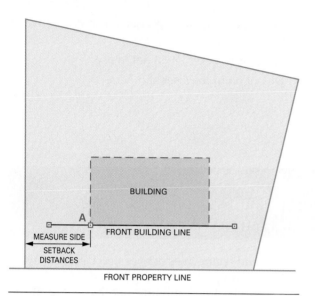

FIGURE 26–3 Measure from the side property line to locate the first building corner stake, Stake A.

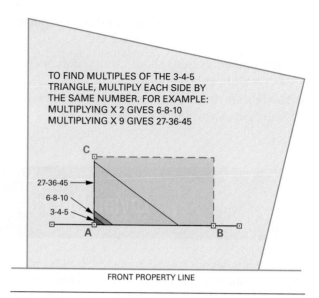

FIGURE 26–5 Use multiples of the 3-4-5 right triangle to place the third building corner stake, Stake C.

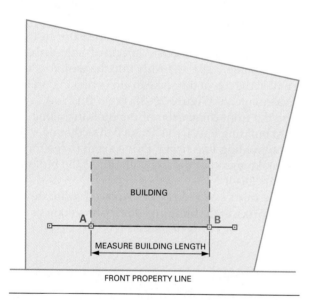

FIGURE 26–4 Measure along the building line to locate the second building corner stake, Stake B.

right angle to the front building line. This stake on the rear building line may be located using one of at least three methods. First, you could use an optical instrument such as a transit-level. It is set up directly over Stake A and sighted to Stake B, aligning the cross-hairs. Then the telescope is rotated exactly 90 degrees. Stake C is located by using a tape measure and the cross-hairs at the same time. (See **Procedure 25–G**). The second method is to use the 3-4-5 method. This process uses multiples of 3-4-5 to create larger right triangles. Multiplying each side of a

3-4-5 triangle by the same number creates a larger triangle that also has a right angle. For example, 3-4-5 multiplied by 9 gives a 27-36-45 right triangle **(Figure 26–5)**.

The third method is faster and more accurate. Two tapes are used to measure the building width from stake A and the **diagonal** of the building from stake B at the same time. To determine the diagonal of the building the **Pythagorean theorem,** $a^2 + b^2 = c^2$, is used. For example, consider a building whose length $a = 40'$ and width $b = 32'$. The diagonal equals c. Using the Pythagorean theorem, $c^2 = 32^2 + 40^2 = 2,624$. Taking the square root of the diagonal C gives us 51.2249939.

To convert to feet-inches to the nearest ¹⁄₁₆th, subtract 51 feet to leave the decimal. Convert the decimal 0.2249939 to inches by multiplying by 12: 0.2249939' × 12 = 2.6999268". Subtract 2 inches and write it down with 51 to make 51'-2". Convert 0.6999268" to a fraction by multiplying by 16, the desired denominator: 0.6999268" × 16 = 11.1988/16th, which rounds off to ¹¹⁄₁₆". Thus, the diagonal of a 32' × 40' rectangle is 51'-2¹¹⁄₁₆". Using two tapes, position Stake C, which is located where the two tapes cross in **Figure 26–6**.

When the locating of Stake C is completed, drive a nail in the top of the stake marking the rear corner exactly. Using two tapes, locate Stake D by measuring the building length from stake C and the width from stake B **(Figure 26–7)**. Secure the stake and drive a nail in its top to mark exactly the other rear corner. Check the accuracy of the work by measuring widths, lengths, and diagonals. The diagonal measurements should be the same **(Figure 26–8)**.

PYTHAGOREAN THEOREM
$c^2 = a^2 + b^2$

FOR EXAMPLE, IF LENGTH = 40' AND WIDTH = 32', THEN
    DIAGONAL EQUALS c.

$c^2 = 32^2 + 40^2 = 2624$
$c = 51.2249939' = 51' + 0.2249939'$
$0.2249939' \times 12 = 2.6999268" = 2" + 0.6999268"$
$0.6999268" \times 16 = 11.1988/16^{th}$ WHICH ROUNDS OFF TO $^{11}/_{16}"$
THUS THE DIAGONAL OF A 32' $\times$ 40' RECTANGLE
    EQUALS $51' - 2^{11}/_{16}"$.

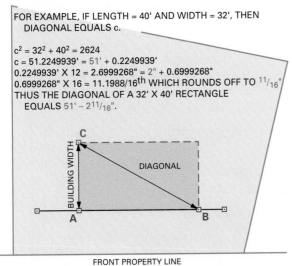

FIGURE 26–6 Use the Pythagorean theorem to place the third building corner stake, Stake C. Two tapes are used to measure the width and diagonal at the same time.

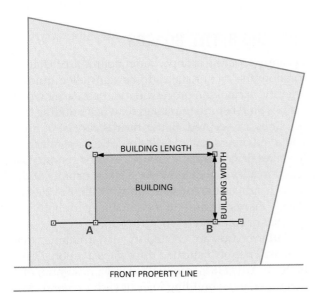

FIGURE 26–7 Measure from established front and rear corners (Stakes B and C) to locate the last building corner stake, Stake D.

All measurements must be made on the level. If the land slopes, the tape is held level with a plumb bob suspended from it **(Figure 26–9).**

Layout of irregularly shaped buildings may seem complicated at first, but they are laid out using the same fundamental principles just outlined. The irregularly shaped building may be staked out

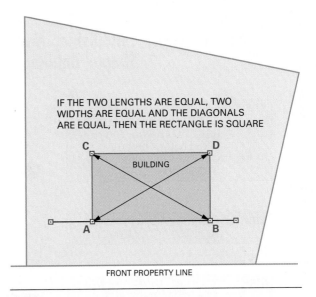

FIGURE 26–8 If the length and width measurements are accurate and the diagonal measurements are equal, then the corners are square.

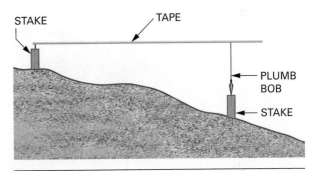

FIGURE 26–9 When layouts are done on sloping land, measurements must be taken on the level.

from a large rectangle **(Procedure 26–A).** The large rectangle corners should align with as many of the building corners as possible. Stake the outermost corners for distance and square, then pull a string to make the large rectangle. The intermediate corner stakes are then located by measuring from the four corners along the strings. Then the final (inside-corner) stakes are measured and located using two tapes.

Placing stakes accurately often requires the builder to move them slightly while installing them. This can sometimes be a nuisance. To speed the process, large nails may be driven into the ground through small squares of thin cardboard **(Figure 26–10).** The cardboard serves to make the nailhead more visible. This process allows for faster erecting of **batter boards** later.

PROCEDURE 26–A Establishing Multiple Corners for an Irregularly Shaped Building

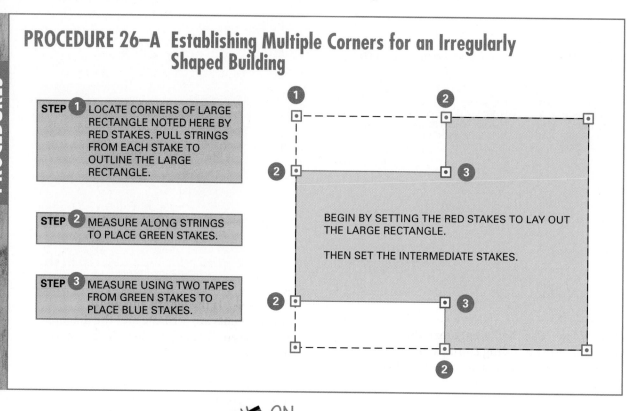

STEP 1 LOCATE CORNERS OF LARGE RECTANGLE NOTED HERE BY RED STAKES. PULL STRINGS FROM EACH STAKE TO OUTLINE THE LARGE RECTANGLE.

STEP 2 MEASURE ALONG STRINGS TO PLACE GREEN STAKES.

STEP 3 MEASURE USING TWO TAPES FROM GREEN STAKES TO PLACE BLUE STAKES.

BEGIN BY SETTING THE RED STAKES TO LAY OUT THE LARGE RECTANGLE.

THEN SET THE INTERMEDIATE STAKES.

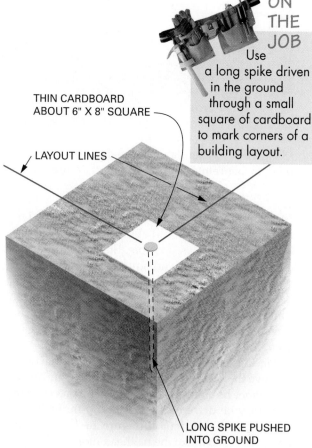

THIN CARDBOARD ABOUT 6" X 8" SQUARE

LAYOUT LINES

LONG SPIKE PUSHED INTO GROUND

FIGURE 26–10 Technique for marking the ground location of stakes.

ON THE JOB

Use a long spike driven in the ground through a small square of cardboard to mark corners of a building layout.

ERECTING BATTER BOARDS

Before excavating is done, batter boards are installed to allow the layout stakes to be reinstalled quickly after the excavation process. Batter boards are wood frames built behind the stakes to which building layout lines are secured. Batter boards consist of horizontal members, called **ledgers.** These are attached to stakes driven into the ground. The ledgers are fastened in a level position to the stakes, usually at the same height as the **foundation wall (Figure 26–11).** Batter boards are built in the same way for both residential or commercial construction.

Batter boards are erected in such a manner that they will not be disturbed during excavation. Drive batter board stakes into the ground a minimum of 4 feet outside the building lines at each corner. When setting batter boards for large construction, increase this distance. This will allow room for heavy excavating equipment to operate without disturbing the batter boards. In loose soil or when stakes are higher than 3 feet, they must be braced **(Figure 26–12).**

Set up the builder's level about center on the building location. Sight to the benchmark, and record the sighting. Determine the difference between the benchmark sighting and the height of the ledgers. Sight and mark each corner stake at the specified elevation. Attach ledgers to the stakes so that the top edge of each ledger is on the mark. Brace the batter boards for strength, if necessary.

MILLWORK CUTTING BILL

Dept. _____

Date: _____

Pieces	DESCRIPTION	Thick	Width	Length	Material	Detail
1						
2						
3						
4						
5						
6						
7						
8						
9						
0						
1						
2						
3						
4						
5						
6						
7						
8						
9						
0						
1						
2						

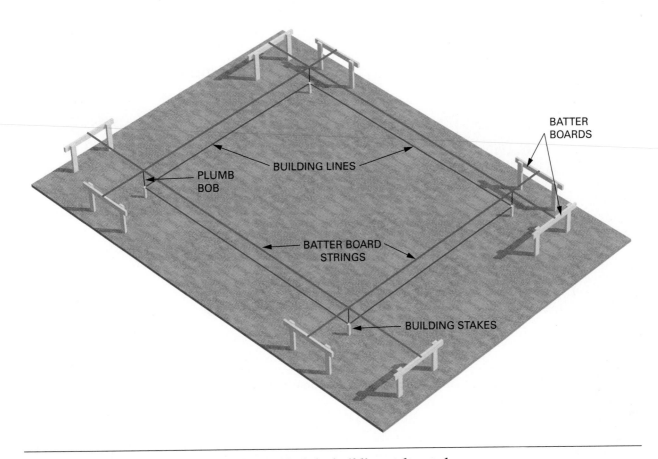

FIGURE 26–11 Batter boards are installed behind the building stakes and nearly level with the top of the foundation.

ON LARGE CONSTRUCTION, STRAIGHT BATTER BOARDS ARE USED IN ORDER TO BE SET BACK FAR ENOUGH TO PROVIDE ROOM FOR HEAVY EXCAVATING EQUIPMENT

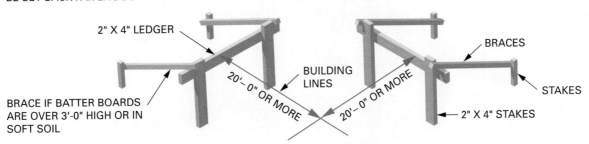

STRAIGHT BATTER BOARDS

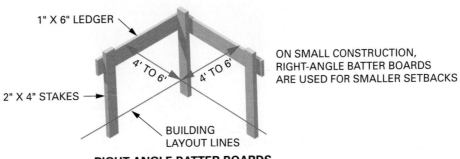

RIGHT-ANGLE BATTER BOARDS

FIGURE 26–12 Batter boards are placed back far enough so they will not be disturbed during excavation operations.

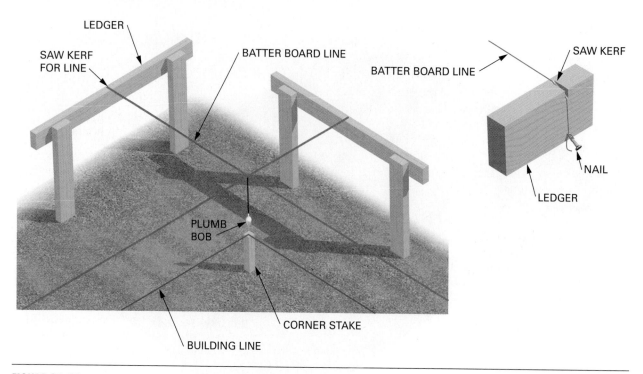

FIGURE 26–13 Saw kerfs are usually made in the ledger board to keep the strings from moving out of line.

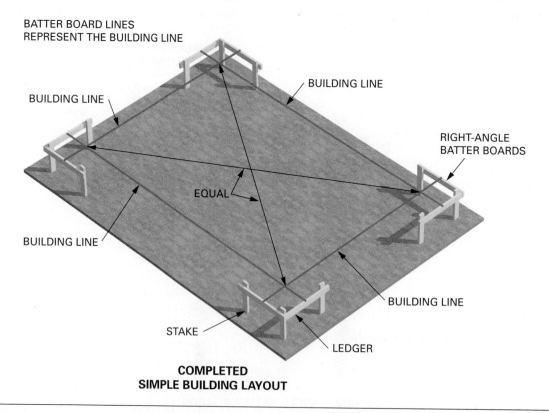

**COMPLETED
SIMPLE BUILDING LAYOUT**

FIGURE 26–14 Check the width and length for accuracy and then check the diagonals.

Stretch lines between batter boards directly over the nailheads in the original corner stakes. Locate the position of the lines by suspending a plumb bob directly over the nailheads. When the lines are accurately located, make a saw cut on the outside corner of the top edge of the ledger. This prevents the layout lines from moving when stretched and secured. Be careful not to make the saw cut below the top edge (**Figure 26–13**). Saw cuts are also often made on batter boards to mark the location of the foundation footing. The footing width usually extends outside and inside of the foundation wall.

Check the accuracy of the layout by again measuring the diagonals to see if they are equal. If not, make the necessary adjustment until they are equal (**Figure 26–14**). Once the excavation is completed, the building stakes may be relocated. Reattach the batter board strings in the appropriate saw kerfs. Use a plumb bob to determine the building stake location (**Figure 26–15**).

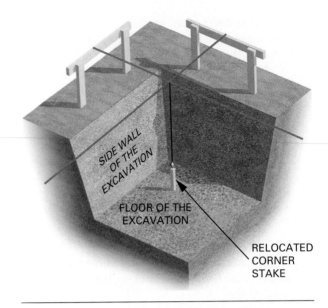

SIDE WALL OF THE EXCAVATION

FLOOR OF THE EXCAVATION

RELOCATED CORNER STAKE

FIGURE 26–15 Building corner stakes are easily repositioned later in the excavation using the batter board strings and a plumb bob.

Key Terms

batter boards	grade rod	horizontal vernier	Pythagorean theorem
benchmark	height of the instrument	laser	transit-level
builder's level	horizontal circle scale	laser level	vertical arc
diagonal		ledgers	
foundation wall		optical levels	

Review Questions

Select the most appropriate answer.

1. The location of the building on the lot is determined from the
   a. foundation plan.   c. architect.
   b. floor plan.   d. plot plan.

2. The reference for establishing elevations on a construction site is called the
   a. starting point.   c. benchmark.
   b. reference point.   d. sight mark.

3. The builder's level is ordinarily used for
   a. laying out straight lines.
   b. reading elevations.
   c. reading vertical angles.
   d. all of these.

4. When leveling an optical level instrument with four leveling screws, turn the opposite leveling screws in
   a. opposite directions, noting the motion of your thumbs.
   b. same direction, noting the motion of your thumbs.
   c. opposite directions, noting the motion of your forefinger.
   d. same direction, noting the motion of your fingers.

5. The cross-hairs of an optical leveling instrument are focused for individual users by rotating the
   a. focus knob on top of the telescope.
   b. eyepiece ring.
   c. tangent screw.
   d. base plate.

6. A tool used with a builder's level to check the level of objects is a
   a. pocket tape backed up by a strip of wood.
   b. grade rod.
   c. piece of wood with graduations marked on it.
   d. all of the above.

7. After a transit-level is moved and set up at a second location, a backsight reading is toward the
   a. previous point and is a plus measurement.
   b. unknown point and is a plus measurement.
   c. previous point and is a minus measurement.
   d. unknown point and is a minus measurement.

8. If the first reading of a rod, at point A is 6' and the second reading at point B is 2', then
   a. point A is higher than point B by 4'.
   b. point A is higher than point B by 8'.
   c. point B is higher than point A by 4'.
   d. point B is higher than point A by 8'.

9. The elevation difference between point A and point D where the backsight rod readings are 36", 48", and 59" and the foresight rod readings are 28", 42", and 60" is such that
   a. point A is 13" higher than point D.
   b. point A is 13" lower than point D.
   c. point A is 23" higher than point D.
   d. point A is 23" lower than point D.

10. When using a vernier scale to measure an angle, read the minutes at the
    a. zero index line.
    b. vernier line that best aligns with the circle scale line.
    c. circle scale.
    d. all of the above.

11. Laser leveling tools are used by
    a. carpenters.
    b. heavy equipment operators.
    c. plumbers.
    d. all of the above.

12. The diagonal of a rectangle whose dimensions are 30 feet × 40 feet is
    a. 45 feet.   c. 60 feet.
    b. 50 feet.   d. 70 feet.

13. The diagonal of a rectangle whose dimensions are 32 feet × 48 feet is
    a. 57'-8⅛".   c. 57'-11".
    b. 57'-8¼".   d. 60'-0".

Concrete formwork is usually the responsibility of the carpenter, whether constructing wood forms or erecting a forming system. The forms must meet specified dimensions and be strong enough to withstand tremendous pressure. The carpenter must know that the forms will be strong enough before the concrete is placed. If a form fails to hold during concrete placement, costly labor and materials are lost. Formwork is unique in construction in that if it is built for a specific purpose then it must be dismantled after the concrete cures. The carpenter must keep this in mind when designing the joints and seams in the formwork so the material can be taken apart easily and salvaged.

SAFETY REMINDER

Concrete is a universal building material used for support of various structures. It must be handled and properly supported during placement. This is particularly important because concrete can only be worked for a short time before it sets up.

Avoid prolonged contact with fresh concrete or wet cement because of possible skin irritation. Wear protective clothing when working with newly mixed concrete. Wash skin areas that have been exposed to wet concrete as soon as possible.

OBJECTIVES

After completing this unit, the student should be able to:

- describe the composition of concrete and factors affecting its strength, durability, and workability.
- explain the reasons for reinforcing concrete and describe the materials used.
- job-mix a batch of concrete and explain the method of and reasons for making a slump test.
- explain techniques used for the proper placement and curing of concrete.
- construct forms for footings, slabs, walks, and driveways.
- construct concrete forms for foundation walls.
- construct concrete column forms.
- lay out and build concrete forms for stairs.
- estimate quantities of concrete.

27 Characteristics of Concrete

U nderstanding the characteristics of concrete is essential for the construction of reliable concrete forms, the correct handling of freshly mixed material, and the final quality of hardened concrete.

PRECAUTIONS ABOUT USING CONCRETE

Avoid prolonged contact with fresh concrete or wet cement because of possible skin irritation. Wear protective clothing when working with newly mixed concrete. Wash skin areas that have been exposed to wet concrete as soon as possible. If any material containing cement gets into the eyes, flush immediately with water and get medical help.

CONCRETE

Concrete is a widely used building material. It can be formed into practically any shape for construction of buildings, bridges, dams, and roads. Improvements over the years have created a product that is strong, durable, and versatile **(Figure 27–1)**.

Composition of Concrete

Concrete is a mixture of **portland cement,** fine and coarse *aggregates,* water, and various *admixtures.* Aggregates are fillers, usually sand, gravel, or stone. Admixtures are materials or chemicals added to the mix to achieve certain desired qualities.

When these concrete ingredients are mixed the portland cement and water form a paste. A chemical reaction, called **hydration,** begins within this paste. This reaction causes the cement to set or harden, referred to as **curing.** As the paste hardens it binds the

FIGURE 27–1   Concrete is widely used in the construction industry.

sand, which in turn binds the small and large aggregates. Together they form a strong, durable, and watertight mass called concrete.

Hydration (hardening) of the concrete is a relatively slow process. Initial setting, to the point of being able to walk on it, takes hours. It often takes a month or more for the concrete to achieve 90 percent of its full strength. Hydration continues for many years.

This process is somewhat fragile. If rapid evaporation or freezing occurs during the curing of concrete, the cement hydration process will stop. If this happens the concrete will never achieve full design strength.

Portland Cement. In 1824, Joseph Aspdin, an Englishman, developed an improved type of cement. He named it portland cement because it produced a concrete that resembled stone found on the Isle of Portland, England. Portland cement is a fine gray powder.

Portland cement is usually sold in plastic-lined paper bags that hold one cubic foot. Each bag weighs 94 pounds or 40 kilograms. The cement can still be used even after a long period of time as long as it remains dry. Eventually, however, it will absorb moisture from the air. It should not be used if it contains lumps that cannot be broken up easily. This condition is known as prehydration.

Types of Portland Cement. New developments have produced several types of portland cement. Type I is the familiar gray kind most generally purchased and used. Type IA is an **air-entrained cement** that is more resistant to freezing and thawing. Types II through V have special uses, such as for low or high heat generation, high early strength, and resistance to severe frost. White portland cement, used for decorative purposes, differs only in color from gray cement. There are many other types, each with its specific purpose—ranging from underwater use to sealing oil wells.

Water. Water is the part of concrete mix that starts the hydration process. It also allows the concrete mixture to flow easily into place. Water used to make concrete must be clean. Other water may be used but should be tested first to make sure it is acceptable. The chemical reaction of hydration can be negatively affected by water impurities. A good general rule is, if it is safe to drink, it is safe to use in concrete.

The amount of water used can affect the quality and strength of the concrete. The amount of water in relation to the amount of portland cement is called the *water-to-cement ratio*. This ratio should be kept as small as possible while allowing for workability. Adding excessive water to a mix simply to increase the flow can weaken the concrete and must be avoided.

Aggregates. Aggregates have no cementing value of their own. They serve only as a filler but are important ingredients. They constitute from 60 to 80 percent of the concrete volume.

Fine aggregate consists of particles ¼ inch or less in diameter. Sand is the most commonly used fine aggregate. *Coarse* aggregate, usually gravel or crushed stone, usually comes in sizes of ⅜, ½, ¾, 1, 1½, and 2 inches. When strength is the only consideration, ¾ inch is the optimum size for most aggregates. Large-size aggregate uses less water and cement. Therefore, it is more economical. However, the maximum aggregate size must not exceed the following:

- ⅕ the smallest dimension of the unreinforced concrete.
- ⅓ the depth of unreinforced slabs on the ground **(Figure 27–2)**.
- ¾ the clear spacing between reinforcing bars or between the bars and the form.

Fine and coarse particles must be in a proportion that allows the finer particles to fill the spaces between the larger particles. The aggregate should be clean and free of dust, loam, clay, or vegetable matter.

Admixtures. Admixtures are available to quicken or retard setting time, develop early strength, inhibit corrosion, retard moisture, control bacteria and fungus, improve pumping, and color concrete, among many other purposes.

Air. An important advance was made with the development of air-entrained concrete. It is produced by using air-entraining portland cement or an admixture. The intentionally made air bubbles are very small. Billions of them are contained in a cubic yard of concrete.

The introduction of air into concrete was designed to increase its resistance to freezing and thawing. It also has other benefits: Less water and sand are required, the workability is improved, separation of the water from the paste is reduced, and the concrete can be finished sooner and is more watertight than ordinary concrete. Air-entrained concrete is now recommended for almost all concrete projects.

Concrete Strength. Concrete strength is measured in pounds per square inch (psi). Test cylinders are compressed in large machines to determine how much stress they will withstand before breaking. Typical strength numbers with concrete are 2,500, 3,500, 4000, and 4,500 psi (ranging from low to high).

The overall strength of the concrete is determined by many things. It is affected by the relative amounts of its ingredients. Water is required for the reaction to begin, but too much water weakens the mix. Also, increasing the amount of portland cement increases strength. Aggregate sizes can also affect the strength.

Strength of concrete begins with the mix, but is also affected by how the concrete is placed and cured. Generally a slow cure is best because the hydration process is slow.

Mixing Concrete

Concrete mixtures are designed to achieve the desired qualities of strength, durability, and workability in the most economical manner. The mix will vary according to the strength and other desired qualities, plus other factors such as the method and time of curing.

For very large jobs, such as bridges and dams, concrete mixing plants may be built on the job site. This is because freshly mixed concrete is needed on a round-the-clock basis. The construction engineer calculates the proportions of the mix and supervises tests of the concrete.

MAXIMUM AGGREGATE SIZE

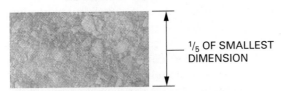

⅕ OF SMALLEST DIMENSION

SECTION THROUGH UNREINFORCED FOOTING

SLAB

⅓ OF DEPTH OF SLAB

SECTION THROUGH SLAB ON GROUND

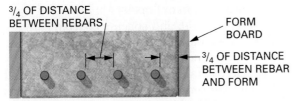

¾ OF DISTANCE BETWEEN REBARS

FORM BOARD

¾ OF DISTANCE BETWEEN REBAR AND FORM

SECTION THROUGH REINFORCED FOOTING

FIGURE 27–2 **Aggregate size depends on the final dimensions of the concrete.**

FIGURE 27–3 Ready-mixed concrete is delivered in trucks.

Ready-Mixed Concrete. Most concrete used in construction is *ready-mixed* concrete. It is sold by the cubic yard or cubic meter. There are 27 cubic feet to the cubic yard. The purchaser usually specifies the amount and the desired strength of the concrete. The concrete supplier is then responsible for mixing the ingredients in the correct proportion to yield the desired strength.

Sometimes the purchasers specify the proportion of the ingredients. They then assume responsibility for the design of the mixture.

The ingredients are accurately measured at the plant with computerized equipment. The mixture is delivered to the jobsite in *transit-mix trucks* **(Figure 27–3).** The truck contains a large revolving drum, capable of holding from 1 to 10 cubic yards. There is a separate water tank with a water measuring device. The drum rotates to mix the concrete as the truck is driven to the construction site.

Specifications require that each batch of concrete be delivered within 1½ hours after water has been added to the mix. If the job site is a short distance away, water is added to the cement and aggregates at the plant. If the job site is farther away, the proper amount of water is added to the dry mix from the water tank on the truck as it approaches or arrives at the job.

Job-Mixed Concrete. A small job may require that the concrete be mixed on the site either by hand or with a powered concrete mixer. All materials must be measured accurately. For measuring purposes, remember that a cubic foot of water weighs about 62½ pounds and contains approximately 7½ gallons.

All ingredients should be thoroughly mixed according to the proportions shown in **Figure 27–4.** A little water should be put in the mixer before the dry materials are added. The water is then added uniformly while the ingredients are mixed from one to three minutes.

Water-to-Cement Ratio	Possible Strength (psi)	Water (gal)	Portland Cement (bags) (94 lb/bag)	Fine Aggregate (sand) (cubic feet)	Coarse Aggregate (size ¾″ max) (cubic feet)
0.45	5500	5.6	1	1.4	2.4
0.50	4500	6.25	1	1.6	2.6
0.65	3000	8	1	2.3	3.4

FIGURE 27–4 Formulas for several concrete mixtures.

Bar #	Bar ∅ (inches)	Metric Sizes (mm)	Bar Weight (lbs per 100 lin ft)
2	¼	6	17
3	⅜	10	38
4	½	13	67
5	⅝	16	104
6	¾	19	150
7	⅞	22	204
8	1	26	267

FIGURE 27–5 Numbers and sizes of commonly used reinforcing steel bars.

Concrete Reinforcement

Concrete has high **compressive strength.** This means that it resists being crushed. It has, however, low **tensile strength.** It is not as resistant to bending or pulling apart. Steel bars, called **rebars,** are used in concrete to increase its tensile strength. Concrete is then called **reinforced concrete.** Fiberglass strands are sometimes used as reinforcement to add tensile strength. They are added to the concrete during the final mixing stage.

Rebars. Rebars used in construction are usually *deformed.* Their surface has ridges that increase the bond between the concrete and the steel. They come in standard sizes, identified by numbers that indicate the diameter in eighths. For instance, a #6 rebar has a diameter of 6/8 or ¾ inch **(Figure 27–5).** Metric rebars are measured in millimeters.

The size, location, and spacing of rebars are determined by engineers and shown on the plans. The rebars are positioned inside the form before the concrete is placed **(Figure 27–6).** Rebars in bridges and roads that experience a salty environment, such as seawater or salting during the winter months, have an epoxy coating. This protects the rebar from rusting, and thus helps prevent premature concrete failure. The tasks of cutting, bending, placing, and tying

FIGURE 27–6 Rebar is placed after the forms are created.

Mesh Size (inches)	Mesh Gauge	Mesh Weight (lbs per 100 sq. ft.)
6 × 6	#6	42
6 × 6	#8	30
6 × 6	#10	21

FIGURE 27–7 Size, gauge, and weight of commonly used welded wire mesh.

rebars require workers who have been trained in that trade.

Wire Mesh. Welded wire mesh is used to reinforce concrete floor slabs resting on the ground, driveways, and walks. It is identified by the gauge and spacing of the wire. Common gauges are #6, #8, and #10. The wire is usually spaced to make 6-inch squares **(Figure 27–7).**

Welded wire mesh is laid in the slab above the vapor barrier, if used, before the concrete is placed. It is spliced by lapping one full square plus 2 inches.

Placing Concrete

Prior to placing the concrete, sawdust, nails, and other debris should be removed from inside the forms. The inside surfaces are brushed or sprayed with oil to make form removal easier. No oil should be allowed to get on the steel reinforcement. This will reduce the steel/cement bond, thereby reducing the tensile strength provided by the steel. Also, before concrete is placed, the forms and subgrade are moistened with water. This is done to prevent rapid absorbing of water from the concrete.

Concrete is *placed*, not poured. Water should never be added so that concrete flows into forms without

FIGURE 27–8 A slump test shows the wetness of a concrete mix.

working it. Adding water alters the water-to-cement ratio on which the quality of the concrete depends.

Slump Test. Slump tests are made by supervisors on the job to determine the consistency of the concrete. The concrete sample for a test should be taken just before the concrete is placed. The *slump cone* is first dampened. It is placed on a flat surface and filled to about one-third of its capacity with the fresh concrete. The concrete is then *rodded* by moving a metal rod up and down twenty-five times over the entire surface. Two more approximately equal layers are added to the cone. Each layer is rodded in a similar manner with the rod penetrating the layer below. The excess is **screeded** from the top. The cone is then turned over onto a flat surface and removed by carefully lifting it vertically in three to seven seconds. The cone is gently placed beside the concrete, and the amount of slump measured between the top of the cone and the concrete **(Figure 27–8).** The test should be completed within two to three minutes.

Changes in slump should be corrected immediately. The table shown in **Figure 27–9** shows recommended slumps for various types of construction. Concrete with a slump greater than 6 inches should not be used unless a slump-increasing admixture has been added.

Placing in Forms. The concrete truck should get as close as possible. Concrete is placed by chutes where needed. It should not be pushed or dragged any more than necessary. It should not be dropped more than 4 to 6 feet, and should be dropped vertically and not angled. Drop chutes should be used in high forms to prevent the buildup of dry concrete on the side of the form or reinforcing bars above the level of the placement. Drop chutes also prevent separation caused by concrete striking and bouncing off the side of the form. Pumps are used on large jobs

Types of Construction	Slump in Inches	
	Maximum	Minimum
Reinforced foundation walls & footings	3	1
Plain footings, caissons & substructure walls	3	1
Beams & reinforced walls	4	1
Building columns	4	1
Pavement & slabs	3	1
Heavy mass concrete	2	1

FIGURE 27–9 Recommended slumps for different kinds of construction.

that need concrete continuously over long distances or to heights up to 500 feet.

Concrete must not be placed at a rapid rate, especially in high forms. The amount of pressure at any point on the form is determined by the height and weight of the concrete above it. Pressure is not affected by the thickness of the wall **(Figure 27–10)**.

A slow rate of placement allows the concrete nearer the bottom to begin to harden. Once concrete hardens it cannot exert more pressure on the forms even though liquid concrete continues to be placed above it **(Figure 27–11)**. The use of stiff concrete with a low slump, which acts less like a liquid, will transmit less pressure. Rapid placing leaves the concrete in the bottom still in a fluid state. It will exert great lateral pressure on the forms at the bottom. This may cause the form ties to fail or the form to deflect excessively.

Concrete should be placed in forms in layers of not more than 12 to 18 inches thick. Each layer should be placed before setting occurs in the previous layer. The layers should be thoroughly consolidated **(Figure 27–12)**. The rate of concrete placement in high forms should be carefully controlled.

RATE (FT/HR)	PRESSURES OF VIBRATED CONCRETE (PSF)			
	50° F		70° F	
	COLUMNS	WALLS	COLUMNS	WALLS
1	330	330	280	280
2	510	510	410	410
3	690	690	540	540
4	870	870	660	660
5	1050	1050	790	790
6	1230	1230	920	920
7	1410	1410	1050	1050
8	1590	1470	1180	1090
9	1770	1520	1310	1130
10	1950	1580	1440	1170

PRESSURE INCREASES AS PLACEMENT RATE INCREASES.
PRESSURE INCREASES AT LOWER TEMPERATURES.

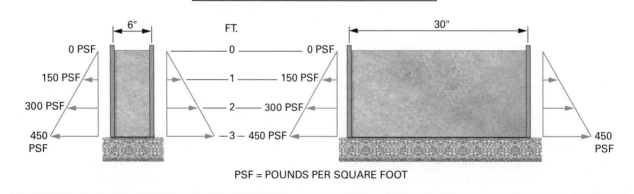

PSF = POUNDS PER SQUARE FOOT

FIGURE 27–10 The height of concrete being poured affects the amount of pressure against the forms. Pressure is not affected by the thickness of the wall.

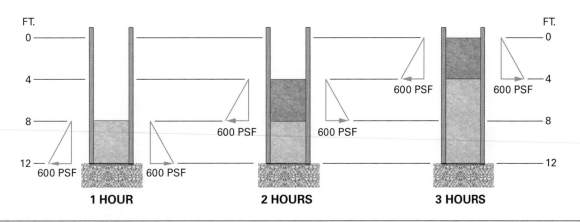

FIGURE 27–11   Once concrete sets, it does not exert more pressure on the form even though liquid concrete continues to be placed above it.

Consolidation of Concrete. To eliminate voids or honeycombs in the concrete, it should be thoroughly worked by hand spading or vibrated after it goes into the form. Vibrators make it possible to use a stiff mixture that would be difficult to consolidate by hand.

An immersion type vibrator, called a **spud vibrator,** has a metal tube on its end. This tube vibrates at a rapid rate. It is commonly used in construction to vibrate and consolidate concrete. Vibration makes the concrete more fluid and able to move, allowing trapped air to escape. This will prevent the formation of air pockets, honeycombs, and cold joints. The operator should be skilled in the use of the vibrator, keeping it moving up and down, uniformly vibrating the entire pour. Over-vibrating should not be done because vibrating increases the lateral pressure on the form **(Figure 27–13).**

Curing Concrete

Concrete hardens and gains strength because of a chemical reaction called hydration. All the desirable properties of concrete are improved the longer this process takes place. A rapid loss of water from fresh concrete can stop the hydration process too soon, weakening the concrete. Curing prevents loss of moisture, allowing the process to continue so that the concrete can gain strength.

Concrete is *cured* either by keeping it moist or by preventing loss of its moisture for a period of time. For instance, if moist-cured for seven days, its strength is up to about 60 percent of full strength. A

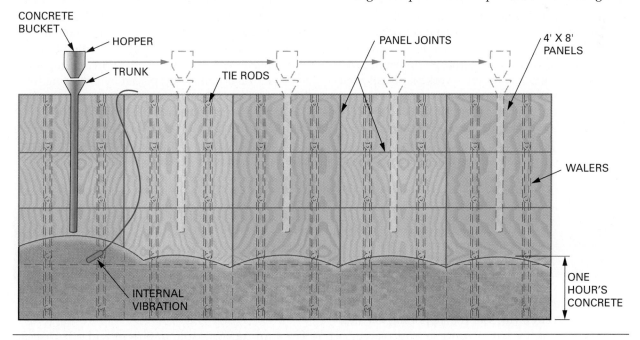

FIGURE 27–12   A safe, consistent pour rate is accomplished using drop chutes to place concrete in internally vibrated layers.

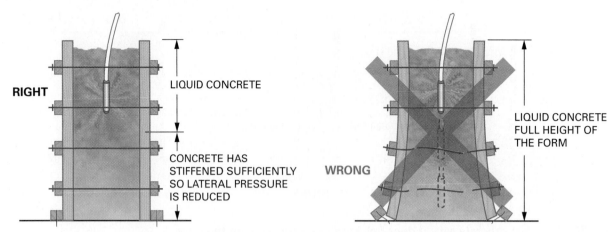

RIGHT

LIQUID CONCRETE

CONCRETE HAS
STIFFENED SUFFICIENTLY
SO LATERAL PRESSURE
IS REDUCED

WRONG

LIQUID CONCRETE
FULL HEIGHT OF
THE FORM

VIBRATE ONLY TO THE DEPTH OF THE FRESHLY PLACED CONCRETE. INSERTING THE VIBRATOR TOO FAR WILL CAUSE
THE CONCRETE AT THE BOTTOM OF THE FORM TO REMAIN IN A LIQUID STATE LONGER THAN EXPECTED. THIS WILL
RESULT IN HIGHER THAN EXPECTED LATERAL FORM PRESSURE AND MAY CAUSE THE FORM TO FAIL. THE DEPTH OF
VIBRATION SHOULD JUST PENETRATE THE PREVIOUS LAYER OF CONCRETE BY A FEW INCHES.

FIGURE 27–13   **Avoid excessive vibration of concrete.** (*Courtesy of Dayton Superior*)

month later the strength is 95 percent, and up to full
strength in about three months. Air-cure will reach
only about 55 percent after three months and will
never attain design strength. In addition, a rapid
loss of moisture causes the concrete to shrink, resulting in cracks. Curing should be started as soon
as the surface is hard enough to resist marring.

Methods of Curing. Flooding or constant sprinkling
of the surface with water is the most effective
method of curing concrete. Curing can also be accomplished by keeping the forms in place, covering
the concrete with burlap, straw, sand, or other material that retains water, and wetting it continuously.

In hot weather, the main concern is to prevent rapid
evaporation of moisture. Sunshades or windbreaks
may need to be erected. The formwork may be allowed
to stay in place or the concrete surface may be covered
with plastic film or other waterproof sheets. The edges
of the sheets are overlapped and sealed with tape or
covered with planks. Liquid curing chemicals may be
sprayed or mopped on to seal in moisture and prevent

evaporation. However, manufacturers' directions for
their use should be carefully followed.

Curing Time. Concrete should be cured for as long as
practical. The curing time depends on the temperature. At or above 70 degrees, curing should take place
for at least three days. At or above 50 degrees, concrete
is cured at least five days. Near freezing, there is practically no hydration and concrete takes considerably
longer to gain strength. There is no strength gain while
concrete is frozen. When thawed, hydration resumes
with appropriate curing. If concrete is frozen within
the first twenty-four hours after being placed, permanent damage to the concrete is almost certain. In cold
weather, *accelerators* that shorten the setting time are
sometimes used. Protect concrete from freezing for at
least four days after being placed by providing insulation or artificial heat, if necessary.

Forms may be removed after the concrete has set
and hardened enough to maintain its shape. This
time will vary depending on the mix, temperature,
humidity, and other factors.

28   Forms for Footings, Slabs, Walks, and Driveways

Concrete for footings, slabs, walks, and driveways is
placed directly on the soil. The supporting soil must
be suitable for the type of concrete work placed on
it. It must be well drained. The concrete should be
placed according to the local environmental conditions.

FOOTING FORMS

The **footing** for a foundation provides a base on
which to spread the load of a structure over a wider
area of the soil. For foundation walls, the most typi-

cal type is a *continuous* or *spread* footing. To provide support for columns and posts, *pier* footings of square, rectangular, circular, or tapered shape are used. Sometimes it is not practical to excavate deep enough to reach load-bearing soil. Then **piles** are driven and capped with a *grade beam* **(Figure 28–1).**

Continuous Wall Footings

In most cases, the footing is formed separately from the foundation wall. In residential construction, often the footing width is twice the wall thickness. The footing depth is often equal to the wall thickness **(Figure 28–2).** However, to be certain, consult local building codes.

For larger buildings, architects or engineers design the footings to carry the load imposed on them. Usually these footings are strengthened by reinforcing rods of specified size and spacing.

Frost Line. In areas where frost occurs, footings must be located below the **frost line.** The frost line is the point below the surface to which the ground typically freezes in winter. Because water expands when frozen, foundations whose footings are above the frost line will heave and buckle when the ground freezes and heaves. In extreme northern climates, footings must be placed as much as 6 feet below the surface **(Figure 28–3).** In tropical climates, footings

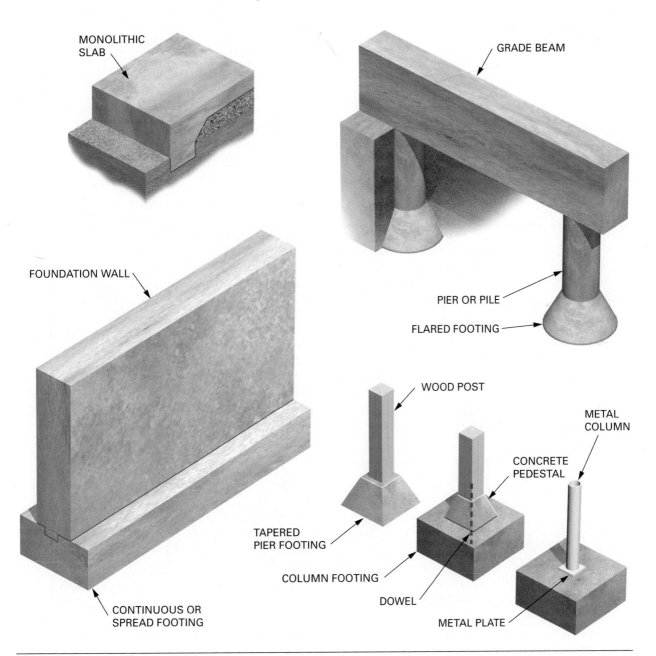

MONOLITHIC SLAB

GRADE BEAM

FOUNDATION WALL

PIER OR PILE

FLARED FOOTING

CONTINUOUS OR SPREAD FOOTING

WOOD POST

METAL COLUMN

CONCRETE PEDESTAL

TAPERED PIER FOOTING

COLUMN FOOTING

DOWEL

METAL PLATE

FIGURE 28–1 **Several types of footings are constructed to support foundations.**

only need to reach solid soil, with no consideration given to frost.

In areas where the soil is stable, footings may be *trench poured* where no formwork is necessary. A trench is dug to the width and depth of the footing. The concrete is carefully placed in the trench **(Figure 28–4)**. In other cases, forms need to be built for the footing.

Locating Footings

To locate the footing forms, stretch lines on the batter boards in line with the outside of the footing.

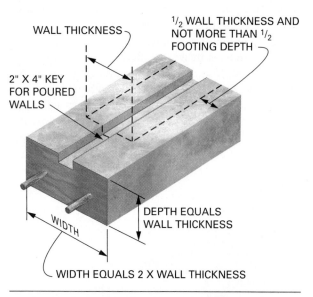

WALL THICKNESS

1/2 WALL THICKNESS AND NOT MORE THAN 1/2 FOOTING DEPTH

2" X 4" KEY FOR POURED WALLS

DEPTH EQUALS WALL THICKNESS

WIDTH

WIDTH EQUALS 2 X WALL THICKNESS

FIGURE 28–2 **Typical footing for residential construction.**

FIGURE 28–4 **In stable soils, the soil acts as the footing form.**

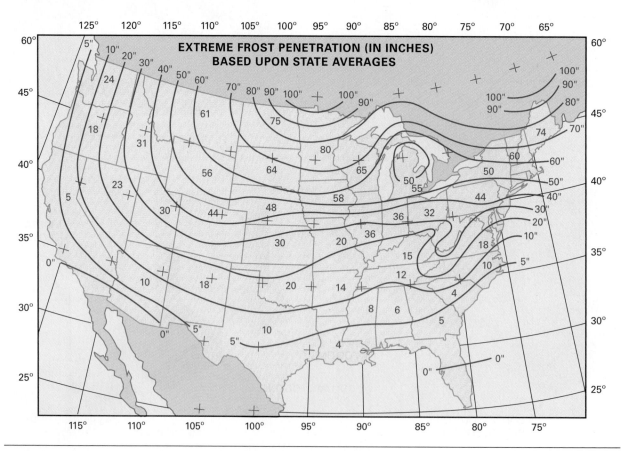

FIGURE 28–3 **Frost line penetration in the United States.** *(Courtesy of U.S. Department of Commerce)*

This is done by noting where the line should be on the batter board for the corner of the building. Then move the strings far enough away from the building to allow for the extra footing width **(Figure 28–5).** Suspend a plumb bob from the batter board lines at each corner. Drive stakes and attach lines to represent the outside surface of the footing.

Steel stakes, **spreaders,** and braces are manufactured for use in building footing forms and other edge formwork. They come with prepunched nail holes for easy fastening **(Figure 28–6).**

Building Wall Footing Forms

Stakes are used to hold the sides in position. Fasten the sides by driving nails through the stakes. Use duplex nails for easy removal. Various methods are used to build wall footing forms. One way is to erect the outside form around the perimeter of the building then return to assemble the inside form. This process will provide a strong, level form system **(Procedure 28–A).**

Set form stakes around the perimeter spaced 4 to 6 feet apart depending on the firmness of the soil. Use a gauge block that is the same thickness as the form board to set stakes. This will ensure that the string does not touch any previous stakes while the current one is being adjusted. Snap a chalk line on the stakes at slightly above the proper elevation. Fasten the form board to the stakes and tap the stakes down until the form is level. Form the outside of the footing in this manner all around.

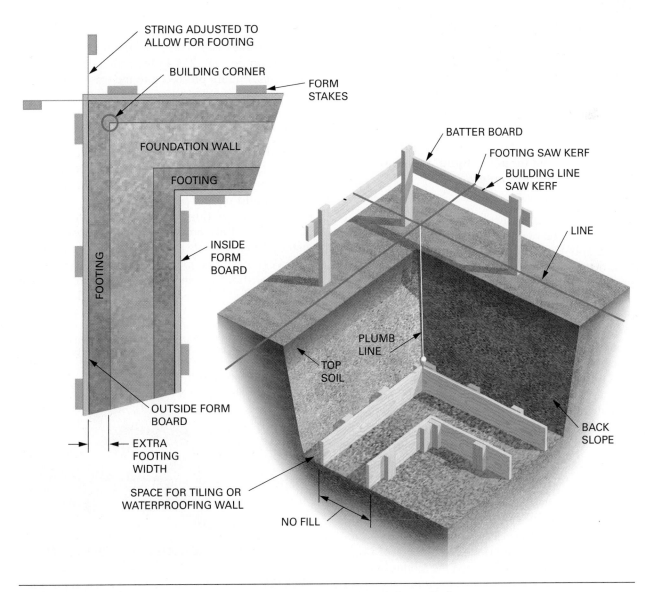

FIGURE 28–5 The footing is located by suspending a plumb bob from the batter board lines.

STEEL STAKES

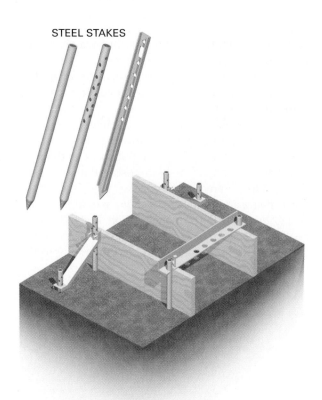

MANUFACTURED FORM SPREADER AND BRACE

FIGURE 28–6   Steel stakes, spreaders, and braces are manufactured for use in building forms. *(Courtesy of Symons Corporation)*

Before erecting the inside forms, cut a number of spreaders. Spreaders serve to tie the two sides together and keep them the correct distance apart. Nail one end to the top edges of the outside form. Erect the inside forms in a manner similar to that used in erecting the outside forms. Place stakes for the inside forms opposite those holding the outside form. Level across from the outside form to determine the height of the inside form. Fasten the spreaders as needed across the form at intervals to hold the form the correct distance apart.

Brace the stakes where necessary to hold the forms straight. In many cases, no bracing is necessary. Footing forms are sometimes braced by shoveling earth or placing large stones against the outside of the forms. **Figure 28–7** shows a typical setup for braced footing forms.

Keyways. A keyway is formed in the footing by pressing 2 × 4 lumber into the fresh concrete **(Figure 28–8)**. The keyway form is beveled on both edges for easy removal after the concrete has set. The purpose of a keyway is to provide a lock between the footing and the foundation wall. This joint helps the foundation wall resist the pressure of the back-filled earth against it. It

FIGURE 28–7   Forms must be braced as necessary.

also helps to prevent seepage of water into the basement. In some cases, where the design of the keyway is not so important, 2 × 4 pieces are not beveled on the edges, but are pressed into the fresh concrete at an angle.

Stepped Wall Footings

When the foundation is to be built on sloped land, it is sometimes necessary to *step* the footing. The footing is formed at different levels, to save material. In building stepped footing forms, the thickness of the footing must be maintained. The vertical and horizontal footing distances are adjusted so that a whole number of blocks or concrete forms can easily be placed into that section of the footing without cutting. The vertical part of each step should not exceed the footing thickness. The

2" X 4" WITH BEVELED EDGES PRESSED INTO FRESH CONCRETE

2" X 4" PRESSED INTO FRESH CONCRETE AT AN ANGLE

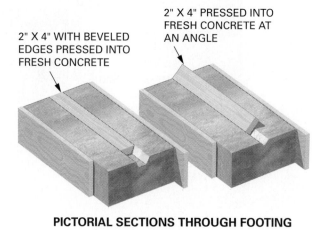

PICTORIAL SECTIONS THROUGH FOOTING

FIGURE 28–8   Methods of forming keyways in the footing.

PROCEDURE 28–A Technique for Setting Footing Forms After Excavation

STEP 1 MOVE BATTER BOARD STRINGS TOWARD THE OUTSIDE TO ALLOW FOR EXTRA WIDTH OF FOOTING.

STEP 2 SUSPEND PLUMB BOB TO LOCATE OUTSIDE EDGES OF FOOTING. DRIVE TWO STAKES AT EACH CORNER AND STRETCH LINES FROM CORNER TO CORNER.

STEP 3 DRIVE INTERMEDIATE STAKES OUTSIDE OF THE STRETCH LINE USING A GAUGE BLOCK OF EQUAL THICKNESS TO THE FORM BOARD.

STEP 4 SNAP A LINE ON THE INSIDE FACE OF THE STAKES AT THE HEIGHT OF THE TOP OF THE FOOTING.

STEP 5 NAIL THE FORM BOARD TO THE STAKE SLIGHTLY HIGHER THAN THE CHALK LINE.

STEP 6 LEVEL THE FORM BOARDS WITH A LASER OR OPTICAL TRANSIT BY TAPPING EACH STAKE DOWN AS NEEDED.

STEP 7 CUT AND ATTACH SPREADERS TO OUTSIDE FORM BOARD.

STEP 8 ATTACH THE INSIDE FORM BOARD TO THE SPREADERS AS INSIDE STAKES ARE PLACED.

STEP 9 USE A LEVEL TO POSITION THE HEIGHT OF THE INSIDE FORM.

STEP 10 BRACE AS NEEDED WITH STAKES DRIVEN AT AN ANGLE AND NAIL TO THE FORMS.

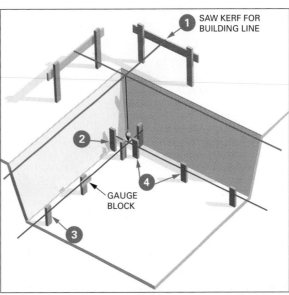

SAW KERF FOR BUILDING LINE

GAUGE BLOCK

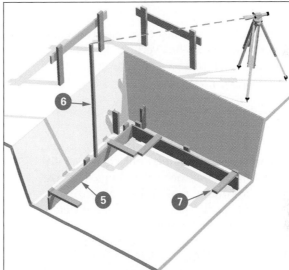

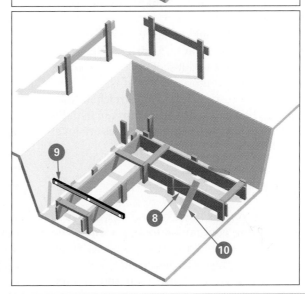

STEP BY STEP PROCEDURES

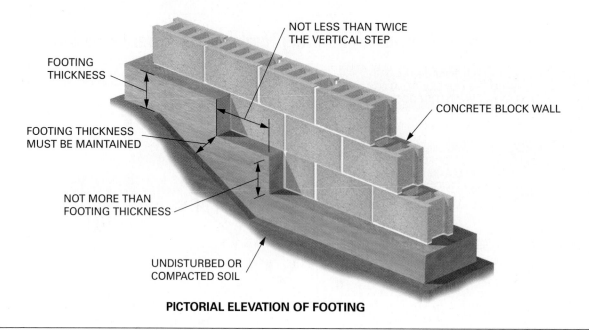

PICTORIAL ELEVATION OF FOOTING

FIGURE 28–9   Stepped footings must be properly dimensioned.

horizontal part of the step must be at least twice the vertical part **(Figure 28–9).** Vertical boards are placed between the forms to retain the concrete at each step.

Column Footings

Concrete for footings, supporting columns, posts, fireplaces, chimneys, and similar objects is usually placed at the same time as the wall footings. The size and shape of the column footing vary according to what it has to support. The dimensions are determined from the foundation plan.

In residential construction, these footing forms are usually built by nailing 2 × 8 pieces together in square, rectangular, or tapered shapes to the specified size **(Figure 28–10).**

Measurements are laid out on the wall footing forms to locate the column footings. Lines are stretched from opposite sides of the wall footing forms to locate the position of the forms. They are laid in position corresponding to the stretched lines **(Figure 28–11).** Stakes are driven. Forms are usually fastened in a position so that the top edges are level with the wall footing forms.

FORMS FOR SLABS

Building forms for slabs, walks, and driveways is similar to building continuous footing forms. The sides of the form are held in place by stakes driven into the ground. Forms for floor slabs are built level. Walks and driveways are formed to shed water.

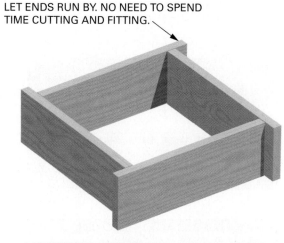

A RECTANGULAR FORM FOR A COLUMN FOOTING

FIGURE 28–10   Construction of column footings.

Usually 2 × 4 or 2 × 6 lumber is used for the sides of the form.

Slab-on-Grade

In warm climates, where frost penetration into the ground is not very deep, little excavation is necessary. The first floor may be a concrete slab placed directly on the ground. This is commonly called **slab-on-grade** construction **(Figure 28–12).** With improvements in the methods of construction, the need for lower construction costs, and the desire to give the structure a lower profile, slabs-on-grade are being used more often in all climates.

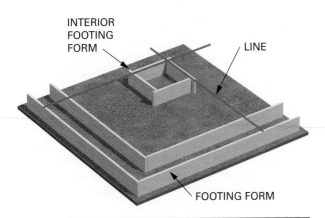

FIGURE 28–11   Locating interior footings using a string.

Basic Requirements

The construction of concrete floor slabs should meet certain basic requirements:

1. The finished floor level must be high enough so that the finish grade around the slab can be sloped away for good drainage. The top of the slab should be no less than 8 inches above the finish grade.

2. All topsoil in the area in which the slab is to be placed must be removed. A base for the slab consisting of 4 to 6 inches of gravel, crushed stone, or other approved material must be well compacted in place.

3. The soil under the slab may be treated with chemicals for control of termites, but caution is advised. Such treatment should be done only by those thoroughly trained in the use of these chemicals.

4. All mechanicals (water and sewer lines, heating ducts, and other utilities) that are to run under the slab must be installed.

5. A vapor barrier must be placed under the concrete slab to prevent soil moisture from rising through the slab. The vapor barrier should be a heavy plastic film, such as 6-mil polyethylene or other material having equal or superior resistance to the passage of vapor. It should be strong enough to resist puncturing during the placing of the concrete. Joints in the vapor barrier must be lapped at least 4 inches and sealed. A layer of sand may be applied to protect the membrane during concrete placement.

6. Where necessary, to prevent heat loss through the floor and foundation walls, waterproof, rigid insulation is installed around the perimeter of the slab.

7. The slab should be reinforced with 6 × 6 inch, #10 welded wire mesh, or by other means to provide equal or superior reinforcing. The concrete slab must be at least 4 inches thick and *haunched* (made thicker) under loadbearing walls **(Figure 28–13).**

Monolithic Slabs

A combined slab and foundation is called a **monolithic slab (Figure 28–14).** This type of slab is also referred to as a *thickened edge slab.* It consists of a shallow footing around the perimeter. The perimeter is placed at the same time as the slab. The slab and footing make up a one-piece integral unit. The

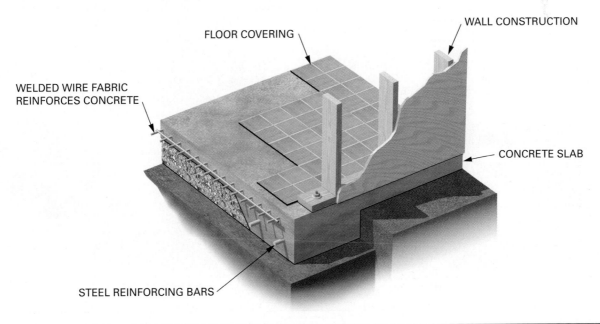

FIGURE 28–12   Slab-on-grade foundation.

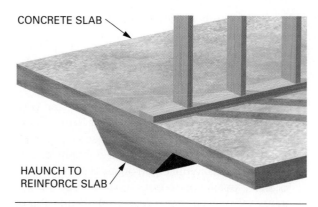

CONCRETE SLAB

HAUNCH TO REINFORCE SLAB

FIGURE 28–13   The slab is haunched under loadbearing walls.

bottom of the footing must be at least one foot below the finish grade, unless local building codes dictate otherwise.

Forms for monolithic slabs are constructed using stakes and edge form boards, plank, or steel manufactured especially for forming slabs and similar objects. The construction procedure is similar to that for wall footing forms (see Procedure 28–A) except the inside form board is omitted **(Figure 28–15).**

- From the batter board lines, plumb down to locate the building corner stakes. Stretch lines from these corner stakes and mark the soil to outline the perimeter of the building. Also mark where

mechanical trenches must be dug. Excavate the trench as needed.

- Reestablish the batter board lines and locate corner stakes. Stretch lines on these corner stakes at the desired elevation.

- Drive intermediate stakes using a gauge block to allow for the form thickness. Snap lines and fasten form boards to the stakes. Level the form with a transit or builder's level. Brace the form as required, remembering that most of the side pressure will be from the concrete. Install reinforcement and mechanicals as required.

Independent Slabs

In areas where the ground freezes to any appreciable depth during winter, the footing for the walls of the structure must extend below the frost line. If slab-on-grade construction is desired in these areas, the concrete slab and foundation wall may be separate. This type of slab-on-grade is called an **independent slab.** It may be constructed in a number of ways according to conditions **(Figure 28–16).**

If the frost line is not too deep, the footing and wall may be an integral unit and placed at the same time. In colder climates, the foundation wall and footing are formed and placed separately. The wall may be set on piles or on a continuous footing, the forming of which has been described previously.

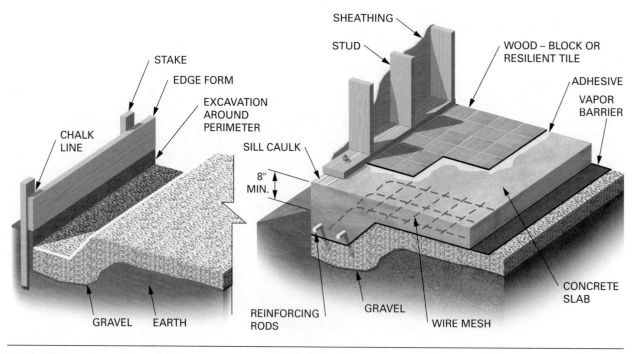

STAKE

EDGE FORM

EXCAVATION AROUND PERIMETER

CHALK LINE

SILL CAULK

8" MIN.

GRAVEL   EARTH

REINFORCING RODS

SHEATHING

STUD

WOOD – BLOCK OR RESILIENT TILE

ADHESIVE

VAPOR BARRIER

CONCRETE SLAB

GRAVEL

WIRE MESH

FIGURE 28–14   Typical monolithic slab-on-grade construction.

FIGURE 28–15 **Thickened edge slabs of two buildings.**

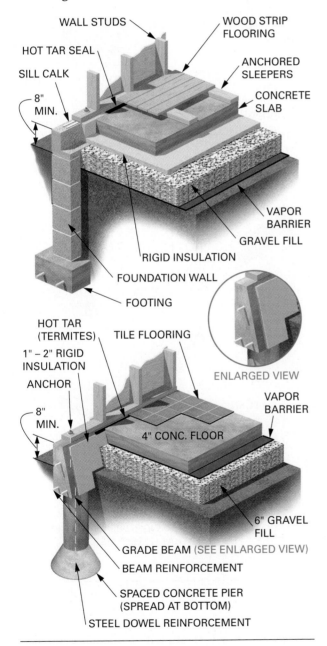

WALL STUDS

HOT TAR SEAL

SILL CALK

8" MIN.

WOOD STRIP FLOORING

ANCHORED SLEEPERS

CONCRETE SLAB

VAPOR BARRIER

GRAVEL FILL

RIGID INSULATION

FOUNDATION WALL

FOOTING

HOT TAR (TERMITES)

1" – 2" RIGID INSULATION

ANCHOR

8" MIN.

TILE FLOORING

ENLARGED VIEW

VAPOR BARRIER

4" CONC. FLOOR

6" GRAVEL FILL

GRADE BEAM (SEE ENLARGED VIEW)

BEAM REINFORCEMENT

SPACED CONCRETE PIER (SPREAD AT BOTTOM)

STEEL DOWEL REINFORCEMENT

FIGURE 28–16 **Independent slabs are constructed in a number of ways.**

Slab Insulation

In colder climates, insulation is required under the slab. The thickness, amount, and location are governed by local energy codes. The only insulation board suited for underground applications is extruded polystyrene. It is produced by various manufacturers in pink, green, blue, and gray colors. It has closed foam cells that keep water from getting into the insulation.

Thin strips of extruded polystyrene may be used between slabs. These strips serve as isolation joints that allow the slabs to expand and contract with temperature and not interfere with each other.

FORMS FOR WALKS AND DRIVEWAYS

Forms for walks and driveways are usually built so water will drain from the surface of the concrete. In these cases, grade stakes must be established and grade lines carefully followed **(Figure 28–17)**.

Establish the grade of the walk or driveway on stakes at both ends. Stretch lines tightly between the end stakes. Drive intermediate stakes. Fasten the edge pieces to the stakes following the line in a manner similar to making continuous footing forms.

Forms for Curved Walks and Driveways

In many instances, walks and driveways are curved. Special metal forms can be purchased to easily form curves, or wood forms can be constructed from ¼-inch plywood or hardboard. They may be used for small-radius curves. If using plywood, install it with the grain vertical for easier bending without breaking. Wetting the stock sometimes helps the bending process.

For curves of a long radius, 1×4 lumber can be used and satisfactorily bent if the curve is not too tight. Lumber can also be curved by making saw **kerfs** spaced close together. The lumber is then bent until the saw kerfs close **(Figure 28–18)**.

The spacing of the kerfs affects the radius of the curve. Closer kerfs yield a tighter bend. The number of kerfs affects the length of the arc—more kerfs equal more curve length.

To determine the kerf spacing, measure the radius distance from the end of the form **(Procedure 28–B)**. Bend the form to close the kerf. The height of the board when the end is raised is the spacing of the kerfs.

The length of the form that has kerfs cut into it can be determined from the circumference of the curve. First calculate the circumference of the whole circle that the curve would make if it were stretched out to a full circle. Then determine what portion of a

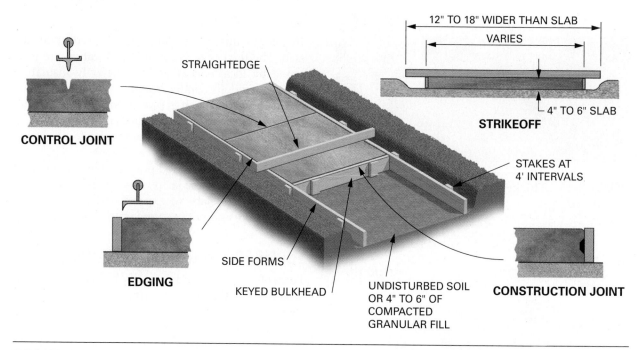

CONTROL JOINT

EDGING

STRAIGHTEDGE

12" TO 18" WIDER THAN SLAB

VARIES

4" TO 6" SLAB

STRIKEOFF

STAKES AT 4' INTERVALS

SIDE FORMS

KEYED BULKHEAD

UNDISTURBED SOIL OR 4" TO 6" OF COMPACTED GRANULAR FILL

CONSTRUCTION JOINT

FIGURE 28–17 Typical way of forming of slabs for walkways.

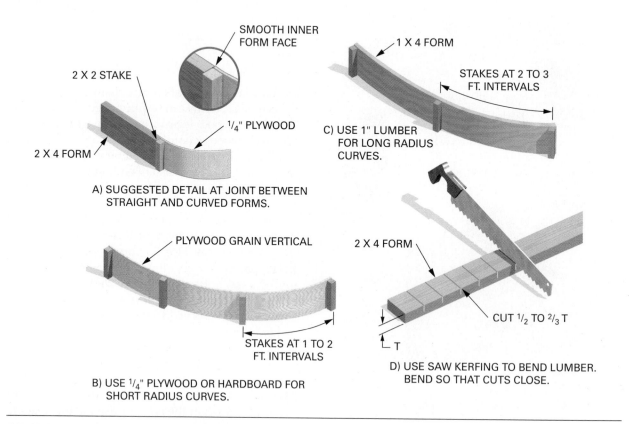

SMOOTH INNER FORM FACE

2 X 2 STAKE

2 X 4 FORM

1/4" PLYWOOD

A) SUGGESTED DETAIL AT JOINT BETWEEN STRAIGHT AND CURVED FORMS.

1 X 4 FORM

STAKES AT 2 TO 3 FT. INTERVALS

C) USE 1" LUMBER FOR LONG RADIUS CURVES.

PLYWOOD GRAIN VERTICAL

STAKES AT 1 TO 2 FT. INTERVALS

B) USE 1/4" PLYWOOD OR HARDBOARD FOR SHORT RADIUS CURVES.

2 X 4 FORM

CUT 1/2 TO 2/3 T

T

D) USE SAW KERFING TO BEND LUMBER. BEND SO THAT CUTS CLOSE.

FIGURE 28–18 Forming curved edge slabs.

whole circle the curve occupies, that is, one-quarter, one-eighth, etc. For example, what is the length of the form if it is a one-quarter bend with a radius of 4 feet? The formula to determine circumference is $C = \pi d$, where π is found with a calculator and $d =$ twice the radius. Thus, $C = \pi(8) = 25.13$ feet. One-quarter of $C = 25.13 \div 4 = 6.28$ feet. The length of form with kerfs cut into it is approximately 6'-3".

PROCEDURE 28–B Making Radius Forms

STEP 1 DETERMINE THE RADIUS OF THE CURVE AND THE AMOUNT OF BEND THE CURVE MAKES, I.E. 4' RADIUS AND ONE-QUARTER BEND.

STEP 2 FROM ONE END OF THE FORM, MEASURE A DISTANCE EQUAL TO THE RADIUS.

STEP 3 MAKE A SAW CUT TO WITHIN 1/4" OF THE BOTTOM OF THE FORM.

STEP 4 FASTEN OR HOLD THE STOCK AND BEND THE END UP TO CLOSE THE KERF.

STEP 5 MEASURE THE DISTANCE THE END HAS RISEN. THIS IS THE KERF SPACING.

STEP 6 CALCULATE THE CIRCUMFERENCE OF A WHOLE CIRCLE WITH THE DESIRED RADIUS.

STEP 7 DIVIDE THE WHOLE BY THE PORTION OF THE CURVE DESIRED.

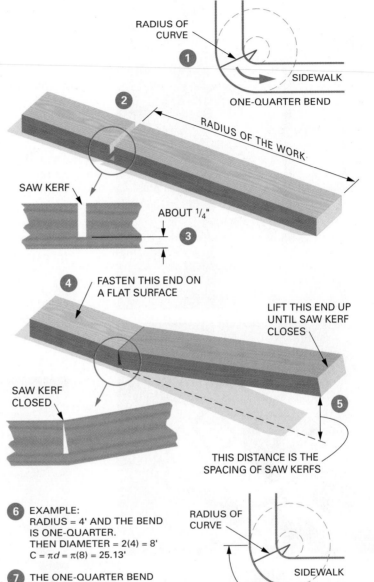

1 RADIUS OF CURVE / SIDEWALK / ONE-QUARTER BEND

2 RADIUS OF THE WORK

SAW KERF

ABOUT 1/4"

3

4 FASTEN THIS END ON A FLAT SURFACE

LIFT THIS END UP UNTIL SAW KERF CLOSES

SAW KERF CLOSED

5 THIS DISTANCE IS THE SPACING OF SAW KERFS

6 EXAMPLE:
RADIUS = 4' AND THE BEND IS ONE-QUARTER.
THEN DIAMETER = 2(4) = 8'
$C = \pi d = \pi(8) = 25.13'$

7 THE ONE-QUARTER BEND HAS A CURVE LENGTH
$(25.13) \div 4 = 6.28'$
APPROXIMATELY 6'-3"

RADIUS OF CURVE / SIDEWALK / LENGTH OF CURVE

STEP BY STEP PROCEDURES

---

29 Wall and Column Forms

Foundation walls and columns are usually formed by using **panels** rather than building forms in place, piece by piece. Panel construction simplifies the erection and stripping of formwork. They also reduce the cost of forming by allowing the material to be used over again.

WALL FORMS

Wall Form Components

Various kinds of panels and panel systems are used. Some concrete panel systems are manufactured of steel, aluminum, or wood. They are designed to be

FIGURE 29–1   Styles of formwork for concrete walls.

used many times. Specially designed hardware is used for joining, spacing, aligning, and bracing the panels **(Figure 29–1).** Care should be taken to keep the form system clean after concrete is placed.

Form Panels. Panels are placed side by side to form the inside and outside of the foundation walls. Panel sizes vary with the manufacturer. Standard sizes for manufactured panels are 2′ × 8′. Wood panels are often 4′ × 8′. Narrower panels of several widths are available to be used as fillers when the space is too narrow for a standard size panel.

Snap Ties. **Snap ties** hold the wall forms together at the desired distance apart. They support both sides against the lateral pressure of the concrete **(Figure 29–2).** They allow the side pressure on the outer panel to be carried or canceled out by the pressure on the inner panel. These ties reduce the need for external bracing and greatly simplify the erection of wall forms. The design of a form for a particular job, including the spacing of the ties and studs, is decided by a structural engineer.

These ties are called snap ties because after removal of the form, the projecting ends are snapped

FIGURE 29–2   A snap tie holds inner and outer forms from spreading when the concrete is placed.

PROCEDURE 29–A   Breaking Snap Ties

STEP 1 SLIDE THE SNAP TIE WRENCH UP AGAINST THE TIE SO THAT THE FRONT OF THE WRENCH IS TOUCHING THE CONCRETE.

STEP 2 KEEPING THE FRONT OF THE WRENCH TIGHT AGAINST THE CONCRETE, PUSH THE HANDLE END TOWARDS THE CONCRETE WALL SO THAT THE TIE IS BENT OVER AT APPROXIMATELY A 90° ANGLE.

STEP 3 ROTATE THE WRENCH AND TIE END 1/4 TO 1/2 TURN BREAKING OFF THE TIE END.

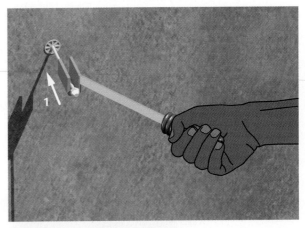

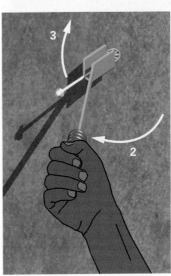

off slightly inside the concrete surface. A special snap tie wrench is used to break back the ties **(Procedure 29–A).** The small remaining holes are easily filled.

Because of the great variation in the size and shape of concrete forms, a large number of snap tie styles are used. For instance, *flat ties* of various styles are used with some manufactured panels. For heavier formwork, *coil ties* and reusable *coil bolts* are used **(Figure 29–3).** For each kind of tie, there are also several sizes and styles. There are hundreds of kinds of form hardware. To become better acquainted with form hardware, study manufacturers' catalogs.

Walers. The snap ties run through and are wedged against form members called **walers.** Walers are doubled 2 × 4 pieces with space between them of about ½". They may be horizontal or vertical. Walers

are spaced at right angles to the panel frame members. The number and spacing depend on the style of the form and the pressure exerted on the form **(Figure 29–4).**

The vertical spacing of the snap ties and walers depends on the height of the concrete wall. The vertical spacing is closer together near the bottom. This is because there is more lateral pressure from the concrete there than at the top **(Figure 29–5).**

For low wall forms less than 4 feet in height, the panel may be laid horizontally with vertical walers spaced as required **(Figure 29–6).**

Care of Forms. After the concrete placed in the footing has hardened sufficiently (sometimes three days), the forms are removed and cleaned. Any concrete clinging to the form should be scraped off. They are easier to clean at this stage rather than later when the concrete has reached full strength.

COIL TIES

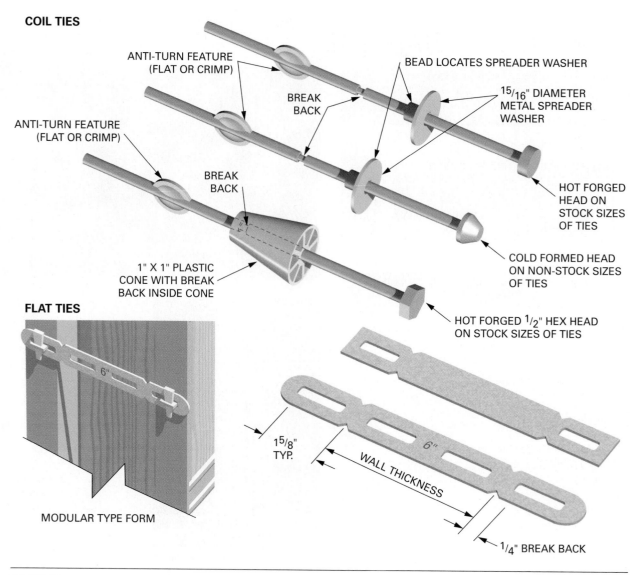

ANTI-TURN FEATURE
(FLAT OR CRIMP)

BEAD LOCATES SPREADER WASHER

BREAK
BACK

$^{15}/_{16}$" DIAMETER
METAL SPREADER
WASHER

ANTI-TURN FEATURE
(FLAT OR CRIMP)

BREAK
BACK

HOT FORGED
HEAD ON
STOCK SIZES
OF TIES

1" X 1" PLASTIC
CONE WITH BREAK
BACK INSIDE CONE

COLD FORMED HEAD
ON NON-STOCK SIZES
OF TIES

FLAT TIES

HOT FORGED $^{1}/_{2}$" HEX HEAD
ON STOCK SIZES OF TIES

6"

$1^{5}/_{8}$"
TYP.

WALL THICKNESS

6"

$^{1}/_{4}$" BREAK BACK

MODULAR TYPE FORM

FIGURE 29–3   **A large variety of snap ties are manufactured.**

FIGURE 29–4   **Walers are easily installed on forming systems when special hardware is used.**

Forms may be oiled to protect them and stored ready for reuse.

The forms last longer if they are handled properly. Place them gently and do not drop them. Damaged corners from dropping the panel will affect the surface appearance of the concrete. They are also easier to use if they are not twisted or broken.

Preparing for Wall Form Assembly

Locating the Forms.   Lines are stretched on the batter boards in line with the outside of the foundation wall. A plumb bob is suspended from the layout lines to the footing. Marks that are plumb with the layout lines are placed on the footing at each corner. A chalk line is snapped on the top of the footing between the corner marks outlining the outside of the foundation wall.

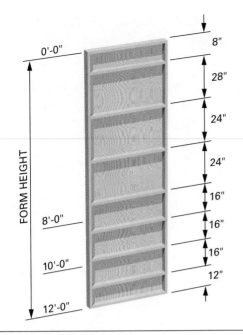

FIGURE 29-5 Horizontal panel stiffeners are placed closer together near the bottom than they are at the top. Dimensions are for purposes of illustration only.

Installing Plates. Sometimes panels are set on 2 × 4 or 2 × 6 lumber plates. Plates provide a positive online wall pattern. They can also level out rough areas on the footing. Plates function to locate the position and size of pilasters and changes in wall thickness, corners, and other variations in the wall **(Figure 29–7).** The outer plate is fastened to the footing using ma-

sonry nails or pins driven by a powder-actuated tool. The inner plate is fastened only to the concrete form. This is done to allow the wall form thickness to swell slightly when the concrete is placed. This movement is due to the slack in the snap ties being taken up.

Manufactured Wall Form Assembly

Erecting Panels. Stack the number of panels necessary to form the inside of the wall in the center of the excavation. Lay the panels needed for the outside of the wall around the walls of the excavation. The face of all panels should be oiled or treated with a chemical releasing agent. This provides a smooth face to the hardened concrete and makes stripping of the forms easy.

Panels may be assembled in inside/outside pairs. But if rebar is required in the wall, then erect the outside wall forms first. This makes installing and tying rebar easier. Set panels in place at all corners first. Set base snap ties into slots while placing the panel and attach the outside corners **(Figure 29–8).** Nail the panel into the plate with duplex nails. Make sure the corners are plumb by testing with a hand level and brace.

Fill in between the corners with panels, keeping the same width panels opposite each other.

Placing Snap Ties. Place snap ties in the dadoes between panels as work progresses. Tie panels together using the wedge bolts or the connection designed by the manufacturer **(Figure 29–9).** Snap ties must be positioned as each panel is placed. Be

FIGURE 29–6 Forms can be laid horizontally or vertically as needed.

IT IS RECOMMENDED THAT THE INSIDE
PLATE NOT BE FASTENED

FASTEN OUTSIDE
PLATE INTO CONCRETE

THIS EDGE
TO CHALK LINE

OUTSIDE EDGE
OF FOOTING

CROSS SECTION A

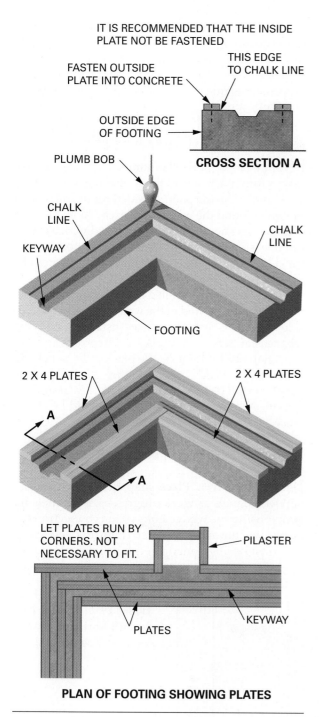

PLUMB BOB

CHALK LINE

CHALK LINE

KEYWAY

FOOTING

2 X 4 PLATES

2 X 4 PLATES

A

A

LET PLATES RUN BY
CORNERS. NOT
NECESSARY TO FIT.

PILASTER

PLATES

KEYWAY

PLAN OF FOOTING SHOWING PLATES

FIGURE 29–7   Attaching plates to the footing before forming walls.

BASE TIE

BASE TIE
WEDGE
BOLT

FIGURE 29–8   Base ties hold the panel bottoms from spreading.

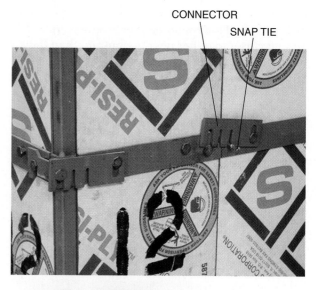

CONNECTOR

SNAP TIE

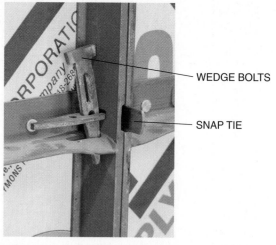

WEDGE BOLTS

SNAP TIE

FIGURE 29–9   Form panels are connected by the snap ties, which are secured by special bolts and connectors.

careful not to leave out any snap ties. Wedge bolts are not hammered tight, only drawn up snug **(Figure 29–10).**

Use filler panels as necessary to complete the outside wall section. Brace the wall temporarily as needed. Install rebar as needed, then erect the panels for the inside of the wall. Keep joints between panels opposite to those for the outside of the wall

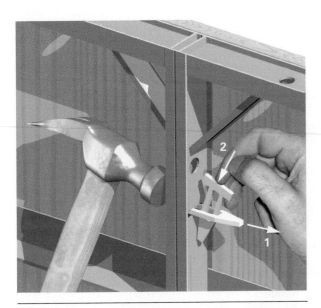

FIGURE 29–10 Tapping the side as the wedge is pushed down adequately tightens the form, yet leaves it loose enough to be easily removed later.

(Figure 29–11). Insert the other ends of the snap ties between panels as they are erected.

Installing Walers. In a typical 8-foot-high wall, walers are only used to help keep the forms straight. Typically the outside panels have one row of walers near the top edge. Special brackets, called waler ties, are used to attach walers to two rows of 2 × 4s **(Figure 29–12).** Let the bracket come between them and wedge into place.

Forming Pilasters. The wall may be formed at intervals for the construction of **pilasters.** These are

FIGURE 29–11 Panels should be assembled with opposing joints so snap ties can be easily installed.

FIGURE 29–12 Special brackets used to attach walers.

thickened portions of the wall that serve to give the wall more lateral (side-to-side) strength. They may also provide support for beams. They are formed with inside and outside corners in the usual manner. If the pilaster is large, longer snap ties are necessary **(Figure 29–13).**

Erecting Wood Wall Forms

Erecting Panels. Stack the number of panels necessary to form the inside of the wall in the center of the excavation. Lay the panels needed for the outside of the wall around the walls of the excavation. The face of all panels should be oiled or treated with a chemical releasing agent. This provides a smooth face to the hardened concrete and makes stripping of the forms easy.

Panels may be assembled in inside/outside pairs. But if rebar is required in the wall then erect

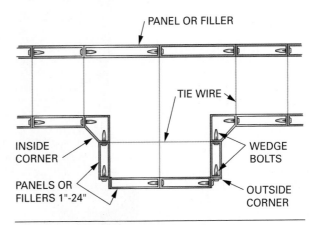

FIGURE 29–13 Pilasters may be formed like any other intersection. Note that inside the pilaster, tie wire is used to hold panel seams.

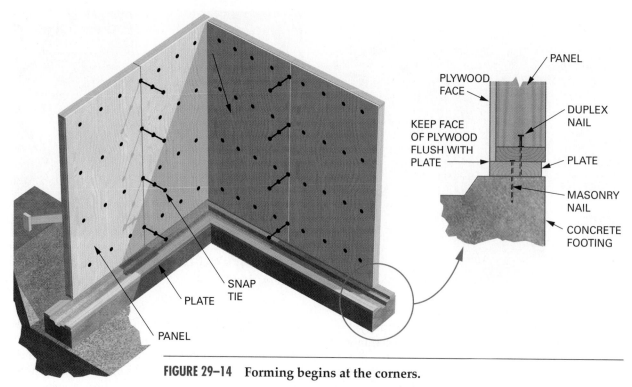

FIGURE 29–14   Forming begins at the corners.

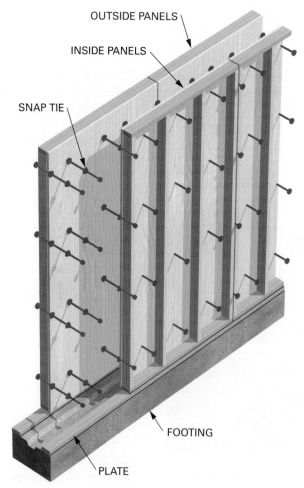

FIGURE 29–15   Inside panels of the form are assembled later with snap ties.

the outside wall forms around the perimeter first. This makes installing and tying rebar easier. Set panels in place at all corners first. Nail the panel into the plate with duplex nails **(Figure 29–14).** Make sure the corners are plumb by testing with a hand level and brace.

Fill in between the corners with panels, keeping the same width panels opposite each other. Place snap ties in the dadoes between panels as work progresses. Tie panels together by driving U-shaped clamps over the edge 2 × 4s or by nailing them together with duplex nails. Use filler panels as necessary to complete each wall section. Brace the wall temporarily as needed.

Placing Snap Ties. After the panels for the outside of the wall have been erected, place snap ties in the in-

termediate holes. Be careful not to leave out any snap ties. Erect the panels for the inside of the wall. Keep joints between panels opposite to those for the outside of the wall. Insert the other end of the snap ties between panels and in intermediate holes as panels are erected **(Figure 29–15).** If the concrete is to be reinforced, the rebars are tied in place before the inside panels are erected.

Installing Walers. When all panels are in place, install the walers. Let the snap ties come through them and wedge into place. Care must be taken when installing and driving snap tie wedges **(Figure 29–16).** Let the ends of the walers extend by the corners of the formwork. Reinforce the corners with vertical 2 × 4s **(Figure 29–17).** This is called **yoking** the corners.

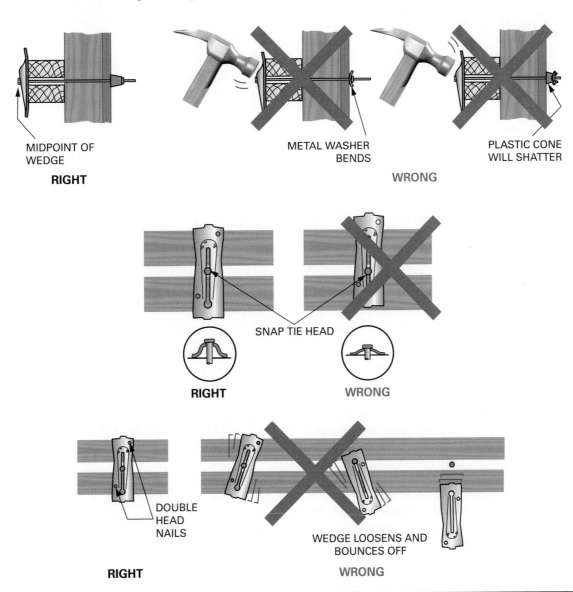

MIDPOINT OF WEDGE

RIGHT

METAL WASHER BENDS

WRONG

PLASTIC CONE WILL SHATTER

SNAP TIE HEAD

RIGHT

WRONG

DOUBLE HEAD NAILS

WEDGE LOOSENS AND BOUNCES OFF

RIGHT

WRONG

FIGURE 29–16 Care must be taken when installing snap tie wedges.

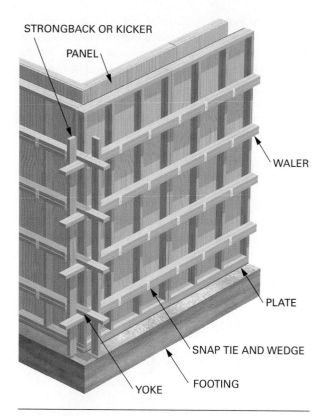

FIGURE 29–17   Yokes are vertical members that hold outside form corners together.

Forming Pilasters. The wall may be formed at intervals for the construction of pilasters. These are thickened portions of the wall that serve to give the wall more lateral (side-to-side) strength. They may strengthen the wall or provide support for beams. They may be constructed on the inside or outside of the wall. In the pilaster area longer snap ties are necessary **(Figure 29–18).**

Completing the Wall Forms

Straightening and Bracing. Brace the walls inside and outside as necessary to straighten them. Wall forms are easily straightened by sighting by eye along the top edge from corner to corner. Another method of straightening is to use line and gauge blocks, stretching a line from corner to corner at the top of the form over two blocks of the same thickness. Move the forms until a test block of equal thickness passes just under the line **(Figure 29–19).**

A special adjustable form brace and aligner are used to position and hold wall forms **(Figure 29–20).** This allows for easy prying when straightening the formwork simply by rotating the turnbuckle. Braces are cut with square ends and nailed into place against **strongbacks.** Strongbacks are placed across walers at right angles wherever braces are needed. The sharp

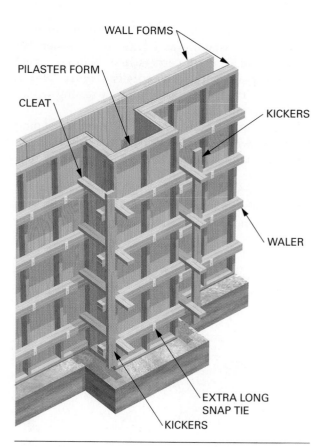

FIGURE 29–18   Formwork for a pilaster.

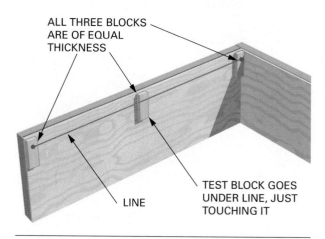

FIGURE 29–19   Straightening the wall form with a line and test block.

corner of the square ends helps hold the braces in place. It also allows easy prying with a bar to tighten the braces and move the forms **(Figure 29–21).** There is no need to make a bevel cut on the ends of braces.

Leveling. After the wall forms have been straightened and braced, chalk lines are snapped on the inside of the form for the height of the foundation

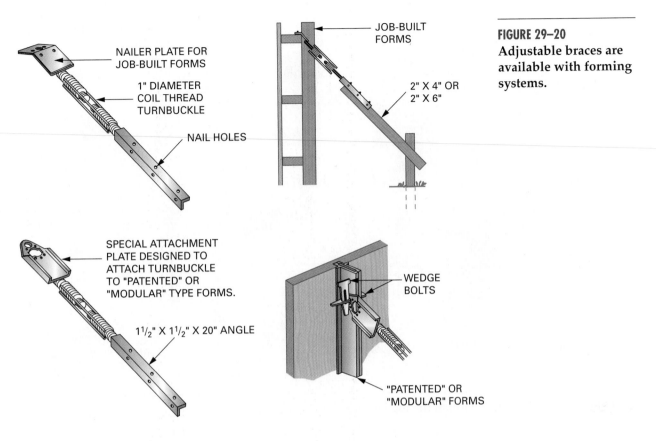

NAILER PLATE FOR JOB-BUILT FORMS

1" DIAMETER COIL THREAD TURNBUCKLE

NAIL HOLES

SPECIAL ATTACHMENT PLATE DESIGNED TO ATTACH TURNBUCKLE TO "PATENTED" OR "MODULAR" TYPE FORMS.

1½" X 1½" X 20" ANGLE

JOB-BUILT FORMS

2" X 4" OR 2" X 6"

WEDGE BOLTS

"PATENTED" OR "MODULAR" FORMS

FIGURE 29–20
Adjustable braces are available with forming systems.

PLYWOOD

WRECKING BAR

SNAP TIE

CUT BRACES WITH SQUARE ENDS

BRACE

STRONGBACK

FIGURE 29–21   Cut braces with square ends. This allows for easy prying when straightening the formwork.

wall. Grade nails may be driven partway in at intervals along the chalk line as a guide for leveling the top of the wall. If the tops of the panels are level with each other, a short piece of stock notched at both ends can be run along the panel tops to screed the concrete. Another method is to fasten strips on the inside walls along the chalk line and use a sim-

ilar screeding board, notched to go over the strips **(Figure 29–22)**.

Setting Anchor Bolts. As soon as the wall is screeded, anchor bolts are set in the fresh concrete. A number of various styles and sizes are manufactured **(Figure 29–23)**. The type is usually specified on the foundation plan. Care must be taken to set the anchor bolts at the correct height and at specified locations. Check local codes for anchor bolt spacing. An anchor bolt *template* is sometimes used to accurately place the bolts **(Figure 29–24)**.

Openings in Concrete Walls

In many cases, openings must be formed in concrete foundation walls for such things as windows, doors, ducts, pipes, and beams. The forms used for providing the larger openings are called **blockouts.**

Constructing Blockouts. Blockouts are also called **bucks.** The blockout is usually made of 2-inch dimension lumber. Its width is the same as the thickness of the foundation wall. Nailing blocks or strips are often fastened to the outside of the bucks. These are beveled on both edges to lock them into the concrete when the form is stripped. They provide for the fastening of window and door frames in the openings. Intermediate pieces may be necessary in

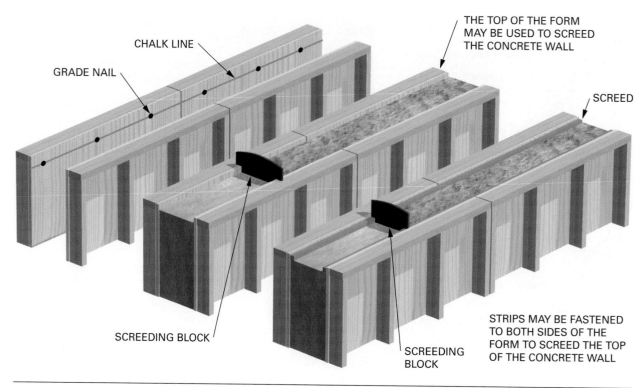

CHALK LINE

GRADE NAIL

THE TOP OF THE FORM MAY BE USED TO SCREED THE CONCRETE WALL

SCREED

SCREEDING BLOCK

SCREEDING BLOCK

STRIPS MAY BE FASTENED TO BOTH SIDES OF THE FORM TO SCREED THE TOP OF THE CONCRETE WALL

FIGURE 29–22 Methods to establish the top surface to the concrete in a wall form.

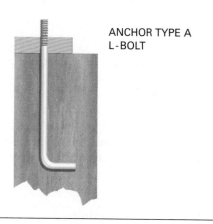

ANCHOR TYPE A L-BOLT

FIGURE 29–23 Typical anchor bolts.

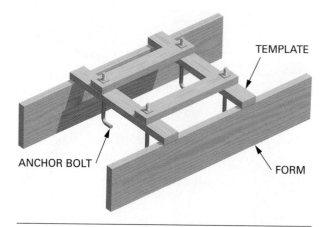

TEMPLATE

ANCHOR BOLT

FORM

FIGURE 29–24 Templates are sometimes used to accurately place anchor bolts in fresh concrete.

bucks for large openings to withstand the pressure of the concrete against them **(Figure 29–25).**

Large blockouts are made to the specified dimension. They are installed against the inside face of the outside panels. Duplex nails through the outside panels hold the blockouts in place. The inside wall panels are then installed against the other side of the blockouts. Nails are driven through the inside wall panels to secure the bucks on the inside.

Girder Pockets. Girder pockets are recesses in the top of the foundation wall. They are sometimes re-

quired to receive the ends of **girders** (beams). A box of the size needed is made and fastened to the inside at the specified location **(Figure 29–26).** Because it is near the top of the form, not much pressure from the liquid concrete is exerted against it.

COLUMN FORMS

To form columns, like all other kinds of formwork, as much use as possible is made of panels, manufac-

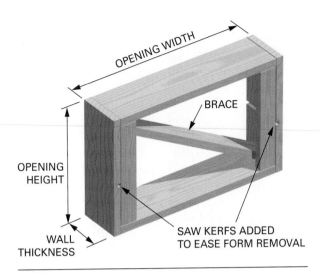

FIGURE 29–25 Construction of a typical window blockout.

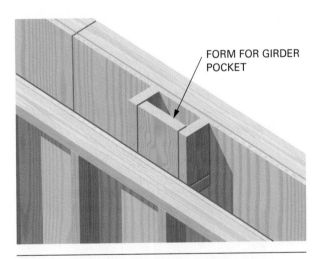

FIGURE 29–26 A small box is attached to the inside form for a girder pocket in a foundation wall.

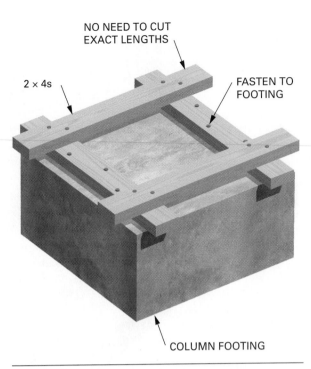

FIGURE 29–27 A yoke is constructed on the column footing to start the forming of a column.

A **quarter-round** molding may be used for a cove shape. Triangular-shaped strips of wood can be used to form a chamfer on the corners of the column **(Figure 29–28)**.

The face of the column may be decorated with **flutes** by fastening vertical strips of **half-round** molding spaced on the panel faces. In addition, *form liners* are often used. They provide various wood, brick, stone, and many other textures in the face of the concrete wall or column.

Build panels for the opposite sides of the column to overlap the previously built panels. Plumb and nail the corners together with duplex nails. Install 2 × 4 yokes around the column forms, letting their ends extend beyond the corners. Nail them together where they overlap. Yoke the column closer together at the bottom. The number and spacing of yokes depend on the height of the column. These details are specified by the form designer. Install vertical 2 × 4s between the overlapping ends of the yokes **(Figure 29–29).** Brace the formwork securely to hold it plumb.

Erecting Manufactured Column Forms

Manufactured forms can be used instead of wood forms. Various width forms can be assembled with outside corners **(Figure 29–30)**. These are erected in the same fashion as for wall forms. Wedge bolts secure the panels to the outside corners. Nail the panels to the plate fastened to the footing. No snap ties

tured or job built, to simplify erection and stripping. Columns may be formed in square, rectangular, circular, or a number of other shapes.

Erecting Wood Column Forms

Stretch lines and mark the location of the column on the footing. Fasten two 2 × 4 pieces to the footing on opposite sides of the column and outside of the line by the panel thickness. Fasten the other two to the overlapping ends of the first pair in a similar location **(Figure 29–27).**

Build and erect two panels to form the thickness and height of the column. A **cove** molding of desired size may be fastened on the edges of this panel to form a radius to the corners of the concrete column.

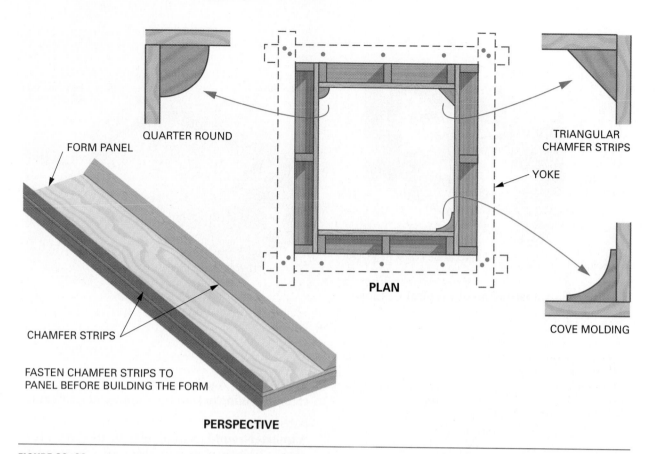

QUARTER ROUND

TRIANGULAR
CHAMFER STRIPS

FORM PANEL

YOKE

PLAN

CHAMFER STRIPS

COVE MOLDING

FASTEN CHAMFER STRIPS TO
PANEL BEFORE BUILDING THE FORM

PERSPECTIVE

FIGURE 29–28   The corners of a concrete column can be formed in several ways.

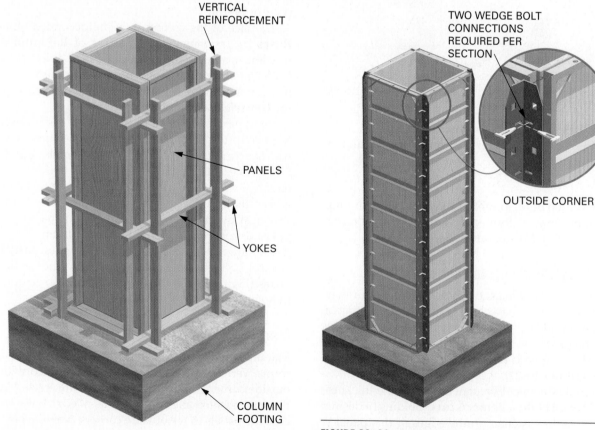

VERTICAL
REINFORCEMENT

TWO WEDGE BOLT
CONNECTIONS
REQUIRED PER
SECTION

PANELS

OUTSIDE CORNER

YOKES

COLUMN
FOOTING

FIGURE 29–29   A completed concrete column form.

FIGURE 29–30   Yokes are not necessary to form columns when forming systems are used.

COLUMN CAPITAL FORM

CIRCULAR COLUMN FORM

FIGURE 29–31 Manufactured circular column and column capital forms simplify forming.

FIGURE 29–32 Some forms are designed to be used only once.

FIGURE 29–33 Insulated concrete forms are made of rigid foam panels.

are needed. Concrete corners can be shaped in the manner previously discussed.

Steel forms are available to form circular columns **(Figure 29–31).** They bolt together and need only to be braced. Another system uses heavy-duty cardboard tubes. These can be assembled on base forms for a monolithic column and footing. After concrete placement the forms are stripped by peeling off the cardboard. This makes this form a one-time-use-only form **(Figure 29–32).**

Insulated Concrete Forms

Insulated concrete forms (ICF) are made of 2-inch-thick expanded polystyrene **(Figure 29–33).** Units are shaped into blocks with a metal or plastic snap tie system. The block sizes vary depending on manufacturer and range from whatever wall thickness needed × 16 to 18 inches high × 3 to 4 feet long. Some blocks have fastening strips included on their surfaces to make it easy to screw on a wall finish later. The benefit to this system is that the finished product is a heavily insulated wall that reduces heating and cooling requirements. Also, no disassembly of forms is necessary.

Assembly begins one course at a time with corner and straight blocks. The next course is set so the vertical seams are broken. This helps interlock the blocks for strength. Rebar is installed as needed after each block layer is installed. Special reinforced tape is sometimes used to help hold the course of blocks together.

The success of this system is in the method of placing the concrete. Placement of the concrete is done in levels or lifts. Only one course is filled at a time, giving the concrete time to hydrate or harden slightly. Typically by the time placement is done for the entire perimeter, the first section is ready to support another lift.

30   Concrete Stair Forms

STAIR FORMS

It may be necessary to refer to Chapters 47 and 48 on stair framing for definition of stair terms, types of stairs, stair layout, and methods of stair construction.

Concrete stairs may be suspended or supported by earth **(Figure 30–1)**. Each type may be constructed between walls or have open ends.

Forms for Earth-Supported Stairs

Before placing concrete, stone, gravel, or other suitable fill is graded to provide proper thickness to the stairs. It should not be overly thick, or concrete will be wasted. It may be necessary to lay out the stairs before the supporting material is placed.

Forming between Existing Walls. When earth-supported stairs are formed between two existing walls, the **rise** and **run** of each step are laid out on the inside of the existing walls. *Rise* is the vertical distance that a step will rise. It is the height of each step, which is also called the *riser. Run* is the horizontal distance of a step. It is roughly the width of each step, which is called the tread. Boards are ripped to width to correspond to the height of each riser. The board is wedged and secured in place with its inside face aligned to the riser layout line. The top and bottom edges of the board are aligned with tread layout lines. After the riser boards are secured in position, they are braced from top to bottom at midspan. This keeps them from bowing outward due to the pressure of the concrete **(Figure 30–2)**.

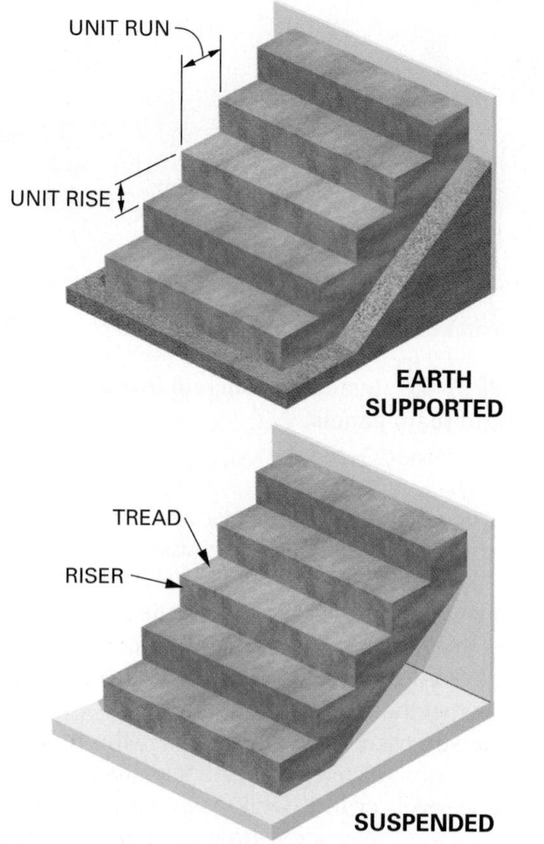

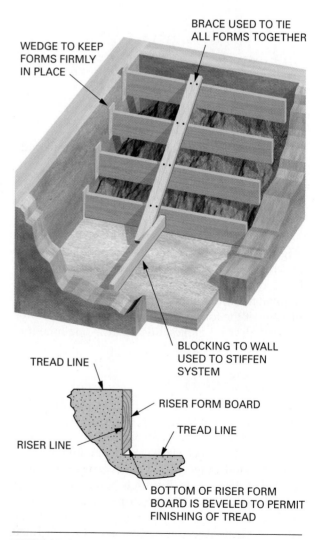

FIGURE 30–1 and accompanying labels:

UNIT RUN

UNIT RISE

EARTH SUPPORTED

TREAD

RISER

SUSPENDED

FIGURE 30–1   Concrete stairs may be formed on supporting soil or suspended.

Figure 30–2 labels:

WEDGE TO KEEP FORMS FIRMLY IN PLACE

BRACE USED TO TIE ALL FORMS TOGETHER

BLOCKING TO WALL USED TO STIFFEN SYSTEM

TREAD LINE

RISER FORM BOARD

TREAD LINE

RISER LINE

BOTTOM OF RISER FORM BOARD IS BEVELED TO PERMIT FINISHING OF TREAD

FIGURE 30–2   Concrete stairs may be formed between existing walls. Note that the bottom edge of the riser form is beveled.

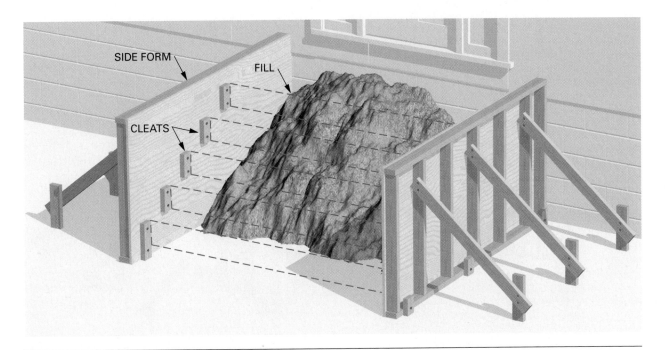

FIGURE 30–3   End panels are braced firmly in position. The stairs are laid out, cleats fastened to them, and the space filled with supporting material.

The riser forms are beveled on the bottom edge. Beveling the bottom edge of the plank permits the mason to trowel the entire surface of the tread. Otherwise, the bottom edge of the riser form will leave its impression in the concrete tread.

Forming Stairs with Open Ends. In cases where the ends of the stairs are to be open, panels are erected on each end. It does no harm if the panels are larger than needed, the panels only need to be plumb. The distance between them is the desired width of the stairs. The end panels are then firmly fastened and braced in position.

The risers and treads are laid out on the inside surfaces of the panels. **Cleats** (short strips of wood) are fastened at each riser location. Allowances must be made for the thickness of the riser form board. Screws or duplex nails should be used to make stripping the form easier. The space should then be filled and compacted with the supporting material to the proper level. Any necessary reinforcing should be installed **(Figure 30–3).**

The boards used to form the risers are ripped to width and cut to length. The bottom edge is beveled to permit finishing of the total tread width. They are then fastened with duplex nails or screws through the side panels in a position against the cleats, with the top and bottom edges aligned to the tread layout lines. The riser forms are then braced from top to bottom at intervals between the two ends **(Figure 30–4).**

Forms for Suspended Stairs

Forms for suspended stairs are more difficult to build. Instead of earth support, a form needs to be built on the bottom to support the stair slab. With proper design and reinforcement, the stairs are strong enough to support themselves in addition to the weight of the traffic. As with earth-supported stairs, suspended stairs may be formed between existing walls, with open ends, or both.

Forming between Existing Concrete Walls. A cross-section of the formwork for suspended stairs between existing walls, with the form members identified, is shown in **Figure 30–5.** First, lay out the treads, risers, and stair slab bottom on both of the walls between which the stairs run. This can be done by using a hand level, ruler, and chalk line **(Figure 30–6).**

The form for the bottom of the stair slab is then laid out. Allow for the thickness of the plywood deck, the width of the supporting joists, and the depth of the *horses.* Snap a line on both walls to indicate their bottom and also the top of the supporting *shores.*

Allowing for a plank sill and wedges at the bottom end of the shores, cut the shores to length with the correct angle at the top. They should be cut a little short to allow for wedging the shore to proper height. Install them at specified intervals on top of the sill. Brace in position, then place and fasten the horses on top with **scabs** or **gusset** plates. Scabs are short lengths of

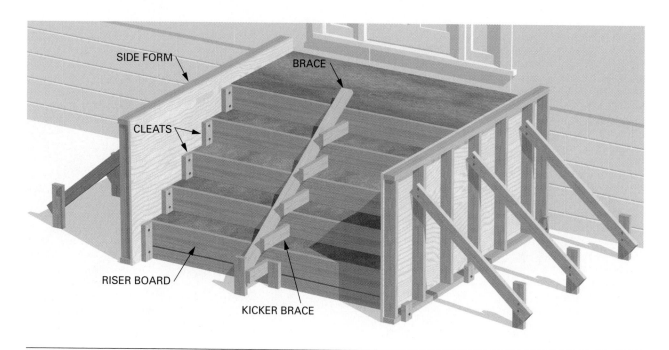

FIGURE 30–4 Typical form construction for concrete stairs having open ends.

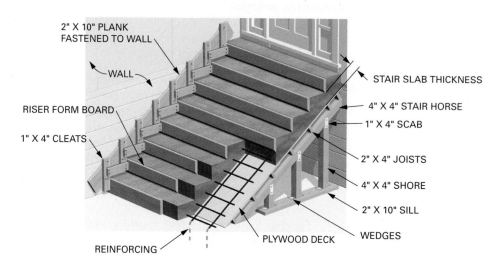

FIGURE 30–5 Cross-section of formwork for suspended stairs between walls.

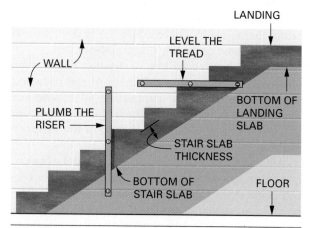

FIGURE 30–6 Make the layout for suspended stairs on the inside of the walls.

narrow boards fastened across a joint to strengthen it. Install joists at right angles across the horses at the specified spacing **(Figure 30–7)**. Fasten the plywood form in position on top of the joists. Use only as many fasteners as needed to hold the plywood in place for easier stripping later. Wedge the shoring as necessary to bring the surface of the plywood to the layout line. Fasten the wedges in place so they will not move. The plywood should now be oiled to facilitate stripping. Install the reinforcing rods.

Rip the riser boards to width. Cut to length, allowing for wedges. Bevel the outside bottom edge of the treads to allow for easy leveling of the concrete at each tread. Install the riser form boards to the layout lines. Wedge in position as shown previously in Figure 30–2.

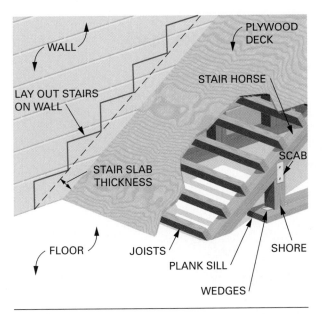

FIGURE 30–7 **The framework for the bottom form of the suspended stairs.**

Another method, shown previously in Figure 30–5, secures the riser form with cleats fastened to a 2 × 10 attached to the wall. The 2 × 10 is attached about an inch above the stair layout. Project the riser layout lines upward on the planks. Fasten cleats to the plank. Then brace the cleats and fasten the riser board to them.

Forming with Open Ends. A completed form for suspended stairs with open ends is shown in **Figure 30–8.** This style of form sets the side panels for the stairs on

the stair horse. The riser layout is made on the side panels and then set in place on top of the stair horse.

Laying Out the Side Forms. First, measure and snap a line on the side form up from the bottom edge a distance equal to the thickness of the slab. The inside corner intersection of the treads and risers lies along this line. Lay out the risers and treads above this line. The stair slab thickness is on the bottom edge of the form panels.

A **pitch board** speeds up the layout process **(Procedure 30–A).** It is a triangular scrap for which the tread width and the riser height are the legs of a right triangle. To mark the tread and riser locations use the pitch board held to the previously snapped slab line.

Laying Out the Stair Horses. The length of the stair horse and shores can be determined from the side form panels. Lay the side panel down on a floor surface. Snap a line that is parallel to the bottom edge of the side form. The distance away from the form bottom is equal to the thickness of the stair plywood deck and supporting joists **(Figure 30–9).**

From the side panel, extend the bottom floor line and the top plumb lines, making a large triangle on the floor. Lay out the thickness of the stair horse and shores including the sill. Measure and cut the length of the stair horse and shore. Cut one stair horse for each side to length at the angles indicated.

Installing Horses and Joists. Temporarily support and brace the horses in position. Install shores in a manner similar to that of closed stairs. Brace shores

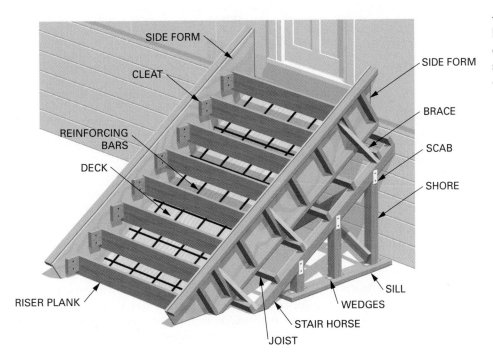

FIGURE 30–8 **A completed form for suspended stairs with open ends.**

STEP BY STEP PROCEDURES

PROCEDURE 30–A Laying Out a Set of Stairs Using a Pitch Board

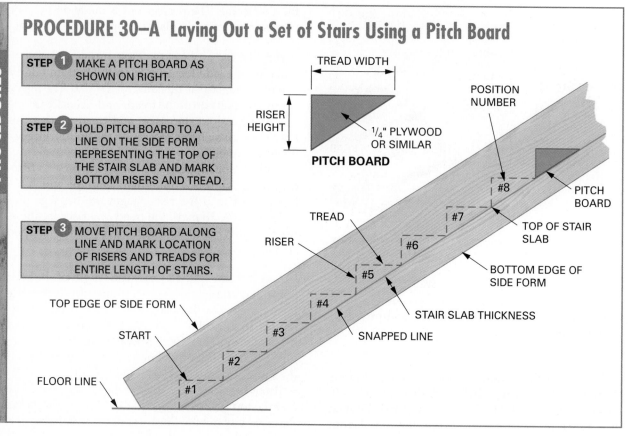

STEP 1 MAKE A PITCH BOARD AS SHOWN ON RIGHT.

STEP 2 HOLD PITCH BOARD TO A LINE ON THE SIDE FORM REPRESENTING THE TOP OF THE STAIR SLAB AND MARK BOTTOM RISERS AND TREAD.

STEP 3 MOVE PITCH BOARD ALONG LINE AND MARK LOCATION OF RISERS AND TREADS FOR ENTIRE LENGTH OF STAIRS.

TREAD WIDTH

RISER HEIGHT

1/4" PLYWOOD OR SIMILAR

PITCH BOARD

POSITION NUMBER

#8

PITCH BOARD

#7

TOP OF STAIR SLAB

TREAD

#6

BOTTOM EDGE OF SIDE FORM

RISER

#5

STAIR SLAB THICKNESS

#4

TOP EDGE OF SIDE FORM

#3

SNAPPED LINE

START

#2

FLOOR LINE

#1

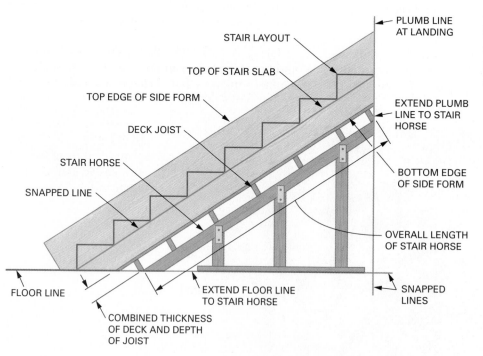

STAIR LAYOUT

PLUMB LINE AT LANDING

TOP OF STAIR SLAB

TOP EDGE OF SIDE FORM

EXTEND PLUMB LINE TO STAIR HORSE

DECK JOIST

STAIR HORSE

SNAPPED LINE

BOTTOM EDGE OF SIDE FORM

OVERALL LENGTH OF STAIR HORSE

FLOOR LINE

EXTEND FLOOR LINE TO STAIR HORSE

SNAPPED LINES

COMBINED THICKNESS OF DECK AND DEPTH OF JOIST

FIGURE 30–9 Method for finding the length and cuts of the stair horses.

and horses firmly in position. Fasten joists at designated intervals to the horses, leaving an adequate and uniform overhang on both sides. Fasten decking to joists using a minimum number of nails.

Installing Side and Riser Forms. Snap lines on the deck for both sides of the stairs. Stand the side forms up with their inside face to the chalk line. Fasten them through the deck into the joists with duplex nails.

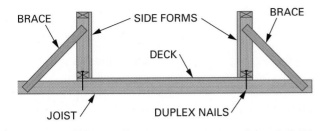

FIGURE 30–10 Cross-section of the side form installed on the deck and braced.

Brace the side forms at intervals so they are plumb for their entire length **(Figure 30–10).** Apply form oil to the deck before reinforcing bars are installed.

Rip the riser boards to width and bevel the bottom edge. Install them on the riser layout line. Fasten them with duplex nails through the side forms into their ends. Install cleats against the riser boards on both ends for additional support. Install intermediate braces to the riser form boards from top to bottom, if needed.

Economy and Conservation in Form Building

Economical concrete construction depends on the reuse of forms. Forms should be designed and built to facilitate stripping and reuse. Use panels to build forms whenever possible. Use only as many nails as necessary to make stripping forms easier.

Care must be taken when stripping forms to prevent damage to the panels so they can be reused. Stripped forms should be cleaned of all adhering concrete and stacked neatly.

 CAUTION CAUTION CAUTION CAUTION

Remove all protruding nails to eliminate the danger of stepping on or brushing against them. ■

Long lengths of lumber can often be used without trimming. Random length boards can extend beyond the forms. There is no need to spend a lot of time cutting lumber to exact length. The important thing is to form the concrete to specified dimensions without spending too much time in unnecessary fitting.

Estimating Concrete Quantities

Sometimes it is the carpenter's responsibility to order concrete for the job. Ready-mix concrete is sold by the cubic yard or cubic meter. To determine the number of cubic yards of concrete needed for a job, first find the number of cubic feet. Because there are 27 cubic feet in one cubic yard, dividing cubic feet by 27 will give cubic yards.

For example, how many cubic yards will be needed for an 8-inch-thick, 8-foot-high, and 36-foot-long wall? First convert the thickness to feet: $8 \div 12 = 0.6666667$ foot. Note here that the 6 is repeated forever but the calculator rounds the last displayed digit to 7. Then calculate the volume by multiplying thickness × width × length or $0.6666667 \times 8 \times 36 = 192$ cubic feet. Dividing by 27 yields 7.1 cubic yards.

The actual amount of concrete needed is often not the same as the amount calculated. Slight variations in the forms may cause large errors in the calculated volume. The quantities from the readymix company are often close, but not perfect. Also some spillage will occur. For these reasons, the amount of concrete ordered should be a little more than calculated. This extra is called the **waste factor.**

The amount of waste factor depends on the type of forms used. Wall forms have smooth, uniform sides and the calculated concrete quantities can be very close to actual quantities. Slabs, on the other hand, have a bottom surface that is irregular, making the actual thickness measurement difficult to determine. Therefore, take several thickness measurements and average them. Also, add a higher waste factor than you would for wall forms. In general, concrete quantities are merely estimates and experience is the best teacher.

Key Terms

admixture	bucks	concrete	footing
aggregates	cleats	cove	frost line
air-entrained cement	compressive strength	curing	girders
blockouts		flutes	gussets

half-round	pilasters	rise	spreaders
hydration	piles	run	spud vibrator
independent slab	pitch board	scabs	strongbacks
joist	portland cement	screed	tensile strength
kerfs	quarter-round	slab-on-grade	walers
monolithic slab	rebars	slump test	waste factor
panels	reinforced concrete	snap ties	yoking

Review Questions

Select the most appropriate answer.

1. Concrete is made of
   a. Portland cement.
   b. large and small aggregates.
   c. water.
   d. all of the above.

2. The steel rods placed in concrete to increase its tensile strength are called
   a. reinforcing bars.
   b. aggregates.
   c. reinforcing nails.
   d. duplex nails.

3. The inside surfaces of forms are oiled to
   a. protect the forms from moisture.
   b. prevent the loss of moisture from concrete.
   c. strip the forms more easily.
   d. prevent honeycombs in the concrete.

4. Keyways are often put in spread footings to
   a. unlock the forms for easy removal.
   b. increase the compressive strength of concrete.
   c. keep the form boards from spreading.
   d. provide a stronger joint between footing and foundation.

5. Rapid placing of concrete
   a. omits the need for vibrating.
   b. may burst the forms.
   c. keeps the aggregate from separating.
   d. reduces voids and honeycombs.

6. Unless footings are placed below the frost line,
   a. the foundation will settle.
   b. the foundation may heave and crack.
   c. excavation will be difficult in winter.
   d. problems with form construction will result.

7. Spreaders for footing forms are used
   a. to allow easy placement of the concrete.
   b. to keep the forms straight.
   c. because they are easier to fasten.
   d. because they maintain the proper footing width.

8. A step in a footing should be dimensioned and sized to match the
   a. concrete block used in the foundation.
   b. form used to pour the foundation wall.
   c. building code requirements.
   d. all of the above.

9. The horizontal surface length of a stepped footing must be at least
   a. 4 feet.
   b. twice the vertical distance.
   c. the vertical distance.
   d. the thickness of the footing.

10. The typical order of installation of a manufactured forming system is
    a. inside forms, outside forms, reinforcing bars, then snap ties.
    b. inside forms, reinforcing bars, snap ties, then outside forms.
    c. outside forms, snap ties, inside forms, then reinforcing bars.
    d. outside forms, snap ties, reinforcing bars, then inside forms.

11. Walers are used on concrete wall forms to ____ the forms.
    a. stiffen
    b. straighten
    c. strengthen
    d. all of the above.

12. A concrete slab should be protected from
    a. freezing before it cures.
    b. curing too fast with a sealer.
    c. moisture after it cures by a sub-slab vapor retarder.
    d. all of the above.

13. While concrete sets it should be protected from
    a. overheating.
    b. freezing.
    c. excessive vibrations.
    d. all of the above.

14. The volume of concrete that should be ordered for a 6″ slab that measures 24′ × 36′ is
    a. 16 cubic yards.   c. 192 cubic yards.
    b. 36 cubic yards.   d. 432 cubic yards.

15. The procedure for erecting footing forms that follows locating the corner stakes with a plumb bob and batter board strings is installation of the
    a. spreaders.
    b. outside form boards.
    c. inside form boards.
    d. reinforcing bars.

16. Overvibrating concrete while placing it in wall forms causes
    a. voids and honeycombs.
    b. the aggregate to rise to the top.
    c. extra side pressure on the forms that could cause form failure.
    d. all of the above.

17. Accessories placed in a foundation wall form to create spaces are called
    a. blockouts.
    b. bucks.
    c. girder pocket forms.
    d. all of the above.

18. To screed the top of a foundation wall,
    a. space nails along a chalk line.
    b. vibrate the top surface so it will flow level.
    c. use a chalk line only.
    d. add enough water to the concrete so that it will flow level.

UNIT 11

Floor Framing

Wood frame construction is used for residential and light commercial construction for important reasons of economy, durability, and variety. The cost for wood frame construction is generally less than for other types of construction. Fuel and air-conditioning expenses are reduced because wood frame construction provides better insulation.

Wood frame homes are very durable. If properly maintained, a wood frame building will last indefinitely. Many existing wood frame structures are hundreds of years old.

Because of the ease with which wood can be cut, shaped, fitted, and fastened, many different architectural styles are possible. In addition to single-family homes, wood frame construction is used for all kinds of lowrise buildings, such as apartments, condominiums, offices, motels, warehouses, and manufacturing plants.

SAFETY REMINDER

Floor framing creates an elevated horizontal plane. Construction of the floor must proceed in a logical and thoughtful manner to reduce the risk of falling.

OBJECTIVES

After completing this unit, the student should be able to:

- describe platform, balloon, and post-and-beam framing, and identify framing members of each.
- describe several energy and material conservation framing methods.
- build and install girders, erect columns, and lay out sills.
- lay out and install floor joists.
- frame openings in floors.
- lay out, cut, and install bridging.
- apply subflooring.
- describe how termites and fungi destroy wood and state some construction techniques used to prevent destruction by wood pests.

31 | Types of Frame Construction

There are several methods of framing a building. Some types are used less often today but still exist, so knowledge of them is necessary when remodeling. Other types are relatively new and knowledge about them is not widespread. Some wood frames are built using a combination of types. New designs utilizing engineered lumber are increasing the height and width to which wood frame structures can be built.

PLATFORM FRAME CONSTRUCTION

The **platform frame,** sometimes called the *western* frame, is most commonly used in residential construction **(Figure 31–1).** In this type of construction, the floor is built and the walls are erected on top of it. When more than one story is built, the second-floor platform is erected on top of the walls of the first story.

A platform frame is easier to erect than a balloon frame. At each floor level a flat surface is provided on which to work. A common practice is to assemble wall framing units on the floor and then tilt the units up into place.

Effects of Shrinkage

Lumber shrinks mostly across width and thickness. A disadvantage of the platform frame is the relatively large amount of settling caused by the shrinkage of the large number of horizontal load-bearing frame members. However, because of the equal amount of horizontal lumber, the shrinkage is more or less equal throughout the building. To reduce shrinkage, only framing lumber with the proper moisture content should be used.

BALLOON FRAME CONSTRUCTION

In **balloon frame** construction, the wall *studs* and first-floor *joists* rest on the **sill.** The second-floor joists rest on a 1 × 4 **ribbon** that is cut in flush with the inside edges of the studs **(Figure 31–2).** This type of construction is used less often today, but a substantial number of structures built with this type of frame are still in use.

Effects of Shrinkage

Shrinkage of lumber along the length of a board is insignificant compared to shrinkage that can occur across the width of the board. Therefore, in the balloon frame, settling caused by shrinkage of lumber is held to a minimum in the exterior walls. This is because the studs are continuous from sill to top **plate.** To prevent unequal settling of the frame due to shrinkage, the studs of **bearing partitions** rest directly on the *girder.*

FIRESTOPS

Firestop blocking is material installed to slow the movement of fire and smoke within smaller cavities of the building frame during a fire. It is sometimes called **draftstop blocking** and *fireblocking.* This allows the occupants more time to get out of a burning building and can reduce the overall damage. They must be installed in many places.

In a wood frame, a firestop in a wall might consist of dimension lumber blocking between studs. In the platform frame, the wall plates act as firestops.

Firestops must be installed in the following locations:

- In all stud walls, partitions, and furred spaces at ceiling and floor levels.
- Between stair *stringers* at the top and bottom. (Stringers are stair framing members. They are sometimes called stair *horses.*)
- Around chimneys, fireplaces, vents, pipes, and at ceiling and floor levels with noncombustible material.
- The space between floor joists at the sill and girder.
- All other locations as required by building codes **(Figure 31–3).**

POST-AND-BEAM FRAME CONSTRUCTION

The *post-and-beam* frame uses fewer but larger pieces. Large timbers, widely spaced, are used for joists, posts, and **rafters. Matched boards** (tongue and grooved) are often used for floors and roof sheathing **(Figure 31–4).**

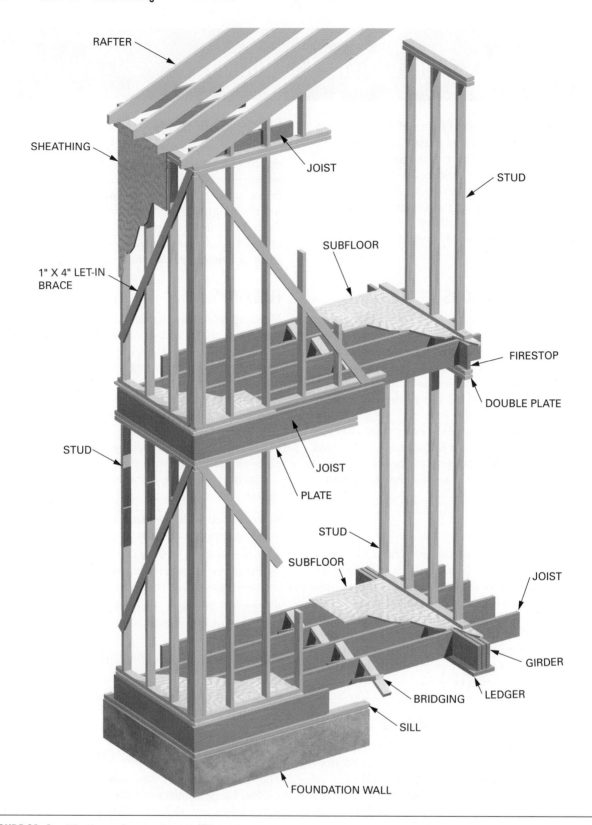

FIGURE 31–1   Platform frame construction.

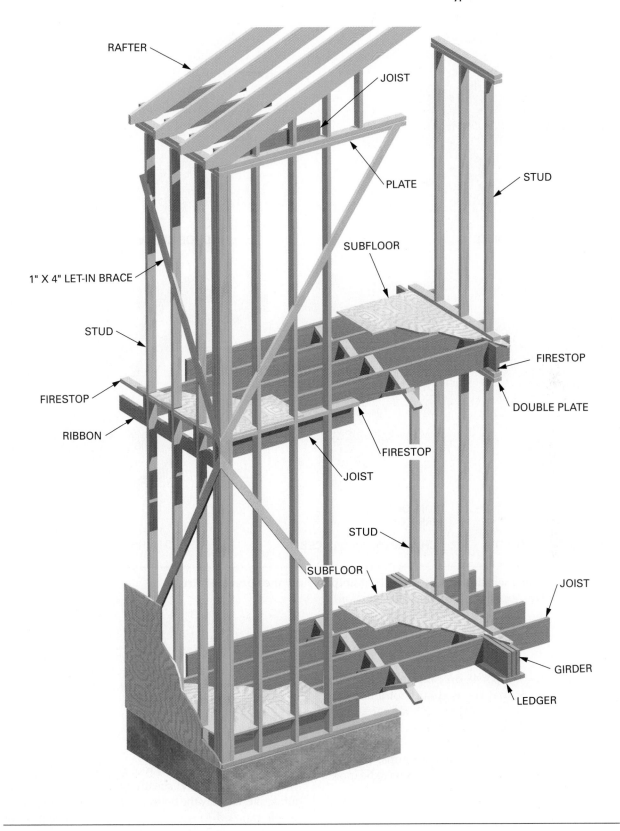

FIGURE 31–2 Balloon frame construction.

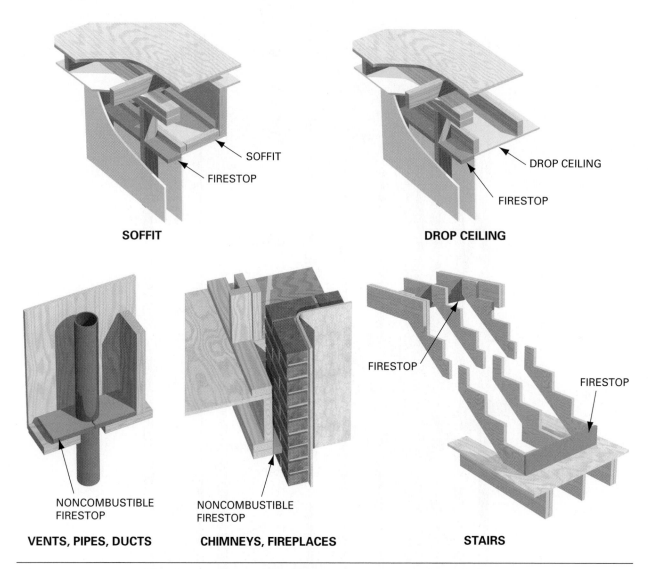

SOFFIT — FIRESTOP

SOFFIT

DROP CEILING — FIRESTOP

DROP CEILING

NONCOMBUSTIBLE FIRESTOP

VENTS, PIPES, DUCTS

NONCOMBUSTIBLE FIRESTOP

CHIMNEYS, FIREPLACES

FIRESTOP — FIRESTOP

STAIRS

FIGURE 31–3 Some locations of firestops that help to slow the spread of fire. *(Courtesy of Western Wood Products Association)*

Floors

APA Rated Sturd-I-Floor 48 **on center (OC),** which is 1¾₂ inches thick, may be used on floor joists that are spaced 4 feet OC instead of matched boards **(Figure 31–5).** In addition to being nailed, the plywood panels are glued to the floor beams with construction adhesive applied with caulking guns. The use of matched planks allows the floor beams to be more widely spaced.

Walls

Exterior walls of a post-and-beam frame may be constructed with widely spaced posts. This allows wide expanses of glass to be used from floor to ceiling. Usually some sections between posts in the wall are studded at close intervals, as in platform fram-

ing. This provides for door openings, fastening for finish, and wall **sheathing.** In addition, close spacing of the studs permits the wall to be adequately braced **(Figure 31–6).**

Roofs

The post-and-beam frame roof is widely used. The exposed roof beams and sheathing on the underside are attractive. Usually the bottom surface of the roof planks is left exposed to serve as the finished ceiling. Roof planks come in 2-, 3-, and 4-inch nominal thicknesses. Some are **end** matched as well as **edge** matched. Some buildings may have a post-and-beam roof, while the walls and floors may be conventionally framed.

The post-and-beam roof may be constructed with a *longitudinal* frame. The beams run parallel to the

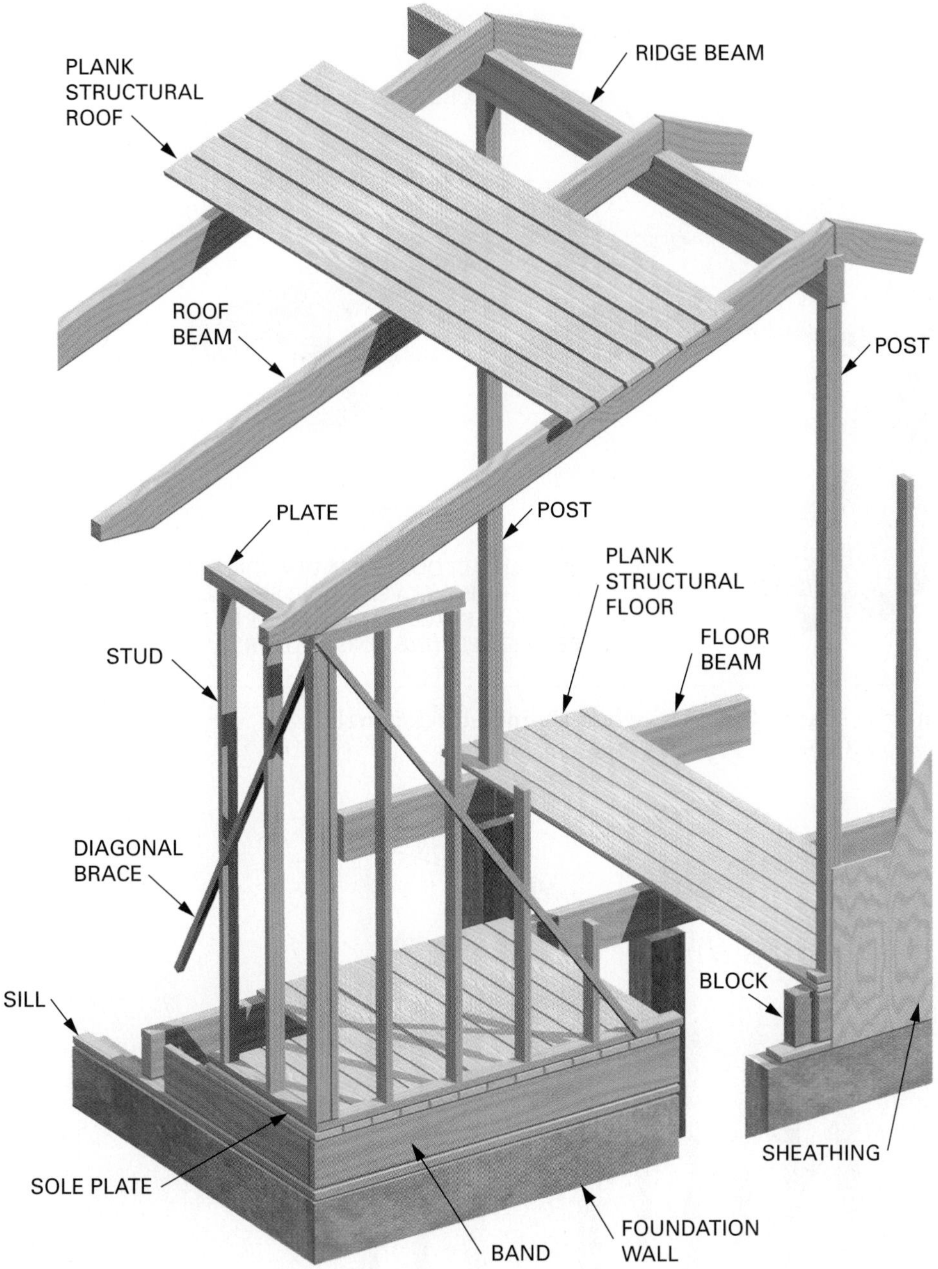

FIGURE 31–4 The post-and-beam frame.

PLANK STRUCTURAL ROOF

RIDGE BEAM

ROOF BEAM

POST

PLATE

POST

PLANK STRUCTURAL FLOOR

FLOOR BEAM

STUD

DIAGONAL BRACE

BLOCK

SILL

SHEATHING

SOLE PLATE

BAND

FOUNDATION WALL

ridge beam. Or they may have a *transverse* frame. The beams run at right angles to the **ridge** beam similar to roof rafters.

The ridge beam and longitudinal beams, if used, are supported at each end by posts in the end walls. They must also be supported at intervals along their length. This prevents the side walls from spreading and the roof from sagging **(Figure 31–7).** One of the disadvantages of a post-and-beam roof is that interior partitions and other interior features must be planned around the supporting roof beam posts.

Because of the fewer number of pieces used, a well-planned post-and-beam frame saves material and labor costs. Care must be taken when erecting the frame to protect the surfaces and make well-

fitting joints on exposed posts and beams. Glulam beams are well suited for and frequently used in post-and-beam construction. A number of metal connectors are used to join members of the frame **(Figure 31–8).**

ENERGY AND MATERIAL CONSERVATION FRAMING METHODS

There has been much concern and thought about conserving energy and materials in building construction. Several systems have been devised that differ from conventional framing methods. They conserve energy and use less material and labor. Check state and local building codes for limitations.

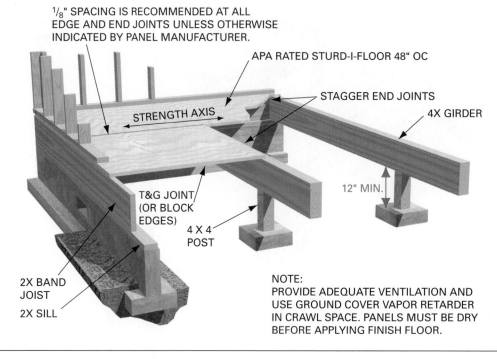

1/8" SPACING IS RECOMMENDED AT ALL EDGE AND END JOINTS UNLESS OTHERWISE INDICATED BY PANEL MANUFACTURER.

APA RATED STURD-I-FLOOR 48" OC

STAGGER END JOINTS

4X GIRDER

STRENGTH AXIS

12" MIN.

T&G JOINT (OR BLOCK EDGES)

4 X 4 POST

2X BAND JOIST

2X SILL

NOTE:
PROVIDE ADEQUATE VENTILATION AND USE GROUND COVER VAPOR RETARDER IN CRAWL SPACE. PANELS MUST BE DRY BEFORE APPLYING FINISH FLOOR.

FIGURE 31–5 Floor beams may be spaced 4 feet OC when 1-3/32-inch-thick panels are used for a floor. *(Courtesy of American Plywood Association)*

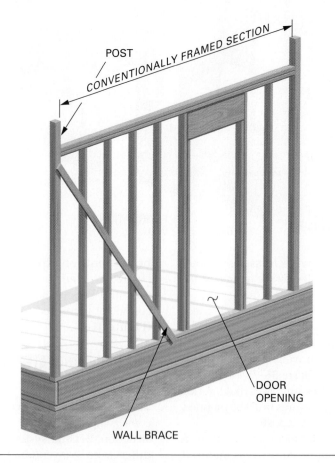

POST

CONVENTIONALLY FRAMED SECTION

DOOR OPENING

WALL BRACE

FIGURE 31–6 Sections of the exterior walls of a post-and-beam wall may need to be conventionally framed.

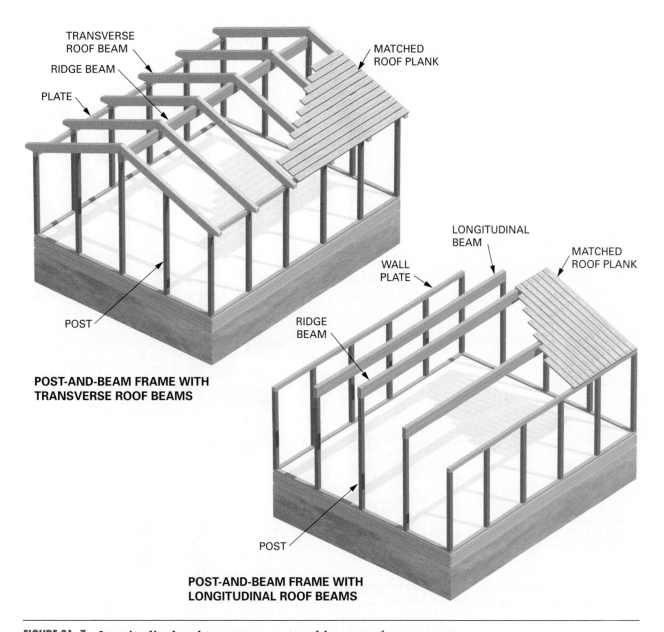

POST-AND-BEAM FRAME WITH
TRANSVERSE ROOF BEAMS

POST-AND-BEAM FRAME WITH
LONGITUDINAL ROOF BEAMS

FIGURE 31–7 Longitudinal and transverse post-and-beam roofs.

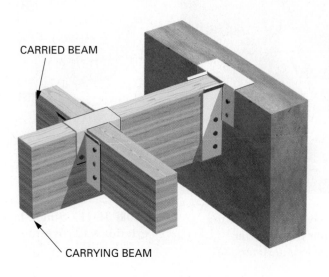

FIGURE 31–8 Metal connectors are specially made to join glulam beams.

Floors

For maximum savings, a single layer of ¾-inch tongue-and-grooved plywood is used over joists. In-line floor joists are used to make installation of the plywood floor easier **(Figure 31–9).** The use of adhesive when fastening the plywood floor is recommended. Gluing increases stiffness and prevents squeaky floors **(Figure 31–10).**

Walls

A single layer of plywood may act as both sheathing and exterior siding. In this case, the plywood must be at least ½-inch thick **(Figure 31–11).** If two-layer construction is used, ⅜-inch plywood is acceptable.

Wall openings are planned to be located so that at least one side of the opening falls on an OC stud.

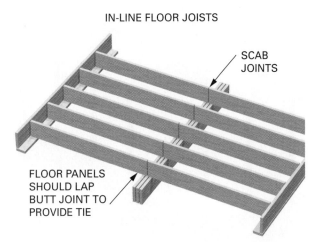

IN-LINE FLOOR JOISTS

SCAB JOINTS

FLOOR PANELS SHOULD LAP BUTT JOINT TO PROVIDE TIE

FIGURE 31–9 **In-line floor joists make installation of plywood subflooring simpler.**

FIGURE 31–10 **Using adhesive when fastening subflooring makes the floor frame stiffer, stronger, and quieter.** *(Courtesy of American Plywood Association)*

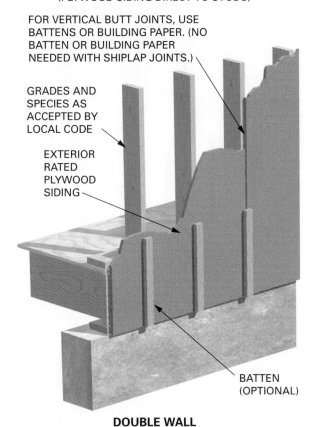

SINGLE WALL
(PLYWOOD SIDING DIRECT TO STUDS)

FOR VERTICAL BUTT JOINTS, USE BATTENS OR BUILDING PAPER. (NO BATTEN OR BUILDING PAPER NEEDED WITH SHIPLAP JOINTS.)

GRADES AND SPECIES AS ACCEPTED BY LOCAL CODE

EXTERIOR RATED PLYWOOD SIDING

BATTEN (OPTIONAL)

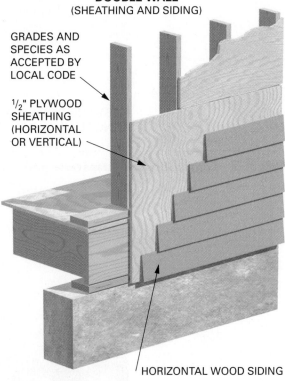

DOUBLE WALL
(SHEATHING AND SIDING)

GRADES AND SPECIES AS ACCEPTED BY LOCAL CODE

½" PLYWOOD SHEATHING (HORIZONTAL OR VERTICAL)

HORIZONTAL WOOD SIDING

FIGURE 31–11 **Single-layer and double-layer exterior wall covering.**

WINDOW OFF MODULE

WINDOW ON MODULE

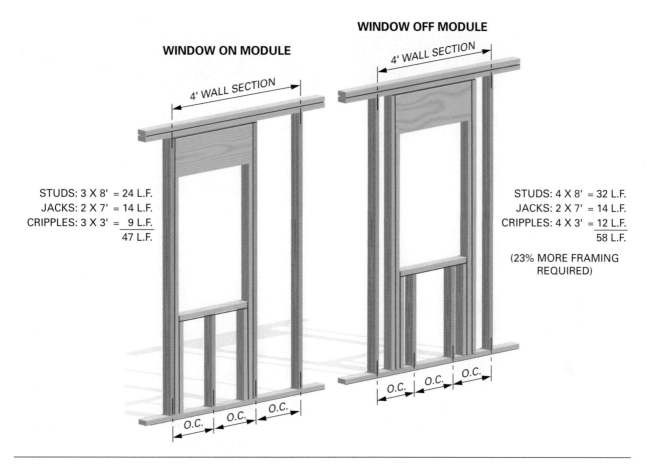

STUDS: 3 X 8' = 24 L.F.
JACKS: 2 X 7' = 14 L.F.
CRIPPLES: 3 X 3' = 9 L.F.
47 L.F.

STUDS: 4 X 8' = 32 L.F.
JACKS: 2 X 7' = 14 L.F.
CRIPPLES: 4 X 3' = 12 L.F.
58 L.F.

(23% MORE FRAMING REQUIRED)

FIGURE 31–12 To conserve materials, locate wall openings so they fall on the OC studs.

Whenever possible, window and door sizes are selected so that the rough opening width is a multiple of the module **(Figure 31–12)**. Also, locate partitions at OC wall stud positions if possible.

Roofs

Roof systems can be modified to improve the insulation over the exterior walls. The raised heel is built into the truss by the manufacturer **(Figure 31–13)**. It raises the roof slightly to allow for full-thickness insulation at the **eaves.** Some areas of the country also use 2 × 6 wall studs to increase the wall insulation.

House Depths

House depths that are not evenly divisible by four waste floor framing and sheathing. Lumber for floor joists is produced in increments of 2 feet. Assuming the girder remains in the center of the building, a house 25 feet wide would require 14-foot-long floor joists. These joists could be used uncut to make a building 28 feet wide. If the girder was installed offset from the center, 12- and 14-foot joists could be used to span 25 feet. Either way material is wasted.

Full-width subfloor panels can be used without cutting on buildings that are 24-, 28- and 32-feet wide. This decreases construction time and saves money.

FULL-WIDTH INSULATION

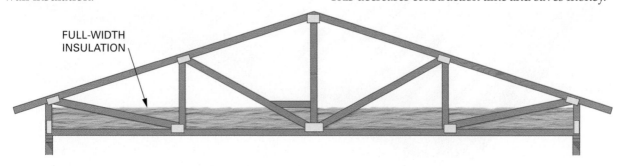

FIGURE 31–13 Modified truss design accommodates thick ceiling insulation without compressing at eaves.

32 Layout and Construction of the Floor Frame

A floor frame consists of members fastened together to support the loads a floor is expected to bear. The floor frame is started after the foundation has been placed and has hardened. A straight and level floor frame makes it easier to frame and finish the rest of the building.

Because platform framing is used more than any other type, this chapter describes how to lay out and construct its floor frame. The knowledge gained in this chapter can be used to lay out and construct any type of floor frame.

DESCRIPTION AND INSTALLATION OF FLOOR FRAME MEMBERS

In the usual order of installation, the floor frame consists of *girders, posts* or *columns, sill plates, joists,* **bridging,** and *subflooring* **(Figure 32–1)**.

Description of Girders

Girders are heavy beams that support the inner ends of the floor joists. Several types are commonly used.

Kinds of Girders. Girders may be made of solid wood or built up of three or more 2-inch planks. Laminated veneer lumber or glulam beams may also be used as girders **(Figure 32–2)**. Sometimes, wide flange, I-shaped steel beams are used.

Determining the Size of Girders and Other Structural Members. The size of solid, built-up, engineered lumber, or steel girders, or any structural component, is best determined by professional architects or engineers. There are many factors to consider when determining structural lumber or steel sizes. Only people specifically trained should select the size of a structural member. Estimating or guessing the size can have serious consequences. If undersized, the members could sag or be too flexible. If oversized, it would not be economical and might take up more room than necessary. If the blueprints do not specify the kind or size, have the design checked by a professional engineer.

Built-Up Girders. If built-up girders of dimension lumber are used, a minimum of three members are fastened together with three 3½-inch or 16d nails at each end. The other nails are staggered not farther than 32 inches apart from end to end **(Figure 32–3)**. Sometimes ½-inch bolts are required. Applying glue between the pieces makes the bond stronger. Laminated veneer lumber may also be built up for use as girders. The end joints of both types are placed directly over supports.

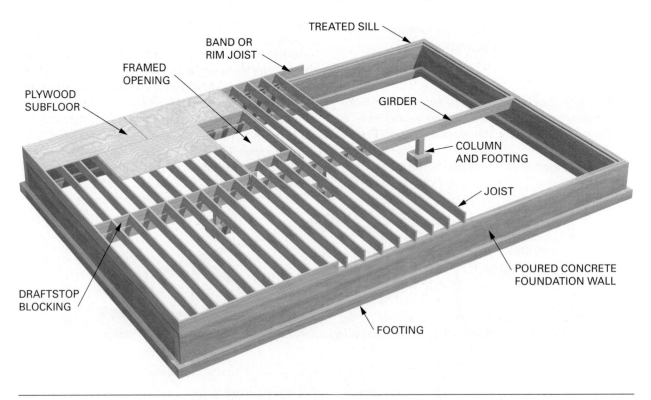

FIGURE 32–1   **A floor frame of platform construction.**

FIGURE 32–2 **Large glulam beams are often used as girders and headers.**

Girder Location. The ends of the girder are usually supported by a *pocket* formed in the foundation wall. The pocket should provide at least a 3-inch bearing for the girder. It should be wide enough to provide ½-inch clearance on both sides and the end. This allows any moisture to be evaporated by circulation of

air. Thus no moisture will get into the girder, which would cause decay of the wood.

The pocket is formed deep enough to provide for shimming the girder to its designated height. A steel bearing plate may be *grouted* in under the girder while it is supported temporarily **(Figure 32–4)**. **Grouting** is the process of filling in small space with a thick paste of cement. Wood **shims** are not usually suitable for use under girders. The weight imposed on them compresses the wood, causing the girder to sink below its designated level.

Installing Girders

Steel girders usually come in one piece and are set in place with a crane. A solid wood girder is often installed in a similar manner but is pieced together with half-lap joints. Joints are made near the posts or columns.

Wood girders are usually built up from dimension lumber and erected in sections. This process begins by building one section at a time. One end is then set in the pocket in the foundation wall. The other end is placed and fastened on a braced temporary support. Continue building and erecting sections on posts until the girder is completed into the opposite pocket (Figure 32–4).

Sight the girder by eye from one end to the other. Place wedges under the temporary supports to straighten the girder. Permanent posts or columns are usually installed after the girder has some weight

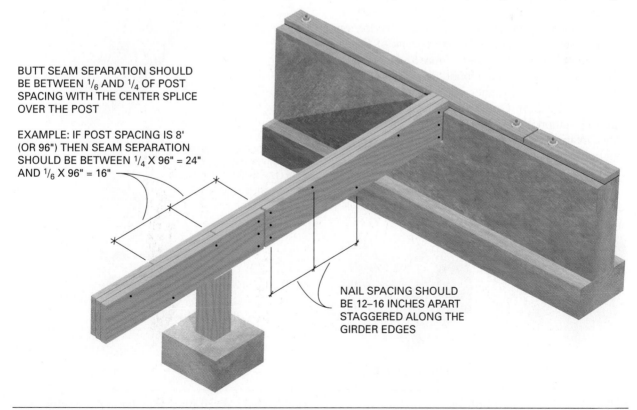

BUTT SEAM SEPARATION SHOULD BE BETWEEN 1/6 AND 1/4 OF POST SPACING WITH THE CENTER SPLICE OVER THE POST

EXAMPLE: IF POST SPACING IS 8' (OR 96") THEN SEAM SEPARATION SHOULD BE BETWEEN 1/4 X 96" = 24" AND 1/6 X 96" = 16"

NAIL SPACING SHOULD BE 12–16 INCHES APART STAGGERED ALONG THE GIRDER EDGES

FIGURE 32–3 **Spacing of fasteners and seams of a built-up girder made with dimension lumber.**

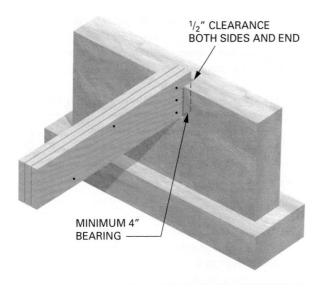

FIGURE 32–4   A girder pocket of a foundation wall should be large enough to provide air space around the end and sides of the girder.

imposed on it by the *floor joists.* Temporary posts should be strong enough to support the weight imposed on them until permanent ones are installed.

SILLS

Sills, also called *mudsills* or *sill plates,* are horizontal members of a floor frame. They lie directly on the foundation wall and provide a bearing for *floor joists.* It is required that the sill be made with a decay-resistant material such as redwood, black locust, cedar, or pressure-treated lumber. Sills may consist of single 2 × 6, or doubled 2 × 6 lumber **(Figure 32–5).**

The sill is attached to the foundation wall with anchor bolts. The size, number, and spacing of bolts is determined by local weather and seismic conditions. This information will be clearly indicated on the set of prints. In any case the maximum spacing betwen anchor bolts is 6 feet and a bolt must be located between 6 and 12 inches from the end of any sill piece.

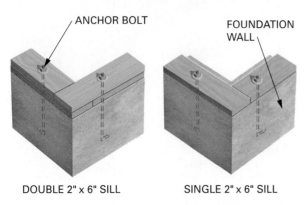

FIGURE 32–5   An anchor bolt should be located between 6 and 12 inches from the end of each sill.

To take up irregularities between the foundation and the sill, a *sill sealer* is used. The sill sealer should be an insulating material used to seal against drafts, dirt, and insects. It comes 6 inches wide and in rolls of 50 feet. It compresses when the weight of the structure is upon it.

Sills must be installed so they are straight, level, and to the specified dimension of the building. The level of all other framing members depends on the care taken with the installation of the sill.

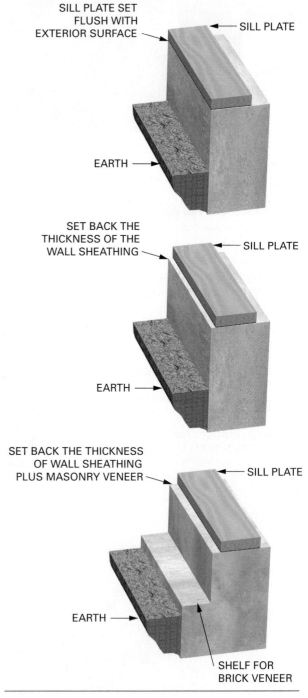

FIGURE 32–6   A sill plate may be located with different setbacks from the foundation edge.

Sometimes the outside edge of the sill is flush with the outside of the foundation wall. Sometimes it is set in the thickness of the wall sheathing, depending on custom or design. In the case of brick-veneered exterior walls, the sill plate may be set back even farther **(Figure 32–6)**.

Installing Sills

Remove washers and nuts from the anchor bolts. Snap a chalk line on the top of the foundation wall in line with the inside edge of the sill.

Cut the sill sections to length. Hold the sill in place against the anchor bolts. Square lines across the sill on each side of the bolts. Measure the distance from the center of each bolt to the chalk line. Transfer this distance at each bolt location to the sill by measuring from the inside edge **(Procedure 32–A)**.

Bore holes in the sill for each anchor bolt. Bore the holes at least ⅛ inch oversize to allow for adjustments. Place the sill sections in position over the anchor bolts after installing the sill sealer. The inside edges of the sill sections should be on the chalk line. Replace the nuts and washers. Be careful not to overtighten the nuts, especially if the concrete wall is still **green.** This may crack the wall.

If the inside edge of the sill plate comes inside the girder pocket, notch the sill plate around the end of the girder. Raise the ends of the girder so it is flush with the top of the sill plate. If steel girders are used, a dimension lumber sill is bolted to the top flanges. This allows for easy fastening of floor joists **(Figure 32–7)**.

Floor Joists

Floor joists are horizontal members of a frame. They rest on and transfer the load to sills and girders. In residential construction, dimension lumber placed on edge has traditionally been used. Wood I-joists, with lengths up to 80 feet, are being specified more often today **(Figure 32–8)**. Steel framing in the form of joists and walls is sometimes used. In general when the price of lumber increases significantly or when termites are a problem steel framing is the solution.

FIGURE 32–8 Engineered lumber makes a strong floor system. *(Courtesy Trus Joist MacMillan)*

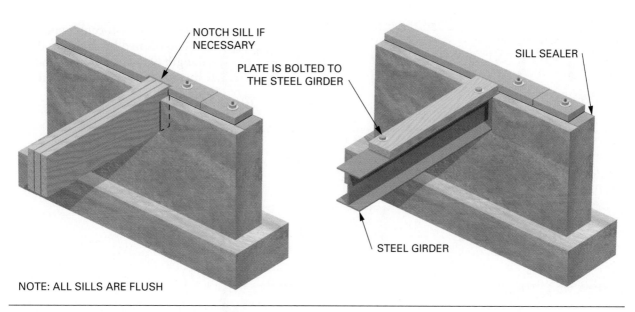

NOTE: ALL SILLS ARE FLUSH

NOTCH SILL IF NECESSARY

PLATE IS BOLTED TO THE STEEL GIRDER

SILL SEALER

STEEL GIRDER

FIGURE 32–7 **Variations in girder and sill installations.**

PROCEDURE 32–A Installing a Sill Plate on Foundation with Anchor Bolts

STEP 1 SNAP CHALK LINE ON FOUNDATION WALL.

STEP 2 ALIGN SILL PLATE AGAINST ANCHOR BOLTS AND PARALLEL TO CHALK LINE.

STEP 3 SQUARE LINES ON SILL FROM BOTH SIDES OF EACH ANCHOR BOLT.

STEP 4 MEASURE EACH BOLT DISTANCE FROM CHALK LINE AND TRANSFER TO SILL.

STEP 5 DRILL HOLES APPROXIMATELY 1/8″ LARGER THAN BOLT DIAMETER.

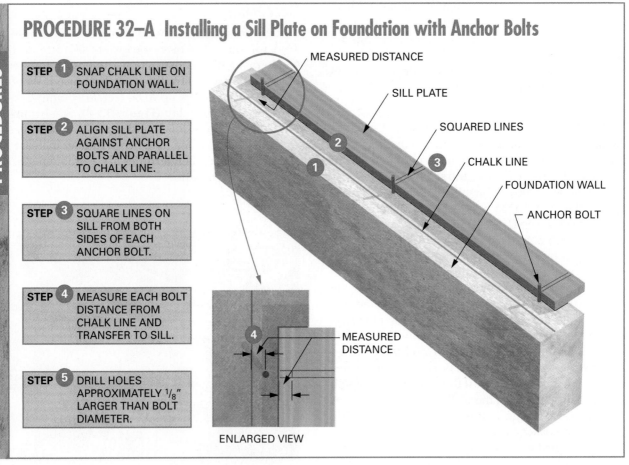

MEASURED DISTANCE

SILL PLATE

SQUARED LINES

CHALK LINE

FOUNDATION WALL

ANCHOR BOLT

MEASURED DISTANCE

ENLARGED VIEW

Joists are generally spaced 16 inches OC in conventional framing. They may be spaced 12, 19.2, or 24 inches OC, depending on the type of construction and the load. The joist spacing is designed so an 8-foot section (96 inches) of plywood is fully supported at its ends. Dividing 96 by the number of joist spaces in that section gives the OC **(Figure 32–9)**. The size of floor joists should be determined from the construction drawings.

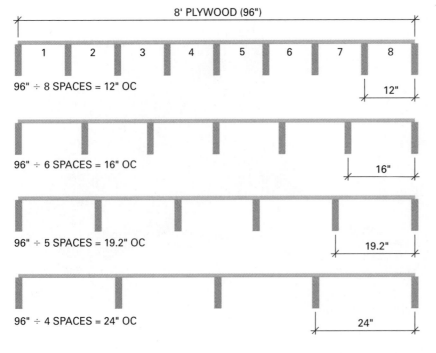

8′ PLYWOOD (96″)

1 2 3 4 5 6 7 8

96″ ÷ 8 SPACES = 12″ OC   12″

96″ ÷ 6 SPACES = 16″ OC   16″

96″ ÷ 5 SPACES = 19.2″ OC   19.2″

96″ ÷ 4 SPACES = 24″ OC   24″

FIGURE 32–9 On-center spacing is determined by dividing the length of a sheet of plywood by the number of joist spaces under it.

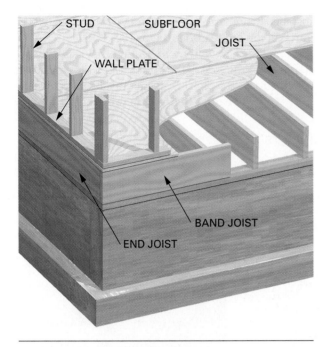

FIGURE 32–10 Typical framing near the sill using dimension lumber.

Joist Framing at the Sill. Joists should rest on at least 1½ inches of bearing on wood and 3 inches on **masonry.** In platform construction, the ends of floor joists are capped with a *band joist,* also called a *rim joist,* **box header,** or **joist header (Figure 32–10).** The use of wood I-joists requires sill construction as recommended by the manufacturer for satisfactory performance of the frame **(Figure 32–11).**

Joist Framing at the Girder. If joists are lapped over the girder, the minimum amount of lap is 3 inches to allow for adequate nailing surface between lapped joists. The maximum overhang of joists at the girder is 12 inches to eliminate floor squeaking. These squeaks are caused when walking on the floor at joist midspan; the overhung joist end raises up rubbing against the

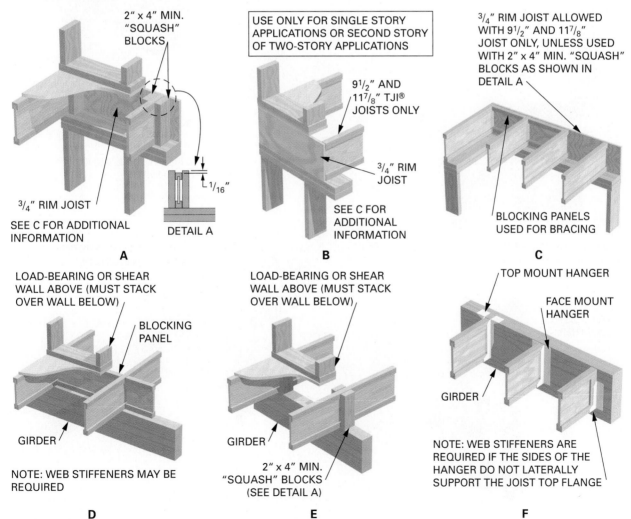

FIGURE 32–11 Selected framing details of wood I-joists.

lapped joist. There is no need to lap wood I-joists. They come in lengths long enough to span the building. However, they may need to be supported by girders depending on the span and size of the wood I-joists. No matter how the joists are framed over the girder, drafted stop blocking is required. It should be

installed using full-width framing lumber between joists on top of the girder **(Figure 32–12)**.

Sometimes, to gain more headroom, joists may be framed into the side of the girder. There are a number of ways to do this. Joist hangers must be used to support wood I-beams. **Web stiffeners** should be ap-

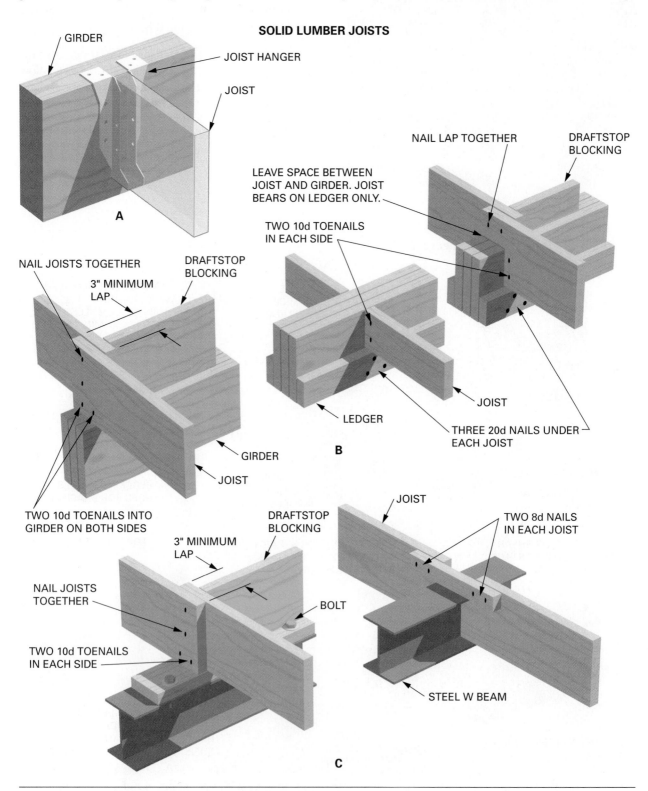

FIGURE 32–12 Various possible framing details at a girder.

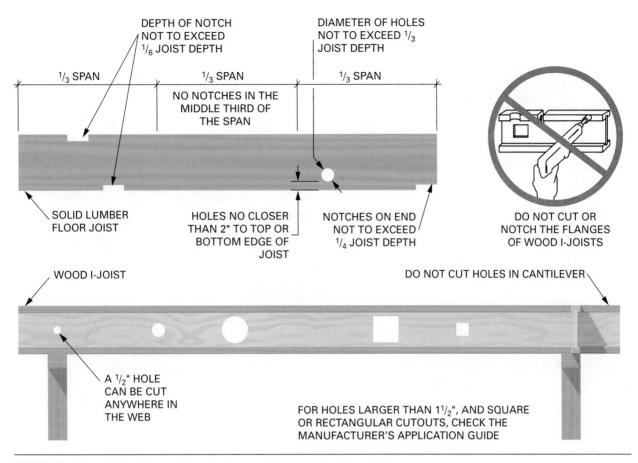

FIGURE 32–13 Allowable notches, holes, and cutouts in floor joists.

plied to the beam ends if the hanger does not reach the top flange of the beam.

Notching and Boring of Joists

Notches in the bottom or top of sawn lumber floor joists should not exceed one-sixth of the joist depth. Notches should not be located in the middle one-third of the joist span. Notches on the ends should not exceed one-fourth of the joist depth.

Holes bored in joists for piping or wiring should not be larger than one-third of the joist depth. They should not be closer than 2 inches to the top or bottom of the joist **(Figure 32–13)**.

Some wood I-joists are manufactured with perforated knockouts in the web along its length. This allows for easy installation of wiring and pipes. To cut other size holes in the web, consult the manufacturer's specifications guide. Do not cut or notch the flanges of wood I-joists.

Laying Out Floor Joists

The locations of floor joists are marked on the sill plate. A squared line marks the side of the joist. An X to one side of the line indicates on which side of the line the joist is to be placed **(Figure 32–14)**.

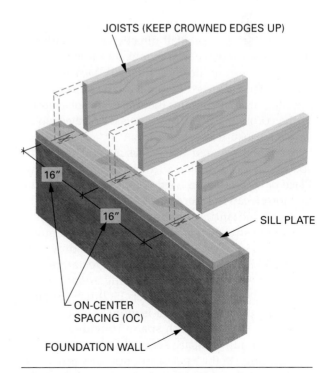

FIGURE 32–14 A line is drawn to mark the edge of a joist; an X is marked to indicate on which side of the line the joist is placed.

PROCEDURE 32–B Laying Out the Sill Plate for Floor Joists

STEP 1 MEASURE THE JOIST SPACING IN FROM THE CORNER.

STEP 2 MEASURE BACK ½ THE JOIST THICKNESS.

STEP 3 SQUARE A LINE ACROSS THE SILL PLATE AND PLACE AN *X* ON THE SIDE OF THE LINE WHERE THE JOIST IS TO BE PLACED.

STEP 4 CONTINUE THE ON-CENTER SPACING ALONG THE LENGTH OF THE BUILDING. USE A STEEL TAPE STRETCHED OVER THE ENTIRE LENGTH.

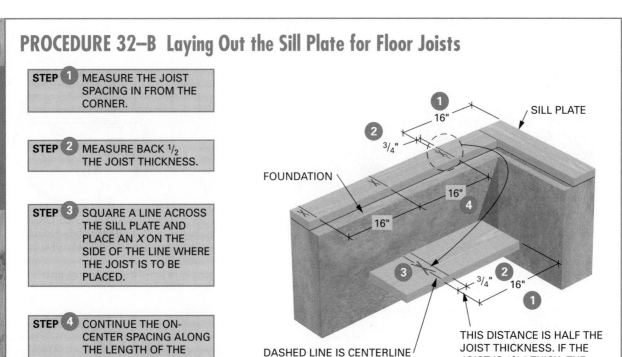

SILL PLATE

FOUNDATION

DASHED LINE IS CENTERLINE OF JOIST

THIS DISTANCE IS HALF THE JOIST THICKNESS. IF THE JOIST IS 1½" THICK, THE DISTANCE IS ¾".

Floor joists must be laid out so that the ends of *plywood subfloor* sheets fall directly on the center of floor joists. Start the joist layout by measuring the joist spacing from the end of the sill. Measure back one-half the thickness of the joist. Square a line across the sill. This line indicates the side of the joist closest to the corner. Place an *X* on the side of the line on which the joist is to be placed.

From the squared line, measure and mark the spacing of the joists along the length of the building. Place an *X* on the same side of each line as for the first joist location **(Procedure 32–B)**.

When measuring for the spacing of the joists, use a tape stretched along the length of the building. Most professional tapes have predominate markings for repetitive layout. Black rectangles for 12- and 24-inch layouts, red rectangles for 16-inch and small black diamonds for 19.2-inch layouts. Using a tape in this manner is more accurate. Measuring and marking each space individually with a rule or framing square generally causes a gain in the spacing. If the spacing is not laid out accurately, the plywood subfloor may not fall in the center of some floor joists. Time will then be lost either cutting the plywood back or adding strips of lumber to the floor joists **(Figure 32–15)**.

Laying Out Floor Openings. After marking floor joists for the whole length of the building, study the plans for the location of floor openings. Mark the sill plate, where joists are to be doubled, on each side of large floor openings. Identify the layout marks that are not for full-length floor joists. Shortened floor joists at the ends of floor openings are called **tail joists.** They are usually identified by changing the *X* to a *T* or a *C* for cripple joist. Lay out for partition supports, or wherever doubled floor joists are required **(Figure 32–16)**. Check the mechanical drawings to make adjustments in the framing to allow for the installation of mechanical equipment.

Lay out the floor joists on the girder and on the opposite wall. If the joists are in-line, *X*s are made on the same side of the mark on both the girder and the sill plate on the opposite wall. If the joists are lapped, a mark is placed on both sides of the line at the girder and on the opposite side of the mark on the other wall **(Figure 32–17)**. These marks may be changed to make it easier to tell which mark is for

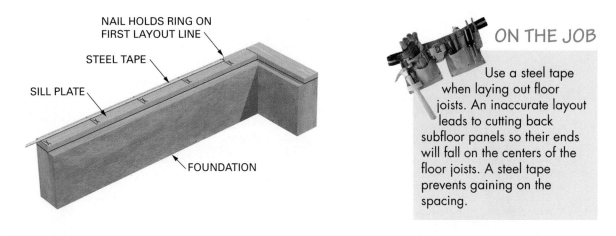

NAIL HOLDS RING ON
FIRST LAYOUT LINE

STEEL TAPE

SILL PLATE

FOUNDATION

FIGURE 32–15 Using a steel tape for layout reduces the possibility of step-off errors.

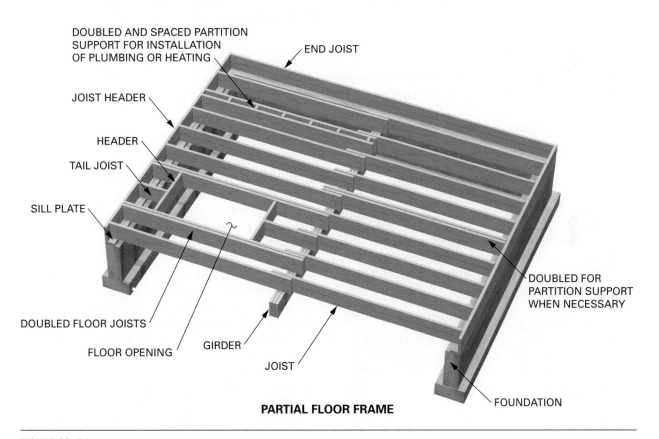

DOUBLED AND SPACED PARTITION SUPPORT FOR INSTALLATION OF PLUMBING OR HEATING

END JOIST

JOIST HEADER

HEADER

TAIL JOIST

SILL PLATE

DOUBLED FLOOR JOISTS

FLOOR OPENING

GIRDER

JOIST

DOUBLED FOR PARTITION SUPPORT WHEN NECESSARY

FOUNDATION

PARTIAL FLOOR FRAME

FIGURE 32–16 Typical framing components of a floor system.

the joist toward the front of the house and which is for the back.

Installing Floor Joists

Stack the necessary number of full-length floor joists at intervals along both walls. Each joist is carefully sighted along its length by eye. Any joist with a severe crook or other warp should not be used. Joists are installed with the crowned edge up.

Keep the end of the floor joist in from the outside edge of the sill plate by the thickness of the band joist. Toenail the joists to the sill and girder with 10d

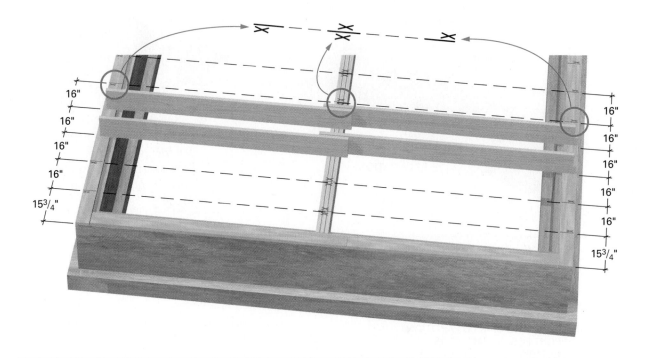

FIGURE 32–17 Floor joist layout lines span the entire width of the building. If the joists are lapped at the girder, then the Xs are marked on different sides of the line.

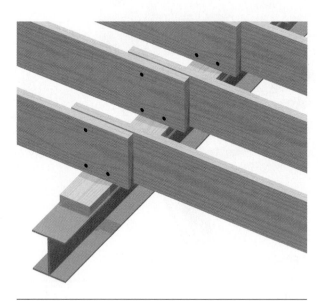

FIGURE 32–18 Floor joists may be lapped at the girder.

or 3-inch common nails. Nail the joists together if they lap at the girder **(Figure 32–18)**. When all floor joists are in position, they are sighted by eye from end to end and straightened. They may be held straight by strips of 1 × 3s tacked to the top of the joists about in the middle of the joist span.

Wood I-joists are installed using standard tools. They can be easily cut to any required length at the jobsite. A minimum bearing of 1¾ inches is required at joist ends and 3½ inches over the girder. The wide, straight wood flanges on the joist make nailing easier, especially with pneumatic framing nailers **(Figure 32–19)**. Nail joists at each bearing with one 8d or 10d nail on each side. Keep nails at least 1½ inches from the ends to avoid splitting.

DOUBLING FLOOR JOISTS

For added strength, doubled floor joists must be securely fastened together. Their top edges must be even. In most cases, the top edges do not lie flush with each other. They must be brought even before they can be nailed together.

To bring them flush, toenail down through the top edge of the higher one, at about the center of their length. At the same time squeeze both together tightly by hand. Use as many toenails as necessary, spaced where needed, to bring the top edges flush **(Procedure 32–C)**. Usually no more than two or

FIGURE 32–19   Typical wood I-joist installation. *(Courtesy Boise Cascade)*

three nails are needed. Then, fasten the two pieces securely together. Drive nails from both sides, staggered from top to bottom, about 2 feet apart. Angle nails slightly so they do not protrude.

Framing Floor Openings

Large openings in floors should be framed before floor joists are installed. This is because room is needed for end nailing. To frame an opening in a floor, first fasten the **trimmer joists** in place. Trimmer joists are full-length joists that run along the inside of the opening. Mark the location of the *headers* on the trimmers. Headers are members of the opening that run at right angles to the floor joists. They should be doubled if they are more than 4′ long.

Cut *headers* to length by taking the measurement at the sill between the trimmers. Taking the measurement at the sill where the trimmers are fastened, rather than at the opening, is standard practice. A measurement between trimmers taken at the opening may not be accurate. There may be a bow in the trimmer joists **(Figure 32–20)**.

Place two headers, one for each end of the opening, on the sill between the trimmers. Transfer the layout of the tail joists on the sill to the headers. Fasten the first header on each end of the opening in position by driving nails through the side of the trimmer into the ends of the headers. Be sure the first header is the header that is farthest from the floor

ON THE JOB

Bring the tops of double floor joists flush by first toenailing the edges before nailing them together.

PROCEDURE 32–C   Aligning the Top Edges of Joists

STEP 1 FASTEN BOTH ENDS OF THE JOISTS TO THEIR BEARINGS.

STEP 2 TOENAIL INTO THE TOP EDGE OF THE JOIST WITH THE GREATER CROWN TO BRING TOP EDGES FLUSH BEFORE FASTENING TOGETHER.

STEP 3 NAIL JOISTS TOGETHER.

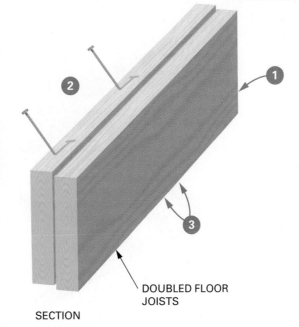

DOUBLED FLOOR JOISTS

SECTION

STEP BY STEP PROCEDURES

opening. Fasten the tail joists in position. Double up the headers. Finally, double up the trimmer joists. **Procedure 32–D** shows the sequence of operations used to frame a floor opening. This particular sequence allows you to end nail the members rather than toenailing them. Use joist hangers as required by local codes. Joist hangers must be installed with the proper amount, size, and type of nails. Roofing nails are not acceptable for joist hangers.

Installing the Band Joist. After all the openings have been framed and all floor joists are fastened, install the band joist. This closes in the ends of the floor joists. Band joists may be lumber of the same size as the floor joists. They also may be a single or double layer of laminated veneer lumber when wood I-joists are used.

Fasten the band joist into the end of each floor joist. If wood I-joists are used as floor joists, drive

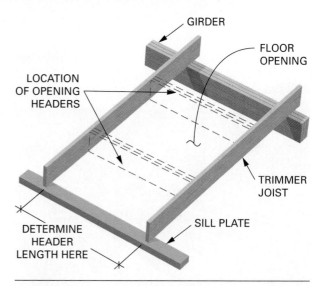

FIGURE 32–20   Length of the header for a floor opening should be measured at the sill, not at the midspan.

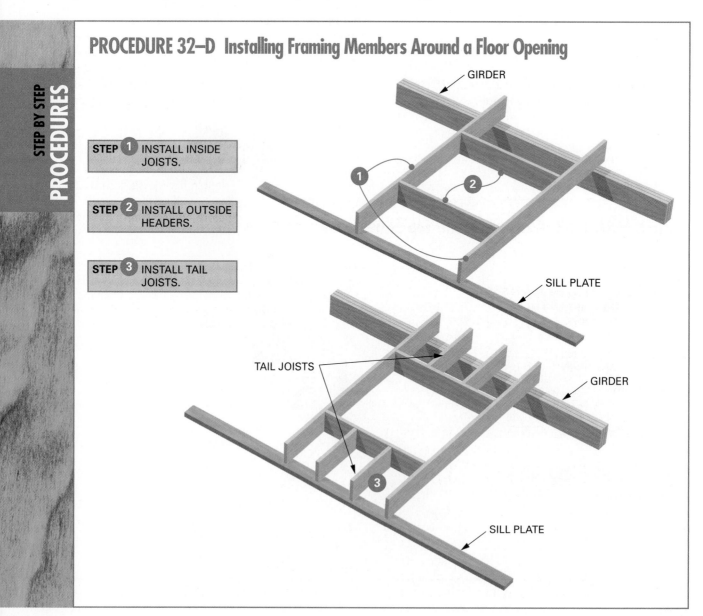

STEP BY STEP PROCEDURES

PROCEDURE 32–D Installing Framing Members Around a Floor Opening

STEP ❶ INSTALL INSIDE JOISTS.

STEP ❷ INSTALL OUTSIDE HEADERS.

STEP ❸ INSTALL TAIL JOISTS.

PROCEDURE 32–D *Continued*

STEP ④ INSTALL INSIDE HEADERS.

STEP ⑤ INSTALL OUTSIDE JOISTS.

STEP ⑥ INSTALL REGULAR SPACED FULL-LENGTH JOISTS.

GIRDER

TAIL JOISTS

JOISTS

SILL PLATE

---

one nail into the top and bottom flange. The band joist is also toenailed to the sill plate at about 6-inch intervals.

BRIDGING

Bridging is installed in rows between floor joists at intervals not exceeding 8 feet. For instance, floor joists with spans 8 to 16 feet need one row of bridging near the center of the span. Its purpose is to distribute a concentrated load on the floor over a wider area. Although codes do not require bridging for 2 × 12 or smaller joists, some builders install it because of customary practice.

Bridging may be solid wood, wood cross-bridging, metal cross-bridging, or 1 × 3 nailed to the bottom of joists **(Figure 32–21)**. Usually solid wood bridging is the same size as the floor joists. It is installed in an off-set fashion to permit end nailing.

Wood cross-bridging should be at least nominal 1 × 3 lumber with two 6d nails at each end. It is placed in double rows that cross each other in the joist space.

Metal cross-bridging is available in different lengths for particular joist size and spacing. It is usually made of 18-gauge steel, and is ¾ inch wide. It comes in a variety of styles. It is applied in a way similar to that used for wood cross-bridging.

Laying Out and Cutting Wood Cross-Bridging

Wood cross-bridging may be laid out using a framing square. Determine the actual distance between floor joists and the actual depth of the joist. For example, 2 × 10 floor joists 24 inches OC measure 22½ inches between them. The actual depth of the joist is 9¼ inches.

Hold the framing square on the edge of a piece of bridging stock. Make sure the 9¼-inch mark of the tongue lines up with the upper edge of the stock. Also make sure the 22½-inch mark of the blade lines up with the lower edge of the stock. Mark lines along the tongue and blade across the stock.

Rotate the square, keeping the same face up. Re-align the square to the previous marks, then mark along the tongue **(Procedure 32–E)**. The bridging may then be cut using a power miter box. Tilt the blade and use a stop set to cut duplicate lengths.

Installing Bridging

Determine the centerline of the bridging. Snap a chalk line across the tops of the floor joist from one

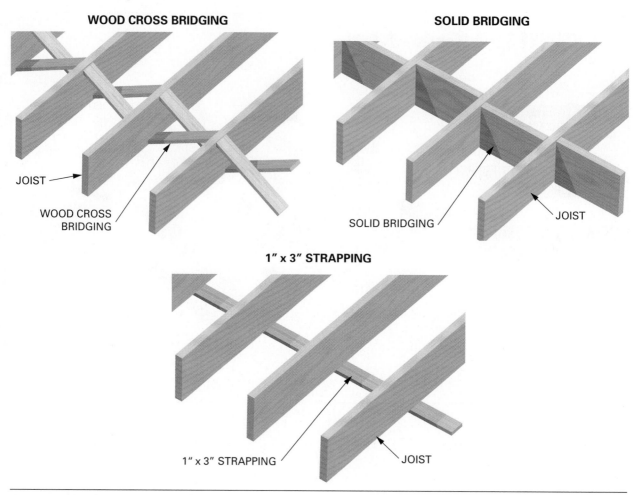

WOOD CROSS BRIDGING

JOIST

WOOD CROSS BRIDGING

SOLID BRIDGING

SOLID BRIDGING

JOIST

1" x 3" STRAPPING

1" x 3" STRAPPING

JOIST

FIGURE 32–21 Types of bridging.

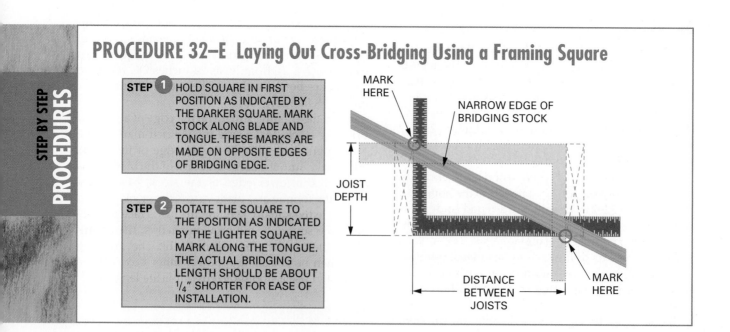

STEP BY STEP PROCEDURES

PROCEDURE 32–E Laying Out Cross-Bridging Using a Framing Square

STEP 1 HOLD SQUARE IN FIRST POSITION AS INDICATED BY THE DARKER SQUARE. MARK STOCK ALONG BLADE AND TONGUE. THESE MARKS ARE MADE ON OPPOSITE EDGES OF BRIDGING EDGE.

STEP 2 ROTATE THE SQUARE TO THE POSITION AS INDICATED BY THE LIGHTER SQUARE. MARK ALONG THE TONGUE. THE ACTUAL BRIDGING LENGTH SHOULD BE ABOUT $1/4$" SHORTER FOR EASE OF INSTALLATION.

MARK HERE

NARROW EDGE OF BRIDGING STOCK

JOIST DEPTH

DISTANCE BETWEEN JOISTS

MARK HERE

PROCEDURE 32–F Installing Solid Bridging

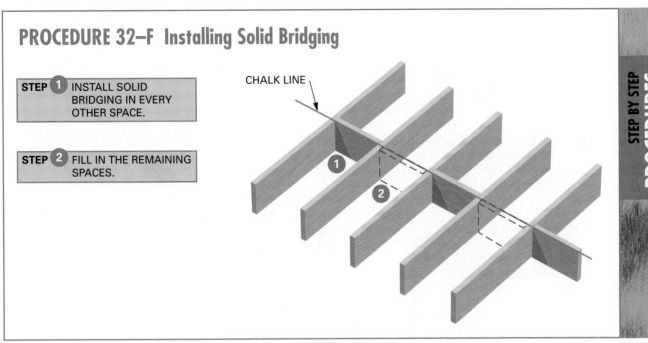

STEP ① INSTALL SOLID BRIDGING IN EVERY OTHER SPACE.

STEP ② FILL IN THE REMAINING SPACES.

CHALK LINE

STEP BY STEP PROCEDURES

---

end to the other. Square down from the chalk line to the bottom edge of the floor joists on both sides.

Solid Wood Bridging. To install solid wood bridging, cut the pieces to length. Install pieces in every other joist space on one side of the chalk line. Fasten the pieces by nailing through the joists into their ends. Keep the top edges flush with the floor joists. Install pieces in the remaining spaces on the opposite side of the line in a similar manner **(Procedure 32–F)**.

Wood Cross-Bridging. To install wood cross-bridging, start two 6d nails in one end of the bridging. Fasten it flush with the top of the joist on one side of the line. Nail only the top end. The bottom ends are not nailed until the subfloor is fastened.

Within the same joist cavity or space, fasten another piece of bridging to the other joist. Make sure it is flush with the top of the joist and positioned on the other side of the chalk line. Also leave a space between the bridging pieces where they form the X to minimize floor squeaks. Continue installing bridging in the other spaces, but alternate so that the top ends of the bridging pieces are opposite each other where they are fastened to same joist **(Procedure 32–G)**.

Metal Cross-Bridging. Metal cross-bridging is fastened in a manner similar to that used for wood cross-bridging. The method of fastening may differ according to the style of the bridging. Usually the bridging is fastened to the top of the joists through predrilled holes in the bridging. Because the metal is thin, nailing to the top of the joists does not interfere with the subfloor.

Some types of metal cross-bridging have steel prongs that are driven into the side of the floor joists. This bridging can be installed from below to layout lines made previous to the installation of the subfloor.

COLUMNS

Girders may be supported by framed walls, wood posts, or steel columns **(Figure 32–22)**. Metal plates are used at the top and bottom of the columns to distribute the load over a wider area. The plates have predrilled holes so that they may be fastened to the girder. Notched sections prevent the columns from slipping off the plates. Column size should be determined from the blueprints.

Installing Columns

After the floor joists are installed and before any more weight is placed on the floor, the temporary posts

STEP BY STEP PROCEDURES

PROCEDURE 32–G Installing Cross-Bridging

STEP 1 SNAP CHALK LINE ACROSS TOPS OF FLOOR JOISTS IN CENTER OF BRIDGING ROW.

STEP 2 SQUARED LINES MAY BE DRAWN DOWN FROM CHALK LINE ON BOTH SIDES OF FLOOR JOISTS.

STEP 3 FASTEN TOP ENDS OF BRIDGING SO THEY OPPOSE EACH OTHER ON THE SAME SIDE OF THE CHALK LINE.

STEP 4 LEAVE BOTTOM ENDS LOOSE UNTIL SUBFLOOR IS APPLIED. THEN FASTEN SO EDGE LINES UP WITH SQUARED LINE. ALSO, LEAVE A SPACE BETWEEN THE BRIDGING PIECES WHERE THEY CROSS.

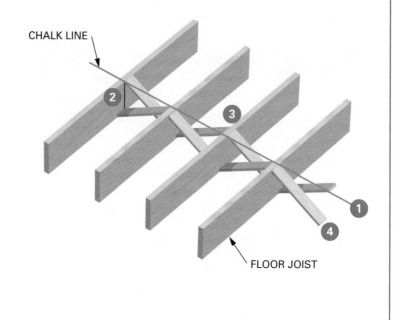

CHALK LINE

FLOOR JOIST

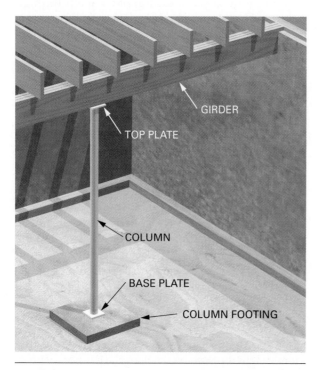

FIGURE 32–22 **Typical column supporting a girder.**

GIRDER

TOP PLATE

COLUMN

BASE PLATE

COLUMN FOOTING

supporting the girder are replaced with permanent posts or columns. Straighten the girder by stretching a line from end to end. Measure accurately from the column footing to the bottom of the girder. Hold a strip of lumber on the column footing. Mark it at the bottom of the girder. Transfer this mark to the column. Deduct the thickness of the top and bottom column plate.

To mark around the column so it has a square end, wrap a sheet of paper around it. Keeping the edges even, mark along the edge of the paper **(Procedure 32–H)**.

Cut through the metal along the line using a hacksaw, reciprocating saw, or circular saw with a metal cutting blade. Install the columns in a plumb position under the girder and centered on the footing. Fasten the top plates to the girder with lag screws. If the girder is steel, then holes must be drilled. The plates are then bolted to the girder, or they may be welded to the girder. The bottoms of the columns are held in place when the finish concrete basement floor is placed around them. If the column is placed on the finished floor, then the bottom plate must be anchored to the footing.

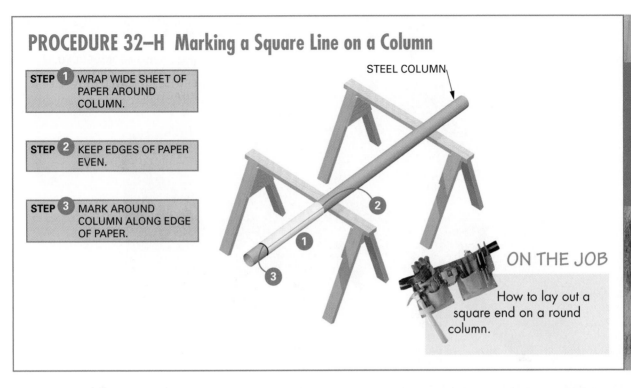

PROCEDURE 32–H Marking a Square Line on a Column

STEP 1 WRAP WIDE SHEET OF PAPER AROUND COLUMN.

STEP 2 KEEP EDGES OF PAPER EVEN.

STEP 3 MARK AROUND COLUMN ALONG EDGE OF PAPER.

STEEL COLUMN

ON THE JOB

How to lay out a square end on a round column.

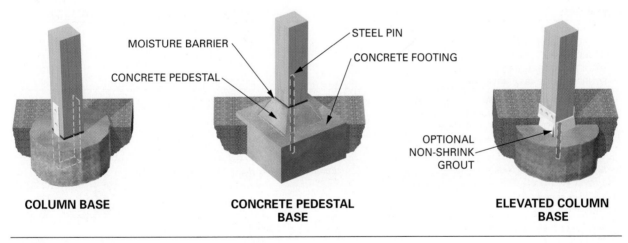

MOISTURE BARRIER

CONCRETE PEDESTAL

STEEL PIN

CONCRETE FOOTING

OPTIONAL NON-SHRINK GROUT

COLUMN BASE

CONCRETE PEDESTAL BASE

ELEVATED COLUMN BASE

FIGURE 32–23 The bottoms of wood posts sometimes rest on pedestal-type footings.

Wood posts are installed in a similar manner, except their bottoms are placed on a pedestal footing **(Figure 32–23)**.

SUBFLOORING

Subflooring is used over joists to form a working platform. This is also a base for finish flooring, such as hardwood flooring, or underlayment for carpet or resilient tiles. APA-Rated Sheathing Exposure 1 is generally used for subflooring in a two-layer floor system. APA-Rated Sturd-I-Floor panels are used when a single-layer subfloor and underlayment system is desired. Blocking is required under the joints of these panels unless tongue-and-groove edges are used. **Figure 32–24** is a selection and fastening guide for APA Sturd-I-Floor panels.

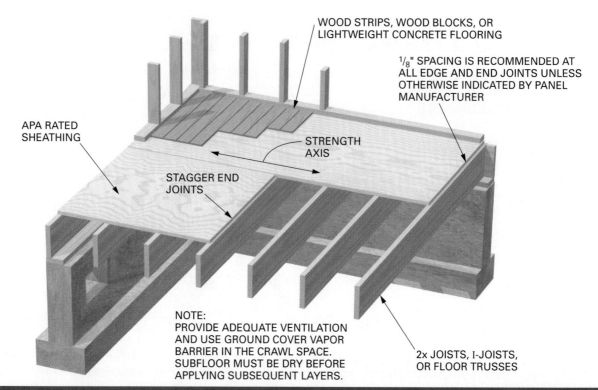

WOOD STRIPS, WOOD BLOCKS, OR LIGHTWEIGHT CONCRETE FLOORING

$\frac{1}{8}$" SPACING IS RECOMMENDED AT ALL EDGE AND END JOINTS UNLESS OTHERWISE INDICATED BY PANEL MANUFACTURER

APA RATED SHEATHING

STRENGTH AXIS

STAGGER END JOINTS

NOTE:
PROVIDE ADEQUATE VENTILATION AND USE GROUND COVER VAPOR BARRIER IN THE CRAWL SPACE. SUBFLOOR MUST BE DRY BEFORE APPLYING SUBSEQUENT LAYERS.

2x JOISTS, I-JOISTS, OR FLOOR TRUSSES

Nailing Recommendations

Type of Panel	Recommended Nail Size & Type	Panel Edges	Intermediate Supports
APA-Rated Sturd-I-Floor—*Glue-nailed installation*	Ring- or screw-shank nails		
STURD-I-FLOOR 16, 20, 24 oc, 3/4" thick or less	6d[a]	12"	12"
STURD-I-FLOOR 24 oc, 7/8" or 1" thick	8d[a]	6"	12"
STURD-I-FLOOR 32, 48 oc, 32" spans	8d[a]	6"	12"
STURD-I-FLOOR 48 oc, 48" spans	8d[b]	6"	6"
APA-Rated Sturd-I-Floor—*Nailed-only installation*	Ring- or screw-shank nails		
STURD-I-FLOOR 16, 20, 24 oc. 3/4" thick or less	6d	6"	12"
STURD-I-FLOOR 24, 32 oc, 7/8" or 1" thick	8d	6"	12"
STURD-I-FLOOR 48 oc, 32" spans	8d[b]	6"	12"
STURD-I-FLOOR 48 oc, 48" spans	8d[b]	6"	6"
APA-Rated Sheathing—*Subflooring*	Common, smooth, ring- or screw-shank[c]		
7/16" to 1/2" thick	6d	6"	12"
7/8" thick or less	8d	6"	12"
Thicker panels	10d	6"	6"
APA-Rated Sheathing—*Wall sheathing*	Common, smooth, ring- or screw-shank or galvanized box[c]		
1/2" thick or less	6d	6"	12"
Over 1/2" thick	8d	6"	12"
APA-Rated Sheathing—*Roof sheathing*	Common, smooth, ring- or screw-shank[c]		
5/16" to 1" thick	8d	6"	12"[d]
Thicker panels	8d ring- or screw-shank or 10d common smooth	6"	12"[d][e]
APA-Rated Siding—*Applied directly to studs or*	Hot dipped galvanized box, siding or casing		
	over nonstructural sheathing		
1/2" thick or less	6d	6"	12"
Over 1/2" thick	8d	6"	12"

(a) 8d common nails may be substituted if ring- or screw-shank nails are not available.

(b) 10d common nails may be substituted if supports are well seasoned.

(c) Other code-approved fasteners may be used.

(d) For spans 48" or greater, space nails 6" at all supports.

(e) Additional fasteners may be required in high wind regions, for further information see APA Data File: Roof Sheathing Fastening Schedules for Wind Uplift, Form T325.

FIGURE 32–24 **Nailing specifications for APA panels.** *(Courtesy of American Plywood Association)*

FIGURE 32–25 **Typical installation of APA-rated panel subfloor.** *(Courtesy Louisiana Pacific Corporation)*

APPLYING PLYWOOD SUBFLOORING

Starting at the corner from which the floor joists were laid out, measure in 4 feet. Note that tongue-and-groove plywood subfloor is only 47½″ wide. Snap a line across the tops of the floor joists from one end to the other. Start with a full panel. Fasten the first row to the chalk line **(Figure 32–25)** and align the joists to the correct spacing before nailing the panel. Leave a ⁄₁₆-inch space at all panel end joints to allow for expansion.

Start the second row with a half-sheet to stagger the end joints. Continue with full panels to finish the row. Leave a ⅛-inch space between panel edges. All end joints are made over joists.

Continue laying and fastening plywood sheets in this manner until the entire floor is covered **(Procedure 32–I)**. Leave out sheets where there are to be openings in the floor. Snap chalk lines across the edges and ends of the building. Trim overhanging plywood with a circular saw.

Estimating Material Quantities for Floor Framing

To estimate the material needed for the floor frame, the length and width of the builing are needed.

Girder. The amount of girder material required is determined by the girder length. If the girder is steel, then the quantity required is merely the actual length of the girder. If, however, the girder is built up, then a calculation must be performed to arrive at the amount of girder material required.

For a built-up girder, multiply the length of the girder by the number of plies in the girder. Then divide by the length of each piece of girder material to get the number of pieces to purchase. For example, if a 45-foot-long girder is made of four plies and made from material that is 12 feet long, then $45 \times 4 = 180$ **linear feet.** Then divide 180 by 12 to arrive at 15 pieces. Add one more waste when seams must be properly aligned, so the total number of pieces to purchase is 16.

Sills. The number of sill pieces depends on the building perimeter and the length of each sill piece. Take the building perimeter and divide by the sill piece length. For example, if a building is 28×48 and sill pieces are 12 feet long, then the number of pieces is calculated as follows: $2(28 + 48) = 152$ feet (perimeter) $\div 12 = 12.667$ or 13 pieces.

Floor Joists. To determine the number of floor joists to order, first determine if they are full length across the building or lapped at the girder. The quantity of lapped floor joists is twice that of full-length joists. To find the number of full-length joists, divide the length of the building by the spacing in terms of feet. For example, if the spacing is 16 inches, then it is $16 \div 12 = 1.333$ feet. So a 40-foot-long building divided by $1.333 = 30$ pieces. Add a joist for every parallel partition above that will need extra support. The band joist material is the building perimeter divided by the length of material used. For example, if the perimeter is 160 feet and the material used is 12 feet long, then the band joist material is $160 \div 12 = 13.333$ or 14 pieces. Add extra for headers as needed.

Bridging. Bridging quantity depends on the style of bridging. The linear feet of solid bridging is simply the length of the building times the number of rows of bridging. Linear feet of cross-bridging is determined by taking the number of joists times 3 feet for 16-inch OC joists. This number is arrived at because 3 feet of bridging is needed for each joist cavity. Four feet is needed for 19.2-inch OC, and 5 feet is needed for 24-inch OC. Then multiply times the number of rows of bridging needed. For example, if two rows of bridging are needed for 30 full-length 16-inch OC joists, then $30 \times 3 \times 2 = 180$ linear feet of bridging must be purchased.

Subfloor. Subfloor is determined from the square footage of the building. Multiply building length times building width to determine the square footage of the building. The number of panel pieces needed is this area divided by the square feet per panel. For example, if the building is 30×50 feet and standard 4×8 panels are used, then the number of pieces is $30 \times 50 = 1,500$ divided by 32 (from $4 \times 8 = 32$ square feet per panel) $= 46.875$ or 47. Note that since tongue-and-groove panels are only 47½ inches wide instead of 48, they do not cover a full 32 square feet. Add 1 percent more to compensate. For example, 46.875 sheets + 1 percent of 46.875 = 47.34 or 48 sheets.

STEP BY STEP
PROCEDURES

PROCEDURE 32–1   Layout Procedure for Installing Plywood Subfloor

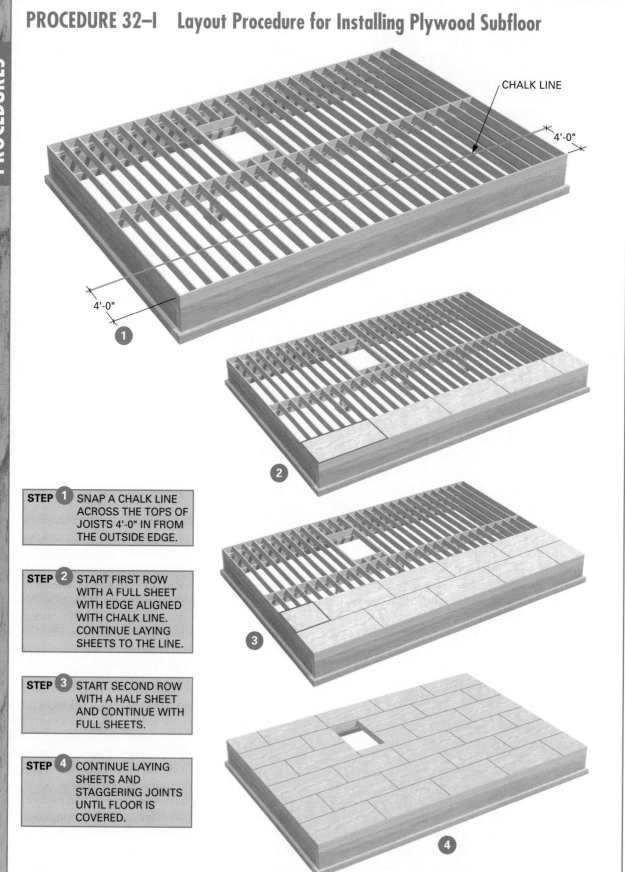

CHALK LINE

4'-0"

4'-0"

①

②

③

④

STEP ① SNAP A CHALK LINE ACROSS THE TOPS OF JOISTS 4'-0" IN FROM THE OUTSIDE EDGE.

STEP ② START FIRST ROW WITH A FULL SHEET WITH EDGE ALIGNED WITH CHALK LINE. CONTINUE LAYING SHEETS TO THE LINE.

STEP ③ START SECOND ROW WITH A HALF SHEET AND CONTINUE WITH FULL SHEETS.

STEP ④ CONTINUE LAYING SHEETS AND STAGGERING JOINTS UNTIL FLOOR IS COVERED.

33 Construction to Prevent Termite Attack

Of all the destructive wood pests, **termites** are the most common. They cause tremendous economic loss annually. They attack wood throughout most of the country, but they are more prevalent in the warmer sections **(Figure 33–1)**. Buildings should be designed and constructed to minimize termite attack. Chemicals should only be applied by trained technicians.

TERMITES

Termites play a beneficial role in their natural habitat. They break down dead or dying plant material to enrich the soil. However, when termites feed on wood structures, they become pests.

Kinds of Termites

There are three kinds of termites: *drywood, dampwood,* and *subterranean.*

Drywood Termites. Drywood termites enter a building usually through attic or foundation vents. They attack sheathing and structural members. They also infest wood door and window frames and furniture. However, the colonies are small. Unchecked, they cause less damage to buildings than other kinds of termites. Treatment is usually tent fumigation of the entire building with a toxic gas.

Dampwood Termites. Dampwood termites infest wood kept wet by lawn sprinklers, leaking toilets, showers, pipes, roofs, and other places where wood is damp. They are capable of doing great damage to a structure if undetected. Entrance into the building is usually where a continual source of moisture keeps wood wet. Discovery of a leak sometimes reveals a dampwood termite infestation. Once the wood is dry, dampwood termites will leave.

Subterranean Termites. Subterranean termites live in the ground. They are the most destructive species because they have such large underground colonies **(Figure 33–2).** For protection against drying out, they must stay in close contact with the soil and its moisture. Above ground, they build earthen *shelter tubes* to protect themselves from the drying effects of

○ REGION I VERY HEAVY
◐ REGION II MODERATE TO HEAVY
◑ REGION III SLIGHT TO MODERATE
○ REGION IV NONE TO SLIGHT

FIGURE 33–1   Degree of subterranean termite hazard in the United States.

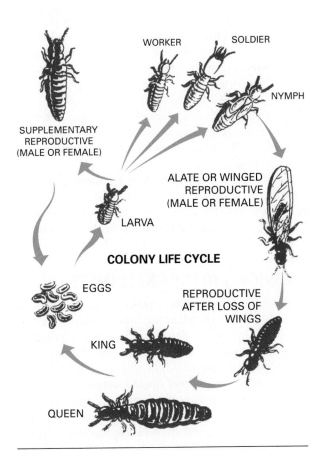

FIGURE 33–2   Typical subterranean termite life cycle.

FIGURE 33–3   **Subterranean termites can build unsupported shelter tubes as high as 12 inches.** (*Courtesy of* **The Termite Report,** *Pear Publishing; Don Pearman, photographer*)

the air. These tubes are usually built over a surface for support. Termites can build unsupported tubes as high as 12 inches in their effort to reach wood **(Figure 33–3).**

Treatment for subterranean termites generally consists of correcting conditions favorable to infestation, installation of ground-to-wood termite barriers, and chemical treatment of the foundation and soil.

TECHNIQUES TO PREVENT TERMITES

Protection against subterranean termites should be considered during planning and construction of a building. Improper design and poor construction practices could lead to termite infestation after completion of the building. Preventive efforts in the planning stage and during construction may save the future owner much anxiety and expense. All of the techniques used for the prevention of termite attack are based on keeping the wood in the structure dry (equilibrium moisture content) and making it as difficult as possible for termites to get

to the wood. In lumber, a moisture content below 20 percent also prevents the growth of fungi, which cause wood to rot.

The Site

All tree stumps, roots, branches, and other wood debris should be removed from the building site. Do not bury it on the site. Footing and wall form planks, boards, stakes, spreaders, and scraps of lumber should be removed from the area before backfilling around the foundation. Lumber scraps should not be buried anywhere on the building site. None should be left on the ground beneath or around the building after construction is completed.

The site should be graded to slope away from the building on all sides. The outside finished grade should always be equal to or below the level of the soil in crawl spaces. This ensures that water is not trapped underneath the building **(Figure 33–4).**

Chemical treatment of the soil before construction is one of the best methods of preventing termite attack. This should be a supplement to, and not a substitute for, proper building practices.

Perforated drain pipe should be placed around the foundation, alongside the footing. This will drain water away from the foundation **(Figure 33–5).** The foundation drain pipe should be sloped so water can drain

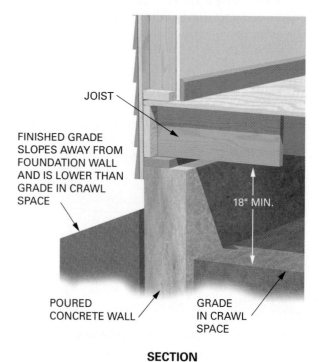

JOIST

FINISHED GRADE
SLOPES AWAY FROM
FOUNDATION WALL
AND IS LOWER THAN
GRADE IN CRAWL
SPACE

18" MIN.

POURED
CONCRETE WALL

GRADE
IN CRAWL
SPACE

SECTION

FIGURE 33–4   **The finished grade should slope away from and be lower than the crawl space floor level.**

FIGURE 33–5 Perforated drain pipe is placed alongside the foundation footing to drain water away from the building. *(Courtesy of Boccia, Inc.)*

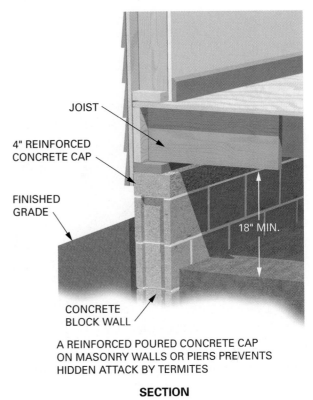

JOIST

4" REINFORCED
CONCRETE CAP

FINISHED
GRADE

18" MIN.

CONCRETE
BLOCK WALL

A REINFORCED POURED CONCRETE CAP
ON MASONRY WALLS OR PIERS PREVENTS
HIDDEN ATTACK BY TERMITES

SECTION

FIGURE 33–6 The top course of concrete block walls should be capped or filled completely with concrete.

to a lower elevation or a **dry well** some distance from the building. A dry well is a pit in the ground filled with stone to absorb the water from the drain pipe. Gutters and downspouts should be installed to lead roof water away from the foundation. Downspouts should be connected to a separate drain pipe to facilitate moving the water quickly.

Crawl Spaces

Solid concrete foundation walls should be properly reinforced and cured to prevent the formation of cracks. Cracks as little as $\frac{1}{32}$ inch wide permit the passage of termites.

Concrete block walls should be either capped with a minimum of 4 inches of reinforced concrete or the top **course** filled completely with concrete **(Figure 33–6).**

Air should be circulated in crawl spaces by means of ventilators placed to leave no pockets of stagnant air. In general, the total area of ventilation openings should be equal to $\frac{1}{150}$th of the ground area of the crawl space. Shrubbery should be kept away from openings to permit free circulation of air. There should be access to the crawl space for inspection of inner wall surfaces for termite tubes.

In crawl spaces and other concealed areas, clearance between the bottom of floor joists and the ground should be at least 18 inches and at least 12 inches for beams and girders **(Figure 33–7).**

Keep all plumbing and electrical conduits clear of the ground in crawl spaces. Suspend them from girders or joists. Do not support them with wood blocks or stakes in the ground. The soil around pipes extending from the ground to the wood above should be treated with chemicals.

Slab-on-Grade

Slab-on-grade is one of the most susceptible types of construction to termite attack. Termites gain access to the building over the edge of the slab, or through isolation joints, openings around plumbing, and

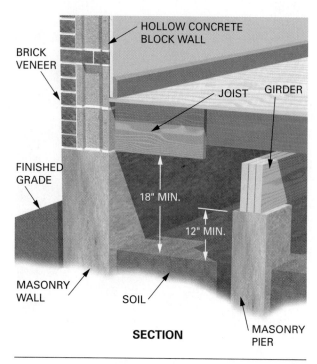

FIGURE 33–7 **Provide adequate clearance between wood and soil in crawl spaces.**

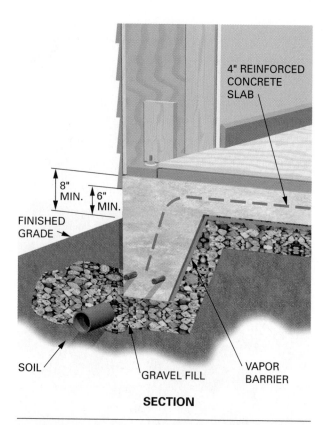

FIGURE 33–8 **In slab-on-grade construction, the monolithic slab provides the best protection against termites.**

cracks in the slab. Termite infestations in this type of construction are difficult to detect and control. For slab-on-grade construction, it is important to have the soil treated with chemicals before placing the concrete slab.

The monolithic slab provides the best protection against termites. The floor and footing are placed in one continuous operation, eliminating joints that permit hidden termite entry **(Figure 33–8).** Proper curing of the slab helps eliminate the development of cracks through which termites can gain access to the wood above.

One type of independent slab extends completely across the top of the foundation. This prevents hidden termite entry. The lower edge of the slab should be open to view from the outside.

The top of the slab should be at least 8 inches above the grade **(Figure 33–9).**

Independent slabs that rest either partway on or against the side of the foundation wall are the least reliable. Termites may gain hidden access to the wood through expansion joints **(Figure 33–10).** Fill the spaces around expansion joints, pipes, conduit, ducts, or steel columns with hot roofing-grade coaltar pitch.

Exterior Slabs. Spaces beneath concrete slabs for porch floors, entrance platforms, and similar units against the foundation should not be filled. Leave

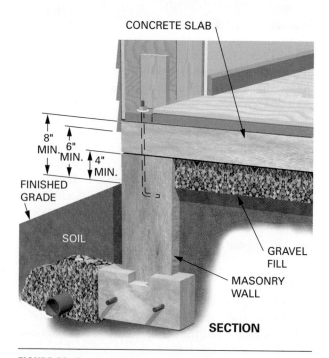

FIGURE 33–9 **An independent slab that extends across the top of the foundation wall prevents hidden termite attack.**

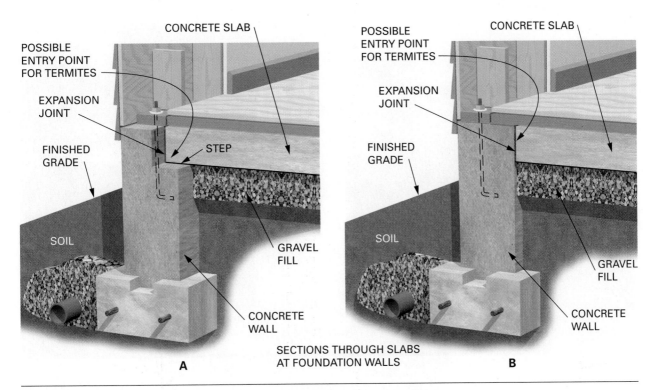

CONCRETE SLAB

POSSIBLE ENTRY POINT FOR TERMITES

EXPANSION JOINT

FINISHED GRADE

STEP

SOIL

GRAVEL FILL

CONCRETE WALL

POSSIBLE ENTRY POINT FOR TERMITES

CONCRETE SLAB

EXPANSION JOINT

FINISHED GRADE

SOIL

GRAVEL FILL

CONCRETE WALL

SECTIONS THROUGH SLABS AT FOUNDATION WALLS

A

B

FIGURE 33–10   Independent slab construction may allow a path for termite attack through isolation joints. (A) Edge of slab rests on ledge of the foundation wall. (B) Slab rests entirely on the ground (floating).

them open with access doors for inspection. If this cannot be done and spaces must be filled, have the soil treated for termites by a professional.

Exterior Woodwork

Wall siding usually extends no more than 2 inches below the top of foundation walls. It should be at least 6 inches above the finished grade.

Porch supports should be placed not closer than 2 inches from the building to prevent hidden access by termites. Wood steps should rest on a concrete base that extends at least 6 inches above the ground.

Door jambs, posts, and similar wood parts should never extend into or through concrete floors.

Termite Shields

If **termite shields** are properly designed, constructed, installed, and maintained, they will force

termites into the open. This will reveal any tubes constructed around the edge and over the upper surface of the shield **(Figure 33–11)**. However, research has shown that termite shields have not been effective in preventing termite infestations. This seems to be due to poor installation, inadvertent damage by homeowner, and infrequent inspections. Check local building codes that may mandate their use.

Use of Pressure-Treated Lumber

All wood decays naturally when not kept dry. To prevent or slow this process some wood species like southern yellow pine for example, are sawed, kiln dried and then treated under pressure with preservatives **(Figure 32–12)**. These preservatives enter the wood cells, virtually poisoning the wood for bacteria, fungi and insects. This process creates **pressure-treated** lumber which is used for foundation sills and structures that touch the ground.

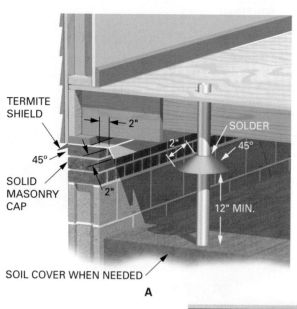

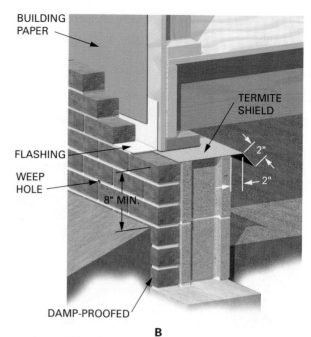

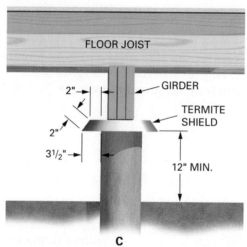

FIGURE 33–11 Typical installation of termite shields: (A) Exterior wall with wood siding, (B) exterior wall with brick veneer, and (C) over interior pier.

FIGURE 33–12 Preservatives are forced into lumber under pressure in large cylindrical tanks. (*Courtesy of Williamette Industries, Inc.*)

Although other grades are available, two are generally used. *Above ground* is used for sill plates, joists, girders, decks, and similar members. *Ground contact* is suitable for contact with soil or freshwater. Typical grade stamps are shown in **Figure 33–13.** Special grades are manufactured for saltwater immersion and wood foundations.

Check building codes for requirements concerning the use of pressure-treated lumber. Generally, building codes require the use of pressure-treated lumber for the following structural members:

■ Wood joists or the bottom of structural floors without joists that are located closer than 18 inches to exposed soil.

■ Wood girders that are closer than 12 inches to exposed soil in crawl spaces or unexcavated areas.

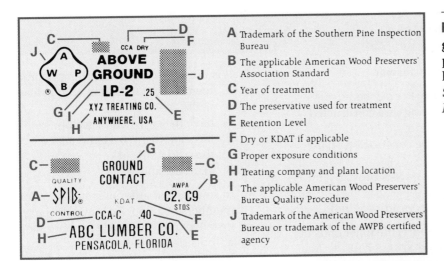

A Trademark of the Southern Pine Inspection Bureau

B The applicable American Wood Preservers' Association Standard

C Year of treatment

D The preservative used for treatment

E Retention Level

F Dry or KDAT if applicable

G Proper exposure conditions

H Treating company and plant location

I The applicable American Wood Preservers' Bureau Quality Procedure

J Trademark of the American Wood Preservers' Bureau or trademark of the AWPB certified agency

FIGURE 33–13 Typical grade stamps for pressure-treated lumber. (*Courtesy of Southern Pine Marketing Council*)

■ Sleeper, sill, and foundation plates on a concrete or masonry slab that is in direct contact with the soil.

Termites generally will not eat treated lumber. They will tunnel over it to reach untreated wood. Their shelter tubes then may be exposed to view and their presence easily detected upon inspection.

Follow these safety rules when handling pressure-treated lumber:

■ Wear eye protection and a dust mask when sawing or machining treated wood **(Figure 33–14).**

■ When the work is completed, wash areas of skin contact thoroughly before eating or drinking.

■ Clothing that accumulates sawdust should be laundered separately from other clothing and before reuse.

■ Dispose of treated wood by ordinary trash collection or burial. Do not burn treated wood. The chemical retained in the ash could pose a health hazard.

FIGURE 33–14 Use safety precautions when handling pressure-treated lumber. (*Courtesy of Southern Pine Marketing Council*)

Key Terms

balloon frame	end	matched boards	sheathing
bearing partitions	firestop blocking	on center (OC)	shims
box header	floor joists	plate	sill
bridging	green	platform frame	tail joist
course	grouting	pressure-treated	termites
draftstop blocking	joist header	rafters	termite shields
dry well	linear feet	ribbon	trimmer joist
eaves	masonry	ridge	web stiffener
edge			

Review Questions

Select the most appropriate answer.

1. A platform frame is easy to erect because
   a. only one-story buildings are constructed with this type of frame.
   b. each platform may be constructed on the ground.
   c. at each level a flat surface is provided on which to work.
   d. fewer framing members are required.

2. A heavy beam that supports the inner ends of floor joists is often called a
   a. pier.   c. stud.
   b. girder.   d. sill.

3. That member of a floor frame that is fastened directly on the foundation wall is called a
   a. pier.   c. stud.
   b. girder.   d. sill.

4. Draftstop blocking is used to
   a. slow the spread of fire.
   b. reduce drafts and improve insulation.
   c. serve as nailing for finish material.
   d. all of the above.

5. To quick and accurately mark a square end on a round column,
   a. use a square.
   b. measure down from the other end several times.
   c. use a pair of dividers.
   d. wrap a piece of paper around it.

6. To protect against termites, keep wood in crawl spaces and other concealed areas above the ground at least
   a. 8 inches.   c. 18 inches.
   b. 12 inches.   d. 24 inches.

7. When the ends of floor joists rest on a supporting member, they should have a bearing of at least
   a. 4 inches.
   b. 3½ inches.
   c. 2½ inches.
   d. 1½ inches.

8. If floor joists lap over a girder, they should have a minimum lap of
   a. 2 inches.   c. 6 inches.
   b. 3 inches.   d. 12 inches.

9. It is important when installing floor joists to
   a. toenail them to the sill with at least two 8d nails.
   b. have the crowned edges up.
   c. face nail them to a band joist with at least three 8d nails.
   d. all of the above.

10. In stating best order of installation for the members of a floor opening, the next member installed after the inside trimmer would be the
    a. outside trimmer.
    b. tail joists.
    c. inside header.
    d. outside header.

11. Holes bored in a floor joist should be no
    a. closer than 2 inches to the edge of the joist.
    b. larger than one-third of the joist width.
    c. larger than necessary.
    d. all of the above.

12. The bearing points of a girder should be at least
    a. as long as the girder is deep (wide).
    b. 4 inches.
    c. 5 inches.
    d. 6 inches.

13. The maximum nail spacing for an engineered panel subfloor on 16 OC floor joists is
    a. 6 inches on the edge and 6 inches on intermediate supports.
    b. 6 inches on the edge and 8 inches on intermediate supports.
    c. 6 inches on the edge and 12 inches on intermediate supports.
    d. 8 inches on the edge and 8 inches on intermediate supports.

14. To lay out the OC joists, the first one is set back a distance equal to
    a. one-half the joist thickness.
    b. the width of a joist.
    c. the thickness of a joist.
    d. ¾ inch always.

15. Pressure treatment is done on lumber to improve its
    a. decay resistance.
    b. pressure resistance
    c. nail holding strength.
    d. all of the above.

UNIT 12

Exterior Wall Framing

Wall framing methods vary across the country and are affected by regional characteristics. These variations are easy to adjust for when the carpenter has an understanding of basic framing methods and practices. Exterior walls must be constructed to the correct height, corners braced plumb, walls straightened from corner to corner, and window and door openings framed to specified size. The techniques described in this unit will enable the apprentice carpenter to frame exterior walls with competence.

Setting the exterior wall frame is the first step in defining the outline of the house. Stack material close enough to the work area, yet not in the way of future work.

OBJECTIVES

After completing this unit, the student should be able to:

- identify and describe the function of each part of the wall frame.
- determine the length of exterior wall studs and the size of rough openings, and lay out a story pole.
- build corner posts and partition intersections, and describe several methods of forming them.
- lay out the wall plates.
- construct and erect wall sections to form the exterior wall frame.
- plumb, brace, and straighten the exterior wall frame.
- apply wall sheathing.
- describe the construction of wood foundation walls.

34 Exterior Wall Frame Parts

The wall frame consists of a number of different parts. The student should know the name, function, location, and usual size of each member. Sometimes the names given to certain parts of a structure may differ according to the geographical area. For that reason, some members may be identified with more than one term.

PARTS OF AN EXTERIOR WALL FRAME

An exterior wall frame consists of *plates, studs, headers, sills, jack studs, trimmers, corner posts, partition intersections, ribbons,* and *braces* **(Figure 34–1).**

Plates

The top and bottom horizontal members of a wall frame are called *plates.* The bottom member is called a **sole plate.** It is also referred to as the *bottom plate* or *shoe.* The top members are called **top plates.** They usually consist of doubled 2-inch stock. In a balloon frame, the sole plate is not used. Instead, the studs rest directly on the sill plate.

Studs

Studs are vertical members of the wall frame. They run full-length between plates. *Jack studs* or *trimmers* are shortened studs that line the sides of an opening. They extend from the bottom plate up to the top of the opening. *Cripple* studs are shorter members above and below an opening, which extend from the top or the bottom plates to the opening.

Studs are usually 2 × 4s, but 2 × 6s are used when 6-inch insulation is desired in exterior walls. Studs are usually spaced 16 inches OC.

Headers

Headers or lintels run at right angles to studs. They form the top of window, door, and other wall openings, such as fireplaces. Headers must be strong enough to support the load above the opening. The depth of the header depends on the width of the opening. As the width of the opening increases, so must the strength of the header. Check drawings, specifications, codes, or manufacturers' literature for header size.

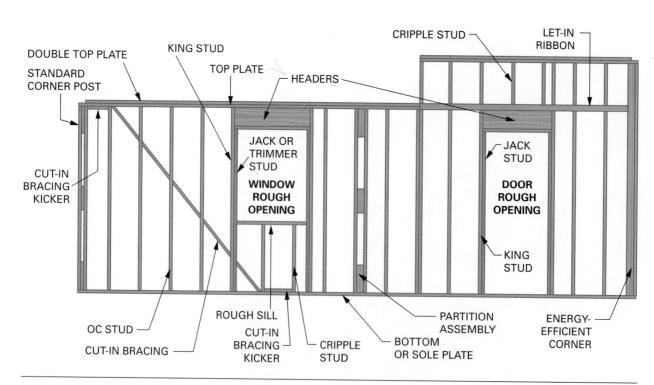

FIGURE 34–1 Typical component parts of an exterior wall frame.

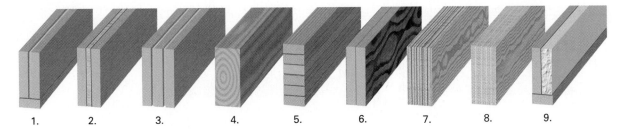

1. A BUILT-UP HEADER WITH A 2 X 4 OR 2 X 6 LAID FLAT ON THE BOTTOM.
2. A BUILT-UP HEADER WITH A ½" SPACER SANDWICHED IN BETWEEN.
3. A BUILT-UP HEADER FOR A 6" WALL.
4. A HEADER OF SOLID SAWN LUMBER.
5. GLULAM BEAMS ARE OFTEN USED FOR HEADERS.
6. A BUILT-UP HEADER OF LAMINATED VENEER LUMBER.
7. PARALLEL STRAND LUMBER MAKES EXCELLENT HEADERS.
8. LAMINATED STRAND LUMBER IS USED FOR LIGHT-DUTY HEADERS.
9. ENERGY-EFFICIENT HEADER WITH RIGID FOAM INSULATION.

FIGURE 34–2 **Types and styles of headers.**

Kinds of Headers. Solid or built-up lumber may be used for headers. For 4-inch walls, two pieces of 2-inch lumber with ½-inch plywood or strand board sandwiched in between them gives the header the full 3½-inch thickness of the wall. In 6-inch walls, three pieces of 2-inch lumber with two pieces of ½-inch plywood or strand board in between makes up the 5½-inch wall thickness **(Figure 34–2).**

Much engineered lumber is now being used for window and door opening headers. Figures 5–6, 6–1, and 8–1 show the use of laminated veneer lum-

ber, parallel strand lumber, and glulam beams as opening headers **(Figure 34–3).**

In many buildings, when the opening must be supported without increasing the header size, the top of a wall opening may be trussed to provide support **(Figure 34–4).** However, when the opening is fairly close to the top plate, the depth of the header is increased. This completely fills the space between the plate and the top of the opening. In this case, the same size header is usually used for all wall openings, regardless of the width of the

GLULAM

PSL (Parallel Strand Lumber)

LSL (Laminated Strand Lumber)

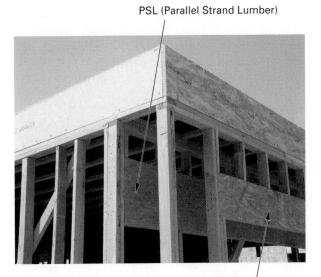

FIGURE 34–3 **Headers may be made of glulam, PSL (parallel strand lumber), and LSL (laminated strand lumber).**

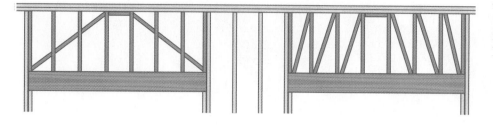

FIGURE 34–4 Two methods of trussing a large opening.

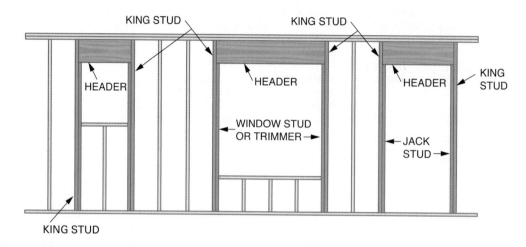

FIGURE 34–5 It is common practice to use the same header height for all wall openings.

opening **(Figure 34–5).** This eliminates the need to install short cripple studs above the header.

Rough Sills

Forming the bottom of a window opening, at right angles to the studs, are members called **rough sills.** They usually consist of a single 2-inch thickness. They carry little load. However, many carpenters prefer to use a double 2-inch thickness rough sill. This provides more surface on which to fasten window trim in the later stages of construction.

Jack Studs and Trimmers

Jack studs and **trimmers** (also called *liners*) are shortened studs that support the headers. They are fastened to the king studs on each side of the opening. In window openings the trimmer should be installed full length from header to bottom plate **(Figure 34–6).** They should not be cut to allow the sill to fit to the king stud. Door jack studs are installed the same as window trimmers **(Figure 34–7).**

Corner Posts

Corner posts are the same length as studs. They are constructed in a manner that provides an outside and an inside corner on which to fasten the exterior

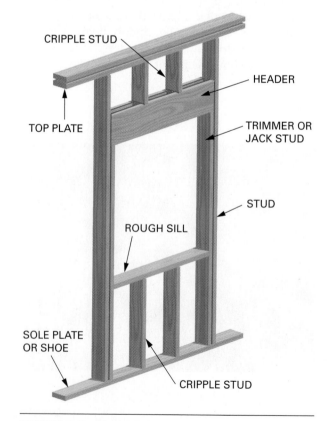

FIGURE 34–6 Typical framing for a window opening.

and interior wall coverings. They may be constructed in several ways.

Corner posts may be full-sized solid lumber such as a 4 × 6 coupled with the 2× dimension lumber

from the intersecting wall **(Figure 34–8).** This may also be parallel strand lumber or three 2× lumber nailed together. A second method uses three blocks in the space between two full pieces of 2× lumber. This method saves material by replacing the middle 2× lumber of the previous method with scraps. The third method is designed to increase the amount of insulation in the corner. One 2× lumber is rotated. This method uses the same amount of full length material as the second method, yet allows for a warmer corner.

Partition Intersections

Wherever interior **partitions** meet an exterior wall, extra studs need to be put in the exterior wall. This provides wood for fastening the interior wall covering in the corner. In most cases, the *partition intersection* is made of two studs nailed to the edge of 2 × 4 blocks about a foot long. One block is placed at the bottom, one at the top, and one about center on the studs **(Figure 34–9).**

Another method is to maintain the regular spacing of the studs. Blocking is then installed between them wherever partitions occur. The block is set back from the inside edge of the stud the thickness of a board. A 1 × 6 board is then fastened vertically on the inside of the wall so that it is centered on the partition.

Another method is to nail a continuous 2 × 6 backer to a full-length stud. The edges of the backer project an equal distance beyond the edges of the stud.

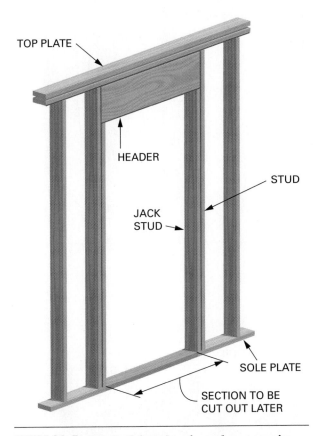

FIGURE 34–7 **Typical framing for a door opening.**

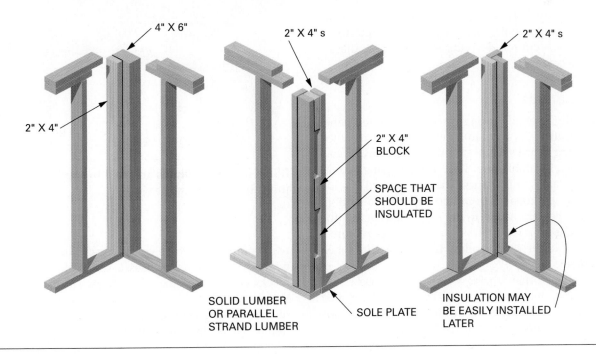

FIGURE 34–8 **Methods of fabricating corner posts.**

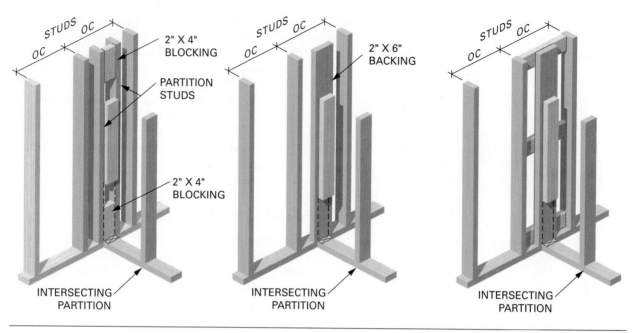

FIGURE 34–9 Methods of fabricating partition intersections.

Ribbons

Ribbons are horizontal members of the exterior wall frame in balloon construction. They are used to support the second-floor joists. The inside edge of the wall studs is notched so that the ribbon lays flush with the edge **(Figure 34–10)**. Ribbons are usually made of 1 × 4 stock. Notches in the stud should be made carefully so the ribbon fits snugly in the notch. This prevents the floor joists from settling. If the notch is cut too deep the stud will be unnecessarily weakened.

Corner Braces

Generally, no wall bracing is required if rated panel wall sheathing is used. In other cases, such as when insulating board sheathing is used, walls are braced with metal wall bracing. They come in gauges of 22 to 16 in flat, T- or L-shapes. They are about 1½″ wide and run diagonally from the top to the bottom plates. They are nailed to the stud edges before the sheathing is applied. The T- and L-shapes require a saw kerf in the stud to allow them to lay flat when installed.

Another corner bracing technique is to use 1 × 4s called *let-in bracing* **(Figure 34–11)**. A 1 × 4 is installed into notches cut out of the inside surface of the studs and plates. This allows the inside surface of all of the

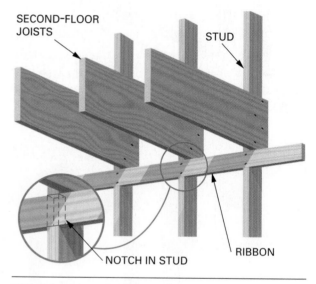

FIGURE 34–10 Ribbons are used to support floor joists in balloon frame walls.

wall components to be continuously flush. The last method, *cut-in bracing,* uses a series of 2× lumber blocks cut between the studs. **Kickers** are nailed to the plates at the ends of the bracing system to spread out the racking load. Both of these methods require tight fits to achieve maximum stiffness.

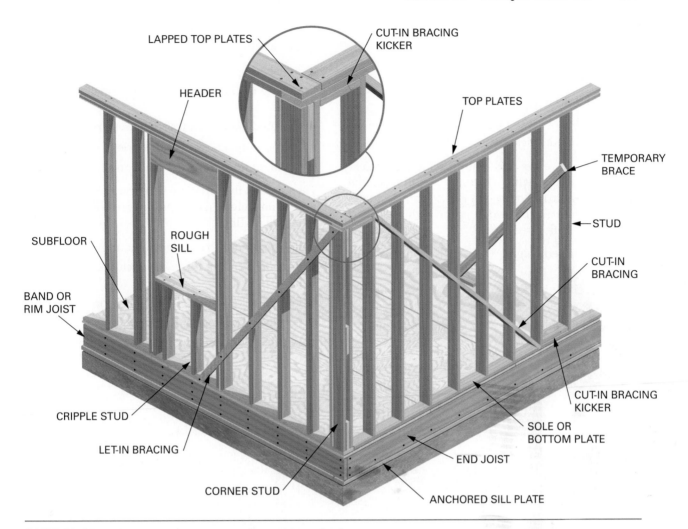

FIGURE 34–11 **Wood wall bracing may be cut-in or let-in.**

35 Framing the Exterior Wall

areful construction of the wall frame makes application of the exterior and interior finish easier. It also reduces problems for those who apply it later.

EXTERIOR WALL FRAMING

The standard height of a rough ceiling is usually 8'-1". Subtracting three plates of a total thickness of 4½ inches, the stud length is 92½ inches. Studs can be purchased precut to length, called precut studs, to save the carpenter time and wasted material. Note, however, that precut studs are usually 92⅝ inches long. The extra ⅛ inch length provides

insurance that the finished material will be easily installed later.

Sometimes the ceiling height is not the standard height. The section view of the house plans will specify the finished floor to finished ceiling height. From this number the length of the stud must be calculated.

Determining the Length of Studs

The stud length must be calculated so that, after the wall is framed, the distance from finish floor to ceiling will be as specified in the drawings. To determine the stud length, the thickness of the finish floor and the finished ceiling thickness below the ceiling joist must be known.

Stud length for platform framing is found by adding these measurements **(Figure 35–1):**

- Finished floor to ceiling height
- Ceiling thickness (includes *furring strips* if used)
- Finished floor thickness.

Then deduct the top and bottom plate total thickness. For example, what is the stud length when the finished floor to ceiling height is 7'–9", ceiling material is ½ inch thick, finished floor is ¾ inch thick, and it is framed with three plates?

7'–9"	Finished height	7'–10¼"	Rough ceiling height
½"	Ceiling thickness	− 4½"	Total plate thickness
+ ¾"	Floor thickness	7'–5¾"	Stud length
7'–10¼"	Rough ceiling height		

Stud length for balloon frame construction is found by adding these measurements **(Figure 35–2):**

- Finished floor to ceiling height of both stories
- Ceiling thickness of both stories (includes *furring strips* if used)
- Finished floor thickness of both stories
- Subfloor thickness of both stories
- Width of floor joists of both stories

Then deduct the top plate total thickness.

Ribbon height to support second-floor joists is found by adding:

- Finished floor to ceiling height of first floor
- First-floor ceiling thickness (includes *furring strips* if used)
- Finished floor thickness of first floor
- Subfloor thickness of first floor
- Width of floor joists of first floor.

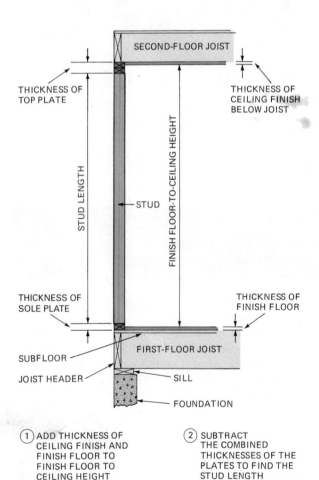

FIGURE 35–1 **Determining stud length in platform construction.**

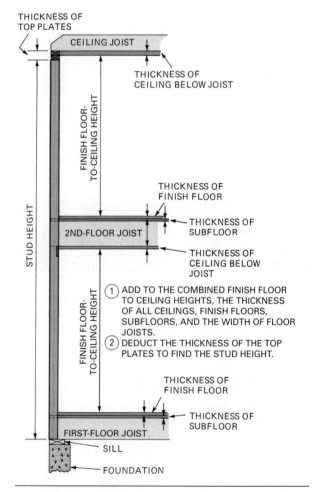

FIGURE 35–2 **Determining stud length in balloon frame construction.**

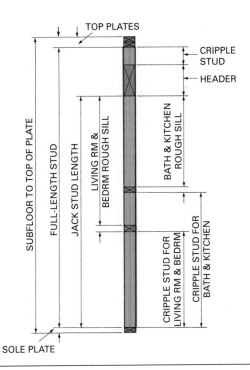

FIGURE 35–3   Layout of a typical story pole.

STORY POLE

The **story pole** is usually a straight strip of 1 × 2 or 1 × 3 lumber. Its length is the distance from the subfloor to the bottom of the joist above. Lines are squared across its width, indicating the height of the sole plates and top plates, headers, and rough sills for all openings.

The story pole is used to determine the length of studs, jack studs, trimmers, and cripple studs even if opening heights differ. The thickness and height of rough sills and the location of headers are also indicated and identified on the story pole **(Figure 35–3).**

A different story pole is made for each floor of the building. As framing progresses, the story pole is used to test the accurate location of framing members in the wall.

Determining Rough Opening Size

A **rough opening** is an opening framed in the wall in which to install doors and windows. The width and height of rough openings are not usually indicated in the plans. It is the carpenter's responsibility to determine the rough opening size for the particular unit from the information given in the door and window schedule. The door and window schedule contains the kind, style, manufacturer's model number, size of each unit, and rough opening dimensions.

Rough Openings for Doors

The rough opening for a door must be large enough to accommodate the door, door frame, and space for shimming the frame to a level and plumb position. Usually ½ inch is allowed for shimming, at the top and both sides, between the door frame and the rough opening. The amount allowed for the door frame itself depends on the thickness of the door frame beyond the door.

Care must be taken not to make the rough opening oversized. If the opening is made too large, the window or door finish may not cover it.

The sides and top of a door frame are called **jambs.** Jambs may vary in thickness. Sometimes *rabbeted* wood jambs are used. The rabbet is that part of the jamb that the door stops up against. At other times nominal 1-inch lumber is used for the jamb. A separate **stop** is applied **(Figure 35–4).** Steel jambs have the door stop built-in similar to a rabbeted wood jamb.

The bottom member of the door frame is called a **sill.** Sills may be hardwood, metal, or a combination of wood and metal. The type of sill and its thickness must be known in order to figure the rough opening height.

Door Rough Opening Height. Rough opening height is determined by adding five dimensions **(Figure 35–5).** The rough opening heights for all openings in a house are usually the same, so only one rough opening height needs to be calculated. Because the wall rests on the subfloor, the subfloor is the starting point:

- Finished floor thickness
- Door sill (or threshold) thickness (if none then add 1 inch for swing clearance under the door)
- Door height
- Head (top) jamb thickness
- Shim space (usually ½ inch)

For example, what is the rough opening height for a 6'-8" (80") door with a ¾-inch finished floor, no threshold, and a ¾-inch jamb?

¾"	Finished floor
1"	Clearance under door
80"	Door height (this includes the small space between the door and jambs that allows opening of the door)
¾"	Jamb thickness
+ ½"	Shim space
83"	Total rough opening height

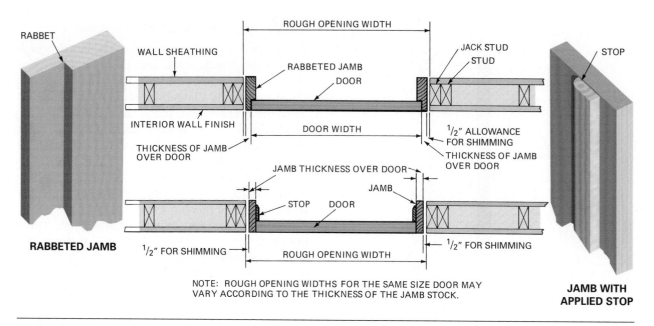

FIGURE 35–4 Determining the rough opening width of a door opening.

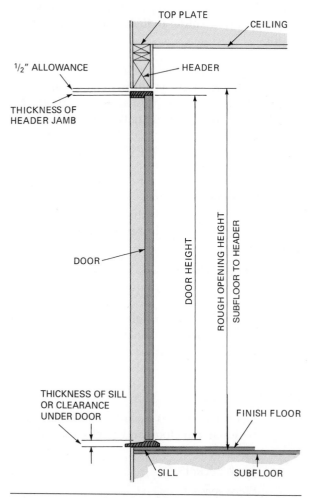

FIGURE 35–5 Determining the rough opening height of a door opening.

Jack stud length can be determined by subtracting bottom plate thickness, which is typically $1\frac{1}{2}$ inches. In this case the jack stud length is $83'' - 1\frac{1}{2}'' = 81\frac{1}{2}''$.

Door Rough Opening Width. Refer back to Figure 35–4 to see that the rough opening width is also found by adding five dimensions. Because door widths vary from room to room, a shorthand method is used. The rough opening width is found by adding $2\frac{1}{2}$ inches to the door width. This number comes from adding $\frac{1}{2}$ inch of shim space and $\frac{3}{4}$ inch of jamb thickness to both sides of the door; or $\frac{1}{2}'' + \frac{3}{4}'' = 1\frac{1}{4}''$ times two = $2\frac{1}{2}''$. Note that if the jam thickness is not $\frac{3}{4}$ inch, then the $2\frac{1}{2}''$ number must be adjusted accordingly. Thus, if the door is 2'–6'' (30'') wide then the rough opening width for this door is $30'' + 2\frac{1}{2}'' = 32\frac{1}{2}''$.

Rough Openings for Windows

Many kinds of windows are manufactured by a number of firms. Because of the number of styles, sizes, and variety of construction methods, it is best to consult the manufacturer's catalog to obtain the rough opening sizes. These catalogs show the style and size of the window unit. They also give the rough opening (RO) for each unit **(Figure 35–6).** Catalogs are available from the lumber company that sells the windows.

WALL LAYOUT

Wall construction begins with careful layout of all wall components. This is usually done on the top and bottom plates. The layout of the walls usually

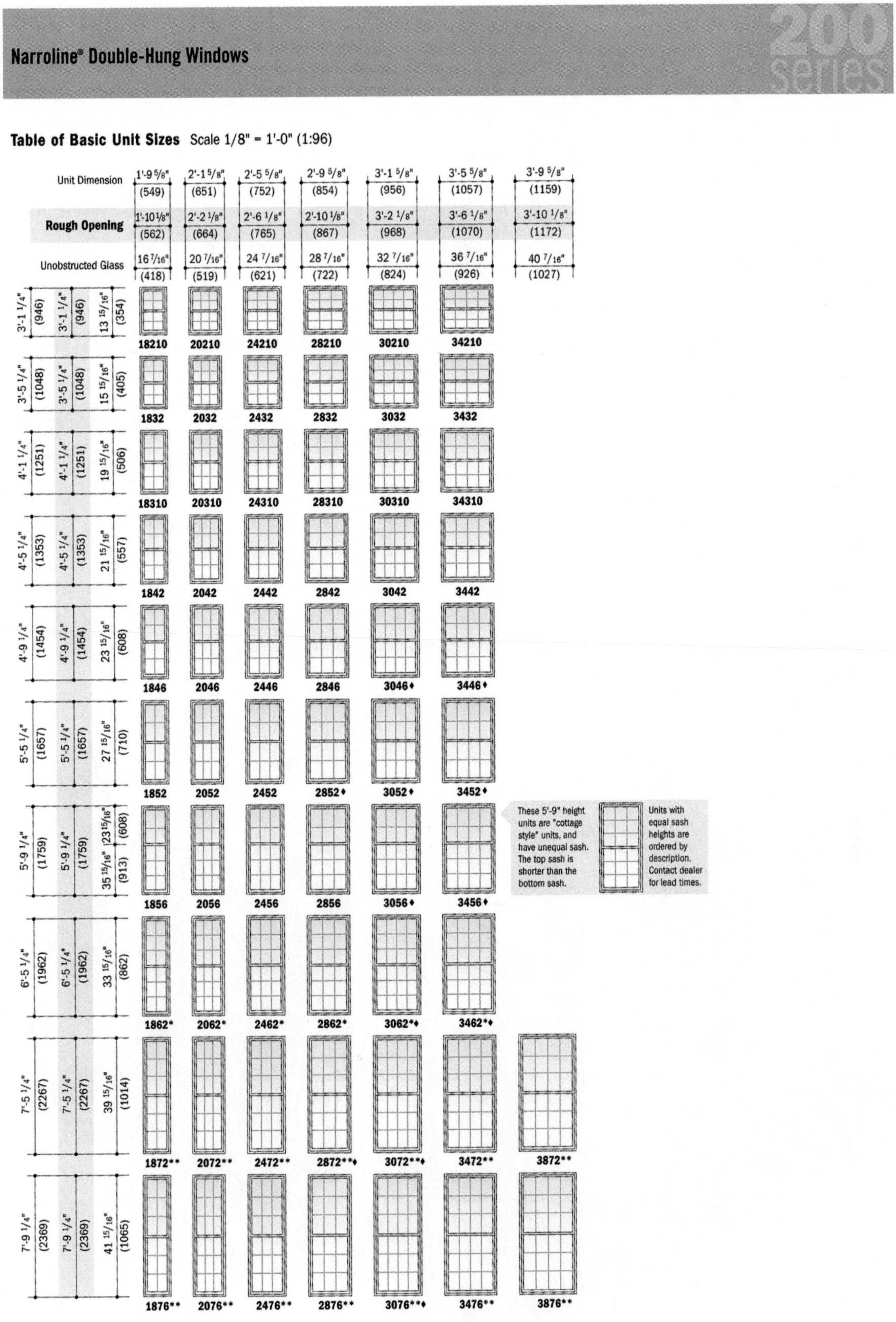
Narroline® Double-Hung Windows
200 series
Table of Basic Unit Sizes Scale 1/8" = 1'-0" (1:96)
Unit Dimension
1'-9 5/8" (549)
2'-1 5/8" (651)
2'-5 5/8" (752)
2'-9 5/8" (854)
3'-1 5/8" (956)
3'-5 5/8" (1057)
3'-9 5/8" (1159)
Rough Opening
1'-10 1/8" (562)
2'-2 1/8" (664)
2'-6 1/8" (765)
2'-10 1/8" (867)
3'-2 1/8" (968)
3'-6 1/8" (1070)
3'-10 1/8" (1172)
Unobstructed Glass
16 7/16" (418)
20 7/16" (519)
24 7/16" (621)
28 7/16" (722)
32 7/16" (824)
36 7/16" (926)
40 7/16" (1027)
3'-1 1/4" (946) 3'-1 1/4" (946) 13 15/16" (354)
18210 20210 24210 28210 30210 34210
3'-5 1/4" (1048) 3'-5 1/4" (1048) 15 15/16" (405)
1832 2032 2432 2832 3032 3432
4'-1 1/4" (1251) 4'-1 1/4" (1251) 19 15/16" (506)
18310 20310 24310 28310 30310 34310
4'-5 1/4" (1353) 4'-5 1/4" (1353) 21 15/16" (557)
1842 2042 2442 2842 3042 3442
4'-9 1/4" (1454) 4'-9 1/4" (1454) 23 15/16" (608)
1846 2046 2446 2846 3046♦ 3446♦
5'-5 1/4" (1657) 5'-5 1/4" (1657) 27 15/16" (710)
1852 2052 2452 2852♦ 3052♦ 3452♦
5'-9 1/4" (1759) 5'-9 1/4" (1759) 35 15/16" (913) 23 15/16" (608)
1856 2056 2456 2856 3056♦ 3456♦
These 5'-9" height units are "cottage style" units, and have unequal sash. The top sash is shorter than the bottom sash.
Units with equal sash heights are ordered by description. Contact dealer for lead times.
6'-5 1/4" (1962) 6'-5 1/4" (1962) 33 15/16" (862)
1862* 2062* 2462* 2862* 3062*♦ 3462*♦
7'-5 1/4" (2267) 7'-5 1/4" (2267) 39 15/16" (1014)
1872** 2072** 2472** 2872**♦ 3072**♦ 3472** 3872**
7'-9 1/4" (2369) 7'-9 1/4" (2369) 41 15/16" (1065)
1876** 2076** 2476** 2876** 3076**♦ 3476** 3876**

FIGURE 35–6 Sample of a manufacturer's catalog showing rough opening sizes for window units. *(Courtesy of Andersen Corporation)*

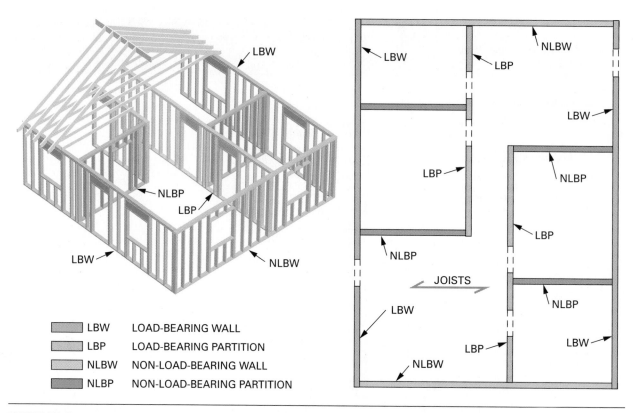

LBW	LOAD-BEARING WALL
LBP	LOAD-BEARING PARTITION
NLBW	NON-LOAD-BEARING WALL
NLBP	NON-LOAD-BEARING PARTITION

FIGURE 35–7   Walls in a building have different functions and characteristics.

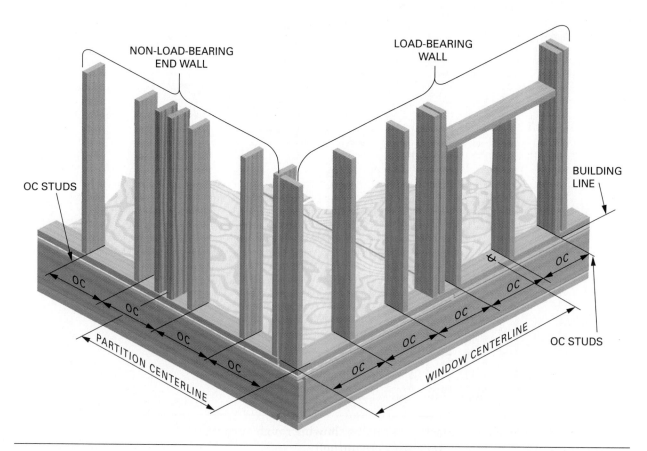

FIGURE 35–8   Layout wall components are measured from the building line.

begins at the same corner of the building where the floor joists began. This will help ensure that the load-bearing studs are directly over the joist below.

Before layout can begin, the carpenter must first determine if the wall to be laid out is load bearing or non-load bearing. Note that exterior walls are referred to as *walls* and interior walls are referred to as *partitions*.

Determining Wall Type

The load-bearing walls (LBW) are usually built first. They support the ceiling joist and rafters and typically run the length of the building. Non-load-bearing walls (NLBW) are end walls and run parallel with the joists. Interior partitions are also load and non-load-bearing. They are load-bearing partitions (LBP) if they run perpendicular to the joists and support the ends of the joists above. All other partitions are considered non-load-bearing partitions (NLBP) **(Figure 35–7).**

Each type of wall has a slightly different layout characteristic. To reduce confusion, remember that all centerline dimensions for openings are measured from the building line, which is on the outside edge of the exterior framing. Layout must take this fact into account **(Figure 35–8). Figure 35–9** notes the similarities and differences of laying out walls and partitions.

Laying Out the Plates

To lay out the plates, measure in on the subfloor, at the corners, the thickness of the wall. Snap lines on the subfloor between the marks. This is done so the wall can be erected to a straight line later. *Tack* the plates in position so their inside edges are to the chalk line. Do not drive nails home (they have to be pulled later). Use only as many as are needed to hold the pieces in place. Plan plate lengths so that joints between them fall in the center of a full-length stud for the convenient erection of wall sections later **(Figure 35–10).**

LAYOUT VARIATIONS FOR WALLS AND PARTITIONS

	MEASURE TO OC STUDS	MEASURE TO CENTERLINES OF OPENINGS
LOAD-BEARING WALL (LBW)	FROM END OF PLATE	FROM END OF PLATE
NON-LOAD-BEARING WALL (NLBW)	INCLUDE WIDTH OF ABUTTING WALL AND SHEATHING THICKNESS	INCLUDE WIDTH OF ABUTTING WALL
LOAD-BEARING PARTITION (LBP)	INCLUDE WIDTH OF ABUTTING WALL	INCLUDE WIDTH OF ABUTTING WALL
NON-LOAD-BEARING PARTITION (NLBP)	FROM END OF PLATE	INCLUDE WIDTH OF ABUTTING WALL

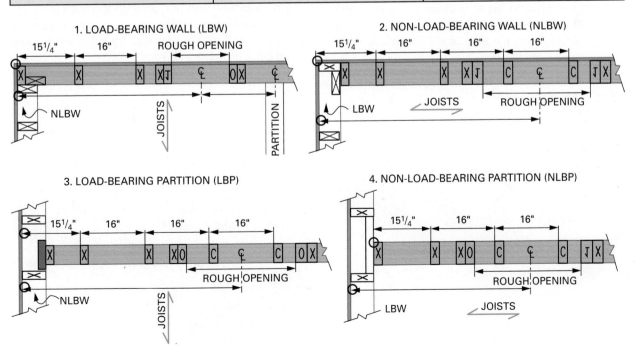

FIGURE 35–9 **Layout details for four types of walls.**

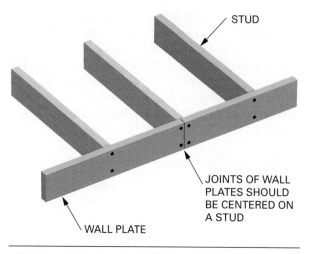

FIGURE 35–10 **Joints in the plates should fall at the center of a stud.**

Wall Openings. From the blueprints, determine the centerline dimension of all the openings in the wall. Lay these out on the plate. Then mark for the king and jack studs by measuring in each direction from the centerline one-half the width of the rough opening **(Figure 35–11)**. Recheck the rough opening measurement to be sure it is correct. Square lines at these points across the plates. Mark an *O* on the side of each line away from the centerline. The *O* represents the jack stud, but a *T* for trimmer or a *J* for jack can also be used. It makes little difference what marks are used as long as the builder understands what they mean.

From the squared lines, measure away from the centerline of the jack stud thickness. Square lines across. Mark *X*s on the side of the line away from center for the king studs on each side of the openings (Figure 35–11).

Partition Intersections

On architectural prints, interior partitions usually are dimensioned to their centerline. Mark on the plates the centerline of all partitions intersecting the wall **(Figure 35–12)**. From the centerlines, measure in each direction one-half the partition stud thickness. Square lines across the plates. Mark *X*s on the side of the lines away from center for the location of partition intersection studs.

Studs and Cripple Studs

After all openings and partitions have been laid out, start laying out all full-length studs and cripple studs. Proceed in the same manner and from the same end as laying out floor joists. This keeps studs directly in line with the joists below.

Measure in from the outside corner the regular stud spacing. From this mark, measure in one direction or the other, one-half the stud thickness. Square a line across the plates. Place an *X* on the side of the line where the stud will be located.

Stretch a tape along the length of the plates from this first stud location. Square lines across the plates at each specified stud spacing. Place *X*s on the same side of the line as the first line.

Where openings occur, mark the OC studs with a *C* instead of an *X*. This will indicate the location of cripple studs **(Procedure 35–A)**. All regular and cripple studs should line up with the floor joists below. When laying out the opposite wall, start from the same end as the first wall to keep all framing lined up with joists below.

MEASURE ½ WIDTH OF ROUGH OPENING IN BOTH DIRECTIONS FROM CENTERLINE

LAY OUT FOR JACK AND KING STUDS ON BOTH SIDES

SUBFLOOR

CORNER POST

DIMENSION TO CENTERLINE OF OPENING

REMEASURE THE ROUGH OPENING WIDTH AS A CHECK

FIGURE 35–11 Laying out a rough opening width.

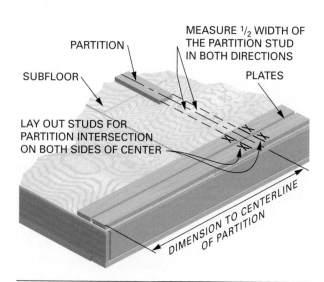

PARTITION

MEASURE ½ WIDTH OF THE PARTITION STUD IN BOTH DIRECTIONS

SUBFLOOR

PLATES

LAY OUT STUDS FOR PARTITION INTERSECTION ON BOTH SIDES OF CENTER

DIMENSION TO CENTERLINE OF PARTITION

FIGURE 35–12 Laying out a partition intersection.

Procedure 35–A Laying Out a Typical Wall Section

STEP 1 MEASURE TO OPENING CENTERLINE. MEASURE ONE-HALF RO IN BOTH DIRECTIONS. MARK *O* BOTH SIDES AWAY FROM CENTER.

STEP 2 MEASURE AND MARK FOR THE KING STUDS.

STEP 3 MEASURE PARTITION CENTERLINE. MARK PARTITION STUDS ON BOTH SIDES.

STEP 4 MEASURE OC SPACING FOR FIRST STUD. DEDUCT ONE-HALF STUD THICKNESS.

STEP 5 LAY OUT AND MARK REMAINING STUDS WITH *X*s MARK A *C* ON OC STUDS WITH THE OPENING SPACE.

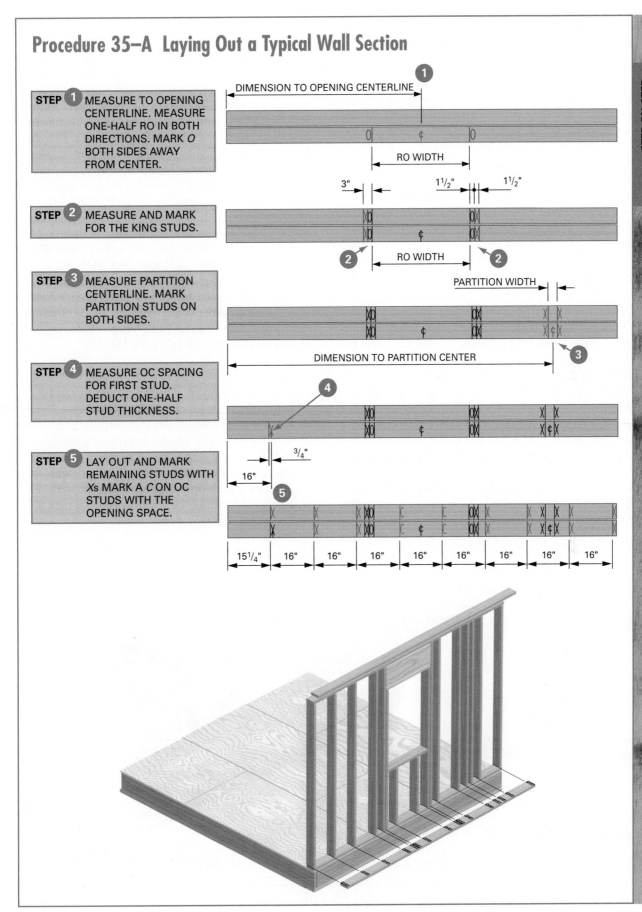

ASSEMBLING AND ERECTING WALL SECTIONS

The usual method of framing the exterior wall is to precut the wall frame members, assemble the wall frame on the subfloor, and erect the frame. With a small crew and without special equipment, the walls are raised section by section. When the frame is erected, the corners are plumbed and braced. Then the walls are straightened between corners. They are also braced securely in position. To prevent problems with the installation of the finish work later, it is important to keep the edges of the frame members flush wherever they join each other.

Precutting Wall Frame Members

Full-length Studs. Studs are often purchased precut to length for a standard 8'–1" wall height. Some builders buy in such high volume that lumber suppliers will also precut headers and jack and cripple studs. Most of the time framing members, other than studs, are cut to length on the job. A power miter saw is an effective tool for cutting studs and other framing to length. Set a stop the desired distance from the saw blade to cut duplicate lengths. If this type of saw is not available, a **jig** can be made for a portable electric circular saw to cut duplicate lengths of framing **(Figure 35–13)**. Reject any studs that are severely warped. They may be cut into shorter lengths for blocking.

Corner Posts and Partition Intersections. Corner posts and partition intersections are often made up ahead of time to speed the assembly. Corners may be made by nailing two full-length studs together where one stud is rotated at a right angle. This detail allows for more insulation in the corner, thereby making the building more energy efficient. Corners may also be made by nailing together two full-length studs with short blocks of the same material laid flat between them **(Figure 35–14)**.

Partition intersections may be made using ladder blocking, which again allows for more insulation in exterior walls. In this case no extra framing layout is needed because the ladder blocking is installed between on-center studs. Another method also allows for insulation to be easily installed later. A 2×6 is nailed at right angles to a stud. A third method is similar to that used for corner posts except for the way the blocks are placed. They are placed in a similar location between two full-length studs, yet with blocks on their edge **(Figure 35–15)**.

Headers, Rough Sills, Jack Studs, and Trimmers. Cut all headers and rough sills. Their length can be determined from the layout on the plates.

Make a story pole. From it determine the length of all trimmers, jack studs, and cripple studs. Cut them accordingly. It may be necessary to place identifying marks on headers, rough sills, jacks, and trimmers if rough openings are different sizes. This will assist in locating the window or door unit to be placed in each rough opening.

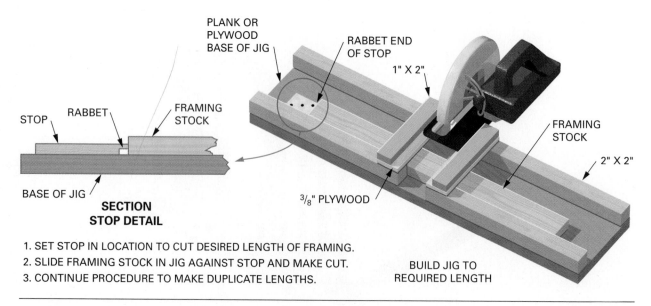

1. SET STOP IN LOCATION TO CUT DESIRED LENGTH OF FRAMING.
2. SLIDE FRAMING STOCK IN JIG AGAINST STOP AND MAKE CUT.
3. CONTINUE PROCEDURE TO MAKE DUPLICATE LENGTHS.

FIGURE 35–13   Techniques for making a cutoff jig for a circular saw.

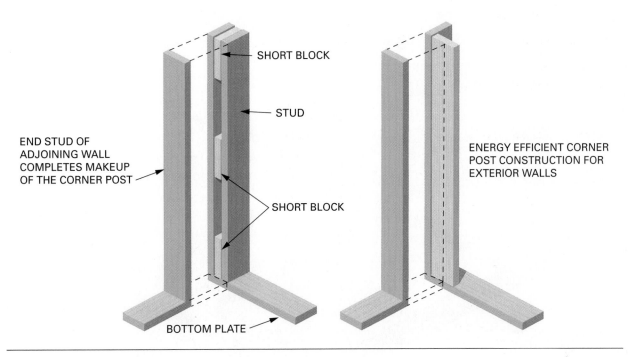

FIGURE 35–14 Construction of corner posts.

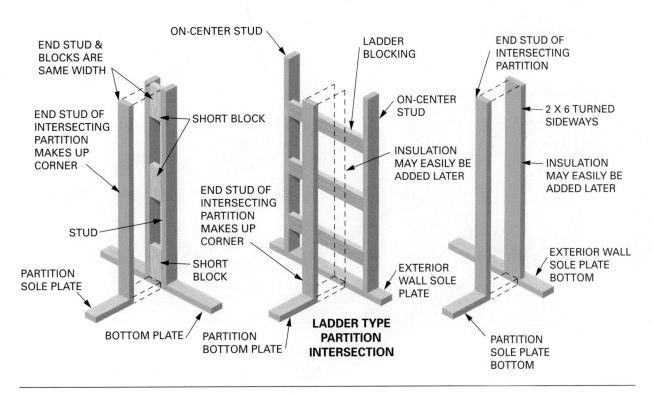

FIGURE 35–15 Construction of partition intersections.

Assembling Wall Sections

A variety of wall assembly procedures are used to build a wall frame. The following technique will quickly create a strong wall **(Procedure 35–B).** Separate the top plate from the bottom plate. Stand them on edge. To avoid a mistake, be careful not to turn one of the plates around. Be certain that the layout lines on top and bottom plates line up. Place all full-length studs, corner posts, and partition intersections in between them.

Procedure 35–B Assembling a Wall Section

STEP 1 PULL TACK NAILS AND SEPARATE PLATES.

STEP 2 ASSEMBLE THE ROUGH OPENING FRAME. NAIL ALL MEMBERS TO EACH OTHER AND THE PLATES.

STEP 3 INSTALL REMAINING STUDS. NAIL ON THE DOUBLED TOP PLATE. LEAVE GAPS WHERE INTERSECTING PLATES WILL FIT.

STEP 4 ALIGN FRAME TO CHALK LINE AND ADJUST IT TO BE SQUARE. TACK IT TO THE SUBFLOOR AND INSTALL PERMANENT BRACING.

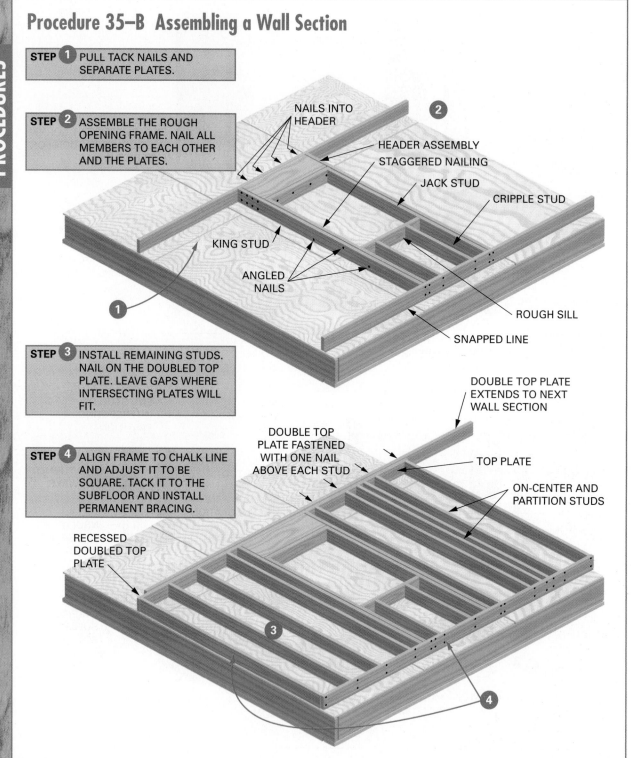

NAILS INTO HEADER

HEADER ASSEMBLY
STAGGERED NAILING

JACK STUD

CRIPPLE STUD

KING STUD

ANGLED NAILS

ROUGH SILL

SNAPPED LINE

DOUBLE TOP PLATE EXTENDS TO NEXT WALL SECTION

DOUBLE TOP PLATE FASTENED WITH ONE NAIL ABOVE EACH STUD

TOP PLATE

ON-CENTER AND PARTITION STUDS

RECESSED DOUBLED TOP PLATE

Very few studs are absolutely straight from end to end so each stud crown will be faced the same way. Sight each full-length stud. It will be difficult for those who apply the interior finish if no attention is paid to the manner in which studs are installed in the wall. A stud that is installed with its crowned edge out next to one with its crowned edge in will certainly present problems later. For this reason some builders come back after the wall is built to flatten and adjust the crowns of the wall studs. Studs

PLANED STUD EDGE   SHIMMED STUD EDGE

FIGURE 35–16 Warped studs can be adjusted to make a flat, straight wall.

FIGURE 35–17 Pneumatic nailers are industry standard tools for framing.

bowing inward are shaved with a power plane and studs bowing outward are shimmed with strips of heavy cardboard **(Figure 35–16)**.

Assemble the window and door openings first to ensure easy nailing. Fasten the headers, jacks, rough sills, and cripples in position, then fasten the king studs. If headers meet the top plate, nail through the plate into the header. Also fasten the jacks to the king studs by driving the nails at an angle. This is a stronger nailing technique and eliminates any protruding nails. A pneumatic framing nailer makes the work easier **(Figure 35–17)**. Fasten each stud, corner post, and partition intersection in the proper position by driving

two 16d nails through the plates into the ends of each member. Nail the doubled top plate to the other top plate. Leave notches where the intersecting walls and partitions are located. Some builders toenail the studs to the bottom plate later as the wall is erected. Toenails are typically 8d or 2½-inch nails.

Nail the doubled top plate to the top plate. Recess or extend the doubled top plate to make a lap joint at the corners and intersections **(Figure 35–18)**. Be sure to nail the doubled top plate into the top plate such that nails are located above the studs. This will ensure that any holes drilled for wiring or plumbing into the top plates will not hit a nail. Where partitions intersect the exterior wall, a space is left for the

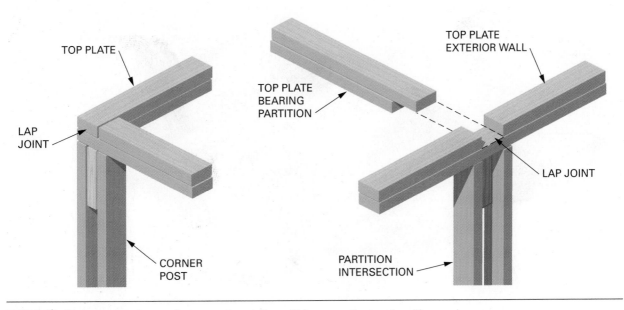

FIGURE 35–18 Doubled top plates are lapped at all intersections of wall sections.

top plate of the partition to lap the plate of the exterior wall. Lapping the plates of the interior partitions with those of the exterior walls ties them together. This results in a more rigid structure.

Bracing Walls

There are several methods of creating a strong wall section that will withstand **racking** loads. Plywood is the most popular material, but let-in and cut-in bracing are also sometimes used. In any case, wall bracing may be applied before the wall is erected.

When plywood is used some builders install it after the wall section is standing. Let-in and cut-in bracing must be installed with tight-fitting seams for maximum strength. For this reason it is much easier to do when the wall is lying down.

Before installing permanent wall bracing, the wall section should be squared while lying on the deck. To do this, align the bottom plate to the previously snapped chalk line on the subfloor where the inside edge of the wall plate will rest. Adjust the ends of the bottom plate into its proper position lengthwise **(Figure 35–19).** Toenail the sole plate to

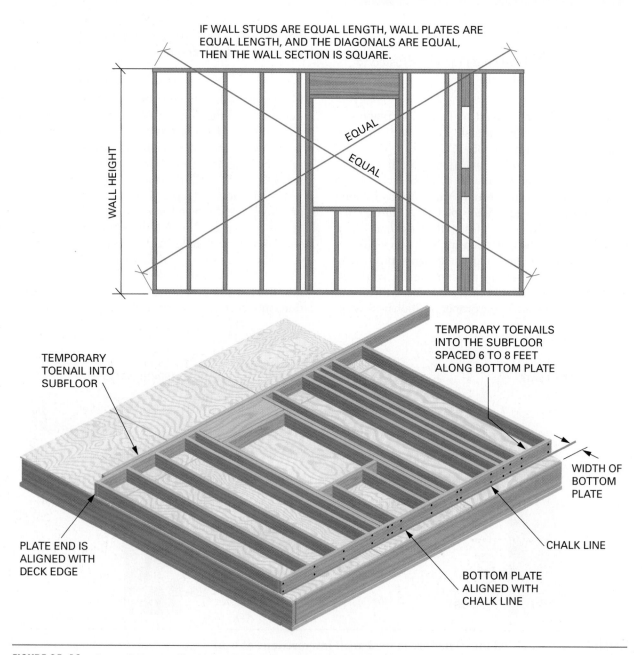

FIGURE 35–19 Squaring a wall section before erecting the frame.

the subfloor with 16d nails spaced about every 6 to 8 feet along what will be the top side of the bottom plate when the wall is in its final position.

Measure both the diagonals from corner to corner. If they are equal, the section is square. Toenail the top plate to the subfloor from the top side using one or two nails. These are used simply to keep the wall square and will be removed before standing the wall. The section is now ready for sheathing or bracing. Sheathing is applied in a manner similar to that used for a subfloor.

If let-in bracing is used, place the brace in position on top of the studs and plates at about a 45-degree angle. Make sure top and bottom plates are covered by the brace. Mark along the side of the brace at each stud and plate. Remove the brace. Using a portable electric circular saw with the blade set for the depth of the notch, make multiple saw cuts between the layout lines. With the claw of a straight-claw hammer, knock out and trim the remaining waste from the notch. Fasten the brace in the notches using two 10d common nails in each framing member **(Figure 35–20).**

Cut-in bracing has two blocks nailed to the plates called kickers at each end of the brace. These serve to transfer the racking load to the plates. First snap a line at about a 45-degree angle on top of the stud edges. Use a speed square to determine the angle to cut the pieces. Fasten them into place with 16d nails, two at each end. The kicker nails should be angled so they do not protrude **(Figure 35–21).**

In regions where seismic activity is severe, such as California, wall bracing takes on a new meaning. Earthquakes occur with such severity that engineers must design the buildings to protect the occupants of

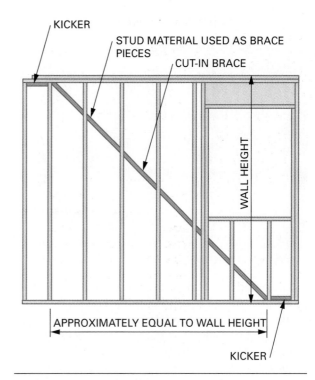

FIGURE 35–21   A cut-in brace.

the house long enough for them to escape the shaking building. **Shear walls** are built into the building **(Figure 35–22).** These are framed of 2 × 6s anchored to the slab with OSB heavily nailed to them. Metal strapping is nailed to the building, anchoring the building to the shear wall. The large wood/metal anchors are bolted to long $^{15}/_{16}$-inch-diameter bolts with $^{5}/_{8}$-inch anchor

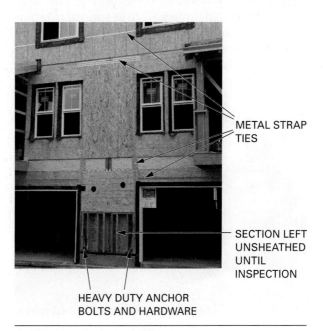

FIGURE 35–22   In earthquake-prone areas, seismic shear walls are constructed using heavy anchors and metal ties.

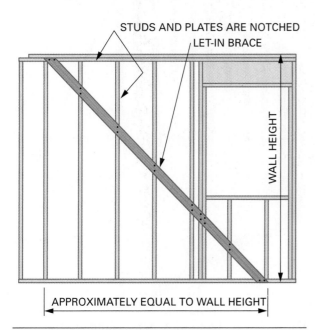

FIGURE 35–20   A let-in corner brace.

1-INCH ANCHOR BOLTS 36 INCHES LONG

3 x 6 PLATE

⅝-INCH ANCHOR BOLTS

FIGURE 35–23   The wall frame can be anchored to the foundation for seismic resistance.

bolts spaced close together. The shear wall bottom plates are sometimes 3 inches thick **(Figure 35–23).**

To provide shear resistance in walls that have large openings such as a garage door, metal frames are used. They are bolted to large 1-inch-diameter bolts anchored 3 feet into concrete. These frames are made with ⅛-inch-thick steel that is bent and folded into the wall thickness **(Figure 35–24).**

Erecting Wall Sections

To erect the wall, remove the toenails from the top plate while leaving the toenails in the bottom plate. The bottom toenails will remain until after the section is erected; they will serve as hinges to keep the bottom plate in position while the frame is raised. Lift the wall section into place, plumb, and temporarily brace. After checking to be sure that the bottom plate is on the chalk line, nail the bottom plate to the band and floor joists or about every 16 inches along the length. In corners, fasten end studs together to complete the construction of the corner

post. A completed corner post provides surfaces to fasten both exterior and interior wall finish.

Brace each section temporarily as erected. Fasten one end of a brace to a 2 × 4 block that has been nailed to the subfloor and the other end to the side of a stud **(Figure 35–25).**

Bracing Walls Temporarily

If the walls have not been previously braced while being framed on the subfloor, all corners must be plumbed and temporarily braced. Install braces for both sides of each corner. Fasten the top end of the brace to the top plate near the corner post on the inside of the wall.

FIGURE 35–25   Temporary braces hold the frame erect during construction.

FIGURE 35–24   Shear strength for walls with large openings can be supplied by metal panels bolted to the foundation.

Temporary braces are fastened to the inside of the wall. They can remain in position until the exterior wall sheathing is applied. Sometimes they remain until it is absolutely necessary to remove them for the application of the interior wall finish. Care must be taken not to let the ends of the braces extend beyond the corner post or top plate. This might interfere with the application of wall sheathing or subsequent ceiling or roof framing.

Methods of Plumbing. Plumb the corner post and fasten the bottom end of the brace. For accurate plumbing of the corner posts, use a 6-foot level with accessory aluminum blocks attached to each end. The blocks keep the level from resting against the entire surface of the corner post. This prevents any bow or irregularity in the surface from affecting the accurate reading of the level.

An alternate method is to use the carpenter's hand level in combination with a long straightedge. On the ends of the straightedge, two blocks of equal thickness are fastened to keep the edge of the straightedge away from the surface of the corner post **(Figure 35–26).** A plumb bob, transit-level, or laser level may also be used to plumb corner posts.

Straightening the Walls

After the corner posts have been plumbed and the top plates doubled, the tops of the walls must be straightened and braced. This can be done with a line and gauge blocks (see Figure 29–19). Nail 2 × 4s at about a 45-degree angle. This will require a length of at least 12 feet for an 8-foot wall. Nail the brace into each plate and twice more into the studs at midspan. Another method is to straighten the brace by eye. After a little practice, eyeing for straightness is fast and surprisingly accurate over distances of less than 40 feet. As one person sights the top plate for straightness by getting the eye as close as possible to the plate, another person nails the brace.

For particularly stubborn wall sections that are difficult to move into plumb, a *spring brace* can be used. Variations of this brace can be used to move the wall top inward or outward.

To create an outward thrust, set a 12- to 16-foot-long 2 × 4 or 2 × 6 against the top plate over a stud **(Procedure 35–C).** The width of the board should be

FIGURE 35–26 Plumbing should be done from the plates and requires a special level or a straightedge with blocks.

Procedure 35–C Plumbing Stubborn Wall Sections with Spring Braces

OUTWARD THRUST SPRING BRACE

STEP 1 SET A 2" X 4" -12' OR 16' AGAINST WALL WITH TOP AGAINST UPPER TOP PLATE.

STEP 2 NAIL BLOCK TO THE SUBFLOOR BEHIND THE BRACE.

STEP 3 PUSH BRACE DOWN AT MIDSPAN SO TOP END SLIDES DOWN AND DIGS INTO STUD.

STEP 4 IF MORE WALL MOVE-MENT IS NEEDED, NAIL TOP WHILE BRACE IS BENT, THEN PICK UP ON BRACE TO STRAIGHTEN IT.

NOTE: DO NOT EXCEED MATERIAL STRENGTH WHERE SOMETHING MIGHT BREAK CAUSING PERSONAL INJURY.

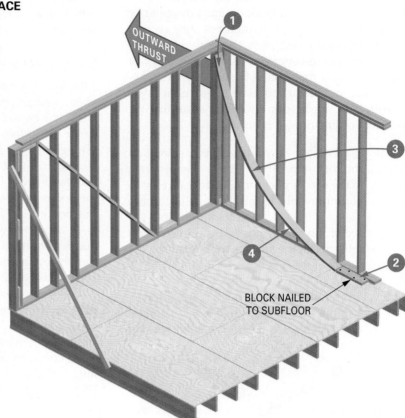

OUTWARD THRUST

BLOCK NAILED TO SUBFLOOR

INWARD THRUST SPRING BRACE

STEP 1 NAIL THE UPPER END OF A 1" X 6" - 16' BRACE TO THE UNDERSIDE OF THE TOP PLATE.

STEP 2 NAIL LOWER END OF BRACE TO FLOOR JOIST WITH SEVERAL 16D NAILS. CAUTION, THIS END MUST BE SECURELY FASTENED SO IT WILL NOT COME LOOSE WHEN UNDER STRESS.

STEP 3 USE A SHORT 2" X 4" AS A POST, ARC THE BRACE AT MIDSPAN. SECURE THE POST WITH A NAIL.

NOTE: DO NOT EXCEED MATERIAL STRENGTH WHERE SOMETHING MIGHT BREAK CAUSING PERSONAL INJURY.

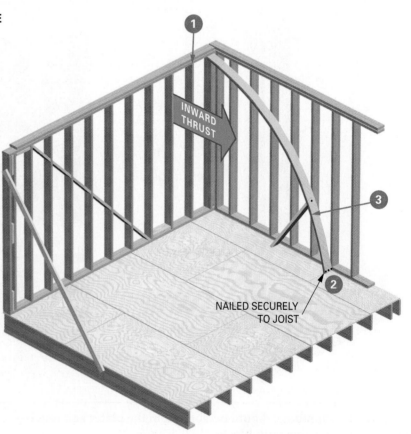

INWARD THRUST

NAILED SECURELY TO JOIST

facing up. Nail a block to the subfloor behind the brace. Push down on the brace to make it arch. This will push the wall out. If more outward push is needed, nail the top of the brace to the stud while it is arched. Then lift the middle of the brace to straighten it. This will create a tremendous outward force. Care should be taken not to break anything.

To bring the wall inward, nail a 16-foot-long 1 × 6 to the top plate and the subfloor. These nails should be set firmly in a floor joist. Lift the brace to an arch. Block the arch with a short piece when the wall is plumb. Care should be taken to watch the brace does not break.

Wall Sheathing

Wall sheathing covers the exterior walls. It may consist of boards, rated panels, fiberboard, **gypsum board,** or rigid foam board.

Boards. Before plywood was created, boards were used predominantly but are seldom used today. Many buildings in existence today have board sheathing. The boards were applied diagonally or horizontally. The diagonally sheathed walls require no other bracing. They made the frame stiffer and stronger than boards applied horizontally. If used today, nail with two 2½-inch or 8d common nails at each stud for 6- and 8-inch boards, or three nails at each stud for 10- and 12-inch boards. End joints must fall over the center of studs. They must be staggered so no two successive end joints fall on the same stud.

Rated Panels. APA-rated wall sheathing panels are used most often today. They may be applied horizontally or vertically. There is no need for corner braces. A minimum ⅜-inch thickness is recommended when the sheathing is to be covered by exterior siding. Greater thicknesses are recommended when the sheathing also acts as the exterior finish siding.

Use 2-inch or 6d nails spaced 6 inches apart on the edges and 12 inches apart on intermediate studs for panels ½ inch thick or less. Use 2½-inch or 8d nails for thicker sheathing panels **(Figure 35–27).**

Rated sheathing panels are sometimes used in combination with rigid foam or gypsum board. When panels are applied vertically on both sides of the corner, no other corner bracing may be necessary.

Other Sheathing Panels. *Fiberboard* sheathing panels are available as wall sheathing **(Figure 35–28).** They are about ⅛"-thick panels with widths and lengths similar to those of plywood. They are made of specially treated, long wood fibers from recycled prod-

ucts. These fibers are pressed into plies that are pressure laminated with a water-resistant adhesive.

They come in three grades of strength. Green panels are for nonstructural applications, red is for permanent wall bracing in the corners, and blue is used where stronger bracing is required. Nails must be 1¼-inch roofing nails or 1-inch crown staples. Nail spacing for the green nonstructural panels is the same as for APA-rated panels, 6" on edges and 12" in the field. Structural panels are nailed 3" on edges and 6" in the field.

It is recommended that panels be nailed along one stud at a time. This serves to help keep ripples out of the sheathing. Nailing begins at one corner, nailing the first stud. Then nail along the plates to the next stud, which is in turn nailed off. Nailing four corners first should be avoided.

Gypsum sheathing consists of a treated gypsum filler between sheets of water-resistant paper. Usually ½ inch thick, 2 feet wide, and 8 feet long, the sheets have matched edges to provide a tighter wall. Because of the soft material, galvanized wall board nails must be used to fasten gypsum sheathing.

Space the nails about 4 inches around the edges and 8 inches in the center. Gypsum board sheathing is used when a more fire-resistant sheathing is required.

Rigid foam sheathing is used when greater wall insulation is desired **(Figure 35–29).** Rated panels are used in the corners to give the building adequate stiffness. It may be applied in thicknesses of ½" or 1". It may be foil faced to increase the thermal barrier. One disadvantage of using these panels as sheathing is that they cannot be used as a nail base for siding. Any siding material applied must be fastened to the studs.

Application of Sheathing Panels. Sheathing panels are installed in a manner similar to that used for installing subfloors. If needed, snap lines to keep the plates and the edges of panels aligned. Panels should be installed with as few seams as possible. Each seam should be as tight as possible. This will make the building more airtight and make it easier to heat or cool the indoor air.

ESTIMATING MATERIALS FOR EXTERIOR WALLS

To estimate the amount of material needed for exterior walls, first determine the total linear feet of exterior wall. Then, figure one stud for every linear foot of wall, if spaced 16 inches on center. This allows for the extras needed for corner posts, partition intersections, trimmers, door jacks, and blocking openings.

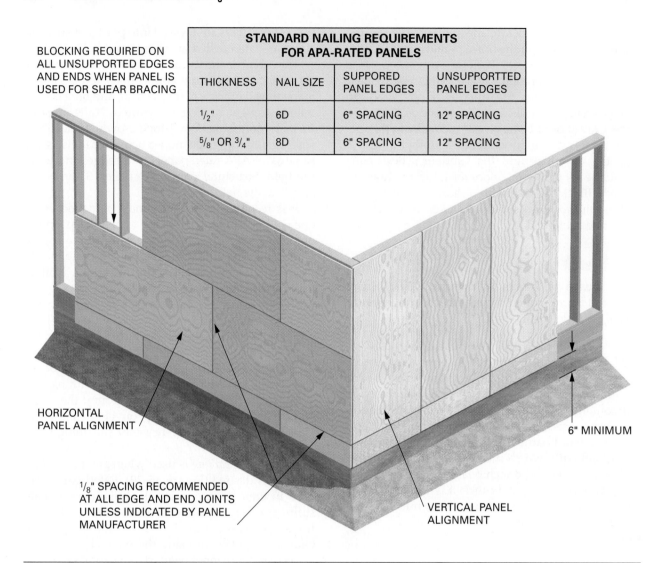

BLOCKING REQUIRED ON
ALL UNSUPPORTED EDGES
AND ENDS WHEN PANEL IS
USED FOR SHEAR BRACING

STANDARD NAILING REQUIREMENTS FOR APA-RATED PANELS			
THICKNESS	NAIL SIZE	SUPPORED PANEL EDGES	UNSUPPORTTED PANEL EDGES
1/2"	6D	6" SPACING	12" SPACING
5/8" OR 3/4"	8D	6" SPACING	12" SPACING

HORIZONTAL
PANEL ALIGNMENT

1/8" SPACING RECOMMENDED
AT ALL EDGE AND END JOINTS
UNLESS INDICATED BY PANEL
MANUFACTURER

VERTICAL PANEL
ALIGNMENT

6" MINIMUM

FIGURE 35–27 Methods of installing APA-rated panel wall sheathing.

FIGURE 35–28 Thin sheets of ⅛" fiberboard may be used as wall sheathing.

FIGURE 35–29 Rigid foam insulation may be used for exterior wall sheathing.

For plates, multiply the total linear feet of wall by three (one sole plate and two top plates). Add 5 percent for waste in cutting.

For headers and rough sills, calculate the size for each opening. Add together the material needed for different sizes.

For wall sheathing, first find the total area to be covered. To find the total number of square feet of wall area, multiply the total linear feet of wall by the wall height. Deduct the area of any large openings. Disregard small openings.

To find the number of sheathing panels, divide the total wall area to be covered by the number of square feet in each sheet. For instance, if a panel measures 4 feet by 8 feet, it contains 32 square feet.

Divide the wall area to be covered by 32 to find the number of panels required. Add about 5 percent for waste. If the answer does not come out even, round it up to the next whole panel.

ALL-WEATHER WOOD FOUNDATION

A wood foundation built of pressure-preservative-treated lumber and plywood is called the *All-Weather Wood Foundation (AWWF)* **(Figure 35–30).** It can be used to support light-frame buildings, such as houses, apartments, schools, and office buildings. It is accepted by major model building codes and by federal agencies such as the VA and FHA.

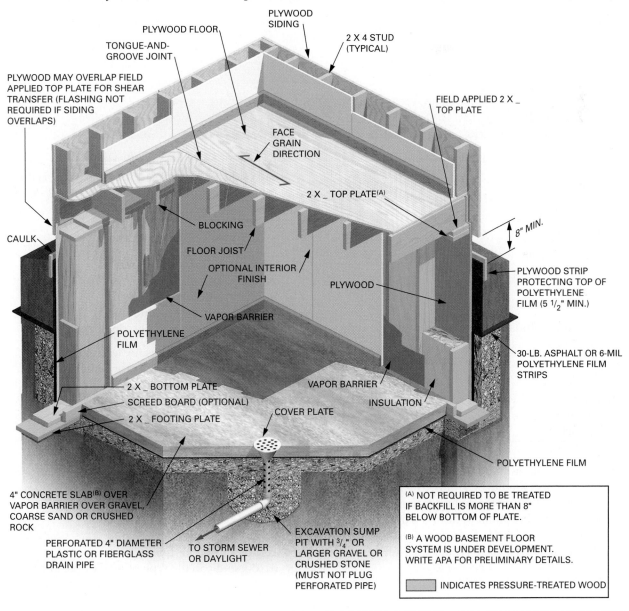

FIGURE 35–30 **Typical details of an all-wood foundation** (*Courtesy of American Plywood Association*)

AWWF is mentioned in this chapter because much of the system is composed of wood exterior walls. The same techniques are used in their assembly and erection as described in this chapter.

For complete design and construction recommendations on the All-Weather Wood Foundation system, contact the American Forest & Paper Association and the Southern Forest Products Association.

Key Terms

corner posts	kicker	ribbons	story pole
gypsum board	lintel	shear walls	studs
headers	partition	sill	top plates
jack studs	racking	sole plate	trimmers
jamb	rough opening	stop	wall sheathing
jig	rough sills		

Review Questions

Select the most appropriate answer.

1. The top and bottom horizontal members of a wall frame are called
   a. headers.   c. trimmers.
   b. plates.   d. sills.

2. The horizontal wall member supporting the load over an opening is called a
   a. header.   c. plate.
   b. rough sill.   d. truss.

3. Shortened studs above and below an opening are called
   a. shorts.   c. cripples.
   b. lame.   d. stubs.

4. Diagonal cut-in bracing requires the installation of
   a. kickers.   c. blocking.
   b. backing.   d. 1 × 4s.

5. The finish floor to ceiling height in a platform frame is specified to be 7′–10″. The finish floor is ¾ inch thick and the ceiling material is ½ inch thick. A single bottom plate and a double top plate are used, each of which has an actual thickness of 1½ inches. What is the stud length?
   a. 7′–5¾″   c. 7′–8¼″
   b. 7′–6¾″   d. 7′–10½″

6. A doorjamb is ¾ inch thick. Allowing ½ inch on each side for shimming the frame, what is the rough opening width for a door that is 2′–8″ wide?
   a. 2′–9½″   c. 2′–11½″
   b. 2′–10½″   d. 3′–0½″

7. A story pole typically shows
   a. the length of headers.
   b. the length of rough sills.
   c. the length of jack studs.
   d. the width of the rough opening.

8. When laying out plates for walls and partitions, measurements for centerlines of openings start from the
   a. end of the plate.
   b. outside edge of the abutting wall.
   c. building line.
   d. nearest intersecting wall.

9. When laying out plates for any OC wall or partition stud, the measurement begins from the
   a. end of the plate.
   b. abutting wall or partition.
   c. opening centerlines.
   d. depends on the type of wall or partition.

10. The first OC wall stud is set back
    a. a distance that is usually ¾ inch.
    b. ½ stud thickness.
    c. to allow the first sheathing piece to be installed flush with the first stud.
    d. all of the above.

11. A corner stud that allows for ample room for insulation in the corner uses
    a. three small blocks.
    b. a stud that is rotated from the others in the wall.
    c. three full studs nailed as a post.
    d. all of the above.

12. Exterior walls are usually straightened before ceiling joists are installed by
    a. using only a carpenter's level.
    b. using a line stretched between two blocks and testing with a gauge block.
    c. using a plumb bob dropped to the bottom plate at intervals along the wall.
    d. by sighting along the length of the wall using a builder's level.

13. Bearing partitions
    a. have a single top plate.
    b. carry no load.
    c. are constructed like bearing walls.
    d. are erected after the roof sheathing is installed.

14. The top plate of the bearing partition
    a. laps the plate of the exterior wall.
    b. is a single member.
    c. butts the top plate of the exterior wall.
    d. is applied after the ceiling joists are installed.

15. Spring braces are typically used
    a. as temporary braces.
    b. as permanent bracing.
    c. only during certain times of the year.
    d. all of the above.

16. What is the rough opening height of a door opening for a 6′–8″ door if the finish floor is ¾ inch thick, ½-inch clearance is allowed between the door and the finish floor, and the jam thickness is ¾ inch?
    a. 6′–9″   c. 6′–10″
    b. 6′–9½″   d. 6′–10½″

17. The type of plywood typically used for wall sheathing is
    a. CDX.   c. BC.
    b. AC.   d. hardwood.

18. The type of permanent wall bracing used most often in construction today is
    a. APA-rated sheathing.
    b. cut-in.
    c. let-in.
    d. all of the above.

19. Estimate the number of 16-inch OC exterior wall studs needed for a rectangular house that measures 28 × 48 by 8 feet high.
    a. 76   c. 1,344
    b. 152   d. 10,752

20. Estimate the number of pieces of wall sheathing needed for a rectangular house that measures 28 × 48 by 8 feet high. Figure an extra foot of material to cover the box header. Neglect the openings and gable end. Add 5 percent for cutting waste.
    a. 42   c. 44
    b. 43   d. 45

UNIT 13

Interior Rough Work

Interior rough work is constructed in the inside of a structure and later covered by some type of finish work. The interior rough work described in this unit includes the installation of *partitions, ceiling joists, furring strips,* and *backing* and *blocking.* The term *rough work* does not imply that the work is crude. It is a kind of work that will eventually be covered by other material. Careful construction of the rough frame makes application of the finish work easier and less complicated.

SAFETY REMINDER

As tools and materials are gathered for interior wall framing, jobsite organization becomes very important for safety and efficiency. Keep waste material outside the work area in organized piles for removal.

OBJECTIVES

After completing this unit, the student should be able to:

- assemble, erect, brace, and straighten bearing partitions.
- determine and make rough openings for doors.
- lay out, cut, and install ceiling joists.
- lay out and erect nonbearing partitions and install backing in walls for fixtures.
- describe various components of light-gauge steel framing.
- lay out and frame light-gauge steel interior partitions.

36 Interior Partitions and Ceiling Joists

Partitions and *ceiling joists* constitute some of the interior framing. Ceiling joists tie the exterior side walls together and support the ceiling finish.

Partitions supporting a load are called *bearing partitions*. Partitions that merely divide the area into rooms are called *nonbearing partitions* **(Figure 36–1).**

PARTITIONS

Load-bearing partitions (LBPs) support the inner ends of ceiling or floor joists. They are placed directly over the girder or the bearing partition in the lower level. If several bearing partitions are used on the same floor, supported girders or walls are placed directly under each.

Non-load-bearing partitions (NLBPs) are built to divide the space into rooms of varying size. They carry no structural load from the rest of the building, only the load of the partition material itself. They may be placed anywhere on the subfloor where joist reinforcement has been provided. They often run parallel to the joists and are nailed to blocking between joists **(Figure 36–2).**

Partitions are erected in a manner similar to that used for exterior walls. A double top plate is used to tie the wall and partition intersections

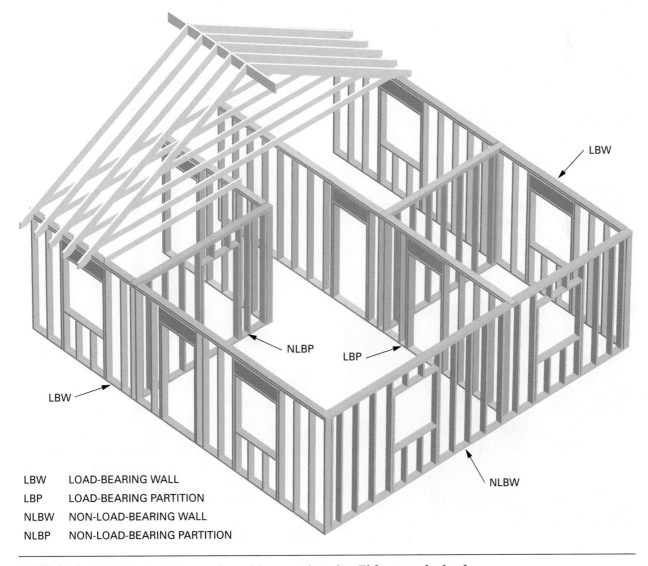

LBW	LOAD-BEARING WALL
LBP	LOAD-BEARING PARTITION
NLBW	NON-LOAD-BEARING WALL
NLBP	NON-LOAD-BEARING PARTITION

FIGURE 36–1 Walls are exterior and partitions are interior. Either may be load bearing or non-load bearing.

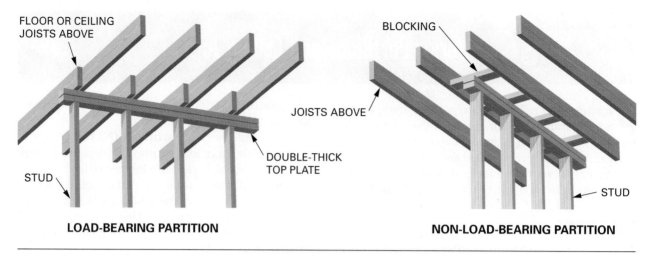

FLOOR OR CEILING
JOISTS ABOVE

STUD

DOUBLE-THICK
TOP PLATE

LOAD-BEARING PARTITION

BLOCKING

JOISTS ABOVE

STUD

NON-LOAD-BEARING PARTITION

FIGURE 36–2 Non-load-bearing partitions merely divide an area into rooms. Load-bearing partitions support the weight of the floor or ceiling above.

together. If *roof trusses* are used, partitions have doubled top plates that are thinner than the other walls **(Figure 36–3).** This is done so the trusses only touch the bearing walls where the roof load is transferred to the foundation. More on roof trusses later in Chapter 46.

Headers on LBPs, as in walls, must be strong enough to support the intended load. Headers on NLBPs may be constructed with 2 × 4s on the flat with cripple studs above **(Figure 36–4).** This saves material since the only load on the partition is the wall finish material.

Bathroom and kitchen walls sometimes must be made thicker to accommodate plumbing. Sometimes 2 × 6 plates and studs are used or a double 2 × 4 partition is erected. Still another wall variation is to use 2 × 6 or 2 × 8 plates with alternated 2 × 4

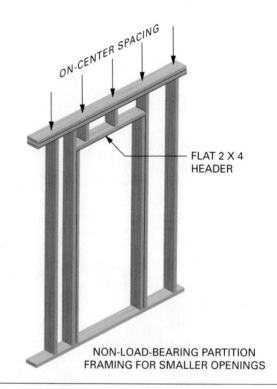

ON-CENTER SPACING

FLAT 2 X 4
HEADER

NON-LOAD-BEARING PARTITION
FRAMING FOR SMALLER OPENINGS

FIGURE 36–4 A method for framing a non-load-bearing header.

studs **(Figure 36–5).** This allows fiberglass insulation to be woven between the studs for increased soundproofing. Also, if the wall thickness needs to be increased only slightly, furring strips may be added to the edges of the studs and plates.

Layout and Framing of Partitions

Partition layout is similar to wall layout. From the floor plan, determine the location of the partitions, noting that the dimensions are usually given to their

FIGURE 36–3 The double top plates of non-load-bearing walls and partitions may be thinner than those of load-bearing walls and partitions.

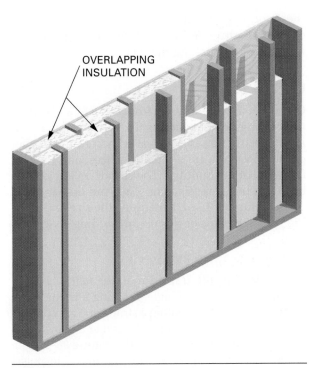

FIGURE 36–5 Sometimes a wide plate is used to build stagger-stud walls for increased soundproofing.

centerlines. To locate all partitions, measure and snap chalk lines on the subfloor.

Lay the edge of the bottom plate to the chalk line, tacking it in position. Lay out the openings and partition intersections then all on-center (OC) studdings including cripple studs. LBP OC studs should line up with floor and ceiling joists above and below. Lay the top plate next to the bottom plate and transfer the layout of the bottom plate to the top plate. Remember that all joints in the plates should end in the center of a stud.

Rough Opening Door Sizes. The rough opening sizes are found in the same way they are found for exterior walls **(Figure 36–6).** The width for a door opening is found by adding to the door width, twice the thickness of the jamb stock, and twice the one-half-inch shim space (one for each side). For example, a 2′-6″ door rough opening width is 2′-8½″ or 32½″.

The rough opening height is found by adding five measurements from the door cross section. They include the following from the subfloor up:

- The thickness of the finish floor
- The 1-inch clearance between the finish floor and the bottom of the door
- The height of the door
- The thickness of the head jamb
- The ½-inch shim space between the head jamb and the rough header.

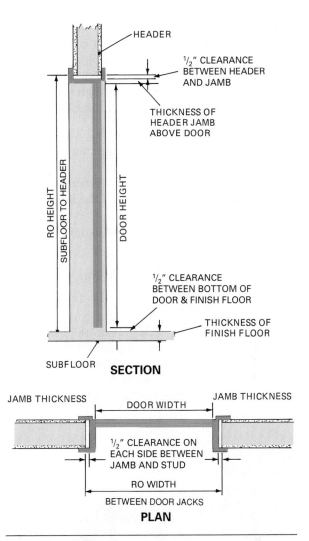

FIGURE 36–6 Figuring the rough opening size for an interior door.

For example a 7′ (84″) door with ¾-inch jambs and ¾-inch finished floor needs a rough opening height of ¾-inch floor + 1-inch clearance + 84-inch door + ¾-inch jamb + ½-inch shim space = 87″.

Usually no threshold is used under interior doors. However, if a threshold is used it will fit within the space allowed for by the 1-inch clearance.

Framing Partitions. For ease of erecting, construct the longer partitions first. Then construct shorter cross-partitions, such as for closets, later. There is no hard and fast rule for constructing partitions; experience will allow the best process to emerge.

Pull the tack nails and separate the top and bottom plates. Place all full-length studs, corner posts, and partition intersections on the floor. Make sure their crowned edges run in the same direction. Nail the framing members around an opening first, then the OC studs to allow fast and easy assembly. Raise the section into position, locking the lapping top plates into position. Plumb and brace the wall section and nail the bottom plate to the floor about 16 inches apart or into floor joists where possible.

The end stud that butts against another wall can be straightened by toenailing through its edge into the center block of the partition intersection **(Figure 36–7).**

Other Openings. Besides door openings, the carpenter must frame openings for heating and air-conditioning ducts, medicine cabinets, electrical panels, and other similar items. If the items do not fit in a stud space, the stud must be cut and a header installed. When ducts run in a wall through the floor, the bottom plate and subfloor must be cut out **(Figure 36–8).** The reciprocating saw is a useful tool for making these cuts.

CEILING JOISTS

Ceiling joists generally run from the exterior walls to the bearing partition across the width of the building. Construction design varies according to geographic location, traditional practices, and the size and style of the building. The size of ceiling joists is based on the span, spacing, load, and the kind and grade of lumber used. Determine the size and spacing from the plans or from local building codes.

Methods of Installing Joists

In a conventionally framed roof, the *rafters* and the ceiling joists form a triangle. Framing a triangle is a common method of creating a strong and rigid building. The weight of the roof and weather is transferred from the roof to the exterior walls **(Figure 36–9).** The rafters are located over the studs and the ceiling joists are fastened to the sides of the rafters **(Figure 36–10).** This binds the rafters and ceiling joists together into a

ON THE JOB

Straighten the edge of end studs in interior partitions that intersect with another wall to frame straight inside corners.

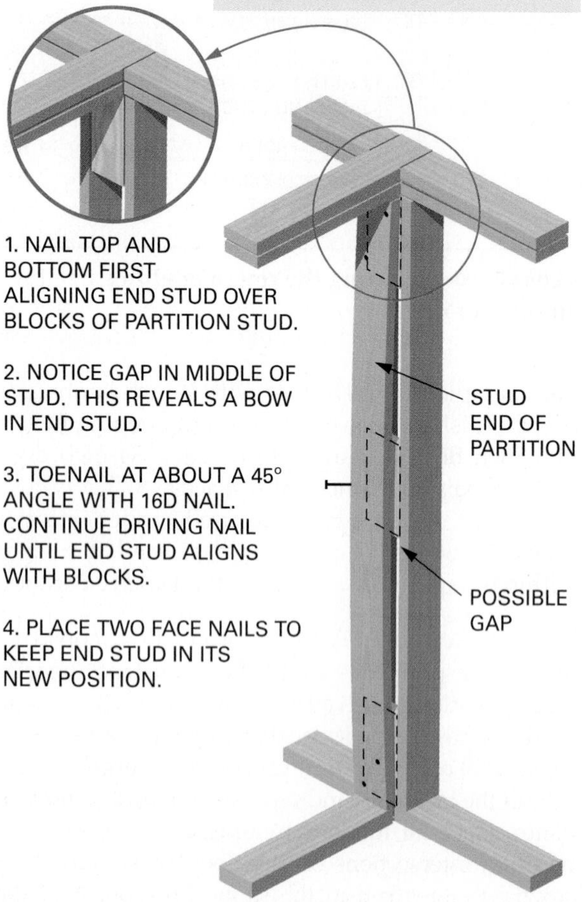

1. NAIL TOP AND BOTTOM FIRST ALIGNING END STUD OVER BLOCKS OF PARTITION STUD.

2. NOTICE GAP IN MIDDLE OF STUD. THIS REVEALS A BOW IN END STUD.

3. TOENAIL AT ABOUT A 45° ANGLE WITH 16D NAIL. CONTINUE DRIVING NAIL UNTIL END STUD ALIGNS WITH BLOCKS.

4. PLACE TWO FACE NAILS TO KEEP END STUD IN ITS NEW POSITION.

STUD END OF PARTITION

POSSIBLE GAP

FIGURE 36–7 Technique for straightening a crowned stud.

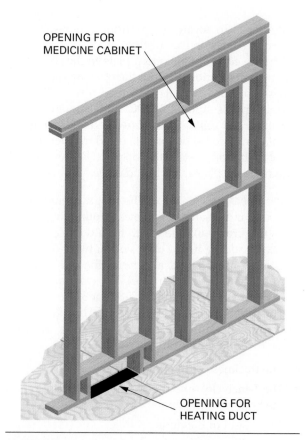

OPENING FOR MEDICINE CABINET

OPENING FOR HEATING DUCT

FIGURE 36–8 Miscellaneous openings in interior partitions are framed with non-load-bearing headers.

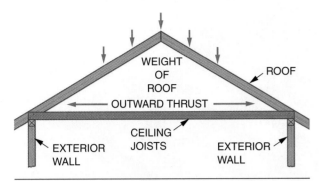

FIGURE 36–9 Ceiling joists tie the roof frame together into a triangle, which resists the outward thrust caused by the rafters.

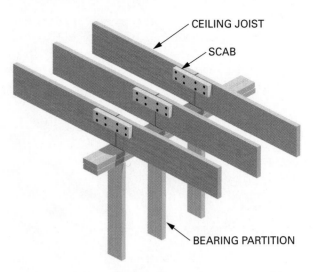

FIGURE 36–11 The joint of in-line ceiling joists must be scabbed at the bearing partition.

Ceiling joists may be made from engineered lumber and purchased in long lengths so that the rafter-ceiling triangle is easily formed. Typically, though, the ceiling joist lengths are half of the building width and therefore must be joined over a beam or bearing partition.

Sometimes the ceiling joists are installed in-line. Their ends butt each other at the centerline of the bearing partition. The joint must be *scabbed* to tie the joints together **(Figure 36–11)**. **Scabs** are short boards fastened to the side of the joist and centered on the joint. They should be a minimum of 24 inches long. In-line ceiling joists are attached to the same side of each rafter pair (front and back rafter).

Another method of joining ceiling joists is to lap them over a bearing partition in the same manner as for floor joists (see Figure 32–18). This puts a stagger in the line of the ceiling joist and consequently in the rafters as well **(Figure 36–12)**. This stagger is visible at the ridgeboard. The layout lines for rafters and ceiling joists are measured from the outside end wall onto the top plate **(Figure 36–13)**. This measurement is exactly the same, with the only difference being the side of the line on which the ceiling joists and rafters are placed.

Cutting the Ends of Ceiling Joists

The ends of ceiling joists on the exterior walls usually project above the top of the rafter. This is especially true when the roof has a low slope. These ends may be cut to the slope of the roof, flush with or slightly below the top edge of the rafter.

Lay out the cut, using a framing square. Cut one joist for a pattern. Use the pattern to mark the rest. Make sure when laying out the joists that you sight each for a crown. Make the cut on the crowned edge

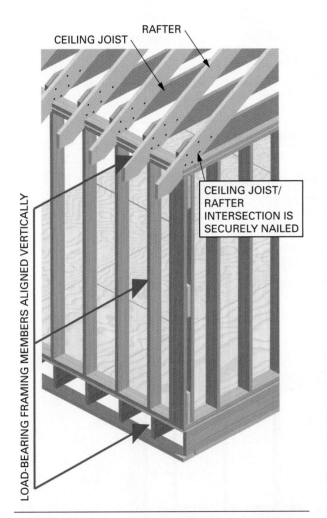

FIGURE 36–10 Ceiling joists are located so they can be fastened to the side of rafters and over the vertically aligned structural frame.

rigid triangle and keeps the walls from spreading outward due to the weight of the roof. The entire roof load is transferred to the foundation through vertically aligned framing members.

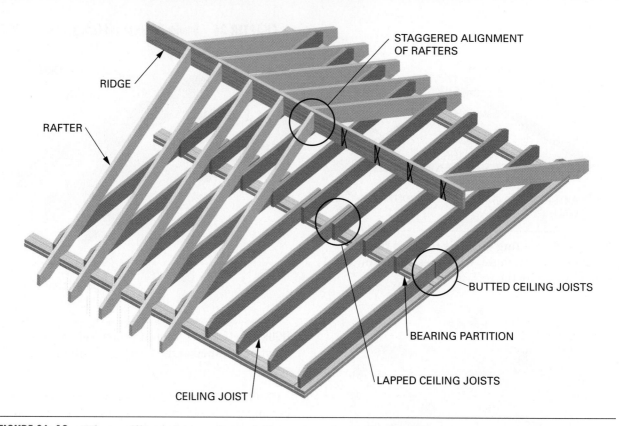

FIGURE 36–12 When ceiling joists are lapped, it causes a stagger in the rafters, which is visible at the ridge.

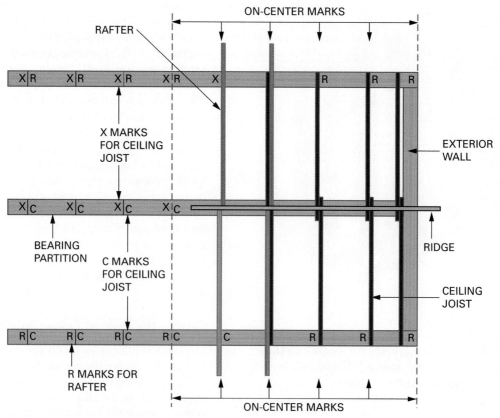

NOTE: ON-CENTER LINES FOR BOTH EXTERIOR WALLS AND BEARING PARTITIONS ARE ALL THE SAME.

FIGURE 36–13 Layout lines on all plates are the same measurements; only the positions of the rafters and ceiling joists vary.

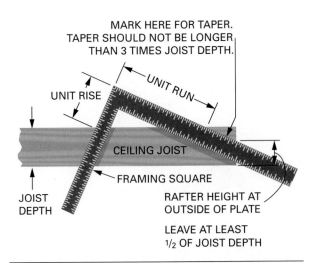

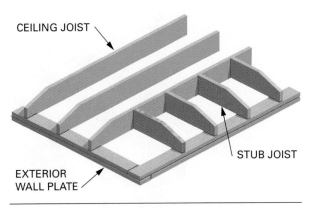

FIGURE 36–15 Stub joists are used for low-pitched hip roofs.

FIGURE 36–14 A framing square is used to mark the slope of a tapered cut on a ceiling joist.

so that edge is up when the joists are installed. Cut the taper on the ends of all ceiling joists before installation. Make sure the length of the taper cut does not exceed three times the depth of the member. Also make sure that the end of the joist remaining after cutting is at least half the member's width **(Figure 36–14)**.

Stub Joists

Usually, ceiling joists run parallel to the end walls and in the same direction as the roof rafters. When low-pitched **hip roofs** are built, **stub joists** are used. These are joists that are shorter and run perpendicular to the normal joists. The use of stub joists allows

clearance for the bottom edge of the rafters that run at right angles to the end wall **(Figure 36–15)**.

Framing Ceiling Joists to a Beam

In many cases, the bearing partition does not run the length of the building because of large room areas. Some type of beam is then needed to support the inner ends of the ceiling joists in place of the supporting wall. Similar in purpose and design to a girder, the beam may be of built-up, solid lumber or engineered lumber.

If the beam is to project below the ceiling, it is installed in the same manner as a header for an opening. The joists are then installed over the beam in the same manner as over the bearing partition **(Figure 36–16)**.

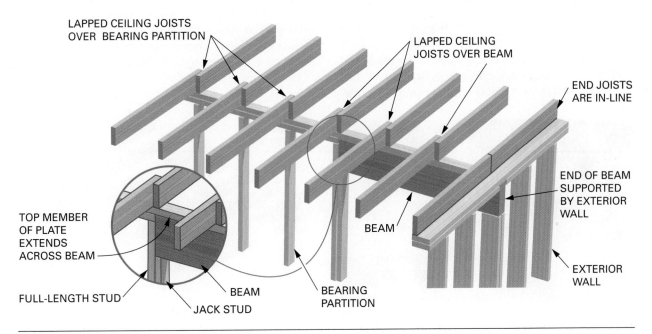

FIGURE 36–16 Support for ceiling joists may be placed as a header in the wall below.

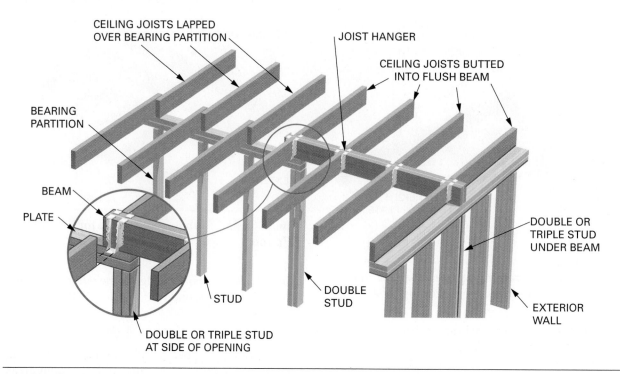

FIGURE 36-17   Support for ceiling joists may be a flush girder. This creates a flush ceiling through the partition below.

If the ceiling is to be flush and continuous through the partition, then the ceiling joists are cut and fitted to a flush beam **(Figure 36–17)**. The beam is supported by the bearing partition and exterior wall. The joists are usually set in **joist hangers.** Adhesive may be used in the joist hanger to eliminate any squeaks that might occur between the joist and ceiling joist **(Figure 36–18).**

Openings

Openings in ceiling joists may need to be made for such things as chimneys, attic access (scuttle), or disappearing stairs. Large openings are framed in the same manner as for floor joists. For small openings, there is no need to double the joists or headers **(Figure 36–19).**

Ribbands and Strongbacks

Ceilings are made stiffer by installing *ribbands* and *strongbacks.* Ribbands are 2 × 4s installed flat on top of the top of ceiling joists. They are placed at midspan to stiffen the joists as well as to keep the spacing uniform. They should be fastened with 16d nails and long enough to be attached to the end

FIGURE 36-18   Floor squeaks created by the joist hanger and joist intersection can be eliminated with caulk.

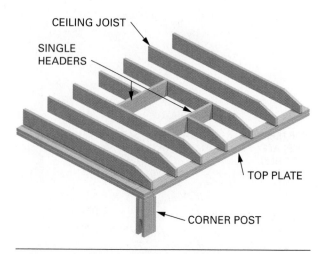

FIGURE 36-19   Joists and headers need not be doubled for small ceiling openings.

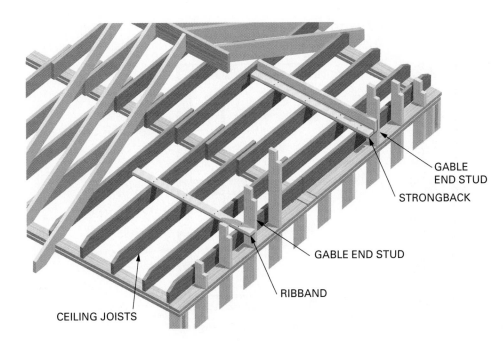

GABLE
END STUD

STRONGBACK

GABLE END STUD

RIBBAND

CEILING JOISTS

walls. With the addition of a 2 × 6 installed on edge, the ribband becomes a strongback. A strongback is used when extra support and stiffness are required on the ceiling joists **(Figure 36–20).**

Layout and Spacing of Ceiling Joists

Roof rafters rest on the plate directly over the regularly spaced studs in the exterior wall. Ceiling joists are installed against the side of the rafters and fastened to them. Spacing of the ceiling joists and rafters should be the same so they can be tied together at the plate line.

- Start the ceiling joist layout from the same corner of the building where the floor joists and wall stud were laid out. Square up from the same side of each regularly spaced stud or cripple stud in the exterior wall and across the top of the plate. Mark an R on the side of the line over top of the stud for the rafter and an X or a C on the other side of the line for the location of the ceiling joists.

- Layout lines on the bearing partition and on the opposite exterior wall are on the same layout as on the first wall. This is similar to floor joist layout. These layout lines should all be the same distance from the end wall. The only difference is the location of the marks for rafters and ceiling joists. They vary depending on whether the ceiling joists are continuous or lapped at the load-bearing partition **(Figure 36–21).**

- Continuous and butted joists are placed on the same side of the layout line. The layout marks for both exterior walls are exactly the same. The load-

bearing partition has only the joist layout line. This places the rafters on the same side of the joist.

- Lapped joist layout marks are reversed, similar to floor joists. Because the joists lap at the load-bearing partition, the layout on the opposite exterior wall is reversed from the first wall. Place on the opposite exterior wall an R and a C or an X on either side of the line but opposite from the first exterior wall. This allows for the lapped joist and creates staggered rafters. The layout for the bearing partition shows where front and back joists lap.

Installing Ceiling Joists

The ceiling joists on each end of the building are placed to allow for installation of **gable end** studs. The last ceiling joist is actually nailed to the gable studs that sit on the exterior wall. It is nailed from the inside of the building. These end joists are installed butted in-line regardless of how the other joists are laid out **(Figure 36–22).**

In addition to other functions, these end joists provide fastening for the ends of the ceiling finish. All other joists are fastened in position with their sides to the layout lines. If the outside ends of the joists have not been tapered, sight each joist for a crown. Install each one with the crowned edge up. If the outside ends have been tapered, install the ceiling joists with the cut edge up. Reject any badly warped joists.

Nailing of the rafter, ceiling joist, and wall plate is critical. This connection establishes the rigid triangle that supports the roof load. The number and size of nails depends on the slope of the roof, intended roof load, and the width of the building. Check local codes for nailing specs.

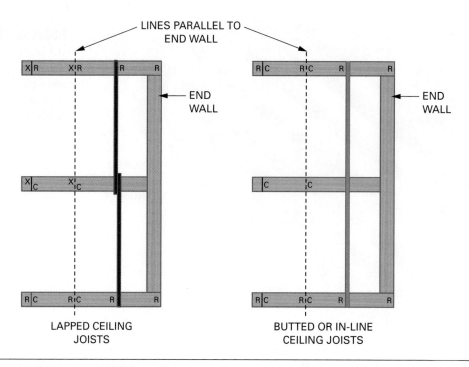

FIGURE 36–21   Layout lines for ceiling joists are the same as for lapped and in-line joists.

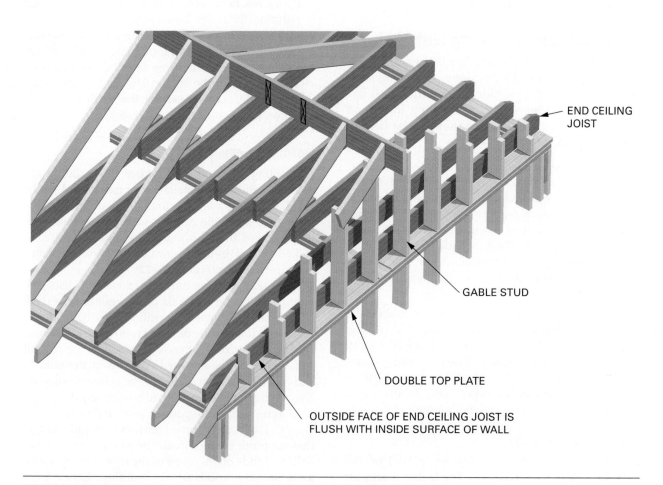

FIGURE 36–22   The end ceiling joist is attached to the gable end studs.

37   Backing and Blocking

This chapter deals with **backing** and **blocking.** A *backing* is, ordinarily, a short block of lumber placed in floor, wall, and ceiling cavities to provide fastening for various parts and fixtures.

Blocking is installed for different purposes, such as providing support for parts of the structure, weathertightness, and firestopping. Sometimes blocking serves as backing.

BACKING AND BLOCKING

There are many places in the structure where blocking and backing should be placed. It would be unusual to find directions for the placement of backing and blocking in a set of plans. It is the wise builder who installs them, much to the delight of those who must install fixtures and finish of all types in later stages of construction.

Backing

It is recommended to install short blocks with their ends against the sole plate and their sides against the studs on inside corners and on both sides of door openings. This provides more fastening surface for the ends of **baseboard (Figure 37–1).**

Much backing is needed in bathrooms. Plumbing rough-in work varies with the make and style of plumbing fixtures. The experienced carpenter will obtain the rough-in schedule from the plumber. He or she then installs backing in the proper location for such things as bathtub faucets, showerheads, lavatories, and water closets **(Figure 37–2).**

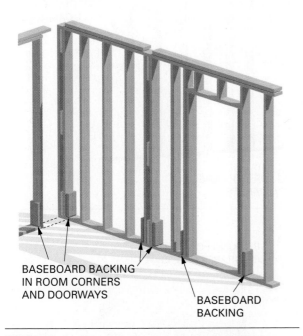

BASEBOARD BACKING IN ROOM CORNERS AND DOORWAYS

BASEBOARD BACKING

FIGURE 37–1   Backing is sometimes installed at corners and door openings for baseboard.

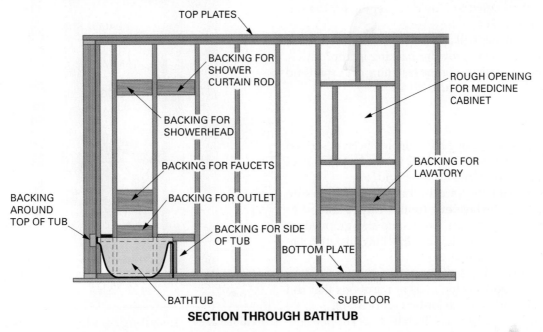

TOP PLATES

BACKING FOR SHOWER CURTAIN ROD

ROUGH OPENING FOR MEDICINE CABINET

BACKING FOR SHOWERHEAD

BACKING FOR FAUCETS

BACKING FOR LAVATORY

BACKING AROUND TOP OF TUB

BACKING FOR OUTLET

BACKING FOR SIDE OF TUB

BOTTOM PLATE

BATHTUB

SUBFLOOR

SECTION THROUGH BATHTUB

FIGURE 37–2   Typical backing needed in bathrooms.

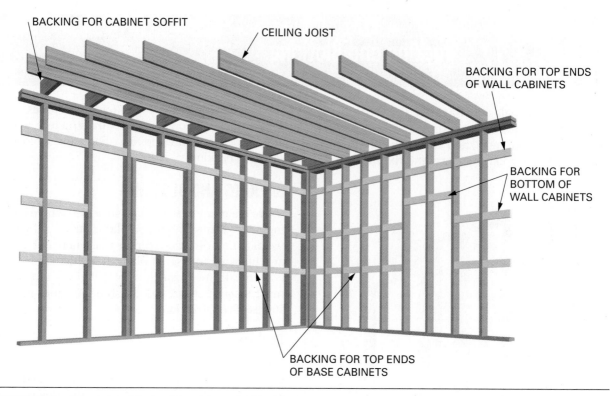

BACKING FOR CABINET SOFFIT

CEILING JOIST

BACKING FOR TOP ENDS OF WALL CABINETS

BACKING FOR BOTTOM OF WALL CABINETS

BACKING FOR TOP ENDS OF BASE CABINETS

FIGURE 37–3   **Considerable backing is needed in kitchens.**

Backing should also be installed around the top of the bathtub.

In the kitchen, backing should be provided for the tops and bottoms of wall cabinets and for the tops of base cabinets. If the ceiling is to be built down to form a *soffit* at the tops of wall cabinets, backing should be installed to provide fastening for the soffit **(Figure 37–3)**.

A homeowner will appreciate the thoughtfulness of the builder who provides backing in appropriate locations in all rooms for the fastening of curtain and drapery hardware **(Figure 37–4)**.

Blocking

Some types of blocking have already been described in earlier units. See Chapter 32 for the installation of solid lumber blocking used for bridging.

When floor panels are used as a combination subfloor and **underlayment** (under carpet and pad), the panel edges must be tongue-and-grooved or supported on 2-inch lumber blocking installed between joists **(Figure 37–5)**.

Ladder-type blocking is needed between ceiling joists to support the top ends of partitions that run parallel to and between joists **(Figure 37–6)**.

Blocks are sometimes installed to support the back edge of bathtubs **(Figure 37–7)**. This stiffens the joint between the tub and the wall finish.

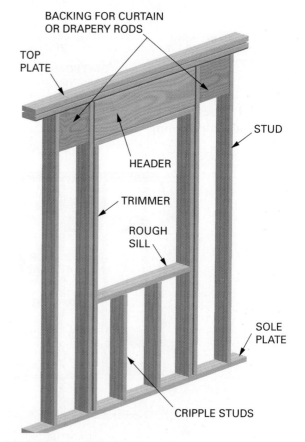

BACKING FOR CURTAIN OR DRAPERY RODS

TOP PLATE

STUD

HEADER

TRIMMER

ROUGH SILL

SOLE PLATE

CRIPPLE STUDS

FIGURE 37–4   **Installing backing around windows allows for easy installation of curtain and drapery hardware.**

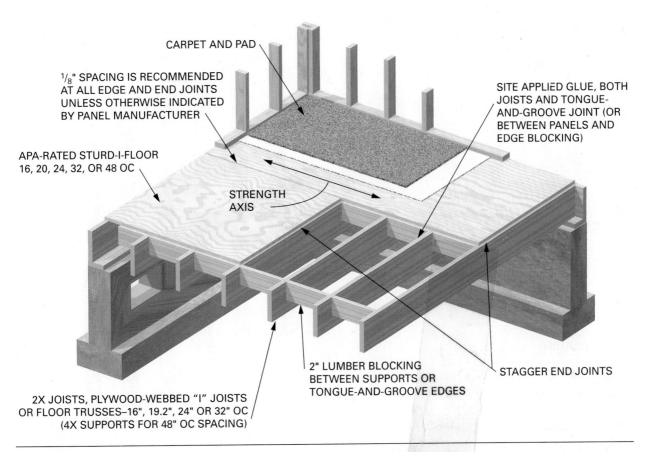

CARPET AND PAD

¹/₈" SPACING IS RECOMMENDED
AT ALL EDGE AND END JOINTS
UNLESS OTHERWISE INDICATED
BY PANEL MANUFACTURER

SITE APPLIED GLUE, BOTH
JOISTS AND TONGUE-
AND-GROOVE JOINT (OR
BETWEEN PANELS AND
EDGE BLOCKING)

APA-RATED STURD-I-FLOOR
16, 20, 24, 32, OR 48 OC

STRENGTH
AXIS

2" LUMBER BLOCKING
BETWEEN SUPPORTS OR
TONGUE-AND-GROOVE EDGES

STAGGER END JOINTS

2X JOISTS, PLYWOOD-WEBBED "I" JOISTS
OR FLOOR TRUSSES–16", 19.2", 24" OR 32" OC
(4X SUPPORTS FOR 48" OC SPACING)

FIGURE 37–5 The edges of APA-Rated Sturd-I-Floor panels must be tongue-and-grooved or supported by blocking.

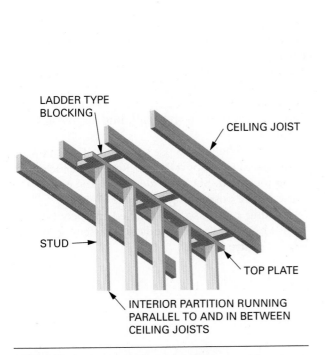

LADDER TYPE
BLOCKING

CEILING JOIST

STUD

TOP PLATE

INTERIOR PARTITION RUNNING
PARALLEL TO AND IN BETWEEN
CEILING JOISTS

FIGURE 37–6 Ladder-type blocking provides support for the top plates of interior partitions.

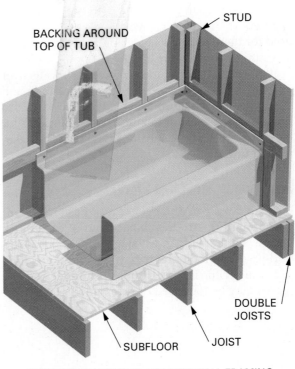

BACKING AROUND
TOP OF TUB

STUD

DOUBLE
JOISTS

SUBFLOOR

JOIST

SUPPORT OF BATHTUB AGAINST WALL FRAMING

FIGURE 37–7 Backing provides in_____ support between tubs and finish material.

FIGURE 37–8 Straight-line blocking is used for structural or exterior wall sheathing.

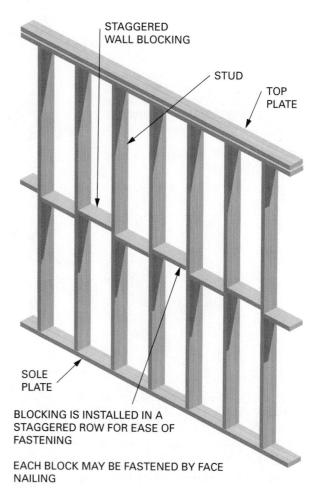

FIGURE 37–9 Blocking used for firestopping may be installed in a staggered fashion for easier nailing.

Wall blocking is required to support the edge of rated panels used in structural shear walls. It is also used in wood foundations and panels permanently exposed to weather for weathertightness. When used for these purposes, blocking must be installed in a straight line (**Figure 37–8**).

Blocking between studs is required in walls over 8'-1" high. The purpose is to stiffen the studs and strengthen the structure. The required blocking also functions as a firestop in stud spaces. Blocking for these purposes may be installed in staggered fashion (**Figure 37–9**).

Installing Backing and Blocking

Install blocking in a staggered row. Fasten by nailing through the studs into each end of each block in the same manner as staggered solid wood bridging described previously. Installing blocking in a straight line is more difficult. The ends of some pieces may be toenailed or face nailed at an angle

(**Procedure 37–A**). Snap a line across the framing. Square lines in from the chalk line on the sides of the studs.

It may be helpful to use a short post on one side of each stud to support the blocking while the end is being toenailed. Start the toenails in the end of the block before positioning it (**Figure 37–10**).

Backing may also be installed in a continuous length by notching the studs and fastening into its edges (**Figure 37–11**).

Procedure 37–A Installing Straight, In-Line Blocking

STEP 1 NAIL FIRST BLOCK WITH FACE NAILS, TOENAIL ONE SIDE IF NECESSARY.

STEP 2 INSTALL SECOND NAIL WITH FACE NAIL TO START. THEN NAIL THE OTHER END, WHERE THE PREVIOUS BLOCK IS IN THE WAY, WITH TOENAIL OR ANGLED FACE NAIL.

STEP 3 INSTALL THE REMAINING BLOCKS IN A SIMILAR MANNER, SKIPPING A BLOCK WHEN THE PRECUT BLOCKS FIT TOO TIGHTLY.

STEP 4 CUSTOM FIT THE MISSING BLOCKS AS NEEDED.

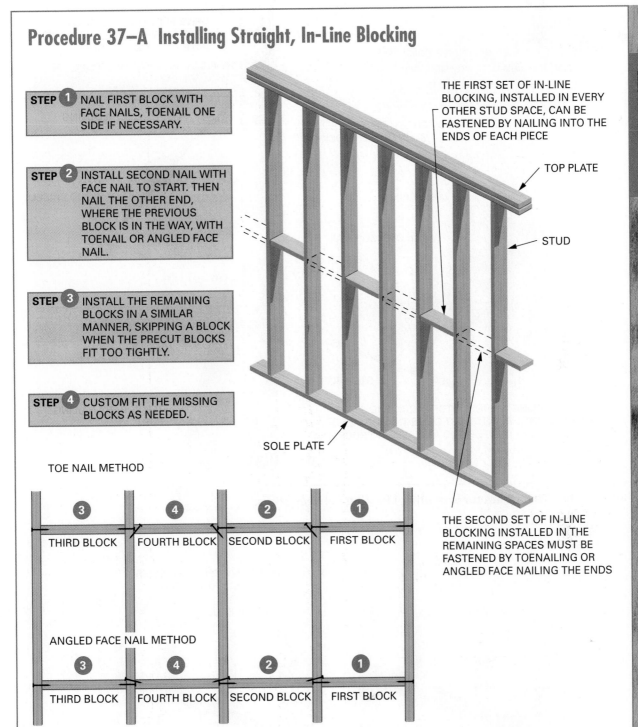

THE FIRST SET OF IN-LINE BLOCKING, INSTALLED IN EVERY OTHER STUD SPACE, CAN BE FASTENED BY NAILING INTO THE ENDS OF EACH PIECE

TOP PLATE

STUD

SOLE PLATE

THE SECOND SET OF IN-LINE BLOCKING INSTALLED IN THE REMAINING SPACES MUST BE FASTENED BY TOENAILING OR ANGLED FACE NAILING THE ENDS

TOE NAIL METHOD

3	4	2	1
THIRD BLOCK	FOURTH BLOCK	SECOND BLOCK	FIRST BLOCK

ANGLED FACE NAIL METHOD

3	4	2	1
THIRD BLOCK	FOURTH BLOCK	SECOND BLOCK	FIRST BLOCK

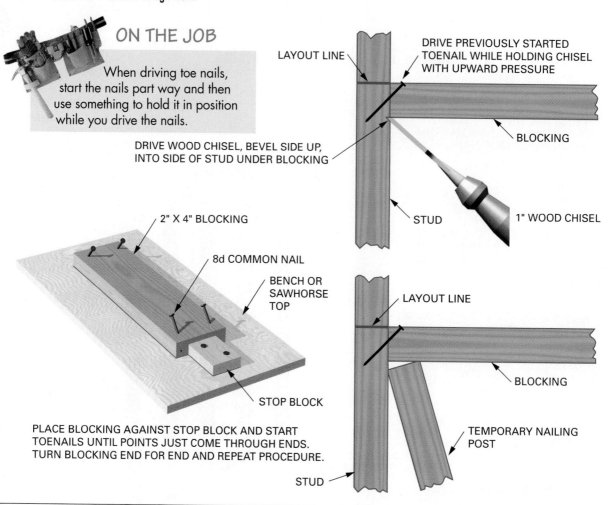

ON THE JOB

When driving toe nails, start the nails part way and then use something to hold it in position while you drive the nails.

DRIVE PREVIOUSLY STARTED TOENAIL WHILE HOLDING CHISEL WITH UPWARD PRESSURE

LAYOUT LINE

BLOCKING

STUD

1" WOOD CHISEL

DRIVE WOOD CHISEL, BEVEL SIDE UP, INTO SIDE OF STUD UNDER BLOCKING

2" X 4" BLOCKING

8d COMMON NAIL

BENCH OR SAWHORSE TOP

STOP BLOCK

PLACE BLOCKING AGAINST STOP BLOCK AND START TOENAILS UNTIL POINTS JUST COME THROUGH ENDS. TURN BLOCKING END FOR END AND REPEAT PROCEDURE.

LAYOUT LINE

BLOCKING

TEMPORARY NAILING POST

STUD

FIGURE 37–10 Techniques for toenailing blocking between studs.

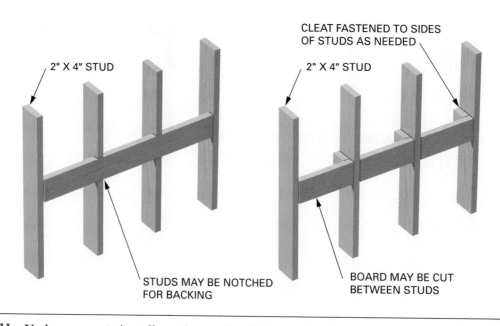

CLEAT FASTENED TO SIDES OF STUDS AS NEEDED

2" X 4" STUD

2" X 4" STUD

STUDS MAY BE NOTCHED FOR BACKING

BOARD MAY BE CUT BETWEEN STUDS

FIGURE 37–11 Various ways to install continuous backing for plumbing and other fixtures.

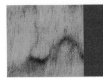

38 Steel Framing

Light steel framing is used for structural framing and interior non-load-bearing partitions **(Figure 38–1)**. Carpenters often frame interior partitions and apply *furring channels* of steel. This chapter is limited to their installation.

The strength of steel framing members of the same design and size may vary with the manufacturer. The size and spacing of steel framing members should be determined from the drawings or by a structural engineer.

INTERIOR STEEL FRAMING

The framing of steel interior partitions is quite similar to the framing of wood partitions. Different kinds of fasteners are used. Some special tools may be helpful.

Steel Framing Components

All steel frame components are cold-formed steel, which is made by one of two methods. The first method is to *press-brake* a steel section into shape. Most shapes are made using this process unless otherwise noted. The other method is to continuously *roll-form* the shape from a coil of steel. This is known as cold-rolled (CR) steel.

Before forming, the steel material is coated with zinc. Steel with this coating is commonly called **galvanized steel.** This coating is designed to protect the steel from corrosion until the building is made weathertight.

The major components of an interior steel frame are *studs, tracks,* and *channels.* Fasteners and accessories are needed to complete the system. The names used to refer to the dimensions of steel framing vary slightly from those used for wood **(Figure 38–2)**. Length is the same for both wood and steel framing. The thickness of wood is typically 1½ inches, whereas the thickness for steel is the thickness of the steel sheet used to make the stud. The steel term that is similar to the *thickness* of a piece of wood is *leg* or **flange size.** A wood stud *width* is similar to steel stud *depth,* which is also referred to as the **web.**

Studs. The thicknesses of steel studs used for non-load-bearing partitions are typically 18, 27, and 33 mil where a mil is 1/1000 of an inch. They are also referred to as 25, 22, and 20 gauge where the larger number denotes the smaller thickness. These studs are used for non-load-bearing partitions. These sizes in metric are 0.46, 0.69, and 0.84 mm. The stud web has punchouts at intervals through which to run pipes and conduit. Studs come in widths of 1⅝, 2½, 3⅝ , 4, and 6 inches, with 1 and 1¼ leg thicknesses.

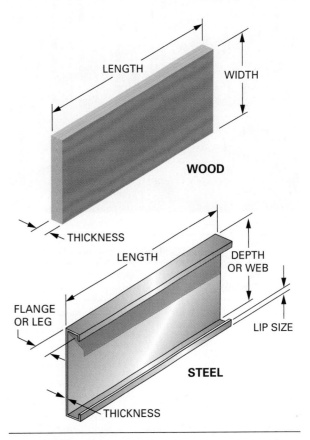

FIGURE 38–1 Steel may be used for load- and non-load-bearing building frames.

FIGURE 38–2 The names for steel stud material are somewhat different from those of wood.

Studs are available in stock lengths of 8, 9, 10, 12, and 16 feet **(Figure 38–3)**. Custom lengths up to 28 feet are also available.

Track. The top and bottom horizontal members of a steel-framed wall are called *track* or *runners*. They are installed on floors and ceilings to receive the studs. They are manufactured by thickness, widths and leg size to match studs **(Figure 38–4)**. Track is available in standard lengths of 10 feet.

Channels. Steel **cold-rolled channels (CRCs)** are formed from 54-mil (16-gauge) steel. They are avail-

able in several widths. They come in lengths of 10, 16, and 20 feet. Channels are used in suspended ceilings and through wall studs. When used for lateral bracing of walls, the channel is inserted through the stud punchouts. It is fastened with welds or clip angles to the studs **(Figure 38–5)**.

Furring Channels. Furring channels (or *hat track*) are hat-shaped pieces made of 18- and 33-mil (25- and 20-gauge) steel. Their overall cross-section size is ⅞ inch by 2⅝ inches. They are available in lengths of 12 feet **(Figure 38–6)**. Furring channels are

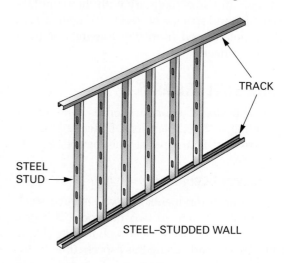

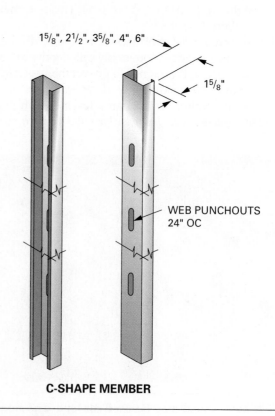

Thickness of Cold-Formed Steel Members			
Mils (1/1000ths inch)	Gauge Number	Metric (mm)	Intended Use
18	25	0.5	Nonstructural
27	22	0.7	Nonstructural
33	20	0.8	Nonstructural
43	18	1.1	Structural
54	16	1.4	Structural
68	14	1.7	Structural
97	12	2.5	Structural

FIGURE 38–3 Steel studs come in several widths, lengths, and thicknesses.

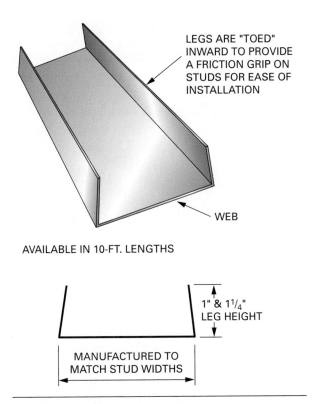

LEGS ARE "TOED" INWARD TO PROVIDE A FRICTION GRIP ON STUDS FOR EASE OF INSTALLATION

WEB

AVAILABLE IN 10-FT. LENGTHS

1" & 1¼" LEG HEIGHT

MANUFACTURED TO MATCH STUD WIDTHS

FIGURE 38–4 Cold-formed channels called *track* are the top and bottom plates of steel-stud walls.

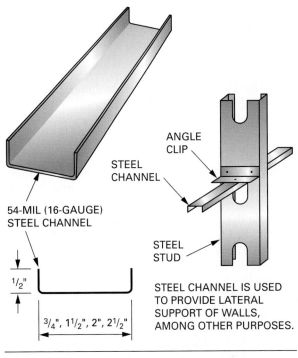

54-MIL (16-GAUGE) STEEL CHANNEL

ANGLE CLIP

STEEL CHANNEL

STEEL STUD

½"

¾", 1½", 2", 2½"

STEEL CHANNEL IS USED TO PROVIDE LATERAL SUPPORT OF WALLS, AMONG OTHER PURPOSES.

FIGURE 38–5 Cold-rolled channels are used to stiffen the framing members of walls and ceilings.

FIGURE 38–6 Furring channels (hat track) are used in both ceiling and wall installations.

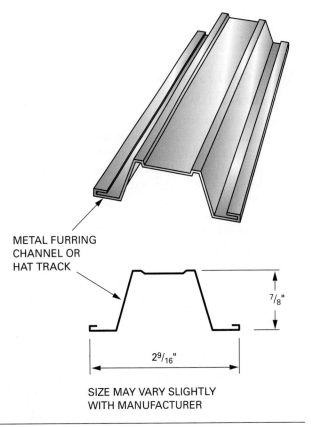

METAL FURRING CHANNEL OR HAT TRACK

⅞"

2⁹⁄₁₆"

SIZE MAY VARY SLIGHTLY WITH MANUFACTURER

SELF-DRILLING

SELF-PIERCING

FIGURE 38–7 Screws to fasten steel members are either self-drilling or self-piercing.

FIGURE 38–8 Bottom track is fastened to the floor next to chalk lines.

applied to walls and ceilings for the screw attachment of gypsum panels. Framing members may exceed spacing limits for various coverings. Furring can then be installed to meet spacing requirements and provide necessary support for the surfacing material.

Fasteners. Steel framing members and components are most commonly fastened with screws. Screws come in a variety of head styles and driving slots. Self-piercing points may be used on lighter gauge studs. Heavy-gauge steel requires self-drilling points **(Figure 38–7).** Screws should be about ½ inch longer than the materials being fastened. A minimum of three threads should penetrate the steel.

Plywood may be attached using specially designed pneumatic nails. Powder-actuated fasteners are often used to attach the framed wall to other support material such as steel or masonry.

LAYOUT AND FRAMING OF STEEL PARTITIONS

Lay out steel-framed partitions as you would wood-framed partitions. Snap chalk lines on the floor. Plumb up from partition ends. Snap lines on the ceiling. Make sure that partitions will be plumb. Using a laser level is an efficient way to lay out floor and ceiling lines for partitions.

Lay out the stud spacing and the wall opening on the bottom track. The top track is laid out after the first stud away from the wall is plumbed and fastened.

Installing Track. Fasten track to floor and ceiling so one edge is to the chalk line **(Figure 38–8).** Make sure both floor and ceiling track are on the same side of the line. Leave openings in floor track for door frames. Allow for the width of the door and thickness of the door frame. Tracks are usually fastened into concrete with powder-driven fasteners. Stub concrete nails or masonry screws may also be used. Fasten into wood with 1¼-inch oval head screws.

Attach the track with two fasteners about 2 inches from each end and a maximum of 24 inches on center in between. At corners, extend one track to the end. Then butt or overlap the other track **(Figure 38–9).** It is not desirable or necessary to make mitered joints.

Install backing, if necessary, between joists or *trusses* where the top track will be attached. Plumb up from the bottom track to the ceiling backing to locate the top track **(Figure 38–10).** Snap lines as needed and fasten with framing screws.

To cut metal framing to length, tin snips may be used on 18-mil (25-gauge) steel. Using tin snips becomes difficult on thicker metal. A power miter box, commonly called a *chop saw,* with a metal-cutting saw blade is the preferred tool **(Figure 38–11).**

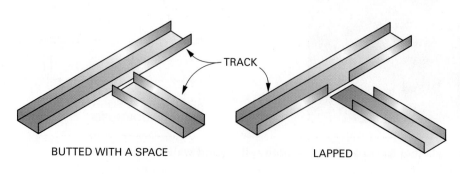
TRACK

BUTTED WITH A SPACE

LAPPED

FIGURE 38–9 Track may be butted or overlapped at intersections.

FIGURE 38–10 Top track may be located on ceiling members or blocking using a level.

⚠ CAUTION CAUTION CAUTION CAUTION

The sharp ends of cut metal can cause serious injury. A pointed end presents an even greater danger. Avoid miter cuts on thin metal. Do not leave short ends of cut metal scattered around the job site. Dispose of them in a container as you cut them. ■

Installing Studs. Cut the necessary number of full-length studs needed. For ease of installation, cut them about ¼ inch short. Install studs at partition intersections and corners. Fasten to bottom and top track. Use ⅜-inch self-drilling pan head screws. If moisture may be present where a stud butts an exterior wall, place a strip of asphalt felt between the stud and the wall.

Place the first stud in from the corner between track. Fasten the bottom in position at the layout line. Plumb the stud. Using a magnetic level can be very helpful. Clamp the top end when plumb, and

FIGURE 38–11 Steel studs may be cut using a chop saw.

fasten to the top track. Lay out the stud spacing on the top track from this stud **(Procedure 38–A).**

Place all full-length studs in position between track with the open side facing in the same direction. The web punchouts should be aligned vertically. This provides for lateral bracing of the wall and the running of plumbing and wiring **(Figure 38–12).**

FIGURE 38–12 Punchouts should line up for bracing and mechanicals.

Procedure 38–A Assembling a Steel Stud Wall

STEP ① LAY OUT STUD LOCATION ON BOTTOM TRACK.

STEP ② PLUMB UP FIRST STUD AWAY FROM WALL.

STEP ③ LAY OUT STUD SPACING ON TOP TRACK FROM PLUMBED STUD.

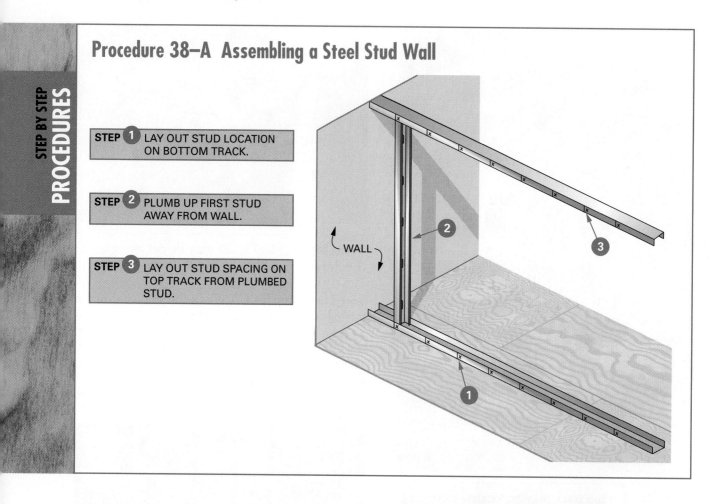

WALL

Fasten all studs except those on each side of door openings securely to top and bottom track.

Wall Openings

Several methods are used to frame around door and window openings. One method involves installing wood jack studs and sills **(Figure 38–13)**. This allows for conventional installation of interior and exterior finishes. The other method uses all steel studs with a metal door frame called a **steel buck.** There are two styles of bucks.

A one-piece metal door frame must be installed before the gypsum board is applied. A three-piece, knocked-down frame is set in place after the wall covering is applied **(Figure 38–14).**

Framing for a Three-Piece Frame. First, place full-length studs on each side of the opening in a plumb position. Fasten securely to the bottom and top plates. Cut a length of track for use as a header. Cut up to 6 inches longer than the width of the opening. Cut the flanges in appropriate places. Bend the web to fit over the studs on each side of the opening. Fasten the fabricated header to the studs at the proper height. Install jack studs over the opening in positions that

FIGURE 38–13 Window and door openings may be lined with wood for easier fastening of finish materials.

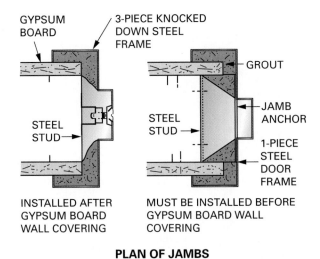

PLAN OF JAMBS

FIGURE 38–14 A one-piece or three-piece knocked-down metal door frame may be used in steel wall framing. *(Courtesy of U.S. Gypsum Corporation)*

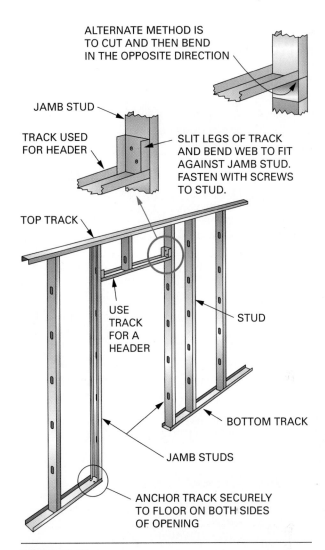

FIGURE 38–15 An opening framed for a three-piece, knocked-down, metal door frame.

continue the regular stud spacing **(Figure 38–15).** Window openings are framed in the same manner. However, a rough sill is installed at the bottom of the opening. Cripples are placed both above and below **(Figure 38–16).**

Framing for a One-Piece Frame. Place the studs on each side of the opening. Do not fasten to the track. Set the one-piece door frame in place. Level the door frame header by shimming under a jamb, if necessary. Fasten the bottom ends of the door jambs to the floor in the proper location. Fasten the studs to the door jambs. Then, fasten the studs to the bottom track. Plumb the door frame. Clamp the stud to the top track, and fasten with screws. Install header and jack studs in the same manner as described previously **(Procedure 38–B).**

The steel framing described above is suitable for average weight doors up to 2'–8" wide. For wider and heavier doors, the framing should be strengthened by using 33-mil (20-gauge) steel framing. Also, double the studs on each side of the door opening **(Figure 38–17).**

Chase Walls

A **chase wall** is made by constructing two closely spaced, parallel walls for the running of plumbing, heating and cooling ducts, and similar items. They are constructed in the same manner as described previously. However, the spacing between the outside edges of the wall frames must not exceed 24 inches.

The studs in each wall should be installed with the flanges running in the same direction. They

FIGURE 38–16 Rough sills are installed similarly to non-load-bearing headers.

Procedure 38–B Assembling a One-Piece Door Frame

STEP 1 FASTEN TOP AND BOTTOM TRACKS IN POSITION. LEAVE SPACE IN BOTTOM TRACK FOR OPENING.

STEP 2 FASTEN ALL STUDS IN POSITION BUT LEAVE JAMB STUDS LOOSE AT THE BOTTOM.

STEP 3 MOVE JAMB STUDS OUT OF THE WAY.

STEP 4 SET DOOR FRAME IN OPENING. LEVEL HEADER AND FASTEN BOTTOM OF FRAME TO FLOOR AT CORRECT WIDTH.

STEP 5 FASTEN JAMB STUDS ON BOTH SIDES TO SIDE JAMBS OF DOOR FRAME.

STEP 6 FASTEN JAMB STUDS ON BOTH SIDES TO BOTTOM TRACK.

STEP 7 PLUMB SIDE JAMB OF DOOR FRAME AND FASTEN JAMB STUDS TO TOP TRACK.

STEP 8 INSTALL HEADER AND JACK STUDS.

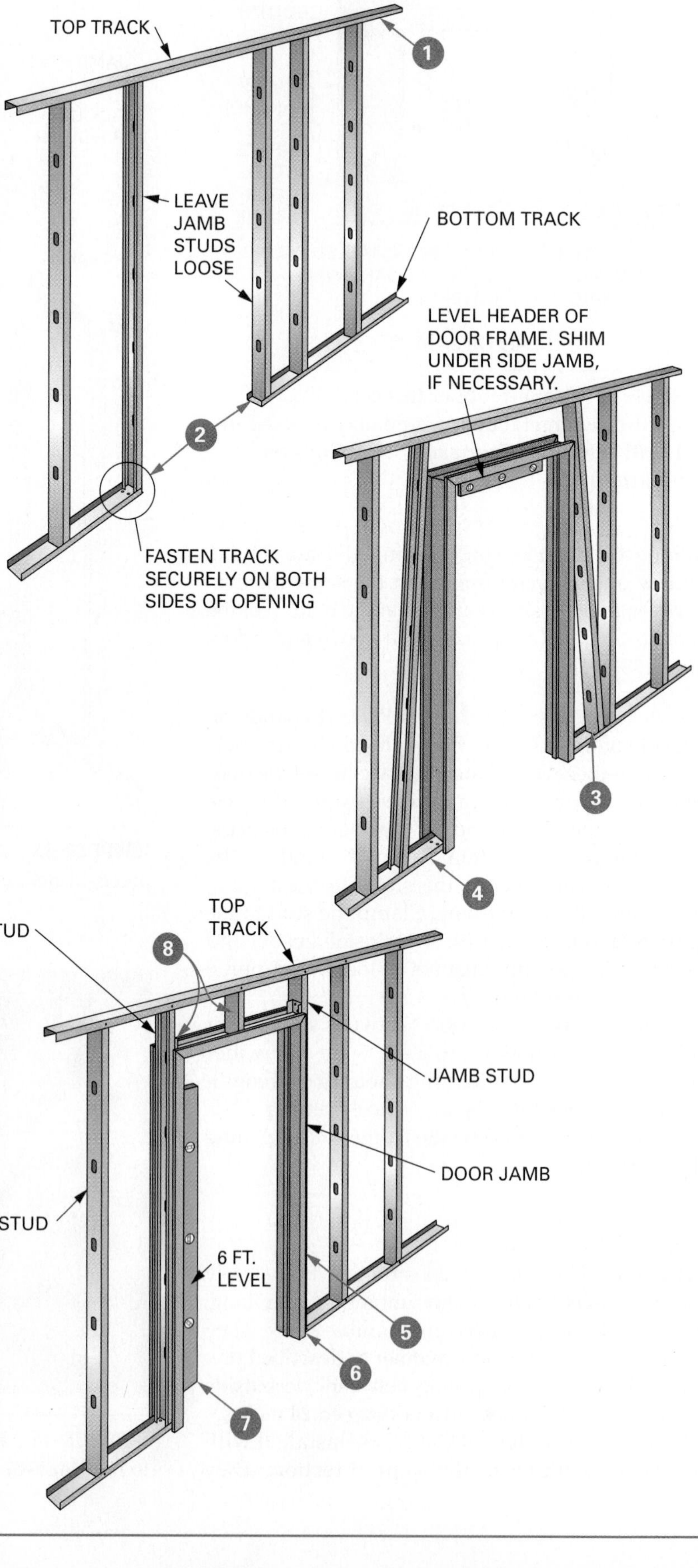

TOP TRACK

LEAVE JAMB STUDS LOOSE

BOTTOM TRACK

LEVEL HEADER OF DOOR FRAME. SHIM UNDER SIDE JAMB, IF NECESSARY.

FASTEN TRACK SECURELY ON BOTH SIDES OF OPENING

JAMB STUD

TOP TRACK

JAMB STUD

DOOR JAMB

STUD

6 FT. LEVEL

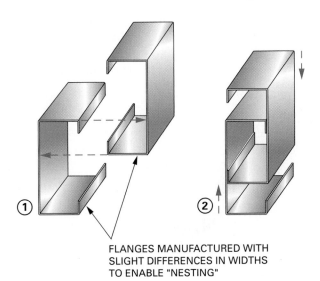

FLANGES MANUFACTURED WITH
SLIGHT DIFFERENCES IN WIDTHS
TO ENABLE "NESTING"

NESTING STUDS

FIGURE 38–17 Steel studs may be nested to create a stronger frame for larger, heavier, doors.

should be directly across from each other. The walls should be tied to each other either with pieces of 12-inch-wide gypsum board or short lengths of steel stud. If the wall studs are not opposite each other, install lengths of steel stud horizontally inside both walls. Tie together with shorter lengths of stud material spaced 24 inches on center **(Figure 38–18).** Wall ties should be spaced 48 inches on center vertically.

Installing Metal Furring

Metal furring may be used on ceilings applied at right angles to joists. They may be applied vertically or horizontally to framed or masonry walls. Space metal furring channels a maximum of 24 inches on center.

Ceiling Furring. Metal furring channels may be attached directly to structural ceiling members or suspended from them. For direct attachment, saddle tie with double-strand 43 mils (18-gauge) wire to each member **(Figure 38–19).** Leave a 1-inch clearance between ends of furring and walls. Metal furring channels may be spliced. Overlap the ends by at least 8 inches. Tie each end with wire. Steel studs may be used with their open side up for furring when supporting framing is widely spaced. Several methods of utilizing metal furring channels or steel studs in suspended ceiling applications are shown in **Figure 38–20.**

Wall Furring. Vertical application of steel furring channels is preferred. Secure the channels by staggering the fasteners from one side to the other not more than 24 inches on center **(Figure 38–21).** For horizontal application on walls, attach furring channels not more than 4 inches from the floor and ceiling. Fasten in the same manner as vertical furring.

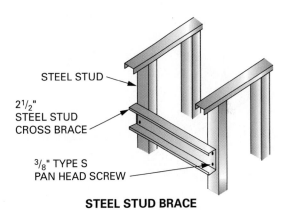

STEEL STUD

2¹⁄₂"
STEEL STUD
CROSS BRACE

³⁄₈" TYPE S
PAN HEAD SCREW

STEEL STUD BRACE

½" OR ⁵⁄₈" GYPSUM
BOARD CROSS BRACE
(12" X WIDTH) SCREW
ATTACHED

STEEL
STUD

24" MAX.

1" TYPE S SCREW

GYPSUM BRACE

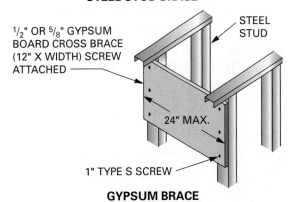

³⁄₈" TYPE S PAN
HEAD SCREWS

STEEL STUD

2¹⁄₂" STEEL STUD
CROSS BRACE

2¹⁄₂"
STEEL TRACK

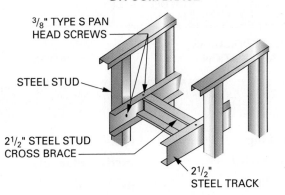

STEEL STUD & TRACK BRACE

FIGURE 38–18 Chase wall construction details.

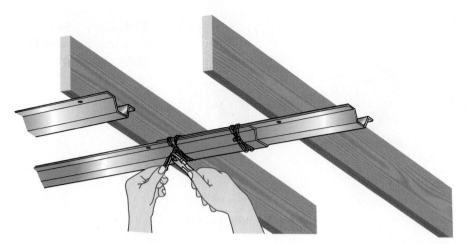

FIGURE 38–19 Method of splicing furring channels.

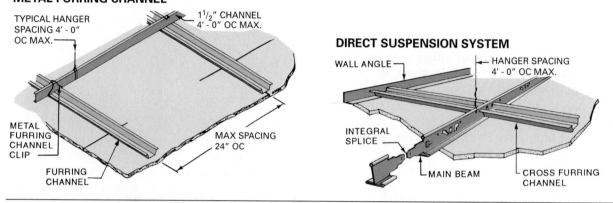

METAL FURRING CHANNEL

TYPICAL HANGER SPACING 4' - 0" OC MAX.

1¹/₂" CHANNEL 4' - 0" OC MAX.

METAL FURRING CHANNEL CLIP

FURRING CHANNEL

MAX SPACING 24" OC

DIRECT SUSPENSION SYSTEM

WALL ANGLE

HANGER SPACING 4' - 0" OC MAX.

INTEGRAL SPLICE

MAIN BEAM

CROSS FURRING CHANNEL

FIGURE 38–20 Metal channels are also used in suspended ceiling applications. *(Courtesy of U.S. Gypsum Corporation)*

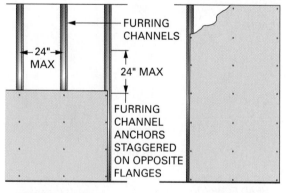

FURRING CHANNELS

24" MAX

24" MAX

FURRING CHANNEL ANCHORS STAGGERED ON OPPOSITE FLANGES

PERPENDICULAR APPLICATION

PARALLEL APPLICATION

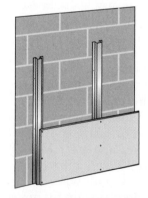

WALL ELEVATION—FURRING

FIGURE 38–21 Furring channels may be attached directly to masonry walls. *(Courtesy of U.S. Gypsum Corporation)*

Key Terms

backing
baseboard
blocking
chase wall
cold-rolled channels (CRCs)

flange size
furring channels
gable end
galvanized steel
hip roof

joist hanger
load-bearing partitions (LBP)
non-load-bearing partitions (NLBP)
scabs

steel buck
stub joists
underlayment
web

Review Questions

Select the most appropriate answer.

1. Wood-framed load-bearing partitions
   a. have a single top plate.
   b. carry no load from the roof.
   c. are constructed like exterior walls.
   d. are erected after the roof is tight.

2. The doubled top plate of a load-bearing partition
   a. is usually a 1× member.
   b. laps the plate of the exterior wall.
   c. butts the top plate of the exterior wall.
   d. is applied after the ceiling joists are installed.

3. What is the rough opening height for a 6′–8″ door if the finish floor is ¾ inch thick, a ½-inch clearance is allowed between the door and the finish floor, and the jamb thickness is ¾ inch?
   a. 6′–9″
   b. 6′–9½″
   c. 6′–10″
   d. 6′–10½″

4. The end stud of partitions that butt against another wall
   a. must be straightened as it is nailed to the intersecting wall.
   b. must be fastened to the intersecting wall with screws.
   c. is usually left out until the wall is erected.
   d. is usually not fastened near its center.

5. A roof load is supported by
   a. the strong triangle frame formed by the ceiling joists and rafters.
   b. framing members aligned vertically.
   c. ceiling joists resisting the outward thrust of rafters.
   d. all of the above.

6. Ceiling joists are typically installed
   a. with their end joints lapped at the bearing partition.
   b. full length along the building width.
   c. by being fastened to the rafter.
   d. all of the above.

7. The ends of ceiling joists are cut to the pitch of the roof
   a. for easy application of the wall sheathing.
   b. so they will not project above the rafters.
   c. to mark the crowned edges up.
   d. after they are fastened to the plate and rafter.

8. Stub joists
   a. run at right angles to regular ceiling joists.
   b. are used on low-pitched common rafter roofs.
   c. are also called blocking and are installed between regular joists.
   d. span from the bearing partition to the exterior wall.

9. Openings in the ceiling may be framed with
   a. single headers and no trimmers.
   b. doubled headers and trimmers.
   c. joist hangers.
   d. all of the above.

10. Non-load-bearing partition headers are usually
    a. stud width material installed on the flat with cripples above.
    b. a doubled 2 × 6 with a plywood spacer.
    c. a doubled 2 × 10 with a plywood spacer.
    d. designed for size by a structural engineer.

11. Blocking and backing are installed
    a. using up scraps of lumber first.
    b. as a nail base for cabinets.
    c. to secure parallel partitions to ceiling joists.
    d. all of the above.

12. The measurement system for sizing steel studs for strength and thickness that uses numbers in a seemingly reverse order is
    a. metric.
    b. mil or 1/1000th.
    c. gauge.
    d. none of the above.

13. The width of a steel stud
    a. is referred to as a web.
    b. is referred to as depth.
    c. is sized similar to wood.
    d. all of the above.

14. When working with steel framing note that
    a. special self-tapping screws are needed.
    b. top plates are usually doubled.
    c. studs are also called track.
    d. all of the above.

15. The top track of non-load-bearing steel partitions is located by
    a. a long level.
    b. by plumbing with a laser transit from the sole plate.
    c. by hanging a plumb bob from the ceiling to the sole plate.
    d. any of the above methods.

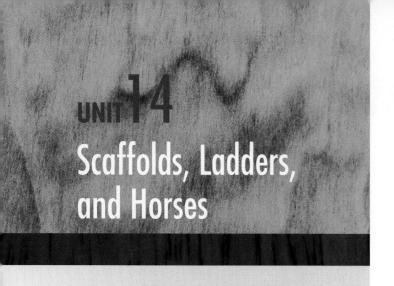

UNIT 14

Scaffolds, Ladders, and Horses

Scaffolding and staging are terms that describe temporary working platforms. They are constructed at convenient heights above the floor or ground. They help workers perform their jobs quickly and safely. This unit includes information on the safe erection of various kinds of working platforms and the building of several construction aids.

SAFETY REMINDER

Scaffolds are involved with most job-site accidents. Know how to safely erect, use and dismantle scaffolding. Always remember where you are when working at elevated levels.

OBJECTIVES

After completing this unit, the student should be able to:

- name the parts of wood single-pole and double-pole scaffolding.
- erect and dismantle metal scaffolding in accordance with recommended, safe procedures.
- build safe staging using roof brackets.
- safely set up, use, and dismantle pump jack scaffolding.
- describe the safe use of ladders, ladder jacks, trestles, and horses.
- build a ladder and sawhorse.

39 Wood, Metal, and Pump Jack Scaffolds

Scaffolds are an essential component of construction, because they allow work to be performed at various elevations. However, they also can create one of the most dangerous working environments. The United States Occupational Safety and Health Administration (OSHA) reports that in construction, falls are the number one killer, and 40 percent of those injured in falls had been on the job less than one year. A survey of scaffold accidents summarizes the problem **(Figure 39–1).** A scaffold fatality and catastrophe investigation conducted by OSHA revealed that the largest percentage, 47 percent, was due to equipment failure. In most instances, OSHA found the equipment did not just break; it was broken due to improper use and erection. Failures at the anchor points, allowing either the scaffold parts or its anchor points to break away, were often involved in these types of accidents. Other factors were improper, inadequate, and improvised construction, and inadequate fall protection. The point of this investigation is that accidents do not just happen; they are caused.

OSHA regulations on the fabrication of frame scaffolds are found in the Code of Federal Regulations 1926.450, 451, 452. These regulations should be thoroughly understood before any scaffold is erected and used. Furthermore, safety codes that are more restrictive than OSHA, such as those in

Canada, California, Michigan, and Washington, should be consulted.

Scaffolds must be strong enough to support workers, tools, and materials. They must also provide an extra safety margin. The standard safety margin requirement is that all scaffolds must be capable of supporting at least four times the maximum intended load.

Those who erect scaffolding must be familiar with the different types and construction methods of scaffolding to provide a safe working platform for all workers. The type of scaffolding depends on its location, the kind of work being performed, the distance above the ground, and the load it is required to support. No job is so important as to justify risking one's safety and life. All workers deserve to be able to return to their families without injury.

The regulations on scaffolding enforced by OSHA make it clear that before erecting or using a scaffold, the worker must be trained about the hazards surrounding the use of such equipment. OSHA has not determined the length of training that should be required. Certainly that would depend on the expertise of the student in training.

Employers are responsible for ensuring that workers are trained to erect and use scaffolding. One level of training is required for workers, such as painters, to work from the scaffold. A higher level of training is required for workers involved in erecting, disassembling, moving, operating, repairing, maintaining, or inspecting scaffolds.

The employer is required to have a **competent person** to supervise and direct the scaffold erection. This individual must be able to identify existing and predictable hazards in the surroundings or working conditions that are unsanitary, hazardous, or dangerous to employees. This person also has authorization to take prompt corrective measures or eliminate such hazards. A competent person has the authority to take corrective measures and stop work if need be to ensure that scaffolding is safe to use.

WOOD SCAFFOLDS

Wood scaffolds are **single pole** or **double pole.** They are used when working on walls. The single-pole scaffold is used when it can be attached to the wall and does not interfere with the work **(Figure 39–2).** The double-pole scaffold is used when the scaffolding must be kept clear of the wall for the application

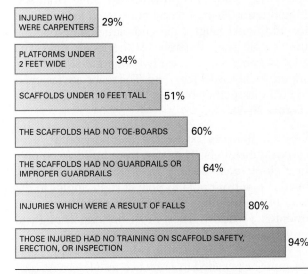

FIGURE 39–1   Accident statistics involving scaffolding.

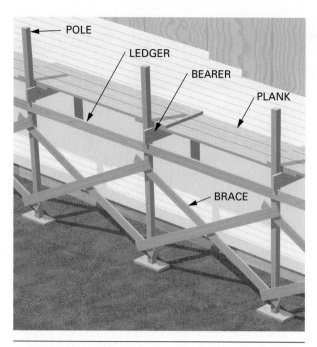

FIGURE 39–2 A light-duty single-pole scaffold. Guardrails are required when the scaffolding is over 10 feet in height.

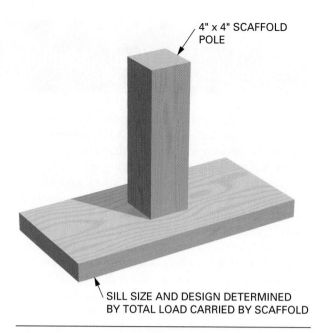

FIGURE 39–4 The bottom ends of scaffold poles are set on footings or pads to prevent them from sinking into the ground.

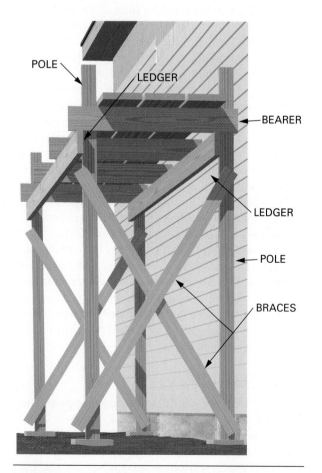

FIGURE 39–3 A double-pole wood scaffold.

of materials or for other reasons **(Figure 39–3).** Wood scaffolds are designated as light-, medium-, or heavy-duty scaffolds, according to the loads they are required to support.

Scaffolding Terms

Poles. The vertical members of a scaffold are called **poles.** All poles should be set plumb. They should bear on a footing of sufficient size and strength to spread the load. This prevents the poles from settling **(Figure 39–4).** If wood poles need to be spliced for additional height, the ends are squared so the upper pole rests squarely on the lower pole. The joint is scabbed on at least two adjacent sides. Scabs should be at least 4 feet long. The scabs are fastened to the poles so they overlap the butted ends equally **(Figure 39–5).**

Bearers. Bearers or *putlogs* are horizontal load-carrying members. They run from building to pole in a single-pole staging. In double-pole scaffolds, bearers run from pole to pole at right angles to the wall of the building. They are set with their width oriented vertically. They must be long enough to project a few inches outside the staging pole.

When placed against the side of a building, bearers must be fastened to a notched *wall ledger*. At each end of the wall, bearers are fastened to the corners of the building **(Figure 39–6).**

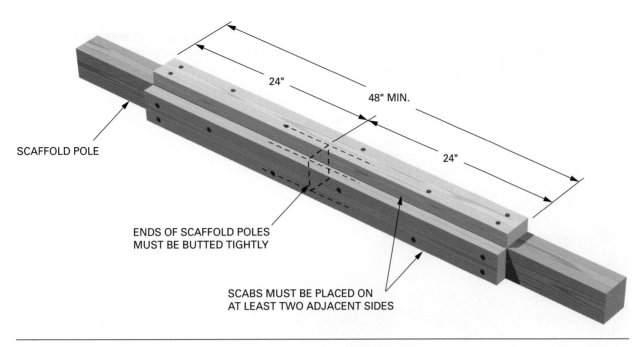

SCAFFOLD POLE

24"

48" MIN.

24"

ENDS OF SCAFFOLD POLES
MUST BE BUTTED TIGHTLY

SCABS MUST BE PLACED ON
AT LEAST TWO ADJACENT SIDES

FIGURE 39–5 Splicing a wooden scaffold pole for additional height.

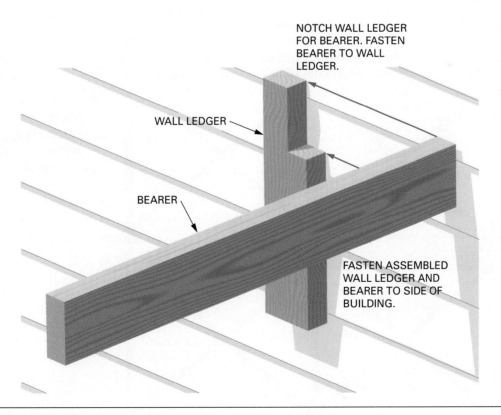

NOTCH WALL LEDGER
FOR BEARER. FASTEN
BEARER TO WALL
LEDGER.

WALL LEDGER

BEARER

FASTEN ASSEMBLED
WALL LEDGER AND
BEARER TO SIDE OF
BUILDING.

FIGURE 39–6 Bearers must be fastened in a notch of a wall ledge for placement
against the side of a building.

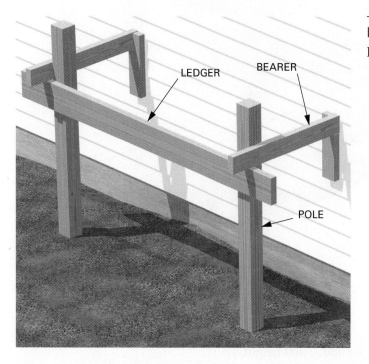

FIGURE 39–7 Ledgers run horizontally from pole to pole and support the bearers.

Ledgers. **Ledgers** run horizontally from pole to pole. They are parallel with the building and support the bearers. Ledgers must be long enough to extend over two pole spaces. They must be overlapped at the pole and not spliced between them **(Figure 39–7).**

Braces. **Braces** are diagonal members. They stiffen the scaffolding and prevent the poles from moving or buckling. Full diagonal face bracing is applied across the entire face of the scaffold in both direc-

tions. On medium- and heavy-duty double-pole scaffolds, the inner row of poles is braced in the same manner. Cross-bracing is also provided between the inner and outer sets of poles on all double-pole scaffolds. All braces are spliced on the poles **(Figure 39–8).**

Plank. **Staging planks** rest on the bearers. They are laid with the edges close together so the platform is tight. There should be no spaces through which

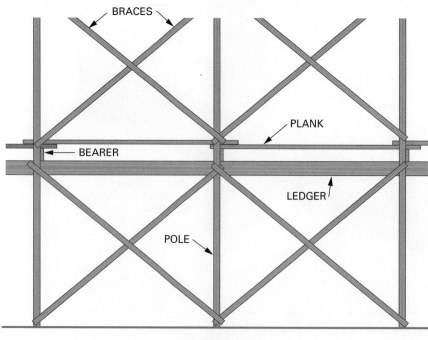

ELEVATION

FIGURE 39–8 Diagonal bracing is applied across the entire face of the scaffold.

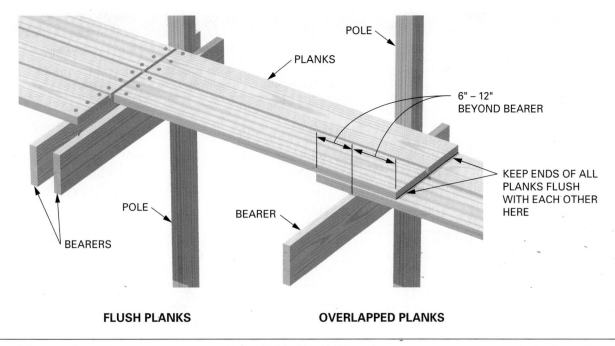

POLE

PLANKS

6" – 12"
BEYOND BEARER

KEEP ENDS OF ALL
PLANKS FLUSH
WITH EACH OTHER
HERE

POLE

BEARER

BEARERS

FLUSH PLANKS

OVERLAPPED PLANKS

FIGURE 39–9 Recommended placement of scaffold plank.

tools or materials can fall. All planking should be scaffold grade or its equivalent. Planking should have the ends banded with steel to prevent excessive checking.

Overlapped planks should extend at least 6 inches beyond the bearer. Where the end of planks butt each other to form a flush floor, the butt joint is placed at the centerline of the pole. Each end rests on separate bearers. The planks are secured to prevent movement. End planks should not overhang the bearer by more than 12 inches **(Figure 39–9)**.

Guardrails. Guardrails are installed on all open sides and ends of scaffolds that are more than 10 feet in height. The top rail is usually of 2×4 lumber. It is fastened to the poles 38 to 45 inches (0.97–1.1 m) above the working platform. A middle rail of 1×6 lumber and a toeboard with a minimum height of 4 inches are also installed **(Figure 39–10)**.

Size and Spacing of Scaffold Members. OSHA requires each member of a wood scaffold to be of a certain size and spacing **(Figure 39–11)**. All members, except planks, are installed on edge (that is, with their width in the vertical position). Also, no wood scaffold should be erected beyond the reach of the local firefighting apparatus.

The OSHA requirements can be found at their website (www.osha.gov/). Care should always be followed when erecting scaffold.

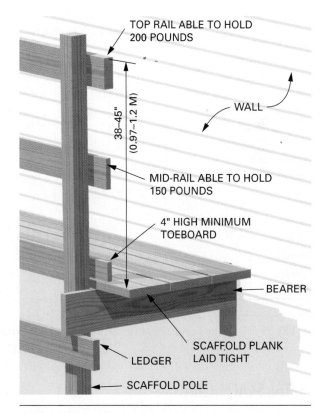

TOP RAIL ABLE TO HOLD
200 POUNDS

WALL

38–45"
(0.97–1.2 M)

MID-RAIL ABLE TO HOLD
150 POUNDS

4" HIGH MINIMUM
TOEBOARD

BEARER

SCAFFOLD PLANK
LAID TIGHT

LEDGER

SCAFFOLD POLE

FIGURE 39–10 Typical wood guardrail specifications.

Single-Pole Wood Scaffolding

Uniformly distributed load	Light Duty		Medium Duty	Heavy Duty
	Not to exceed 25 pounds/square foot	Not to exceed 25 pounds/square foot	Not to exceed 50 pounds/square foot	Not to exceed 75 pounds/square foot
Maximum height of scaffold	20 ft	60 ft	60 ft	60 ft
Poles or uprights	2 × 4 in.	4 × 4 in.	4 × 4 in.	4 × 4 in.
Pole spacing (longitudinal)	6 ft 0 in.	10 ft 0 in.	8 ft 0 in.	6 ft 0 in.
Maximum width of scaffold	5 ft 0 in.	5 ft 0 in.	5 ft 0 in.	5 ft 0 in.
Bearers or putlogs	2 × 6 in. or 3 × 4 in.	2 × 6 in. or 3 × 4 in.	2 × 9 in. or 3 × 4 in.	2 × 9 in. or 2 × 5 in. (rough)
Spacing of bearers or putlogs	6 ft 0 in.	10 ft 0 in.	8 ft 0 in.	6 ft 0 in.
Ledgers	1 × 4 in.	1¼ × 9 in.	2 × 9 in.	2 × 9 in.
Vertical spacing of horizontal members	7 ft 0 in.	7 ft 0 in.	9 ft 0 in.	6 ft 6 in.
Bracing, horizontal and diagonal	1 × 4 in.	1 × 4 in.	1 × 6 in. or 1¼ × 4 in.	2 × 4 in.
Planking	1¼ × 9 in.	2 × 9 in.	2 × 9 in.	2 × 9 in.
Toeboards	4 in. high (min)	4 in. high (min)	4 in. high (min)	4 in. high (min)
Guardrail	2 × 4 in.	2 × 4 in.	2 × 4 in.	2 × 4 in.

Independent (Double-Pole) Wood Scaffolding

Uniformly distributed load	Light Duty		Medium Duty	Heavy Duty
	Not to exceed 25 pounds/square foot	Not to exceed 25 pounds/square foot	Not to exceed 50 pounds/square foot	Not to exceed 75 pounds/square foot
Maximum height of scaffold	20 ft	60 ft	60 ft	60 ft
Poles or uprights	2 × 4 in.	4 × 4 in.	4 × 4 in.	4 × 4 in.
Pole spacing (longitudinal)	6 ft 0 in.	10 ft 0 in.	8 ft 0 in.	6 ft 0 in.
Pole spacing (transverse)	6 ft 0 in.	10 ft 0 in.	8 ft 0 in.	8 ft 0 in.
Ledgers	1¼ × 4 in.	1¼ × 9 in.	2 × 9 in.	2 × 9 in.
Vertical spacing of horizontal members	7 ft 0 in.	7 ft 0 in.	6 ft 0 in.	4 ft 6 in.
Spacing of bearers	6 ft 0 in.	10 ft 0 in.	8 ft 0 in.	8 ft 0 in.
Bearers	2 × 5 in. or 3 × 4 in.	2 × 9 in. (rough) or 3 × 8 in.	2 × 9 in. (rough)	2 × 9 in. (rough)
Bracing, horizontal	1 × 4 in.	1 × 4 in.	1 × 6 in. or 1¼ × 4 in.	2 × 4 in.
Bracing, diagonal	1 × 4 in.	1 × 4 in.	1 × 4 in.	2 × 4 in.
Planking	1¼ × 9 in.	2 × 9 in.	2 × 9 in.	2 × 9 in.
Toeboards	4 in. high (min)	4 in. high (min)	4 in. high (min)	4 in. high (min)
Guardrail	2 × 4 in.	2 × 4 in.	2 × 4 in.	2 × 4 in.

FIGURE 39–11 Sizes and spacing requirements of wood scaffold members.

METAL SCAFFOLDING

Metal Tubular Frame Scaffold

Metal tubular frame scaffolding consists of manufactured *end frames* with folding *cross braces, adjustable screw legs, baseplates, platforms,* and *guardrail hardware* (**Figure 39–12**). They are erected in sections that consist of two end frames and two cross braces and typically come in 5 × 7 modules. Frame scaffolds are easy to assemble, which can lead to carelessness. Because untrained erectors may think scaffolds are just stacked up, serious injury and death can result from a lack of training.

End frames consist of posts, horizontal bearers, and intermediate members. End frames come in a number of styles depending on the manufacturer.

FIGURE 39–12   A typical metal tubular frame scaffold.

Frames can be wide or narrow, and some are designed for rolling tower scaffolds, while other frames have an access ladder built into the end frame **(Figure 39–13)**.

Cross braces rigidly connect one scaffold member to another member. Cross braces connect the bottoms and tops of frames. This diagonal bracing keeps the end frames plumb and provides the rigidity that allows them to attain their designed strength. The braces are connected to the end frames using a variety of locking devices **(Figure 39–14)**.

OSHA regulations require the use of baseplates on all supported or ground-based scaffolds **(Figure 39–15)** in order to transfer the load of scaffolding, material, and workers to the supporting surface. It is extremely important to distribute this load over an area large enough to reduce the pounds per square foot load on the ground. If the scaffold sinks into the ground when it is being used, accidents could occur. Therefore, baseplates should sit on and be nailed to a **mud sill (Figure 39–16)**. A mud still is typically a 2 × 10 (5 × 25 cm)

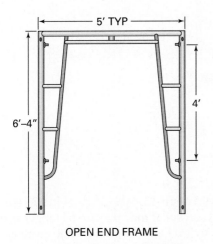

OPEN END FRAME

WALK THROUGH FRAME
WITH BUILT IN LADDER

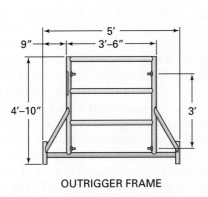

OUTRIGGER FRAME

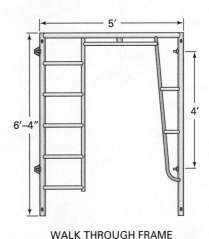

SIDEWALK CANOPY FRAME

FIGURE 39–13   Examples of typical metal tubular end frames.

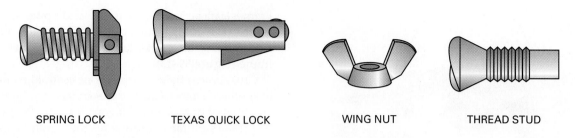

SPRING LOCK   TEXAS QUICK LOCK   WING NUT   THREAD STUD

FIGURE 39–14   Typical locking devices used to connect cross braces to end frames.

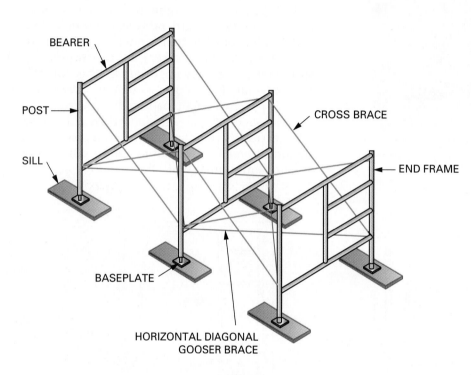

BEARER

POST

SILL

BASEPLATE

CROSS BRACE

END FRAME

HORIZONTAL DIAGONAL
GOOSER BRACE

FIGURE 39–15   Typical baseplate setup for a metal tubular frame scaffold.

FIGURE 39–16   Baseplates should be nailed to the mud sill.

board approximately 18 to 24 inches (45–60 cm) long. On soft soil it may need to be longer and/or thicker.

To level an end frame while erecting a frame scaffold, screwjacks may be used. At least one-third of the screwjack must be inserted in the scaffold leg. Lumber may be used to crib up the legs of the scaffold **(Figure 39–17)**. Cribbing height is restricted to equal the length of the mud sill. Therefore, using a 19-inch (48-cm)-long, 2 × 10-inch (5 × 25-cm) mud sill, the crib height is limited to 19 inches (48 cm). OSHA also prohibits the use of concrete blocks to level scaffolding.

A guardrail system is a vertical fall-protection barrier consisting of, but not limited to, toprails, midrails, toeboards, and posts **(Figure 39–18)**. It prevents employees from falling off a scaffold platform or walkway. A guardrail system is required when the working height is 10 feet or more. Guardrail systems must have a top rail capacity of 200 pounds applied downward or horizontally. The top rail must be between 38 and 45 inches (0.97–1.2 m) above the work deck, with the midrail installed midway be-

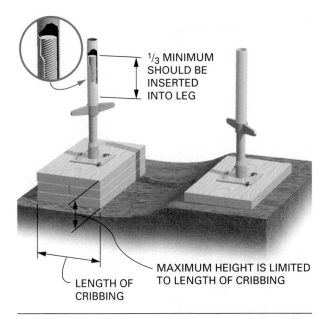

FIGURE 39–17 Cribbing, interlocked blocking may be used to level the ground under the scaffold.

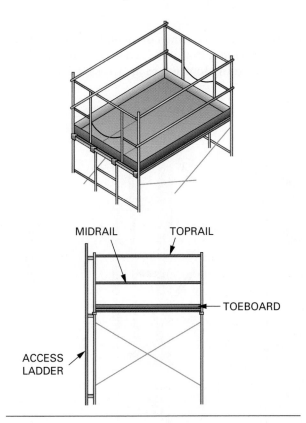

FIGURE 39–18 Typical guardrail system for a metal tubular frame scaffold.

tween the upper guardrail and the platform surface. The midrail must have a capacity of 150 pounds applied downward or horizontally.

If workers are on different levels of the scaffold, toeboards must be installed as an overhead pro-

tection for lower level workers. Toeboards are typically 1 × 4-inch boards installed touching the platform. If materials or tools are stacked up higher than the toeboards, screening must be installed. Moreover, all workers on the scaffold must wear hard hats.

Coupling pins are used to stack the end frames on top of each other **(Figure 39–19).** They have holes in them that match the holes in the end-frame legs; these holes allow locking devices to be installed. Workers must ensure that the coupling pins are designed for the scaffold frames in use.

The scaffold end frames and platforms must have uplift protection installed when a potential for uplift exists. Installing locking devices through the legs of the scaffold and the coupling pins provides this protection **(Figure 39–20).** If the platforms are not equipped with uplift protection devices, they can be tied down to the frames with number nine steel-tie wire.

FIGURE 39–19 Coupler pins to join end frames.

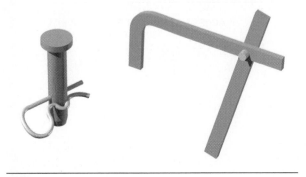

FIGURE 39–20 Coupler locking devices to prevent scaffold uplift.

OSHA requires safe access onto the scaffold for both erectors and users of the scaffolds. Workers can climb end frames only if they meet OSHA regulations. Frames may only be used as a ladder if they are designed as such. Frames meeting such design guidelines must have level horizontal members that are parallel and are not more than 16¾ inches (43 cm) apart vertically **(Figure 39–21)**. Scaffold erectors may climb end-frame rungs that are spaced up to 22 inches (46 cm). Platform planks should not extend over the end frames where end-frame rungs are used as a ladder access point. The cross braces should never be used as a means of access or egress. Attached ladders and stair units may be used **(Figure 39–22)**. A rest platform is required for every 35 feet (10.7 m) of ladder.

Manually propelled mobile scaffolds use wheels or casters in the place of baseplates **(Figure 39–23)**. Casters have a designed load capacity that should never be exceeded. Mobile scaffolds made from tubular end frames must use diagonal horizontal braces, or gooser braces (see Figure 39–15) to keep the mobile tower frame square.

Side brackets are light-duty (35 pounds per square foot maximum) extension pieces used to increase the working platform **(Figure 39–24)**. They are designed to hold personnel only and are not to be used for material storage. When side brackets

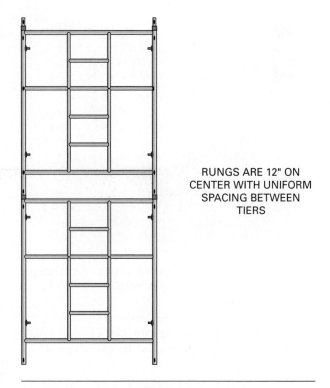

RUNGS ARE 12" ON CENTER WITH UNIFORM SPACING BETWEEN TIERS

FIGURE 39–21   The rungs of an end frame must be spaced no more than 16¾ inches (43 cm) apart if users are going to use it as an access ladder.

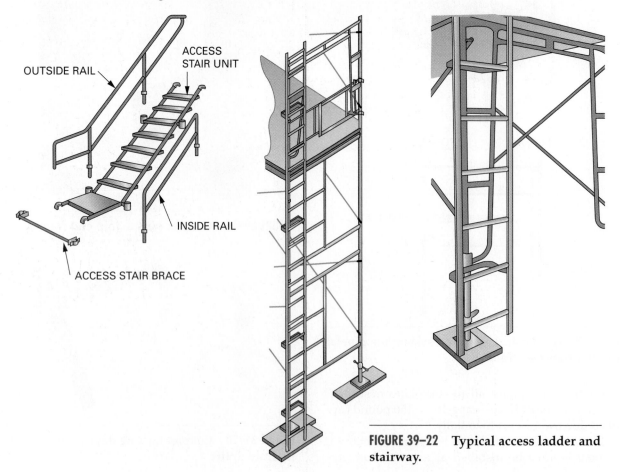

OUTSIDE RAIL

ACCESS STAIR UNIT

INSIDE RAIL

ACCESS STAIR BRACE

FIGURE 39–22   Typical access ladder and stairway.

FIGURE 39–23 Casters replace a baseplate to transform a metal tubular frame scaffold into a mobile scaffold.

FIGURE 39–25 A hoist that attaches to the top of a scaffold is used to raise material and equipment.

are used, the scaffold must have tie-ins, braces, or outriggers to prevent the scaffold from tipping.

Hoist arms and wheel wells are sometimes attached to the top of the scaffold to hoist scaffold parts to the erector or material to the user of the scaffold **(Figure 39–25).** The load rating of these hoist arms and wheel wells is typically no more than 100 pounds. The scaffold must be secured from overturning at the level of the hoist arm, and workers should never stand directly under the hoist arm when hoisting a load. They should stand a slight distance away, but not too far to the side, because this will increase the lateral or side-loading force on the scaffold.

Scaffold Inspection

Almost half of all scaffold accidents, according to the U.S. Bureau of Labor Statistics, involve defective scaffolds or defective scaffold parts. This statistic means ongoing visual inspection of scaffold parts must play a major role in safe scaffold erection and use. OSHA requires that a competent person inspect all scaffolds at the beginning of every work shift.

Visual inspection of scaffold parts should take place at lease five times: before erection, during erection, during scaffold use, during dismantling, and before scaffold parts are put back in storage. All damaged parts should be red-tagged and removed from service and then repaired or destroyed as required. Things to look for during the inspection process include the following:

Broken and excessively rusted welds
Split, bent, or crushed tubes
Cracks in the tube circumference
Distorted members
Excessive rust
Damaged brace locks
Lack of straightness
Excessively worn rivets or bolts on braces
Split ends on cross braces
Bent or broken clamp parts
Damaged threads on screwjacks

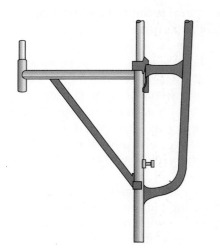

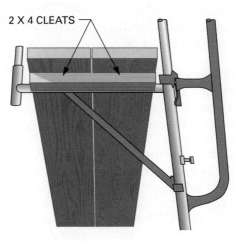

2 X 4 CLEATS

FIGURE 39–24 Side brackets are used to extend a scaffold work platform. These brackets should only be used for workers and never for material storage.

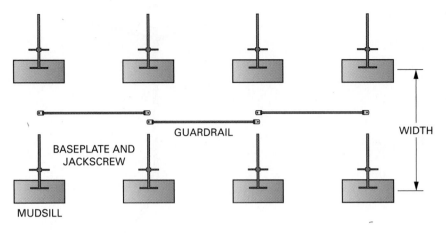

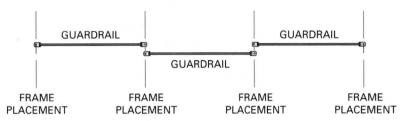

FIGURE 39–26 During scaffold erection, the baseplates are spaced according to guardrail braces.

Damaged caster brakes
Damaged swivels on casters
Corrosion of parts
Metal fatigue caused by temperature extremes
Leg ends filled with dirt or concrete.

Scaffold Erection Procedure

The first thing to be done during the erection procedure is to inspect all scaffold components delivered to the jobsite. Defective parts must not be used.

The foundation of the scaffold must be stable and sound, able to support the scaffold and four times the maximum intended load without settling or displacement.

Always start erecting the scaffold at the highest elevation, which will allow the scaffold to be leveled without any excavating by installing cribbing, screwjacks, or shorter frames under the regular frames. The scaffold must always be level and plumb. Lay out the baseplates and screwjacks on mud sills so the guardrails and end frames with cross braces can be properly installed **(Figure 39–26).**

Stand one of the end frames up and attach the cross braces to each side, making sure the correct length cross braces have been selected for the job. Connect the other end of the braces to the second end frame. All scaffold legs must be braced to at least one other leg **(Figure 39–27).** Make sure that all brace connections are secure. If any of these mechanisms are not in good working order, replace the frame with one that has properly functioning locks.

Use a level to plumb and level each frame **(Figure 39–28).** Remember that OSHA requires

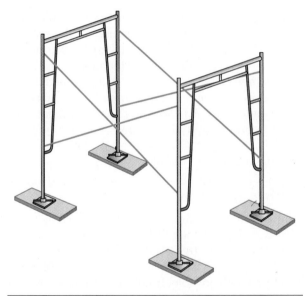

FIGURE 39–27 Cross braces connect end frames, keeping them rigid and plumb.

that all tubular welded frame scaffolds be plumb and level. Adjust screwjacks or cribbing to level the scaffold. As each frame is added, keep the scaffold bays square with each other. Repeat this procedure until the scaffold run is erected. Remember, if the first level of scaffolding is plumb and level, the remaining levels will be more easily assembled.

The next step is to place the planks on top of the end frames. All planking must meet OSHA requirements and be in good condition. If planks that do not have hooks are used, they must extend over

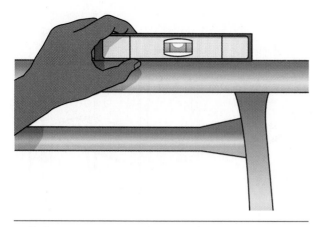

FIGURE 39–28 End frames must be level. Level and plumb begins on the first row of scaffolding.

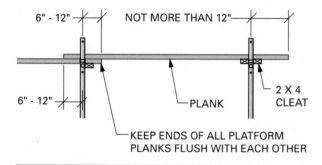

FIGURE 39–29 Recommended placement for scaffold planks.

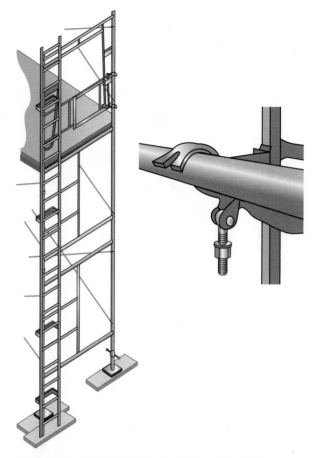

FIGURE 39–30 Attachable ladders are connected so that the bolt attaches from the bottom.

their end supports by at least 6 inches (15 cm) and not more than 12 inches (30 cm). A cleat should be nailed to both ends of wood planks to prevent plank movement **(Figure 39–29)**. Platform laps must be at least 12 inches, and all platforms must be secured from movement. Hooks on planks also have uplift protection installed on the ends.

It is a good practice to plank each layer fully as the scaffold is erected. If the deck is only to be used for erecting, then a minimum of two planks can be used. However, full decking is preferred, because it is a safer method for the erector.

Before the second lift is erected, the erector must provide an access ladder. Access may be on the end frame, if it is so designed, or an attached ladder. If the ladder is bolted to a horizontal member, the bolt must face downward **(Figure 39–30)**. Next, the second level of frames may be hung temporarily over the ends of the first frames and then installed onto the coupling pins of the first-level frames **(Figure 39–31)**. Special care must be taken to ensure proper footing and balance when lifting and placing frames. OSHA requires erector fall protection—a full body harness attached to a proper anchor point on the structure—when it is feasible and not a greater hazard to do so. Never attach fall protection to the scaffold.

After the end frames have been set in place and braced, they should have uplift protection pins installed through the legs and coupling pins. Wind, side brackets, and wheel wells can cause uplift so it is a good practice to pin all scaffold legs together.

The remaining scaffolding is erected in the same manner as the first. Remember, all work platforms must be fully decked and have a guardrail system or personal fall-arrest system installed before it can be turned over to the scaffold users. If the scaffold is higher than four times its minimum base dimension, it must be restrained from tipping by guying, tying, bracing, or equivalent means **(Figure 39–32)**. The scaffold is not allowed to tip into or away from the structure.

After the scaffold is complete, it is inspected again to make sure that it is plumb, level, and square before turning it over for workers to use. The inspection should also include checking that all legs are on baseplates or screwjacks and mud sills (if required), ensuring the scaffolding is properly braced with all brace connections secured, and making sure all tie-ins are properly placed and secured, both at the scaffold and at the structure. All platforms must be fully planked with proper decking

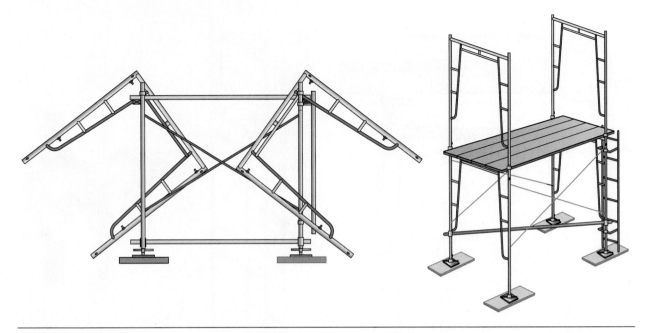

FIGURE 39–31 The next lift of end frames may be preloaded by hanging them on the previous end frame.

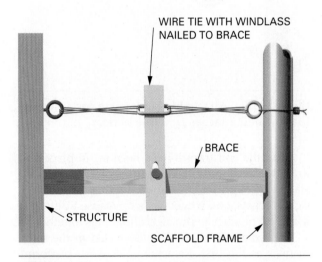

WIRE TIE WITH WINDLASS NAILED TO BRACE

BRACE

STRUCTURE

SCAFFOLD FRAME

FIGURE 39–32 A wire tie and a windlass may be used to secure the scaffold tightly to the building.

and in good shape. Toeboards and/or screening should be installed as needed. Check that end and/or side brackets are fully secured and that any overturning forces are compensated for. All access units are inspected to ensure they are correctly installed and ladders and stairs are secured. Again, workers on the scaffold must wear hard hats.

After the scaffolding passes all inspections, it is ready to be turned over to the workers. Remember that this scaffolding must be inspected by a competent person at the beginning of each work shift and after any occurrence, such as a high wind or a rainstorm, that could affect its structural integrity.

Scaffold Capacity

All scaffolds and their components must be capable of supporting, without failure, their own weight and at least four times the maximum intended load applied or transmitted to them. Erectors and users of scaffolding must never exceed this safety factor.

Erectors and users of the scaffold must know the maximum intended load and the load-carrying capacities of the scaffold they are using. The erector must also know the design criteria, maximum intended load-carrying capacity, and intended use of the scaffold.

When erecting a frame scaffold, the erector should know the load-carrying capacities of its components. The rated leg capacity of a frame may never be exceeded on any leg of the scaffold. Also, the capacity of the top horizontal member of the end frame, called the bearer, may never be exceeded. Remember, it is possible to overload the bottom legs of the scaffold without overloading the bearer or top horizontal member of any frame. It is also possible to overload the bearer or top horizontal member of the frame scaffold and not overload the leg of that same scaffold. Erectors must pay careful attention to the load capacities of all scaffold components.

If the scaffold is covered with weatherproofing plastic or tarps, the lateral pressure applied to the scaffold will dramatically increase. Consequently, the number of tie-ins attached to prevent overturning must be increased. Additionally, any guy wires added for support will increase the downward pressure and weight of the scaffold.

OSHA regulations state that supported scaffolds with a ratio larger than four-to-one (4:1) of the height to narrow base width must be restrained from tipping by guying, tying, bracing, or equivalent means. Guys, ties, and braces must be installed at locations where horizontal members support both inner and outer legs. Guy, ties, and braces must be installed according to the scaffold manufacturer's recommendations or at the closest horizontal member to the 4:1 height. For scaffolds greater than 3 feet (0.91 m) wide, the vertical locations of horizontal members are repeated every 26 feet (7.9 m). The top guy, tie, or brace of completed scaffolds must be placed no further than the 4:1 height from the top. Such guys, ties, and braces must be installed at each end of the scaffold and at horizontal intervals not to exceed 30 feet (9.1 m). The tie or standoff should be able to take pushing and pulling forces so the scaffold does not fall into or away from the structure.

The supported scaffold poles, legs, post, frames, and uprights should bear on baseplates, mud sills, or other adequate, firm foundation. Because the mud sills have more surface area than baseplates, sills distribute loads over a larger area of the foundation. Sills are typically wood and come in many sizes. Erectors should choose a size according to the load and the foundation strength required. Mud sills made of 2 × 10-inch (5 × 25-cm) full-thickness or nominal lumber should be 18 to 24 inches (46 to 60 cm) long and centered under each leg **(Figure 39–33)**.

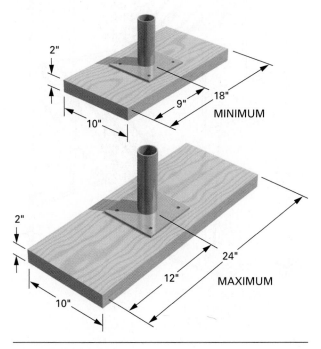

FIGURE 39–33 Baseplates should be centered on the mud sills.

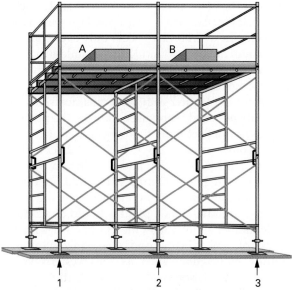

THE CENTER END FRAME LABELED #2 CARRIES TWICE THE LOAD OF EACH OF THE END FRAMES LABELED #1 AND #3

FIGURE 39–34 The inner frames, such as #2 shown here, often carry twice the load of the frames located at the end of the scaffold.

The loads exerted onto the legs of a scaffold are not equal. Consider a scaffold with two loads on two adjacent platforms **(Figure 39–34)**. Half of load A is carried by end frame #1 and the other half is carried by #2. Half of load B is carried by end frames #2 and #3. End frame #2 carries two half loads or one full load, which is twice the load of end frames #1 and #3. At no time should the manufacturer's load rating for its scaffolding be exceeded.

Scaffold Platforms

The scaffolding's work area must be fully planked between the front uprights and the guardrail supports in order for the user to work from the scaffold. The planks should not have more than a 1-inch gap between them unless it is necessary to fit around uprights such as a scaffold leg. If the platform is planked as fully as possible, the remaining gap between the last plank and the uprights of the guardrail system must not exceed 9½ inches.

Scaffold platforms must be at least 18 inches wide with a guardrail system in place. In areas where they cannot be 18 inches wide, they will be as wide as is feasible. The platform is allowed to be as much as 14 inches away from the face of the work.

Planking for the platforms, unless cleated or otherwise restrained by hooks or equivalent means, should extend over the centerline of their support at least 6 inches (15 cm) and no more than 12 inches (30 cm). If the platform is overlapped to create a long platform, the overlap shall occur only over supports and should not be less than 12 inches unless the platforms are nailed together or otherwise restrained to prevent movement.

When fully loaded with personnel, tools, and/or material, the wood plank used to make the platform must never deflect more than one-sixtieth (⅟₆₀) of its span. In other words, a 2 × 10-inch (5 × 25-cm) plank that is 12 feet (3.7 m) long and is sitting on two end frames spaced 10 feet (3 m) apart should not deflect more than 2 inches (5 cm) or one-sixtieth of the span.

Any solid sawn wood planks should be scaffold-grade lumber as set out by the grading rules for the species of lumber being used. A recognized lumber-grading association, such as the Western Wood Products Association (WWPA) or the National Lumber Grades Authority (NLGA), establishes these grading rules. A grade should be stamped on the scaffold-grade plank, indicating that it meets OSHA and industry requirements for scaffold planks. Two of the most common wood species used for scaffold planks are southern yellow pine and Douglas fir. OSHA does not require wood scaffold planks to bear grade stamps. The erector may use "equivalent" planks, which are determined to be equivalent by visually inspecting or test loading the wood plank in accordance with grading rules.

Scaffold platforms are usually rated for the intended load. Light-duty scaffolds are designed at 25 pounds per square foot, medium-duty scaffolds are rated at 50 pounds per square foot, and heavy-duty scaffolds at 75 pounds per square foot. The maximum span of a plank is tabulated in **Figure 39–35.** Using this chart, the maximum load that could be put on a nominal thickness plank (1½ inch or 3.8 cm) with a span of 7 feet (2.1 m) is 25 pounds per square foot. Note that a load of 50 pounds per square foot would require a span of no more than 6 feet (1.8 m).

Fabricated planks and platforms are often used in lieu of solid sawn wood planks. These planks and platforms include fabricated wood planks that use a pin to secure the lumber sideways, oriented strand board

planks, fiberglass composite planks, aluminum-wood decked planks, and high-strength galvanized steel planks. The loading of fabricated planks or platforms should be obtained from the manufacturer and never exceeded. Scaffold platforms must be inspected for damage before each use.

Scaffold Access

A means of access must be provided to any scaffold platform that is 2 feet above or below a point of access. Such means include a hook-on or attachable ladder, a ramp, or a stair tower and are determined by the competent person on the job.

If a ladder is used, it should extend 3 feet above the platform and be secured both at the top and bottom. Hook-on and attachable ladders should be specifically designed for use with the type of scaffold used, have a minimum rung length of 11½ inches, and have uniformly spaced rungs with a maximum spacing between rung length of 16¾ inches. Sometimes a stair tower can be used for access to the work platform, usually on larger jobs **(Figure 39–36).** A ramp can also be used as access to the scaffold or the work platform. When using a ramp, it is important to remember that

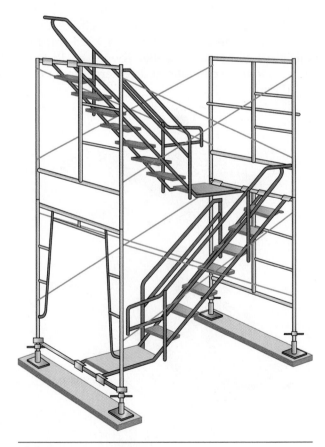

FIGURE 39–36   Scaffold access may be provided by a stair tower.

Maximum Permissible Plank Span				
Maximum Intended Load	Rough 2× Full Thickness		2× Nominal Thickness	
Lbs/sq ft	Feet	Meters	Feet	Meters
25	10	3	8	2.4
50	8	2.4	6	1.8
75	6	1.8	—	—

FIGURE 39–35   Maximum spacing of planks based on the load rating of the scaffold.

a guardrail system or fall protection is required at 6 feet above a lower level.

The worker using the scaffold can sometimes access the work platform using the end frames of the scaffold itself. According to regulations, the end frame must be specifically designed and constructed for use as ladder rungs. The rungs can run up the center or to one side of the end frame; some have the rungs all the way across the end frame. Scaffold users should never climb any end frame unless the manufacturer of that frame designated it to be used for access.

Scaffold Use

Scaffolds must not be loaded in excess of their maximum intended load or rated capacities, whichever is less. Workers must know the capacity of scaffolds they are erecting and/or using. Before the beginning of each work shift, or after any occurrence that could affect a scaffold's structural integrity, the competent person must inspect all scaffolds on the job.

Employees must not work on scaffolds covered with snow or ice except to remove the snow or ice. Generally, work on or from scaffolds is prohibited during storms or high winds. Debris must not be allowed to accumulate on the platforms. Makeshift scaffold devices, such as boxes or barrels, must not be used on the scaffold to increase workers' working height. Step ladders should not be used on the scaffold platform unless they are secured according to OSHA regulations.

Fall Protection

Current OSHA standards on scaffolding require **fall protection** when workers are working at heights above 10 feet. This regulation applies to both the user of the scaffold and the erector or dismantler of the scaffold. These regulations allow the employer the option of a guardrail system or a personal fall protection system. The fall protection system most often used is a complete guardrail system. A guardrail system has a toprail 38 to 45 inches (0.97–1.2 m) above the work deck, with a midrail installed midway between the toprail and the platform. The work deck should also be equipped with a toeboard. These requirements are for all open sides of the scaffold, except for those sides of the scaffold that are within 14 inches of the face of the building.

A typical personal fall protection system consists of five related parts: the harness, lanyard, lifeline, rope grab, and anchor **(Figure 39–37).** The failure of any one part means failure of the system. Therefore, constant monitoring of a lifeline system is a critical responsibility. It is easy for a system to lose its integrity almost immediately, even on first use.

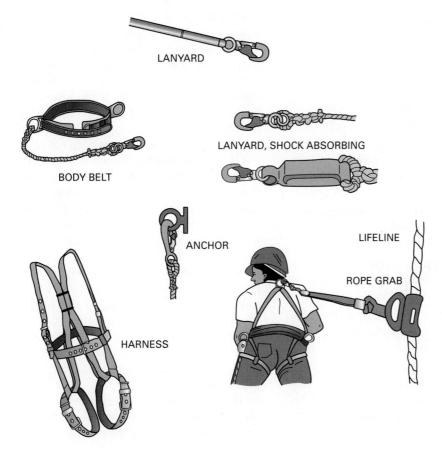

FIGURE 39–37
Components of a personal protection system.

LANYARD

BODY BELT

LANYARD, SHOCK ABSORBING

ANCHOR

HARNESS

LIFELINE

ROPE GRAB

OSHA recognizes that sometimes fall protection may not be possible for erectors. As the scaffold increases in length, the personal fall-arrest system may not be feasible because of its fixed anchorage and the need for employees to traverse the entire length of the scaffold. Additionally, fall protection may not be feasible due to the potential for lifelines to become entangled or to create a tripping hazard for erectors or dismantlers as they traverse the scaffold. Do not use the scaffold components as an anchor point of the fall-protection harness. OSHA puts the responsibility of when to use fall protection, both for the user of the scaffold and the erector, on the competent person.

Falling Object Protection

According to industry standards and OSHA requirements, workers must wear hard hats during the process of erecting a scaffold. In addition to hard hats, protection from potential falling objects may be required. When material on the scaffold could fall on workers below, some type of barricade must be installed to prevent that material from falling. OSHA lists toeboards as part of the falling object protection for the workers below the scaffold. The toeboard can serve two functions: It keeps material on the scaffold and keeps the workers on the scaffold platform if they happen to slip.

Dismantling Scaffolds

Many guidelines and rules for erection also apply to scaffold dismantling. However, dismantling requires additional precautions to ensure the scaffold will come down in a controlled, safe, and logical manner. Important factors to consider include the following:

1. Check every scaffold before dismantling. Any loose or missing ties or bracing must be corrected.

2. If a hoist is to be used to lower the material, the scaffold must be tied to the structure at the level of the hoist arm to dispel any overturning effect of the wheel and rope.

3. The erector should be tied off for fall protection, as required by the regulations, unless it is infeasible or a greater hazard to do so.

4. Start at the top and work in reverse order, following the step-by-step procedures for erection. Leave the work platforms in place as long as possible.

5. Do not throw planks or material from the scaffold. This practice will damage the material and presents overhead hazards for workers below.

6. Building tie-ins and bracing can only be removed when the dismantling process has reached that level or location on the scaffold. An improperly removed tie can cause the entire scaffold to overturn.

7. Remove the ladders or the stairs only as the dismantling process reaches that level. Never climb or access the scaffold by using the cross braces.

8. As the scaffold parts come off the scaffold, they should be inspected for any wear or damage. If a defective part is found, it should be tagged for repair and not used again until inspected by the competent person.

9. Dismantled parts and materials should be organized, stacked, and placed in bins or racks out of the weather.

10. Secure the disassembled scaffold equipment to ensure that no unauthorized, untrained employees use it. All erectors must be trained, experienced, and under the supervision and direction of a competent person.

11. Always treat the scaffold components as if a life depended on them, because the next time the scaffold is erected, someone's life will indeed be depending on it being sound.

MOBILE SCAFFOLDS

The rolling tower, or **mobile scaffold,** is widely used for small jobs, generally not more than 20 feet in height **(Figure 39–38).** The components of the mobile scaffold are the same as those for the stationary frame scaffold, with the addition of casters and horizontal diagonal bracing. There are additional restrictions on rolling towers as well.

The height of a rolling tower must never exceed four times the minimum base dimension. For example, if the frame sections are 5×7, the rolling tower can only be 20 feet high. If the tower exceeds this height-to-base ratio, it must be secured to prevent overturning. When outriggers are used on a mobile tower, they must be used on both sides.

Casters on mobile towers must be locked with positive wheel swivel locks or the equivalent to prevent movement of the scaffold while it is stationary. Casters typically have a load capacity of 600 pounds each, and the legs of a frame scaffold can hold 2,000 to 3,000 pounds each. Care must be taken not to overload the casters.

Never put a cantilevered work platform, side bracket, or well wheels on the side or end of a mobile tower. Mobile towers can tip over if used incorrectly. Mobile towers must have horizontal, diagonal, or gooser braces at the base to prevent racking of the tower during movement **(Figure 39–39).** Metal hook planks also help prevent racking if they are secured to the frames.

The force to move the scaffold should be applied as close to the base as practicable, but not more than 5 feet (1.5 m) above the supporting surface. The cast-

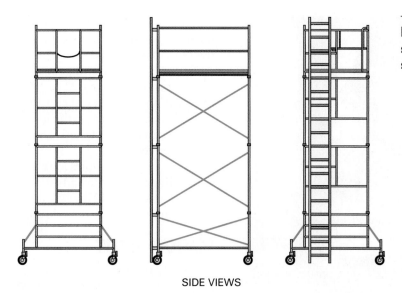

SIDE VIEWS

FIGURE 39–38 Typical setup for a mobile scaffold.

ers must be locked after each movement before beginning work again. Employees are not allowed to ride on rolling tower scaffolds during movement unless the height-to-base width ratio is two-to-one or less. Before the scaffold is moved, each employee on the scaffold must be made aware of the move. Caster and wheel stems shall be pinned or otherwise secured in scaffold legs or adjustment screws. The surface that the mobile tower rolls on must be free of holes, pits, and obstructions and must be within 3 degrees of level. Only use a mobile scaffold on firm floors.

PUMP JACK SCAFFOLDS

Pump jack scaffolds consist of 4 × 4 poles, a pump jack mechanism, and metal braces for each pole **(Figure 39–40).** The braces are attached to the pole at intervals and near the top. The arms of the bracket extend from both sides of the pole at 45-degree angles. The arms are attached to the sidewall or roof to hold the pole steady.

The scaffold is raised by pressing on the foot pedal of the pump jack **(Figure 39–41).** The mechanism has brackets on which to place the scaffold plank. Other brackets hold a guardrail or platform. Reversing a lever allows the staging to be pumped downward.

Pump jack scaffolds are used widely for siding, where staging must be kept away from the walls,

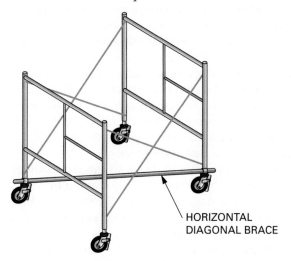

FIGURE 39–39 The horizontal diagonal brace (or gooser) is used to keep the tower square when it is rolled.

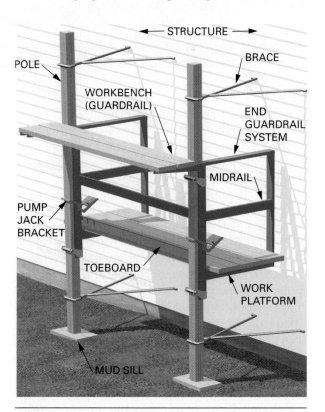

FIGURE 39–40 Components of a pump jack system.

FIGURE 39–41 Pump jacks are raised by pressing the foot lever.

and when a steady working height is desired. However, pump jack scaffolds have their limitations. They should not be used when the working load exceeds 500 pounds. No more than two persons are permitted at one time between any two supports.

Wood poles must not exceed 30 feet in height. Braces must be installed at a maximum vertical spacing of not more than 10 feet.

To pump the scaffold past a brace location, temporary braces are used. The temporary bracing is installed about 4 feet above the original bracing. Once the scaffold is past the location of the original brace, it can be reinstalled. The temporary brace is then removed.

Wood pump jack poles are constructed of two 2 × 4s nailed together. The nails should be 3-inch or 10d, and no less than 12 inches apart, staggered uniformly from opposite outside edges.

SCAFFOLD SAFETY

The safety of those working at a height depends on properly constructed scaffolds and proper use of scaffolds. Those who have the responsibility of constructing scaffolds must be thoroughly familiar with the sizes, spacing, and fastening of scaffold members and other scaffold construction techniques. Those who use the scaffold must be aware of where they are working at all times. They must watch their step and the material they are using to prevent accidents. They should also make it a habit to visually inspect the scaffold before each use.

40 Brackets, Horses, and Ladders

ROOFING BRACKETS

Roofing brackets are used when the pitch of the roof is too steep for carpenters to work on without slipping **(Figure 40–1).** Usually any roof with more than 4 on 12 **slope** requires roof brackets. They keep the worker from slipping and also hold the roofing materials. Roofing brackets are made of metal. Some are adjustable for roofs of different pitches.

A metal plate at the top of the bracket has slots in which to drive nails to fasten the bracket to the roof **(Figure 40–2).** The slots are open angled or *key holed*. This enables removal of the bracket from the fasteners without pulling the nails. The bracket is simply tapped upward from the bottom, and then lifted over the nailheads. The nails that remain are then driven all the way to the surface or flush.

Applying Roof Brackets

Roof brackets are usually used when the roof is being shingled. Apply roof brackets in rows. Space them out so that they can be reached without climbing off the roof bracket staging below.

On asphalt-shingled roofs, place the brackets at about 6- to 8-foot (1.8- to 2.4-m) horizontal intervals. The nails for the brackets should be installed so they land where a shingle nail is located (see Figure 40–2). The bottom nail of the bracket should be at or slightly above the nailing line of the shingle. (See Chapter 51.) This will place the bracket so it will be in centered under the next shingle tab, ensuring that the bracket nails are covered and won't cause leaks. No joint or cutout in the course above should fall in line with the nails holding the bracket. Otherwise, the roof will leak. Use three 3¼-inch or 12d common nails driven flush. Try to get at least one nail into a rafter.

Open adjustable brackets so the top member is approximately level or slightly leaning toward the roof. Place a staging plank on the top of the brackets. Overlap them as in wall scaffolds. Keep the inner edges against the roof for greater support. A toeboard made of 1 × 6 or 1 × 8 lumber is usually placed flat on the roof with its bottom edge on top of the brackets. This protects the new roofing from damage by the workers' toes during construction **(Figure 40–3).** After the shingles are applied, tap

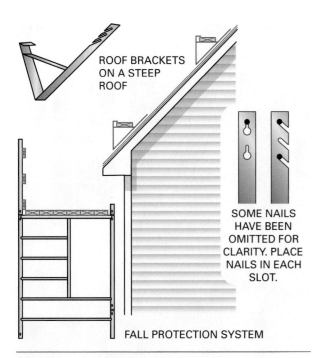

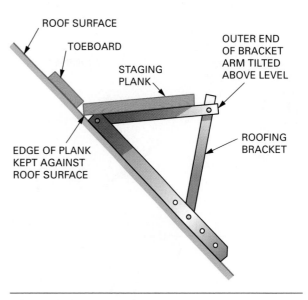

FIGURE 40–3 A toeboard protects the roofing material from damage from workers' toes.

end. This type of scaffold may be used in the building interior for working on ceilings.

Trestle Jacks

Manufactured metal trestle supports, called *trestle jacks,* are available for frequent and prolonged use. Trestle jacks are adjustable in height at about 3-inch intervals. They are clamped to a 2-inch wood support on which the scaffold planks are placed. The size of the support depends on how far apart the trestle jacks are placed. Metal braces hold the trestle rigid **(Figure 40–4).**

FIGURE 40–1 Roof brackets are used when the roof pitch is too steep for carpenters to work on without slipping.

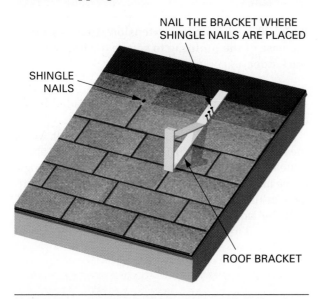

FIGURE 40–2 Roof brackets are nailed where a shingle nail is located.

the bottom of the bracket upward along the slope of the roof to release it from the nails. Raise the shingle tab and drive the nails flush so they do not stick up and damage the shingles.

TRESTLES

Trestles are used when a low scaffold on a level surface is desired. A trestle is a low working platform supported by a bearer with spreading legs at each

FIGURE 40–4 Adjustable metal trestle jacks are manufactured for prolonged and repetitive use.

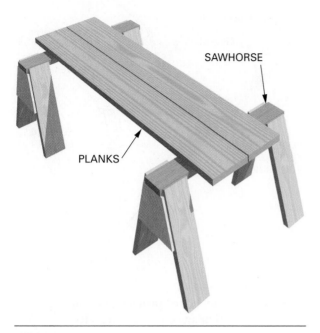

FIGURE 40–5   **Sawhorses are sometimes used as supports for a trestle scaffold.**

Sawhorses are used for trestle supports when occasional use is required and their height meets the requirements of the job **(Figure 40–5).** For light-duty work, horses and trestle jacks should not be spaced more than 8 feet apart. Do not use horses that have become weak or defective. The horse should be strong enough to support four times the intended load.

If a horse scaffold is arranged in tiers, no more than two tiers should be used. The legs of the horses in the upper tier should be nailed to the planks. Each tier should be cross-braced.

LADDERS

Carpenters must often use *ladders* to work from or to reach working platforms above the ground. Most commonly used ladders are the **stepladder** and the **extension ladder.** They are usually made of wood, aluminum, or fiberglass. Make sure all ladders are in good condition before using them.

Extension Ladders

To raise an extension ladder, the bottom of the ladder must be secured. This is done either with the help of another person or by placing the bottom against a solid object, such as the base of the wall. Pick up the other end. Walk forward under the ladder, pushing upward on each rung until the ladder is upright **(Figure 40–6).** With the ladder vertical, and leaning toward the wall, extend the ladder by pulling on the rope with one hand while holding the ladder upright with the other. Raise the ladder to the

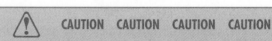

FIGURE 40–6   **Raising an extension ladder using the base of the building to secure the bottom of the ladder.**

desired height. Make sure the spring-loaded hooks are over the rungs on both sides.

Lean the top of the ladder against the wall. Move the base out until the distance from the wall is about ¼ the vertical height. This will give the proper angle to the ladder. The proper angle for climbing the ladder can also be determined, as shown in **Figure 40–7.**

If the ladder is used to reach a roof or working platform, it must extend above the top support by at least 3 feet.

> ⚠️ **CAUTION   CAUTION   CAUTION   CAUTION**
>
> Be careful of overhead power lines when using ladders. Metal ladders conduct electricity. Contact with power lines could result in electrocution. ∎

When the ladder is in position, shim one leg, if necessary, to prevent wobbling, and secure the top of the ladder to the building. Check that the feet of the ladder are secure and will not slip. Then face the ladder to climb, grasping the side rails or rungs with both hands **(Figure 40–8).**

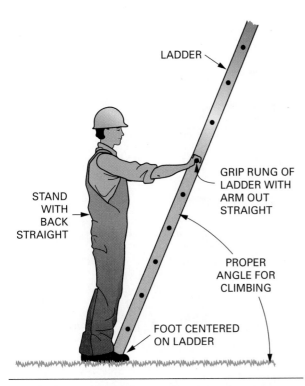

FIGURE 40-7 Techniques for finding the proper ladder angle before climbing.

Stepladders

When using a stepladder, open the legs fully so the brackets are straight and locked. Make sure the ladder does not wobble. If necessary, place a shim under the leg to steady the ladder. Never work above the recommended top step indicated by the manufacturer. This is usually the second step down from the top, not including the top as a step. Do not use the ledge in back of the ladder as a step. The ledge is used to hold tools and materials only. Move the ladder as necessary to avoid overreaching. Make sure all materials and tools are removed from the ladder before moving it.

Ladder Jacks

Ladder jacks are metal brackets installed on ladders to hold scaffold planks. At least two ladders and two jacks are necessary for a section. Ladders should be heavy duty, free from defects, and placed no more than 8 feet apart. They should have devices to keep them from slipping.

The ladder jack should bear on the side rails in addition to the ladder rungs. If bearing on the rungs only, the bearing area should be at least 10 inches on each rung. No more than two persons should occupy any 8 feet of ladder jack scaffold at any one time. The platform width must not be less than 18 inches. Planks must overlap the bearing surface by at least 10 inches **(Figure 40–9).**

FIGURE 40-8 Face the ladder when climbing. Hold on with both hands.

FIGURE 40-9 Ladder jacks are used to support scaffold plank for short-term, light repair work.

Procedure 40–A Constructing a Typical Sawhorse

STEP 1 CUT 2" X 6" SAWHORSE TOP 36" LONG AND BEVEL BOTH EDGES OF EACH END AS SHOWN.

STEP 2 CUT FOUR LEGS TO 24" LENGTH OR AS DESIRED, WITH BEVEL ON EACH END AT SAME ANGLE AS TOP.

STEP 3 FASTEN ALL FOUR LEGS TO SAWHORSE TOP.

STEP 4 HOLD PLYWOOD BRACE AS SHOWN AND MARK ITS LENGTH AT THE TOP EDGE.

FROM EACH MARK, LAY OUT SAME ANGLE AS TOP AND LEGS. CUT, MAKE DUPLICATE, AND FASTEN ONE ON EACH END OF HORSE FLUSH WITH OUTSIDE FACE OF LEGS. MOVE LEGS TO FIT THE PIECE.

2" X 6" TOP

TOP VIEW

36"

1" X 6" LEGS

SIDE VIEW

3/8" PLYWOOD BRACE

24" OR AS DESIRED

10"

END VIEW

SAWHORSE MULTI-VIEW

1   6"   3/8"   3/8"

DETAIL OF END

3

TOP

2

LEG

DETAIL OF BEVEL

4

3/8" PLY.

END VIEW

PERSPECTIVE

CONSTRUCTION AIDS

Sawhorses, work stools, ladders, and other construction aids are sometimes custom built by the carpenter on the job or in the shop.

Sawhorses

Sawhorses are used on practically every construction job. They support material that is being laid out or cut to size. They may be built of various materials depending on the desired strength of the sawhorse. Light-duty horses may be made with a 2 × 4 top and 1 × 4 legs, while heavy-duty horses are made with a 2 × 6 top and 2 × 4 legs. Both use plywood for leg braces.

Sawhorses are constructed in a number of ways according to the preference of the individual **(Procedure 40–A).** However, they should be of sufficient width, a comfortable working height, and light enough to be moved easily from one place to another. A typical sawhorse is 36 inches wide with 24-inch legs. A tall person may wish to make the leg 26 inches long.

Procedure 40–B Constructing a Job-Built Ladder

STEP ❶ TEMPORARILY FASTEN RAILS OF LADDER SIDE BY SIDE AND LAY OUT DADOES 12" OC AS SHOWN BELOW.

STEP ❷ CUT DADOES, CUTTING ONLY $^3/_8$" DEEP. ANY DEEPER WILL WEAKEN THE RAIL.

STEP ❸ CUT RUNGS 15–20" LONG. FASTEN RUNGS IN DADOES WITH 16d NAILS KEEPING ENDS OF RUNGS FLUSH WITH OUTSIDE FACE OF RAILS.

STEP ❹ CUT CLEATS TO FIT BETWEEN RUNGS.

15–20" WIDE

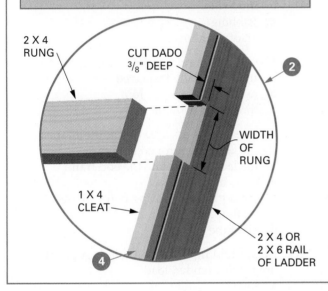

2 X 4 RUNG

CUT DADO $^3/_8$" DEEP

WIDTH OF RUNG

1 X 4 CLEAT

2 X 4 OR 2 X 6 RAIL OF LADDER

Job-Made Ladders

At times it is necessary to build a ladder on the job **(Procedure 40–B).** These are usually straight ladders no more than 24 feet in length. The side rails are made of clear, straight-grained 2 × 4 or 2 × 6 stock spaced 15 to 20 inches apart. *Cleats* or *rungs* are cut from 2 × 4 stock and inset into the edges of the side rails not more than ¾ inch. Filler blocks are sometimes used on the rails between the cleats. Cleats are uniformly spaced at 12 inches top to top.

Key Terms

bearer	double pole	guardrails	mobile scaffold
braces	extension ladder	ladder jacks	mud sill
competent person	fall protection	ledgers	poles

pump jack	**scaffolds**	**slope**	**step ladder**
roofing brackets	**single pole**	**staging planks**	**trestles**
sawhorses			

Review Questions

Select the most appropriate answer.

1. The vertical members of a scaffold are called
   a. columns.   c. poles.
   b. piers.   d. uprights.

2. Bearers support
   a. ledgers.   c. rails.
   b. planks.   d. braces.

3. Scaffold planks should be at least
   a. 2 × 6.   c. 2 × 10.
   b. 2 × 8.   d. 2 × 12.

4. Overlapped planks should extend beyond the bearer at least ___ inches and no more than ___ inches.
   a. 3, 6   c. 6, 8
   b. 3, 8   d. 6, 12

5. Metal tubular frame scaffolding is held rigidly plumb by
   a. end frames.   c. cross braces.
   b. goosers.   d. cribbing.

6. The part of a scaffold that protects workers below from falling objects is a
   a. toeboard.   c. top rail.
   b. midrail.   d. posts.

7. The workers allowed to climb an access ladder for a metal tubular scaffold that has its rungs spaced 18 inches apart are the scaffold
   a. users only.
   b. erectors only.
   c. erector and dismantlers only.
   d. all of the above.

8. Wood scaffold planks, when loaded, should deflect no more than
   a. $\frac{1}{8}$ of the span.
   b. $\frac{1}{16}$ of the span.
   c. $\frac{1}{20}$ of the span.
   d. $\frac{1}{60}$ of the span.

9. The height of a mobile scaffold must not exceed the minimum base dimension by
   a. three times.   c. five times.
   b. four times.   d. six times.

10. Guardrails must be installed on all scaffolds more than
    a. 10 feet in height.
    b. 16 feet in height.
    c. 20 feet in height.
    d. 24 feet in height.

11. Tubular scaffold end frames should be installed level and plumb to sit on top of
    a. base plates.
    b. mud sills.
    c. cribbing.
    d. all of the above.

12. The minimum number of times a scaffold should be visually inspected is
    a. two.   c. four.
    b. three.   d. five.

13. To access the work area of a scaffold, the user should use an approved ladder or
    a. the ladder built into the end frame.
    b. cross braces.
    c. the horizontal bearing points of the scaffold.
    d. all of the above.

14. All scaffolding should be able to support ___ times the intended load.
    a. four
    b. five
    c. ten
    d. twenty

15. The responsibility for safety on a jobsite relies on
    a. the local OSHA inspector.
    b. the general contractor.
    c. scaffold erectors.
    d. every worker on the job.

16. Pump jacks differ from metal tubular scaffolding in that metal tubular scaffolding
    a. often requires toeboards.
    b. is usually installed on a mud sill.
    c. has braces.
    d. none of the above.

Roof Framing

The ability to lay out rafters and frame all types of roofs is an indication of an experienced carpenter. On most jobs, the lead carpenter lays out the different rafters, while workers make duplicates of them. Those persons aspiring to supervisory positions in the construction field must know how to frame various kinds of roofs.

SAFETY REMINDER

Roof framing involves sloped and often slippery surfaces while also requiring greater mental energy to perform more complicated framing details. Maintain a constant awareness of your surroundings and what your feet are doing as you focus on roofing details.

OBJECTIVES

After completing this unit, the student should be able to:

- describe several roof types and define roof framing terms.
- describe the members of gable, gambrel, hip, intersecting, and shed roofs.
- lay out a common rafter and erect gable, gambrel, and shed roofs.
- lay out and install gable studs.
- lay out the members of and frame hip and equal-pitch intersecting roofs.
- erect a trussed roof.
- apply roof sheathing.

41 | Roof Types and Terms

oofs may be framed in a stick-built fashion using rafters and ridgeboards, or they may be framed with trusses. Trusses will be discussed later in Chapter 46. Careful thought and patience are required for a carpenter who wants to become proficient at roof framing. Knowledge of diverse roof types and the associated terms are essential to framing roofs. Carpenters demonstrate their craftsmanship when constructing a roof frame.

ROOF TYPES

Several roof styles are in common use. These roofs are described in the following material and are illustrated in **Figure 41–1.**

Gable Roof. The **gable roof** is the most common roof style. Two sloping roof surfaces meet at the top. They form triangular shapes at each end of the building called *gables.*

Shed Roof. The **shed roof** slopes in one direction. It is sometimes referred to as a *lean-to.* It is commonly used on additions to existing structures. It is also used extensively on contemporary homes.

Hip Roofs. The **hip roof** slopes upward to the ridge from all walls of the building. This style is used when the same overhang is desired all around the building. The hip roof eliminates maintenance of gable ends.

Intersecting Roof. An **intersecting roof** is required on buildings that have wings. Where two roofs intersect, valleys are formed. This requires several types of rafters.

Gambrel Roof. The **gambrel roof** is a variation of the gable roof. It has two slopes on each side instead of one. The lower slope is much steeper than the upper slope. It is framed somewhat like two separate gable roofs.

Mansard Roof. The **mansard roof** is a variation of the hip roof. It has two slopes on each of the four sides. It is framed somewhat like two separate hip roofs.

Butterfly Roof. The **butterfly roof** is an inverted gable roof. It is used on many modern homes. It resembles two shed roofs with their low ends placed against each other.

Other Roofs. Other roof styles are a combination of the styles just mentioned. The shape of the roof can be one of the most distinctive features of a building.

ROOF FRAME MEMBERS

A roof frame may consist of a ridgeboard and common, hip, or valley rafters. It may also have hip jacks, valley jacks, cripple jack rafters, collar ties, and gable end studs **(Figure 41–2).** Each of these components may be laid out and cut using similar mathematical principles and theory. Rafters are the sloping members of the roof that support the roof covering.

Ridgeboard. The **ridgeboard** is the uppermost member of a roof. Although not absolutely necessary, the ridgeboard simplifies the erection of the roof. It provides a place for the upper ends of rafters to be secured before the sheathing is applied.

Common Rafters. **Common rafters** are so named because they are the most common rafter. They make up the major portion of the roof, spanning from the ridgeboard to the wall. They extend at right angles from the plate to the ridge. They are used as a basis or starting point for all other rafters.

GABLE ROOF   HIP ROOF   INTERSECTING ROOF

GAMBREL ROOF   MANSARD ROOF   BUTTERFLY ROOF

SHED ROOF

FIGURE 41–1   Many roof styles can be used for residential buildings.

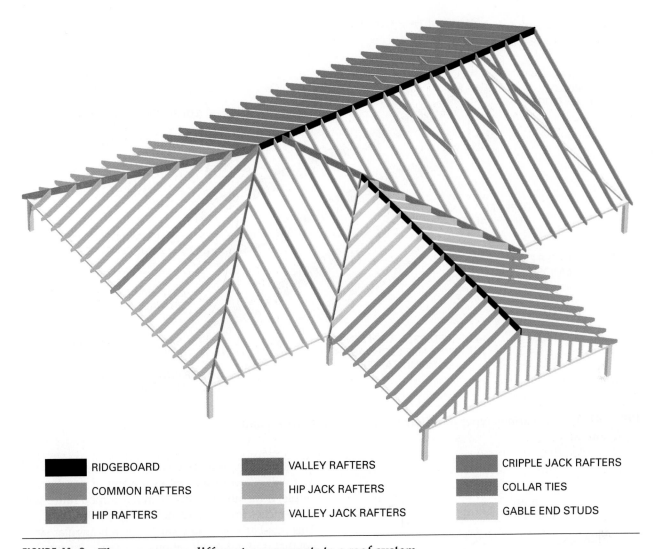

	RIDGEBOARD		VALLEY RAFTERS		CRIPPLE JACK RAFTERS
	COMMON RAFTERS		HIP JACK RAFTERS		COLLAR TIES
	HIP RAFTERS		VALLEY JACK RAFTERS		GABLE END STUDS

FIGURE 41–2 **There are many different components to a roof system.**

Hip Rafters. **Hip rafters** form an intersection of two roof sections. They project out from the roof plane forming an outside corner. They usually run at a 45-degree angle to the plates.

Valley Rafters. **Valley rafters** also create the intersection of two roof sections but project inward, forming an inside corner. Valley and hip rafters are similar in theory and layout.

Jack Rafters. **Jack rafters** come in three types: **hip jacks, valley jacks,** and **cripple jacks.** They are essentially common rafters that are cut shorter to land on a hip, valley, or both.

Collar Ties. **Collar ties** are horizontal members that add strength to the common rafter. They are raised and shortened ceiling joists.

Gable End Studs. **Gable end studs** form the wall that closes in the triangular wall area under a gable roof.

Understanding rafters can be confusing at first, and it may take some time to become comfortable with rafters. The various types have similarities and differences that make them unique **(Figure 41–3).** All rafters except the valley jack start from the wall plate. All rafters except the hip jack land on the ridge. Hip and valley rafters have longer lengths and tails because they are not parallel to the other rafters. They run at an angle from the plate to the ridge.

ROOF FRAMING TERMS

Roof framing theory has many terms. Successful roof construction begins with understanding these terms. They are defined as follows and most are illustrated in **Figure 41–4.** The unit run is always a fixed number. A carpenter can always expect the total span and unit rise to be given. The others terms and measurements are marked with a square or calculated. These terms will also be discussed in more detail in later chapters.

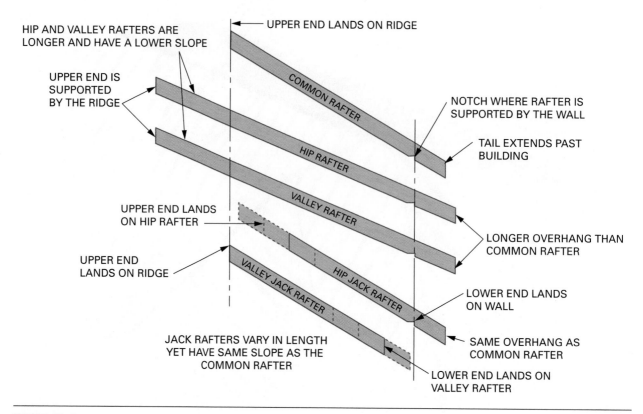

HIP AND VALLEY RAFTERS ARE LONGER AND HAVE A LOWER SLOPE

UPPER END LANDS ON RIDGE

UPPER END IS SUPPORTED BY THE RIDGE

COMMON RAFTER

NOTCH WHERE RAFTER IS SUPPORTED BY THE WALL

HIP RAFTER

TAIL EXTENDS PAST BUILDING

VALLEY RAFTER

UPPER END LANDS ON HIP RAFTER

LONGER OVERHANG THAN COMMON RAFTER

UPPER END LANDS ON RIDGE

VALLEY JACK RAFTER

HIP JACK RAFTER

LOWER END LANDS ON WALL

JACK RAFTERS VARY IN LENGTH YET HAVE SAME SLOPE AS THE COMMON RAFTER

SAME OVERHANG AS COMMON RAFTER

LOWER END LANDS ON VALLEY RAFTER

FIGURE 41–3 The various types of rafters have many similarities and differences.

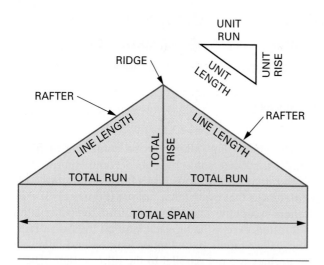

UNIT RUN

UNIT RISE

UNIT LENGTH

RIDGE

RAFTER

RAFTER

LINE LENGTH

LINE LENGTH

TOTAL RISE

TOTAL RUN

TOTAL RUN

TOTAL SPAN

FIGURE 41–4 The terms associated with roof theory.

Unit Run. The **unit run** is a standardized horizontal distance. It is the number that is used as a base for the roof angle, and it is a horizontal distance under a rafter. This distance is always 12 inches (or 200 mm in metric measure) for a common rafter. It is 16.97 inches (or 283 mm) for hip and valley rafters.

Total Span. The **total span** of a roof is the horizontal distance covered by the roof. This is usually the width of the building measured from the outer faces of the framing.

Unit Rise. The **unit rise** is the number of inches the roof will rise vertically for every unit of run. For example, if the unit rise is 6 then a common rafter will rise 6 inches for every 12 inches it covers horizontally. This number is typically shown on a triangular symbol that is found on elevations and section views.

Total Run. The **total run** of a rafter is the total horizontal distance over which the rafter slopes. This is usually one-half the span of the roof.

Total Rise. The **total rise** is the vertical distance that the roof rises from plate to ridge. Total rise may be calculated by multiplying the unit rise by the total run of the rafter. For example, if the unit rise is 6 and the run is 13, then the total rise is $6 \times 13 = 78$ inches. This measurement is not to the top of the rafter, but rather to some point inside the rafter.

Theoretical Line Length. The **line length** of a rafter is the length of the rafter from the plate to the ridge. It is the hypotenuse (longest side) of the right triangle formed by the total run as the base and the total rise as the vertical leg. Line length gives no consideration to the thickness or width of the framing stock.

Unit Length. **Unit length** is the length of rafter needed to cover a horizontal distance of one unit of run. It is the hypotenuse of a right triangle formed by the unit run and unit rise.

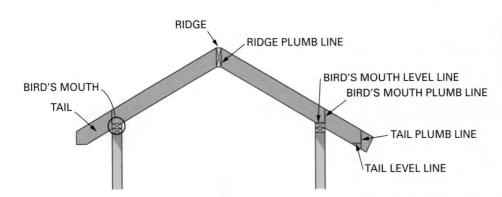

RIDGE
RIDGE PLUMB LINE
BIRD'S MOUTH LEVEL LINE
BIRD'S MOUTH PLUMB LINE
BIRD'S MOUTH
TAIL
TAIL PLUMB LINE
TAIL LEVEL LINE

FIGURE 41–5 A bird's mouth is a notch for the rafter to sit on the wall.

Pitch. The **pitch** of a roof is a fraction. It indicates the amount of incline of a roof. The pitch is found by dividing the rise by the span. For example, if the span of the building is 32 feet and total rise is 8 feet, then 8 divided by 32 is $\frac{8}{32}$, which reduces to ¼. The roof is then said to have a ¼ pitch.

Slope. **Slope** is the common term used to express the steepness of a roof. It is stated using the unit rise and the unit run. For example, if the unit rise is 4 inches and the unit run is always 12 inches, then the slope is said to be 4 on 12.

Bird's Mouth. **Bird's mouth** is the term used to refer to the notch cut in a rafter. This is done to make the rafter sit securely on the wall plate so that it can be adequately fastened **(Figure 41–5)**.

Plumb Line. A **plumb line** is any line on the rafter that is vertical when the rafter is in position. There is a plumb line at the ridge, at the wall plate, and usually at the end of the tail. They are marked using a square. A framing square is marked along the tongue. A speed square is marked along the edge of the square where the inch ruler is located.

Level Line. A **level line** is any line on the rafter that is horizontal when the rafter is in position. It is marked along the blade of the framing square when laying out level cuts on rafters. A speed square is marked along the long edge of the square where the degree scale is located after lining up the alignment guide with the plumb line.

DETERMINING RAFTER LENGTHS

Three methods are available for finding the length of a rafter: the estimated, step-off, and calculated methods. They are used for different reasons and have varying degrees of speed and accuracy.

Estimated Rafter Length

With the estimated method, the carpenter uses a framing square and a tape measure to scale a dis-

FIGURE 41–6 Scaling the rough length of a common rafter.

tance that the rafter will cover. It serves only as an estimate for ordering the proper length material **(Figure 41–6)**.

The blade of the square represents the total run of the rafter. The tongue represents the total rise. Using a scale of one inch to one foot, locate the run and rise. For example, a run of 10 feet becomes 10 inches and a rise of 5 feet becomes 5 inches. Then measure the shortest distance between the blade and tongue. This distance in inches is easily converted to feet. Add extra if the rafter overhangs the wall of the building.

EXAMPLE If the span of a building is 28 feet and the unit rise is 6 inches, what is the estimated length? The run of the rafter is $28 \div 2 = 14$. The total rise of a rafter is $14 \times 6 = 84$ inches. Divide this by 12 to convert it to feet, $84 \div 12 = 7$ feet. Locate seven inches on the tongue of the square and 14 inches on the blade. Measure the distance between these points. This is approximately 15⅝ inches. Converting back to feet, the rough rafter length from plate to ridge is 15⅝ feet. A stock length of 16 feet would have to be used for the rafter with no overhang.

Step-Off Rafter Length Method

The step-off method uses a framing square to step off the length of a rafter. It can be used for most types of rafters. The step-off method is based on the unit of run (12 inches for the common rafter). The rafter stock is stepped off for each unit of run until the desired number of units or parts of units are stepped-off **(Figure 41–7)**. If the rafter has a total run of 16 feet, for example, the square is moved 16 times. This method works well in most situations but it can cause errors in the length because small incremental errors that add up occur during each step. More on the step-off method in Chapter 42.

Calculated Rafter Length

The calculated rafter length method uses the rafter tables located on a framing square and a calculator. It is the most accurate way of determining rafter length. The procedure can be confusing at first, but once it is understood, any and all rafter lengths can be determined. This includes hip, valley, and jack rafters. The following chapters of this book will focus mainly on the calculated method.

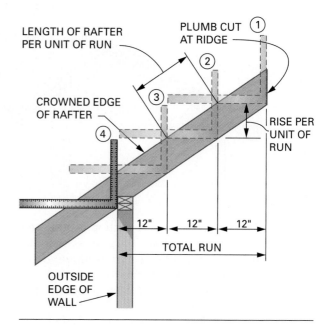

FIGURE 41–7   **Step-off method of determining rafter length.**

---

42   Gable and Gambrel Roofs

The equal-pitched gable roof is the most commonly used roof style. Gable roofs have an equal slope on both sides intersecting the ridge in the center of the span **(Figure 42–1)**. This roof is the simplest to frame. Only one type of rafter, the common rafter, needs to be laid out.

Like the gable roof, the gambrel roof is symmetrical (see Figure 42–1). Each side of the building has

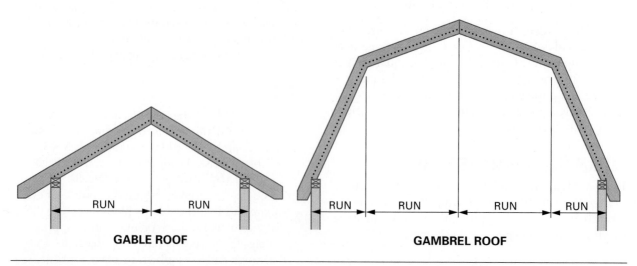

FIGURE 42–1   **The gable and gambrel roofs have opposing rafters that are symmetrical.**

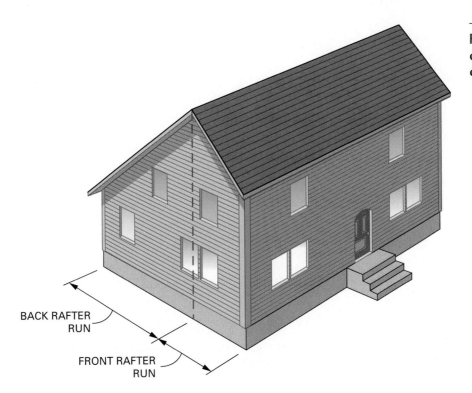

FIGURE 42–2 The ridge of the saltbox roof is off-center.

rafters with two different slopes. They have different unit rises. Their runs may or may not be the same depending on the desired roof style.

The saltbox is a gable roof with rafters having different slopes on each side of the ridge **(Figure 42–2).** One rafter slopes faster to reach to the ridge from a lower floor.

COMPONENTS OF A COMMON RAFTER

The common rafter requires several cuts, which have to be laid out. The cuts and lines of a rafter have several names **(Figure 42–3).** The cut at the top is called the *plumb cut* or *ridge cut*. It fits against the ridgeboard. The notch cut at the plate is called the *bird's mouth, seat cut,* or *heel.* It fits against the top and outside edge of the wall plate. It consists of a plumb line and a level line layout. These lines are named according to the notch name used. For example, if the notch is referred to as a seat cut, then it is laid out with a seat plumb line and a seat level line.

The distance between the ridge cut and the seat cut is referred to as the *rafter length.* It is also called the line length or theoretical line length of the rafter. This distance must be determined by the carpenter.

At the bottom end of the rafter is the tail or overhang, which extends beyond the building. It supports the **fascia** and **soffit.** The plumb line is called the *tail plumb line* or *fascia cut.* The *tail level line* is also referred to as the *soffit cut.*

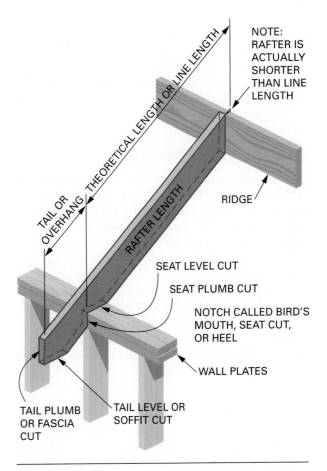

FIGURE 42–3 Names of the cuts and lines of a common rafter.

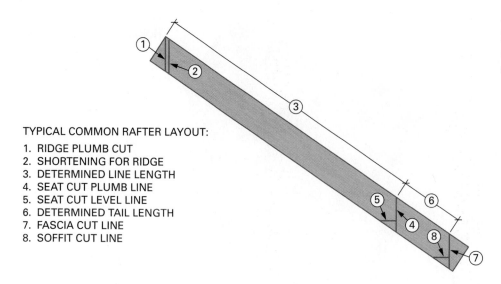

<figure id="2" />

FIGURE 42–4 Rafters are laid out in a stepwise fashion.

TYPICAL COMMON RAFTER LAYOUT:

1. RIDGE PLUMB CUT
2. SHORTENING FOR RIDGE
3. DETERMINED LINE LENGTH
4. SEAT CUT PLUMB LINE
5. SEAT CUT LEVEL LINE
6. DETERMINED TAIL LENGTH
7. FASCIA CUT LINE
8. SOFFIT CUT LINE

COMMON RAFTER LAYOUT

Even though rafters may look different, they are laid out using a similar process. The principles for framing rafters, whether equally or unequally sloped, common or hip, are the same. Layout follows a logical flow **(Figure 42–4)**. The ridge cut is marked and the rafter length is adjusted for the ridge thickness. The line length is determined and a plumb line is made for the seat cut. A level line is drawn to complete the bird's mouth. The tail length is determined and a fascia line is drawn. The layout is finished with a soffit cut.

COMMON RAFTER PATTERN

Because common rafters are all cut the same way, only one layout is required. Once a rafter is laid out and cut, it becomes the pattern for the rest of the rafters. It should be tested and verified to fit before cutting additional rafters.

Select the straightest piece of stock for the pattern. Lay it across two sawhorses. Sight the stock along the edge for straightness and mark the crowned edge. This edge will become the top side where the roof sheathing is attached. Further discussion of rafters will assume the carpenter is standing near the rafter's lower edge, opposite of the crowned edge.

Laying Out the Ridge Cut

Place the square down on the side of the stock at its left end. Hold the tongue of the square with the left hand and the blade with the right hand. Move the square until the outside edge of the tongue and the edge of the stock line up with the specified rise in inches.

Make sure the blade of the square and the edge of the stock line up with the unit run or 12 inches. Slide the square to the left until it reaches the end of the rafter stock. Recheck the alignment and mark along the outside edge of the tongue for the plumb cut at the ridge **(Figure 42–5)**.

When using a speed square, place the pivot point of the square on the top edge of the rafter. Rotate the square with the pivot point touching the rafter. Looking in the rafter scale window, align the edge of the rafter with the number that corresponds to the rise per unit of run desired. Note that there are two scales, one for a common rafter and one for hips and valleys. Mark the plumb line along the edge of the square that has the inch ruler marked on it.

FIGURE 42–5 Laying out the plumb cut of the common rafter at the ridge.

EXAMPLE Assume the slope of the roof is 6 on 12. Hold the square so the 6-inch mark on the tongue and the 12-inch mark on the blade line up with the top edge of the rafter stock. Mark along the outside edge of the tongue of the square. Make a second plumb line for practice using a speed square. Place the speed square on the rafter stock and rotate the square, keeping the pivot point touching the edge of the rafter. Align the edge of the rafter with the number 6 on the common scale in the rafter scale window. Mark the plumb line along the edge of the square that has the ruler marked on it. Both layouts should look like that shown in Figure 42–4.

Rafter Shortening Due to Ridge

The length of the rafter must be shortened when a ridgeboard is inserted between abutting rafters **(Procedure 42–A).** The total shortening is equal to the thickness of the ridgeboard, thus each rafter will be shortened one-half the ridgeboard's thickness.

To shorten the rafter, measure at a right angle from the ridge plumb line a distance equal to one-half the thickness of the ridgeboard. Lay out another plumb line at this point. This will be the cut line. Note that shortening is always measured at right angles to the ridge cut, regardless of the slope of the roof.

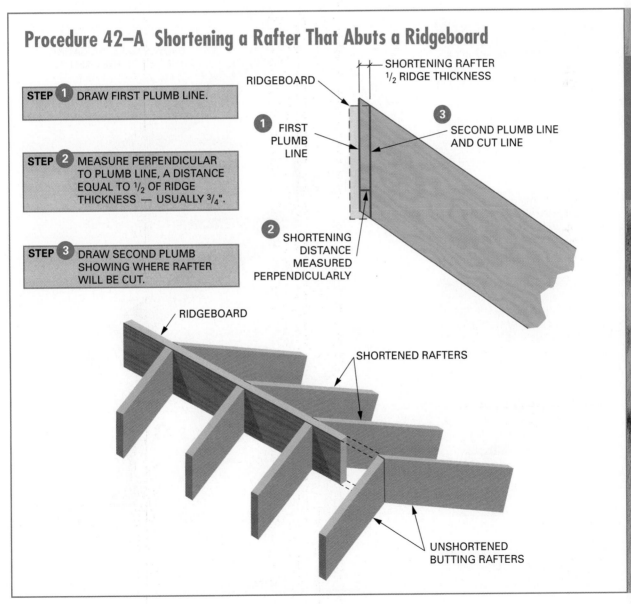

Procedure 42–A Shortening a Rafter That Abuts a Ridgeboard

STEP 1 DRAW FIRST PLUMB LINE.

STEP 2 MEASURE PERPENDICULAR TO PLUMB LINE, A DISTANCE EQUAL TO 1/2 OF RIDGE THICKNESS — USUALLY 3/4".

STEP 3 DRAW SECOND PLUMB SHOWING WHERE RAFTER WILL BE CUT.

RIDGEBOARD

SHORTENING RAFTER 1/2 RIDGE THICKNESS

1 FIRST PLUMB LINE

3 SECOND PLUMB LINE AND CUT LINE

2 SHORTENING DISTANCE MEASURED PERPENDICULARLY

RIDGEBOARD

SHORTENED RAFTERS

UNSHORTENED BUTTING RAFTERS

STEP BY STEP PROCEDURES

COMMON RAFTER LENGTH

The length of a common rafter can be determined in two ways, via the step-off or calculation method. The step-off method requires close and fine measuring as each increment is made. The calculation method uses a calculator and the **rafter tables** found on a rafter square. Both methods can be used to determine the length of any kind of a rafter, but the calculation method is faster and more accurate.

Step-Off Method

The step-off method is based on the unit of run (12 inches or 200 mm for the common rafter). The rafter stock is stepped off for each unit of run until

Procedure 42–B Using the Step-Off Method for Rafter Layout

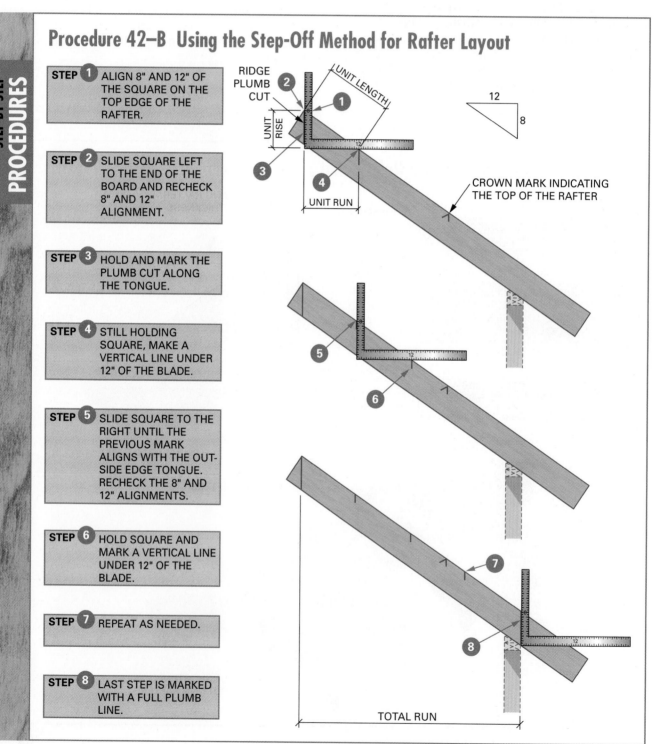

STEP 1 ALIGN 8" AND 12" OF THE SQUARE ON THE TOP EDGE OF THE RAFTER.

STEP 2 SLIDE SQUARE LEFT TO THE END OF THE BOARD AND RECHECK 8" AND 12" ALIGNMENT.

STEP 3 HOLD AND MARK THE PLUMB CUT ALONG THE TONGUE.

STEP 4 STILL HOLDING SQUARE, MAKE A VERTICAL LINE UNDER 12" OF THE BLADE.

STEP 5 SLIDE SQUARE TO THE RIGHT UNTIL THE PREVIOUS MARK ALIGNS WITH THE OUTSIDE EDGE TONGUE. RECHECK THE 8" AND 12" ALIGNMENTS.

STEP 6 HOLD SQUARE AND MARK A VERTICAL LINE UNDER 12" OF THE BLADE.

STEP 7 REPEAT AS NEEDED.

STEP 8 LAST STEP IS MARKED WITH A FULL PLUMB LINE.

RIDGE PLUMB CUT

UNIT LENGTH

UNIT RISE

UNIT RUN

CROWN MARK INDICATING THE TOP OF THE RAFTER

12

8

TOTAL RUN

the desired number of units or parts of units have been stepped off.

First, lay out the ridge cut. Keeping the square in place, mark where the blade intersects with the top edge of the rafter. Hold the square at the same angle and slide it to the right until the tongue lines up with this mark. Move the square in a similar manner until the total run of the rafter is laid out **(Procedure 42–B).** Mark a plumb line along the tongue of the square at the last step. This line will be parallel to the ridge cut.

Fractional Step-Off. First, step off for the total whole units of run. Move the square as if to step off one more time. Instead of marking 12 inches on the blade of the square, measure the fractional part of the unit of run along the blade. Mark it on the rafter. Holding the square in the same position, move it so the outside of the tongue lines up with the mark. Lay out a plumb line along the tongue of the square **(Figure 42–6).**

> **EXAMPLE** If the rafter has a total run of 16 feet, 7 inches, step off 16 times. Hold the square for the 17th step. Mark along the blade at the 7-inch mark. With the square held at the same angle, move it to the mark. Lay out the plumb cut of the bird's mouth by marking along the tongue.

Framing Square Gauges. *Framing square* gauges can be attached to the square. They are used as stops against the top edge of the rafter. These gauges are attached to the tongue of the square for the desired

FIGURE 42–7 Framing square gauges are attached to the square to hold it in the same position for easy alignment.

rise per foot of run and to the blade at the unit of run **(Figure 42–7).** They speed up the alignment part of the step-off process and greatly increase the accuracy of the overall layout.

Calculation Method

The calculation method uses the idea that each step of the step-off method has a unit length. Unit length is the hypotenuse of the right triangle included in each step-off **(Figure 42–8).** Because the total run is the number of step-offs, the rafter length is found by multiplying total run by unit length. For example, if the unit length is 13 inches and the number of step-offs is 4 then the rafter length is $13 \times 4 = 52$ inches.

This may seem strange to multiply inches times feet. Normally in arithmetic, this will give a result that has no meaning. But in this case the run is 4, a number of units, not 4 feet. Total run must always look like feet but it is always simply the number of unit runs (step-offs) under the rafter.

Rafter tables provide rafter information in six rows for slopes of 2 through 18 inches of unit rise. The table is stamped on the blade of a rafter (framing) square or sometimes printed in small booklets that come with the square **(Figure 42–9).** The inch marks 2 through 18 represent the unit rises available with the information directly below each number.

The first row of the rafter table is the common rafter length per unit run or, simply stated, unit length. This number comes from using the Pythagorean theorem, $a^2 + b^2 = c^2$ on the unit triangle. The unit rise and unit run are the a and b with the unit length being c. All unit lengths are calculated this way. Knowing this, the unit lengths of usual unit rises, such as 6 ¾ inches, can be calculated.

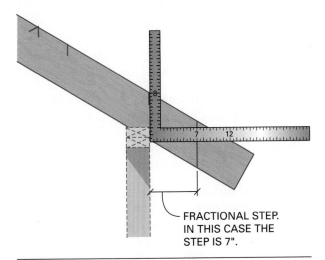

FRACTIONAL STEP. IN THIS CASE THE STEP IS 7".

FIGURE 42–6 Laying out a fractional part of a run for a common rafter.

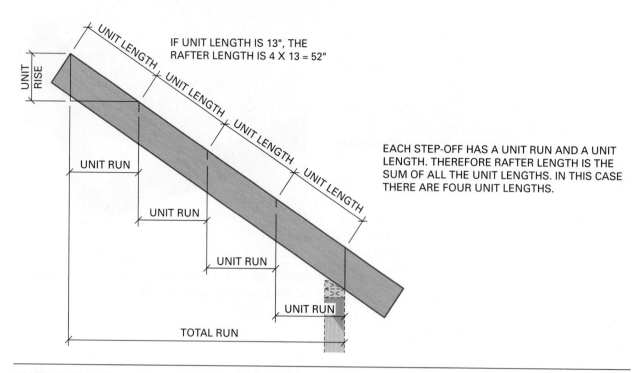

FIGURE 42–8 Unit length is the hypotenuse of the right triangle included in each step-off.

FIGURE 42–9 Rafter tables are found on the framing square.

EXAMPLE Suppose the unit rise of a rafter is 7 inches. What is the unit length? Unit run is always 12 inches for a common rafter.

$$a^2 + b^2 = c^2$$
$$7^2 + 12^2 = c^2$$
$$49 + 144 = 193 = c^2$$
$$c = \sqrt{193} = 13.892443 \text{ inches}$$

Unit length = 13.89 inches rounded to the nearest hundredths.

Using Rafter Tables. The unit length for a rafter with a unit rise of 7 is found on the rafter tables. First, locate which line of the table will have the unit lengths for a common rafter. They are placed in the top line of the rafter table. Looking at the first row under the 7-inch mark on the edge of the square will show the number 13 89 or 13.89 inches. The space represents a decimal point.

EXAMPLE: **Example of Using the Calculation Method to Determine Common Rafter Line Length**

Find the line length of a common rafter with a unit rise of 8 inches for a 28-foot-wide building.

STEP 1: Read below the 8-inch mark in the first row to find 14 42 or 14.42 inches unit length.

STEP 2: Divide 28 by 2 to determine the run.
$$28 \div 2 = 14 \text{ units of run.}$$

STEP 3: Multiply unit length times total run.
$$14.42 \times 14 = 201.88 \text{ inches.}$$
That is 201 inches plus a fraction.

STEP 4: Convert decimal to sixteenths.
$$0.88 \times 16 = 14.08$$

STEP 5: Round off to nearest whole sixteenths.
$14.08 \Rightarrow {}^{14}\!/_{16}$, which reduces to $\frac{7}{8}$ inches.

STEP 6: Add to determine rafter length.
$$201 + \frac{7}{8}'' = 201\frac{7}{8}''$$

Note that the calculated line length must be measured from the first ridge plumb line. If it is located at the end of the rafter, a tape may be hooked on the end of the board. Measure from the first ridge cut along the top edge of the rafter, the length of the rafter **(Figure 42–10).** The measurement mark should be on the edge of the rafter. Mark the plumb line for the seat.

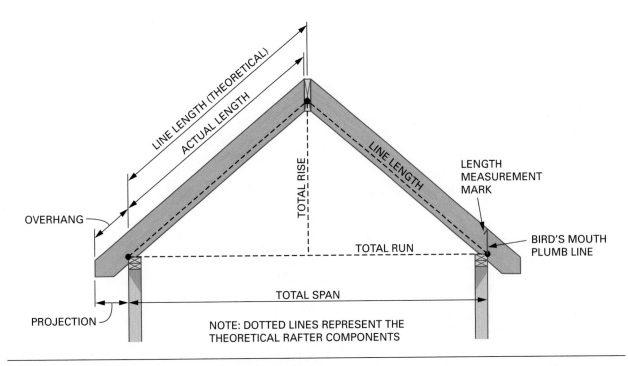

FIGURE 42–10 The line length is theoretical and measured from the first plumb line drawn.

Mark a plumb cut at the ridge. From the ridge cut, measure along the top edge of the rafter, the length of the rafter as determined by calculations. Mark the length and make a plumb line at the seat. At this point, the rafter length is theoretical because the rafter must be shortened due to the thickness of the ridge (see Figure 42–10).

Seat Cut Layout

The **seat cut** or *bird's mouth* of the rafter is a combination of a level cut and a plumb cut. The level cut rests on top of the wall plate. The plumb cut fits snugly against the outside edge of the wall.

The seat cut level line is not a precise measurement. The main concern is that it be perpendicular to the plumb line and not cut too deeply. Position the level so there is sufficient stock, about two-thirds of the rafter remaining above the seat cut, to ensure there is enough strength to support any overhang. Typically, the seat level line is 2 to 4 inches long. This line should not be longer than the width of the wall plate.

To mark the seat level line, hold the framing square in the same manner as for marking plumb lines. Slide it along the stock until the blade lines up with the plumb line. Mark the appropriate level line along the blade.

With a speed square, hold the alignment guide of the square in line with the plumb line previously drawn. Mark along the long edge of the square to achieve level lines for seat cuts **(Figure 42–11)**. On roofs with moderate slopes, the length of the level cut of the seat is usually the width of the wall plate. For steep roofs, the level cut is shorter.

Sometimes it is desirable to cut the seat cut to allow for wall sheathing **(Figure 42–12)**. Extend the level line a distance equal to the wall sheathing thickness. Draw a new plumb line down from the end of the line.

Common Rafter Tails

The **tail cut** is the cut at the end of the rafter tail. It may be a plumb cut, a combination of cuts, or a square cut **(Figure 42–13)**. Sometimes the rafter tails are left to run wild. This means they are slightly longer than needed. They are cut off later after the roof frame is erected.

On most plans the projection is given, not the overhang. The projection is a level measurement, whereas the overhang is a sloping measurement along the rafter.

Stepping-Off Projection.
To mark the fascia cut, place a framing square on the rafter to mark a plumb cut. Align the square using the unit rise on the tongue and unit run on the blade. Larger projections will require the square to be moved to the lower edge of the rafter. **Figure 42–14** shows a fascia cut example when the unit rise is 8 inches and the projection is 18 inches.

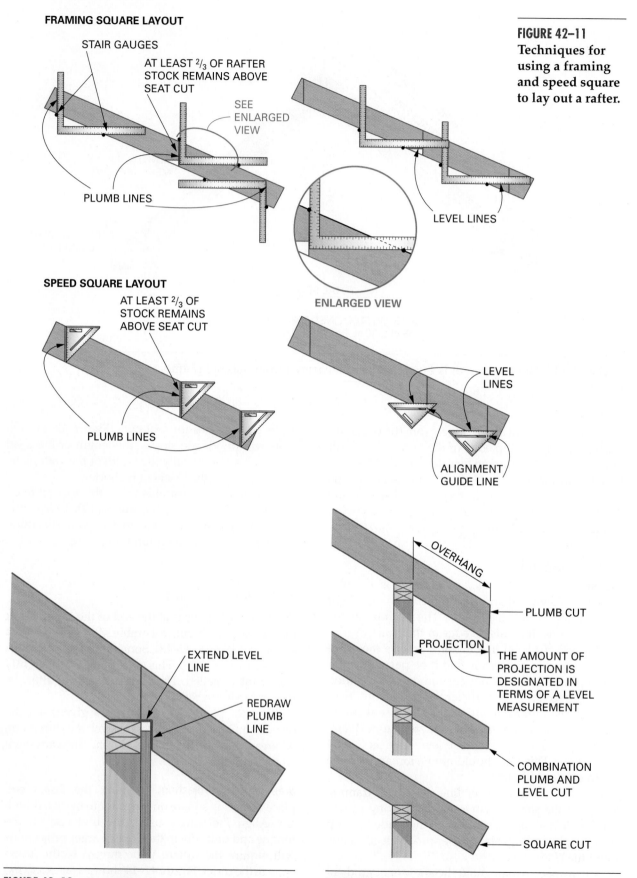

FRAMING SQUARE LAYOUT

STAIR GAUGES

AT LEAST ²/₃ OF RAFTER STOCK REMAINS ABOVE SEAT CUT

SEE ENLARGED VIEW

PLUMB LINES

LEVEL LINES

ENLARGED VIEW

SPEED SQUARE LAYOUT

AT LEAST ²/₃ OF STOCK REMAINS ABOVE SEAT CUT

PLUMB LINES

LEVEL LINES

ALIGNMENT GUIDE LINE

FIGURE 42–11 Techniques for using a framing and speed square to lay out a rafter.

EXTEND LEVEL LINE

REDRAW PLUMB LINE

OVERHANG

PLUMB CUT

PROJECTION

THE AMOUNT OF PROJECTION IS DESIGNATED IN TERMS OF A LEVEL MEASUREMENT

COMBINATION PLUMB AND LEVEL CUT

SQUARE CUT

FIGURE 42–12 Sometimes seat cuts are made deeper to allow for wall sheathing.

FIGURE 42–13 Various tail cut styles for the common rafter.

OVERHANG VS. PROJECTION

STEP-OFF METHOD

8 AND 12 ALIGNED
ON THE LOWER
EDGE OF RAFTER

21⁵⁄₈ " OVERHANG

18"
PROJECTION

CALCULATION METHOD

OVERHANG LENGTH
STEP 1 18" ÷ 2 = 1.5 UNITS OF RUN
STEP 2 8" UNIT RISE HAS UNIT
        LENGTH OF 14.42"
STEP 3 UNIT LENGTH X RUN = RAFTER
        LENGTH
STEP 4 1.5 X 14.42" = 21.63"
STEP 5 0.63 X 16 = 10.08 SIXTEENTHS
        = ¹⁰⁄₁₆ = ⁵⁄₈
STEP 6 OVERHANG = 21⁵⁄₈"

FIGURE 42–14 Laying out the tail of the common rafter using the projection or overhand calculation.

Calculating Overhang. The overhang may also be calculated and measured using the same formula for determining rafter length: unit length times run. Convert the projection to a run by changing inches to feet. In this case, 18 inches is 1.5 units of run and the unit length is 14.42 inches (see Figure 42–14). Thus the overhang measurement is 1.5 × 14.42 = 21.63 inches or 21⅝ inches. This number is measured along the rafter edge.

Soffit Cut. The soffit cut line varies according to the finish material. It should be made so the plumb line is long enough to adequately support the fascia **(Figure 42–15).**

Cutting Wild Rafter Tails. To cut wild rafter tails after the rafters are installed, measure and mark the two rafters on either end for the amount of overhang. This is usually a level measurement from the outside of the wall studs to the tail plumb line **(Procedure 42–C).** Plumb the marks up to the top edge of the rafters with a level. Stretch a line across the top edges of all the rafters. Using a T-bevel plumb down on the side of each rafter from the chalk line. Use a circular saw to cut each rafter. Cutting down along the line is the easiest way to make the cut.

If the tail cut is a combination of plumb and level cuts, make the plumb cuts first. Then snap a line across the cut ends as desired. Level each soffit cut in from the chalk line. Working from the outside toward the wall is the easiest way to accomplish this.

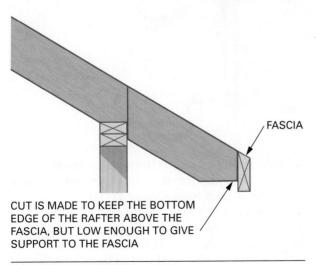

CUT IS MADE TO KEEP THE BOTTOM
EDGE OF THE RAFTER ABOVE THE
FASCIA, BUT LOW ENOUGH TO GIVE
SUPPORT TO THE FASCIA

FASCIA

FIGURE 42–15 Soffit cuts are positioned to allow for adequate support of fascia.

⚠ **CAUTION   CAUTION   CAUTION   CAUTION**

On some cuts that are at a sharp angle with the edge of the stock (i.e., the level cut of the seat), the guard of the circular saw may not retract. In this case, retract the guard by hand until the cut is made a few inches into the stock. Never wedge the guard in an open position to overcome this difficulty. ■

Wood I-Joist Roof Details

In addition to solid lumber, wood I-joists may be used for rafters **(Figure 42–16).** Layout is the same for both dimension lumber and wood I-joists. Some roof framing details and variations for wood I-joists

Procedure 42–C Optional Method for Cutting Rafter Tails

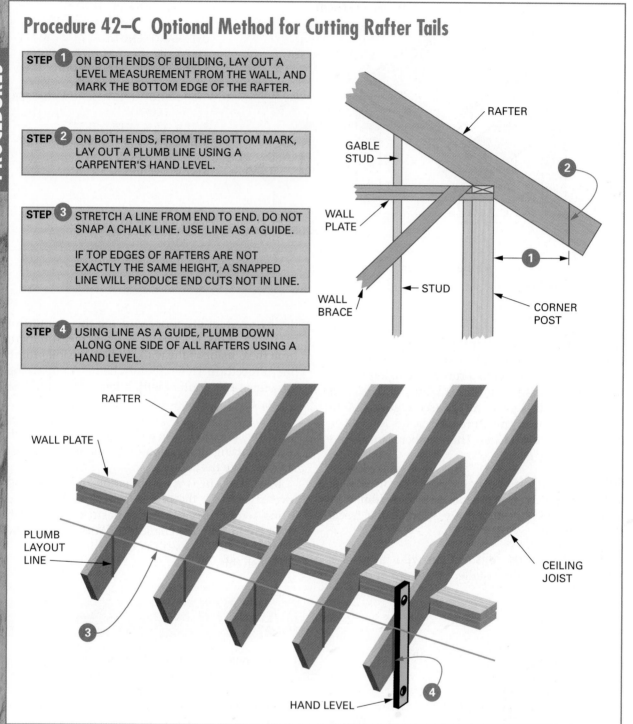

STEP 1 ON BOTH ENDS OF BUILDING, LAY OUT A LEVEL MEASUREMENT FROM THE WALL, AND MARK THE BOTTOM EDGE OF THE RAFTER.

STEP 2 ON BOTH ENDS, FROM THE BOTTOM MARK, LAY OUT A PLUMB LINE USING A CARPENTER'S HAND LEVEL.

STEP 3 STRETCH A LINE FROM END TO END. DO NOT SNAP A CHALK LINE. USE LINE AS A GUIDE.

IF TOP EDGES OF RAFTERS ARE NOT EXACTLY THE SAME HEIGHT, A SNAPPED LINE WILL PRODUCE END CUTS NOT IN LINE.

STEP 4 USING LINE AS A GUIDE, PLUMB DOWN ALONG ONE SIDE OF ALL RAFTERS USING A HAND LEVEL.

RAFTER

GABLE STUD

WALL PLATE

WALL BRACE

STUD

CORNER POST

RAFTER

WALL PLATE

PLUMB LAYOUT LINE

CEILING JOIST

HAND LEVEL

are shown in **Figure 42–17.** To determine sizes for various spans and more specific information, consult the manufacturer's literature.

LAYING OUT THE RIDGEBOARD

The ridgeboard must be laid out with the same spacing as was used on the wall plate. Use only the straightest lumber for the ridge. The total length of

the ridgeboard should be the same as the length of the building plus the necessary amount of gable overhang on both ends **(Figure 42–18).**

From the end of the ridgeboard, measure and mark the required amount of gable end overhang. This distance will vary with the style of the house. The second rafter is nailed on this line. Continue the layout of remaining rafters, marking on both sides of the ridgeboard.

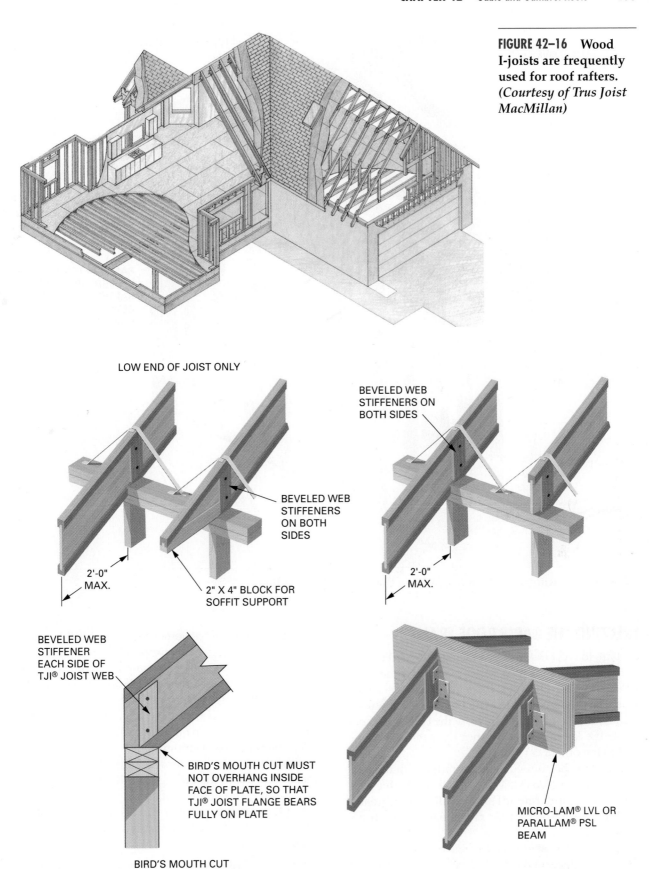

FIGURE 42–16 Wood I-joists are frequently used for roof rafters. *(Courtesy of Trus Joist MacMillan)*

LOW END OF JOIST ONLY

BEVELED WEB STIFFENERS ON BOTH SIDES

BEVELED WEB STIFFENERS ON BOTH SIDES

2'-0" MAX.

2" X 4" BLOCK FOR SOFFIT SUPPORT

2'-0" MAX.

BEVELED WEB STIFFENER EACH SIDE OF TJI® JOIST WEB

BIRD'S MOUTH CUT MUST NOT OVERHANG INSIDE FACE OF PLATE, SO THAT TJI® JOIST FLANGE BEARS FULLY ON PLATE

BIRD'S MOUTH CUT

MICRO-LAM® LVL OR PARALLAM® PSL BEAM

FIGURE 42–17 Some wood I-joist roof framing details. *(Courtesy of Trus Joist MacMillan)*

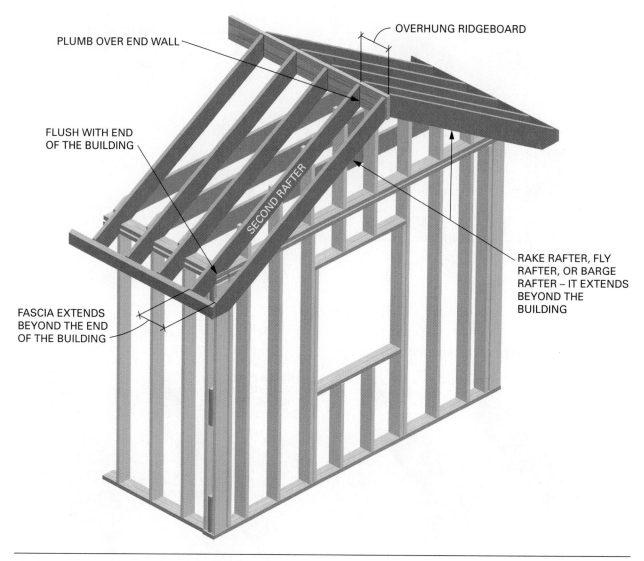

OVERHUNG RIDGEBOARD

PLUMB OVER END WALL

FLUSH WITH END
OF THE BUILDING

SECOND RAFTER

RAKE RAFTER, FLY
RAFTER, OR BARGE
RAFTER – IT EXTENDS
BEYOND THE
BUILDING

FASCIA EXTENDS
BEYOND THE END
OF THE BUILDING

FIGURE 42–18   Ridgeboards may overhang the end of the building.

ERECTING THE GABLE ROOF FRAME

Prepare to erect the roof by placing plywood on top of the ceiling joists to serve as a safe work surface **(Procedure 42–D).** Tack them in place for safety. Place the ridgeboard on top in the direction it will be installed. Take care not to reverse the ridgeboard because the layout is usually not the same from one end to the other.

Align the bottom edge of the rafter with the bottom of the ridgeboard. This will allow greater support of the entire rafter. This also allows for greater airflow over the top of the ridgeboard through the ridge vent.

Erection is most efficiently done by three workers, one at the ridge and one each at the bearing walls. Nail two rafters to the ridge at each end of the ridge. Raise and hold the ridge with rafters attached in position as the lower ends are nailed at the bird's mouth.

A temporary brace sometimes is helpful to support the ridge and two rafters while nailing takes place. Position and nail opposing pairs of rafters to complete the outline of the roof frame. Next, plumb and brace the ridge to the end wall. Attach the brace to the ridge and load-bearing wall or end wall. Fill in remaining rafters in opposing pairs. This will help keep the ridgeboard straight.

Rafters should be toenailed into the plate and faced nailed into the ceiling joist. Roof slope, width of building, and amount of snow or wind load will affect the number and size of nails needed to fasten the ceiling joist/rafter connection. Check local code requirements.

Collar Ties.   Collar ties are horizontal members fastened to opposing pairs of rafters, which effectively reduce the span of the rafter. As a load is placed on one side of the roof, it is transferred to the other side

Procedure 42–D Assembling Gable Rafters

STEP 1 PLACE PLYWOOD ON CEILING JOISTS FOR A SAFE WORK SURFACE.

STEP 2 POSITION RIDGEBOARD ON THE WORK SURFACE.

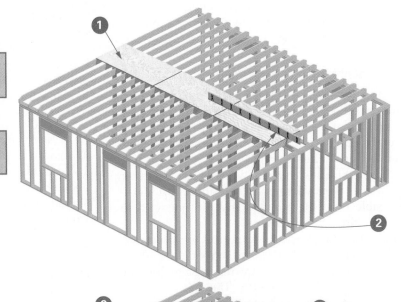

STEP 3 NAIL TWO RAFTERS TO THE RIDGEBOARD.

STEP 4 RAISE THE RIDGE AND RAFTERS INTO POSITION AND NAIL AT THE BIRD'S MOUTH.

STEP 5 NAIL TEMPORARY POSTS TO RIDGE IF DESIRED.

STEP 6 RAISE AND NAIL THE OPPOSING RAFTER PAIRS INTO POSITION.

STEP 7 PLUMB THE RIDGE OVER THE END WALL. BRACE RIDGE TO THE WALL. FILL IN THE REMAINING RAFTERS IN PAIRS TO KEEP THE RIDGEBOARD STRAIGHT.

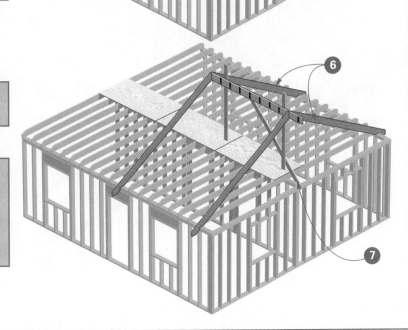

through the collar tie. In effect, the load on one rafter supports the load on the other rafter and vice versa.

Install collar ties to every third rafter pair or as required by drawings or codes. The length of a collar tie varies, but they are usually about one-third to one-half of the building span.

Constructing a Rake Overhang

The first and last rafters of a gable roof are called **rake rafters,** *fly rafters,* or *barge rafters.* They support the finish material. They may be constructed using several methods, depending on the amount of gable overhang and rafter length.

They should always be made using straight rafter pieces since they will form the roof edge. This is easily visible from the ground. They do not need to have the bird's mouth cut out.

Rake rafters are often supported by **lookouts.** Lookouts are framed in before the roof is sheathed. One method uses 2 × 4s on the flat notched into the second rafter **(Figure 42–19).** The other lookout method uses rafter width material and is strongest. This method looks similar to a cantilevered joist system. Lookouts may be spaced 16 to 32 inches on center (OC) according to the desired strength.

INSTALLING GABLE STUDS

The triangular areas formed by the rake rafters and the wall plate at the ends of the building are called **gables.** They must be framed with studs. These studs are called **gable studs.** The bottom ends are cut square. They fit against the top of the wall plate. The top ends fit snugly against the bottom edge and inside face of the end rafter. They are cut off flush with or below the top edge of the rafter **(Figure 42–20).**

Laying Out Gable Studs

The end wall plate is laid out for the location of the gable studs. Studs should be positioned directly above the end wall studs. This allows for easier installation of the wall sheathing. Square a line up from the wall studs over to the top of the wall plate.

Finding Gable Stud Length

Gable end stud lengths can be found using either of two methods.

Cut-and-Fit Method. To use this method, stand each stud plumb in place on the layout line and then mark it along the bottom and top edge of the rafter

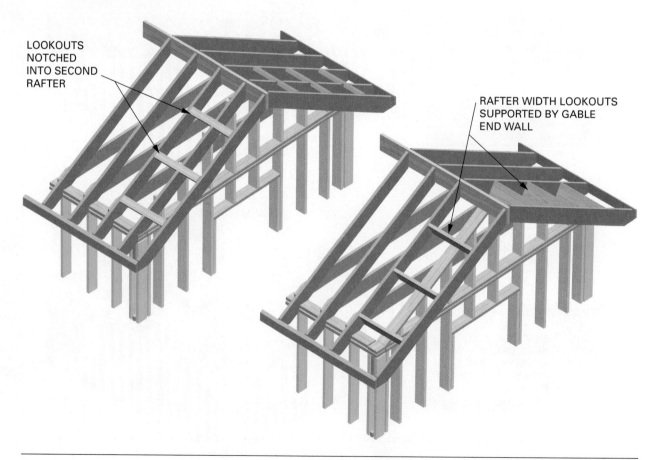

LOOKOUTS NOTCHED INTO SECOND RAFTER

RAFTER WIDTH LOOKOUTS SUPPORTED BY GABLE END WALL

FIGURE 42–19   **Two styles of lookouts can support the rake overhang.**

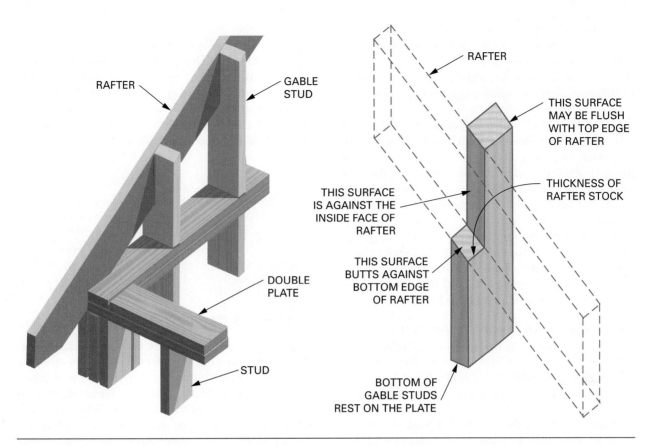

FIGURE 42–20 The cut at the top end of the gable stud slopes with the rafter.

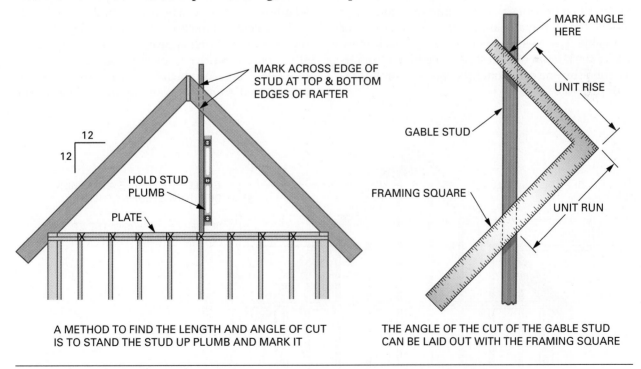

A METHOD TO FIND THE LENGTH AND ANGLE OF CUT IS TO STAND THE STUD UP PLUMB AND MARK IT

THE ANGLE OF THE CUT OF THE GABLE STUD CAN BE LAID OUT WITH THE FRAMING SQUARE

FIGURE 42–21 Cut-and-fit method of finding the length and cut of a gable stud.

(Figure 42–21). Remove the stud. Use a scrap piece of rafter stock to mark the depth of cut on the stud. Mark and cut all studs in a similar manner.

The studs are fastened by toenailing to the plate and by nailing through the rafter into the edge of the stud. Care must be taken when installing gable studs not to force a bow in the end rafters. This creates a crown in them. Sight the top edge of the end rafters for straightness as gable studs are installed. After all gable studs are installed, the end ceiling joist is nailed to the inside edges of the studs.

Common Difference Method. Gable studs that are spaced equally have a common difference in length. Each stud is shorter than the next one by the same amount **(Figure 42–22).** Once the length of the first stud and the common difference are known, gable studs can be easily laid out and gang cut on the ground.

To find the common difference in the length of gable studs, multiply the spacing, in terms of unit run, by the unit rise of the roof. For example, if the stud spacing is 16 inches OC and the unit rise is 6, what is the common difference in length?

STEP 1: Convert OC spacing to units of run.

$$16 \div 12 = 1\tfrac{1}{3} \text{ feet or units of run.}$$

STEP 2: Multiply unit rise by units of run.

$$6 \times 1\tfrac{1}{3} = 8 \text{ inches common difference.}$$

Next, determine the length of a stud. A framing square works well to measure the distance plumb and accurately. Mark it with the angle for the notch as described previously using a framing square. Assemble the stud material necessary for one side of the gable. Lay them on horses next to each other where the bottom of each stud is separated from its neighbor by one common difference in length **(Procedure 42–E).**

Using the first measured stud as a reference, square a line across the ganged studs. Mark the angle for each stud the same as the first. Be sure to measure and mark to the same point for each, that is, to the long point or short point of each notch.

Toenail each stud to the plate and face nail into the rafter. Check to be sure the rafter does not bow upward as each stud is installed.

> **EXAMPLE** Gable studs are spaced 16 inches OC. The roof rises 8 inches per foot of run. Change 16 inches to 1⅓ feet. Multiply 1⅓ by 8 to get 10.666. The common difference in length for gable end studs here is 10⅔ inches or roughly 10¹¹⁄₁₆ inches.

FRAMING A GAMBREL ROOF

A gambrel roof is one where each side of the ridgeboard has two sloping rafters. The rafters of a true gambrel roof are chords of a semicircle whose diameter is the width of the building **(Figure 42–23).** To frame a true gambrel roof, the rafter lengths are calculated using the Pythagorean theorem or determined from a full-scale layout on a large, flat surface, such as the subfloor. From the layout, rafter lengths and angles can be determined.

The calculated method involving the Pythagorean theorem may at first appear intimidating, but it really is only repetitive steps of the same process. Before calculations can begin, some information must be obtained from the plans or the architect. The

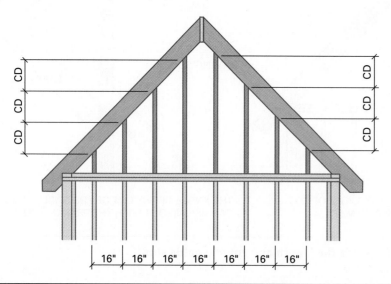

GABLE STUDS HAVE A COMMON DIFFERENCE (CD) IN LENGTH

IF THE UNIT RISE IS 6" THE CD = 8" FOR BOTH SIDES

FIGURE 42–22   Equally spaced gable studs have a common difference in length.

Procedure 42–E Cutting Gable Studs

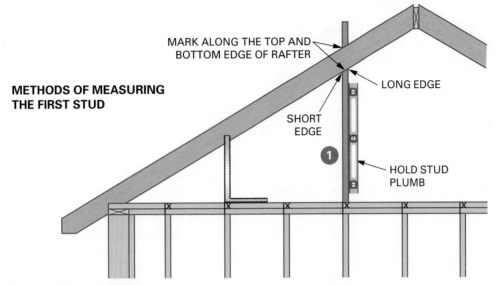

MARK ALONG THE TOP AND
BOTTOM EDGE OF RAFTER

LONG EDGE

**METHODS OF MEASURING
THE FIRST STUD**

SHORT
EDGE

1

HOLD STUD
PLUMB

STEP 1 MEASURE ANY STUD TO BEGIN LAYOUT. A FRAMING SQUARE MAY BE USED TO MEASURE A SMALL STUD.

STEP 2 ASSEMBLE STUD MATERIAL.

STEP 3 SPACE OUT STUDS SO THE BOTTOMS ARE SEPARATED BY THE COMMON DIFFERENCE.

STEP 4 SQUARE A LINE ACROSS STUD MATERIAL FROM THE STUD PREVIOUSLY MEASURED.

STEP 5 MARK ANGLED CUT ON EACH STUD.

STEP 6 MARK AND LAY OUT EXTRA FOR TOP TO FASTEN TO RAFTER.

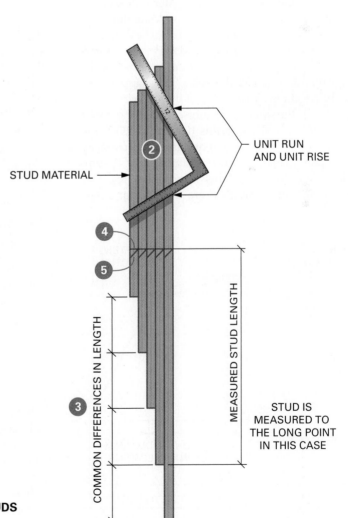

2

UNIT RUN
AND UNIT RISE

STUD MATERIAL

4

5

MEASURED STUD LENGTH

COMMON DIFFERENCES IN LENGTH

3

STUD IS
MEASURED TO
THE LONG POINT
IN THIS CASE

GANG CUTTING GABLE STUDS

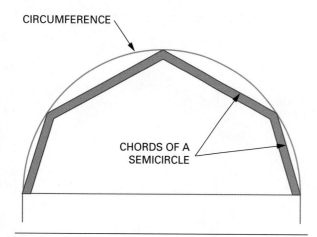

FIGURE 42–23 The slopes of a true gambrel roof form chords of a semicircle.

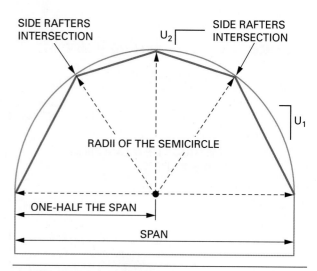

FIGURE 42–25 Rafter intersections are located a radius from the center of the span.

measurements needed are the building span and the horizontal distances that the side rafter intersections are from the building's center and building line (**Figure 42–24**). Another method, that will not be described here, uses the ceiling height created at the rafter intersections. It is helpful to note that the total rise of the roof is equal to one-half the span. This happens because the height is also a radius of the semicircle. The distance from the side rafters intersection to the center of the building is also a radius (**Figure 42–25**). Note that there are only two different rafters because they come in pairs. This is because a gambrel roof is symmetrical. Bringing all of this information together shows that right triangles are formed and the

Pythagorean theorem can be used to solve for rafter lengths (**Figure 42–26**).

EXAMPLE Find the lengths of the rafters labeled R_1 and R_2 for a gambrel roof system, given that the span is 26 feet and the horizontal distances of the side intersections are 6 and 7 feet (**Figure 42–27**).

1. $a^2 + b^2 = c^2$ therefore $a^2 = c^2 - b^2$
2. From triangle 1
$a^2 = 13^2 - 7^2 = 169 - 49 = 120$
$a = \sqrt{120} = 10.954'$

3. From triangle 2
$R_1 = \sqrt{a^2 + 6^2} = \sqrt{10.952^2 + 6^2} = \sqrt{120 + 36}$
$R_1 = \sqrt{156} = 12.490' = 12'\text{-}5\frac{7}{8}''$

4. From triangle 3
$b = \text{radius} - a = 13' - 10.954' = 2.046'$
5. $R_2 = \sqrt{b^2 + 7^2} = \sqrt{2.046^2 + 7^2} = \sqrt{4.186 + 49}$
$R_2 = \sqrt{53.186} = 7.293' = 7'\text{-}3\frac{1}{2}''$

If the rafters are to be framed to a **purlin knee wall** as shown in **Figure 42–28**, the unit rise for each rafter may be found by dividing the total rise of each rafter, in inches, by its total run. For example, in Figure 42–27, U_1 is found by first converting (a) to inches, 10.954 feet times 12 = 131.448 inches, then dividing it by 6, the run, to get 21.908 or $21\frac{15}{16}$ inches unit rise. U_2 is found similarly, (b), in inches, divided by 7. That is, 2.046 feet times 12 = 24.552 inches, then divide by 7 to get 3.507 or $3\frac{1}{2}$ inches unit rise.

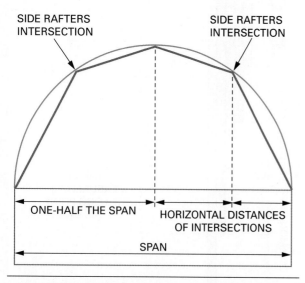

FIGURE 42–24 Gambrel rafter calculations begin with determining the horizontal distances of the intersections and the span.

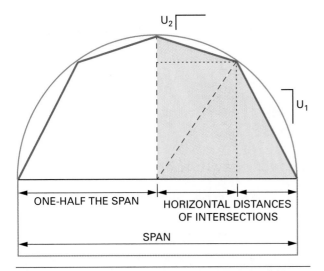

FIGURE 42–26 Various right triangles are formed by horizontal and vertical lines and radii of the semicircle.

Sometimes the slope of gambrel rafters is given in the drawings. The slopes, then, may not actually be chords of a semicircle **(Figure 42–29).** In this case, the rafter lengths and cuts can also be found with a full-size layout. They are usually laid out with a framing square similar to the layout considerations of gable roof rafters.

Gambrel Rafter Layout

Determine the construction at the intersection of the two slopes. Usually gambrel roof rafters meet at a continuous member, similar to the ridge, called a *purlin*. Any structural member that runs at right angles to and supports rafters is called a purlin. Sometimes, the rafters of a gambrel roof meet at the top of a *knee wall*. In the lower and steeper slope, the rafters may be sized as wall members rather than roof members if the rafter is within 30 degrees of vertical (Figure 42–29).

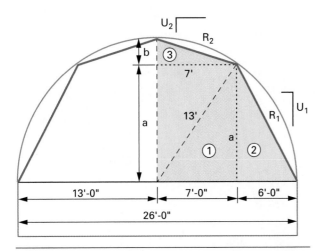

FIGURE 42–27 Starting information gathered for a gambrel calculation.

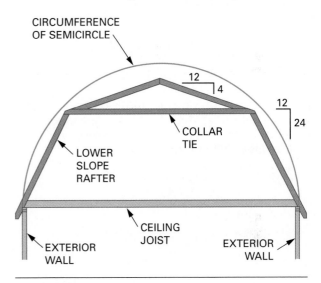

FIGURE 42–29 In some cases, the slope of the gambrel rafters is given in the drawings.

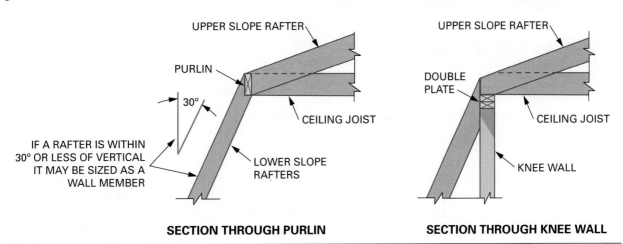

SECTION THROUGH PURLIN

SECTION THROUGH KNEE WALL

FIGURE 42–28 Gambrel roof rafters may intersect each other at a purlin or knee wall.

Procedure 42–F Laying Out Gambrel Rafters

UPPER SLOPE RAFTER

STEP ① LAY OUT PLUMB LINE.

STEP ② SHORTEN ½ THICKNESS OF RIDGE.

STEP ③ LAY OUT LINE LENGTH.

STEP ④ MARK PLUMB LINE.

STEP ⑤ DRAW LEVEL LINE THE WIDTH OF KNEE WALL PLATE.

LOWER SLOPE RAFTER

STEP ① LAY OUT PLUMB LINE— NO SHORTENING.

STEP ② LAY OUT LINE LENGTH.

STEP ③ DRAW PLUMB LINE.

STEP ④ LAY OUT LEVEL LINE OF SEAT CUT. LEAVE ON MINIMUM OF ⅔ WIDTH OF RAFTER STOCK.

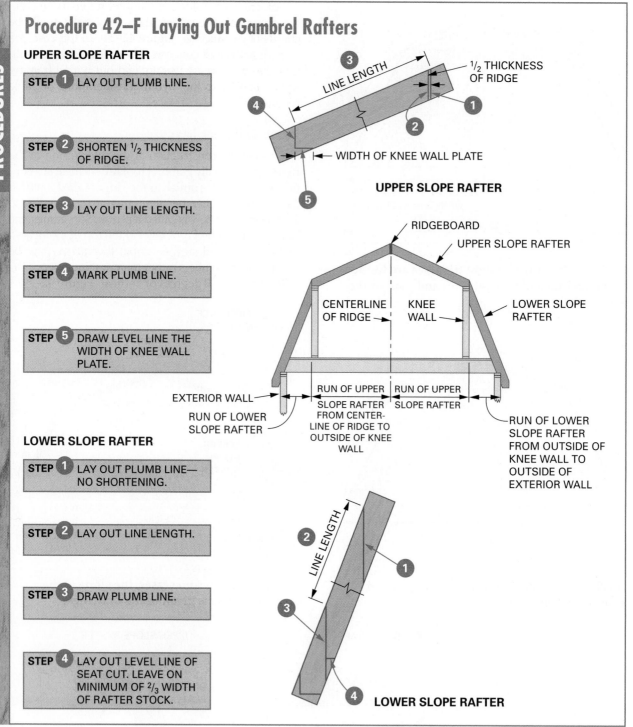

Determine the run and rise of the rafters for both slopes. Find their line length in the same way as for rafters in a gable roof. Lay out plumb lines at the top and bottom of rafters for each slope. Make the seat cut on the lower rafter. Notch all rafters where they intersect at the purlin **(Procedure 42–F)**. Because of the steep slope of the lower roof, the level cut of the seat cannot be made the full width of the plate. At least ⅔ of the width of the rafter stock must remain.

Erecting the Gambrel Roof

If a purlin is used, fasten a rafter to each end of a section of purlin. Raise the assembly. Fasten the lower end of the rafters against ceiling joists and on the top plate of the wall. Brace the section under each end of the purlin. Continue framing sections until the other end of the building is reached. Place temporary and adequate bracing under the purlin where needed. If a knee wall is used, build, straighten, and brace the wall.

Frame the roof by installing both lower and upper slope rafters opposite each other. This maintains equal stress on the frame. Plumb and brace the roof after a few rafters have been installed. Sight along the ridge and purlin or knee wall for straightness as framing progresses. After all rafters have been erected, install ceiling joists.

The open ends formed by the gambrel roof are framed with studs in the same manner as installing studs in gable ends.

43 Hip Roofs

The hip roof is a little more complicated to frame than the gable roof. Two additional kinds of rafters need to be laid out.

To frame the hip roof it is necessary to lay out not only common rafters and a ridge, but also **hip rafters** and **hip jack rafters (Figure 43–1).** Hip rafters form the outside corners where two sloping roof planes meet. They are longer than common rafters and their

slope is at a lower angle. Hip jacks are common rafters that must be cut shorter to land on a hip rafter. They all have a common difference in length.

HIP RAFTER THEORY

The theory behind fashioning a hip rafter is quite similar to that used for a common rafter, yet slight

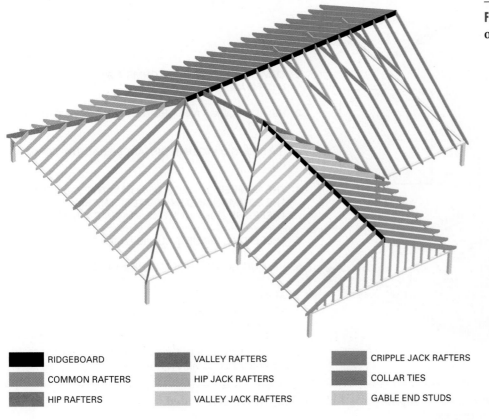

FIGURE 43–1 Members of a hip roof frame.

	RIDGEBOARD		VALLEY RAFTERS		CRIPPLE JACK RAFTERS
	COMMON RAFTERS		HIP JACK RAFTERS		COLLAR TIES
	HIP RAFTERS		VALLEY JACK RAFTERS		GABLE END STUDS

differences allow the hip rafter to take a different shape in the roof. The length of the hip rafter can be estimated using a framing square.

Unit Run

Both hip and common rafters have a total run that always runs horizontal under the rafters. A common rafter run forms a 90-degree angle with the plates and the hip run forms a 45-degree angle. Therefore, a hip rafter run covers more distance than does a common run, but the number of units of run is the same.

The hip unit run forms the diagonal of a square created by the common unit run (**Figure 43–2**). Using the Pythagorean theorem with 12 as a and b, c turns out to be 16.97 inches. This amount is rounded off to 17 inches when using a framing square to lay out the rafter.

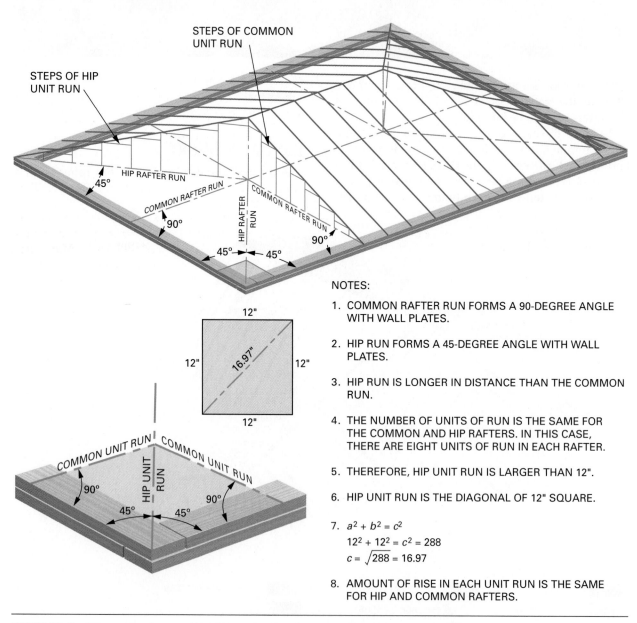

NOTES:

1. COMMON RAFTER RUN FORMS A 90-DEGREE ANGLE WITH WALL PLATES.

2. HIP RUN FORMS A 45-DEGREE ANGLE WITH WALL PLATES.

3. HIP RUN IS LONGER IN DISTANCE THAN THE COMMON RUN.

4. THE NUMBER OF UNITS OF RUN IS THE SAME FOR THE COMMON AND HIP RAFTERS. IN THIS CASE, THERE ARE EIGHT UNITS OF RUN IN EACH RAFTER.

5. THEREFORE, HIP UNIT RUN IS LARGER THAN 12".

6. HIP UNIT RUN IS THE DIAGONAL OF 12" SQUARE.

7. $a^2 + b^2 = c^2$
   $12^2 + 12^2 = c^2 = 288$
   $c = \sqrt{288} = 16.97$

8. AMOUNT OF RISE IN EACH UNIT RUN IS THE SAME FOR HIP AND COMMON RAFTERS.

FIGURE 43–2 The unit of run of the hip rafter is 16.97 inches, which is rounded to 17 inches for layout purposes.

Unit Rise

Both hip and common rafters start at the wall plate and slope up to meet at the ridge, which is at the same height. Each rafter has the same number of unit runs under it. Therefore, the rise for each unit run must be the same.

A framing square can be used to lay out a hip rafter. The square is held the same as for a common rafter. The unit rise is held on the tongue and the unit run on the blade. The difference is the unit run is now 17 inches, not 12.

Estimating the Rough Length of the Hip Rafter

The rough length of the hip rafter can be found in the manner previously described for the common rafter. Let the tongue of the square represent the length of common rafter. Let the blade of the square represent its total run. Measure across the square the distance between the two points. A scale of one inch equals one foot gives the length of the hip rafter from plate to ridge **(Figure 43–3)**.

EXAMPLE The common rafter length is 15 feet. Its total run is 12 feet. A measurement across these points scales off to 19'–2½". It will be necessary to order 20-foot lengths for the hip rafters unless extra is needed for overhang.

Laying Out the Hip Rafter

The steps to lay out a hip rafter are similar to those for a common rafter with some differences. The differences are caused by the fact that the hip runs at a 45-degree angle. There are at least 11 lines to draw in order to lay out a hip rafter **(Figure 43–4)**.

Ridge Cut of the Hip

The ridge plumb line on a hip is drawn similar to that for a common rafter. The squares are held in the same manner except different numbers or scales are used **(Figure 43–5)**. The hip ridge cut is also a compound angle called a **cheek cut** or *side cut*. A *single cheek* cut or a *double cheek* cut may be made on the hip rafter according to the way it is framed at the ridge **(Figure 43–6)**.

Shortening the Hip

The rafter must be shortened before the cheek cuts are laid out. The hip rafter is shortened by one half the 45-degree thickness of the ridge board. The rafter must be shortened due to the ridge before the cheek cuts are laid out **(Figure 43–7)**. This is due to the fact that the hip run is at a 45-degree angle.

To lay out the ridge cut, select a straight length of stock for a pattern. Lay it across two sawhorses. Mark a plumb line at the left end. Hold the tongue of the square at the rise and the blade of the square at 17 inches, the unit of run for the hip rafter. Shorten the rafter by measuring at right angles to the plumb line. This measurement is 1¹⁄₁₆ inches for dimension lumber. Lay out another plumb line at that point **(Procedure 43–A)**.

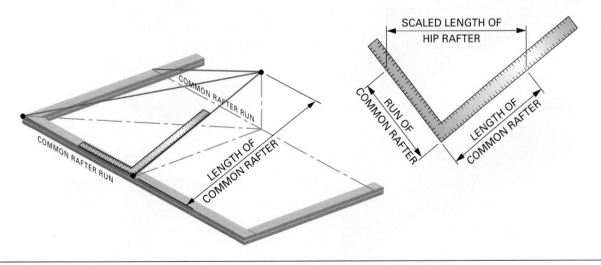

FIGURE 43–3 An estimated length of the hip rafter by scaling across the framing square.

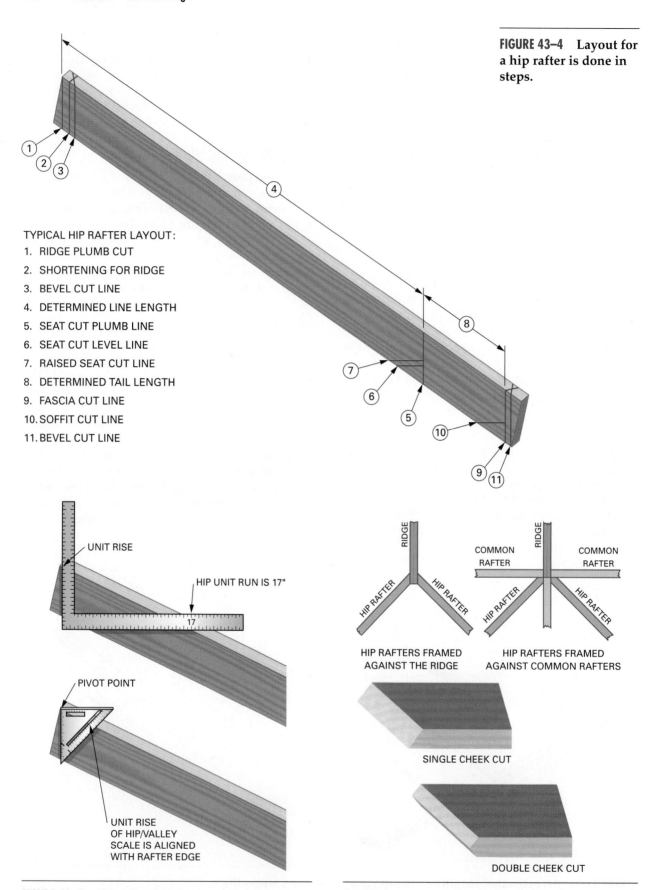

FIGURE 43–4 Layout for a hip rafter is done in steps.

TYPICAL HIP RAFTER LAYOUT:

1. RIDGE PLUMB CUT
2. SHORTENING FOR RIDGE
3. BEVEL CUT LINE
4. DETERMINED LINE LENGTH
5. SEAT CUT PLUMB LINE
6. SEAT CUT LEVEL LINE
7. RAISED SEAT CUT LINE
8. DETERMINED TAIL LENGTH
9. FASCIA CUT LINE
10. SOFFIT CUT LINE
11. BEVEL CUT LINE

UNIT RISE

HIP UNIT RUN IS 17"

17

PIVOT POINT

UNIT RISE
OF HIP/VALLEY
SCALE IS ALIGNED
WITH RAFTER EDGE

RIDGE

HIP RAFTER

HIP RAFTER

HIP RAFTERS FRAMED
AGAINST THE RIDGE

COMMON
RAFTER

RIDGE

COMMON
RAFTER

HIP RAFTER

HIP RAFTER

HIP RAFTERS FRAMED
AGAINST COMMON RAFTERS

SINGLE CHEEK CUT

DOUBLE CHEEK CUT

FIGURE 43–5 Hip plumb lines are drawn in a fashion similar to that for common rafters.

FIGURE 43–6 Single or double cheek cuts are used depending on the method of framing at the ridge.

Procedure 43–A   Marking a Rafter Plumb Line

STEP **1** NOTE HOW THE SQUARE IS SET UP ON THE RAFTER EDGE.

STEP **2** ALIGN THE SQUARE AND DRAW A PLUMB LINE FROM THE CORNER OF THE BOARD.

STEP **3** MEASURE AT A RIGHT ANGLE ONE-HALF THE DIAGONAL THICKNESS OF THE RIDGE. IT IS 1$\frac{1}{16}$" FOR DIMENSION LUMBER.

STEP **4** DRAW SECOND PLUMB LINE.

UNIT RISE

1

HIP UNIT RUN IS 17"

17

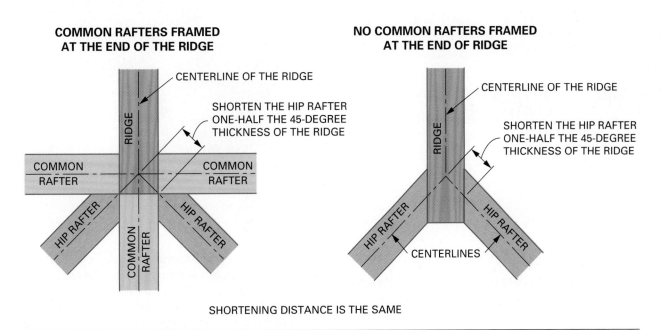

COMMON RAFTERS FRAMED AT THE END OF THE RIDGE

CENTERLINE OF THE RIDGE

SHORTEN THE HIP RAFTER ONE-HALF THE 45-DEGREE THICKNESS OF THE RIDGE

RIDGE

COMMON RAFTER

COMMON RAFTER

HIP RAFTER

HIP RAFTER

COMMON RAFTER

NO COMMON RAFTERS FRAMED AT THE END OF RIDGE

CENTERLINE OF THE RIDGE

SHORTEN THE HIP RAFTER ONE-HALF THE 45-DEGREE THICKNESS OF THE RIDGE

RIDGE

HIP RAFTER

HIP RAFTER

CENTERLINES

SHORTENING DISTANCE IS THE SAME

FIGURE 43–7   Amount to shorten the hip rafter for either method of framing at the ridge.

Procedure 43–B Making Cheek Cuts Using the Measurement Method

CHEEK CUTS USING MEASUREMENT METHOD

SINGLE CHEEK LINE

STEP 1 SQUARE A LINE ACROSS THE TOP EDGE FROM THE SECOND PLUMB LINE.

STEP 2 MARK THE CENTER OF THE TOP EDGE.

STEP 3 FROM THE SECOND LINE, MEASURE AT A RIGHT ANGLE ONE-HALF THE THICKNESS OF THE HIP RAFTER. IT IS $3/4$" FOR DIMENSION LUMBER.

STEP 4 DRAW THIRD PLUMB LINE.

STEP 5 DRAW A DIAGONAL LINE ACROSS THE TOP FROM THE THIRD LINE THROUGH THE CENTERLINE.

DOUBLE CHEEK LINES

STEP 1 SQUARE ACROSS FROM THE THIRD PLUMB LINE.

STEP 2 DRAW A DIAGONAL LINE FROM SQUARED LINE THROUGH THE CENTERLINE.

STEP 3 A FOURTH PLUMB LINE MAY BE DRAWN AS A CUT LINE.

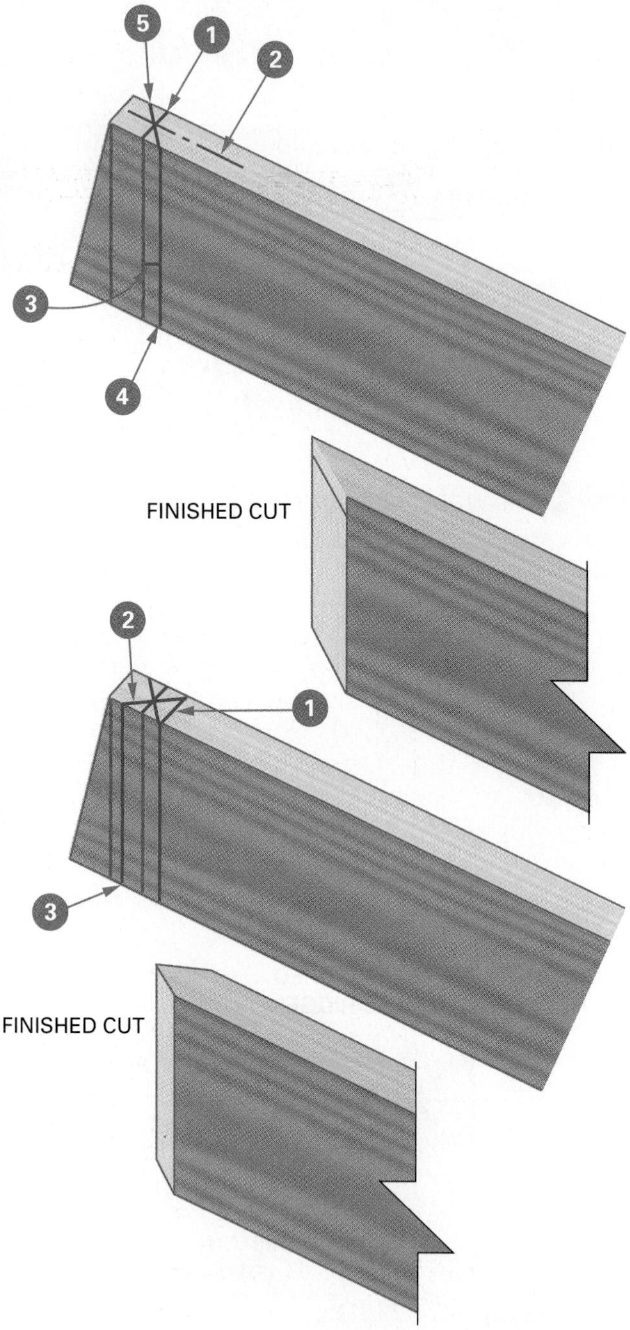

FINISHED CUT

FINISHED CUT

Laying Out Cheek Cuts

To complete the layout of the ridge cut, mark lines for a single or double cheek cut as required. The method of laying out cheek cuts shown in **Procedure 43–B** gives accurate results.

The bottom line of the rafter tables on the framing square can be used to determine the angle of the side cuts of hip rafters. The number found is used with 12 on a square to position it for marking the angle. Use the number from the tables on the tongue with 12 used on the blade. Align these numbers along the edge of the rafter. The angle is laid out by marking along the blade **(Figure 43–8).**

CHEEK CUTS USING RAFTER TABLES

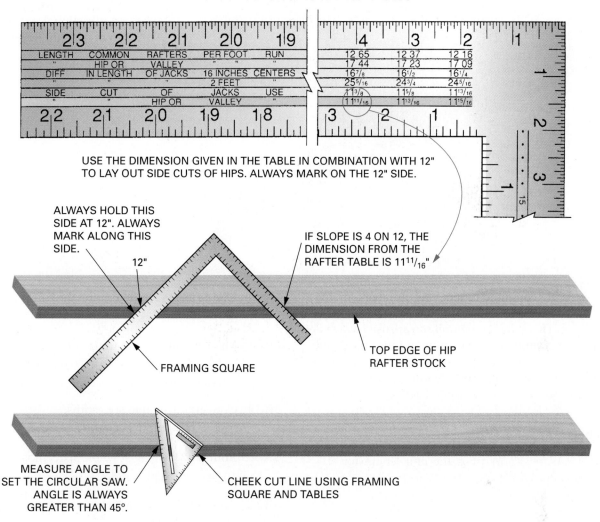

USE THE DIMENSION GIVEN IN THE TABLE IN COMBINATION WITH 12"
TO LAY OUT SIDE CUTS OF HIPS. ALWAYS MARK ON THE 12" SIDE.

ALWAYS HOLD THIS
SIDE AT 12". ALWAYS
MARK ALONG THIS
SIDE.

12"

IF SLOPE IS 4 ON 12, THE
DIMENSION FROM THE
RAFTER TABLE IS 11¹¹/₁₆"

TOP EDGE OF HIP
RAFTER STOCK

FRAMING SQUARE

MEASURE ANGLE TO
SET THE CIRCULAR SAW.
ANGLE IS ALWAYS
GREATER THAN 45°.

CHEEK CUT LINE USING FRAMING
SQUARE AND TABLES

FIGURE 43–8   Cheek cuts using the rafter table method.

Laying Out the Hip-Rafter Length

The length of the hip rafter is laid out in a manner similar to that used for the common rafter. It may be found by the step-off method or calculated. Remember always to start any layout for length from the first ridge plumb line, the one before any shortening.

Stepping Off the Hip. In the step-off method, the number of steps for the hip rafter is the same as for the common rafter in the same roof. The same rise is used, but the unit of run for the hip is 17 not 12. For example, for a roof with a rise of 6 inches per foot of run, the square is held at 6 and 12 for the common rafter, and 6 and 17 for the hip rafter of the same roof.

If the total run of the common rafter contains a fractional part of 12, then the hip run must contain the same fractional part of 17. In other words, if the common run is 12½ steps of 12 inches then the hip run is 12½ steps of 17 inches.

EXAMPLE   If the common run distance is 15'–9", then the run is 15⁹⁄₁₂ or 15¾ steps. The hip run is the same number of steps. In this case, 15 steps are made. The last step is ¾ of 17 or 12¾ inches along the blade **(Figure 43–9).**

Using Rafter Tables. Finding the length of the hip using the tables found on the framing square is similar to the process used for finding the length of the common rafter. However, the numbers from the second line are used instead of the first line. These numbers are the unit length of the hip rafter in inches or length per unit of run.

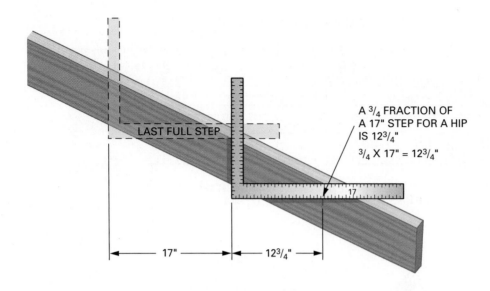

FIGURE 43–9 **Stepping off a fractional unit of run on the hip rafter.**

As with common rafters, the numbers in the rafter tables can be calculated using the Pythagorean theorem. The unit rise and unit run are used as *a* and *b* to find *c*, unit length. A common rafter uses unit rise and 12 inches, where a hip uses unit rise and 16.97 inches. It is important not to use the rounded off number of 17 in these calculations.

EXAMPLE   What is the unit length for a hip rafter with a unit rise of 7 inches?

$$a^2 + b^2 = c^2$$
$$7^2 + 16.97^2 = c^2$$
$$49 + 287.9809 = 336.9809 = c^2$$
$$c = \sqrt{336.9809} = 18.35704 \text{ inches}$$

Unit length = 18.36 inches rounded to the nearest hundredth. This number is also found in the rafter table: On the second line under the 7 is 18 36, which is 18.36.

As with the common rafter, the hip length is then found by multiplying the unit length by the total run of a common rafter. This is because the number of units of run under a common rafter is the same as for a hip.

EXAMPLE: Example of Using the Calculation Method to Determine Hip Rafter Line Length

Find the line length of a hip rafter with a unit rise of 8 inches for a 28-foot-wide building.

STEP 1:   Read below the 8-inch mark in the first row to find 18 76 or 18.76 inches unit length.

STEP 2:   Divide 28 by 2 to determine the run.
$$28 \div 2 = 14 \text{ units of run.}$$

STEP 3:   Multiply unit length times total run.
$$18.76 \times 14 = 262.64 \text{ inches. That is } 262 \text{ inches plus a fraction.}$$

STEP 4:   Convert decimal to sixteenths.
$$0.64 \times 16 = 10.24 \text{ sixteenths.}$$

STEP 5:   Round off to nearest whole sixteenths.
$$10.24 \Rightarrow {}^{10}\!/_{16}\text{, which reduces to } \frac{5}{8} \text{ inches.}$$

STEP 6:   Collect information into rafter length.
$$262 + \frac{5}{8}'' = 262\frac{5}{8}''$$

If the total run of the rafter contains a fractional part of a foot, multiply the figure in the tables by the whole number of feet plus the fractional part changed to a decimal. For instance, if the total run of the rafter is 15'–6", multiply the figure in the tables by 15½ changed to 15.5.

Lay out the length obtained along the top edge of the hip rafter stock. This must be done from the first plumb line before shortening and cheek cuts are made **(Figure 43–10)**.

Laying Out the Seat Cut of the Hip Rafter

Like the common rafter, the seat cut of the hip rafter is a combination of plumb and level cuts. The height above the bird's mouth on a hip rafter is the same distance as the common rafter. To locate where the seat level line should be, first measure down along the plumb line from the top of the common rafter. Then mark that same measurement on the hip. No

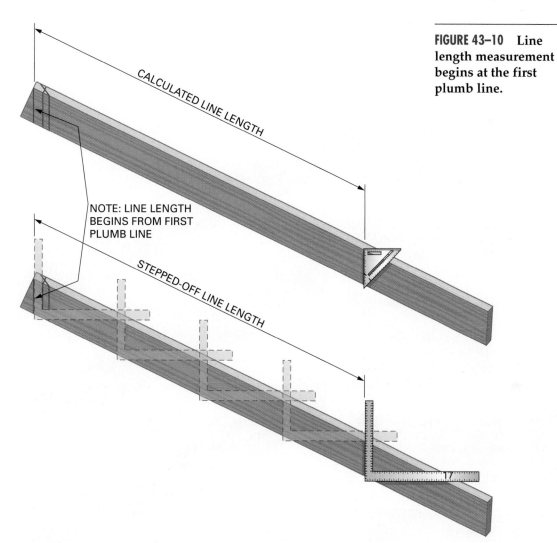

FIGURE 43–10 Line length measurement begins at the first plumb line.

CALCULATED LINE LENGTH

NOTE: LINE LENGTH BEGINS FROM FIRST PLUMB LINE

STEPPED-OFF LINE LENGTH

17

consideration needs to be given to fitting it around the corner of the wall.

Backing and Dropping the Hip. When making the level cut of the seat, special consideration must be given. The hip rafter is at the intersection of the two roof slopes. The center of the top edge is in perfect alignment with the other rafters, but the top outside corners of the top edge surface stick above the sheathing. To remedy this situation, the seat level line is cut deeper, a process called **dropping the hip** rafter, or the rafter top edge is bevel planed, a process called **backing the hip (Figure 43–11).**

Dropping the hip is much easier to do, so it is done more frequently than backing the hip. The amount of dropping must be determined for each roof, because this amount changes with the slope of the roof. As steepness increases, the amount of drop increases. The process for determining the amount of drop is shown in **Procedure 43–C.**

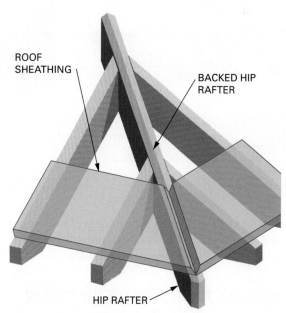

ROOF SHEATHING

BACKED HIP RAFTER

HIP RAFTER

FIGURE 43–11 The hip rafter must be adjusted to allow for the roof sheathing.

STEP BY STEP
PROCEDURES

Procedure 43–C Dropping or Backing a Hip Rafter

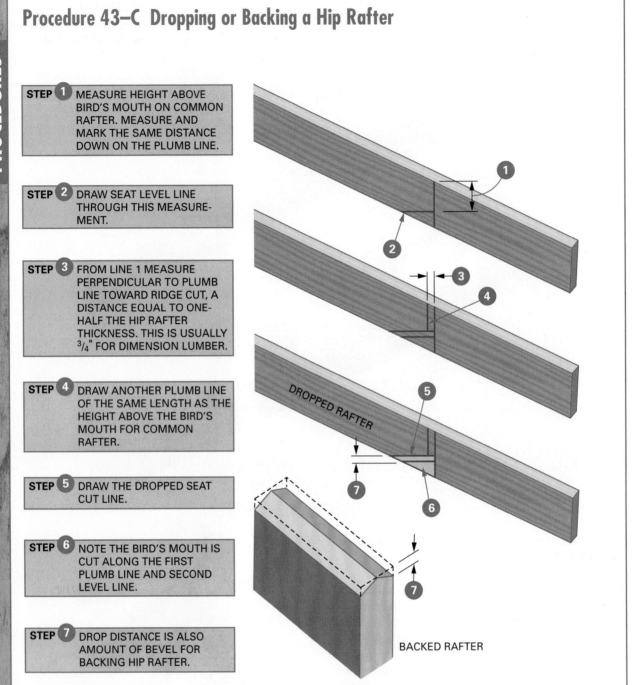

STEP 1 MEASURE HEIGHT ABOVE BIRD'S MOUTH ON COMMON RAFTER. MEASURE AND MARK THE SAME DISTANCE DOWN ON THE PLUMB LINE.

STEP 2 DRAW SEAT LEVEL LINE THROUGH THIS MEASUREMENT.

STEP 3 FROM LINE 1 MEASURE PERPENDICULAR TO PLUMB LINE TOWARD RIDGE CUT, A DISTANCE EQUAL TO ONE-HALF THE HIP RAFTER THICKNESS. THIS IS USUALLY ³/₄" FOR DIMENSION LUMBER.

STEP 4 DRAW ANOTHER PLUMB LINE OF THE SAME LENGTH AS THE HEIGHT ABOVE THE BIRD'S MOUTH FOR COMMON RAFTER.

STEP 5 DRAW THE DROPPED SEAT CUT LINE.

STEP 6 NOTE THE BIRD'S MOUTH IS CUT ALONG THE FIRST PLUMB LINE AND SECOND LEVEL LINE.

STEP 7 DROP DISTANCE IS ALSO AMOUNT OF BEVEL FOR BACKING HIP RAFTER.

DROPPED RAFTER

BACKED RAFTER

To back the rafter, the measurement of the drop is used to determine the bevel. This amount is the distance the top outside corners are beveled.

LAYING OUT THE TAIL OF THE HIP RAFTER

The hip tail can be found in one of two ways: Calculate the overhang or calculate the projection. Hip tail projection is determined with the fractional step-off method as shown earlier in Figure 43–9. For exam-ple, if the common projection is 15 inches, then the run is 15 ÷ 12 = 1.25 or 1¼ units of run. The hip tail has the same run. To step this off, mark one full step plus a quarter step or ¼ × 16.97 = 4.25 inches, which is 4¼ inches. The total hip projection here is 16.97 or 17 + 4¼ = 21¼ inches.

The overhang calculation uses the same formula as rafter length: unit length times run (tail) = line length (tail). For example, if the unit rise is 6, then the unit length of a hip is 18 inches. If the common

Procedure 43–D Laying Out a Hip Rafter Tail

STEP 1 UNIT RISE = 6, THEN UNIT LENGTH IS 18 FROM THE RAFTER TABLES.

STEP 2 PROJECTION IS 15".
15 ÷ 12 = 1.25 UNITS OF RUN.

STEP 3 OVERHANG LENGTH:
18 X 1.25 = 22$\frac{1}{2}$".

STEP 4 MEASURE THE OVERHANG LENGTH.

STEP 5 DRAW FASCIA PLUMB LINE.

STEP 6 SQUARE A LINE ACROSS THE TOP OF THE RAFTER FROM THIS LINE AND MARK THE CENTER.

STEP 7 FROM LINE 6 MEASURE PERPENDICULAR TOWARD THE RIDGE A DISTANCE EQUAL TO ONE-HALF THE HIP THICKNESS.

STEP 8 DRAW A SECOND PLUMB LINE AND SQUARE A LINE ACROSS THE TOP.

STEP 9 DRAW DIAGONALS ACROSS THE TOP FROM THE ENDS OF THE SQUARED LINE THROUGH THE CENTER MARK.

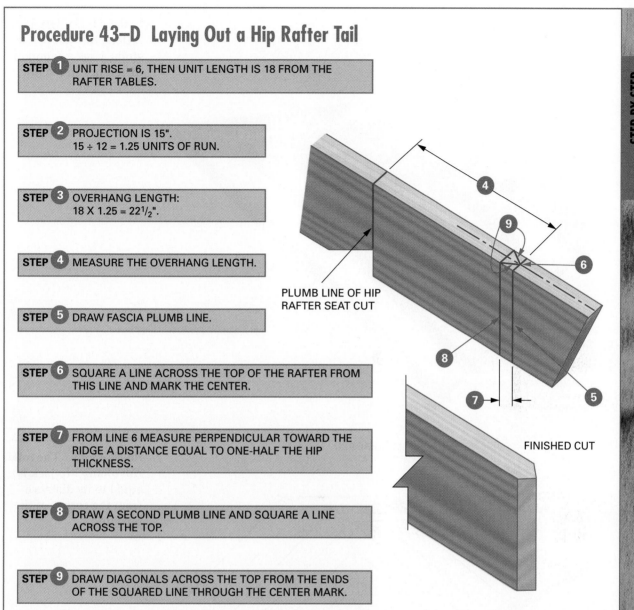

PLUMB LINE OF HIP
RAFTER SEAT CUT

FINISHED CUT

projection is 1.25 units of run, then the overhang is 18 × 1.25 = 22½ inches **(Procedure 43–D).**

HIP JACK RAFTERS

The hip jack rafter is a shortened common rafter. It is parallel to the common rafter, but is simply cut shorter to meet the hip. The seat cut and the tail are exactly like those of a common rafter, only the length varies as shown earlier in Figure 43–1.

Hip Jack Length

Each jack is shorter or longer than the next set by the same amount. This is called the *common difference* **(Figure 43–12).** The common difference is found in the rafter tables on the framing square for jacks 16 and 24 inches on center (OC).

Once the length of one jack is determined, the length of all others can be found by making each set shorter or longer by the common difference. To find the length of any jack, its total run must be known.

Finding the Total of the Run. The total run of any jack rafter and its distance from the corner along the outside edge of the wall plate form a square. Since all sides of a square are equal, the total run of any hip jack rafter is equal to its distance from the corner of the building. To determine the jack run, measure the distance from the corner to the center of where the jack will sit on the wall plate **(Figure 43–13).**

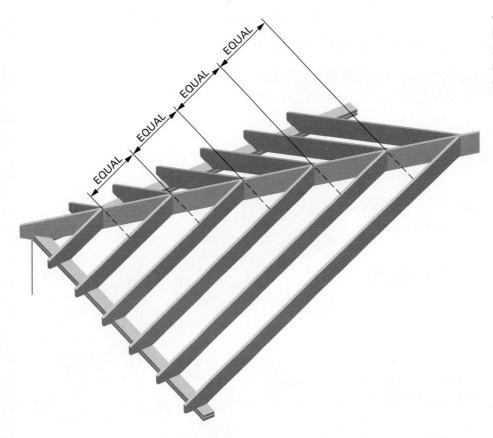

FIGURE 43–12 Hip jack rafters, like gable end studs, have a common difference in length.

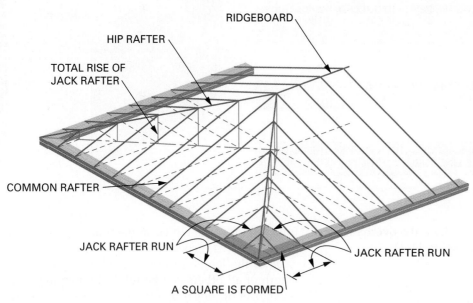

FIGURE 43–13 The total run of any hip jack is equal to its distance from the corner of the building.

Find the length of the hip jack rafter by the step-off method, or use the rafter tables in the same way as for common rafters. The only difference is that the total run of the jack found above is used. Use the common rafter pattern to lay out the jack tail.

Hip Jack Shortening. Since the hip jack rafter meets the hip rafter at a 45-degree angle, it must be short-

ened one-half the 45-degree angle or diagonal thickness of the hip rafter stock. Measure the distance at right angles to the original plumb line toward the tail. Draw another plumb line **(Procedure 43–E).**

Hip Jack Cheek Cuts. The hip jack rafter has a single cheek cut where it lands on the hip rafter. Square the shortened plumb line across the top of

Procedure 43–E Laying Out a Hip Jack Cheek Cut

STEP 1	LAY OUT PLUMB LINE.
STEP 2	MEASURE AT RIGHT ANGLE TO PLUMB LINE ONE-HALF 45-DEGREE THICKNESS OF HIP RAFTER AND DRAW SHORTENED PLUMB LINE.
STEP 3	SQUARE SHORTENED PLUMB LINE ACROSS TOP EDGE OF RAFTER STOCK.
STEP 4	MEASURE AT RIGHT ANGLE FROM SHORTENED PLUMB LINE ONE-HALF THE THICKNESS OF THE JACK RAFTER STOCK AND DRAW ANOTHER PLUMB LINE.
STEP 5	DRAW CENTERLINE ALONG TOP EDGE OF RAFTER.
STEP 6	DRAW DIAGONAL FROM LAST PLUMB LINE THROUGH CENTERLINE.

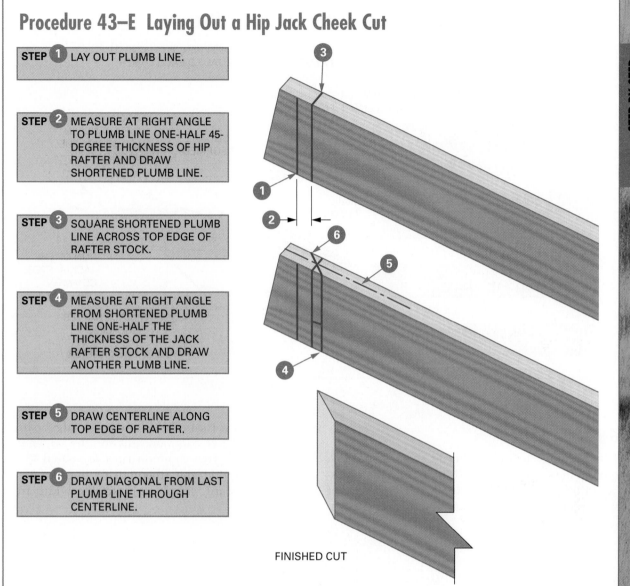

FINISHED CUT

the rafter. Draw an intersecting centerline along the top edge. Measure back toward the tail at right angles from the second plumb line a distance equal to one-half the jack rafter stock. Draw another plumb line. On the top edge, draw a diagonal from the third plumb line through the intersection of the centerline. Take care when drawing the diagonal line on the top edge. The direction of the diagonal depends on which side of the hip the jack rafter is framed.

Determine the common difference of hip jack rafters from the rafter tables under the inch mark that coincides with the slope of the roof.

The common difference may also be calculated. For rafters spaced 24 inches OC, the common difference is equal to the common rafter length for two units of run. For rafters spaced 16 inches OC, the common difference is equal to the common rafter length for 1⅓ units of run **(Procedure 43–F).**

Once the common difference is determined, measure the distance successively along the top edge of the pattern for the longest jack rafter. This pattern is then used to cut all the jack rafters necessary to frame that section of roof. This process is repeated for each section on jacks. Some jacks are framed in pairs depending on the width and length dimensions of the house.

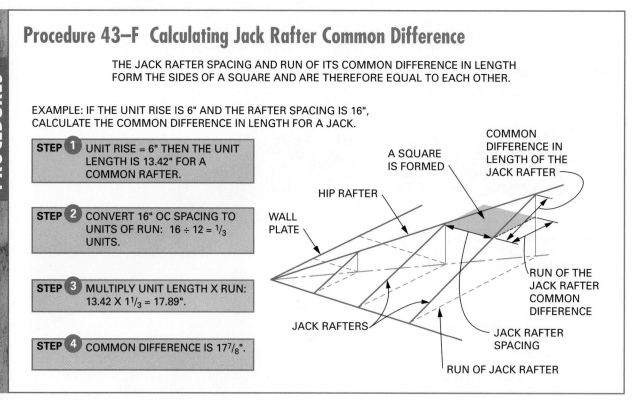

Procedure 43–F Calculating Jack Rafter Common Difference

THE JACK RAFTER SPACING AND RUN OF ITS COMMON DIFFERENCE IN LENGTH FORM THE SIDES OF A SQUARE AND ARE THEREFORE EQUAL TO EACH OTHER.

EXAMPLE: IF THE UNIT RISE IS 6" AND THE RAFTER SPACING IS 16", CALCULATE THE COMMON DIFFERENCE IN LENGTH FOR A JACK.

STEP 1 UNIT RISE = 6" THEN THE UNIT LENGTH IS 13.42" FOR A COMMON RAFTER.

STEP 2 CONVERT 16" OC SPACING TO UNITS OF RUN: 16 ÷ 12 = $\frac{1}{3}$ UNITS.

STEP 3 MULTIPLY UNIT LENGTH X RUN: 13.42 X $1\frac{1}{3}$ = 17.89".

STEP 4 COMMON DIFFERENCE IS $17\frac{7}{8}$".

COMMON DIFFERENCE IN LENGTH OF THE JACK RAFTER

A SQUARE IS FORMED

HIP RAFTER

WALL PLATE

RUN OF THE JACK RAFTER COMMON DIFFERENCE

JACK RAFTERS

JACK RAFTER SPACING

RUN OF JACK RAFTER

Hip Roof Ridge Length

The hip ridge length is shorter than for a gable roof on the same sized building. To find the ridgeboard length, it is helpful to visualize the run of the major components of a roof **(Figure 43–14)**.

The length of the building under the ridgeboard is made up of two common runs and the ridgeboard length. Also two common runs are equal to the width. To find the (theoretical) line length of the ridgeboard, simply subtract the width from the length.

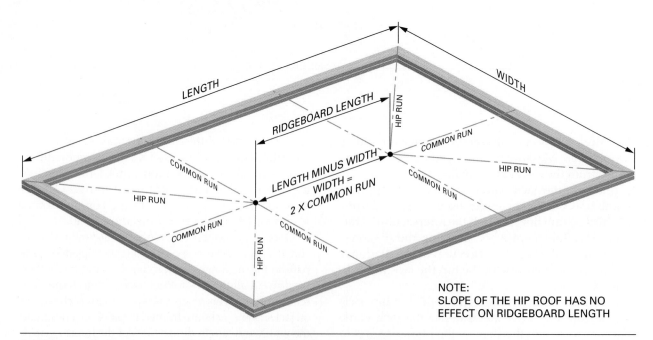

LENGTH

WIDTH

RIDGEBOARD LENGTH

HIP RUN

COMMON RUN

COMMON RUN

HIP RUN

LENGTH MINUS WIDTH WIDTH = 2 X COMMON RUN

HIP RUN

COMMON RUN

COMMON RUN

COMMON RUN

HIP RUN

COMMON RUN

NOTE: SLOPE OF THE HIP ROOF HAS NO EFFECT ON RIDGEBOARD LENGTH

FIGURE 43–14 Determining the line length of the hip roof ridgeboard.

The actual length is longer than the line length. The amount to add depends on two styles of framing. One style has a common rafter framed at the end of the ridgeboard. The other has no common rafter at the end **(Figure 43–15)**.

The amount to add in the first case is one-half the thickness of the common rafter at each end. This is ¾ inch for dimension lumber. The amount to add when no common rafter is framed at the end is one-half the thickness of the common rafter plus one-half the diagonal thickness of the hip. For dimension lumber this is ¾ + 1¹⁄₁₆ = 1¹³⁄₁₆ inches added to each end.

Raising the Hip Roof

Before raising the hip roof, the ridgeboard must first be laid out for the location of the common rafters **(Procedure 43–G)**. Measure from the outside corner the common run distance. Mark this length on the plate. Deduct the distance that the ridgeboard was lengthened on one end. This is ¾ or 1¹³⁄₁₆ inches for dimension lumber. Measure from this second mark to the next common rafter. This distance can then be transferred to the ridgeboard, measured from the end. Mark an X on the appropriate side.

Erect the common roof section in the same manner as described previously for a gable roof. If the hip rafters are framed against the ridge, install them next. If they are framed against the common rafters, install the common rafters against the end of the ridge. Then install the hip rafters.

Fasten jack rafters to the plate and to hip rafters in pairs. As each pair of jacks is fastened, sight the hip rafter along its length. Keep it straight. Any *bow* in the hip is straightened by driving the jack a little tighter against the bowed-out side of the hip as the roof is framed **(Procedure 43–H)**.

FIGURE 43–15
Determining the actual length of the hip roof ridge.

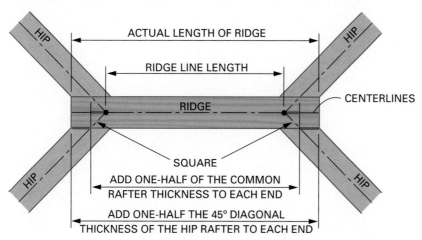

COMMON RAFTERS FRAMED AT THE END OF THE RIDGE

ACTUAL RIDGE LENGTH

RIDGE LINE LENGTH

HIP · COMMON · COMMON · HIP

COMMON · RIDGE · COMMON — CENTERLINES

HIP · COMMON · COMMON · HIP

SQUARE

ADD ONE-HALF OF THE THICKNESS OF THE COMMON RAFTER AT EACH END

NO COMMON RAFTERS FRAMED AT THE END OF THE RIDGE

ACTUAL LENGTH OF RIDGE

RIDGE LINE LENGTH

HIP · RIDGE · HIP — CENTERLINES

HIP · HIP

SQUARE

ADD ONE-HALF OF THE COMMON RAFTER THICKNESS TO EACH END

ADD ONE-HALF THE 45° DIAGONAL THICKNESS OF THE HIP RAFTER TO EACH END

Procedure 43–G Laying Out the Hip Ridgeboard

STEP 1 MEASURE ALONG WALL FROM THE CORNER THE DISTANCE EQUAL TO THE COMMON RUN. MARK ON THE WALL PLATE.

STEP 2 DEDUCT BACK TOWARD THE CORNER THE AMOUNT ADDED TO THE RIDGEBOARD (DETERMINED FROM FIGURE 43–15).

STEP 3 MEASURE FROM THIS MARK TO THE SIDE OF THE NEXT COMMON RAFTER.

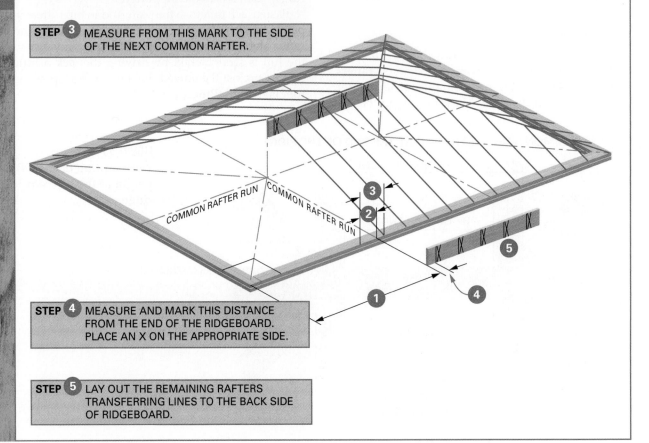

STEP 4 MEASURE AND MARK THIS DISTANCE FROM THE END OF THE RIDGEBOARD. PLACE AN X ON THE APPROPRIATE SIDE.

STEP 5 LAY OUT THE REMAINING RAFTERS TRANSFERRING LINES TO THE BACK SIDE OF RIDGEBOARD.

Procedure 43–H Erecting the Hip Roof Frame

STEP 1 ERECT THE RIDGE WITH ONLY AS MANY COMMON RAFTERS AS NEEDED.

STEP 2 INSTALL THE HIP RAFTERS. THE HIP RAFTERS EFFECTIVELY BRACE THE ASSEMBLY.

NOTE: THIS PROCEDURE SHOWS THE HIPS FRAMED TO THE RIDGE. IF HIPS ARE FRAMED TO THE COMMON RAFTERS, INSTALL THE COMMONS TO THE END OF THE RIDGE BEFORE THE HIPS.

STEP 3 INSTALL THE REMAINING COMMON RAFTERS IN PAIRS OPPOSING EACH OTHER. SIGHT THE TOP EDGE OF THE RIDGE FOR STRAIGHTNESS AS FRAMING PROGRESSES.

STEP 4 INSTALL THE HIP JACK RAFTERS IN PAIRS OPPOSING EACH OTHER. SIGHT THE HIP FOR STRAIGHTNESS AS JACKS ARE INSTALLED.

NOTE: IT IS BEST TO INSTALL A PAIR OF JACKS ABOUT HALFWAY UP THE HIP TO STRAIGHTEN IT. THE HIP CAN BE STRAIGHTENED BY DRIVING THE JACK, ON THE CROWN SIDE OF THE BOW, DOWN TIGHTER.

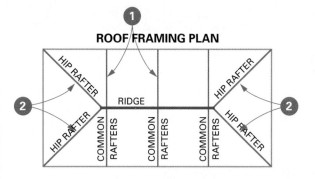

ROOF FRAMING PLAN

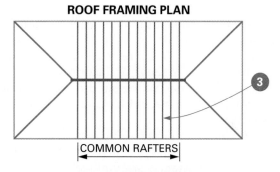

ROOF FRAMING PLAN

COMMON RAFTERS

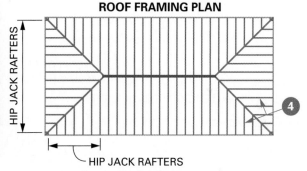

ROOF FRAMING PLAN

HIP JACK RAFTERS

HIP JACK RAFTERS

STEP BY STEP PROCEDURES

44 The Intersecting Roof

Buildings of irregular shape, such as L-, H-, or T-shaped buildings, require a roof for each section. These roofs meet at an inside corner intersection, where **valleys** are formed. They are called *intersecting roofs*. The roof may be a gable, a hip, or a combination of types.

THE INTERSECTING ROOF

The intersecting roof requires the layout and installation of several kinds of rafters not previously described **(Figure 44–1)**. Some buildings have sections of different widths referred to as **major spans** and **minor spans.** A major span is the width of the main part of the building, whereas a minor span refers to the width of any extension to the building.

Valley rafters form the inside corner intersection of two roofs. If the heights of both roofs are different, two kinds of valley rafters are required.

Supporting valley rafters run from the plate to the ridge of the main roof. Their run is one-half of the major span.

Shortened valley rafters run from the plate to the supporting valley rafter. Their run is one-half of the minor span.

Valley jack rafters run from the ridge to the valley rafter.

Valley cripple jack rafters run between the supporting and shortened valley rafter.

Hip-valley cripple jack rafters run between a hip rafter and a valley rafter.

Confusion concerning the layout of so many different kinds of rafters can be eliminated by remembering the following.

- The length of any kind of rafter can be found using its run.

- The amount of shortening is always measured at right angles to the plumb cut.

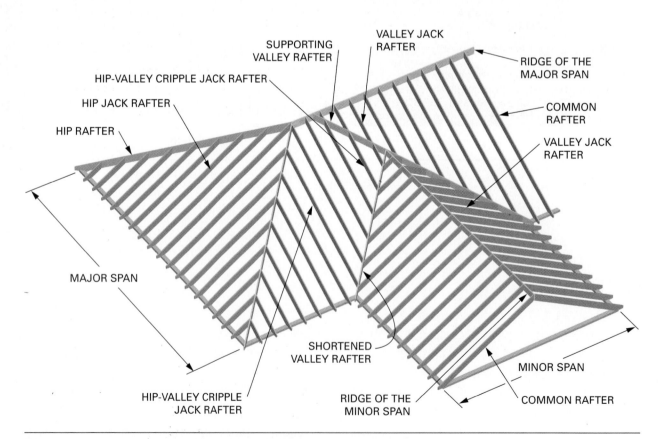

FIGURE 44–1 Members of the intersecting roof frame.

Procedure 44–A Laying Out a Valley Rafter

STEP 1 LAY OUT A PLUMB LINE.

STEP 2 MEASURE PERPENDICULARLY TO PLUMB LINE, ONE–HALF THE 45° THICKNESS OF THE RIDGE.

STEP 3 LAY OUT A SECOND PLUMB LINE. SQUARE THE PLUMB LINE ACROSS THE TOP EDGE.

STEP 4 MEASURE PERPENDICULARLY TO SECOND PLUMB LINE THE THICKNESS OF THE SUPPORTING VALLEY RAFTER STOCK.

STEP 5 DRAW THIRD PLUMB LINE.

STEP 6 DRAW A SQUARE LINE AND CENTERLINE ACROSS THE TOP FROM SECOND PLUMB LINE.

STEP 7 DRAW A DIAGONAL FROM THIRD PLUMB LINE THROUGH THE CENTERLINE.

STEP 8 DRAW A SQUARE LINE ACROSS THE TOP FROM THIRD PLUMB LINE.

STEP 9 DRAW SECOND DIAGONAL.

STEP 10 DRAW A FOURTH PLUMB LINE ON BACK SIDE OF RAFTER.

WHEN RIDGE HEIGHTS ARE DIFFERENT, THE SUPPORTING VALLEY RAFTER RUNS CONTINUOUS TO THE RIDGE OF THE MAJOR SPAN AND SUPPORTS THE SHORTENED VALLEY RAFTER.

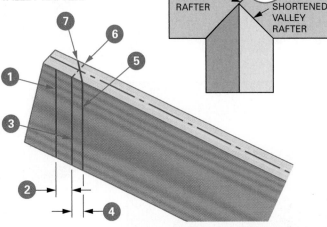

RIDGE OF MAJOR SPAN

SUPPORTING VALLEY RAFTER

SHORTENED VALLEY RAFTER

RIDGE HEIGHTS ARE THE SAME WHEN BOTH SPANS WITH THE SAME ROOF PITCH ARE EQUAL. BOTH VALLEY RAFTERS MEET AT THE RIDGE.

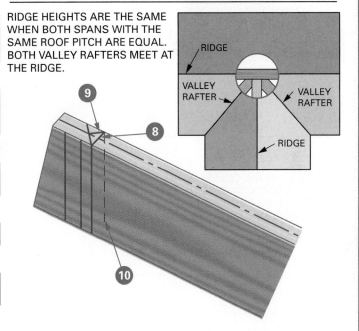

RIDGE

VALLEY RAFTER

VALLEY RAFTER

RIDGE

- Hip and valley rafters are similar. Common, jack, and cripple jack rafters are similar.
- The method previously described for laying out cheek cuts works on all rafters for any slope.

Supporting Valley Rafter Layout

The layout of a supporting valley rafter is more similar than not to that of a hip rafter. The unit of run for both is 17 inches. The total run for both is the same, one-half the major span. The valley rafter is shortened by half the 45-degree or diagonal thickness of the ridge board **(Procedure 44–A).** Cheek cuts are made at the ridge depending on how the building is framed.

The line length of either valley is found by the step-off method or by the calculation method, using the formula unit length × run.

Procedure 44–B Laying Out a Valley Seat Cut

STEP 1 MEASURE LINE LENGTH OF RAFTER AND DRAW PLUMB LINE.

STEP 2 MEASURE DOWN THE SAME DISTANCE AS HEIGHT ABOVE BIRD'S MOUTH OF COMMON RAFTER.

STEP 3 DRAW LEVEL LINE FOR THE SEAT CUT.

STEP 4 MEASURE PERPENDICULAR TO PLUMB LINE TOWARD THE TAIL, A DISTANCE THAT IS ONE-HALF THE VALLEY THICKNESS. THIS IS $3/4$" FOR DIMENSION LUMBER.

STEP 5 DRAW SECOND PLUMB LINE.

STEP 6 DRAW TWO SQUARED LINES ACROSS BOTTOM FROM THE TWO PLUMB LINES.

STEP 7 FIND THE CENTER OF THE FIRST SQUARED LINE AND DRAW DIAGONALS FOR CHEEK CUTS.

NOTE: THE BIRD'S MOUTH MAY BE CUT SQUARE FROM THE SECOND PLUMB LINE.

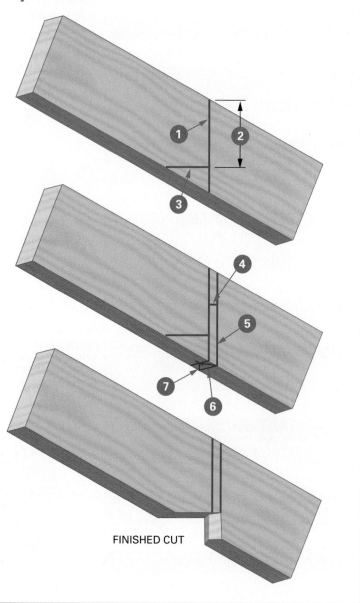

FINISHED CUT

Laying Out a Valley Rafter Seat Cut

Valley rafters are not dropped or backed like hip rafters. Instead, the seat cut line is made with side cuts or lengthened. This is done so the valley will clear the inside corner of the wall plate **(Procedure 44–B).**

To lay out a valley bird's mouth, first measure the line length and draw the seat plumb line. Then measure and mark down along this line, from the top edge of the rafter, a distance equal to the height above the bird's mouth on the common rafter. This makes the height above the bird's mouth the same

for all rafters. Draw a level line at this point to make a bird's mouth. From the plumb line, measure at a right angle toward the tail a distance equal to one-half the valley rafter thickness. Draw a new plumb line and extend to it to the previously drawn seat cut line. On the bottom edge of the rafter, draw two squared lines from the two plumb lines. Find the center of the first line and draw the appropriate diagonals.

There is no reason to drop or back a valley rafter as is done for hip rafters. The top corners of a valley are actually below the roof sheathing. The centerline of the top edge of the valley is aligned with the roof sheathing.

However, one edge of a supporting valley must be backed. This portion is located between the ridgeboards of the major and minor spans **(Figure 44–2)**. Only the lower edge facing the wall plate is backed.

Laying Out A Valley Rafter Tail

The length of the valley rafter tail is found in the same manner as that used for the hip rafter. It may be stepped off as shown in Figure 43–9 or calculated as shown in Procedure 43–D. Step off the same number of times, from the plumb line of the seat cut, as stepped off for the common rafter. However, use a unit run of 17 instead of 12.

The tail cut of the valley rafter is a double cheek cut. It is similar to that of the hip rafter, but angles outward instead. From the tail plumb line, measure at right angles to the plumb line toward the tail, one-half the thickness of the rafter stock. Lay out another plumb line. Square both plumb lines across the top of the rafter. Draw diagonals from the center **(Procedure 44–C)**.

Shortening a Valley

The shortened valley differs from the supporting valley only in length and ridge cut. The length of the shortened valley is found by using the run of the common rafter of the minor span.

The plumb cut at the top end is different from that of the supporting valley. Since the two valley rafters meet at right angles, the cheek cut of the shortened valley is a squared miter, not a compound miter. It looks similar to the ridge plumb line of a common rafter. To shorten this rafter, measure back, at right angles from the plumb line at the top end, a distance that is half the thickness of the supporting valley rafter stock. Lay out another plumb line **(Procedure 44–D)**.

Valley Jack Layout

The valley jack is a common rafter that is shortened at its bottom end where it meets the valley. The length of the valley jack is found, like any rafter, by multiplying unit length × run. The total run of any jack is the horizontal distance measured along the ridge to that rafter from the upper end of its valley rafter **(Figure 44–3)**. Remember to use unit run and length of a common rafter for the valley jack.

The ridge cut of the valley jack is the same as a common rafter. It is shortened in the same way. The

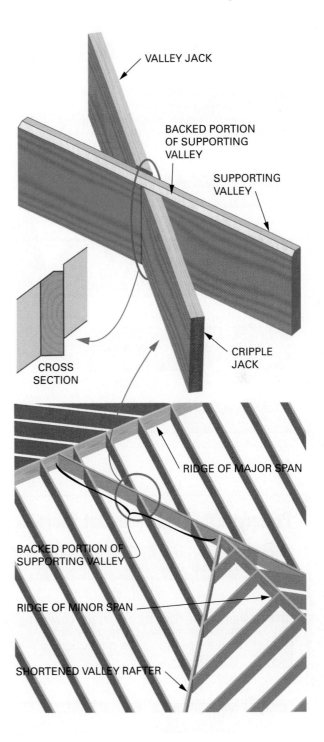

FIGURE 44–2   Backing the upper section of supporting valley.

cheek cut against the valley rafter is a single cheek cut that is shortened, like other rafters meeting at a diagonal, by deducting one-half the 45-degree thickness of the valley rafter.

Procedure 44–C Laying Out a Valley Rafter Tail

STEP ① MEASURE TAIL LENGTH OR STEP-OFF PROJECTION.
NOTE: THIS IS DONE FROM THE FIRST BIRD'S MOUTH PLUMB LINE.

STEP ② DRAW PLUMB LINE.

STEP ③ MEASURE PERPENDICULAR TO PLUMB LINE ONE-HALF THICKNESS OF VALLEY. THIS IS ³/₄" FOR DIMENSION LUMBER.

STEP ④ DRAW TWO SQUARED LINES ACROSS TOP OF RAFTER AND MARK CENTERLINE.

STEP ⑤ DRAW DIAGONALS.

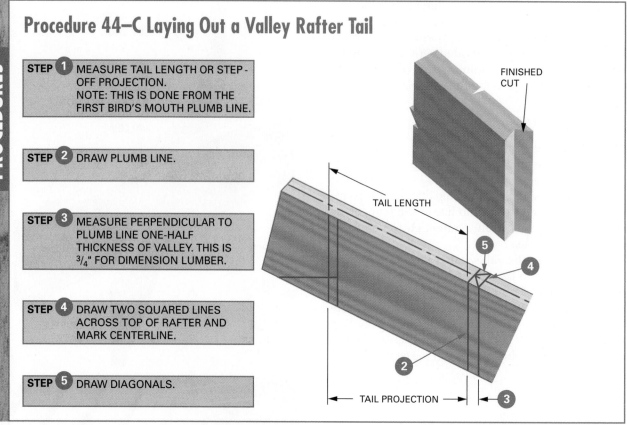

FINISHED CUT

TAIL LENGTH

TAIL PROJECTION

Procedure 44–D Laying Out Shortened Valley Rafter Ridge Plumb Cut

STEP ① DRAW PLUMB LINE USING SQUARE SETUP FOR A HIP OR VALLEY.

STEP ② MEASURE PERPENDICULAR TO PLUMB LINE ONE-HALF THICKNESS OF SUPPORTING VALLEY. THIS IS ³/₄" FOR DIMENSION LUMBER.

STEP ③ DRAW SECOND PLUMB LINE.

NOTE: CUT IS SQUARED ACROSS TOP OF RAFTER, SIMILAR TO COMMON RAFTERS.

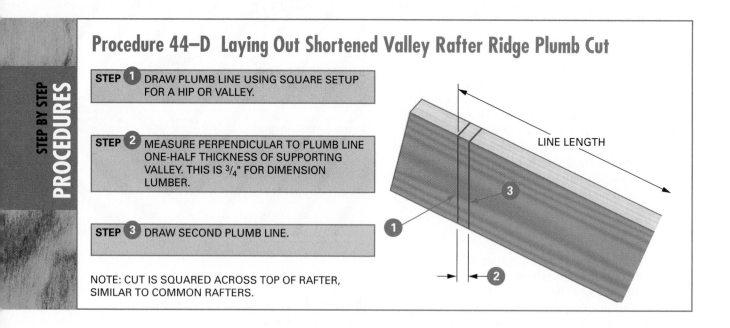

LINE LENGTH

A valley jack has a common difference in length, similar to hip jacks. The layout for other valley jacks is made by adding or deducting the common difference from the jack length. This number is found on the fifth line of the rafter tables or may be calculated as shown in Procedure 43–F.

Valley Cripple Jack Layout

As stated before, the length of any rafter can be found if its total run is known. The run of the valley cripple jack is always twice its horizontal distance from the intersection of the valley rafters

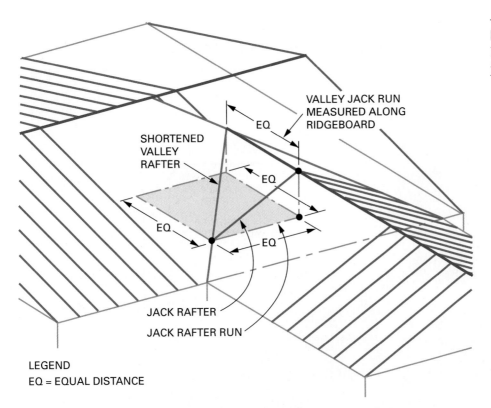

FIGURE 44–3
Determining the valley jack rafter run.

LEGEND
EQ = EQUAL DISTANCE

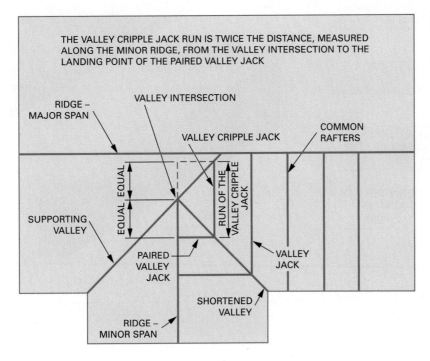

FIGURE 44–4
Determining a valley cripple jack rafter run.

(Figure 44–4). Use the common rafter tables or step off in a manner similar to that used for common rafters to find its length. Shorten each end by one-half the 45-degree thickness of the valley rafter stock **(Procedure 44–E).** Noting that they angle in opposite directions, make a single cheek cut.

Hip-Valley Cripple Jack Rafter Layout

All hip-valley cripple jacks cut between the same hip and valley rafters are the same. This is because the hip and valley rafters are parallel. To determine the length of the rafter, first find its total run.

Procedure 44–E Laying Out a Valley Cripple Jack Rafter

STEP 1 DRAW COMMON PLUMB LINE AT UPPER END.

STEP 2 SHORTEN BY ONE-HALF 45-DEGREE THICKNESS OF SUPPORTING VALLEY. THIS IS 1¹/₁₆" FOR DIMENSION LUMBER. DRAW SECOND PLUMB LINE.

STEP 3 SQUARE AND CENTER A LINE OVER TOP OF RAFTER.

STEP 4 MEASURE PERPENDICULAR ONE-HALF CRIPPLE JACK THICKNESS. THIS IS ³/₄" FOR DIMENSION LUMBER. DRAW THIRD PLUMB LINE.

STEP 5 DRAW DIAGONAL ACROSS TOP FROM THIRD PLUMB LINE THROUGH CENTERLINE.

STEP 6 MEASURE LINE LENGTH OF CRIPPLE JACK.

STEP 7 REPEAT PROCEDURE, BUT SHORTEN AND MEASURE TOWARD UPPER END.

NOTE: DIAGONAL ON TOP OF RAFTER MAY BE DRAWN IN TWO DIRECTIONS. MAKE SURE THEY ARE APPROPRIATE AND IN OPPOSITE DIRECTIONS.

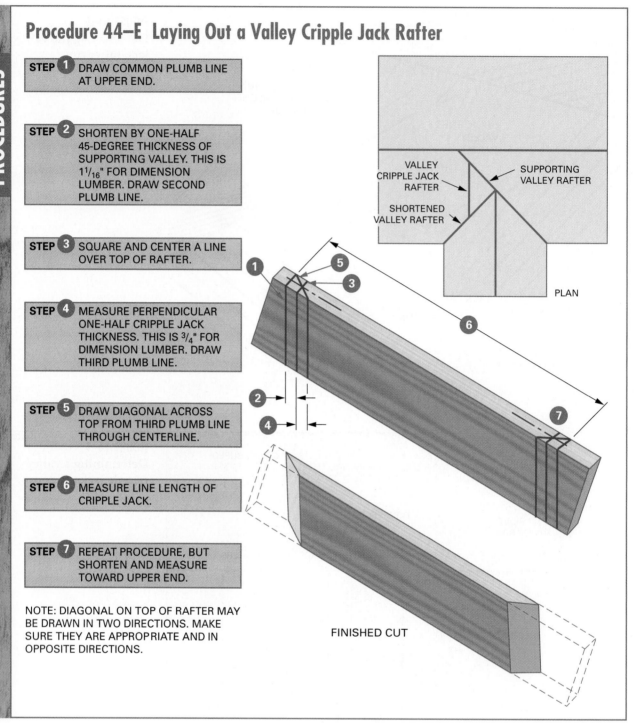

VALLEY CRIPPLE JACK RAFTER

SUPPORTING VALLEY RAFTER

SHORTENED VALLEY RAFTER

PLAN

FINISHED CUT

The run of a hip-valley cripple jack rafter is equal to the plate line distance between the seat cuts of the hip and valley rafters **(Figure 44–5)**. Remember to lay out all jack plumb lines the same as for common rafters.

Draw a plumb line at the top end. Shorten by measuring perpendicular toward the lower end, one-half 45-degree thickness of the hip rafter. This is 1¹⁄₁₆ inches for dimension lumber **(Procedure 44–F)**. Draw a second plumb line. Square a line over the top edge of the

Procedure 44–F Laying Out a Hip-Valley Cripple Jack Rafter

STEP 1 DRAW COMMON PLUMB LINE AT UPPER END.

STEP 2 SHORTEN ONE-HALF 45 DEGREE THICKNESS OF SUPPORTING VALLEY. THIS IS 1$^1/_{16}$" FOR DIMENSION LUMBER. DRAW SECOND PLUMB LINE SQUARE AND CENTER A LINE OVER TOP OF RAFTER.

STEP 3 MEASURE PERPENDICULAR ONE-HALF CRIPPLE JACK THICKNESS. THIS IS $^3/_4$" FOR DIMENSION LUMBER. DRAW THIRD PLUMB LINE.

STEP 4 DRAW DIAGONAL ACROSS TOP FROM THIRD PLUMB LINE THROUGH CENTERLINE.

STEP 5 MEASURE LINE LENGTH OF HIP - VALLEY CRIPPLE JACK.

STEP 6 REPEAT PROCEDURE, BUT SHORTEN USING ONE-HALF 45-DEGREE THICKNESS OF VALLEY. THIS IS 1$^1/_{16}$" FOR DIMENSION LUMBER.

NOTE: DIAGONALS ON TOP OF RAFTER MAY BE DRAWN IN TWO DIRECTIONS. MAKE SURE THEY ARE APPROPRIATE AND IN SAME DIRECTION.

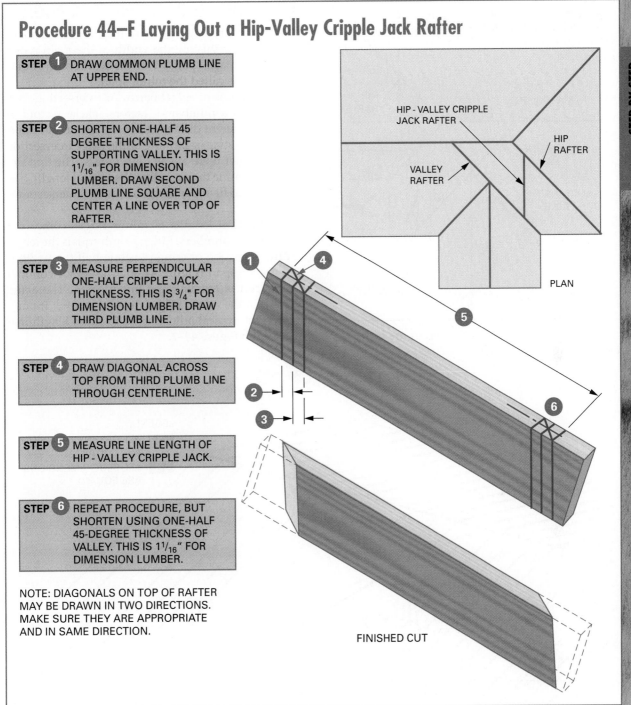

HIP - VALLEY CRIPPLE JACK RAFTER

HIP RAFTER

VALLEY RAFTER

PLAN

FINISHED CUT

cripple also marking the center of the line. Measure and mark perpendicular toward the lower end, one-half the thickness of the cripple rafter. This is ¾ inch for dimension lumber. Draw a third plumbline. Across the top edge, draw a diagonal from the top of the third line through the centerline.

Measure the line length of the cripple and repeat the exact layout of the upper end. This is done because the cuts are parallel. The only exception is that the shortening is one-half 45-degree thickness of the valley rafter which is 1¹⁄₁₆ inches for dimension lumber.

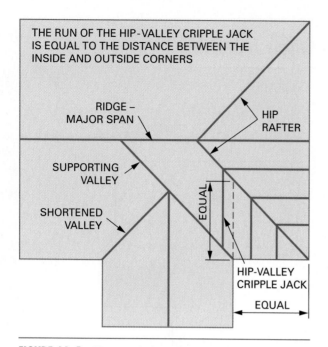

THE RUN OF THE HIP-VALLEY CRIPPLE JACK IS EQUAL TO THE DISTANCE BETWEEN THE INSIDE AND OUTSIDE CORNERS

RIDGE – MAJOR SPAN

HIP RAFTER

SUPPORTING VALLEY

SHORTENED VALLEY

EQUAL

HIP-VALLEY CRIPPLE JACK

EQUAL

EQUAL

FIGURE 44–5 Determining the run of the hip-valley cripple jack rafter.

RIDGE LENGTHS OF INTERSECTING ROOFS

Intersecting roofs have more than one ridgeboard. To reduce confusion, the width of the main roof is called the major span, while the width of the building wings is called the minor span.

The lengths of ridgeboards for intersecting roofs depend on a number of factors. These factors include the height differences of the major and minor ridges, the style of framing at the ridge intersection, and whether there are any hip rafters. The first step is to determine the overall or theoretical length, then modify it to fit depending on how the intersection is framed.

Theoretical lengths of minor ridges come from adding two numbers. Ridge length equals the length of the extension or wing plus one-half the width of the minor span **(Figure 44–6).** The minor common rafter run is one-half the minor span. The actual length of the minor ridges depends on the framing style. The typical adjustments to ridge length may be seen in **Figure 44–7.**

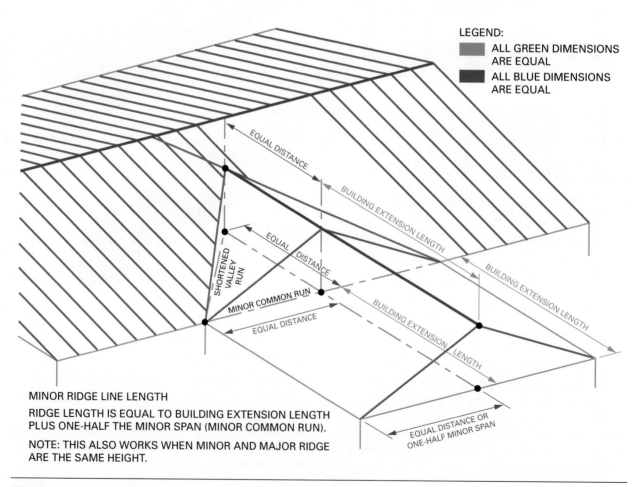

LEGEND:

ALL GREEN DIMENSIONS ARE EQUAL

ALL BLUE DIMENSIONS ARE EQUAL

EQUAL DISTANCE

BUILDING EXTENSION LENGTH

EQUAL DISTANCE

BUILDING EXTENSION LENGTH

SHORTENED VALLEY RUN

MINOR COMMON RUN

BUILDING EXTENSION LENGTH

EQUAL DISTANCE

EQUAL DISTANCE OR ONE-HALF MINOR SPAN

MINOR RIDGE LINE LENGTH

RIDGE LENGTH IS EQUAL TO BUILDING EXTENSION LENGTH PLUS ONE-HALF THE MINOR SPAN (MINOR COMMON RUN).

NOTE: THIS ALSO WORKS WHEN MINOR AND MAJOR RIDGE ARE THE SAME HEIGHT.

FIGURE 44–6 Determining the line length of the minor ridge.

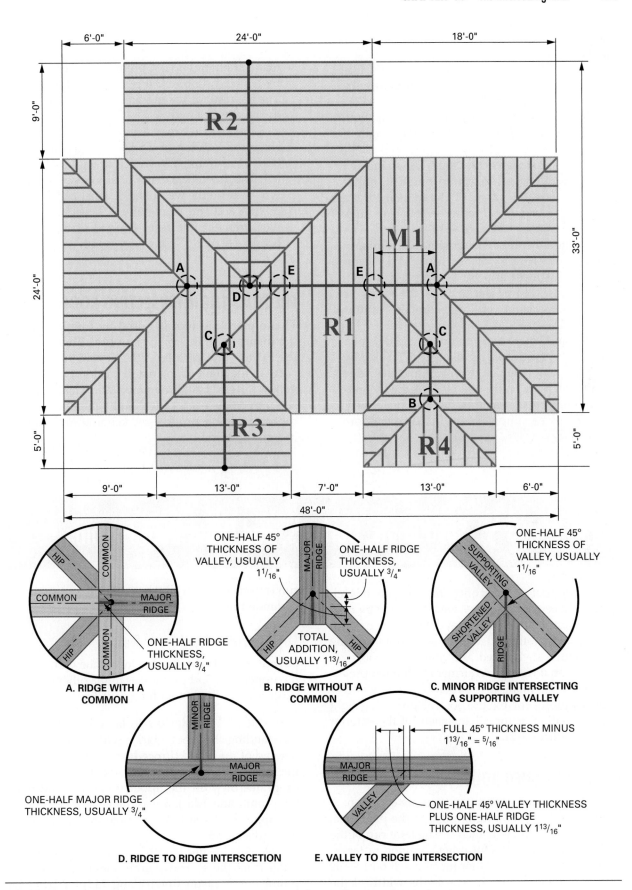

FIGURE 44–7 Ridgeboard line lengths are adjusted depending on the style of framing.

EXAMPLE   Using the information in Figure 44–7 and dimension lumber framing, the theoretical length of Ridge 1 is 48' − 24' = 24'. Adjustments including adding ¾" at both points A give an actual Ridge 1 (R1) length of 24'–1½".

Ridge 2 theoretical length is found by adding the wing length to one-half minor span. This is 9' + 12' = 21'. The adjustment is a deduction at point B of one-half major ridge thickness or 21 − ¾" = 20'–11¼".

Ridge 3 theoretical length is found similarly to Ridge 2. Thus 5' + 6.5' = 11'–6". The necessary adjustment at point C is to deduct one-half the 45-degree thickness of the valley or 1⁄₁₆". Actual length of Ridge 3 is 11'–6" minus 1⁄₁₆" = 11'–4¹⁵⁄₁₆".

Ridge 4 length looks complicated but is not. Ridge 4 is a side of a parallelogram formed by the hip and valley on two sides and the ridge and wall plate on the others. Thus the theoretical length is the same as the wall plate measurement or simply 5 feet. Adjustments are

made at either end. At point C it is shortened by 1⁄₁₆" and at point D it is lengthened by 1¹³⁄₁₆". The net result is to lengthen Ridge 4 by ¾" giving an actual length of 5'–¾".

Sometimes we may want to measure on the ridge the location of a valley/ridge intersection. Measurement 1 is noted in Figure 44–7 as M1. The theoretical distance of Measurement 1 is 6'–0" from the wall plate distance. Add ¾" at point A and add ⁵⁄₁₆" at point E. Thus the actual measurement line from the end of the ridge board is 6' plus ¾" plus ⁵⁄₁₆" = 6'–1¹⁄₁₆".

ERECTING THE INTERSECTING ROOF

The intersecting roof is raised by framing opposing members of the main span first. Then, the valley rafters are installed. To prevent bowing the ridge of the main span, install rafters to oppose the valley rafters. Install common and jack rafters in sets opposing each other. Sight members of the roof as framing progresses, keeping all members in a straight line.

45 Shed Roofs, Dormers, and Other Roof Framing

The shed roof slopes in only one direction. It is relatively easy to frame. A shed roof may be freestanding or one edge may rest on an exterior wall while the other edge butts against an existing wall **(Figure 45–1).** Shed and other type roofs are also used on **dormers.** A dormer is a framed projection above the plane of the roof. It contains one or more windows for the purpose of providing light and ventilation or for enhancement of the exterior design **(Figure 45–2).**

FRAMING A SHED ROOF

A shed roof is made of rafters that are common. A shed rafter run is at a right angle to the plate line, and the unit of run is 12 inches. The total run of the rafter is determined as for other common rafters—from where it bears on other members. If both ends rest on exterior walls, the run is the width of the building minus the thickness of one of the walls **(Figure 45–3).** If one end rests on another wall, the

run is from wall to wall. These are laid out similar to common rafters.

Laying Out a Shed Roof Rafter

Layout for a shed rafter begins at one end, such as the upper one, and progresses to the other end. The process is similar to that for any rafter.

Freestanding Shed Rafter. To lay out a rafter for a freestanding shed roof, mark a plumb line at the upper end of the board **(Procedure 45–A).** This will be the fascia cut line. Deduct for fascia thickness, if appropriate. Step off the upper projection and mark the plumb line. Mark a reasonable seat cut line for the upper bird's mouth and note the height above the bird's mouth.

Calculate the line length and measure it along the rafter from the second plumb line. Draw the third plumb line. Measure down along this plumb line the height above the bird's mouth. Draw another seat cut. Step off the lower projection and draw the final

FIGURE 45–1 A building with shed roofs butting each other.

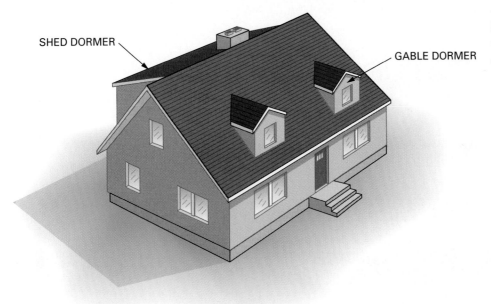

FIGURE 45–2 A roof with gable and shed dormers.

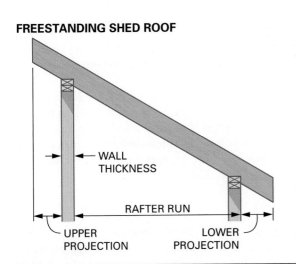

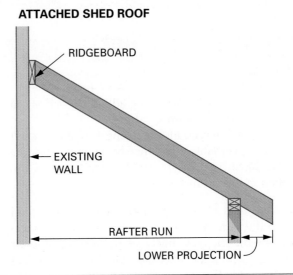

FIGURE 45–3 Styles of shed roofs.

STEP BY STEP PROCEDURES

Procedure 45–A Laying Out a Freestanding Shed Rafter (Double Overhang)

STEP 1 MARK A PLUMB LINE AT THE UPPER END OF THE BOARD.

STEP 2 STEP OFF THE UPPER PROJECTION AND MARK THE SECOND PLUMB LINE.

STEP 3 MARK A REASONABLE SEAT CUT LINE FOR THE UPPER BIRD'S MOUTH AND NOTE THE HEIGHT ABOVE THE BIRD'S MOUTH.

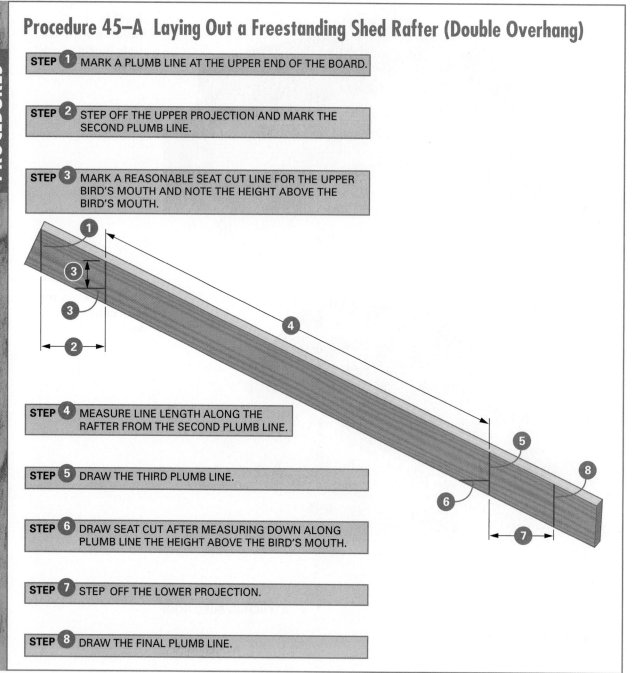

STEP 4 MEASURE LINE LENGTH ALONG THE RAFTER FROM THE SECOND PLUMB LINE.

STEP 5 DRAW THE THIRD PLUMB LINE.

STEP 6 DRAW SEAT CUT AFTER MEASURING DOWN ALONG PLUMB LINE THE HEIGHT ABOVE THE BIRD'S MOUTH.

STEP 7 STEP OFF THE LOWER PROJECTION.

STEP 8 DRAW THE FINAL PLUMB LINE.

plumb line. Level cuts may be added to each fascia cut as is done for common rafters.

The rafters may be installed long with the tails left wild. The overhang would then be trimmed after the rafters have been installed. This procedure is described in Procedure 42–C.

Attached Shed Rafter. Attached shed rafters are laid out in the same way as for common rafters **(Procedure 45–B).** The only difference is the shortening for the ridge. A common rafter shortens by deducting one-half the ridge thickness. For an attached shed rafter, however, the shortening is a full ridge thickness.

Shed Rafter Erection. Shed rafters are toenailed to the plates or ridge at the designated spacing. It is important to keep the bird's mouth snug up against the walls. In addition to nailing the rafters, metal framing connectors may be required. This includes all types of framed roofs.

Dormers

Dormers typically have either a gable or a shed roof. A gable dormer roof is framed like an intersecting gable roof with valleys and jacks. A shed dormer is simpler in design and often framed to

Procedure 45–B Laying Out an Attached Shed Rafter (Single Overhang)

STEP 1 MARK A PLUMB LINE AT THE UPPER END OF THE BOARD.

STEP 2 MEASURE PERPENDICULAR THE FULL THICKNESS OF RIDGEBOARD.

STEP 3 MARK THE SECOND PLUMB LINE.

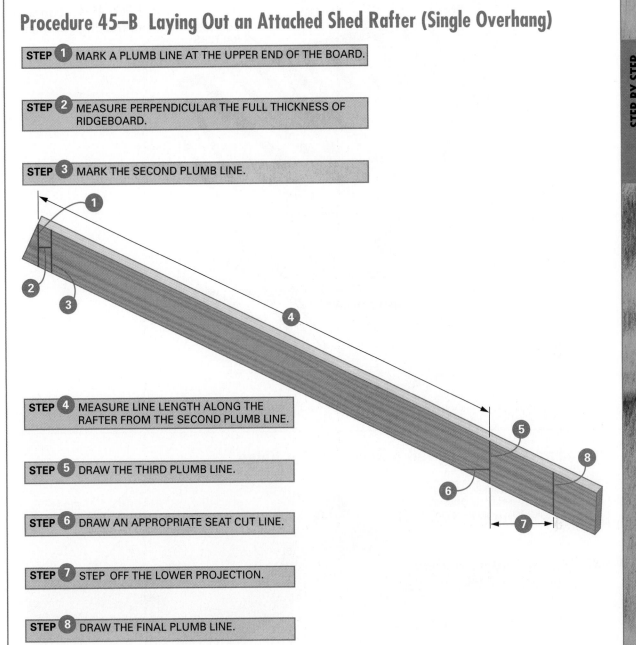

STEP 4 MEASURE LINE LENGTH ALONG THE RAFTER FROM THE SECOND PLUMB LINE.

STEP 5 DRAW THE THIRD PLUMB LINE.

STEP 6 DRAW AN APPROPRIATE SEAT CUT LINE.

STEP 7 STEP OFF THE LOWER PROJECTION.

STEP 8 DRAW THE FINAL PLUMB LINE.

the ridge of the main roof. This is typically done to gain sufficient rafter slope **(Figure 45–4).**

When framing openings in the main roof for dormers, the rafters on both sides of the opening are doubled. Some dormers have their front walls directly over the exterior wall below. In this case much of the load of the dormer is transferred into the load-bearing wall below. If dormers are placed such that the front wall is recessed or if the dormers are partway up the main roof, the main roof must be strengthened. Double headers are also added at the top and bottom of the opening.

OTHER ROOF FRAMING

A number of other types of roof framing exist, but they are all related, in some way, to the framing theory previously described. To solve these various roof framing situations, begin by returning to the basics of understanding rafter length as compared with run and rise.

Sometimes a gable roof or dormer is framed to an already existing building. Ridgeboards and valley jack rafters are cut to fit to the existing sheathing line and not to valley rafters. Shed rafters may also be cut to existing roofs as well.

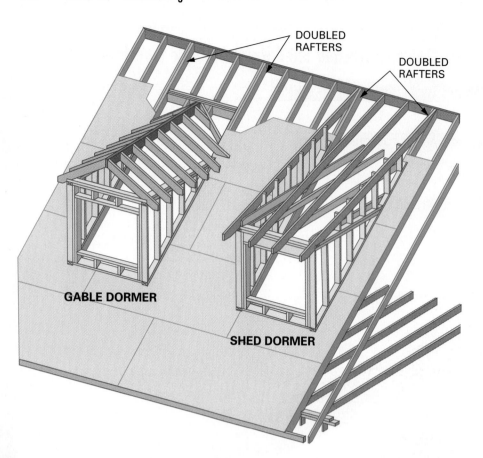

DOUBLED RAFTERS

DOUBLED RAFTERS

GABLE DORMER

SHED DORMER

FIGURE 45–4   Two styles of framing a dormer.

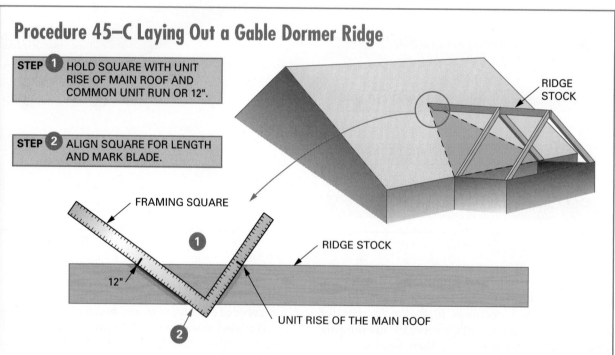

Procedure 45–C Laying Out a Gable Dormer Ridge

STEP 1 HOLD SQUARE WITH UNIT RISE OF MAIN ROOF AND COMMON UNIT RUN OR 12".

STEP 2 ALIGN SQUARE FOR LENGTH AND MARK BLADE.

RIDGE STOCK

FRAMING SQUARE

1

RIDGE STOCK

12"

UNIT RISE OF THE MAIN ROOF

2

Fitting a Gable Dormer to Ridge Roof Sheathing

The ridge must be fitted level into the existing roof. The simplest way to determine the ridge length is to first cut a slope on the end of the ridge, then install it with common rafters and adjust it until it fits. The ridge is later cut for length after plumbing up from the end wall.

The ridge slope cut layout uses the unit rise for the existing roof **(Procedure 45–C).** Hold the unit rise and 12 on a framing square and adjust to the end of the board. Mark the blade of the square.

Fitting a Valley Jack Rafter to Roof Sheathing

After the ridge is placed against the existing roof, a 1 × 6 is nailed where a valley would be. It provides a place for the new valley jacks to land. The upper end of the valley jack is laid out like a common rafter. The lower end is a compound miter.

Jack rafter run is measured along the ridge from the center of the rafter where it lands on the ridge to the intersection of the ridge and the 1 × 6. No shortening need be done at the bottom miter because the run is measured as actual.

The cross-sectional rise of the jack rafter must be determined to cut the compound miter at the proper angle. This amount of rise is related to the slope of the main roof and the thickness of the jack rafter. Layout is done from the center of the jack since the run is measured from its center. Therefore the cross-sectional run is one-half of the jack thickness, usually ¾". Converting ¾" to feet is $0.75" \div 12 = 0.0625'$. The cross-sectional rise is unit rise times 0.0625 (**Procedure 45–D**).

Fitting a Shed Roof Rafter to Roof Sheathing

The top ends of shed dormer rafters may be fit against the main roof which has a steeper slope. The cuts are laid out by using a framing square as outlined in **Procedure 45–E**.

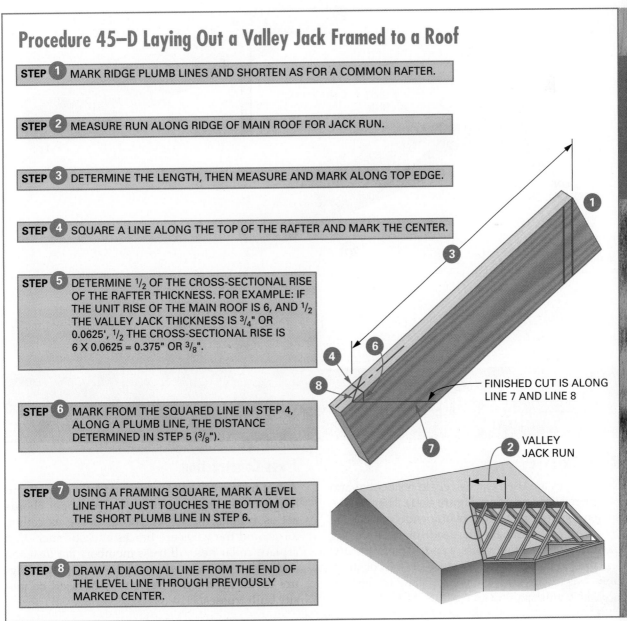

Procedure 45–D Laying Out a Valley Jack Framed to a Roof

STEP 1 MARK RIDGE PLUMB LINES AND SHORTEN AS FOR A COMMON RAFTER.

STEP 2 MEASURE RUN ALONG RIDGE OF MAIN ROOF FOR JACK RUN.

STEP 3 DETERMINE THE LENGTH, THEN MEASURE AND MARK ALONG TOP EDGE.

STEP 4 SQUARE A LINE ALONG THE TOP OF THE RAFTER AND MARK THE CENTER.

STEP 5 DETERMINE ½ OF THE CROSS-SECTIONAL RISE OF THE RAFTER THICKNESS. FOR EXAMPLE: IF THE UNIT RISE OF THE MAIN ROOF IS 6, AND ½ THE VALLEY JACK THICKNESS IS ³/₄" OR 0.0625', ½ THE CROSS-SECTIONAL RISE IS 6 X 0.0625 = 0.375" OR ³/₈".

STEP 6 MARK FROM THE SQUARED LINE IN STEP 4, ALONG A PLUMB LINE, THE DISTANCE DETERMINED IN STEP 5 (³/₈").

STEP 7 USING A FRAMING SQUARE, MARK A LEVEL LINE THAT JUST TOUCHES THE BOTTOM OF THE SHORT PLUMB LINE IN STEP 6.

STEP 8 DRAW A DIAGONAL LINE FROM THE END OF THE LEVEL LINE THROUGH PREVIOUSLY MARKED CENTER.

FINISHED CUT IS ALONG LINE 7 AND LINE 8

VALLEY JACK RUN

STEP BY STEP PROCEDURES

Procedure 45–E Laying Out a Shed Dormer Rafter

STEP ① DRAW A LEVEL LINE TO THE SLOPE OF THE SHED ROOF.

STEP ② REVERSE THE SQUARE AND HOLD ON THE LEVEL LINE TO THE SLOPE OF THE MAIN ROOF.

STEP ③ DRAW CUTTING LINE ALONG THE BLADE OF THE SQUARE. EXTEND THE LINE ACROSS ENTIRE SIDE OF RAFTER.

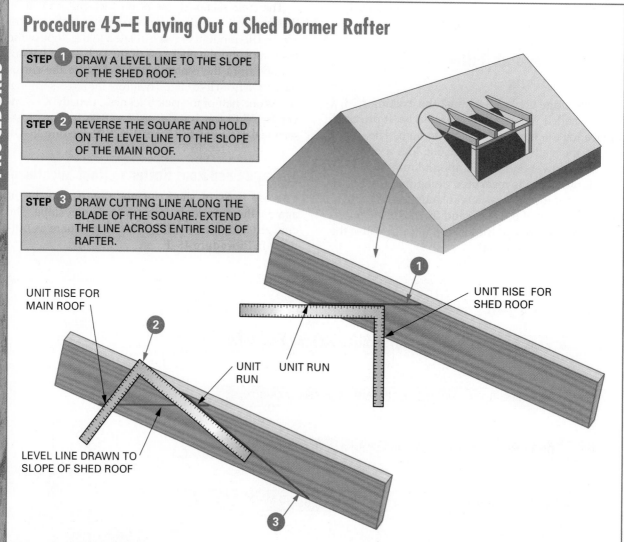

UNIT RISE FOR MAIN ROOF

UNIT RISE FOR SHED ROOF

UNIT RUN

UNIT RUN

LEVEL LINE DRAWN TO SLOPE OF SHED ROOF

46 Trussed Roofs

ROOF TRUSSES

Roof trusses are used extensively in residential and commercial construction **(Figure 46–1).** They are designed to support the roof over a wide span, up to 100 feet, and eliminate the need for load-bearing partitions. They require less wood to manufacture, less time to install, and can reduce overall construction costs. Some styles, however, allow for little or no usable attic space.

Truss Construction

A roof truss consists of upper and lower *chords* and diagonal members called *webs*. The upper chords act as rafters, the lower chords serve as ceiling joists, and the webs are braces and stiffeners that replace collar ties. All truss members are fastened and joined securely with metal gusset plates **(Figure 46–2).** Plywood gussets may be used on jobsite built trusses.

FIGURE 46–1   Trusses are used extensively in residential construction.

Trusses are designed with smaller member sizes than are rafter-ceiling joist systems. This causes higher stresses in the roof system members. Never cut any webs or chords of a truss unless directed by an engineer. Also, installing trusses can be very dangerous. Lives have been lost while installing trusses improperly. For these reasons, care must be employed by engineers in their designs and carpenters in their installation practices. ■

FIGURE 46–2   The members of roof trusses are securely fastened with metal gussets.

Truss Design

Most trusses are made in fabricating plants. They are transported to the job site. Trusses are designed by engineers to support prescribed loads. Trusses may also be built on the job, but approved designs must be used. The carpenter should not attempt to design a truss. Approved designs and instructions for job-built trusses are available from the American Plywood Association and the Truss Plate Institute.

The most common truss design for residential construction is the *Fink* truss **(Figure 46–3).** Other truss shapes are designed to meet special requirements **(Figure 46–4).**

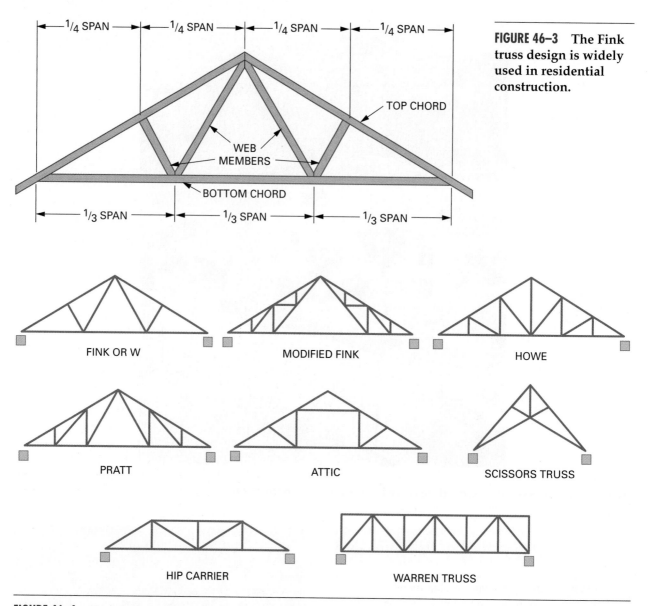

FIGURE 46–3   The Fink truss design is widely used in residential construction.

FIGURE 46–4   Various truss designs for special requirements.

Erecting a Trussed Roof

Carpenters are more involved in the erection than the construction of trusses. Trusses are delivered to the jobsite using specially designed trucks. They should be unloaded and stored on a flat dry surface. A print is provided showing the location of all trusses. A drawing of each truss is also provided that outlines important installation points **(Figure 46–5).**

The erection and bracing of a trussed roof is a critical stage of construction. Failure to observe recommendations for erection and bracing could cause a collapse of the structure. This could result in loss of life or serious injury, not to mention the loss of time and material.

The recommendations contained herein are technically sound. However, they are not the only method of bracing a roof system. They serve only as a guide. The builder must take necessary precautions during handling and erection to ensure that trusses are not damaged, which might reduce their strength.

Small trusses, which can be handled by hand, are placed upside down, hanging on the wall plates, toward one end of the building. Trusses for wide spans require the use of a crane to lift them into position. One at a time the truss is lifted, fastened, and braced in place. The end truss is installed first by swinging it up into place and bracing it securely in a plumb position **(Figure 46–6).**

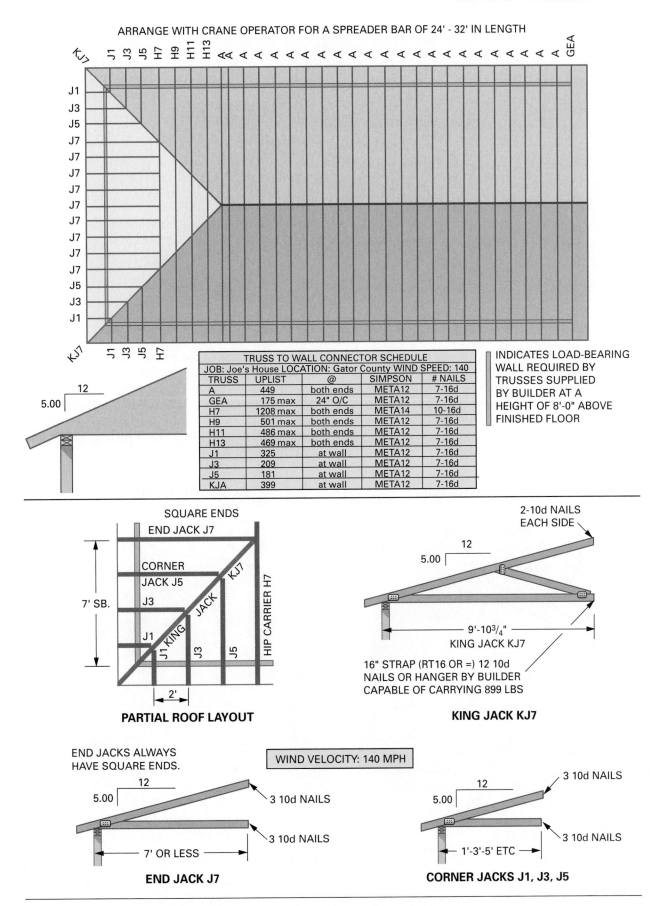

FIGURE 46–5 Trusses are often delivered with an engineered set of prints showing truss labels and locations.

FIGURE 46-6 Trusses are often lifted into place with a crane for speed and safety.

First Truss Installation

The end truss is usually installed first, but not always. In either case installation of the first truss requires great care. Trusses must be adequately braced in position, held secure, and plumb. They must be temporarily braced with enough strength to support the tip-over force of the trusses to follow. This may be achieved by bracing to securely anchored stakes driven into the ground or by bracing to the inside floor under the truss **(Figure 46-7).** These braces should be located directly in line with all rows of continuous top chord **lateral bracing,** which will

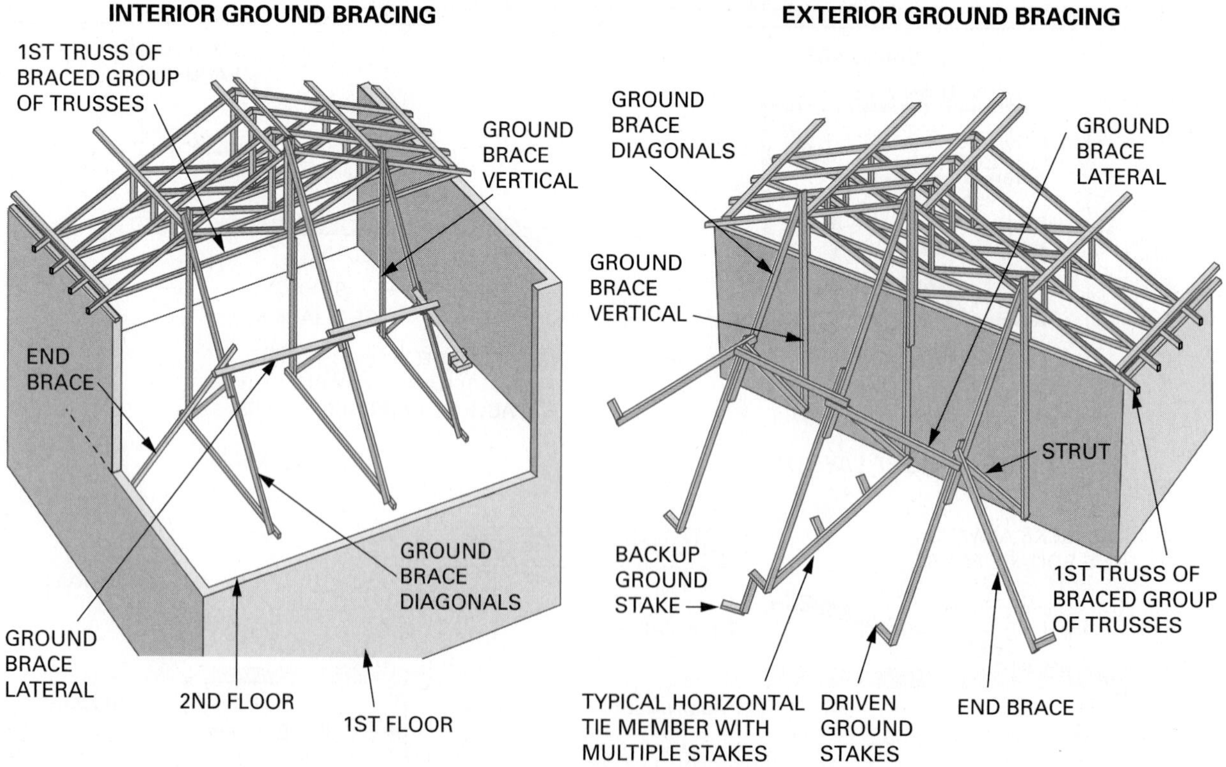

INTERIOR GROUND BRACING

1ST TRUSS OF
BRACED GROUP
OF TRUSSES

GROUND
BRACE
VERTICAL

END
BRACE

GROUND
BRACE
DIAGONALS

GROUND
BRACE
LATERAL

2ND FLOOR

1ST FLOOR

EXTERIOR GROUND BRACING

GROUND
BRACE
DIAGONALS

GROUND
BRACE
LATERAL

GROUND
BRACE
VERTICAL

STRUT

1ST TRUSS OF
BRACED GROUP
OF TRUSSES

BACKUP
GROUND
STAKE →

TYPICAL HORIZONTAL
TIE MEMBER WITH
MULTIPLE STAKES

DRIVEN
GROUND
STAKES

END BRACE

FIGURE 46-7 The first set of trusses must be well braced before the erection of other trusses.

be installed later. All bracing should be securely fastened with the appropriate size and quantity of nails remembering that lives depend on the bracing doing its intended job.

Temporary Bracing of Trusses

As trusses are set in place, they are nailed to the plate. Metal framing connectors are usually applied **(Figure 46–8).** Information on these connectors may be found on the connector schedule of the set of prints. Sufficient temporary bracing must be applied to secure trusses until the finish material is applied and/or until permanent bracing is installed. Temporary bracing should be no less than 2 × 4 lumber as long as practical, with a minimum length of 8 feet. The 2 × 4s should be fastened with two 16d nails at every intersection and should be overlapped by two trusses **(Figure 46–9).** Exact spacing of the trusses should be maintained. Adjusting trusses later, while possible, is time consuming and risky.

Temporary bracing must be applied to three planes of the truss assembly: the top chord or sheathing plane, the bottom chord or ceiling plane, and the vertical web plane at right angles to the bottom chord **(Figure 46–10).**

Top Chord Bracing. Continuous lateral bracing should be installed within 6 inches of the ridge and at about 8- to 10-foot intervals between the ridge and wall plate. Diagonal bracing should be set at approximately 45-degree angles between the rows of lateral bracing. It forms triangles that provide stability to the plane of the top chord **(Figure 46–11).**

Web Plane Bracing. Temporary bracing in the plane of the web members is made up of diagonals placed at right angles to the trusses from top to bottom chords **(Figure 46–12).** They usually become permanent braces of the web member plane.

Bottom Chord Bracing. To maintain the proper spacing on the bottom chord, continuous lateral bracing for the full length of the building must be applied. The bracing should be nailed to the top of the bottom chord at intervals no greater than 8 to 10 feet along the width of the building.

Diagonal bracing should be installed, at least, at each end of the building **(Figure 46–13).** In most cases, temporary bracing of the plane of the bottom chord is left in place as permanent bracing.

Permanent Bracing

Permanent bracing is designed by the structural engineer when designing the truss, and all bracing

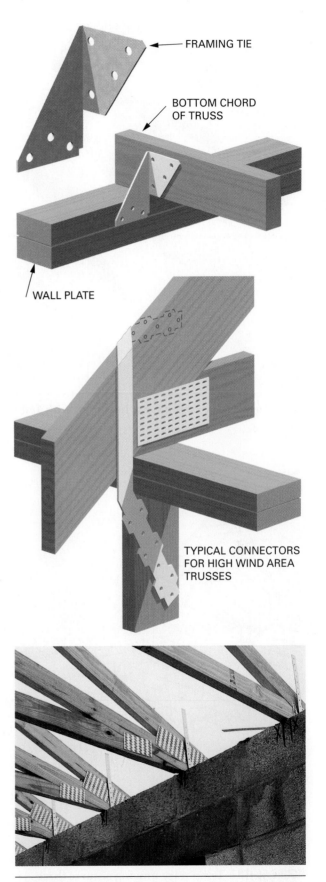

FRAMING TIE

BOTTOM CHORD OF TRUSS

WALL PLATE

TYPICAL CONNECTORS FOR HIGH WIND AREA TRUSSES

FIGURE 46–8 Metal framing ties are used to fasten trusses to the wall plate.

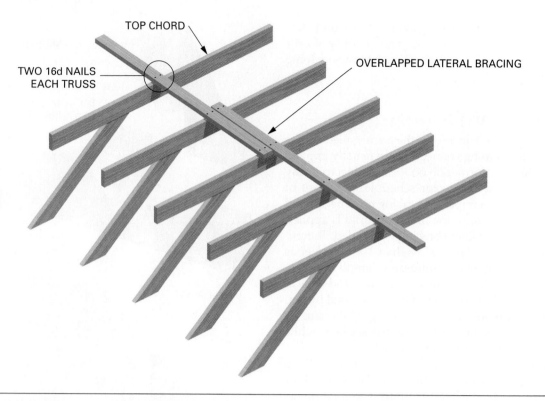

FIGURE 46–9   All bracing must be nailed with 16d (3½-inch) common nails.

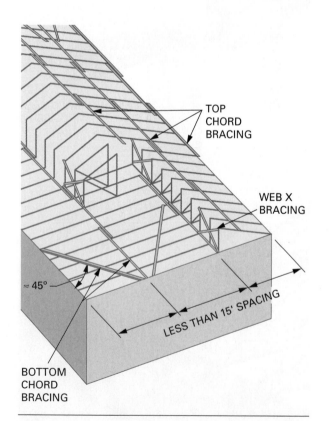

FIGURE 46–10   Temporary bracing secures three planes of trusses.

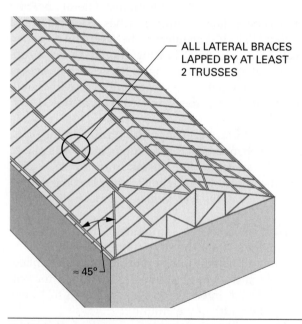

FIGURE 46–11   Typical temporary bracing of the top chord plane.

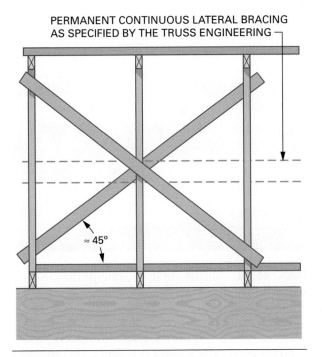

PERMANENT CONTINUOUS LATERAL BRACING
AS SPECIFIED BY THE TRUSS ENGINEERING

≈ 45°

FIGURE 46–12 Bracing of the web member plane prevents lateral movement of the trusses.

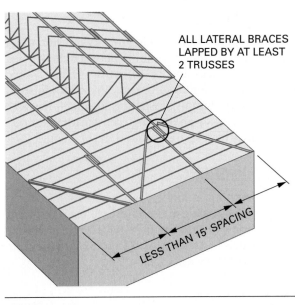

ALL LATERAL BRACES
LAPPED BY AT LEAST
2 TRUSSES

LESS THAN 15' SPACING

FIGURE 46–13 Typical bracing of the top chord plane.

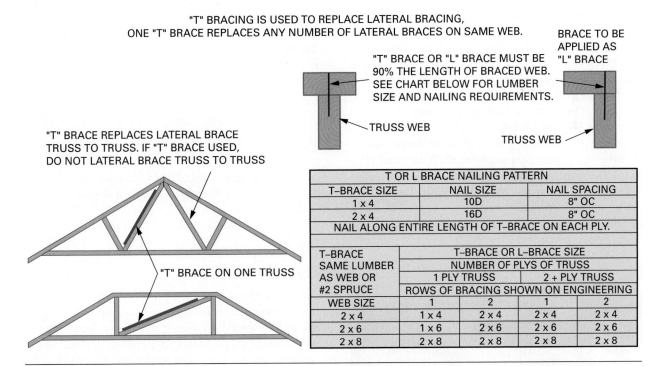

"T" BRACING IS USED TO REPLACE LATERAL BRACING,
ONE "T" BRACE REPLACES ANY NUMBER OF LATERAL BRACES ON SAME WEB.

"T" BRACE OR "L" BRACE MUST BE
90% THE LENGTH OF BRACED WEB.
SEE CHART BELOW FOR LUMBER
SIZE AND NAILING REQUIREMENTS.

BRACE TO BE
APPLIED AS
"L" BRACE

TRUSS WEB

TRUSS WEB

"T" BRACE REPLACES LATERAL BRACE
TRUSS TO TRUSS. IF "T" BRACE USED,
DO NOT LATERAL BRACE TRUSS TO TRUSS

"T" BRACE ON ONE TRUSS

T OR L BRACE NAILING PATTERN		
T–BRACE SIZE	NAIL SIZE	NAIL SPACING
1 x 4	10D	8" OC
2 x 4	16D	8" OC
NAIL ALONG ENTIRE LENGTH OF T–BRACE ON EACH PLY.		

T–BRACE SAME LUMBER AS WEB OR #2 SPRUCE	T–BRACE OR L–BRACE SIZE			
	NUMBER OF PLYS OF TRUSS			
	1 PLY TRUSS		2 + PLY TRUSS	
	ROWS OF BRACING SHOWN ON ENGINEERING			
WEB SIZE	1	2	1	2
2 x 4	1 x 4	2 x 4	2 x 4	2 x 4
2 x 6	1 x 6	2 x 6	2 x 6	2 x 6
2 x 8	2 x 8	2 x 8	2 x 8	2 x 8

FIGURE 46–14 Some webs require stiffeners to be added by the carpenter.

designs should be carefully followed. The top chord permanent bracing is often provided by the roof sheathing. Web bracing may have X braces as well as lateral bracing and web stiffeners **(Figure 46–14).** These are usually installed after the trusses, but before the sheathing is installed.

Framing Openings

Openings in the roof or ceiling for skylights or access ways must be framed within or between the trusses. The chords, braces, and webs of a truss system should never be cut or removed unless directed to do so by an engineer. Simply installing headers between trusses will create an opening. The sheathing or ceiling finish is then applied around the opening.

ROOF SHEATHING

Roof sheathing is applied after the roof frame is complete. Sheathing gives rigidity to the roof frame. It also provides a nailing base for the roof covering. Rated panels of plywood and strand board are commonly used to sheath roofs.

Panel Sheathing

Plywood and other rated panel roof sheathing is laid with the face grain running perpendicular to the rafter or top chord for greater strength **(Figure 46–15)**. End joints are made on the framing member and staggered, similar to subfloor. Nails are typically spaced 6 inches apart on the ends and 12 inches apart on intermediate supports.

Additional nails are added according to local codes for increased strength. Some areas require extra nails in the sheathing to protect from uplift caused by high winds. These nailing zones put more nails along the perimeter of the roof and at the corners of the roof. Other areas require more nails on special trusses that are designed to improve shear resistance **(Figure 46–16)**.

Adequate blocking, tongue-and-groove edges, or other suitable edge support such as *panel clips* must be used when spans exceed the indicated value of the plywood roof sheathing. Panel clips are small metal pieces shaped like a capital H. They are used between adjoining edges of the plywood

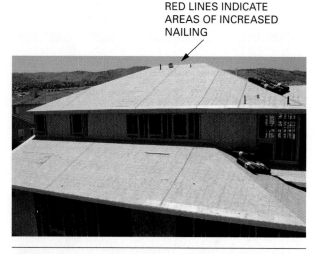

RED LINES INDICATE AREAS OF INCREASED NAILING

FIGURE 46–16 **Sheathing in seismic-prone areas has extra nails.**

between rafters **(Figure 46–17)**. One clip is typically used for 24-inch spans and two panel clips are used for 32- and 48-inch spans. If OSB (oriented strand board) is used as sheathing, two clips are required for 24-inch spans.

The ends of sheathing may be allowed to slightly and randomly overhang at the *rakes* until sheathing is completed. A chalk line is then snapped between the ridge and plate at each end of the roof. The sheathing ends are trimmed straight to the line with a circular saw.

> ⚠️ CAUTION CAUTION CAUTION CAUTION
>
> Sawdust from a circular saw on roof sheathing can be very slippery. Make sure of your footing as you cut. ■

Plank Sheathing

Plank decking is used in post-and-beam construction where the roof supports are spaced farther apart. Plank roof sheathing may be of 2-inch nominal thickness or greater, depending on the span. Usually both edges and ends are matched. The plank roof often serves as the finish ceiling for the rooms below.

ESTIMATING

Common Rafters for Gable Roof. Divide the length of the building by the spacing of the rafters. Add one, as a starter, and then multiply the total by two.

FIGURE 46–15 **Sheathing a trussed roof with plywood.**

APA PANEL ROOF SHEATHING

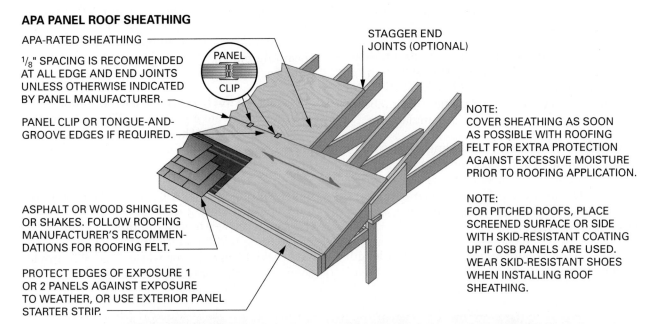

APA-RATED SHEATHING

1/8" SPACING IS RECOMMENDED AT ALL EDGE AND END JOINTS UNLESS OTHERWISE INDICATED BY PANEL MANUFACTURER.

PANEL CLIP OR TONGUE-AND-GROOVE EDGES IF REQUIRED.

ASPHALT OR WOOD SHINGLES OR SHAKES. FOLLOW ROOFING MANUFACTURER'S RECOMMEN-DATIONS FOR ROOFING FELT.

PROTECT EDGES OF EXPOSURE 1 OR 2 PANELS AGAINST EXPOSURE TO WEATHER, OR USE EXTERIOR PANEL STARTER STRIP.

PANEL CLIP

STAGGER END JOINTS (OPTIONAL)

NOTE:
COVER SHEATHING AS SOON AS POSSIBLE WITH ROOFING FELT FOR EXTRA PROTECTION AGAINST EXCESSIVE MOISTURE PRIOR TO ROOFING APPLICATION.

NOTE:
FOR PITCHED ROOFS, PLACE SCREENED SURFACE OR SIDE WITH SKID-RESISTANT COATING UP IF OSB PANELS ARE USED. WEAR SKID-RESISTANT SHOES WHEN INSTALLING ROOF SHEATHING.

RECOMMENDED MINIMUM FASTENING SCHEDULE FOR APA PANEL ROOF SHEATHING
(INCREASED NAIL SCHEDULES MAY BE REQUIRED IN HIGH WIND ZONES)

PANEL THICKNESS (b) (in.)	NAILING [c] [d]		
		MAXIMUM SPACING (in.)	
	SIZE	PANEL EDGES	INTERMEDIATE
5/16 – 1	8D	6	12 [a]
1 – 1/8	8D OR 10D	6	12 [a]

A) FOR SPANS 48 INCHES OR GREATER, SPACE NAILS 6 INCHES AT ALL SUPPORTS.
B) FOR STAPLING ASPHALT SHINGLES TO 5/16-INCH AND THICKER PANELS, USE STAPLES WITH A 15/16-INCH MINIMUM CROWN WIDTH AND A 1-INCH LEG LENGTH. SPACE ACCORDING TO SHINGLE MANUFACTURER'S RECOMMENDATION.
C) USE COMMON SMOOTH OR DEFORMED SHANK NAILS WITH PANELS TO 1 INCH THICK. FOR 1 1/8-INCH PANELS, USE 8D RING- OR SCREW-SHANK OR 10D COMMON SMOOTH-SHANK NAILS.
D) OTHER CODE-APPROVED FASTENERS MAY BE USED.

FIGURE 46–17 Recommendations for the application of APA panel roof sheathing. *(Courtesy of American Plywood Association)*

EXAMPLE A building is 42 feet long. The rafter spacing is 16 inches on center (OC). Divide 42 by 1⅓ (16 inches divided by 12 equals 1⅓ feet) to get 31½ spaces. Change 31⅓ to the next whole number to get 32. Add 1 to make 33. Multiplying by 2 equals 66 rafters.

Common Rafters with Hip or Valley Rafters. Use the same procedure as for a gable roof, but add two rafters for each hip and valley rafter. This will allow material for the common and jack rafters.

Hip and Valley. Count the number of hips and valleys.

Ridge Board for Gable Roof. Take the length of the building plus the rake overhang. Divide this sum by the length of the material to be used for the ridge. Typically 12- or 16-foot boards are used because there is minimal cutting waste with the various OCs. Round up to the next whole number.

Ridge Board for Hip Roof. Subtract the width of the building from the length of the building. The actual length could be as much as 3⅝ inches longer, so add enough to compensate.

Gable End Studs. Divide the width of the building by the OC spacing. Add two extra studs to allow for two gable ends.

Trusses. Divide the building length by the OC spacing, usually 2 feet. Subtract one. This is the number of common trusses. Two additional gable end trusses will be needed for ends. *Note:* The number of hip trusses are determined by the manufacturer.

Bracing Material. Divide the building width by four to get the number of rows of top, bottom, and web bracing. Round this number up to the nearest whole number. Divide by the bracing length, typically 16 feet. Round to the nearest whole number. Add extra for ground bracing of the first truss.

Sheathing Gable Roof. Round up the rafter length to the nearest 2 feet. Round up the ridgeboard length to the nearest even number of feet (i.e., 14.7 rounds up to 16). Multiply these numbers together and then double it to account for the other side. Divide the total by 32 (the number of square feet in a sheathing panel). Round up to the nearest whole number.

Sheathing Hip Roof. Calculate the number of panels as if this roof were a gable roof. Then add another 5 percent for waste. There is some waste when cutting triangles.

Gable End Sheathing. Calculate the total rise of the roof by multiplying the unit rise times the rafter run (one-half the building width), then divide by 12 to change the answer to feet. Multiply this number by the building width and divide by 32 (panel square feet). Add 10 percent for waste. This will provide enough material for two gable ends.

Key Terms

backing the hip	gambrel roof	minor span	supporting valley rafter
bird's mouth	hip jack rafter	pitch	tail cut
butterfly roof	hip rafter	plumb line	total rise
cheek cut	hip roof	purlin knee wall	total run
collar ties	hip-valley cripple jack rafter	rafter tables	total span
common rafter	intersecting roof	rake rafter	unit length
cripple jack rafter	jack rafter	ridgeboard	unit rise
dormer	line length	seat cut	unit run
dropping the hip	lateral bracing	shed roof	valley
fascia	level line	shortened valley rafter	valley cripple jack rafter
gable end stud	lookouts	slope	valley jack rafter
gable roof	major span	soffit	valley rafter
gable studs	mansard roof		
gables			

Review Questions

Select the most appropriate answer.

1. The type of roof that has the fewest number of different sized members is a
   - **a.** gable.
   - **b.** hip.
   - **c.** gambrel.
   - **d.** mansard.

2. The term used to represent horizontal distance under a rafter is
   - **a.** run.
   - **b.** span.
   - **c.** line length.
   - **d.** pitch.

3. The rafter that spans from a hip rafter to the ridge is a
   a. hip jack.
   c. valley.
   b. valley jack.
   d. cripple jack.

4. The rafter that spans from a wall plate to the ridge and whose run is perpendicular to the ridge is a
   a. hip.
   c. both a and b.
   b. valley.
   d. none of the above.

5. The minimum amount of stock left above the seat cut of the common rafter to ensure enough strength to support the overhang is usually _____ of the rafter stock.
   a. one-quarter
   c. two-thirds
   b. one-half
   d. three-quarters

6. What is the line length of a common rafter from the centerline of the ridge to the plate with a unit rise of 5 inches, if the building is 28'–0" wide?
   a. 70 inches
   c. 186 inches
   b. 140 inches
   d. 364 inches

7. The common difference in the length of gable studs spaced 24 inches OC for a roof with a slope of 8 on 12 is
   a. 8 inches.
   c. 16 inches.
   b. 12 inches.
   d. 20 inches.

8. What is the hip projection, in inches, if the common rafter projection is 6 inches?
   a. 6 inches
   b. ⁶/₁₇ foot
   c. 8½ inches
   d. 17 inches minus 6 inches

9. The line length of a hip rafter with a unit rise of 6 inches and a total run of 12 feet is
   a. 72 inches.
   c. 216 inches.
   b. 144 inches.
   d. 224 inches.

10. The jack rafter is most similar to a
    a. common rafter.
    c. valley rafter.
    b. hip rafter.
    d. gable end stud.

11. The total run of any hip jack rafter is equal to its
    a. distance along the plate from the outside corner.
    b. distance along the plate from the inside corner.
    c. line length.
    d. common difference in length.

12. The total run of the shortened valley rafter is equal to the total run of the
    a. minor span.
    b. major span.
    c. common rafter of minor roof.
    d. hip rafter.

13. The total run of the hip-valley cripple jack rafter is equal to
    a. one-half the run of the hip jack rafter.
    b. one-half the run of the longest valley jack rafter.
    c. the distance between seat cuts of the hip and valley rafters.
    d. the difference in run between the supporting and shortened valley rafters.

14. The length of any rafter in a roof of specified slope may be found if its _____ is known.
    a. total run
    c. unit run
    b. unit rise
    d. ridgeboard length

15. The typical amount that a dimension lumber hip rafter is shortened because of the ridge is
    a. ¾ inches.
    c. 1½ inches.
    b. 1¹/₁₆ inches.
    d. none of the above.

16. The member that provides the most resistance of trusses to tipping over is called a
    a. diagonal brace.
    c. web.
    b. lateral brace.
    d. gusset.

17. The part of a truss that may be cut if necessary is the
    a. web.
    c. gusset.
    b. chord.
    d. none of the above.

18. The length of a ridgeboard for a hip roof installed on a rectangular building measuring 28 × 48 feet is slightly more than
    a. 20 feet.
    c. 48 feet.
    b. 28 feet.
    d. 76 feet.

19. Neglecting the rake rafters, the estimated number of gable rafters for a rectangular building measuring 28 × 48 feet is
    a. 28.
    c. 72.
    b. 48.
    d. 74.

20. Neglecting the roof overhangs, the estimated number of pieces of sheathing needed for a hip roof installed on a rectangular building that measures 28 × 48 feet that has a unit rise of 6 inches is
    a. 24.
    c. 48.
    b. 25.
    d. 51.

UNIT 16
Stair Framing

Staircases can be a showcase for carpenters to demonstrate their skill and talent. They are often intricate and ornate, requiring close cutting and fitting. Stairs must be carefully designed and laid out to ensure safe passage and ease of use. Also many stair dimensions must comply with national and local codes. Building a set of stairs challenges the carpenter to work at his or her best while proving a place for great reward in pride of workmanship.

SAFETY REMINDER

Stairways must obliviously be built strong enough to support all intended loads. They must also be built to exact standards to ensure they are safe to use by many different sized people.

OBJECTIVES

After completing this unit, the student should be able to:

- describe several stairway designs.
- define terms used in stair framing.
- determine the rise and tread run of a stairway.
- determine the length of and frame a stairwell.
- lay out a stair carriage and frame a straight stairway.
- lay out and frame a stairway with a landing.
- lay out and frame a stairway with winders.
- lay out and frame service stairs.

47 Stairways and Stair Design

A set of stairs or a **staircase** can be an outstanding feature of an entrance. A staircase provides beauty and grace to a room, generally affecting the character of the entire interior.

A set of stairs generally refers to one or more flights of steps leading from one level of a structure to another. *Staircase* is a term usually saved to refer to a finished set of stairs that has architectural appeal. Stairs are further defined as finish or service stairs. Finish stairs extend from one habitable level of a house to another. Service stairs extend from a habitable to a non-habitable level, typically a basement.

Stairs are also governed closely by national and local building codes. Codes set limits on total stair height, step width, step rise, step depth, and the acceptable amount of variation between steps. These codes are designed to ensure a set of stairs will be safe for use by anyone.

Stairs, like rafters, are built by the most experienced carpenter on the job. Many design concepts and terms must be understood to successfully build a set of stairs. In addition to the fine carpentry of a staircase, consideration must be given to framing of the stairwell, or the opening in the floor through which a person must pass when climbing and going down the stairs **(Figure 47–1)**. The stairwell is framed at the same time as the floor.

TYPES OF STAIRWAYS

A straight stairway is continuous from one floor to another. There are no turns or landings. Platform stairs have intermediate landings between floors. Platform-type stairs sometimes change direction at the landing. An L-type platform stairway changes direction 90 degrees. A U-type platform stairway changes direction 180 degrees.

Platform stairs are installed in buildings that have a high floor-to-floor level. They also provide a temporary resting place. They are a safety feature in case of falls. The landing is usually constructed at roughly the middle of the staircase.

A winding staircase gradually changes direction as it ascends from one floor to another. In many cases, only a part of the staircase winds. Winding stairs may solve the problem of a shorter straight horizontal run. However, their use is not recommended due to the potential danger associated with climbing tapered steps **(Figure 47–2)**.

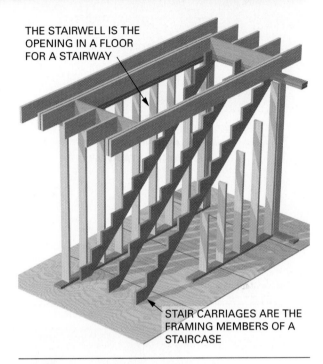

THE STAIRWELL IS THE OPENING IN A FLOOR FOR A STAIRWAY

STAIR CARRIAGES ARE THE FRAMING MEMBERS OF A STAIRCASE

FIGURE 47–1 Typical framing details for stairs.

Stairways constructed between walls are called **closed stairways.** Closed stairways are more economical to build, but they add little charm or beauty to a building. Stairways that have one or both sides open to the room are called **open stairways.** Open stairways have more parts and pieces, adding to the charm and beauty of the stairs.

The terms used in stair framing have similarities with those used for rafters **(Figure 47–3)**.

Total Rise. The total rise of a stairway is the vertical distance between finish floors.

Total Run. The total run is the total horizontal distance that the stairway covers.

Riser. The riser is the finish material used to cover the unit rise distance.

Unit Rise. The unit rise is the vertical distance from one step to another.

Tread. The tread is the horizontal finish material used to make up the step on which the feet are placed when ascending or descending the stairs.

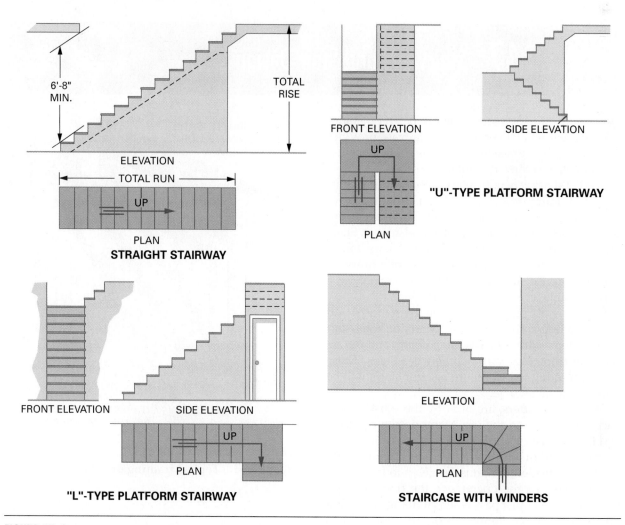

STRAIGHT STAIRWAY

6'-8" MIN.

ELEVATION

TOTAL RUN

UP

PLAN

TOTAL RISE

FRONT ELEVATION

SIDE ELEVATION

"U"-TYPE PLATFORM STAIRWAY

UP

PLAN

FRONT ELEVATION

SIDE ELEVATION

UP

PLAN

"L"-TYPE PLATFORM STAIRWAY

ELEVATION

UP

PLAN

STAIRCASE WITH WINDERS

FIGURE 47–2 Various types of stairways.

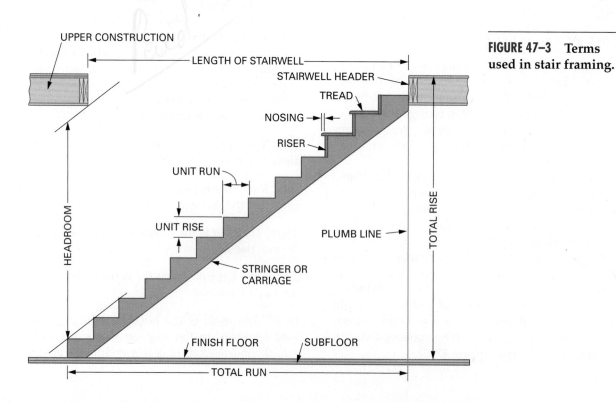

UPPER CONSTRUCTION

LENGTH OF STAIRWELL

STAIRWELL HEADER

TREAD

NOSING

RISER

UNIT RUN

UNIT RISE

HEADROOM

PLUMB LINE

TOTAL RISE

STRINGER OR CARRIAGE

FINISH FLOOR

SUBFLOOR

TOTAL RUN

FIGURE 47–3 Terms used in stair framing.

Unit Run. The unit run is the horizontal distance between the faces of the risers.

Nosing. The **nosing** is that part of the tread that extends beyond the face of the riser. It is not part of the calculations for stairs, but rather an add-on to treads.

Stair Carriage. A **stair carriage,** sometimes called a **stair horse,** provides the main strength for the stairs. It is usually a nominal 2 × 10, 2 × 2, or 2 × 14 framing member cut to support the treads and risers.

Stair Stringer. The **stringer** is the finish material applied to cover the stair carriage. It can also take the place of the stair carriage when side walls are used to support the stairs.

Stairwell. A **stairwell** is an opening in the floor for the stairway to pass through. It provides adequate headroom for persons using the stairs.

Headroom. The headroom of a set of stairs is the vertical distance above the stairs from a line drawn from nosing to nosing to the upper construction.

STAIR DESIGN

Stairs in residential construction are at least 36 inches wide **(Figure 47–4).** If more than 50 people occupy the building the width must be 44 inches. The maximum height of a single flight of stairs is 12 feet, unless a platform is built in to break up the continuous run. This platform must be as long as the stair is wide but need not be longer than 48 inches.

Staircases must be constructed at a proper angle for maximum ease in climbing and for safe descent. The relationship of the rise and run determines this angle **(Figure 47–5).** The preferred angle is between 30 and 38 degrees. Stairs with a slope of less than 20 degrees waste a lot of valuable space. Stairs with a slope that is excessively steep (50 degrees or over) are difficult to climb and dangerous to descend.

The international residential code specifies that the height of a riser shall not exceed 7¾ inches and that the width of a tread not be less than 10 inches. Some local codes modify these limits to 8¼ inches maximum rise and 9 inches minimum tread. However, a rise of between 7 and 7½ inches and a tread run of between 10 and 12 inches is recommended to maximize safety and ease of use.

Riser Height

Because every building presents a different situation, the unit rise and unit run must be determined for each building. The unit rise is determined first. To determine the individual rise, first measure the

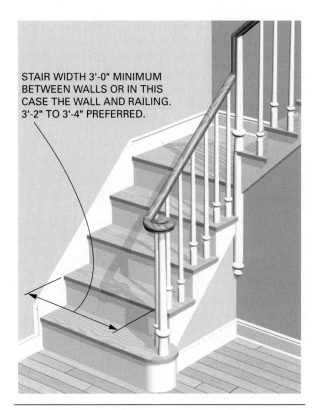

STAIR WIDTH 3'-0" MINIMUM BETWEEN WALLS OR IN THIS CASE THE WALL AND RAILING. 3'-2" TO 3'-4" PREFERRED.

FIGURE 47–4 Recommended stair widths.

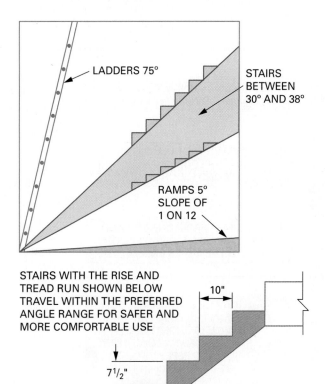

LADDERS 75°

STAIRS BETWEEN 30° AND 38°

RAMPS 5° SLOPE OF 1 ON 12

STAIRS WITH THE RISE AND TREAD RUN SHOWN BELOW TRAVEL WITHIN THE PREFERRED ANGLE RANGE FOR SAFER AND MORE COMFORTABLE USE

10"

7½"

37°

FIGURE 47–5 Appropriate angles and dimensions for stairs.

total rise of the stairway from finished floor to finished floor. Next find the number of risers that will fit in the opening. This is done by dividing the total rise by an assumed unit rise. The result is then rounded up to a whole number of risers and divided again into the total rise to give the actual unit rise.

EXAMPLE   A total rise of 8'–10" (106 inches) is measured. Assume the unit rise to be 7¾" at first.

- Divide total rise by 7¾".
  106 ÷ 7.75 = 13.677 risers.
- Round up to the next whole number.
  13.677 rounds up to 14 risers.
- Divide total rise by number of risers.
  106 ÷ 14 = 7.57 or 7⁹⁄₁₆ inches actual unit rise.

The unit rise height may be reduced by increasing the number of risers. For example, if the number of risers is increased to 15 the unit rise is 106 divided by 15, which equals about 7¹⁄₁₆ inches.

Another method of determining the riser height involves repetitive step-offs using wing dividers. Stand a *story pole* vertically. This could be a 1 × 2 strip of lumber. Mark the total rise allowing for the finish floor to finish floor height. Set the dividers to 7¾-inch rise spacing. Step-off the rod over the entire total rise. If the last step is not even, then adjust the dividers slightly smaller. Repeat this process as often as necessary **(Procedure 47–A).**

Tread Run

The tread run (width) is measured from the face of one riser to the next. It does not include the nosing **(Figure 47–6).** It needs to be adjusted to create a proper stair angle. There are two rules to follow to accomplish this.

17–18 Method. First, the sum of one rise and one tread should equal between 17 and 18. For example, if the riser height is 7⅜ inches, then the minimum tread width may be 17 inches minus 7⅜ inches. This

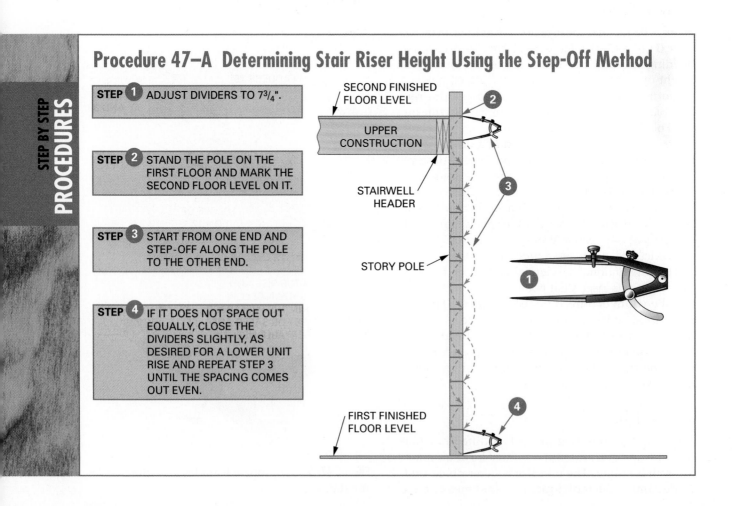

Procedure 47–A Determining Stair Riser Height Using the Step-Off Method

STEP ① ADJUST DIVIDERS TO 7³⁄₄".

STEP ② STAND THE POLE ON THE FIRST FLOOR AND MARK THE SECOND FLOOR LEVEL ON IT.

STEP ③ START FROM ONE END AND STEP-OFF ALONG THE POLE TO THE OTHER END.

STEP ④ IF IT DOES NOT SPACE OUT EQUALLY, CLOSE THE DIVIDERS SLIGHTLY, AS DESIRED FOR A LOWER UNIT RISE AND REPEAT STEP 3 UNTIL THE SPACING COMES OUT EVEN.

SECOND FINISHED FLOOR LEVEL

UPPER CONSTRUCTION

STAIRWELL HEADER

STORY POLE

FIRST FINISHED FLOOR LEVEL

STEP BY STEP PROCEDURES

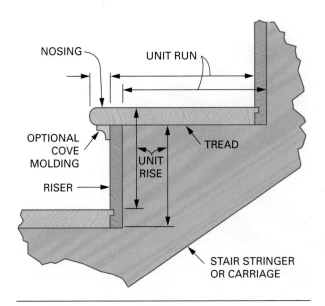

FIGURE 47–6 The tread unit run does not include the nosing.

equals 9⅝ inches. The maximum tread width may be 18 inches minus 7⅜ inches. This equals 10⅝ inches.

24–25 Method. The second rule for determining the tread width is found in many building codes, which state that the sum of two risers and one tread shall not be less than 24 inches nor more than 25 inches **(Figure 47–7).** With this formula, a rise of 7⅜ inches calls for a minimum tread width of 9¼ inches and a maximum of 10¼ inches.

These numbers are further restricted with a minimum run of 10 inches by some codes. Thus, the run should be between 10 and 10¼ inches wide.

Variations in Stair Steepness

Decreasing the rise increases the run of the stairs. This makes the stairs easier to climb but uses up more space. Increasing the rise height decreases the run. This makes the stairs steeper and more difficult to climb but uses less space. The carpenter must use good judgment when adapting the rise and run dimensions to fit the space in which the stairway is constructed. In general, a rise height of 7½ inches and a tread run of 10 inches create a safe, comfortable stairway.

FRAMING A STAIRWELL

A stairwell is framed in the same manner as any large floor opening, as discussed in Unit 11, Floor Framing. Several methods of framing stairwells are illustrated in **Figure 47–8.**

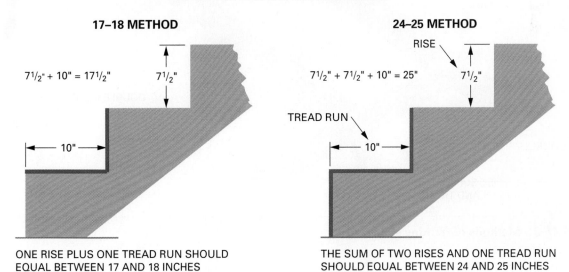

TWO FORMULAS ARE USED TO DETERMINE THE UNIT RUN FOR STAIRS
AFTER UNIT RISE HAS BEEN DETERMINED

17–18 METHOD

7½" + 10" = 17½"   7½"

10"

ONE RISE PLUS ONE TREAD RUN SHOULD
EQUAL BETWEEN 17 AND 18 INCHES

24–25 METHOD

RISE

7½" + 7½" + 10" = 25"   7½"

TREAD RUN

10"

THE SUM OF TWO RISES AND ONE TREAD RUN
SHOULD EQUAL BETWEEN 24 AND 25 INCHES

NOTE: CHECK LOCAL CODES FOR MAXIMUM ALLOWABLE
RISE AND MINIMUM ALLOWABLE RUN.

FIGURE 47–7 Two techniques for determining the unit run with the desired unit rise.

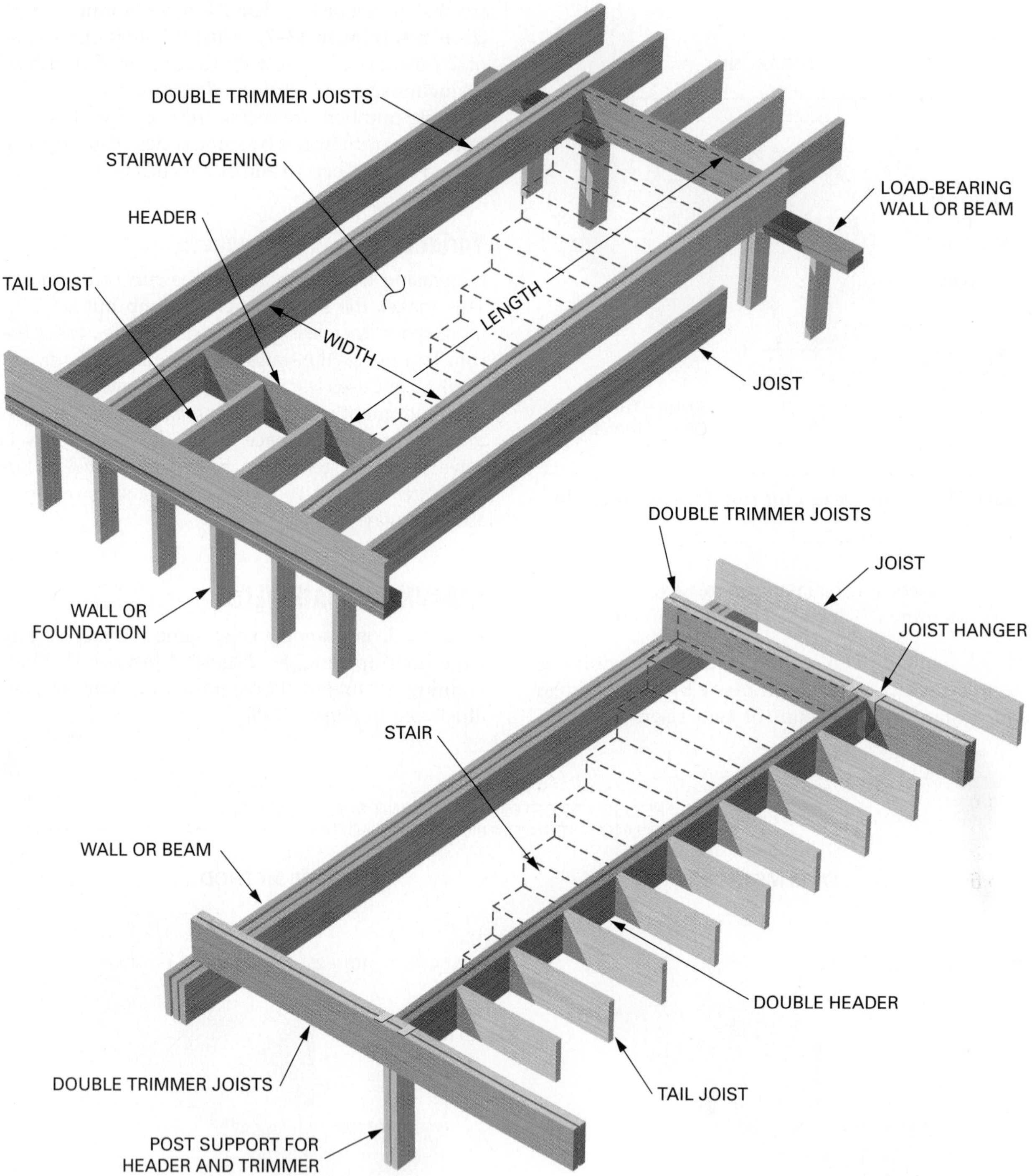

FIGURE 47–8 Methods of framing stairwells.

Stairwell Width

The width of the stairwell depends on the width of the staircase. The drawings show the finish width of the staircase. However, the stairwell must be made wider than the staircase to allow for wall and stair fin-

ish **(Figure 47–9)**. Extra width will be required for a handrail and other finish parts of an open staircase that makes a U-turn on the landing above. The carpenter must be able to determine the width of the stairwell by studying the prints for size, type, and placement of the stair finish before framing the stairs.

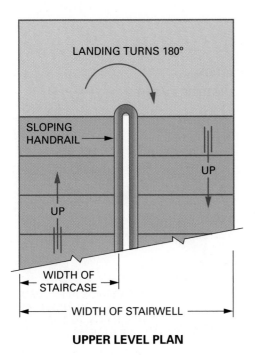

UPPER LEVEL PLAN

FIGURE 47–9 **The stairwell must be made wider than the staircase.**

Length of the Stairwell

The length of the stairwell depends on the slope of the staircase. Stairs with a low angle require a longer stairwell to provide adequate headroom **(Figure 47–10).** Building codes require a minimum of 6'–8" for headroom; however this is minimum where more headroom is preferred.

To find the minimum length of the stairwell, add the headroom to the thickness of the upper floor construction. The upper floor construction includes subfloor and floor joists of the second floor as well as the first floor ceiling thickness. Divide this number by the riser height. The result is the number of risers needed to create that rise and will probably be a whole number and a fraction. Round up to the next larger whole number. Multiply the number of risers by the tread width to find the length of the stairwell **(Procedure 47–B).**

EXAMPLE A stairway has a riser height of 7½ inches, a tread width of 10 inches, and a total thickness of upper floor construction of 11¾ inches. What is the stairwell length to allow for 7'–0" headroom?

- Add upper floor construction to headroom.
  11¾ + 84 = 95¾" or 95.75"
- Divide by unit rise to find number of risers.
  95.75 ÷ 7.5 equals 12.77 risers
- Round up to nearest whole number.
  12.77 rounds up to 13
- Multiplying number of risers by unit run.
  13 × 10 = 130 inches = minimum length of the stairwell

This process assumes the carriage is framed to the upper construction where the header of the stairwell acts as the top riser. If the top tread is framed flush with the second floor, add another tread width to the length of the stairwell **(Figure 47–11).**

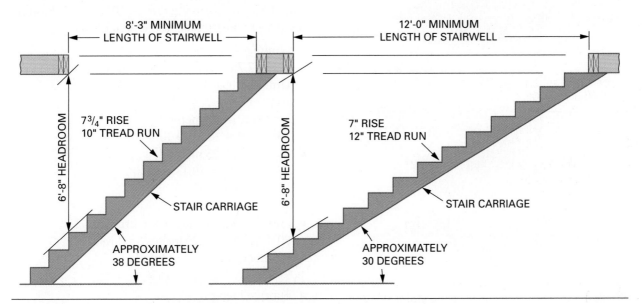

FIGURE 47–10 **Stair angles affect the stairwell opening length.**

Procedure 47–B Determining Stairwell Length

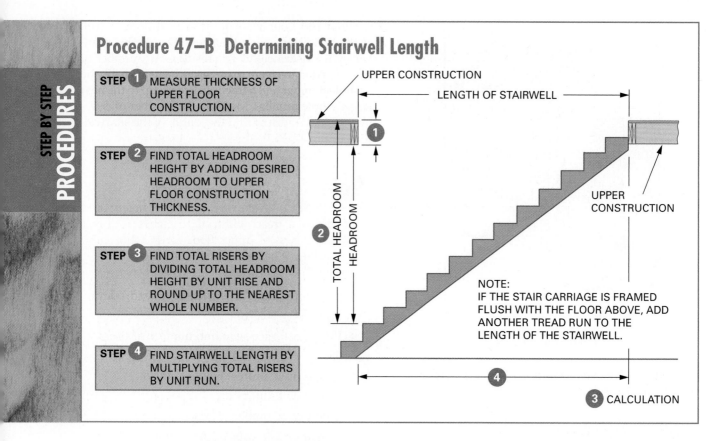

STEP 1	MEASURE THICKNESS OF UPPER FLOOR CONSTRUCTION.
STEP 2	FIND TOTAL HEADROOM HEIGHT BY ADDING DESIRED HEADROOM TO UPPER FLOOR CONSTRUCTION THICKNESS.
STEP 3	FIND TOTAL RISERS BY DIVIDING TOTAL HEADROOM HEIGHT BY UNIT RISE AND ROUND UP TO THE NEAREST WHOLE NUMBER.
STEP 4	FIND STAIRWELL LENGTH BY MULTIPLYING TOTAL RISERS BY UNIT RUN.

NOTE:
IF THE STAIR CARRIAGE IS FRAMED FLUSH WITH THE FLOOR ABOVE, ADD ANOTHER TREAD RUN TO THE LENGTH OF THE STAIRWELL.

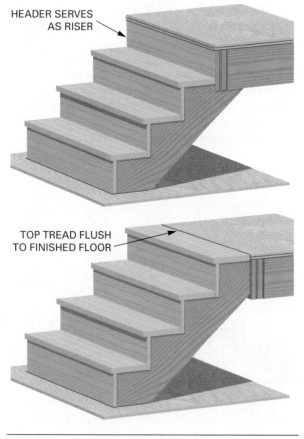

FIGURE 47–11 Methods of framing stairs at the stairwell header.

Additional floor space may be obtained by framing the subfloor past the header. This addition is framed at the same angle as the stairs (**Figure 47–12**).

Stair Design Example

First, what would be the largest riser (unit rise) and tread (unit run) allowed for a set of stairs that has a total rise of 102 inches? Second, calculate the appropriate stairwell length if the headroom is 80 inches and the upper construction is 11¼ inches. Assume the upper construction serves as a riser.

- Number of risers
  $102 \div 7.75 = 13.16129 \to 14$ risers
- Unit rise
  $102 \div 14 = 7.29 \to 7\frac{5}{16}''$
- Unit run 17–18 method
  $18 - 7.29 = 10.71 \to 10\frac{11}{16}''$
- Unit run 24–25 method
  $25 - 2(7.29) = 10.43 \to 10\frac{7}{16}''$
- Maximum tread
  $10\frac{11}{16}''$
- Headroom plus upper construction
  $80 + 11.25 = 91.25$ inches total rise
- Number of risers and treads
  $91.25 \div 7.29 = 12.5 \to 13$
- Stairwell length
  $13 \times 10.71 = 139.23 \to 139\frac{1}{4}''$

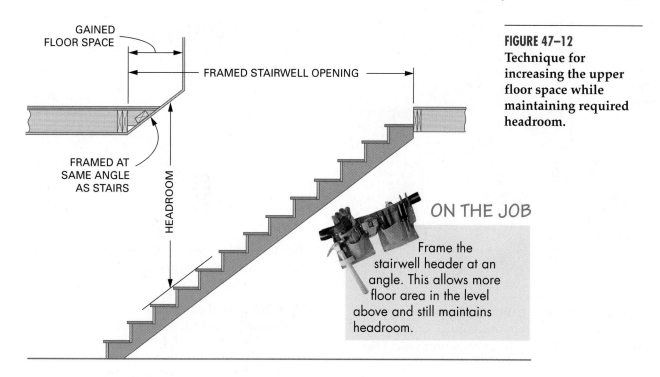

GAINED FLOOR SPACE

FRAMED STAIRWELL OPENING

FRAMED AT SAME ANGLE AS STAIRS

HEADROOM

FIGURE 47–12
Technique for increasing the upper floor space while maintaining required headroom.

ON THE JOB

Frame the stairwell header at an angle. This allows more floor area in the level above and still maintains headroom.

48 Stair Layout and Construction

All stairs are laid out in approximately the same way. The use, location, and cost of stairs determine the way they are built. Regardless of the kind of stairs, where they are located, or how much they cost, care should be taken in their layout and construction.

METHODS OF STAIR CONSTRUCTION

There are two principal methods of stair construction. The **housed stringer** is laid out and cut to be a finished product. It is often fabricated off site in a shop and installed near the conclusion of the construction process. The job-built staircase uses stair carriages that are usually built on site. These are finished later as the construction process comes to an end.

Housed Stringer Staircase

For the housed stringer staircase, the framing crew frames only the stairwell. The staircase is installed when the house is ready for finishing. Dadoes are routed into the sides of the finish stringer. They *house* (make a place for) and support the risers and treads **(Figure 48–1)**.

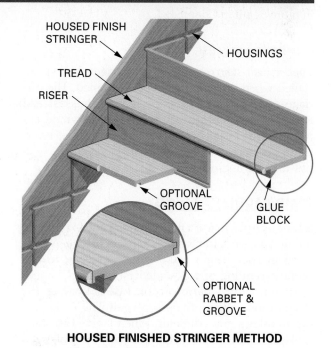

HOUSED FINISH STRINGER

HOUSINGS

TREAD

RISER

OPTIONAL GROOVE

GLUE BLOCK

OPTIONAL RABBET & GROOVE

HOUSED FINISHED STRINGER METHOD

FIGURE 48–1   A housed finish stringer has dadoed sides to accept treads and risers.

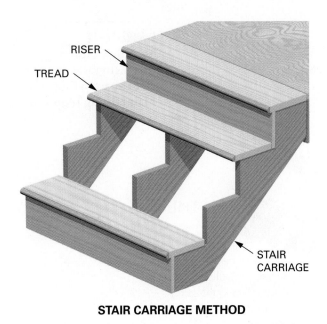

STAIR CARRIAGE METHOD

FIGURE 48–2 Stair carriages have notches cut to support treads and risers.

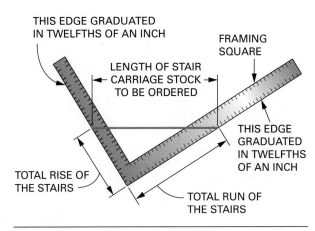

FIGURE 48–3 The framing square may be used to estimate the rough length of a stair carriage.

Occasionally, the finish carpenter builds a housed stringer staircase on the jobsite. A router and jig are used to dado the stringers. Then, the treads, risers, and other stair parts are cut to size. The staircase is then assembled either in place or as a unit and then installed into place. Stair carriages are not required when the housed stringer method of construction is used. Building of housed stringer stairs is described in more detail in Chapter 81.

Job-Built Staircase

The *job-built staircase* uses stair carriages that are installed when the structure is being framed. The carriage is laid out and cut with risers and treads fastened to the cutouts **(Figure 48–2)**. This style uses temporary rough treads installed for easy access to upper levels during construction. Later the carriage is fitted with finish treads and risers with other stair trim.

STAIR CARRIAGE LAYOUT

It is dangerous to use a flight of stairs and experience an unexpected variation in stair dimensions. A variation in riser height could cause someone to trip while ascending or fall while descending. Changes in tread width, narrower or wider, change the rhythm and pattern of a person's gait. This makes using the stairs more difficult. When laying out stairs make sure that all riser heights are equal and all tread widths are equal.

This problem is addressed by building codes, which require all dimensions in stair layout to be ac-

curate. The height from one riser to the next must be within ³⁄₁₆ inch. Total variation between the largest and smallest riser and largest and smallest tread is not to exceed ³⁄₈ inch. Note that these variations can be substantially reduced with reasonable care and skill.

Scaling Rough Carriage Length

The length of lumber needed for the stair carriage is often determined using the Pythagorean theorem. It also can be found by scaling across the framing square. Use the edge of the square that is graduated in 12ths of an inch. Mark the total rise on the tongue. Then mark the total run on the blade. Scale off in between the marks **(Figure 48–3)**.

EXAMPLE A stairway has a total rise of 8'–9" and a total run of 12'–3". What is the length of material needed to build the carriage?

Pythagorean Theorem
- Change dimensions to decimals.

$$8'-9'' = 8.75'; 12'-3'' = 12.25'$$

- Substitute into $a^2 + b^2$ and c^2 and solve.

$$8.75^2 + 12.25^2 = c^2 = 226.625$$
$$c = \sqrt{226.625} = 15.05'$$

- Round up to nearest even number.

$$15.05 \rightarrow 16 \text{ feet}$$

Scaling
Locate 8–³⁄₁₂ths and 12–³⁄₁₂ths on framing square. Measuring across the square between these dimensions results in a reading of a little over 15. At a scale of 1 inch = 1 foot, board length ordered should be 16 feet.

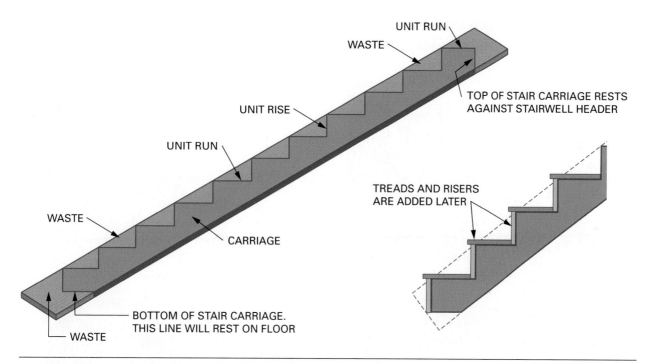

FIGURE 48–4 A completed stair carriage layout.

Stepping Off the Stair Carriage

Place the stair carriage stock on a pair of saw-horses. Sight the stock for a crowned edge. Set gauges on the framing square with the rise on the tongue and the tread run on the blade. Lines laid out along the tongue will be plumb lines. Those laid out along the blade will be level lines when the stair carriage is in its final position.

Using the framing square gauges against the top edge of the carriage, step off on the carriage the necessary number of times. Mark both rise and run along the outside of the tongue and blade. Lay out enough level and plumb lines to include the top tread and the lower finished floor line **(Figure 48–4).** The layout lines represent the bottom of the tread and backside of the riser.

Equalizing the Bottom Riser

The bottom of the carriage must be adjusted to allow for the finished floor. This amount depends on how the finish material is applied. This will make the bottom riser equal in height to all of the other risers when the staircase is finished. This process is known as dropping the stair carriage.

If the carriage rests on the finish floor, then the first riser is cut shorter by the thickness of the tread stock **(Figure 48–5).** If the carriage rests on the subfloor the riser height must be adjusted using the tread and finished floor thicknesses. To achieve the first riser height dimension, take the riser height minus the tread thickness plus finished floor thickness.

Sometimes the finish floor and tread stock are the same thickness. In this special case nothing is cut off the bottom end of the stair carriage.

Equalizing the Top Riser

The top of the carriage must be properly located against the upper construction to maintain the unit rise. This location depends on the style of the stairs. When the header of the upper construction acts as the top riser, the top of the carriage is roughly one riser height below the subfloor **(Figure 48–6).** This distance is equal to riser height plus tread thickness minus the finished floor thickness.

If the top tread is flush with the finished floor, the carriage elevation is only slightly below the subfloor. This distance is the tread thickness minus the finished floor thickness. If the finished floor thickness is larger than the tread, a negative number will result. In this case raise the carriage this calculated distance above the subfloor.

Attaching the Stair Carriage to Upper Construction

When the upper construction is part of the top riser, most of the carriage is below the header. This offers poor support and room for fastening. Two methods of

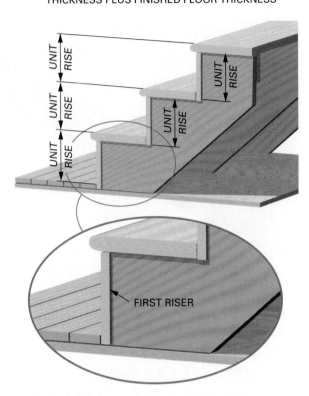

CARRIAGE LANDS ON FINISHED FLOOR

FIRST RISER EQUALS UNIT RISE MINUS TREAD THICKNESS

CARRIAGE LANDS ON SUBFLOOR

FIRST RISER EQUALS UNIT RISE MINUS TREAD THICKNESS PLUS FINISHED FLOOR THICKNESS

FIRST RISER

FIRST RISER

FIGURE 48–5 Adjusting the stair carriage bottom to equalize the first riser.

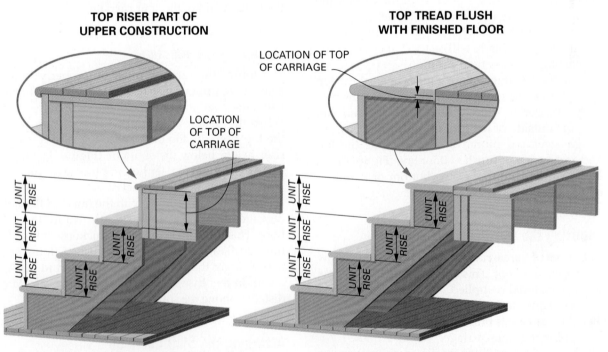

TOP RISER PART OF UPPER CONSTRUCTION

LOCATION OF TOP OF CARRIAGE

TOP TREAD FLUSH WITH FINISHED FLOOR

LOCATION OF TOP OF CARRIAGE

LOCATION OF TOP OF CARRIAGE

TOP OF CARRIAGE IS LOCATED BELOW SUBFLOOR. DISTANCE EQUALS UNIT RISE PLUS TREAD THICKNESS MINUS FINISHED FLOOR THICKNESS.

TOP OF CARRIAGE IS LOCATED BELOW SUBFLOOR. DISTANCE EQUALS TREAD THICKNESS MINUS FINISHED FLOOR THICKNESS.

FIGURE 48–6 Locating the top of the carriage on the header.

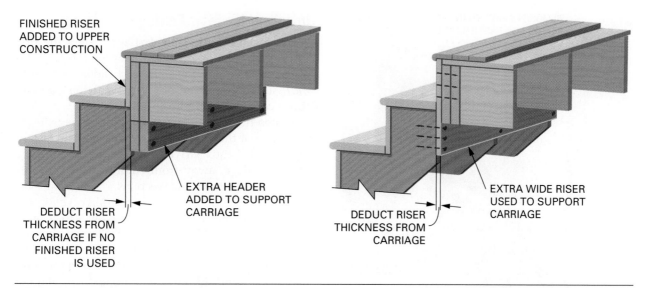

FINISHED RISER
ADDED TO UPPER
CONSTRUCTION

EXTRA HEADER
ADDED TO SUPPORT
CARRIAGE

DEDUCT RISER
THICKNESS FROM
CARRIAGE IF NO
FINISHED RISER
IS USED

EXTRA WIDE RISER
USED TO SUPPORT
CARRIAGE

DEDUCT RISER
THICKNESS FROM
CARRIAGE

FIGURE 48–7 **Methods of framing the stair carriage to the stairwell header.**

securing the carriage are shown in **Figure 48–7.** One method uses an extra header under the existing one. The carriage can then be attached in a variety of ways. The second method uses a wider riser on the top. It is long enough to fasten into the header and the carriage. It should be made with plywood for strength.

If no finished riser is desired, the carriage must be adjusted. With no riser board, the plumb line needs to be cut back a distance equal to one riser's thickness. This will allow the back edge of the top tread to rest against the header.

Cutting the Stair Carriages

After the first carriage is laid out, cut it. Follow the layout lines carefully because this will be a pattern for others.

 CAUTION CAUTION CAUTION CAUTION

When making a cut at a sharp angle to the edge, the guard of the saw may not retract. Retract the guard by hand until the cut is made a few inches into the stock. Then release the guard and continue the cut. Never wedge the guard in an open position. ■

Finish the cuts at the intersection of the riser and tread run with a handsaw. Use the first carriage as a pattern. Lay out and cut as many other carriages as needed. Three carriages are often used for residential staircases of average width. For wider stairs, the number of carriages depends on such factors as whether or not risers are used and the thickness of

the tread stock. Check the drawings or building code for the spacing of carriages for wider staircases.

FRAMING A STRAIGHT STAIRWAY

If the stairway is either completely closed or closed on one side, the walls must be prepared before the stair carriages are fastened in position.

Preparing the Walls of the Staircase

Gypsum board (drywall) is sometimes applied to walls before the stair carriages are installed against them. This procedure saves time. It eliminates the need to cut the drywall around the cutouts of the stair carriage. This method also requires no blocking between the studs to fasten the ends of the drywall panels.

However, blocking between the studs in back of the stair carriage provides backing for fastening the stair trim. Lack of it may cause difficulty for those who apply the finish.

If the drywall is to be applied after the stairs are framed, blocking is required between studs in back of the stair carriage to fasten the ends of the gypsum board. Snap a chalk line along the wall sloped at the same angle as the stairs. Be sure the top of the blocking is sufficiently above the stair carriage to be useful. Install 2 × 6 or 2 × 8 blocking, on edge, between and flush with the edges of the studs. Their top edge should be to the chalk line **(Figure 48–8).**

Housed staircases may be installed after the walls are finished. Sometimes they are installed before the walls are finished. In such a case, they must be furred out away from the studs. This allows the wall finish to extend below the top edge of the finished stringer.

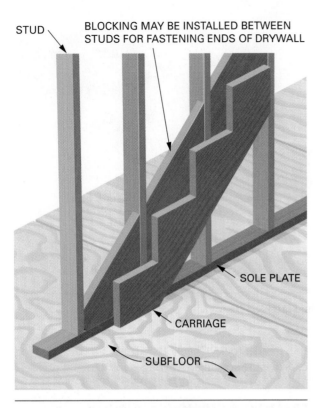

FIGURE 48–8   **Preparation of the wall for application of drywall before stair carriages are installed.**

Installing the Stair Carriage

When installing the stair carriage, fasten the first carriage in position on one side of the stairway. Attach it at the top to the stairwell header. Make sure the distance from the subfloor above to the top tread run is correct (see Figure 48–6).

Fasten the bottom end of the stair carriage to the sole plate of the wall and with intermediate fastenings into the studs. Drive nails near the bottom edge of the carriage, away from cutouts. This prevents splitting of the triangular sections.

Fasten a second carriage on the other wall in the same manner as the first. If the stairway is to be open on one side, fasten the carriage at the top and bottom of the staircase. The location of the stair carriage on the open end of a stairway is in relation to the position of the **handrail.** First, determine the location of the centerline of the handrail. Then, position the stair carriage on the open side of a staircase. Make sure its outside face will be in a line plumb with the centerline of the handrail when it is installed **(Figure 48–9).**

Fasten intermediate carriages at the top into the stairwell header and at the bottom into the subfloor. Test the tread run and riser cuts with a straightedge placed across the outside carriages **(Figure 48–10).** About halfway up the flight, or where necessary, fas-

FIGURE 48–9   **The outside stair carriage is located plumb under the handrail.**

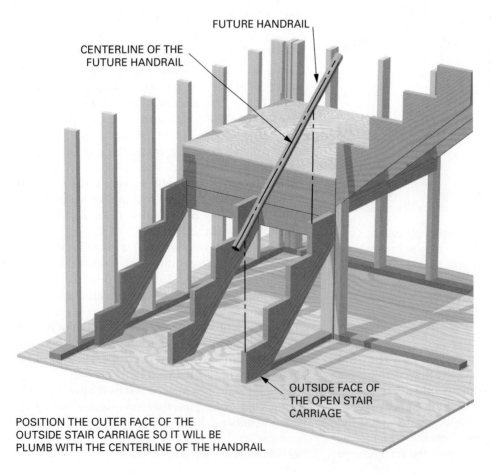

POSITION THE OUTER FACE OF THE
OUTSIDE STAIR CARRIAGE SO IT WILL BE
PLUMB WITH THE CENTERLINE OF THE HANDRAIL

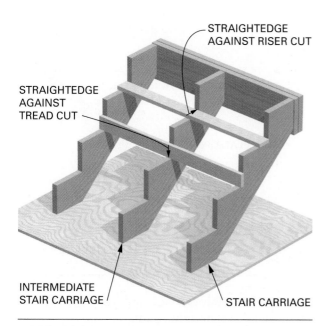

FIGURE 48–10 The alignment of tread and riser cuts on carriages must be checked with a straightedge.

ten a temporary riser board. This straightens and maintains the spacing of the carriages **(Figure 48–11)**.

If a wall is to be framed under the stair carriage at the open side, fasten a bottom plate to the subfloor plumb with the outside face of the carriage. Lay out the studs on the plate. Cut and install studs under the carriage in a manner similar to that used to install

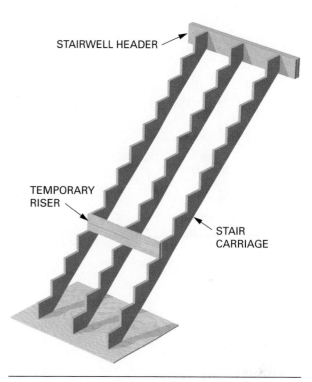

FIGURE 48–11 A temporary riser about halfway up the flight straightens and maintains the carriage spacing.

gable studs. Be careful to keep the carriage straight. Do not crown it up in the center **(Figure 48–12)**. Install rough lumber treads on the carriages until the stairway is ready for the finish treads.

FIGURE 48–12 Studs are typically installed under the stair carriage on the open side of the stairway.

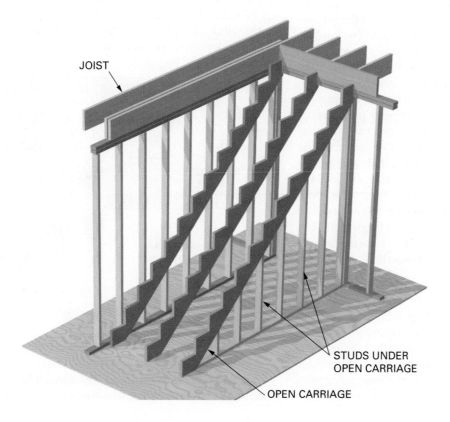

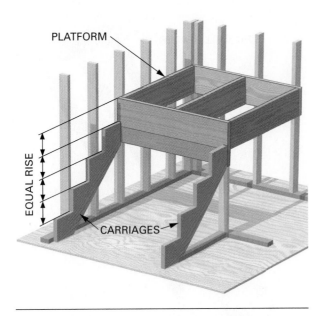

FIGURE 48–13 The top side of the stair platform is located as if it were a tread.

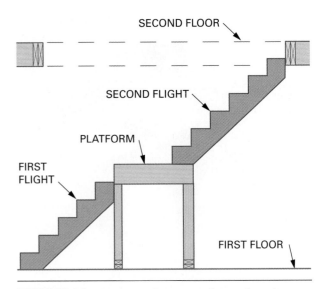

FIGURE 48–14 Stair carriages are framed to the platform as two straight flights of stairs.

STAIRWAY LANDINGS

A stair landing is an intermediate platform between two flights of stairs. A landing is designed for changing the direction of the stairs and as a resting place for long stair runs. The landing usually is floored with the same materials as the main floors of the structure. Some codes require that the minimum length of a landing be not less than 2′–6″. Other codes require the minimum dimension to be the width of the stairway.

L-type stairs may have the landing near the bottom or the top. U-type stairs usually have the landing about midway on the flight. Many codes state that no flight of stairs shall have a vertical rise of more than 12 feet. Therefore, any staircase running between floors with a vertical distance of more than 12 feet must have at least one intermediate landing or platform.

Platform stairs are built by first erecting the platform. The finished floor of the platform may be thought of as an extra wide tread. It should be the same height as if it were a finish tread in the staircase. This allows an equal riser height for both flights **(Figure 48–13)**. The stairs are then framed to the platform as two straight flights **(Figure 48–14)**. Either the stair carriage or the housed stringer method of construction may be used.

LAYING OUT WINDING STAIRS

Winding stairs change direction without a conventional landing. They often will allow for two extra risers in the space normally occupied by the landing. This is particularly useful for stairwells that are too small for normal stair construction. **Winders,** however, are not recommended for safety reasons. Some codes state

that they may be used in individual dwelling units, if the required tread width is provided along an arc. This is called the *line of travel.* It is a certain distance from the side of the stairway where the treads are narrower. The IRC (International Residential Code) states the line of travel is 12 inches from the narrow edge of the tread. The minimum tread width at the line of travel is 11 inches and the minimum width at the narrow edge is 6 inches **(Figure 48–15)**. Other codes may permit a narrower tread. Check the building code for the area in which the work is being performed.

To lay out a winding turn of 90 degrees, draw a full-size winder layout on the floor directly below where the winders are to be installed **(Procedure 48–A)**. On a

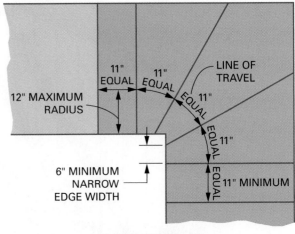

PLAN OF WINDERS

IRC (INTERNATIONAL RESIDENTIAL CODE)
REQUIREMENTS FOR WINDERS

FIGURE 48–15 International Residential Building Code specifications for winding stairs.

Procedure 48–A Laying Out Winding Stairs

STEP 1 BEGIN WITH LAYOUT OF STAIR WIDTH AND THE LOWER TREADS. MARK THEM TOWARD THE PLATFORM.

STEP 2 FROM POINT A OF LAST TREAD, MEASURE 6" TOWARD PLATFORM AND MARK IT AS POINT B.

STEP 3 MEASURE FROM POINT A ALONG TREAD LINE 12" TO POINT C.

STEP 4 SWING 11" ARC FROM POINT C TOWARD WHERE POINT D WILL BE. SWING 12" ARC FROM POINT B TO LOCATE POINT D.

STEP 5 DRAW LINE TO WALL FROM POINT B THROUGH POINT D. THIS IS TREAD LINE FOR NEXT STEP.

STEP 6 SWING 6" ARC FROM POINT B TO LOCATE POINT E ON THE STAIR WIDTH LINE.

STEP 7 SWING 11" ARC FROM POINT D TOWARD WHERE POINT F WILL BE. SWING 12" ARC FROM POINT E TO LOCATE POINT F.

STEP 8 DRAW LINE TO WALL FROM POINT E THROUGH POINT F.

STEP 9 REPEAT THIS PROCEDURE FOR THE REMAINING WINDERS.

STEP 10 FIRST NORMAL TREAD WILL BE 11" FROM LAST WINDER MEASURED 12" OUT AS SEEN BETWEEN G AND H.

STEP 11 LAY OUT REMAINING NORMAL TREADS.

STEP 12 DRAW PLUMB LINES UP THE WALL TO LOCATE RISER LINES FOR EACH STEP.

PLAN

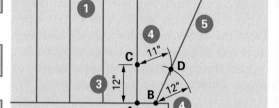

PLAN

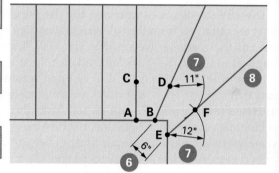

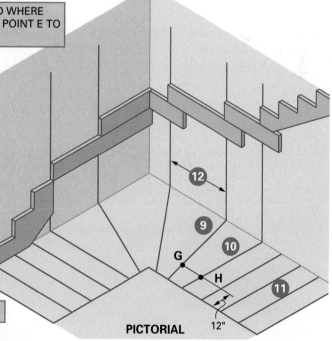

PICTORIAL

closed staircase, the walls at the floor line represent its sides. For stairs open on one side, lay out lines on the floor representing the outside of the staircase. The wall represents the inside. Swing an arc, showing the line of travel. Use the outside corner of the wall under the outer carriage or layout lines as center. The radius of the arc may be 12 to 18 inches, as the codes permit.

From the same center, lay out the width of the narrow end of the treads in both directions. Square lines from the end points to the opposite side of the staircase. Divide the arc into equal parts. Project lines from the narrow end of the tread, through the intersections at the arc, to the wide end at the wall. These lines represent the faces of the risers. Draw lines parallel to these to indicate the riser thickness. Plumb these lines up the wall to intersect with the tread run for each winder.

The cuts on the stair carriage for the winding steps are obtained from the full-size layout. Lay out and cut the carriage. Fasten it to the wall. Install rough treads until the stairs are ready for finishing.

If one side of the staircase is to be open, a **newel post** is installed. Then, the risers are mitered to or **mortised** into the post **(Figure 48–16).** A mortise is a rectangular cavity in which the riser is inserted. Newel posts are part of the stair finish. They are described in more detail in Chapters 80 and 82.

SERVICE STAIRS

Service stairs, typically used for basement stairs, are built as a quick set of stairs. They are not considered finish stairs and are often built without risers. Two

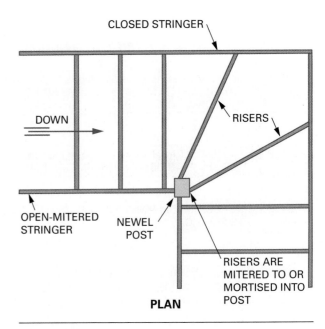

PLAN

FIGURE 48–16 Risers of open-sided winders are mitered against or mortised into a newel post.

carriages are used with nominal 2×10 treads cut between them. The carriages are not always cut out, like those previously described. They may be dadoed to receive the treads. An alternative method is to fasten **cleats** to the carriages to support the treads **(Figure 48–17).**

Lay out the carriages in the usual manner. Cut the bottoms on a level line to fit the floor. Cut the tops along a plumb line to fit against the header of the stairwell. "Drop" the carriages as necessary to pro-

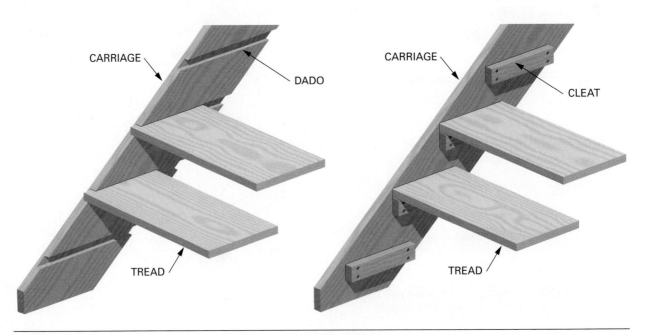

FIGURE 48–17 Service stair carriages may be dadoed or cleated to support the treads.

vide a starting riser with a height equal to the rest of the risers.

Dadoed Carriages

If the treads of rough stairs are to be dadoed into the carriages, lay out the thickness of the tread on the stringer below the layout line. The top of the tread is to the original layout line. Mark the depth of the dado on both edges of the carriage. Set the circular saw to cut the depth of the dado. Make cuts along the layout lines for the top and bottom of each tread. Then, make a series of saw cuts between those just made.

Chisel from both edges toward the center, removing the excess to make the dado. Nail all treads into the dadoes. Assemble the staircase. Fasten the assembled staircase in position. Locate the top tread at a height to obtain an equal riser at the top. The lower end of a basement stairway is sometimes anchored by installing a **kicker** plate, which is fastened to the floor **(Figure 48–18)**.

Cleated Carriages

If the treads are to be supported by cleats, measure down from the top of each tread a distance equal to its thickness. Draw another level line. Fasten 1 × 3 cleats to the carriages with screws. Make sure their

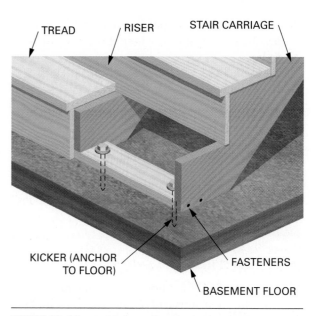

FIGURE 48–18   A kicker plate may be used to anchor a carriage to floor.

top edges are to the bottom line. Fasten the carriages in position. Cut the treads to length. Install the treads between the carriages so the treads rest on the cleats. Fasten the treads by nailing the stair carriage into their ends.

Key Terms

cleats	kicker	open stairways	stairwell
closed stairways	mortise	stair carriage	stringer
handrail	newel post	stair horse	winders
housed stringer	nosing	staircase	

Review Questions

Select the most appropriate answer.

1. The rounded outside edge of a tread that extends beyond the riser is called a
   a. housing.   c. turnout.
   b. coving.   d. nosing.

2. The unit rise of a stair refers to the
   a. vertical part of a step.
   b. riser.
   c. calculation from total rise and number of risers.
   d. all of the above.

3. The unit run of a stair refers to the
   a. tread width without nosing.
   b. horizontal portion of each step plus the nosing.
   c. stairwell opening.
   d. all of the above.

4. Stairways in residential construction should have a minimum width of
   a. 30 inches.   c. 32 inches.
   b. 36 inches.   d. 40 inches.

5. The IRC (International Residential Code) calls for the maximum riser height for a residence to be
   a. 8¼ inches.
   b. 7¾ inches.
   c. 7½ inches.
   d. 7¼ inches.

6. The largest riser height for a stairway with a total vertical rise of 8'–11" is
   a. 8 inches.
   b. 7¾ inches.
   c. 7⅝ inches.
   d. 7½ inches.

7. Using the 17–18 method, what is the range of unit run for a stair with a unit rise of 7¼ inches?
   a. 8¾ to 9¾ inches
   b. 9 to 10 inches
   c. 9¾ to 10¾ inches
   d. 10 to 11 inches

8. Using the 24–25 method, what is the range of unit run for a stair with a unit rise of 7¼ inches?
   a. 8½ to 9½ inches
   b. 9¾ to 10¾ inches
   c. 9½ to 10½ inches
   d. 10 to 11 inches

9. Many building codes specify a minimum unit run of
   a. 8½ inches.
   b. 9½ inches.
   c. 9 inches.
   d. 10 inches.

10. Most building codes specify a minimum headroom clearance of
    a. 6'–6".
    b. 6'–8".
    c. 7'–0".
    d. 7'–6".

11. The stairwell length for a set of stairs with a unit rise of 7½ inches, a unit run of 10 inches, a desired headroom of 82 inches, and an upper construction dimension of 11½ inches is
    a. 93¼".
    b. 110.75".
    c. 124".
    d. 130".

12. The stair carriage with a unit rise of 7½ inches rests on the finish floor. What is the riser height of the first step if the tread thickness is ¾ inch?
    a. 6¾ inches
    b. 7¾ inches
    c. 7½ inches
    d. 8½ inches

13. A stairway has a unit rise of 7½ inches and the stairwell header acts as the top riser. The tread stock thickness is 1¼ inches. The finish floor thickness of the upper floor is ¾ inch. What distance down from the top of the subfloor must the rough carriage be fastened?
    a. 7³⁄₁₆ inches
    b. 7½ inches
    c. 7¾ inches
    d. 7¹³⁄₁₆ inches

14. The line of travel for a set of winding stairs
    a. is the location of the minimum tread width for the stairs.
    b. are drawn for layout of treads.
    c. is at least 12 inches from the narrow edge.
    d. all of the above.

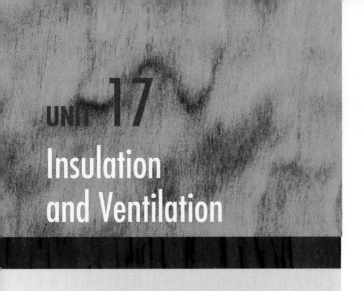

UNIT 17

Insulation and Ventilation

Thermal insulation prevents the loss of heat in buildings during cold seasons. It also resists the passage of heat into air-conditioned buildings in hot seasons. Moisture in the air may cause severe problems for a building. Vapor retarders are essential to a long-lived building. Adequate ventilation must be provided in the living environment and within the building materials. Ventilation encourages evaporation of harmful moisture formed in living spaces and within the insulation. Acoustical insulation reduces the passage of sound from one area to another.

Insulation and ventilation should be thought of together. If a house has insulation it must also have allowances for ventilation. This will ensure the insulation performance is maximized while also making the house pleasant to live in.

OBJECTIVES

After completing this unit, the student should be able to:

- describe how insulation works and define insulating terms and requirements.
- describe the commonly used insulating materials and state where insulation is placed.
- properly install various insulation materials.
- explain relative humidity and moisture migration.
- explain the need for ventilating a structure.
- explain the kinds and purpose of vapor retarders and how they are applied.
- describe various methods of construction to reduce the transmission of sound.

49 Thermal and Acoustical Insulation

All materials used in construction have some insulating value. **Insulation** is a material that interrupts or slows the transfer of heat. Heat transfer is a complex process involving three mechanisms: conduction, convection, and radiation.

Conduction is the transfer of heat by contact. Warmer particles touch cooler ones, causing them to vibrate with more energy, thus transferring energy into the cooler material. Heat energy is thought to move from warmer materials into the cooler ones in an attempt to reach **equilibrium.**

Convection involves a fluid, such as a gas or a liquid. The fluids affecting a house are usually air and water. When a portion of the fluid is warmed, it becomes less dense than the surrounding fluid. The warmer fluid requires more space than the cooler fluid. Gravity causes the cooler, more dense fluid to force the warmer, less dense fluid to rise. This action can be seen in the streams of fluid in a pan of hot oil on a cooking stove. A hot-air balloon encapsulates less dense hot air, which creates lift, causing the craft to rise.

Radiation is a general word referring to electromagnetic radiation, which includes microwaves, radio waves, infrared, visible light, ultraviolet, x rays, and cosmic rays. These can be thought of as particles of energy with varying wavelengths and vibration frequencies that all travel at the same speed, the speed of light. Protection from radiation is best achieved by reflecting it with clean shiny surfaces such as aluminum foil.

The best way to slow heat transfer within a house depends on where the house is located. In southern climes, protection from solar radiation is important, whereas in northern climes this importance is low. In both climes, protection from conduction and convection are important.

HOW INSULATION WORKS

Insulating materials create a space between two surfaces, thereby breaking contact and reducing conduction. They trap air and slow convection. They also can either reflect or absorb some radiation.

Air is an excellent insulator if confined to a small space and kept still. Insulation materials are designed to trap air into small unmoving pockets. Insulation effectiveness increases as the air spaces become smaller in size and greater in number. Millions of tiny air cells, trapped in its unique cellular structure, make wood a better insulator than concrete, steel, or aluminum.

All insulating materials are manufactured from materials that are themselves poor insulators. For example, fiberglass insulation is made from glass, which is a good conductor of heat. The improved insulation value comes from air trapped within the insulating material. Insulation also provides resistance to sound travel.

THERMAL INSULATION

Among the materials used for insulating are glass fibers, mineral fibers (rock), organic fibers (paper), and plastic. Aluminum foil is also used. It works by reflecting heat instead of stopping air movement.

Resistance Value of Insulation

The **R-value** of insulation is a number that indicates its measured resistance to the flow of heat. The higher the R-value, the more efficient the material is in retarding the passage of heat **(Figure 49–1).** R-values are clearly printed on insulation packages **(Figure 49–2).**

Insulation Requirements

The rising costs of energy coupled with the ecological need to conserve have resulted in higher R-value recommendations for new construction than in previous years. Average winter low-temperature zones of the United States are shown in **Figure 49–3.** This information is used to determine the R-value of insulation installed in walls, ceilings, and floors. Insulation requirements vary according to the average low temperature.

In warmer climates, less insulation is needed to conserve energy and provide comfort in the cold season. However, air-conditioned homes should also receive more insulation in walls, ceilings, and floors. This ensures economy in the operation of air-conditioning equipment in hot climates.

Comfort and operating economy are dual benefits. Insulating for maximum comfort automatically provides maximum economy of heating and cooling operations.

Foundation Materials	
8" concrete block (2-hole core)	1.11
12" concrete block (2-hole core)	1.28
8" lightweight aggregate block	2.18
12" lightweight aggregate block	2.48
Common brick	0.20/inch
Sand or stone	0.08/inch
Concrete	0.08/inch

Structural Materials	
Softwood	1.25/inch
Hardwood	0.91/inch
Steel	0.0032/inch
Aluminum	0.00070/inch

Sheathing Materials	
½" plywood	0.63
⅝" plywood	0.78
¾" plywood	0.94
½" aspenite, OSB	0.91
¾" aspenite, OSB	1.37

Insulating Materials	
Batts and Blankets	
3½" fiberglass	11
6" fiberglass	19
8" fiberglass	25
12" fiberglass	38
3½" high-density fiberglass	15
5½" high-density fiberglass	21
8½" high-density fiberglass	30
10" high-density fiberglass	38
fiberglass	3.17/inch

Loose Fill	
Fiberglass	3.17/inch
Cellulose	3.70/inch

Reflective Foil	
Foil-faced bubble pack	1.0
1-layer foil	0.22
4-layer foil	11

Rigid Foam	
Expanded polystyrene foam (bead board)	4.0/inch
Extruded polystyrene foam	5.0/inch
Polyisocyanurate/urethane foam	7.2/inch

Spray Foams	
Low-density polyurethane	3.60/inch
High-density polyurethane	7.0/inch

Finish Materials	
Wood shingles	0.87
Vinyl siding	Negligible
Aluminum siding	Negligible
Wood siding (½" × 8")	0.81
½" hardboard siding	0.36
Polyethylene film	Negligible
Builder's felt (15#)	0.06
²⁵⁄₃₂" hardwood flooring	0.68
Vinyl tile (1/8")	0.05
Carpet and pad	1.23
¼" ceramic tile	0.05
⅜" gypsum board	0.32
½" gypsum board	0.45
⅝" gypsum board	0.56
½" plaster	0.09
Asphalt shingles	0.27

FIGURE 49–1 Insulating R-values of various building materials. (*Courtesy of Richard Harrington*)

FIGURE 49–2 The R-value is broadly stamped on insulation packaging.

Where Heat Is Lost

The amount of heat lost from the average house varies with the type of construction. The principal areas and approximate amount of heat loss for a typical house with moderate insulation are shown in **Figure 49–4.**

Houses of different architectural styles vary in their heat loss characteristics. A single-story house, for example, contains less exterior wall area than a two-story house, but has a proportionately greater ceiling area. Greater heat loss through floors is experienced in homes erected on concrete slabs or unheated crawl spaces unless these areas are well insulated.

The transfer of heat through uninsulated ceilings, walls, or floors can be reduced almost any desired amount by installing insulation. Maximum quantities

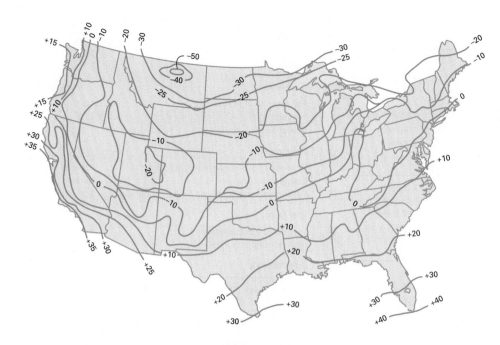

FIGURE 49–3 Average low-temperature zones of the United States. *(Courtesy U.S. Department of Agriculture, Forest Service)*

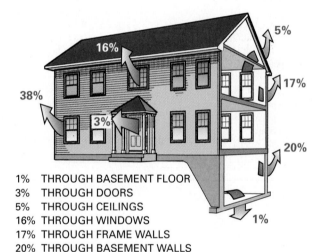

1%   THROUGH BASEMENT FLOOR
3%   THROUGH DOORS
5%   THROUGH CEILINGS
16% THROUGH WINDOWS
17% THROUGH FRAME WALLS
20% THROUGH BASEMENT WALLS
38% AIR LEAKAGE THROUGH CRACKS IN WALLS, WINDOWS, AND DOORS

FIGURE 49–4 Typical heat loss for a house built with moderate insulation. *(Courtesy of Dow Chemical Company)*

in these areas can cut heat losses by up to 90 percent. The use of 2 × 6 studs in exterior walls permits installation of 6-inch-thick insulation. This achieves an R-19 value, or, with improved insulation, an R-21 value.

Windows and doors are generally sources of great heat loss. **Weatherstripping** around windows and doors reduces heat loss. Heat loss through glass surfaces can be reduced 50 percent or more by installing double- or triple-glazed windows. This is referred to as **insulated glass.** Improved insulated glass design uses argon gas between two panes of

glass. The trapped gas improves the R-value significantly. A **low-emissivity coating** on the inner glass surface also improves the window performance. This coating allows ultraviolet rays from the sun to pass but reflects the inside infrared (radiant) rays back into the building. Windows having both of these features are called low E argon windows.

Adding a **storm window** or storm door is also effective. Adding additional layers of trapped air creates additional thermal resistance.

Constructing homes to be more airtight is an effective method of reducing energy needs. Since heat is lost through air leakage, anything that stops air movement will improve energy efficiency. An additional benefit of airtightness includes reducing moisture migration into building components. Moisture migration is covered further in Chapter 50.

TYPES OF INSULATION

Insulation is manufactured in a variety of forms and types. These are commonly grouped as *flexible, loose-fill, rigid, reflective,* and *miscellaneous.*

Flexible Insulation

Flexible insulation is manufactured in *blanket* or *batt* form. Blanket insulation **(Figure 49–5)** comes in rolls. Widths are suited to 16- and 24-inch stud and joist spacing. The usual thickness is from 3½ inches and 6 inches. The body of the blanket is made of fluffy rockwool or glass fibers. Blanket insulation is unfaced or faced with asphalt-laminated Kraft paper or aluminum foil with flanges on both edges for fastening to studs or joists. These facings may be

FIGURE 49–5 Blanket insulation comes in rolls.

considered a **vapor retarder,** also called a **vapor barrier,** but installation of the flanges must be airtight to function as a vapor retarder. These facings should always face the warm side of the wall.

Batt insulation **(Figure 49–6)** is made of the same material as blanket insulation. It comes in thicknesses up to 12 inches. Widths are for standard stud and joist spacing. Lengths are either 48 or 93 inches. Batts come faced or unfaced.

Loose-Fill Insulation

Loose-fill insulation is usually composed of materials in bulk form. It is supplied in bags or bales. It is placed by pouring, blowing, or packing by hand. Materials include rockwool, fiberglass, and cellulose.

Loose-fill insulation is suited for use between ceiling joists and hollow-core masonry walls. It is also used in the sidewalls of existing houses that were not insulated during construction.

FIGURE 49–6 Batt insulation is made up to 12 inches thick.

FIGURE 49–7 Types of rigid insulation boards: (A) Foil-faced polyurethane. (B) Extruded polystyrene. (C) Expanded polystyrene.

Rigid Insulation

Rigid insulation is usually a foamed plastic material in sheet or board forms **(Figure 49–7).** The material is available in a wide variety of sizes, with widths up to 4 feet and lengths up to 12 feet. The most common types are made from polystyrene and polyurethane.

Expanded polystyrene or white bead board has good insulating qualities. It is sometimes installed on the inside of dry basement walls. If expanded polystyrene comes in contact with water it will absorb moisture. The absorbed moisture will significantly reduce the R-value. Therefore, this foam should not be installed in damp areas.

Extruded polystyrene comes in a variety of colors depending on the manufacturer. It is often installed as insulation for footings and basement walls. Extruded polystyrene is a closed-cell foam that will not absorb moisture, keeping its R-value while wet. It may be installed in wet locations and is sometimes used for flotation in docks.

Polyurethane or polyisocyanurate foam is usually made with a facing of foil or building paper. It has a superior R-value of 7.2 per inch. It is used under roof decks and as wall sheathing. Polyurethane foam should only be used on dry areas. It absorbs moisture when in contact with water.

Reflective Insulation

Foil conducts heat well, but reflects radiant energy. It is typically used in warm climates to protect the building from the sun and reflected heat. Reflective insulation usually consists of foil bonded to a surface of some other material such as drywall. It is only effective when installed with a minimum of ¾ inch of space.

Miscellaneous Insulation

Foamed-in-place insulation is sometimes used. A urethane foam is produced by mixing two chemicals together. It is injected into place and expands on contact with the surface. This product requires special equipment provided by an insulating contractor. It not only adds superior insulation value but also adds rigidity to the overall structure. It forms a structural bond between the sheathing and studs, giving racking strength. It also has superior soundproofing qualities that result from its air sealing properties.

Spray foams are available in aerosol cans. They are used to seal and insulate gaps between doors and window units and the wall frame. They are also used to seal mechanical penetrations such as those made by electricians. Cans are typically inverted and sprayed into gaps. As the foam reacts and expands, it fills the space, creating an airtight seal.

Other types of insulating materials include lightweight *vermiculite* and *perlite* aggregates. Vermiculite is made by superheating mica, which causes it to expand quickly, creating air spaces. It is able to withstand high temperatures and can be used around chimneys. Perlite is made from heating volcanic glass, causing it to expand. It is sometimes used in plaster and concrete to make it lighter and more thermal resistant.

WHERE AND HOW TO INSULATE

Any building surfaces that separate conditioned air, heated or cooled, from unconditioned air must be insulated. This will save energy and money over the life of the building. Insulation should be placed in all outside walls and top-floor ceilings. In houses with unheated crawl spaces, insulation should be placed between the floor joists. Collectively these surfaces are called the **thermal envelope.**

Great care should be exercised when installing insulation. Insulation that is not properly installed can render the material useless and a waste of time and money. Pay attention to details around outlets, pipes, and any obstructions. Make the insulation conform to irregularities by cutting and piecing without bunching or squeezing. Keep the natural fluffiness of the insulation intact at all times. Voids in insulation of only 5 percent can create overall efficiency reduction of 25 percent. This is caused by the fact that heat will move to the colder areas. Insulation should be installed neatly. Generally, if insulation looks neat, it will perform well.

In houses with flat or low-pitched roofs, insulation should be used in the ceiling area only if sufficient space is left above the insulation for ventilation. Insulation is used along the perimeter of houses built on slabs when required **(Figure 49–8).**

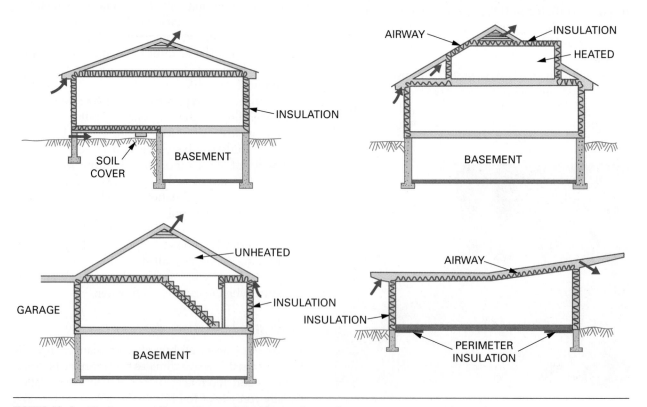

FIGURE 49–8 **Various configurations of the thermal envelope.**

Installing Flexible Insulation

Cut the batts or blankets with a knife. Make lengths slightly oversize by about an inch at the top and bottom. Measure out from one wall a distance equal to the desired lengths of insulation. Draw a line on the floor. Roll out the material from the wall. Compress the insulation with a straightedge on the line. Cut it with a sharp knife on a cutting board to protect the floor **(Figure 49–9).** Cut the necessary number of lengths.

 CAUTION CAUTION CAUTION CAUTION

Always protect lungs, sinuses, eyes, and skin from insulation fibers. Long-term effects could be severe. ∎

Place the batts or blankets between the studs. Staple the flanges of the facing either to the stud thickness edge or to the inside surface of the studs as well as the top and bottom plates. Use a hammer tacker or a hand stapler to fasten the insulation in place **(Figure 49–10).**

Fill any spaces around windows and doors with spray-can foam. Foam will fill the voids with an airtight seal and protect the house from air leakage **(Figure 49–11).** After the foam cures, flexible insulation may be added to fill the remaining space. Do not pack the insulation tightly. Squeezing or compressing it reduces its effectiveness. If the insulation has no covering or stapling tabs, it is friction-fitted between the framing members.

FIGURE 49–10 A hand stapler can be used to fasten wall insulation.

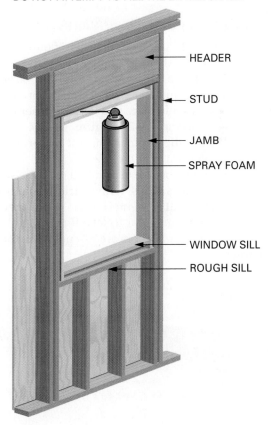

SPRAY A BEAD OF FOAM JUST LARGE ENOUGH TO SEAL DOOR OR WINDOW UNIT TO STUD FRAMING. DO NOT ATTEMPT TO FILL THE ENTIRE SPACE.

HEADER

STUD

JAMB

SPRAY FOAM

WINDOW SILL

ROUGH SILL

FIGURE 49–11 Spaces around doors and windows are filled with spray-foam insulation to seal them to the rough opening.

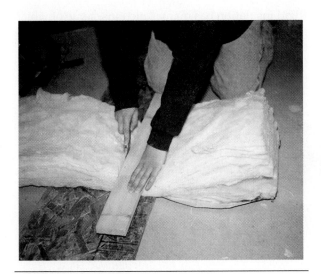

FIGURE 49–9 Flexible insulation is compressed with a straightedge as it is cut.

Ceiling Insulation

Ceiling insulation is installed by stapling it to the ceiling joists or by friction-fitting it between them. If furring strips have been applied to the ceiling joists, the insulation is simply laid on top of the furring strips. In most cases, the use of unfaced insulation is recommended. This makes it easier to determine proper fit of the insulation as well as lowering the cost of materials. Extend the insulation across the top plate.

In most areas of North America, attics should be well ventilated. This ensures that any moisture and heat that escapes the building will not be trapped in the attic space. As warm/moist air is vented out, cool air from the outside replaces it. This has a drying effect on the house. This venting usually begins at the eaves and ends with ridge or attic vents. Therefore, the ceiling insulation must not block this venting.

It may be necessary to compress the insulation against the top of the wall plate to permit the free flow of air. An air-insulation dam should also be installed to protect the insulation from air movement inside the fibers. Otherwise, the insulation's R-value and performance will be reduced **(Figure 49–12).**

Insulating Floors over Crawl Spaces

Flexible insulation installed between floor joists over a crawl space should be unfaced because the facing could act as a vapor retarder and introduce the risk of trapped moisture. Unfaced batts or blankets may be secured in place with an air barrier material stapled to the joists **(Figure 49–13).** This supports the insulation from falling out and protects it from air movement.

A vapor barrier should be installed over the soil under crawl spaces. This will restrict moisture from leaving the soil and entering the crawl space cavity. The easiest material to install is a sheet of polyethylene. Vapor barriers are discussed further in Chapter 50.

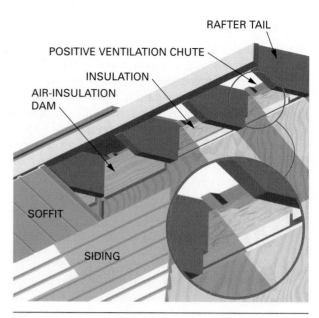

FIGURE 49–12   Air-insulation dams protect the insulation from air infiltration.

Installing Loose-Fill Insulation

Loose-fill insulation is typically blown into place with special equipment. The surface of the material between ceiling joists is leveled to the desired depth **(Figure 49–14).** Care should be taken to evenly distribute the required amount of insulation. This amount is determined from the instructions and proportions printed on the bails of insulation. Take care not to overfill the eave areas. This will restrict the attic ventilation and might allow insulation to fill the soffits.

Installing Rigid Insulation

Rigid insulation has many functions, some of which are not interchangeable. Take care when selecting the material if it will be installed near water. All foams are easily cut with a knife or saw. A table saw can also be used to cut rigid foam.

FIGURE 49–13   Air barrier material protects insulation between floor joists in a crawl space from air infiltration.

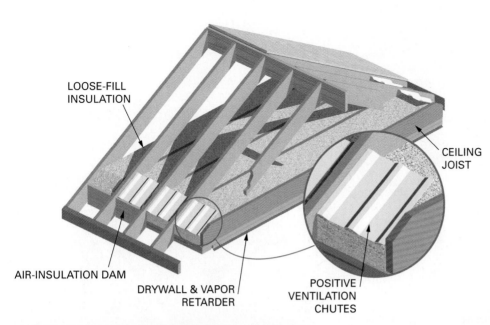

LOOSE-FILL INSULATION

CEILING JOIST

AIR-INSULATION DAM

DRYWALL & VAPOR RETARDER

POSITIVE VENTILATION CHUTES

FIGURE 49–14 Loose-fill insulation is leveled to the desired depth.

 CAUTION CAUTION CAUTION CAUTION

Wear a respirator to protect lungs and sinuses from the fine airborne particles released from insulation when it is cut with a power saw. ■

Rigid foams may be attached with fasteners or adhesives or applied by friction-fitting between the framing members. Fasteners include plastic capped nails that are ring shanked with large plastic washers. They hold the foam in place without pulling through the foam. Adhesives are used to glue the sheet to a surface. Some adhesives will dissolve the plastic foam, so be sure to use the correct adhesive as listed on the adhesive tube.

Polyurethane foam is sometimes used on roofs of cathedral ceilings **(Figure 49–15).** Various thicknesses may be used depending on the desired R-value. Roof sheathing should be installed on furring strips above the foam. Since the foam is an insulator, installing sheathing and shingles directly to it will cause the shingles to overheat. This will cause the shingle life span to be reduced. The air space allows for removal of moisture and excess heat.

Insulating Masonry Walls

Masonry walls can be insulated on the inside or outside of the building **(Figure 49–16).** Interior applications require the masonry wall to be furred. Furring can be wood or metal hat track. Fasten furring strips 16 or 24 inches on center (OC). Dry walls may be insulated with fiberglass batts, expanded polystyrene, or polyurethane sheets. For damp walls, only extruded polystyrene boards should be used.

Insulating the exterior has advantages and disadvantages over interior applications. An advantage is that the masonry wall becomes part of the thermal envelope. It serves as a heat battery helping to reduce the high and low swings of temperature within the room.

The disadvantage is that a protective layer must be installed over the foam. This layer should be a masonry mastic troweled on to protect the foam from physical abuse and the sun, which will degrade it over time. Also, exterior application is not recommended in regions where termites are a problem. Termites are more difficult to detect behind the insulation layer.

Installing Reflective Insulation

Reflective insulation usually is installed between framing members in the same manner as blanket insulation. It is attached to either the face or side of the studs. However, an air space of at least ¾ inch must be maintained between its surface and the inside edge of the stud.

ACOUSTICAL INSULATION

Acoustical or *sound* insulation resists the passage of noise through a building section. The reduction of sound transfer between rooms is important in offices, apartments, motels, and homes. Excessive noise is not only annoying but harmful. It not only causes fatigue and irritability, it can damage the sensitive hearing nerves of the inner ear.

Sound insulation between active areas and quiet areas of the home is desirable. Sound insulation between the bedroom area and the living area is important. Bathrooms also should be insulated.

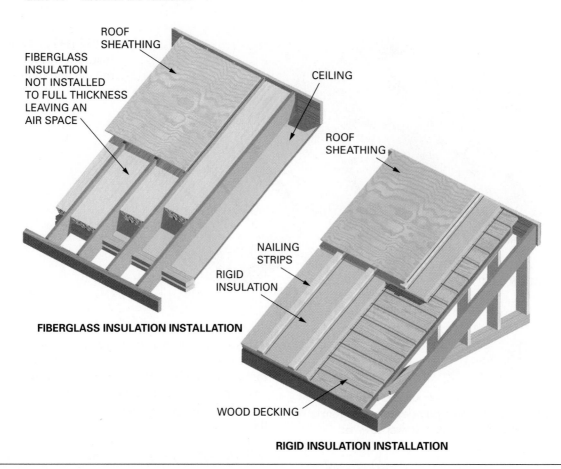

FIBERGLASS INSULATION INSTALLATION

RIGID INSULATION INSTALLATION

FIGURE 49–15 Methods of installing rigid insulation under roof sheathing.

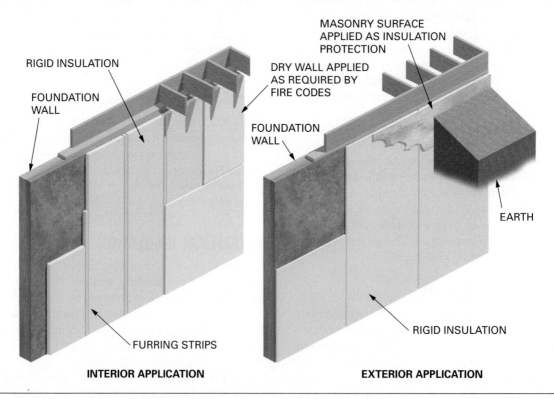

INTERIOR APPLICATION

EXTERIOR APPLICATION

FIGURE 49–16 Extruded polystyrene rigid insulation may be applied to the exterior or interior of basement walls.

Sound Transmission

Noises create sound waves. These waves radiate outward from the source until they strike a wall, floor, or ceiling. These surfaces then begin to vibrate by the pressure of the sound waves in the air. Because the wall vibrates, it conducts sound to the other side in varying degrees, depending on the wall construction.

Sound Transmission Class. The resistance of a building section, such as a wall, to the passage of sound is rated by its **Sound Transmission Class (STC).** The higher the number, the better the sound barrier. The approximate effectiveness of walls with varying STC numbers is shown in **Figure 49–17.**

Sound travels readily through the air and through some materials. When airborne sound strikes a wall, the studs act as conductors unless they are separated in some way from the covering material. Electrical outlet boxes placed back to back in a wall easily transmit sound. Faulty construction, such as poorly fitted doors, often allows sound to pass through. Therefore, good, airtight construction practices are the first line of defense in controlling sound.

Wall Construction

A wall that provides sufficient resistance to airborne sound should have an STC rating of 45 or greater. At one time, the resistance usually was provided only

25	Normal speech can be understood quite easily
30	Loud speech can be understood fairly well
35	Loud speech audible but not intelligible
42	Loud speech audible as a murmur
45	Must strain to hear loud speech
48	Some loud speech barely audible
50	Loud speech not audible

This chart from the Acoustical and Insulating Materials Association illustrates the degree of noise control achieved with barriers having different STC numbers.

FIGURE 49–17 Approximate effectiveness of sound reduction in walls with varying STC ratings.

by double walls, which resulted in increased costs. However, a system of using **sound-deadening insulating board** with a gypsum board outer covering has been developed. This system provides good sound resistance. Resilient steel channels placed at right angles to the studs isolate the gypsum board from the stud. **Figure 49–18** shows various types of wall construction and their STC rating.

Floor and Ceiling Construction

Sound insulation between an upper floor and the ceiling below involves not only the resistance of airborne sounds, but also that of **impact noise.** Impact

SOUND INSULATION OF SINGLE WALLS			SOUND INSULATION OF DOUBLE WALLS		
DETAIL	DESCRIPTION	STC RATING	DETAIL	DESCRIPTION	STC RATING
A — 16" — 2" X 4"	1/2" GYPSUM WALL-BOARD	32	A — 16" — 2" X 4"	1/2" GYPSUM WALLBOARD	38
	5/8" GYPSUM WALL-BOARD	37			
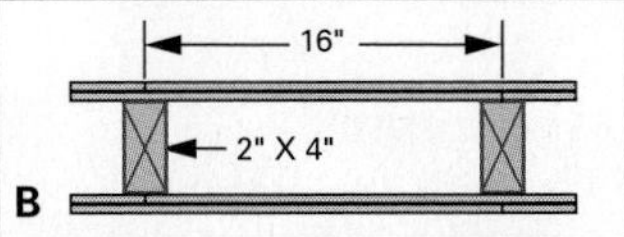 B — 16" — 2" X 4"	3/8" GYPSUM LATH (NAILED) PLUS 1/2" GYPSUM PLASTER WITH WHITECOAT FINISH (EACH SIDE)	39	B — 16" — 2" X 4"	5/8" GYPSUM WALLBOARD (DOUBLE LAYER EACH SIDE)	45
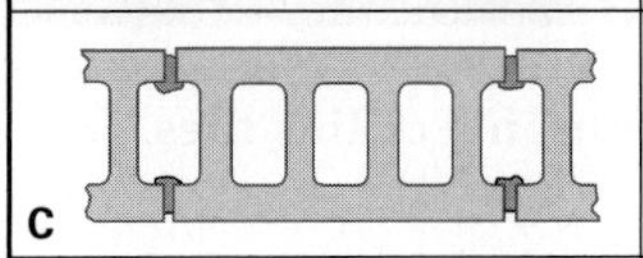 C	8" CONCRETE BLOCK	45	C — 16" — BETWEEN OR WOVEN 2" X 4"	1/2" GYPSUM WALLBOARD 1 1/2" FIBROUS INSULATION	49
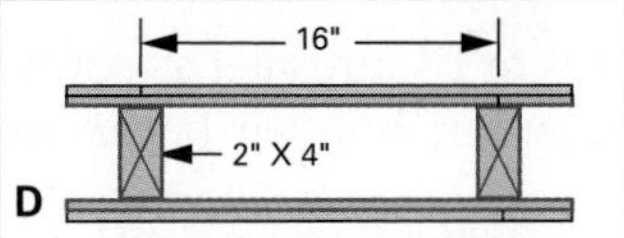 D — 16" — 2" X 4"	1/2" SOUND DEADENING BOARD (NAILED) 1/2" GYPSUM WALLBOARD (LAMINATED) (EACH SIDE)	46	D — 16" — 2" X 4"	1/2" SOUND DEADENING BOARD (NAILED) 1/2" GYPSUM WALLBOARD (LAMINATED)	50
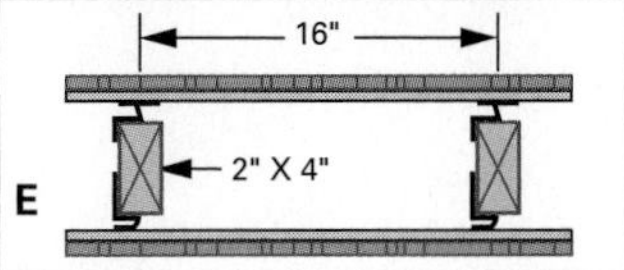 E — 16" — 2" X 4"	RESILIENT CLIPS TO 3/8" GYPSUM BACKER BOARD 1/2" FIBERBOARD (LAMINATED) (EACH SIDE)	52			

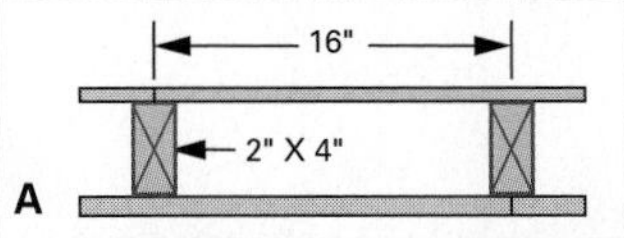

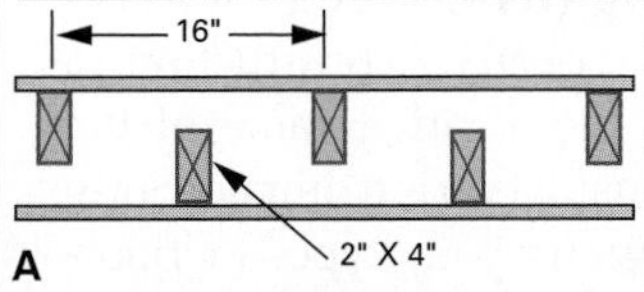

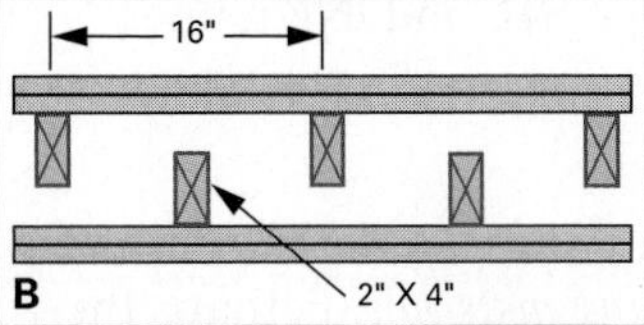

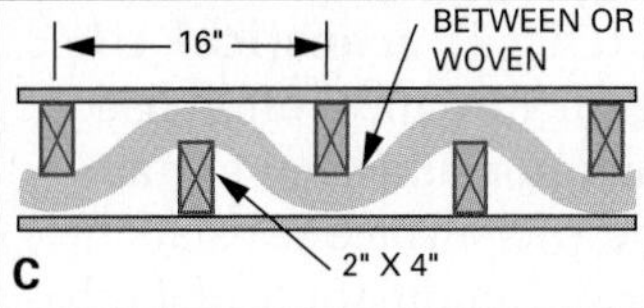

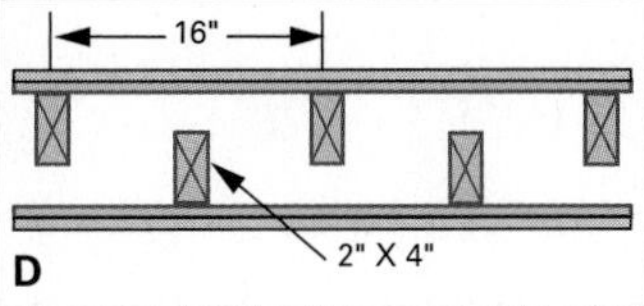

FIGURE 49–18 STC ratings of various types of wall construction.

RELATIVE IMPACT AND SOUND TRANSFER IN FLOOR-CEILING COMBINATIONS (2" X 10" JOISTS)		ESTIMATED VALUE		RELATIVE IMPACT AND SOUND TRANSFER IN FLOOR-CEILING COMBINATIONS (2" X 8" JOISTS)		ESTIMATED VALUE	
DETAIL	DESCRIPTION	STC RATING	APPROX. INR	DETAIL	DESCRIPTION	STC RATING	APPROX. INR
A	FLOOR 3/4" SUBFLOOR BUILDING PAPER 3/4" FINISH FLOOR CEILING GYPSUM LATH AND SPRING CLIPS 1/2" GYPSUM PLASTER	52	–2	D	FLOOR 7/8" T. & G. FLOORING CEILING 3/8" GYPSUM BOARD	30	–18
B	FLOOR 5/8" PLYWOOD SUBFLOOR 1/2" PLYWOOD UNDERLAYMENT 1/8" VINYL - ASBESTOS TILE CEILING 1/2" GYPSUM WALLBOARD	31	–17	E	FLOOR 3/4" SUBFLOOR 3/4" FINISH FLOOR CEILING 3/4" FIBERBOARD	42	–12
C	FLOOR 5/8" PLYWOOD SUBFLOOR 1/2" PLYWOOD UNDERLAYMENT FOAM RUBBER PAD 3/8" NYLON CARPET CEILING 1/2" GYPSUM WALLBOARD	45	+5	F	FLOOR 3/4" SUBFLOOR 3/4" FINISH FLOOR CEILING 1/2" FIBERBOARD LATH 1/2" GYPSUM PLASTER 3/4" FIBERBOARD	45	–4

FIGURE 49–19   STC and INR for floor and ceiling constructions.

noise is caused by such things as dropped objects, footsteps, or moving furniture. The floor is vibrated by the impact. Sound is then radiated from both sides of the floor. Impact noise control must be considered as well as airborne sounds when constructing floor sections for sound insulation.

An **Impact Noise Rating (INR)** shows the resistance of various types of floor-ceiling construction to impact noises. The higher the positive value of the INR, the more resistant the insulation is to impact noise transfer. **Figure 49–19** shows various types of floor-ceiling construction with their STC and INR ratings.

Sound Absorption

The amount of noise in a room can be minimized by the use of *sound-absorbing materials*. Perhaps the most commonly used material is **acoustical tile** made of fiberboard. These tiles are most often used in the ceiling where they are not subjected to damage. The tiles are soft. The tile surface consists of small holes or fissures or a combination of both

FIGURE 49–20   Sound-absorbing ceiling tiles.

(Figure 49–20). These holes or fissures act as sound traps. The sound waves enter, bounce back and forth, and finally die out.

50 Condensation and Ventilation

Energy use and costs are reduced when buildings are insulated. Unfortunately, when a building is made more airtight and more energy efficient, a negative side effect occurs. This effect is caused by water.

Water is the enemy of a building. Roofing and siding are installed to protect the building from water. Water can also cause the interior to be musty and moldy, the insulation to degrade and perform poorly, and potentially cause structural failures. Moisture in and around a building must be understood and dealt with for the building to function properly and last a long time.

Energy-efficient construction is desired to reduce energy costs and make the building more comfortable. But the problems of excess moisture within a building must be addressed. Solutions include controlling air leakage, using vapor retarders, also called vapor barriers and building drying potential into building systems.

CONDENSATION

Water exists in solid, liquid, and gaseous forms. Most familiar are the solid and liquid forms, which we all know as ice and water. Gaseous water is called steam when it is very hot and **vapor** when it is cool **(Figure 50–1).** Vapor is normally invisible. **Condensation** is when the vapor falls out of the air on to cooler surfaces. **Dew point** is a term used to identify the air temperature when condensation has occurred. Fog is an example of air that has reached its dew point.

Relative Humidity

The amount of moisture held in the air is referred to as **relative humidity.** This amount can vary with the temperature of the air. Warm air can hold more moisture than cool air. Consider four identical 1-cubic-foot containers of air, each having about 5 drops (0.00047 pound) of water suspended in the

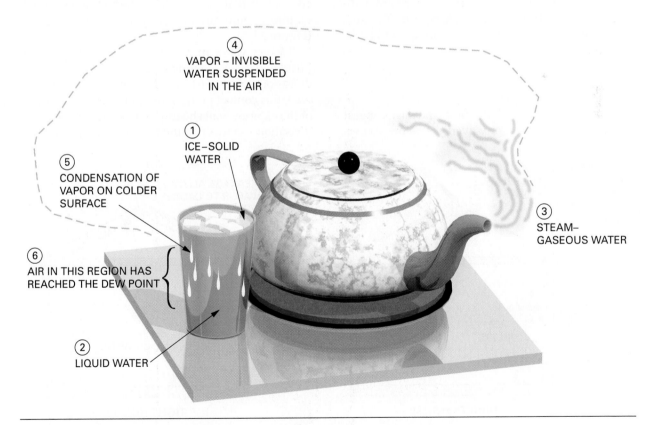

FIGURE 50–1 Water can exist in three forms or states. Moisture in warm air condenses when it comes in contact with a cold surface.

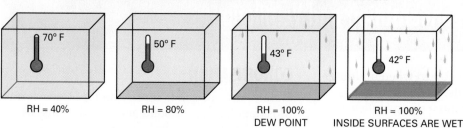

IDENTICAL BOXES, IDENTICAL AMOUNTS OF WATER VAPOR INSIDE.
TEMPERATURE DIFFERENCE CAUSES RELATIVE HUMIDITY TO DIFFER.

70° F   50° F   43° F   42° F

RH = 40%   RH = 80%   RH = 100%   RH = 100%
                                     DEW POINT   INSIDE SURFACES ARE WET

⟵ INCREASING TEMPERATURE

INCREASING RELATIVE HUMIDITY (RH) ⟶

FIGURE 50–2 Relative humidity changes with air temperature.

air as vapor **(Figure 50–2)**. The relative humidity (RH) in each container depends on the temperature of the air. At 70°, the RH would be 40 percent, or put another way, the air would be holding 40 percent of the moisture it could hold at that temperature.

As the temperature drops, the air is less able to hold moisture. When the temperature reaches 43°, the air can hold no more than the 5 drops, thus the relative humidity is 100 percent. The air is completely saturated with moisture. A slight lowering of temperature will cause the air to be at the dew point and the container walls begin to feel moist.

Dew point is not always at 43°; it can be any temperature where the RH reaches 100 percent. The dew point occurs when the air can no longer hold any more moisture.

Moisture in Buildings

Building materials perform best and last longest when the relative humidity averages below 50 per-

cent, but the RH in the living environment is often much higher.

Moisture in the form of water vapor inside a building comes from many activities. It is produced by cooking, bathing, washing, drying, and cleaning as well as many other sources. Reducing the moisture production within the building is one step in solving the problem, but can only go so far.

Older uninsulated homes did not have the interior moisture problems that we have today. They were drafty enough to dry the house during the heating season. Often humidifiers were operated to add moisture to the air. Today tighter homes might require dehumidifiers during the heating seasons **(Figure 50–3)**.

If water vapor moves into the thermal envelope (insulated walls, ceilings, and floors) it will condense on cooler surfaces **(Figure 50–4)**. Inside the wall, this contact point may be on the inside surface of the exterior wall sheathing. In a crawl space, condensation can form on the floor frame and subfloor.

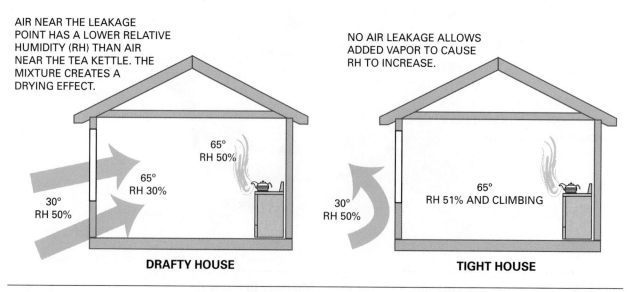

AIR NEAR THE LEAKAGE POINT HAS A LOWER RELATIVE HUMIDITY (RH) THAN AIR NEAR THE TEA KETTLE. THE MIXTURE CREATES A DRYING EFFECT.

65°
RH 50%

65°
RH 30%

30°
RH 50%

DRAFTY HOUSE

NO AIR LEAKAGE ALLOWS ADDED VAPOR TO CAUSE RH TO INCREASE.

65°
RH 51% AND CLIMBING

30°
RH 50%

TIGHT HOUSE

FIGURE 50–3 Comparison of drafty and airtight houses with respect to relative humidity.

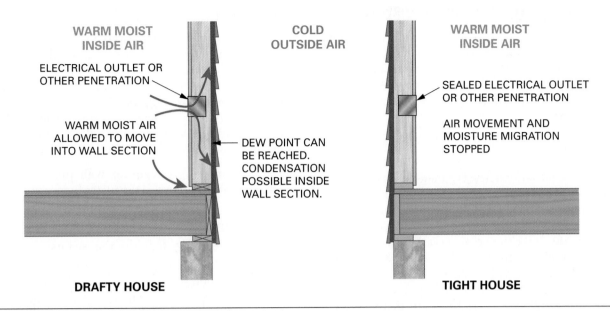

WARM MOIST INSIDE AIR

ELECTRICAL OUTLET OR OTHER PENETRATION

WARM MOIST AIR ALLOWED TO MOVE INTO WALL SECTION

COLD OUTSIDE AIR

DEW POINT CAN BE REACHED. CONDENSATION POSSIBLE INSIDE WALL SECTION.

DRAFTY HOUSE

WARM MOIST INSIDE AIR

SEALED ELECTRICAL OUTLET OR OTHER PENETRATION

AIR MOVEMENT AND MOISTURE MIGRATION STOPPED

TIGHT HOUSE

FIGURE 50–4   Moisture migration into building materials is caused mostly by air leakage.

In attics, the ceiling joists and roof frame can become saturated with moisture.

Controlling moisture in the building is essential. Condensation of water vapor inside walls, attics, roofs, and floors could lead to serious problems.

Reducing the production of moisture within the house is ultimately the responsibility of the homeowner. Homeowners must be educated about the problems of moisture. Proper maintenance of exhaust piping for clothes dryers and using bathroom and kitchen fans regularly are important. If they are defective, constricted, or unused, the moist air will not be removed.

Moisture Problems

- High relative humidity and a warm environment will allow mold and mildew to grow. This can be seen in bathrooms and basements.

- When insulation absorbs water, the dead air spaces in the insulation may become filled with water. Insulation may compress when it gets wet and will not return to its original shape. This significantly reduces the insulation R-value.

- Uncontrolled moisture can move through the wall. This may cause exterior paint to blister and peel.

- A warm attic will cause the formation of ice dams at the **eaves.** After a heavy snowfall, lost heat from the building causes the snow next to the roof to melt. Water then runs down the roof and

freezes on the cold roof overhang, forming ice buildup. As this continues an **ice dam** is formed. This causes water to back up on the roof, under the shingles, and into the walls and ceiling **(Figure 50–5).**

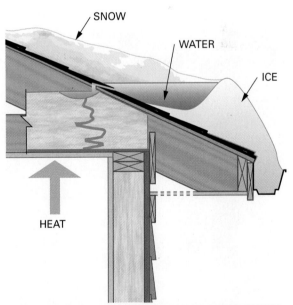

SNOW

WATER

ICE

HEAT

HEAT ESCAPING FROM CEILING MELTS SNOW. WATER FLOWS TO THE OVERHANG WHERE IT FREEZES INTO AN ICE DAM. WATER BUILDUP BEHIND DAM BACKS UP UNDER ROOFING MATERIAL.

FIGURE 50–5   A properly constructed and ventilated attic will keep ice dams from forming.

PREVENTION OF MOISTURE PROBLEMS

The goal is not to remove all moisture, because this would be virtually impossible and undesirable from the standpoint of human comfort. The goal is to reduce moisture migration into the building components and remove excess moisture through ventilation.

Vapor Retarder

A vapor retarder (barrier) is a material used to slow the flow of airborne moisture from passing through building materials. Polyethylene sheeting installed behind the drywall or interior finish serves as a vapor retarder. It is a transparent plastic sheet. It comes in rolls of usually 100 feet in length and 8, 10, 12, 14, 16, and 20 feet in width. The most commonly used thicknesses are 4 mils and 6 mils.

Blanket and batt insulations are manufactured with facings that also may serve as a vapor retarder. In order for these facings to perform as a vapor barrier, the stapling flanges must be installed in a tight manner. The flanges should be overlapped and stapled to the stud thickness, not the side **(Figure 50–6)**. The drywall or finish material will help hold the flanges tight.

Polyethylene sheeting, while effective, is not the only type of vapor retarder. Vapor retardant paints are also effective.

FIGURE 50–6 The vapor retarder material of faced insulation must be overlapped and stapled to the stud thickness to be effective.

Air Leakage

Most excess moisture migration into building components occurs by air leakage. If air is allowed to move into a wall section, moisture will be included. Thus, if air movement into the wall section is minimized, so is moisture migration.

Many methods and techniques are available to reduce air leakage. Some involve only a little extra time and money:

- Apply caulk under exterior wall plates before they are stood up to seal the plates to the subfloor.
- Install wall sheathing panels with adhesive applied to studs and plates to seal the sheet perimeter.
- Apply sheathing tape to the seams of wall sheathing.
- Seal all penetrations in the framing members. These include those created by plumbing, electrical, heating, air conditioning, and ventilating installations.
- Install foamed-in-place insulation.
- Install drywall panels with construction adhesive applied to studs and plates to seal the sheet perimeter.

It is not necessary to use every technique to reduce air leakage; each will make a difference. For example, an effective approach could be to seal the wall plates to the subfloor, seal holes and penetrations, and seal wall sheathing seams with sheathing tape.

Ventilation

Ventilation is the exchange of air to allow for drying and improved air quality. This often must take place both inside the living environment and within the building components.

Some believe buildings can be built too airtight. This is not an accurate statement. More to the point, buildings must control unwanted air leakage and ventilate unwanted moisture. With airtight construction techniques, the energy costs of the building are reduce. With proper ventilation, the building is comfortable and will last a long time.

Ventilation can be achieved by either allowing the natural flow of air or by using a fan. The interior living areas are vented with fans installed in rooms where air quality needs to be improved. These areas include bathrooms, kitchens, and laundry rooms. Using them daily exhausts the moisture-laden air to the outside.

Drying Potential

Materials of a building are exposed to severe weather. The building should be constructed such that building materials can dry easily. This is referred to as **drying**

potential. Drying will allow condensed moisture and wind-driven water to be removed by evaporation. It can be achieved by constructing natural ventilation in the building.

Proper ventilation depends on where the building is built. The relative humidity of the region must be considered. In cool climes where the building will require heat, the inside air usually has a higher RH than the outside air. In warm moist climes, the outside air usually has the higher RH.

If ventilation takes place, the outside air brought in is adjusted to the same temperature as the inside air. This may be accomplished heating or cooling. When air is warmed its ability to hold moisture increases and its RH is reduced. This has a drying effect. If the air must be cooled, its ability to hold moisture goes down and its RH goes up. This will possibly cause dampness.

In crawl spaces under floors where the air is usually cooler than the outside air, no ventilation should be installed. Here warmer outside air entering the crawl space area would be cooled. This would raise the relative humidity, causing more moisture problems.

Drying would happen more effectively when the outside air was cooler. The best solution for crawl spaces is to install the vapor retarder on the ground. A sheet of polyethylene will inhibit moisture from leaving the ground and getting into the floor system above. A layer of sand over the sheet will protect it from accidental perforations.

Attic Ventilation

With a well-insulated ceiling and adequate ventilation, attic temperatures are lower and excess moisture is removed.

On roofs where the ceiling finish is attached to the roof rafters, insulation is usually installed between the rafters. An adequate air space of at least 1 inch must be maintained between the insulation and the roof sheathing. The air space must be connected to air inlets in the soffit and outlets at the ridge **(Figure 50–7).** Failure to do so may result in reduced shingle life, formation of ice dams at the eaves, and possible decay of the roof frame.

Types of Ventilators

There are many types of ventilators. Their location and size are factors in providing adequate ventilation.

Ventilating Gable Roofs. The best way to vent an attic is with a combination of ridge and soffit vents **(Figure 50–8).** Each rafter cavity is vented from soffit to ridge. The roof sheathing is cut back about 1 inch from the ridge on each side, and the vent material is nailed over this slot.

Cap shingles then can be nailed directly to the vent installed over the vent space. Perforated material or screen vents are installed in the soffits to provide the entry point for the ventilation. Positive-ventilation chutes should be installed to prevent any air obstructions caused by the ceiling insulation near the eaves **(Figure 50–9).** This system can adequately vent the attic space of a house that is up to 50 feet wide.

Triangularly shaped louver vents are sometimes installed in both end walls of a gable roof. They come in various shapes and sizes and are installed as close to the roof peak as possible. Their effectiveness depends on the prevailing wind direction.

The minimum free-air area for attic ventilators is based on the ceiling area of the rooms below. The

FIGURE 50–7 Method of providing ventilation when the entire rafter cavity is filled with insulation.

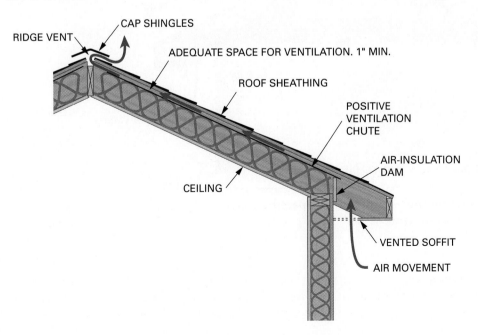

CAP SHINGLES

RIDGE VENT

ADEQUATE SPACE FOR VENTILATION. 1" MIN.

ROOF SHEATHING

POSITIVE VENTILATION CHUTE

AIR-INSULATION DAM

CEILING

VENTED SOFFIT

AIR MOVEMENT

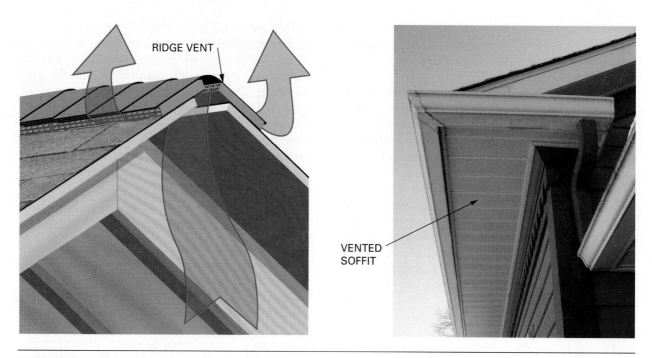

FIGURE 50–8 Ridge and soffit vents work together to provide adequate attic ventilation.

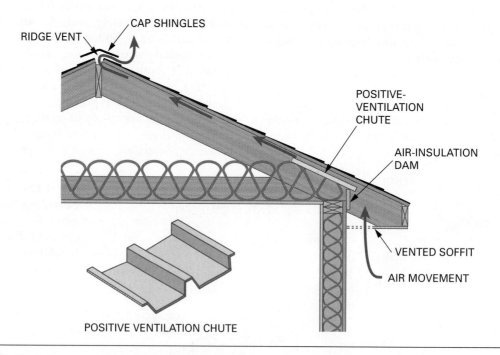

FIGURE 50–9 Positive-ventilation chutes maintain air space between compressed insulation and roof sheeting.

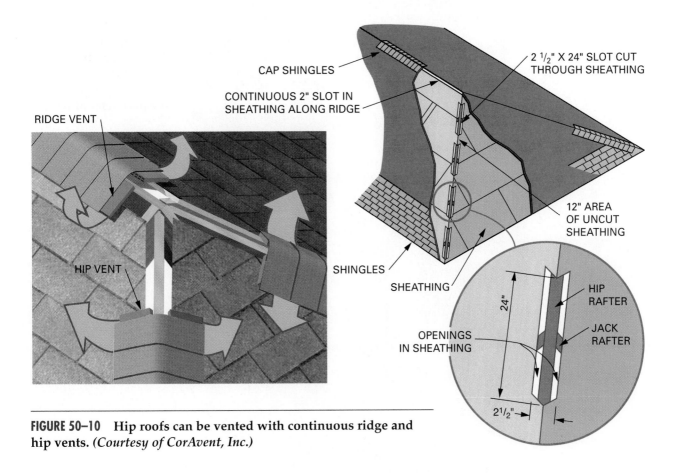

FIGURE 50–10   Hip roofs can be vented with continuous ridge and hip vents. *(Courtesy of CorAvent, Inc.)*

free-air area for the openings should be ¹⁄₁₅₀th of the ceiling area. For example, if the ceiling area is 1,200 square feet, the minimum total free-air area of the ventilators should be 1,200 divided by 150 = 8 square feet of open area.

Ventilating Hip Roofs.

Hip roofs should have additional continuous venting along each hip rafter. This allows each hip-jack rafter cavity to be vented. When cutting a 2½-inch-wide slot for the vents, it is recommended to leave a 1-foot section of sheathing uncut between every 2 feet of slot section **(Figure 50–10)**. This allows for adequate ventilation while maintaining the integrity sheathing for the hip roof.

Reducing Air Leakage

Airtight sheathing is installed by adding construction adhesive. Apply a continuous bead on the studs and plates where the perimeter of the panel will fit.

Nail the panel as required by nailing codes. Another technique uses sheathing tape that is installed such that it covers every panel seam. Either technique is completed with a bead of adhesive applied under the bottom plate before the wall is raised.

Airtight drywall is installed in a similar fashion as airtight sheathing **(Figure 50–11)**. Adhesive is applied continuously to the studs and plates along the sheet perimeter. The plate should also be sealed to the subfloor before the wall is erected.

Foamed-in-place insulation creates an airtight thermal envelope. It can only be installed by insulation contractors with special equipment **(Figure 50–12)**.

All penetrations should be sealed before the interior finish is applied. No crack or hole is too small to be sealed **(Figure 50–13)**. Interior and exterior walls are equally important. Any air movement in the structure can cause air leakage and moisture migration.

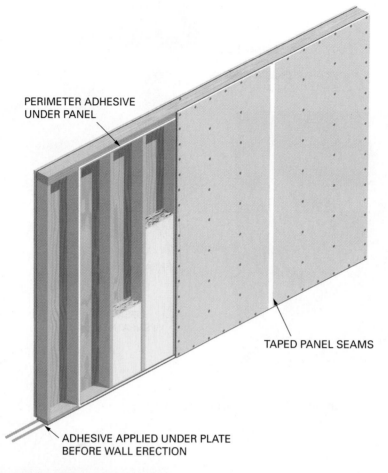

PERIMETER ADHESIVE
UNDER PANEL

TAPED PANEL SEAMS

ADHESIVE APPLIED UNDER PLATE
BEFORE WALL ERECTION

FIGURE 50–11 Airtight drywall may be achieved with adhesive placed on the perimeter of panel.

FIGURE 50–12 Foamed-in-place insulation provides an excellent R-value and airtightness. *(Courtesy of Richard Harrington)*

FIGURE 50–13 All penetrations within a house should be sealed.

Procedure 50–A Installing a Polyethylene Film Vapor Retarder

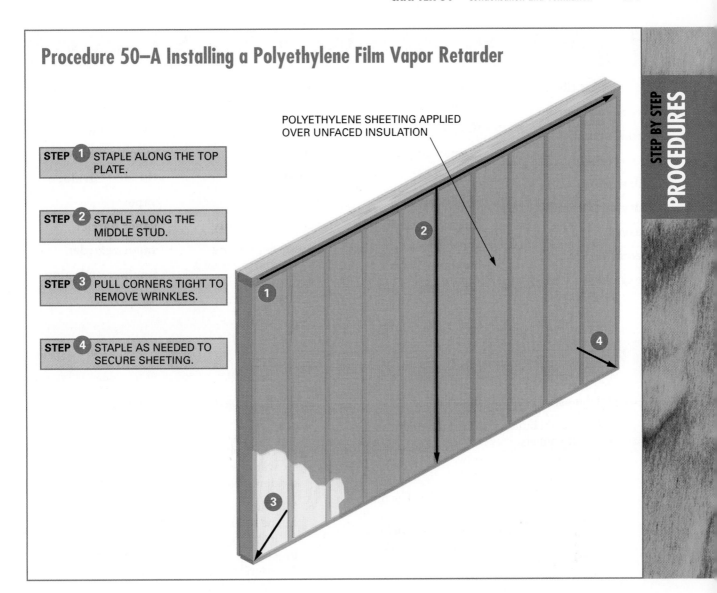

POLYETHYLENE SHEETING APPLIED
OVER UNFACED INSULATION

STEP 1 STAPLE ALONG THE TOP PLATE.

STEP 2 STAPLE ALONG THE MIDDLE STUD.

STEP 3 PULL CORNERS TIGHT TO REMOVE WRINKLES.

STEP 4 STAPLE AS NEEDED TO SECURE SHEETING.

STEP BY STEP PROCEDURES

Polyethylene sheeting is installed by unrolling a length long enough to cover the wall. Add extra length and width to ensure proper coverage. All seams should be along a nailing surface such as a plate or stud.

Partially unfold the section to expose the long edge. Staple it along the top plate about 6 to 12 inches apart, letting the rest drape down to the floor. Next begin along the stud in the middle of the sheeting. Smooth out wrinkles downward as it is stapled. The result is that the staples will form a large T shape in the sheeting. Have someone pull each unfastened corner snug to smooth wrinkles. Staple as needed **(Procedure 50–A).**

Carefully fit it around all openings. Lap all joints by several inches, keeping them on a surface nailing. Repair any tears with sheathing tape. Cut off the excess at the floor line.

Some carpenters cut the film out of openings after the drywall finish is applied. This assures a more positive seal. The retarder should be fitted tightly around outlet boxes. Add a ribbon of sealing compound around outlets and switch boxes.

Key Terms

acoustical tile

condensation

conduction

convection

dew point

drying potential

eaves

equilibrium

expanded
   polystyrene

extruded
   polystyrene

flexible insulation

foamed-in-place

ice dam

impact noise

Impact Noise Rating
   (INR)

insulated glass

insulation

loose-fill insulation

low-emissivity
   coating

polyurethane

radiation

R-value

relative humidity

sound-deadening
   insulating board

Sound Transmission
   Class (STC)

spray foams

storm window

thermal envelope

vapor

vapor barrier

vapor retarder

weather-stripping

Review Questions

Select the most appropriate answer.

1. The structural building material with the greater R-value is
   a. concrete.
   b. steel.
   c. stone.
   d. wood.

2. The insulating term *R-value* is defined as the measure of
   a. resistance of a material to the flow of heat.
   b. relative amount of the heat lost through a building section.
   c. the conductivity of a material.
   d. the total heat transfer through a building section.

3. The boundary between conditioned and unconditioned air is called the thermal
   a. envelope.
   b. resistance.
   c. retarder.
   d. dam.

4. The material used to protect insulation from air infiltration at the eaves is called an air
   a. barrier.
   b. dam.
   c. retarder.
   d. stopper.

5. The term used to identify air when water droplets form on cooler surfaces is called
   a. condensation point.
   b. vapor point.
   c. water point.
   d. dew point.

6. Moisture migration into the insulation layer can be reduced by
   a. installing a vapor retarder.
   b. placing sheathing tape on the seams of exterior sheathing.
   c. airtight construction techniques.
   d. all of the above.

7. A vapor barrier in a crawl space should be installed
   a. just below the subfloor.
   b. just under the joists.
   c. on top of the ground.
   d. all of the above.

8. If air temperature suddenly increases, the relative humidity would
   a. increase.
   b. decrease.
   c. remain the same.
   d. all of the above.

9. Squeezing or compressing flexible insulation tightly into spaces
   a. reduces its effectiveness.
   b. increases its efficiency by creating more air spaces.
   c. is necessary to hold it in place.
   d. helps prevent air leakage by sealing cracks.

10. When insulation is placed between roof framing members, there should be an air space between the insulation and the roof sheathing of at least
    a. ½ inch.
    b. 1 inch.
    c. 2 inches.
    d. 3 inches.

11. Ice dams on top of the roofing can be eliminated by properly installed
    a. attic ventilation.
    b. roofing material.
    c. vapor retarder.
    d. air dam.

12. The best method of venting an attic space is with
    a. gable vents.
    b. hip vents.
    c. roof windows.
    d. continuous ridge and soffit vents.

Success Story

Tanya Marquette

Title: General Contractor

Company: Tanya Marquette Construction, New Paltz, New York

EDUCATION

When Tanya re-entered the workforce after raising her children, she began by doing odd carpentry jobs. "Using a hammer and a drill was very empowering—I loved it!" she recalls. When there were fewer jobs to be had for unskilled workers, Tanya decided training would help put her ahead of the pack. "I realized I didn't know things, like how to bid a job," she says. So she started attending a carpentry program at the Hudson Valley Occupational Industrial Center (HVOIC), which helped her hone the skills she'd taught herself.

HISTORY

A teacher from HVOIC asked her to be his assistant carpenter, and she accepted. He was formative in making her the carpenter she is today, Tanya says. "I learned a tremendous amount from him," she recalls. "He was a very old-style craftsperson who didn't like new-fangled things, so I learned how to do things with hand tools." In the late 1970s, she decided to go into business for herself, and started Tanya Marquette Construction. Word of mouth brought her more jobs, and as her workload would swell and shrink, she'd add or subtract crew members. By the 1990s, her work shifted away from hands-on tasks and she became more of a manager. Today she finds it easier to work with subcontractors rather than a crew, because she can more accurately serve the divergent needs of each job, as well as better control her income.

ON THE JOB

Each day, Tanya gets up around 6:30 a.m.: "I figure out the goal of the day and put my running shoes on." For the last fourteen months, she's been working as a general contractor on an eight-acre property that included a 150-year-old barn remodeled into a house. The job kept growing: She first built a two-car garage; then a four-stall barn for a horse; clear-cut four acres of property to make a pasture; put up a 1,500-foot fence for it; and moved a driveway. She also did finish work inside the home, including gutting and rebuilding bathrooms, turning a walk-in closet into a library, installing new ceramic tile floors, redoing other floors, putting in skylights, and painting the basement.

BEST ASPECTS

Tanya loves working with people. And she also likes how so many different kinds of people must unite to complete a project. "The process of working on a project is very Zen," she says. "On my last job I was dealing with twenty-one subcontractors, and I felt like the leader of an orchestra." And, of course, she loves the sense of accomplishment of creating beautiful projects for people.

CHALLENGES

"This is one of the most stressful professions there is in trades," says Tanya. "Everyone is stressed, and the physical conditions are rarely the most favorable. I've found that humor can be a great survival strategy on those especially tough days."

IMPORTANCE OF EDUCATION

According to Tanya, it's important for each person to take responsibility for getting himself or herself the training that is needed, whether through unions, vocational schools, apprenticeships, or books. And then you must apply any education to the real world. "If you want to be a good craftsperson, you must develop a good set of building standards and practices. You must also learn how to do business, how to best communicate, what's important in labor laws, and what new materials are out there, among many other things," she says. "I'm happy to still be learning things every day."

FUTURE OPPORTUNITIES

Her latest interest is in property development; she'd like to buy several properties, fix them up, and turn them over for profit. Right now, she's remodeling a pre–Civil War wood frame house, and hopes to invest in a second property soon. "This way, I won't have customers' needs to worry about or as much pressure with time scheduling, which will limit the stress in my life," she says. Another thing she'd like to complete? Her own house, which she says "will never get finished!"

WORDS OF ADVICE

Don't worry if you don't know *everything* once you graduate from a program. You'll continue learning—and gaining confidence—on the job.

UNIT 18
Roofing

Materials used to cover a roof and make it tight are part of the exterior finish called *roofing*. Roofing adds beauty to the exterior and protects the interior. Before roofing is applied, the roof deck must be securely fastened. There must be no loose or protruding nails and the deck must be clean of all debris. Properly applied roofing gives years of dependable service.

SAFETY REMINDER

A variety of roofing materials is available to protect a building from weather. Application of materials is usually straight forward and designed to be installed with speed. Care must be taken to remember the dangers of falling.

OBJECTIVES

After completing this unit, the student should be able to:

- define roofing terms.
- describe and apply roofing felt, organic or fiberglass asphalt shingles, tile roofing and roll roofing.
- describe various grades and sizes of wood shingles and shakes and apply them.
- flash valleys, sidewalls, chimneys, and other roof obstructions.
- estimate needed roofing materials.

51 | Asphalt Shingles

Asphalt shingles are the most commonly used roof covering for residential and light commercial construction. They are designed to provide protection from the weather for a period ranging from 20 to 30 years. They are available in many colors and styles.

ROOFING TERMS

An understanding of the terms most commonly used in connection with roofing is essential for efficient application of roofing material.

- A **square** is the amount of roofing required to cover 100 square feet of roof surface. There are usually three bundles of shingles per square or about 80 three-tab shingle strips **(Figure 51–1)**. One square of shingles can weigh between 235 and 400 pounds depending on shingle quality.
- **Deck** is the wood roof surface to which roofing materials are applied.
- **Coverage** is the number of overlapping layers of roofing and the degree of weather protection offered by roofing material. Roofing may be called single, double, or triple coverage.
- A **shingle butt** is the bottom exposed edge of a shingle.
- **Courses** are horizontal rows of shingles or roofing.

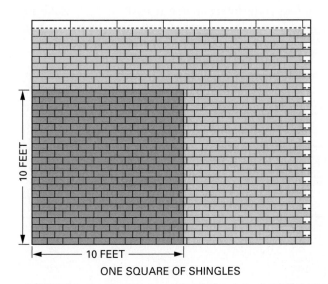

FIGURE 51–1   One square of shingles will cover 100 square feet.

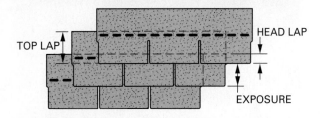

FIGURE 51–2   Asphalt strip exposure and lap.

- **Exposure** is the distance between courses of roofing. It is the amount of roofing in each course exposed to the weather **(Figure 51–2)**.
- The **top lap** is the height of the shingle or other roofing minus the exposure. In roll roofing this is also known as the **selvage.**
- The **head lap** is the distance from the bottom edge of an overlapping shingle to the top of a shingle two courses under, measured up the slope.
- **End lap** is the horizontal distance that the ends of roofing in the same course overlap each other.
- **Flashing** is strips of thin roofing material. It is usually made of lead, zinc, copper, vinyl, or aluminum. It may also be strips of roofing material used to make watertight joints on a roof. Metal flashing comes in rolls of various widths that are cut to the desired length.
- *Asphalt cements* and *coatings* are manufactured to various consistencies depending on the purpose for which they are to be used. **Cements** are classified as *plastic*, *lap*, and *quick-setting*. They will not flow at summer temperatures. They are used as adhesives to bond asphalt roofing products and flashings. They are usually troweled on the surface. **Coatings** are usually thin enough to be applied with a brush. They are used to resurface old roofing or metal that has become weathered.
- **Electrolysis** is a reaction that occurs when unlike metals come in contact with water. This contact causes one of the metals to corrode. A simple way to prevent the disintegration caused by electrolysis is to secure metal roofing material with fasteners of the same metal.

PREPARING THE DECK

In preparation for installing roofing materials, the deck should be clean and clear of debris. This will prevent any damage to the roofing material and

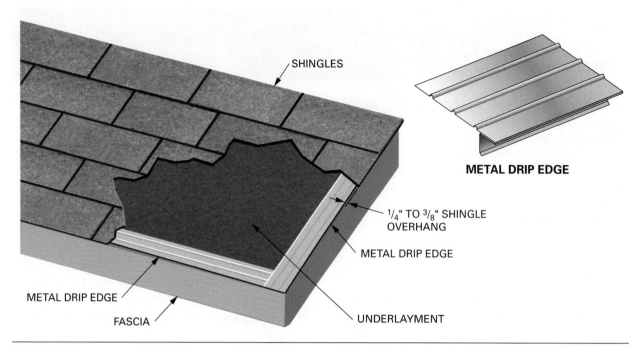

FIGURE 51–3 A metal drip edge may be used to support the shingle edge overhang.

make a safer working surface. A metal **drip edge** is often installed along the roof edges. The metal drip edge is usually made of aluminum or galvanized steel **(Figure 51–3)**. The drip edge is used to support the asphalt shingle overhang. Otherwise the shingles would droop from the heat of the sun.

Install the metal drip edge by using roofing nails of the same metal spaced 8 to 10 inches along its inner edge. End joints may be lapped by about 2 inches.

Underlayment

The deck should next be covered with an asphalt shingle **underlayment.** The underlayment protects the sheathing from moisture until the roofing is applied. It also gives additional protection to the roof afterward.

Asphalt Felts

Asphalt felts consist of heavy felt paper saturated with asphalt or coal tar. They are usually made in various weights of pounds per square **(Figure 51–4)**. Asphalt felt comes in 36-inch wide rolls. The rolls are 72 or 144 feet long covering 2 or 4 squares. Usually the lightest-weight felt is used as an underlayment for asphalt shingles.

Apply a layer of asphalt felt underlayment over the deck starting at the bottom. Lay each course of felt over the lower course at least 2 inches. Make any end laps at least 4 inches. Lap the felt 6 inches from both sides over all hips and ridges **(Figure 51–5)**.

Nail or staple through each lap and through the center of each layer about 16 inches apart. Roofing nails driven through the center of metal discs or specially designed, large head felt fasteners hold the un-

	Approx. Weight Per Roll	Approx. Weight Per Square	Squares Per Roll	Roll Length	Roll Width	Side or End Laps	Top Lap	Exposure
SATURATED FELT	60 #	15 #	4	144'	36"	4"–6"	2"	34"
	60 #	30 #	2	72'	36"	4"–6"	2"	34"
	60 #	60 #	1	36'	36"	4"–6"	2"	34"

FIGURE 51–4 Sizes and weights of asphalt-saturated felts.

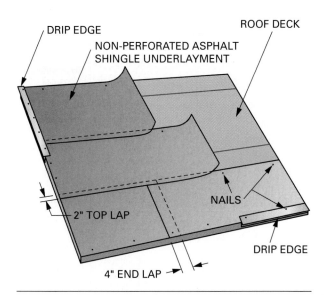

FIGURE 51–5 **Application of asphalt shingle underlayment felt.**

FIGURE 51–6 **Asphalt felt underlayment is applied to the roof deck before shingling.** *(Courtesy of APA—The Engineered Wood Association)*

derlayment securely in strong winds until shingles are applied **(Figure 51–6).** A metal drip edge is often installed along the rakes after the application of the underlayment.

Ice and Water Shield

As an added measure of protection to roofs, an ice and water shield may be applied. An ice and water shield is a roofing membrane composed of two waterproofing materials bonded into one layer. It is comprised of a rubberized asphalt adhesive backed by a layer of polyethylene. It comes in 3-foot × 75-foot rolls. The rubberized asphalt surface is backed with a release paper to protect the sticky side. During installation, the release paper is peeled off, allowing the asphalt to bond to the roof deck.

This material is used in the trouble spots of a roof such as along the eaves, in valleys, and in unique areas where leaks are more likely. Low-slope roofs and roofs exposed to severe blowing weather are at increased risk of leaks. An ice and water shield is often used during re-roofing applications of houses that experience ice damming at the eaves. (See Chapter 50.)

The ice and water shield is installed directly on the roof deck. The deck should be clean and clear of debris. There should be no voids in the deck. Laps should be 6 inches at the ends and 4 inches at the sides. Cut pieces to the desired size. After positioning the sheet, fold back enough membrane to peel off some release paper. Reposition the membrane and press the membrane to the deck. Pull remaining release paper off and press the membrane to the deck. This will ensure proper adherence to the roof deck. This membrane is not recommended for use in the hot climes of intense sun such as the desert areas of the Southwest.

KINDS OF ASPHALT SHINGLES

Two types of **asphalt shingles** are manufactured. *Organic* shingles have a base made of heavy asphalt-saturated paper felt coated with additional asphalt. *Fiberglass* shingles have a base mat of glass fibers. The mat does not require the saturation process and only requires an asphalt coating. Both kinds of shingles are surfaced with selected mineral granules. The asphalt coating provides weatherproofing qualities. The granules protect the shingle from the sun, provide color, and protect against fire.

Fiberglass-based asphalt shingles have an Underwriters Laboratories Class A fire resistance rating. The Class A rating is the highest standard for resistance to fire. The Class C rating for organic shingles, while not as high, will meet most residential building codes.

Asphalt shingles come in a wide variety of colors, shapes, and weights **(Figure 51–7).** They are applied in the same manner. Shingle quality and longevity are generally determined by the weight per square. Asphalt shingles can weigh anywhere from 235 to 400 pounds per square. Most asphalt shingles are manufactured with factory-applied adhesive. This increases their resistance to the wind.

APPLYING ASPHALT SHINGLES

Before applying strip shingles, make sure that the roof deck is properly prepared. The underlayment and drip edge should be applied. Asphalt roofing products become soft in hot weather. Be careful not to damage them by digging in with heavily cleated shoes during application or by unnecessary walking on the surface after application. The slope of a roof should not be less than 4 inches per unit of run when

| Product | Configuration | Per Square | | | Size | | Exposure | Underwriters' Listing |
		Approx. Shipping Weight	Shingles	Bundles	Width	Length		
Wood appearance strip shingle more than one thickness per strip 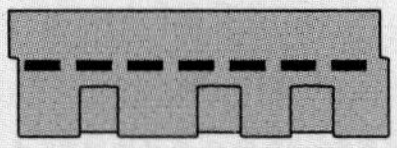 Laminated or job applied	Various edge, surface texture, & application treatments	285# to 390#	67 to 90	4 or 5	11½" to 15"	36" or 40"	4" to 6"	A or C—many wind resistant
Wood appearance strip shingle single thickness per strip 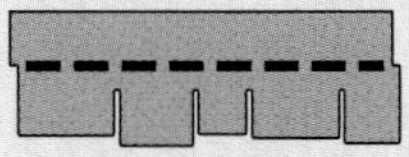	Various edge, surface texture, & application treatments	Various 250# to 350#	78 to 90	3 or 4	12" or 12¼"	30" or 40"	4" to 5⅛"	A or C—many wind resistant
Self-sealing strip shingle 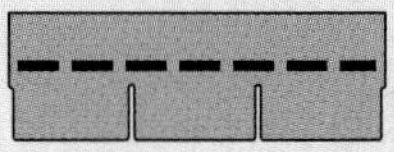	Conventional 3 tab	205#-240#	78 or 80	3	12" or 12¼"	36"	5" or 5⅛"	A or C—all wind resistant
	2 or 4 tab	Various 215# to 325#	78 or 80	3 or 4	12" or 12¼"	36"	5" or 5⅛"	
Self-sealing strip shingle 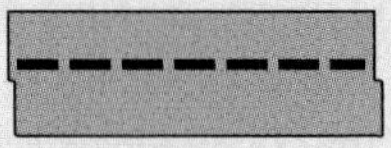 No cut out	Various edge and texture treatments	Various 215# to 290#	78 to 81	3 or 4	12" or 12¼"	36" or 36¼"	5"	A or C—all wind resistant
Individual lock down 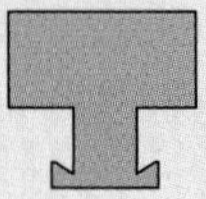 Basic design	Several design variations	180# to 250#	72 to 120	3 or 4	18" to 22¼"	20 to 22½"	—	C—many wind resistant

FIGURE 51–7   Asphalt shingles are available in a wide variety of sizes, shapes, and weights. (*Courtesy of Asphalt Roofing Manufacturers' Association*)

conventional methods of asphalt shingle application are used.

Asphalt Shingle Layout

On small roofs, strip shingles are applied by starting from either rake. On long buildings, a more accurate vertical alignment is ensured by starting at the center and working both ways. Mark the center of the roof at the eaves and at the ridge. Snap a chalk line between the marks. Snap a series of chalk lines 6 inches apart on each side of the centerline if the shingle tab cutouts are to break on the halves. Snap lines 4 inches apart if the cutouts are to break on the thirds. When applying the shingles, start the course with the end of the shingle to the vertical chalk line. Start succeeding courses in the same manner. Break the joints as necessary. Pyramid the shingles up in the center. Work both ways toward the rakes **(Figure 51–8)**.

If it is decided to start at the rakes and cutouts are to break on the halves, start the first course with a whole tab. The second course is started with a shingle from which 6 inches have been cut; the third course, with a strip from which the entire first tab is removed; the fourth, with one and one-half tabs removed, and so on **(Procedure 51–A).** These starting strips are precut for faster application. Waste from these strips is used on the opposite rake.

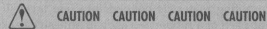

| ⚠ | CAUTION   CAUTION   CAUTION   CAUTION |

No butt seams of any three consecutive courses should line up. This maximizes the life of the shingles. ▪

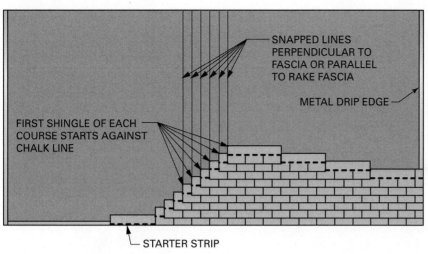

FIGURE 51-8 On long roofs, start shingling from the center working toward the rakes.

Procedure 51-A Laying Out Stagger Cutouts on the Halves Starting from the Rake

STEP 1 STARTER — BEGIN WITH A FULL STARTER SHINGLE MINUS 3" SO BUTT SEAMS DO NOT ALIGN WITH FIRST COURSE.

STEP 2 START FIRST COURSE WITH FULL STRIP.

STEP 3 START SECOND COURSE WITH FULL STRIP MINUS 1/2 TAB.

STEP 4 START THIRD COURSE WITH FULL STRIP MINUS FIRST TAB.

STEP 5 START FOURTH COURSE WITH $1/2$ STRIP. REPEAT STARTING WITH A FULL TAB OR CONTINUE CUTTING $1/2$ TAB FOR EACH COURSE UNTIL A $1/2$ TAB IS MADE. THEN REPEAT WITH FULL STRIPS.

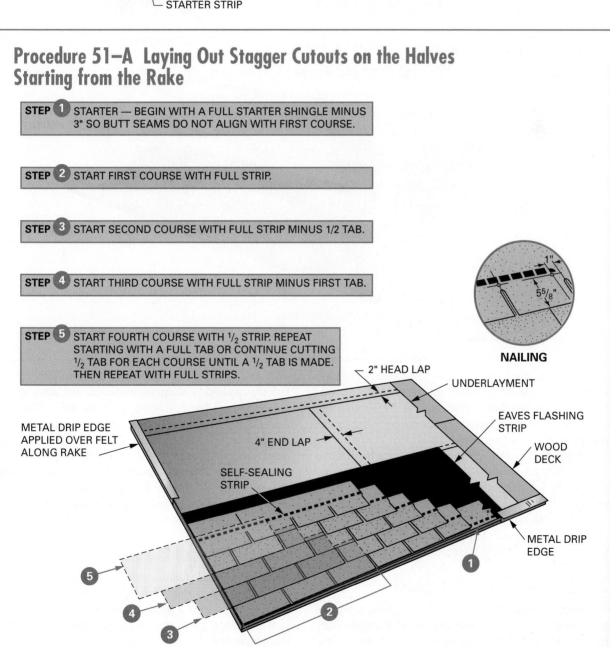

NAILING

Procedure 51–B Laying Out Stagger Cutouts on the Thirds Starting from the Rake

STEP ① STARTER — BEGIN WITH FULL STARTER SHINGLE WITH ABOUT 3" CUT OUT SO BUTT SEAMS DO NOT ALIGN WITH FIRST COURSE BUTTS.

STEP ② START FIRST COURSE WITH FULL SHINGLE.

STEP ③ START SECOND COURSE WITH FULL SHINGLE MINUS 4".

STEP ④ START THIRD COURSE WITH FULL SHINGLE MINUS 8". REPEAT WITH FULL STRIP.

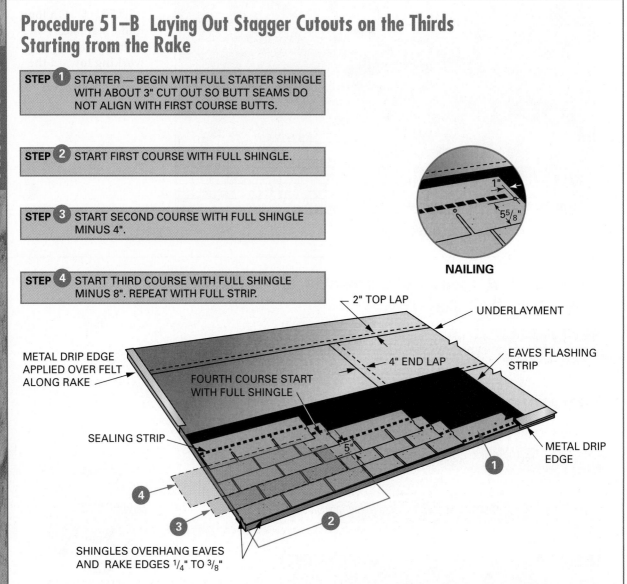

NAILING

2" TOP LAP

UNDERLAYMENT

4" END LAP

EAVES FLASHING STRIP

METAL DRIP EDGE APPLIED OVER FELT ALONG RAKE

FOURTH COURSE START WITH FULL SHINGLE

SEALING STRIP

METAL DRIP EDGE

SHINGLES OVERHANG EAVES AND RAKE EDGES ¼" TO ³⁄₈"

If the cutouts are to break on the thirds, cut the starting strip for the second course by removing 4 inches. Remove 8 inches from the strip for the third course, and so on **(Procedure 51–B)**.

Cut the shingles by scoring them on the back side with a utility knife or use a hook blade. Use a square as a guide for the knife. Bend the shingle. It will break on the scored line.

The layouts may have to be adjusted so that tabs on opposite rakes will be of approximately equal widths. No rake tab should be less than 3 inches in width.

Starter Course of Asphalt Shingles

The **starter course** backs up and fills in the spaces between tabs of the first regular course of shingles. Cut the exposed tabs off or about 5 inches of exposure from a regular shingle. Apply these pieces so the factory-applied sealing strip is near the eave edge. This will seal the first shingle course tabs. The first starter strip applied must be shorter in length by a minimum of 3 inches. This step will keep the butt seams or end joints from aligning with the first course. Overhang the shingles past the drip edge ¼ to ⅜ inches.

Some roofers install the starter strips along the perimeter of the roof, keeping the sealing strips closest to the edge of the roof.

Fastening Asphalt Shingles

Selecting suitable fasteners, using the recommended number, and putting them in the right places are important steps in the application of asphalt shingles. Lay the first regular course of shingles on top of the starter course. Keep their bottom edges flush with each other. Use a minimum of four fasteners in each strip shingle. Do not nail into or above the factory-applied adhesive **(Figure 51–9)**.

The fastener length should be sufficient to penetrate the sheathing at least ¾ inch, or through approved panel sheathing. Roofing nails should be 11- or 12-gauge galvanized steel or aluminum with barbed shanks. They should have ⅜- to ⁷⁄₁₆-inch heads. Roofing nails may be driven by hand or with power nailers. In some locales power-driven staples may be used in place of nails. However, their use is limited to shingles with factory-applied adhesive **(Figure 51–10)**. Staples must be at least 16-gauge. They should have a minimum crown width of ¹⁵⁄₁₆ inch.

Align each shingle of the first course carefully. Fasten each shingle from the end nearest the shingle just laid. This prevents buckling. Drive fasteners straight so that the nailheads or staple crowns will not cut into the shingles. The entire crown of the staple or nailhead should bear tightly against the shingle. It should not penetrate its surface **(Figure 51–11)**. Continue applying shingles until the first course is complete.

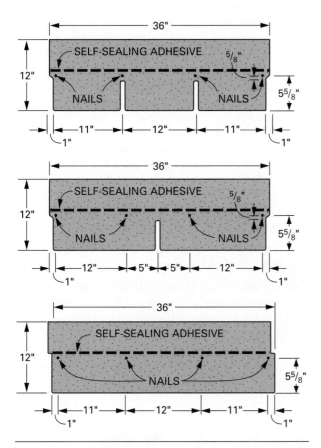

FIGURE 51–9 Recommended fastener locations for asphalt strip shingles. *(Courtesy of Asphalt Roofing Manufacturers' Association)*

FIGURE 51–10 Pneumatic staplers and nailers are often used to fasten asphalt shingles. *(Courtesy of Senco Products, Inc.)*

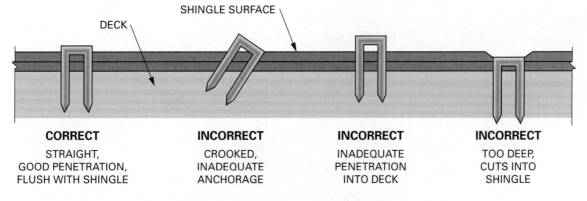

FIGURE 51–11 It is important to fasten asphalt shingles correctly.

Shingle Exposure

The maximum *exposure* of asphalt shingles to the weather depends on the type of shingle. Recommended maximum exposures range from 4 to 6 inches. Most commonly used asphalt shingles have a maximum exposure of 5 inches. Less than the maximum recommended exposure may be used, if desired.

LAYING OUT SHINGLE COURSES

When laying out shingle courses, space the desired exposure up each rake from the top edge of the first course of shingles. Snap lines across the roof for five or six courses. When snapping a long line, it may be necessary to thumb the line down against the roof at about center. Then snap the lines on both sides of the thumb.

Lay succeeding courses so their top edges are to the chalk line. Start each course so the cutouts are staggered in the desired manner. Continue snapping lines and applying courses until a point 3 or 4 feet below the ridge is reached. Some carpenters snap a line to straighten out the course after using the top of the cutout for the shingle exposure for a number of courses.

Spacing Shingle Courses to the Ridge

The last course of shingles below the *ridge cap* should be exposed by about the same amount as all the other shingle courses. This can be achieved by changing the exposure of the last 3 to 4 feet of roof near the ridge. To determine the new exposure first cut a full tab from a shingle. Center it on the ridge. Then bend it over the ridge. Do this at both ends of the building. Mark the bottom edges on the roof.

Measure up 2 inches. Snap a line between the marks. This line will be the top edge of the next to last course of shingles. It should be about 3½ inches down from the ridge. Divide the distance between this line and the top of the last course of shingles applied into spaces as close as possible to the exposure of previous courses. Do not exceed the maximum exposure **(Procedure 51–C).** Snap lines across the roof and shingle up to the ridge.

The line for the last course of shingles is snapped on the face of the course below. Lay out the exposure from the bottom edge of the course on both ends of the roof. Snap a line across the roof. Fasten the last course of shingles by placing their butts to the line. If the ridge is to be vented, cut the top edge of the shingle at the sheathing. If the ridge is not vented, bend the top shingle edges over the ridge. Fasten their top edges to the opposite slope.

Applying the Ridge Cap

Cut hip and ridge shingles from shingle strips to make approximately 12-inch × 12-inch squares. Cut shingles from the top of the cutout to the top edge on a slight taper. The top edge should be narrower than the bottom edge **(Figure 51–12).** Cutting the shingles in this manner keeps the top half of the shingle from protruding when it is bent over the ridge.

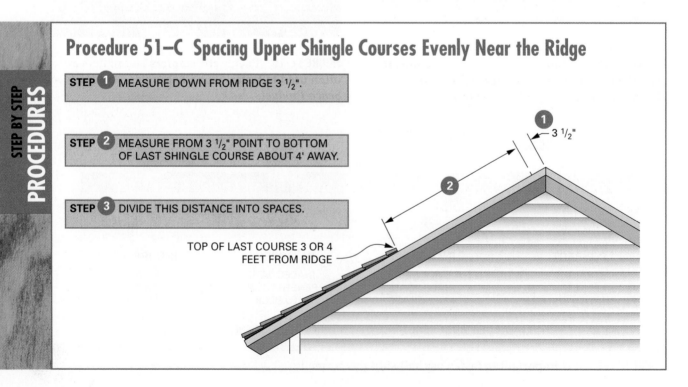

STEP BY STEP PROCEDURES

Procedure 51–C Spacing Upper Shingle Courses Evenly Near the Ridge

STEP 1 MEASURE DOWN FROM RIDGE 3 ½".

STEP 2 MEASURE FROM 3 ½" POINT TO BOTTOM OF LAST SHINGLE COURSE ABOUT 4' AWAY.

STEP 3 DIVIDE THIS DISTANCE INTO SPACES.

TOP OF LAST COURSE 3 OR 4 FEET FROM RIDGE

3 ½"

CUT ALONG DOTTED LINE, TAPERING
TOP PORTION SLIGHTLY

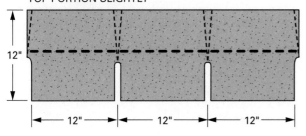

FIGURE 51–12   Ridge and hip cap shingles are cut from strip shingles.

The ridge cap is applied after both sides of the roof have been shingled. At each end of the roof, center a shingle on the ridge. Bend it over the ridge. Mark its bottom edge on the front slope or the one most visible. Snap a line between the marks.

Beginning at the bottom of a hip or at one end of the ridge, apply the shingles over the hip or ridge.

FIGURE 51–13   Applying ridge cap shingles to a vented ridge.

Expose each 5 inches. In cold weather, ridge cap shingles may have to be warmed in order to prevent cracking when bending them over the ridge. On the ridge, shingles are started from the end away from prevailing winds. The wind should blow over the shingle butts, not against them. Keep one edge, from the butt to the start of the tapered cut, to the chalk line. Secure each shingle with one fastener on each side, 5½ inches from the butt and one inch up from each edge **(Figure 51–13).** Apply the cap across the ridge until 3 or 4 feet from the end. Then space the cap to the end in the same manner as spacing the shingle course to the ridge. The last ridge shingle is cut to size. It is applied with one fastener on each side of the ridge. The two fasteners are covered with asphalt cement to prevent leakage.

52   Roll and Tile Roofing

Roll roofing can be used on roof slopes with as little as one inch of rise per foot of run. On steeper roofs, roll roofing is used when economy is the major factor and appearance is not as important. Cement tile roofing is used in warmer climes. The material is well suited for hot and wet areas. It is used when a long roof life span is desired.

TYPES OF ROLL ROOFING

Roll roofing is made of the same materials as asphalt shingles. Various types are made with a base sheet of organic felt or glass fibers in a number of weights **(Figure 52–1).** Some types are applied with exposed or concealed fasteners. They have a top lap of 2 to 4 inches. A concealed-nail type, called **double coverage** roll roofing, has a top lap of 19 inches. All kinds come in rolls that are 36 inches wide.

Roll roofing is recommended for use on roofs with slopes with less than 4 inches of rise per foot. However, the exposed-nail type should not be used on pitches with less than 2 inches of rise per foot. Roll roofing applied with concealed nails and having a top lap of at least 3 inches may be used on pitches as low as 1 inch per foot. Use the same type and length of nails as for asphalt shingles.

Typical Asphalt Rolls								
1	2		3	4		5	6	7
	Approximate Shipping Weight							
Product	Per Roll	Per Sq.	Sqs. Per Package	Length	Width	Side or End Lap	Top Lap	Exposure	Underwriters' Listing
 Mineral surface roll	75# to 90#	75# to 90#	one	36' 38'	36" 36"	6"	2" 4"	34" 32"	C
Mineral surface roll Double coverage	55# to 70#	55# to 70#	one half	36'	36"	6"	19"	17"	C

FIGURE 52–1 Details of roll roofing types.

General Application Methods

Apply all roll roofing when the temperature is above 45°F (7°C). This prevents cracking of the coating. Cut the roll into 12- to 18-foot lengths. Spread in a pile on a smooth surface to allow the pieces to flatten out.

Use only the lap or quick-setting cement recommended by the manufacturer. Store cement in a warm place until ready for use. To warm it rapidly, place the unopened container in hot water.

> ⚠ **CAUTION CAUTION CAUTION CAUTION**
>
> These materials are flammable. Never warm them over an open fire or place them in direct contact with a hot surface. ■

Apply roll roofing only on a solid, smooth, well-seasoned deck. Make sure the area below has sufficient ventilation to prevent the deck from absorbing condensation. This would cause the roofing to warp and buckle. A felt underlayment is not used with roll roofing.

APPLYING ROLL ROOFING

Apply 9-inch wide strips of the roofing along the eaves and rakes overhanging about ⅜ inch. Fasten the strips with two rows of nails one inch from each edge. Space them about 4 inches apart (**Figure 52–2**).

Apply the first course with its edges and ends flush with the strips. Secure the upper edge with nails staggered about 4 inches apart. Do not fasten within 18 inches of the rake edge. Apply cement only to that part of the edge strips covered by the first course. Press the lower edge and rake edges of the first course firmly in place over the edge strips. Finish nailing the upper edge out to the rakes. Apply succeeding courses in like manner. Make all end laps 6 inches wide. Apply cement the full width of the lap.

After all courses are in place, lift the lower edge of each course. Apply the cement in a continuous layer over the full width and length of the lap. Press the lower edges of the upper courses firmly into the cement. A small bead should appear along the entire edge of the sheet. Care must be taken to apply the correct amount of cement. Avoid getting cement on the roof deck because this will make subsequent re-roofing more difficult.

Covering Hips and Ridges

Cut strips of 12 × 36 roofing. Bend the pieces lengthwise through their centers. Snap a chalk line on both sides down 5½ inches from the hip or ridge. Apply cement from one line over the top to the other line.

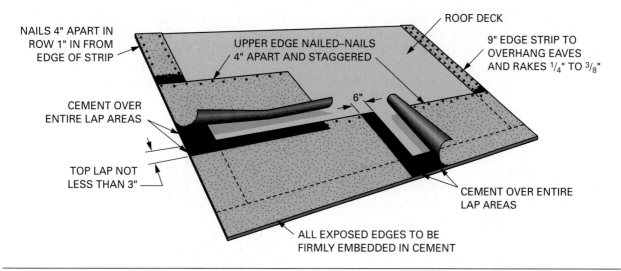

NAILS 4" APART IN ROW 1" IN FROM EDGE OF STRIP

UPPER EDGE NAILED–NAILS 4" APART AND STAGGERED

ROOF DECK

9" EDGE STRIP TO OVERHANG EAVES AND RAKES $1/4$" TO $3/8$"

CEMENT OVER ENTIRE LAP AREAS

6"

TOP LAP NOT LESS THAN 3"

CEMENT OVER ENTIRE LAP AREAS

ALL EXPOSED EDGES TO BE FIRMLY EMBEDDED IN CEMENT

FIGURE 52–2 Recommended details for applying roll roofing.

Fit the first strip over the hip or ridge. Press it firmly into place. Start at the lower end of a hip and at either end of a ridge. Lap each strip 6 inches over the preceding one. Nail each strip only on the end that is to be covered by the overlapping piece.

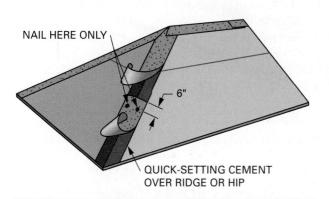

NAIL HERE ONLY

6"

QUICK-SETTING CEMENT OVER RIDGE OR HIP

FIGURE 52–3 Finishing hips and ridges using roll roofing.

Spread cement on the end of each strip that is lapped before the next one is applied. Continue in this manner until the end is reached **(Figure 52–3)**.

APPLYING DOUBLE COVERAGE ROLL ROOFING

Cut the 19-inch *selvage* portion from enough double coverage roll roofing to cover the length of the roof. Save the surfaced portion for the last course. Apply the selvage portion parallel to the eaves **(Figure 52–4)**. It should overhang the drip edge by $3/8$ inch. Secure it to the roof deck with three rows of nails. Place the top row $4\frac{1}{2}$ inches below the upper edge. Put the bottom row one inch above the bottom edge. Place the other row halfway between. Place the nails in the upper and middle rows slightly staggered about 12 inches apart. Place the nails in the lower row about 6 inches apart and slightly staggered. Nail along rakes in the same manner.

FIGURE 52–4 Details for nailing the selvage starter strip of roll roofing.

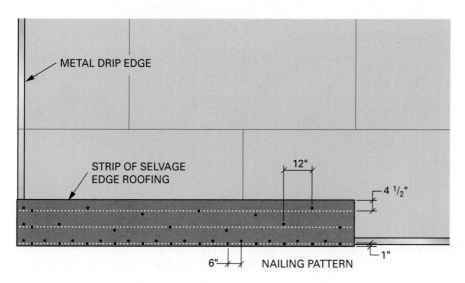

METAL DRIP EDGE

STRIP OF SELVAGE EDGE ROOFING

12"

$4\frac{1}{2}$"

1"

6"

NAILING PATTERN

Apply the first course. Secure it with two rows of nails in the selvage portion. Place one row about 4¾ inches below the upper edge. Put the second row about 8½ inches below the first. Space the nails about 12 inches apart in each row and stagger them **(Procedure 52–A)**.

Apply succeeding courses in the same manner. Lap the full width of the 19-inch selvage each time. Make all end laps 6 inches wide. End laps are made in the manner shown in **Procedure 52–B**. Stagger end laps in succeeding courses.

Procedure 52–A Applying Double Coverage Roll Roofing

STEP 1 STARTER STRIP OF 19" SELVAGE CUT FROM A FULL SHEET LAID TO OVERHANG EAVE AND RAKE ¼" TO ³/₈".

STEP 2 FIRST COURSE.

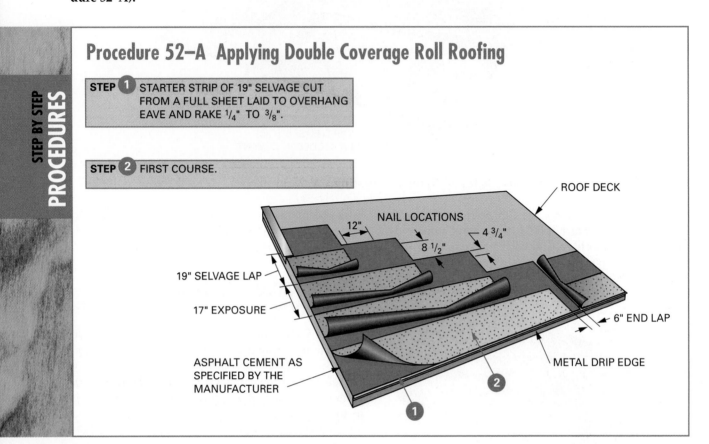

NAIL LOCATIONS

ROOF DECK

12"

8 ½"

4 ³/₄"

19" SELVAGE LAP

17" EXPOSURE

6" END LAP

ASPHALT CEMENT AS SPECIFIED BY THE MANUFACTURER

METAL DRIP EDGE

Procedure 52–B Making End Laps on Double Coverage Roll Roofing

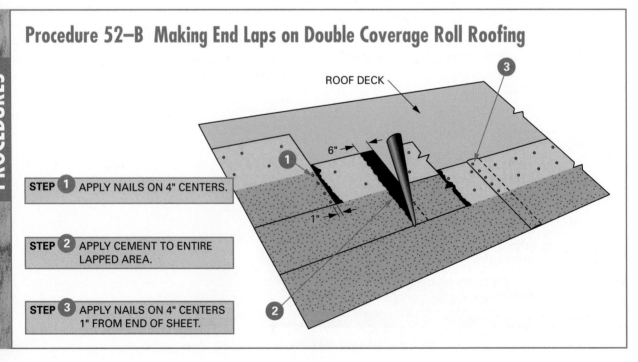

ROOF DECK

6"

1"

STEP 1 APPLY NAILS ON 4" CENTERS.

STEP 2 APPLY CEMENT TO ENTIRE LAPPED AREA.

STEP 3 APPLY NAILS ON 4" CENTERS 1" FROM END OF SHEET.

STEP BY STEP PROCEDURES

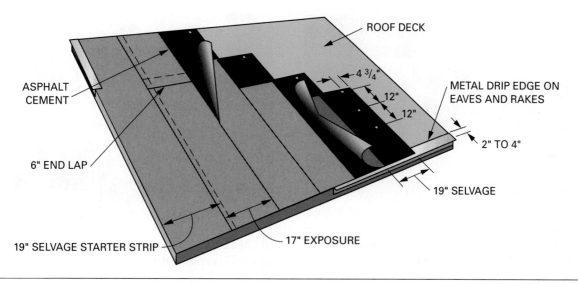

ROOF DECK

ASPHALT CEMENT

6" END LAP

19" SELVAGE STARTER STRIP

17" EXPOSURE

4 ¾"

12"

12"

METAL DRIP EDGE ON EAVES AND RAKES

2" TO 4"

19" SELVAGE

FIGURE 52–5   Vertical application of double coverage roll roofing.

Lift and roll back the surface portion of each course. Starting at the bottom, apply cement to the entire selvage portion of each course. Apply it to within ¼ inch of the surfaced portion. Press the overlying sheet firmly into the cement. Apply pressure over the entire area using a light roller to ensure adhesion between the sheets at all points.

Apply the remaining surfaced portion left from the first course as the last course. This type roofing may also be applied in like manner parallel to the rakes **(Figure 52–5).** Lap the seams away from the prevailing wind to keep wind blown water from being forced into the seam. Hips and ridges are covered in the same manner shown in **Figure 52–6.**

It is important to follow specific application instructions because of differences in the manufacture of roll roofing. Some instructions call for hot asphalt. Others call for cold cement. Others give the option of either. Specific requirements for quantities and types of adhesive must be followed.

CEMENT ROOF TILES

Another roofing option is cement roof tiles **(Figure 52–7),** which are typically used in hot locales. They are designed for various architectural styles and will last 30 to 60 years. The lifetime range is determined by the amount of sun and windblown rain the roof receives.

Concrete and clay tiles are manufactured in a variety of styles and thicknesses. Consequently, their weight can range from 500 to 1,200 pounds per square. Regular tiles range between 700 and 800 pounds per square. This increased weight affects the strength required from the roof system. Many truss manufacturers will require the roof to be loaded with the tiles before any finishes are applied **(Figure 52–8).** Only weeks later will the interior framing surfaces be shimmed and trimmed flush. This time allows the trusses to shift and adjust to the load.

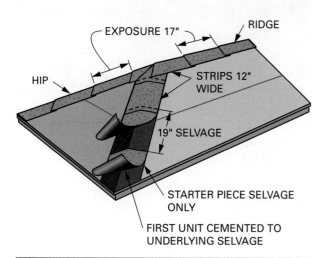

EXPOSURE 17"

RIDGE

HIP

STRIPS 12" WIDE

19" SELVAGE

STARTER PIECE SELVAGE ONLY

FIRST UNIT CEMENTED TO UNDERLYING SELVAGE

FIGURE 52–6   Method of covering hips and ridges with double coverage roofing.

FIGURE 52–7   Cement tile roofing is applied to many homes in warm climates.

FIGURE 52–8 Cement tile must be preloaded on roofs to allow the roof system to settle.

FIGURE 52–10 Areas of windblown rain have a second layer of hot-mopped felt.

Underlayment requirements vary with local codes. Most areas use at least one layer of 30# felt fastened with nails and metal caps **(Figure 52–9)**. Coastal areas prone to wind blown rain typically use another layer comprised of 90# asphalt roll installed in hot, mopped in place, bitumen **(Figure 52–10)**.

The tile pieces range in sizes from 7 to 13 inches wide by 15 to 18 inches long. They typically have two prepunched holes along the top edge that are used for fastening with screws. Tiles are installed to chalk lines with screws. At corners and hips, the tiles are cut with a circular saw with a diamond or composite blade **(Figure 52–11)**.

FIGURE 52–9 First step in tile roofing is felt applied with capped nails.

FIGURE 52–11 Tiles screwed in place.

53 Wood Shingles and Shakes

Wood shingles and shakes are common roof coverings **(Figure 53–1).** Most shingles and shakes are produced from western red cedar. Cedar logs are first cut into desired lengths. They are then split into sections from which shingles and shakes are sawn or split. All shingles are sawed. Most shakes are split. Shingles, therefore, have a relatively smooth surface. Most shakes have at least one highly textured, natural grain surface. Most shakes also have thicker butts than shingles **(Figure 53–2).**

Most wood shingles and shakes are produced by mills that are members of the Cedar Shake and Shingle Bureau. Their product label **(Figure 53–3)** is assurance that the products meet quality standards.

DESCRIPTION OF WOOD SHINGLES AND SHAKES

Wood shingles are available for use on roofs in four standard grades. Shakes are manufactured by different methods to produce four types. Both shingles and shakes may be treated to resist fire or premature decay in areas of high humidity **(Figure 53–4).**

Sizes and Coverage

Shingles come in lengths of 16, 18, and 24 inches. The butt thickness increases with the length. Shakes are available in lengths of 15, 18, and 24 inches. Their

FIGURE 53–2   All shingles are sawed from the log. Most shakes are split and have a rough surface. *(Courtesy of Cedar Shake and Shingle Bureau)*

FIGURE 53–1   Wood shingles and shakes may be used as siding and roof covering.

FIGURE 53–3 **The product label ensures that quality requirements are met.** (*Courtesy of Cedar Shake and Shingle Bureau*)

butt thicknesses are from ⅜ to ¾ inch. The 15-inch length is used for starter and finish courses. **Figure 53–5** shows the sizes and the amount of roof area that one square, laid at various exposures, will cover.

Maximum Exposures

The area covered by one square of shingles or shakes depends on the amount exposed to the weather. The maximum amount of shingle exposure depends on the length and grade of the shingle or shake and the pitch of the roof. Shakes are not generally applied to

roofs with slopes of less than 4 inches rise per foot. Shingles, with reduced exposures, may be used on slopes down to 3 inches rise per foot. **Figure 53–6** shows the maximum recommended roof exposure for wood shingles and shakes.

SHEATHING AND UNDERLAYMENT

Shingles and shakes may be applied over spaced or solid roof sheathing. **Spaced sheathing** or *skip sheathing* is usually 1 × 4 or 1 × 6 boards. **Solid sheathing** is usually APA-rated panels. Solid Sheathing may be required in regions subject to frequent earthquakes or under treated shingles and shakes. It is also recommended for use with shakes in areas where wind-driven weather is common.

Spaced Sheathing

Solid wood sheathing is applied from the eaves up to a point that is plumb with a line 12 to 24 inches inside the wall line. An eaves flashing is installed, if required. Spaced sheathing may then be used above the solid sheathing to the ridge.

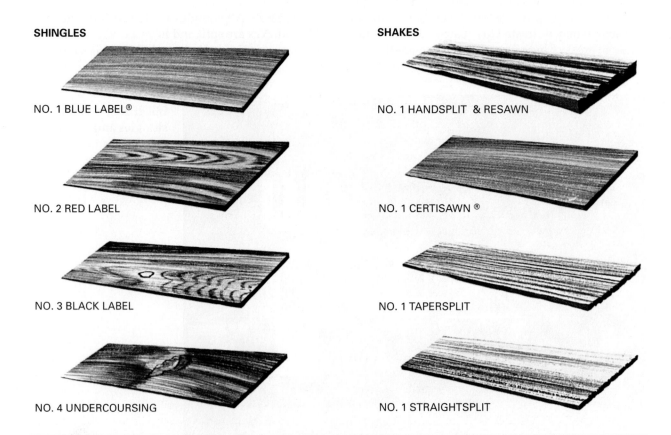

FIGURE 53–4   Description of grades and kinds of wood shingles and shakes. (*Courtesy of Cedar Shake and Shingle Bureau*)

Coverage and Exposure Tables

Shingle Coverage Table

Length and Thickness	Approximate Coverage of One Square (4 bundles) of Shingles Based on Following Weather Exposures								
	3½″	4″	4½″	5″	5½″	6″	6½″	7″	7½″
16″ × 5/2″	70	80	90	100*	—	—	—	—	—
18″ × 5/2¼″	—	72½	81½	90½	100*	—	—	—	—
24″ × 4/2″	—	—	—	—	73½	80	86½	93	100*

NOTE *Maximum exposure recommended for roofs.

Shake Coverage Table

Shake Type, Length and Thickness	Approximate Coverage (in sq. ft.) of One Square, When Shakes are Applied with an Average ½″ Spacing, at Following Weather Exposures, in Inches (d):				
	5	5½	7½	8½	10
18″ × ½″ Handsplit-and-Resawn Mediums(a)	—	55(b)	75(c)	—	—
18″ × ¾″ Handsplit-and-Resawn Heavies(a)	—	55(b)	75(c)	—	—
18″ × ⅝″ Tapersawn —	55(b)	75(c)	—		
24″ × ⅜″ Handsplit 50(e)	—	75(b)	—		
24″ × ½″ Handsplit-and-Resawn Mediums	—	—	75(b)	85	100(c)
24″ × ¾″ Handsplit-and-Resawn Heavies	—	—	75(b)	85	100(c)
24″ × ⅝″ Tapersawn —		75(b)	85	100(c)	
24″ × ½″ Tapersplit —		75(b)	85	100(c)	
18″ × ⅜″ Straight-Split	—	65(b)	90(c)	—	—
24″ × ⅜″ Straight-Split	—	—	75(b)	85	100(c)
15″ Starter-Finish course	Use supplementary with shakes applied not over 10″ weather exposure				

(a) 5 bundles will cover 100 sq. ft. roof area when used as starter-finish course at 10″ weather exposure; 7 bundles will cover 100 sq. ft. roof area at 7½″ weather exposure; see footnote (d).

(b) Maximum recommended weather exposure for 3-ply roof construction.

(c) Maximum recommended weather exposure for 2-ply roof construction.

(d) All coverage based on an average ½″ spacing between shakes.

(e) Maximum recommended weather exposure.

FIGURE 53–5   Tables show the sizes and coverage of wood shingles and shakes. *(Courtesy of Cedar Shake and Shingle Bureau)*

For shingles, either 1 × 4 or 1 × 6 spaced sheathing may be used. Space 4-inch boards the same amount as the shingles are exposed to the weather. In this method of application, each course of shingles is nailed to the center of the board. If 6-inch boards are used, they are spaced two exposures. Two courses of shingles are nailed to the same board when courses are exposed up to, but not exceeding, 5 ½ inches. For shingles with greater exposures, the sheathing is spaced a distance of one exposure (**Figure 53–7**).

In shake application, spaced sheathing is usually 1 × 6 boards spaced the same distance, on center, as the shake exposure (**Figure 53–8**). The spacing should never be more than 7 ½ inches for 18-inch shakes and 10 inches for 24-inch shakes installed on roofs.

Shingle Exposure Table

Pitch	Maximum Exposure Recommended for Roofs								
	Length								
	No. 1 Blue Label			No. 2 Red Label			No. 3 Black Label		
	16″	18″	24″	16″	18″	24″	16″	18″	24″
3/12 to 4/12	3¼″	4¼″	5⅝″	3½″	4″	5½″	3″	3½″	5″
4/12 and steeper	5″	5½″	7½″	4″	4½″	6½″	3½″	4″	5½″

Shake Exposure Table

Pitch	Maximum exposure recommended for roofs	
	Length	
	18″	24″
4/12 and steeper	7½″	10″(a)

(a) 24″ × ⅜″ handsplit shakes limited to 5″ maximum weather exposure, per UBC.

FIGURE 53–6 Maximum exposures of wood shingles and shakes for various roof pitches. (*Courtesy of Cedar Shake and Shingle Bureau*)

Underlayment

No underlayment is required under wood shingles. A breather-type roofing felt may be used over solid or spaced sheathing. Underlayment is typically applied under shakes.

APPLICATION TOOLS AND FASTENERS

Shingles and shakes are usually applied with a shingling hatchet. Recommendations for the type and size of fasteners should be closely followed.

Shingling Hatchet

A **shingling hatchet (Figure 53–9),** should be lightweight. It should have both a sharp *blade* and a *heel*. A **sliding gauge** is sometimes used for fast and accurate checking of shingle exposure. The gauge permits several shingle courses to be laid at a time without snapping a chalk line. A power nailer may be used. However, shingles and shakes often need to be trimmed or split with the hatchet. More time may be lost than gained by using a power nailer.

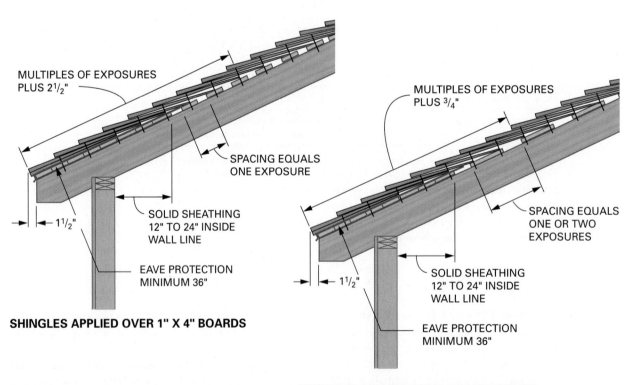

SHINGLES APPLIED OVER 1" X 4" BOARDS

SHINGLES APPLIED OVER 1" X 6" BOARDS

FIGURE 53–7 Application of wood shingles on spaced sheathing. (*Courtesy of Cedar Shake and Shingle Bureau*)

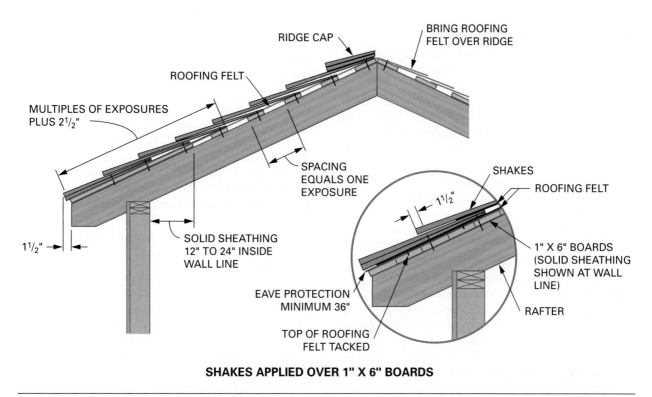

SHAKES APPLIED OVER 1" X 6" BOARDS

FIGURE 53–8 Method of applying shakes on spaced sheathing. (*Courtesy of Cedar Shake and Shingle Bureau*)

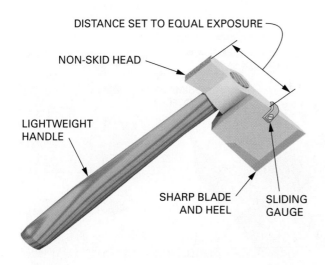

FIGURE 53–9 A shingling hatchet is commonly used to apply wood shingles and shakes.

Type of Shingle or Shake	Nail Type and Minimum Length	
Shingles—New Roof	**Type**	**(in.)**
16" and 18" Shingles	3d Box	1¼
24" Shingles	4d Box	1½
Shakes—New Roof	**Type**	**(in.)**
18" Straight-Split	5d Box	1¾
18" and 24" Handsplit-and-Resawn	6d Box	2
24" Tapersplit	5d Box	1¾
18" and 24" Taper-sawn	6d Box	2

FIGURE 53–10 Recommended nail types and sizes for wood shingles and shakes. (*Courtesy of Cedar Shake and Shingle Bureau*)

Fasteners

Apply each shingle with only two corrosion-resistant nails, such as stainless steel, hot-dipped galvanized, or aluminum nails. Box nails usually are used because their smaller gauge minimizes splitting. Minimum nail lengths for shingles and shakes are shown in **Figure 53–10.** Staples should be 16-gauge aluminum or stainless steel with a minimum crown of 7/16 inch. The staple should be driven with its crown across the grain of the shingle or shake. Staples should be long enough to penetrate the sheathing at least ½ inch.

APPLYING WOOD SHINGLES

With a string pulled along the bottom edge of shingles, install the first layer of a starter course of wood shingles. If a gutter is used, overhang the shingles plumb with the center of the gutter. If no gutter is installed, overhang the starter course 1½ inches beyond the fascia. Space adjacent shingles between ¼ and ⅜ inch apart. Place each fastener about ¾ inch in from the edge of the shingle and not more than one inch above the exposure line. Drive fasteners flush with the surface. Make sure that the head does not crush the surface.

Apply another layer of shingles on top of the first layer of the starter course. The starter course may be tripled, if desired, for appearance. This procedure is recommended in regions of heavy snowfall.

Lay succeeding courses across the roof. Apply several courses at a time. Stagger the joints in adjacent courses at least 1½ inches. There should be no joint in any three adjacent courses in alignment. Joints should not line up with the centerline of the heart of wood or any knots or defects. **Flat grain** or *slash grain* shingles wider than 8 inches should be split in two before fastening. Trim shingle edges with the hatchet to keep their butts in line **(Figure 53–11).**

After laying several courses, snap a chalk line to straighten the next course. Proceed shingling up the roof. When 3 or 4 feet from the ridge, check the distance on both ends of the roof. Divide the distance as close as possible to the exposure used. A full course should show below the ridge cap. Shingle tips are cut flush with the ridge. Tips can be easily cut across the grain with a hatchet if cut on an angle to the side of the shingle.

On intersecting roofs, stop shingling the roof a few feet away from the valley. Select and cut a shingle at a proper taper. Apply it to the valley. Do not break joints in the valley. Do not lay shingles with their grain parallel to the valley centerline. Work back out. Fit a shingle to complete the course **(Procedure 53–A).**

Hips and Ridges

After the roof is shingled, 4- to 5-inch-wide hip and ridge caps are applied. Measure down on both sides of the hip or ridge at each end for a distance equal to the exposure. Snap a line between the marks.

Lay a shingle so its bottom edge is to the line. Fasten with two nails. Use longer nails so that they penetrate at least ½ inch into or through the sheathing.

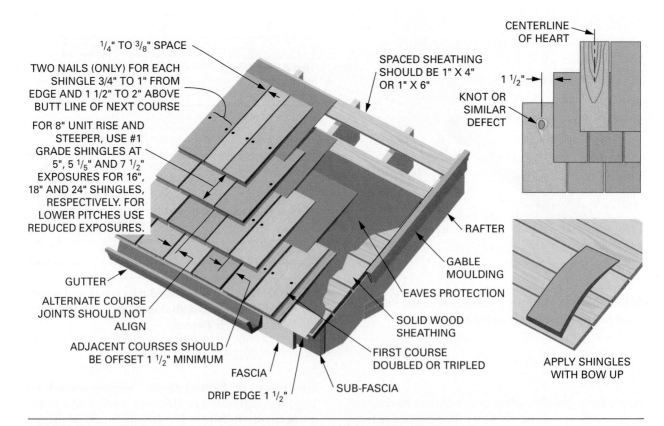

FIGURE 53–11   Details of wood shingle application. (*Courtesy of Cedar Shake and Shingle Bureau*)

Procedure 53–A Applying Shingles or Shakes Along an Open Valley

(Courtesy of Cedar Shake and Shingle Bureau)

ORDER OF APPLYING SHINGLES OR SHAKES AT VALLEY

STEP 1 TRIM AND FIT FIRST PIECE AT THE VALLEY CHALK LINE.

STEP 2 PLACE PRE-CUT PIECE SO THAT CUT ANGLE IS POSITIONED ON CHALK LINE WITH BUTT ON COURSE LINE.

STEP 3 SELECT A SHINGLE OR SHAKE OF THE REQUIRED WIDTH TO COMPLETE THE COURSE.

KEEP NAILS WELL AWAY FROM THE CENTER OF THE VALLEY

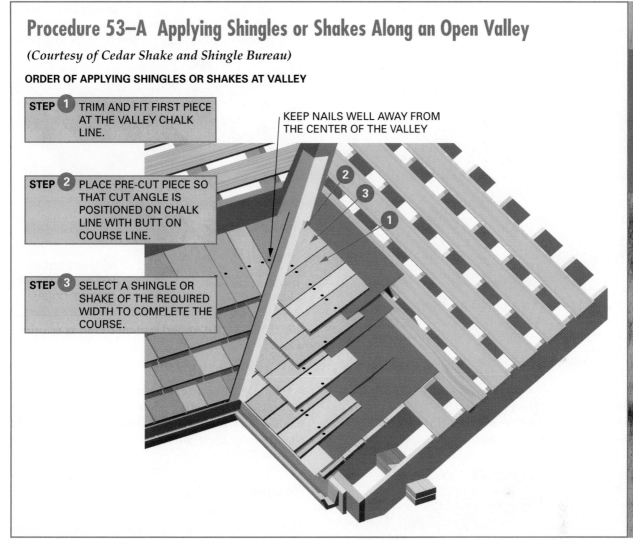

STEP BY STEP PROCEDURES

With the hatchet, trim the top edge flush with the opposite slope.

Lay another shingle on the other side with the butt even and its upper edge overlapping the first shingle laid. Trim its top edge at a bevel and flush with the side of the first shingle. Double this first set of shingles, alternating the joint. Apply succeeding layers of cap, with the same exposure as used on the roof. Alternate the overlap of each layer **(Figure 53–12).** Space the exposure when nearing the end so all caps are about equally exposed.

APPLYING WOOD SHAKES

Shakes are applied in much the same manner as shingles **(Figure 53–13).** Mark the handle of the shingling hatchet at 7½ and 10 inches from the top of the head. These are the exposures that are used most of the time when applying wood shakes.

Underlayment

A full width of felt underlayment is first applied under the starter course. Butts of the starter course should project 1½ inches beyond the fascia. Next, lay an 18-inch-wide strip of #30 roofing felt over the upper end of the starter course. Its bottom edge should be positioned at a distance equal to twice the shake exposure above the butt line of the first exposed course of shakes **(Figure 53–14).** For example, 24-inch shakes, laid with 10 inches of exposure, would have felt applied 20 inches above the butts of the shakes. The felt will cover the top 4 inches of the shakes. It will extend up 14 inches on the sheathing. The top edge of the felt must rest on the spaced sheathing, if used.

Nail only the top edge of the felt. Fasten successive strips on their top edge only. Their bottom edges should be one shake exposure from the bottom of the

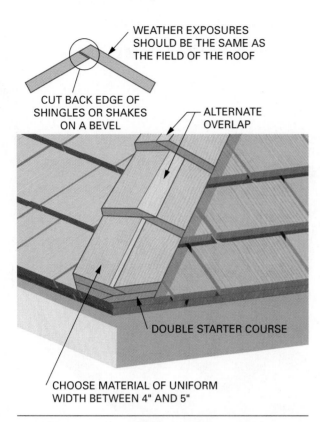

FIGURE 53–12 **When applying hip and ridge shingles, alternate the overlap.** *(Courtesy of Cedar Shake and Shingle Bureau)*

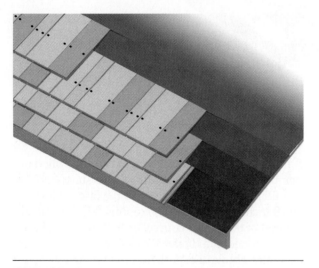

FIGURE 53–13 **Three or four shake courses at a time are carried across the roof.** *(Courtesy of Cedar Shake and Shingle Bureau)*

previous strip. It is important to lay the felt straight. It serves as a guide for applying shakes. After the roof is felted, the tips of the shakes are tucked under the felt. The bottom should be exposed by the distance of twice the exposure.

Apply the second and successive courses with joints staggered and fasteners placed the same as

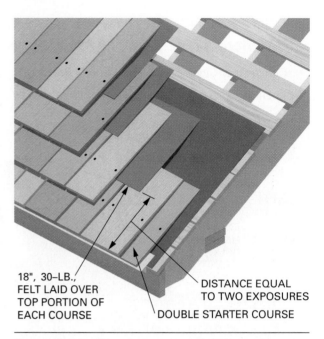

FIGURE 53–14 **An underlayment of felt is required when laying shakes.**

for shingles. The spacing between shakes should be at least ⅜ inch, but not more than ⅝ inch to allow wood to dry properly after rain. Lay straight-split shakes with their smooth end toward the ridge **(Figure 53–15)**.

Maintaining Shake Exposure

There is a tendency to angle toward the eave. Therefore, check the exposure regularly with the hatchet handle. An easy way to be sure of correct exposure is to look through the joint between the edges of the shakes in the course below the one being nailed. The bottom edge of the felt will be visible. The butt of the shake being nailed is positioned directly above it **(Figure 53–16)**.

Ridges and Hips

Adjust the exposure so that tips of shakes in the next-to-last course just come to the ridge. Use economical 15-inch starter-finish shakes for the last course. They save time by eliminating the need to trim shake tips at the ridge. Cap ridges and hips with shakes in the same manner as that used for wood shingles.

ESTIMATING ROOFING MATERIALS

Find the area of the roof in square feet. Divide the total by 100 to determine the number of squares needed. Add about 5 to 10 percent for waste. A simple roof with no dormers, valleys, or other obstruc-

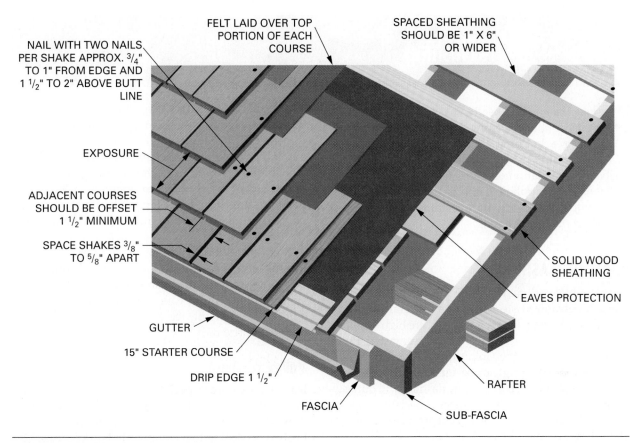

NAIL WITH TWO NAILS PER SHAKE APPROX. $3/4$" TO 1" FROM EDGE AND $1 1/2$" TO 2" ABOVE BUTT LINE

FELT LAID OVER TOP PORTION OF EACH COURSE

SPACED SHEATHING SHOULD BE 1" X 6" OR WIDER

EXPOSURE

ADJACENT COURSES SHOULD BE OFFSET $1 1/2$" MINIMUM

SPACE SHAKES $3/8$" TO $5/8$" APART

SOLID WOOD SHEATHING

EAVES PROTECTION

GUTTER

15" STARTER COURSE

DRIP EDGE $1 1/2$"

FASCIA

SUB-FASCIA

RAFTER

FIGURE 53–15 **Shake application details.** *(Courtesy of Cedar Shake and Shingle Bureau)*

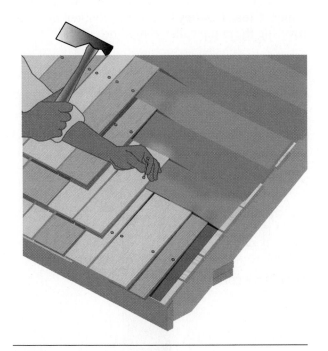

FIGURE 53–16 **The tip of the shake is inserted between the layers of underlayment.** *(Courtesy of Cedar Shake and Shingle Bureau)*

tions requires less allowance for waste. A complicated roof requires more.

For wood shingles, add one square for every 240 linear feet of starter course. Add one square of shakes for 120 linear feet of starter course.

Add one extra square of shingles for every 100 linear feet of valleys and about two squares for shakes.

Add an extra bundle of shakes or shingles for every 16 feet of hip and ridge to be covered.

Figure 2 pounds of nails per square at standard exposure.

Remember that a square of roofing will cover 100 square feet of roof surface only when applied at standard exposures. Allow proportionally more material when these exposures are reduced.

54 Flashing

*F*lashing is a material used in various locations that are susceptible to leaking. It prevents water from entering a building **(Figure 54–1)**. The words *flash, flashed,* and *flashing* are also used as verbs to describe the installation of the material. Various kinds of flashing are applied at the eaves, valleys, chimneys, vents, and other roof projections. They prevent leakage at the intersections.

KINDS OF FLASHING

Flashing material may be sheet copper, zinc, aluminum, galvanized steel, vinyl, or mineral-surfaced asphalt roll roofing. Copper and zinc are high-quality flashing materials, but they are more expensive. Roll roofing is less expensive. Colors that match or contrast with the roof covering can be used. If properly applied, roll roofing used as a valley flashing will outlast the main roof covering. Sheet metal, especially copper, may last longer. However, it is good practice to replace all flashing when reroofing.

Eaves Flashing

Whenever there is a possibility of ice dams forming along the eaves and causing a backup of water, **ice and water shield flashing** is needed. Apply the flashing such that it overhangs the drip edge by ¼ to ⅜ inch. The flashing should extend up the roof far enough to cover a point at least 12 inches inside the wall line of the building. If the overhang of the eaves requires that the flashing be wider than 36 inches, the necessary horizontal lap joint is located on the portion of the roof that extends outside the wall line **(Figure 54–2)**.

For a slope of at least 4 inches rise per foot of run, install a single course of 36-inch-wide flashing covering the underlayment and drip edge. On lower slopes greater protection against water leakage is gained by applying ice and water shield flashing over the entire roof.

Valley Flashing

Roof valleys are especially vulnerable to leakage. This is because of the great volume of water that flows down through them. Valleys must be carefully flashed according to recommended procedures. Valleys are flashed in two ways: *open* or *closed*. **Open valleys** are constructed with no shingle or roofing material installed within several inches of the valley center. **Closed valleys** have the roofing material covering the entire valley centerline.

FIGURE 54–1 Flashings are used to seal against leakage where the roofing butts against adjoining surfaces. (*Courtesy of Cedar Shake and Shingle Bureau*)

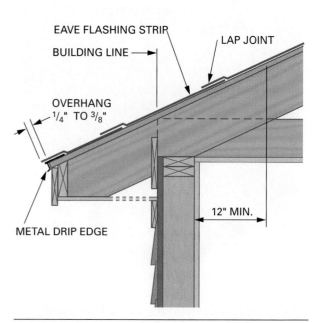

FIGURE 54–2 If there is danger of ice dams forming along the eaves, eave flashing is installed with seams away from the building line.

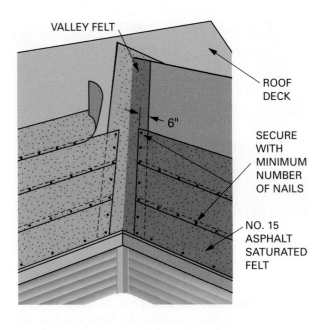

FIGURE 54–3　Felt underlayment is applied in the valley before roof underlayment.

Open Valley Flashing

Begin valley flashing by applying a 36-inch-wide strip of asphalt felt centered in the valley **(Figure 54–3)**. Fold or crease the roll along the length and seat it well into the valley. Be careful not to cause any break in the felt. Fasten it with only enough nails along its edges to hold it in place. Let the courses of felt underlayment applied to the roof overlap the valley underlayment by not less than 6 inches. The eave flashing, if required, is then applied.

Using Roll Roofing Flashing

Lay an 18-inch-wide layer of mineral-surfaced roll roofing centered in the valley **(Figure 54–4)**. Its mineral-surfaced side should be down. Use only enough nails spaced 1 inch in from each edge to hold the strip smoothly in place. Press the roofing firmly in the center of the valley when nailing the opposite edge. Next, lay a 36-inch-wide strip with its surfaced side up. Center it in the valley. Fasten it in the same manner as the first strip pressing firmly into the valley center.

Snap a chalk line on each side of the valley. Use them as guides for trimming the ends of the shingle courses. These lines are spaced 6 inches apart at the ridge. Because more water may be present in the valley at the eave than at the ridge, these lines are spread ⅛ inch per foot as they approach the eave. Thus, a valley 16 feet long will be 8 inches wide at the eaves.

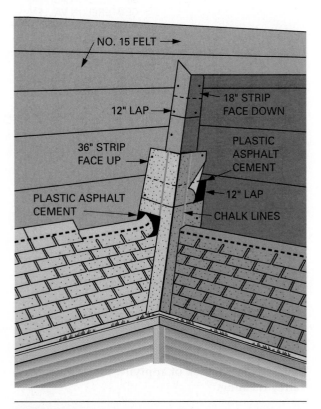

FIGURE 54–4　Method of applying roll roofing open valley flashing.

The upper corner of each end asphalt shingle is clipped. This helps keep water from entering between the courses. Each roof shingle is cemented to the valley flashing with plastic asphalt cement.

Metal Flashing

Prepare the valley with underlayment in the same manner as described previously. Next, lay a strip of sheet metal flashing centered in the valley **(Figure 54–5)**. The metal should extend at least 10 inches on each side of the valley centerline for slopes with a 6-inch rise or less and 7 inches on each side for a steeper slope. Carefully press and form it into the valley. Fasten the metal with nails of similar material spaced close to its outside edges. Use only enough fasteners to hold it smoothly in place.

Snap lines on each side of the valley, as described previously. Use them as guides for cutting the ends of the shingle courses. Trim the last shingle of each course to fit on the chalk line. Clip 1 inch from its upper corner at a 45-degree angle. To form a tight seal, cement the shingle to the metal flashing with a 3-inch width of asphalt plastic cement.

If a valley is formed by the intersection of a low-pitched roof and a much steeper one, a 1-inch-high,

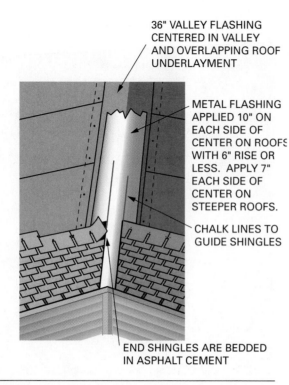

36" VALLEY FLASHING CENTERED IN VALLEY AND OVERLAPPING ROOF UNDERLAYMENT

METAL FLASHING APPLIED 10" ON EACH SIDE OF CENTER ON ROOFS WITH 6" RISE OR LESS. APPLY 7" EACH SIDE OF CENTER ON STEEPER ROOFS.

CHALK LINES TO GUIDE SHINGLES

END SHINGLES ARE BEDDED IN ASPHALT CEMENT

FIGURE 54–5   Method of applying metal open valley flashing.

crimped standing seam should be made in the center of the metal flashing of an open valley. The seam will keep heavy rain water flowing down the steeper roof from overrunning the valley and possibly being forced under the shingles of the lower slope.

Closed Valley Flashing

A closed valley protects the valley flashing. It thus adds to the weather resistance at vulnerable points. Several methods are used to flash closed valleys.

The first step for any method is to apply the asphalt felt underlayment as previously described for open valleys. Then, center a 36-inch width of smooth or mineral surface roll roofing, 50-pound per square, or heavier, in the valley over the underlayment. Form it smoothly in the valley. Secure it with only as many nails as necessary. Another method uses a strip of wide metal flashing on top of the felt underlayment.

Woven Valley Method

Valleys may be flashed by applying asphalt shingles on both sides of the valley and alternately weaving each course over and across the valley. This is called a **woven valley (Figure 54–6).**

Lay the first course of shingles along the edge of one roof up to and over the valley for a distance of

FIGURE 54–6   Valleys are sometimes flashed by weaving shingles together.

at least 12 inches. Lay the first course along the edge of the adjacent roof. Extend the shingles over the valley on top of the previously applied shingles **(Figure 54–7).**

Succeeding courses are then applied. Weave the valley shingles alternately, first on one roof and then on the other. When weaving the shingles, make sure they are pressed tightly into the valley. Also make sure no nail is closer than 6 inches to the valley centerline. Fasten the end of the woven shingle with two nails. Most carpenters prefer to cover each roof

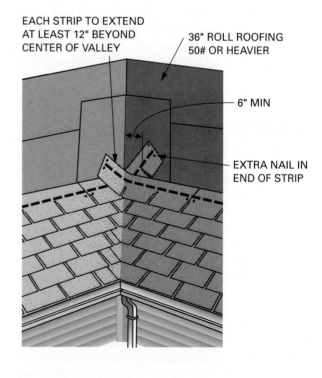

EACH STRIP TO EXTEND AT LEAST 12" BEYOND CENTER OF VALLEY

36" ROLL ROOFING 50# OR HEAVIER

6" MIN

EXTRA NAIL IN END OF STRIP

FIGURE 54–7   Details for applying a woven valley.

area with shingles to a point approximately 3 feet from the valley. They weave the valley shingles in place later.

No end joints should occur within 6 inches of the center of the valley. Therefore, it may be necessary to occasionally cut a strip short that would otherwise end near the center. Continue from this cut end with a full-length strip over the valley.

Closed Cut Valley Method

Apply the shingles to one roof area. Let the end shingle of every course overlap the valley by at least 12 inches **(Figure 54–8)**. Make sure no end joints occur within 6 inches of the center of the valley. Place fasteners no closer than 6 inches from the center of the valley. Form the end shingle of each course snugly in the valley. Secure its end with two fasteners.

Snap a chalk line 2 inches short of and parallel to the center of the valley on top of the overlapping shingles. Apply shingles to the adjacent roof area. Cut the end shingle to the chalk line. Clip the upper corner of each shingle as described previously for open valleys. Bed the end of each shingle that lies in the valley in about a 3-inch-wide strip of asphalt cement. Make sure that no fasteners are located closer than 6 inches to the valley centerline.

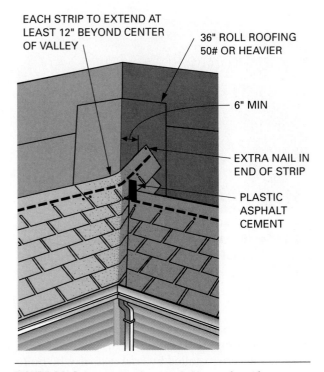

FIGURE 54–8 Details for applying a closed cut valley.

Step Valley Flashing Method

Step flashings are individual metal pieces tucked between courses of shingles (see Figure 54–1). When applying step flashings in valleys, first estimate the number of shingle courses required to reach the ridge. Cut a piece of metal flashing for each course of shingles. Each piece should be at least 18 inches wide for slopes with a 6-inch rise or greater and 24 inches wide for slopes with less pitch. The height of each piece should be at least 3 inches more than the shingle exposure.

Prepare the valley with underlayment as described previously. Snap a chalk line in the center of the valley. Apply the starter course on both roofs. Trim the end shingle of each course to the chalk line. Fit and form the first piece of flashing to the valley. Trim its bottom edge flush with the drip edge. Fasten it in the valley over the first layer of the starter course. Use fasteners of like material to prevent electrolysis. Fasten the upper corners of the flashing only.

Apply the first regular course of shingles to both roofs on each side of the valley. Trim the valley shingles so their ends lay on the chalk line. Bed them in plastic asphalt cement. Do not drive nails through the metal flashing.

Apply the next piece of flashing in the valley over the first course of shingles. Keep its bottom edge about ½ inch above the butts of the next course of shingles. Apply the second course of shingles in the same manner as the first. Secure a flashing over the second course. Apply succeeding courses and flashings in this manner **(Figure 54–9)**. Remember, flashing is placed over each course of shingles. Do not leave any flashings out. When the valley is completely flashed, no metal flashing surface is exposed. If the valley does not extend all the way to the ridge of the main roof, a *saddle* is applied over the ridge of the minor roof **(Figure 54–10)**.

Flashing against a Wall

When roof shingles butt up against a vertical wall, the joint must be made watertight. The usual method of making the joint tight is with the use of metal step flashings **(Figure 54–11)**.

The flashings are purchased or cut about 8 inches in width. They are bent at right angles in the center so they will lay about 4 inches on the roof and extend about 4 inches up the sidewall. The length of the flashings is about 3 inches more than the exposure of the shingles. When used with shingles exposed 5 inches to the weather, they are made 8 inches in length. Cut and bend the necessary number of metal flashings.

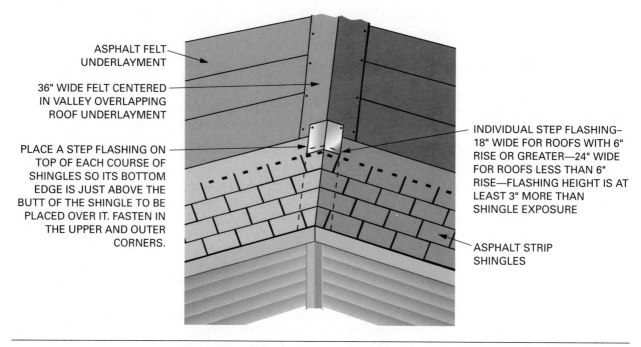

ASPHALT FELT UNDERLAYMENT

36" WIDE FELT CENTERED IN VALLEY OVERLAPPING ROOF UNDERLAYMENT

PLACE A STEP FLASHING ON TOP OF EACH COURSE OF SHINGLES SO ITS BOTTOM EDGE IS JUST ABOVE THE BUTT OF THE SHINGLE TO BE PLACED OVER IT. FASTEN IN THE UPPER AND OUTER CORNERS.

INDIVIDUAL STEP FLASHING– 18" WIDE FOR ROOFS WITH 6" RISE OR GREATER—24" WIDE FOR ROOFS LESS THAN 6" RISE—FLASHING HEIGHT IS AT LEAST 3" MORE THAN SHINGLE EXPOSURE

ASPHALT STRIP SHINGLES

FIGURE 54–9   Details for applying metal step flashings in a valley.

The roofing is applied and flashed before the siding is applied to the vertical wall. First, apply an underlayment of asphalt felt to the roof deck. Turn the ends up on the vertical wall by about 3 to 4 inches.

Apply the first layer of the starter course, working toward the vertical wall. Fasten a metal flashing on top of the starter course of shingles. Its bottom edge should be flush with the drip edge. Use one fastener in each top corner. Lay the first regular course with its end shingle over the flashing and against the sheathing of the sidewall. Do not drive any fasteners through the flashings. It is usually not necessary to bed the shingles to the flashings with asphalt cement. The step flashing holds down the end of the shingle below it.

Apply a flashing over the upper side of the first course and against the wall. Keep its bottom edge at a point that will be about ½ inch above the butt of the next course of shingles. Continue applying shingles and flashings in this manner until the ridge is reached. Some carpenters prefer to cut the shingles back if a tab cutout occurs over a flashing. This prevents metal from being exposed to view.

Flashing a Chimney

In many cases, especially on steep pitch roofs, a **cricket** or **saddle** is built between the upper side of the chimney and roof deck. The cricket, although not

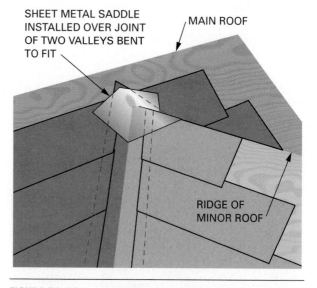

SHEET METAL SADDLE INSTALLED OVER JOINT OF TWO VALLEYS BENT TO FIT

MAIN ROOF

RIDGE OF MINOR ROOF

FIGURE 54–10   A saddle is installed over the ridge of a minor roof where it intersects with the main roof.

a flashing in itself, prevents accumulation of water behind the chimney **(Figure 54–12).**

Flashings are installed by *brick masons* who build the chimney. The upper ends of the flashing are bent around and **mortared** between the courses of brick as the chimney is built. The flashings are long

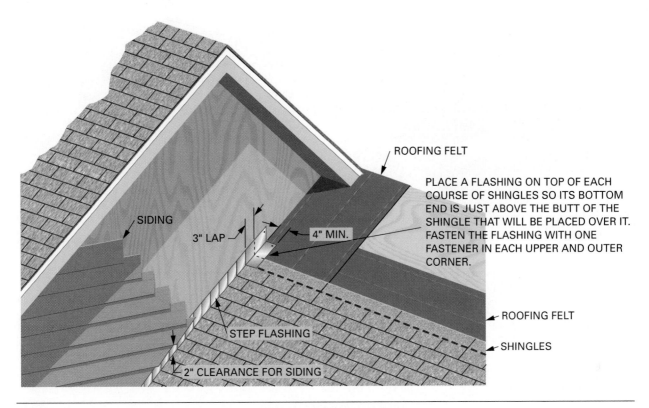

SIDING

3" LAP

4" MIN.

ROOFING FELT

PLACE A FLASHING ON TOP OF EACH COURSE OF SHINGLES SO ITS BOTTOM END IS JUST ABOVE THE BUTT OF THE SHINGLE THAT WILL BE PLACED OVER IT. FASTEN THE FLASHING WITH ONE FASTENER IN EACH UPPER AND OUTER CORNER.

ROOFING FELT

SHINGLES

STEP FLASHING

2" CLEARANCE FOR SIDING

FIGURE 54–11   Using metal step flashing where a roof butts a wall.

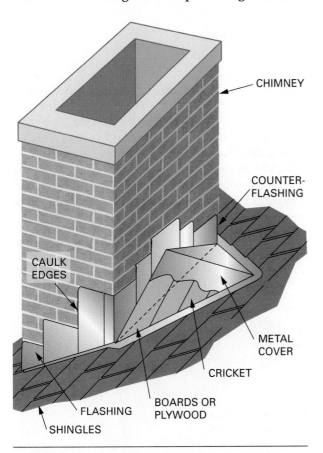

CHIMNEY

COUNTER-FLASHING

CAULK EDGES

METAL COVER

CRICKET

BOARDS OR PLYWOOD

FLASHING

SHINGLES

FIGURE 54–12   A cricket is built to prevent the accumulation of water behind the chimney.

enough to be bent at and over the roof sheathing for tucking between shingles. These flashings are usually in place before the carpenter applies the roof covering.

The underlayment is applied and tucked under the existing flashings. The shingle courses are brought up to the chimney. They are applied under the flashing on the lower side of the chimney. This is called the **apron flashing.** Shingles are tucked under the **apron.** The top edges of the shingles are cut as necessary, until the shingle exposure shows below the apron. The apron is then pressed into place on top of the shingles in a bed of plastic cement. Its projecting ends are carefully and gently formed up around the sides of the chimney and under the lowest side flashings.

Along the sides of the chimney, the flashings are tucked in between the shingles in the same manner as in flashing against a wall. No nails are used in the flashings. The standing portions of the **side flashings** are bedded to the chimney with asphalt cement. The roof portion is bedded to the shingle. The projecting edges of the lowest side flashings are carefully formed around the corner. They are folded against the low side on the chimney. The top edges of the highest side flashings are also folded around the corner and under the **head flashing** on the upper side of the chimney.

The head flashing is cemented to the roof. Shingles are applied over it. They are bedded to it with asphalt

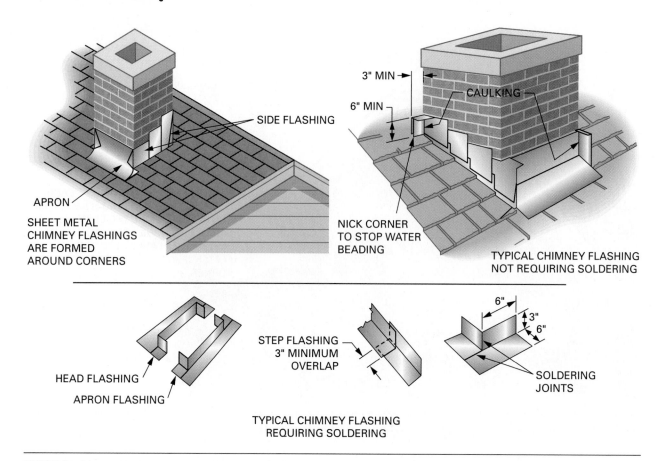

SIDE FLASHING

APRON

SHEET METAL
CHIMNEY FLASHINGS
ARE FORMED
AROUND CORNERS

3" MIN

6" MIN

CAULKING

NICK CORNER
TO STOP WATER
BEADING

TYPICAL CHIMNEY FLASHING
NOT REQUIRING SOLDERING

HEAD FLASHING

APRON FLASHING

STEP FLASHING
3" MINIMUM
OVERLAP

6"

3"

6"

SOLDERING
JOINTS

TYPICAL CHIMNEY FLASHING
REQUIRING SOLDERING

FIGURE 54–13 **Chimney flashings wrap the entire chimney in overlapping pieces.**

cement. Its projecting ends are also carefully formed around the corner on top of the side flashings.

The projecting ends of chimney flashings are carefully formed and folded around the corners of the chimney. Gently and carefully tap the metal with a hammer handle. Care must be taken not to break through the flashings. Chimneys may be flashed by other methods and materials, other than described above, depending on the custom of certain geographical areas **(Figure 54–13).** Other rectangular roof obstructions, such as skylights, are flashed in a similar manner.

Flashing Vents

Flashings for round pipes, such as *stack vents* for plumbing systems and *roof ventilators,* usually come as *flashing collars* made for various roof pitches. They fit around the stack. They have a wide flange on the bottom that rests on the roof deck. The flashing is installed over the stack vent, with its flange on the roof sheathing. It is fastened in place with one fastener in each upper corner.

Shingle up to the lower end of the stack vent flashing. Lift the lower part of the flange. Apply shingle

courses under it. Cut the top edge of the shingles, where necessary, until the shingle exposure, or less, shows below the lower edge of the flashing. Apply asphalt cement under the lower end of the flashing. Press it into place on top of the shingle courses.

Apply shingles around the stack and over the flange. Do not drive nails through the flashing. Bed shingles to the flashing with asphalt cement, where necessary **(Figure 54–14).**

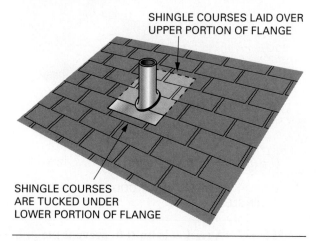

SHINGLE COURSES LAID OVER
UPPER PORTION OF FLANGE

SHINGLE COURSES
ARE TUCKED UNDER
LOWER PORTION OF FLANGE

FIGURE 54–14 **A vent stack flashing boot.**

Key Terms

apron	deck	ice and water shield flashing	side flashing
apron flashing	double coverage	mortared	sliding gauge
asphalt felt	drip edge	open valley	solid sheathing
asphalt shingles	electrolysis	roll roofing	spaced sheathing
cements	end lap	saddle	square
closed valley	exposure	selvage	starter course
coatings	flashing	shakes	top lap
courses	flat grain	shingle butt	underlayment
coverage	head flashing	shingling hatchet	wood shingles
cricket	head lap		woven valley

Review Questions

Select the most appropriate answer.

1. A square is the amount of roofing required to cover
   a. 1 square foot.
   b. 100 square feet.
   c. 150 square feet.
   d. 200 square feet.

2. One roll of #15 asphalt felt will cover about
   a. 1 square.
   b. 2 squares.
   c. 3 squares.
   d. 4 squares.

3. When applying asphalt felt on a roof deck as underlayment, lap each course over the lower course by at least
   a. 2 inches.
   b. 4 inches.
   c. 6 inches.
   d. 12 inches.

4. Laying out three tab shingles "on the thirds" requires starter tabs to be
   a. 3 inches shorter.
   b. 4 inches shorter.
   c. 6 inches shorter.
   d. 12 inches shorter.

5. Nailing for asphalt shingles should be located _____ the self-sealing strip.
   a. above
   b. on
   c. below
   d. above or below

6. Checking and adjusting shingle exposure should be done when top shingle course is _____ from ridge.
   a. 1–2 feet
   b. 3–4 feet
   c. 5–6 feet
   d. 7–8 feet

7. Selvage is a term used to describe
   a. recycling building materials.
   b. the portion of roofing material that is not to be exposed.
   c. leftover unused roofing material.
   d. all of the above.

8. Most wood shingles and shakes are made from
   a. cypress.
   b. redwood.
   c. eastern white cedar.
   d. western red cedar.

9. The longest available length of wood shingles and shakes is
   a. 16 inches.
   b. 18 inches.
   c. 24 inches.
   d. 28 inches.

10. Wood shingles normally overhang the fascia by
    a. ⅜ inch.
    b. 1 inch.
    c. 1½ inches.
    d. 2 inches.

11. Prior to applying shakes, it is important to lay underlayment straight
    a. so there are no wrinkles in the felt.
    b. to obtain the proper lap.
    c. to improve its looks after the roof is completed.
    d. because the felt serves as an installation guide for shingles.

12. The first step in installation of roll roof valley flashing is
    a. installing an 18-inch-wide flashing piece face up.
    b. installing an 18-inch-wide flashing piece face down.
    c. installing a full-width flashing piece face up.
    d. installing a full-width underlayment.

13. When nailing shingles, do not locate any nails closer to the valley centerline than
    a. 6 inches.   c. 10 inches.
    b. 8 inches.   d. 12 inches.

14. Step flashings used against a vertical wall are cut about 8 inches wide. They should be bent so that

    a. 3 inches lay on the wall and 5 inches lay on the roof.
    b. 4 inches lay on the wall and 4 inches lay on the roof.
    c. 5 inches lay on the wall and 3 inches lay on the roof.
    d. 2 inches lay on the wall and 6 inches lay on the roof.

15. A built-up section between the roof and the upper side of chimney is called a
    a. cricket.   c. furring.
    b. dutchman.   d. counter flashing.

UNIT 19
Windows

Windows are normally installed prior to the application of exterior *siding*. Care must be taken to provide easy-operating, weathertight, attractive units. Quality workmanship results in a more comfortable interior, saves energy by reducing fuel costs, minimizes maintenance, gives longer life to the units, and makes application of the exterior siding easier.

SAFETY REMINDER

Windows are made in many styles, sizes and shapes. They are often a focal point of large rooms. Follow manufacturer's recommendations and local code requirements carefully when installing them to avoid injury.

OBJECTIVES

After completing this unit, the student should be able to:

- describe the most popular styles of windows and name their parts.
- select and specify desired sizes and styles of windows from manufacturers' catalogs.
- install various types of windows in an approved manner.
- cut glass and glaze a sash.

55   Window Terms and Types

W ood *windows* are one of many types of **mill-work (Figure 55–1).** Millwork is a term used to describe products, such as windows, doors, and cabinets, fabricated in woodworking plants that are used in the construction of a building. Windows are usually fully assembled and ready for installation when delivered to the construction site. Windows are also made of aluminum and steel. Windows made with exposed wood parts encased in vinyl are called **vinyl-clad** windows. The names given to various parts of a window are the same, in most cases, regardless of the window type.

PARTS OF A WINDOW

When shipped from the factory, the window is complete except for the interior trim. It is important that the installer know the names, location, and functions of the parts of a window in order to understand, or to give, instructions concerning them.

The Sash

The **sash** is a frame in a window that holds the glass. The type of window is generally determined by the way the sash operates. The sash may be installed in a fixed position, move vertically or horizontally, or swing outward or inward.

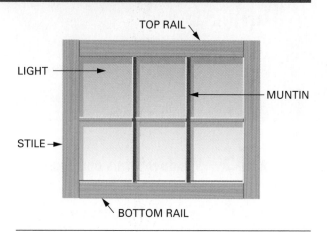

FIGURE 55–2   A window sash and its parts.

Sash Parts. Vertical edge members of the sash are called **stiles.** Top and bottom horizontal members are called **rails.** The pieces of glass in a sash are called **lights.** There may be more than one light of glass in a sash. Small strips of wood that divide the glass into smaller lights are called **muntins.** Muntins may divide the glass into rectangular, diamond, or other shapes **(Figure 55–2).**

Many windows come with false muntins called **grilles.** Grilles do not actually separate or support the glass. They are applied as an overlay to simulate small lights of glass. They are made of wood or plastic. They snap in and out of the sash for easy cleaning of the lights **(Figure 55–3).** They may also be

FIGURE 55–1   Windows of many types and sizes are fully assembled in millwork plants and ready for installation. (*Courtesy of Andersen Windows, Inc.*)

FIGURE 55–3   Removable grilles simulate true divided-light muntins.

preinstalled between the layers of glass in double- or triple-glazed windows.

WINDOW GLASS

Several qualities and thicknesses of sheet glass are manufactured for **glazing** and other purposes. The installation of glass in a window sash is called glazing. **Single-strength (SS) glass** is about ³⁄₃₂ inch thick. It is used for small lights of glass. For larger lights, **double-strength (DS) glass** about ⅛ inch thick may be used. *Heavy sheet glass* about ³⁄₁₆ inch thick is also manufactured. Many other kinds of glass are made for use in construction.

Safety Glass

Most residential windows are not glazed with safety glass, so if they break, they could fragment into large pieces and cause injury. Care must be taken to handle windows in a manner to prevent breaking the glass. Some codes, however, do require a type of **safety glass** in windows with low sill heights or located near doors. Skylights and roof windows are generally required to be glazed with safety glass.

Safety glass is constructed, treated, or combined with other materials to minimize the possibility of injuries resulting from contact with it. Several types of safety glass are manufactured.

Laminated glass consists of two or more layers of glass with inner layers of transparent plastic bonded together. **Tempered glass** is treated with heat or chemicals. When broken at any point, the entire piece immediately disintegrates into a multitude of small granular pieces. *Transparent plastic* is also used for safety glazing material.

Insulated Glass

To help prevent heat loss, and to avoid condensation of moisture on glass surfaces, **insulated glass,** or *thermal pane windows,* are used frequently in place of single-thickness glass.

Insulated glass consists of two or three (generally two) layers of glass separated by a sealed air space ³⁄₁₆ to 1 inch in thickness **(Figure 55–4).** Moisture is removed from the air between the layers before the edges are sealed. To raise the R-value of insulated glass, the space between the layers is filled with **argon gas.** Argon conducts heat at a lower rate than air. Additional window insulation may be provided with the use of *removable glass panels* or *combination storm sash.*

Solar Control Glass. The R-value of windows may also be increased by using special *solar-control insulated glass,* called *high performance* or *Low-E glass.* Low-E is an abbreviation for **low emissivity.** It is used to

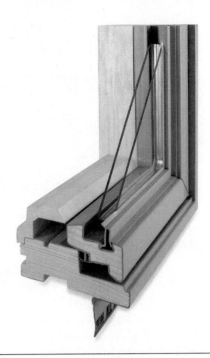

FIGURE 55–4 Cutaway of insulated glass used to increase the R-value in a window. (*Courtesy of Andersen Windows, Inc.*)

designate a type of glazing that reflects heat back into the room in winter and blocks heat from entering in the summer **(Figure 55–5).** An invisible, thin, metallic coating is bonded to the air space side of the inner glass. This lets light through, but reflects heat.

THE WINDOW FRAME

The sash is hinged to, slides, or is fixed in a **window frame.** The frame usually comes with the exterior trim applied. It consists of several distinct parts **(Figure 55–6).**

The Sill

The bottom horizontal member of the window frame is called a **sill.** It is set or shaped at an angle to shed water. Its bottom side usually is grooved so a weathertight joint can be made with the wall siding.

Jambs

The vertical sides of the window frame are called **side jambs.** The top horizontal member is called a **head jamb.**

Extension Jambs. The inside edge of the jamb should be flush with the finished interior wall surface when the window is installed. In some cases, windows can be ordered with jamb widths for standard wall thicknesses. In other cases, jambs are

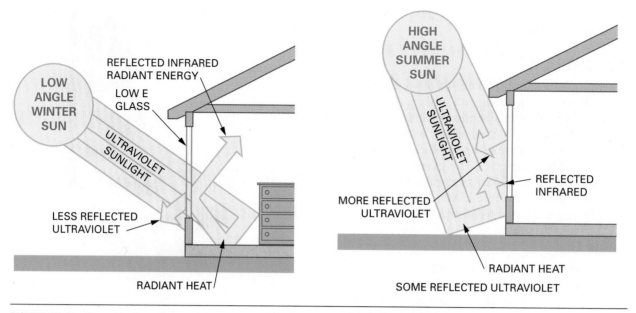

FIGURE 55–5 Low-E glass is used in windows to keep heat in during cold weather and out during hot weather.

made narrow. **Extension jambs** are then provided with the window unit. The extensions are cut to width to accommodate various wall thicknesses. They are applied to the inside edge of the jambs of the window frame **(Figure 55–7)**. The extension jambs are installed at a later stage of construction when the interior trim is applied. They should be stored for safekeeping until needed.

Windows may also be purchased with extension jambs already installed. Care should always be

taken to protect the jambs throughout the construction process.

Blind Stops

Blind stops are sometimes applied to certain types of window frames. They are strips of wood attached to the outside edges of the jambs. Their inside edges project about ½ inch inside the frame. They provide a weathertight joint between the outside casings and

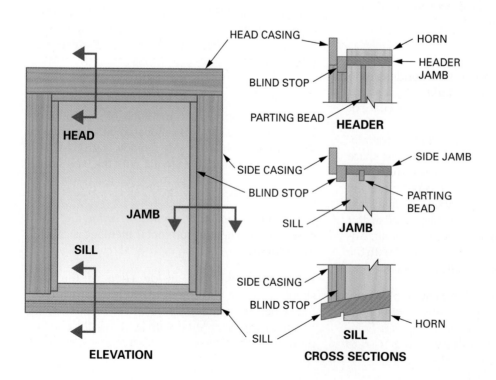

FIGURE 55–6 A window frame consists of parts with specific terms.

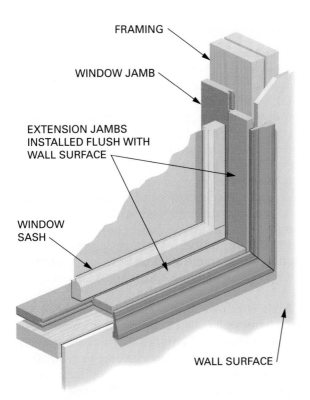

FRAMING

WINDOW JAMB

EXTENSION JAMBS
INSTALLED FLUSH WITH
WALL SURFACE

WINDOW
SASH

WALL SURFACE

EXTENSION JAMBS MAY BE INSTALLED ON ALL FOUR
SIDES AS SHOWN OR MAY EXCLUDE THE BOTTOM WHERE
A SILL IS INSTALLED

FIGURE 55–7 To compensate for varying wall thicknesses, extension jambs are provided with some window units.

the frame. They also act as stops for screens and storm sash. They make the outer edge of the channel for top sash of double-hung windows.

Casings

Window units usually come with exterior casings applied. The side members are called **side casings.**

In most windows, the lower ends are cut at a bevel and rest on the sill. The top member is called the **head casing.** On flat casings, a weathertight *rabbeted* or *tongue-and-grooved* joint is made between them. When molded casings are used, the mitered joints at the head are usually bedded in compound **(Figure 55–8).**

When windows are installed or manufactured, side by side, in multiple units, a **mullion** is formed where the two side jambs are joined together. The casing covering the joint is called a mullion casing **(Figure 55–9).**

The Drip Cap

A **drip cap** comes with some windows. It is applied on the top edge of the head casing. It is shaped with a sloping top to carry rain water out over the window unit. Its inside edge is rabbeted. The wall siding is applied over it to make a weathertight joint.

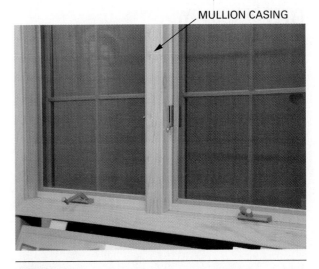

MULLION CASING

FIGURE 55–9 Window units that are joined create a mullion.

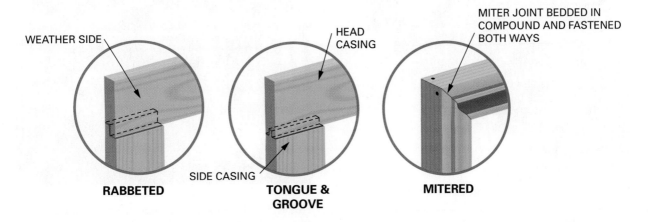

WEATHER SIDE

HEAD
CASING

MITER JOINT BEDDED IN
COMPOUND AND FASTENED
BOTH WAYS

SIDE CASING

RABBETED

**TONGUE &
GROOVE**

MITERED

FIGURE 55–8 A weathertight joint is made between side and header casings.

Window Flashing

In some cases, a **window flashing** is also provided. This is a piece of metal as long as the head casing, which is also called a *drip cap*. It is bent to fit over the head casing and against the exterior wall **(Figure 55–10).** The flashing prevents the entrance of water at this point. Flashings are usually made of aluminum or zinc. The vinyl flanges of vinyl-clad wood windows are usually formed as an integral part of the window. No additional head flashings are required.

Protective Coatings

Most window units with wood exterior casings are primed with a first coat of paint applied at the factory. Priming should be done before installation. Store the units under cover and protected from the weather until installed. Additional protective coats should be applied as soon as practical. **Vinyl-clad** wood windows are designed to eliminate painting.

TYPES OF WINDOWS

Common types of windows are fixed, single- or double-hung, casement, sliding, awning, and hopper windows.

Fixed Windows

Fixed windows consist of a frame in which a sash is fitted in a fixed position. They are manufactured in many shapes **(Figure 55–11).**

Oval and *circular* windows are usually installed as individual units. *Elliptical, half rounds,* and *quarter rounds* are widely used in combination with other types. In addition, fixed windows are manufactured in other *geometric* shapes (*squares, rectangles, triangles, parallelograms, diamonds, trapezoids, pentagons, hexagons* and *octagons*). They may be assembled or combined with other types of windows in a great variety of shapes **(Figure 55–12).** In addition to factory-assembled units, lengths of the frame stock may be purchased for cutting and assembling odd shape or size units on the job.

Arch windows have a curved top or head that make them well suited to be joined in combination with a number of other types of windows or doors. All of the windows mentioned come in a variety of sizes. With so many shapes and sizes, hundreds of interesting and pleasing combinations can be made. Arched windows may be made as part of the sash **(Figure 55–13).**

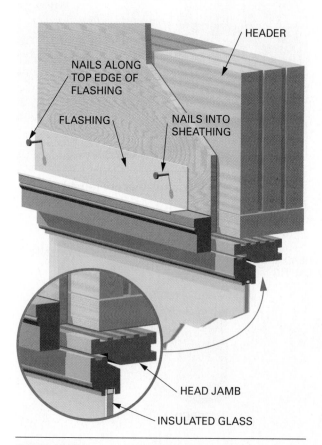

FIGURE 55–10 A window flashing covers the top edge of the header casing.

FIGURE 55–11 Fixed windows are often used in combinations. (*Courtesy of Andersen Windows, Inc.*)

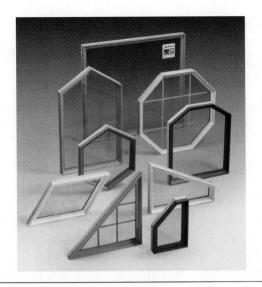

FIGURE 55-12 Windows come in a variety of shapes and sizes. *(Courtesy of Andersen Windows, Inc.)*

FIGURE 55-13 Arched windows can be part of a window sash.

Single- and Double-Hung Windows

The **double-hung window** consists of an upper and a lower sash that slide vertically by each other in separate channels of the side jambs **(Figure 55-14).** The **single-hung window** is similar except the up-

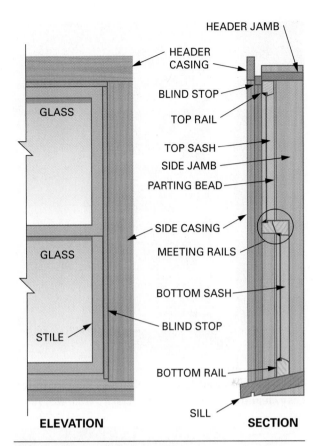

FIGURE 55-14 The double-hung window and its parts.

per sash is fixed. A strip separating the sash is called a **parting strip.**

In most units, the sash slides in metal channels that are installed in the frames. Each sash is provided with *springs, sash balances,* or *compression weatherstripping* to hold it in place in any position. Compression weatherstripping prevents air infiltration, provides tension, and acts as a counterbalance. Some types provide for easy removal of the sash for painting, repair, and cleaning.

When the sashes are closed, specially shaped *meeting rails* come together to make a weathertight joint. *Sash locks* located at this point not only lock the window, but draw the rails tightly together. Other hardware consists of *sash lifts* that are fastened to the bottom rail of the bottom sash. They provide an uplifting force to make raising the sash easier and help keep the sash in the position it is placed.

Most double-hung windows are also designed to be removed via a tilt-in action **(Figure 55-15).** This makes cleaning both inside and outside surfaces easy. Double-hung windows can be arranged in a number of ways. They can be installed side by side in multiple units or in combination with other types. **Bow window** units project out from the building often creating more floor space **(Figure 55-16).** The look is

FIGURE 55–15 Double-hung windows may tilt in for easy cleaning. *(Courtesy of Andersen Windows, Inc.)*

FIGURE 55–16 Bow window units are made by joining smaller units into a curved wall. *(Courtesy of Andersen Windows, Inc.)*

of a smooth curve. A **bay window** unit is similar to a bow except the sides are straight with corners at the window intersections.

Casement Windows

The **casement window** consists of a sash hinged at the side. It swings outward by means of a crank or lever. Most casements swing outward. The inswinging type is very difficult to make weathertight. An advantage of the casement type is that the entire sash can be opened

FIGURE 55–17 Casement windows swing outward. *(Courtesy of Andersen Windows, Inc.)*

for maximum ventilation. **Figure 55–17** shows the use of casement windows in a bow window unit.

Sliding Windows

Sliding windows have sashes that slide horizontally in separate tracks located on the header jamb and sill **(Figure 55–18)**. When a window-wall effect is desired, many units can be placed side by side. Most units come with all necessary hardware applied.

FIGURE 55–18 The sashes in sliding windows move horizontally by each other. *(Courtesy of Andersen Windows, Inc.)*

Awning and Hopper Windows

An **awning window** unit consists of a frame in which a sash hinged at the top swings outward by means of a crank or lever. A similar type, called the **hopper window,** is hinged at the bottom and swings inward.

Each sash is provided with an individual frame so that many combinations of width and height can be used. These windows are often used in combination with other types **(Figure 55–19).**

FIGURE 55–19 Awning windows are often used in stacks or in combination with other types of windows. *(Courtesy of Andersen Windows, Inc.)*

Skylight and Roof Windows

Skylights provide light only. **Roof windows** contain operating sashes to provide light and ventilation **(Figure 55–20).** One type of roof window comes with a tilting sash that allows access to the outside surface for cleaning. Special flashings are used when multiple skylights or roof windows are ganged together.

FIGURE 55–20 Skylights and roof windows are made in a number of styles and sizes. *(Courtesy of Andersen Windows, Inc.)*

56 Window Installation and Glazing

There are numerous window manufacturers that produce hundreds of kinds, shapes, and sizes of windows. Because of the tremendous variety and design differences, follow the manufacturer's instructions closely to ensure a correct installation. Directions in this unit are basic to most window installations. They are intended as a guide to be supplemented by procedures recommended by the manufacturer.

SELECTING AND ORDERING WINDOWS

The builder must study the plans to find the type and location of the windows to be installed. The floor plan shows the location of each unit. Each unit is usually identified by a number or a letter next to the window symbol.

Those responsible for designing and drawing plans for building or selecting windows must be aware of, and comply with, building codes that set certain stan-

dards in regard to windows. Most codes require minimum areas of natural light. Codes also require minimum ventilation by windows unless provided by other means. Some codes stipulate minimum window sizes in certain rooms for use as emergency egress.

Outswinging windows, such as awning and casement windows, should not swing out over decks, patios, and similar areas unless they are high enough to permit persons to travel under them. When lower, the projecting sash could cause serious injury.

Window Schedule

The numbers or letters found in the floor plan identify the window in more detail in the **window schedule.** This is, usually, part of a set of plans **(Figure 56–1).** This schedule normally includes window style, size, manufacturer's name, and unit number. Rough opening sizes may or may not be shown.

Window Schedule

Ident.	Quan.	Manufacturer	Size	Remarks
A	6	Andersen	TW28310	D.H. Tiltwash
B	1	Andersen	WDH2442	Woodwright D.H.
C	2	Andersen	3062	Narrowline D.H.
D	1	Andersen	CW24	Casement Single
E	1	Andersen	C34	Casement Triple
F	1	Andersen	C23	Casement Double

FIGURE 56–1 Typical window schedule found in a set of plans.

Manufacturers' Catalogs

Sometimes a window schedule is not included. Units are identified only by the manufacturer's name and number on the floor plan. To get more information, the builder must refer to the window manufacturer's catalog.

The catalog usually includes a complete description of the manufactured units and optional accessories, such as insect screens, glazing panels, and grilles. For a particular window style, the catalog typically shows overall unit dimensions, rough opening widths and heights, and glass sizes of manufactured units. Large-scale, cross-section details of the window unit also usually are included so the builder can more clearly understand its construction **(Figure 56–2).**

Table of Basic Unit Sizes Scale 1/8" = 1'-0" (1:96)

Unit Dimension	1'-5" (432)	1'-8 1/2" (521)	2'-0 1/8" (613)	2'-4 3/8" (721)	2'-11 15/16" (913)	2'-9 3/4" (857)	3'-4 3/4" (1035)	4'-0" (1219)	4'-8 1/2" (1435)	5'-11 5/8" (1819)	5'-11 7/8" (1826)
Rough Opening	1'-5 1/2" (445)	1'-9" (533)	2'-0 5/8" (625)	2'-4 7/8" (733)	3'-0 1/2" (927)	2'-10 1/4" (870)	3'-5 1/4" (1048)	4'-0 1/2" (1232)	4'-9" (1448)	6'-0 1/8" (1832)	6'-0 3/8" (1838)
Unobstructed Glass*	12 5/8" (321)	16 1/8" (410)	19 3/4" (502)	24" (610)	31 9/16" (802)	12 5/8" (321)	16 1/8" (410)	19 3/4" (502)	24" (610)	31 9/16" (802)	19 3/4" (502)

Row Dim.											
2'-0 1/8" (613) / 2'-0 5/8" (625) / 19 5/16" (491)	CR12	CN12	C12	CW12							
2'-4 3/8" (721) / 2'-4 7/8" (733) / 23 9/16" (598)	CR125	CN125	C125	CW125							
2'-11 15/16" (913) / 3'-0 1/2" (927) / 31 1/8" (791)	CR13	CN13	C13	CW13	CXW13	CR23	CN23	C23	CW23	CXW23	C33
3'-4 13/16" (1037) / 3'-5 3/8" (1051) / 36" (914)	CR135	CN135	C135	CW135†♦	CXW135♦	CR235	CN235	C235	CW235†♦	CXW235†♦	C335
4'-0" (1219) / 4'-0 1/2" (1232) / 43 3/16" (1097)	CR14	CN14	C14	CW14†♦	CXW14♦	CR24	CN24	C24	CW24†♦	CXW24†♦	C34
4'-4 13/16" (1341) / 4'-5 5/8" (1356) / 48" (1219)	CR145	CN145	C145	CW145†♦	CXW145♦	CR245	CN245	C245	CW245†♦	CXW245†♦	C345

* "Unobstructed Glass" measurement is for single sash only.

** These units have straight arm operators, see opening specifications.

† CW series units (except CW2, CW25 and CW3 height) open to 20" clear opening width using sill hinge control bracket. Bracket can be pivoted allowing for cleaning position. CW series units are also available with a 22" clear opening width.

♦ These units meet or exceed the following dimensions: Clear Openable Area of 5.7 sq. ft., Clear Openable Width of 20" and Clear Openable Height of 24", when appropriate hardware (straight arm or split arm) is specified.

• Andersen® art glass panels are available for all sizes on this page.

• "Unit Dimension" always refers to outside frame to frame dimension.

• Dimensions in parentheses are in millimeters.

• When ordering, be sure to specify color desired: White, Sandtone, Terratone® or Forest Green.

Left   Right   Stationary

Venting Configuration

Hinging shown on size table is standard. Specify left, right or stationary, as viewed from the outside. For other hinging of multiple units, contact your local supplier.

FIGURE 56–2 Typical page from a window manufacturer's catalog. *(Courtesy of Andersen Windows, Inc.)*

Order window units giving the type and identification letters and/or numbers found in the window schedule or manufacturer's catalog. The size of all existing rough openings should be checked to make sure they correspond to the size given in the catalog before windows are ordered.

INSTALLING WINDOWS

All rough window openings should be prepared to ensure weathertight window installations.

Housewraps and Building Paper

Exterior walls are sometimes covered with a building paper, 15# asphalt felt, prior to the application of siding. This prevents the infiltration of air into the structure. Yet, it also allows the passage of water vapor to the outside. In place of building paper, exterior walls are often covered with a type of air infiltration barrier commonly called **housewrap.**

Housewrap is a very thin, tough plastic material. It is used to cover the sheathing on exterior walls for the same purpose as building paper **(Figure 56–3).** Housewraps are commonly known by the brand names of Typar and Tyvek. They are more resistant than building paper to air leakage and are virtually tearproof. Yet they are also breathable to allow water vapor to escape. Building paper comes in 36-inch wide rolls. Housewrap rolls are 1.5, 3, 4.5, 5, 9, and 10 feet wide.

Housewrap gets its name because it is completely wrapped around the building. It covers corners, window and door openings, plates, and sills. Housewrap is designed to survive prolonged periods of exposure to the weather. It can and usually is applied immediately after framing is completed, but before doors and windows have been installed. The

wrap, then, serves also as a flashing for the sides of windows and doors.

Applying Housewrap. Begin at the corner, holding the roll vertically on the wall. Unroll it a short distance. Make sure the roll is plumb. Secure the sheet to the corner, leaving about a foot extending beyond the corner to overlap later. Continue to unroll. Make sure the sheet is straight, with no buckles. Fasten every 12 to 18 inches **(Figure 56–4).**

FIGURE 56–4   Housewrap is unrolled as it is applied to a sidewall. (*Courtesy of Reemay, Inc.*)

FIGURE 56–3
Housewrap is widely used as an air infiltration barrier on sidewalls. (*Courtesy of Reemay, Inc.*)

FIGURE 56–5 Housewrap is cut and folded in wall openings. *(Courtesy of Reemay, Inc.)*

Unroll directly over window and door openings and around the entire perimeter of the building. Overlap all joints by at least 3 inches. Secure them with a special housewrap tape. On horizontal joints, the upper layer should overlap the lower layer.

Make cuts in the housewrap from corner to corner of rough openings. Fold the triangular flaps in and around the opening **(Figure 56–5)**. Secure the flaps on the inside with fasteners spaced about every 6 inches.

⚠ CAUTION   CAUTION   CAUTION   CAUTION

Housewraps are slippery. They should not be used in any application where they will be walked on. ■

Fastening Windows

Remove all protection blocks from the window unit. Do not remove any diagonal braces applied at the factory. Close and lock the sash. If windows are stored inside, they can easily be moved through the openings and set in place **(Figure 56–6)**. It is important to center the unit in the opening on the

rough sill with the window casing overlapping the wall sheathing.

⚠ CAUTION   CAUTION   CAUTION   CAUTION

Have sufficient help when setting large units. Handle them carefully to avoid damaging the unit or breaking the glass. Broken glass can cut through protective clothing and cause serious injury. ■

Place a level on the window sill. If not level, determine which side of the window is the highest. Check to see if the high side of the unit is at the desired height. If not, bring it up by shimming with a wood shim between the rough sill and the bottom end of the window's side jamb **(Figure 56–7)**. Remove the window unit from the opening and caulk the backside of the casing or nailing flange. This will seal the unit to the housewrap. Replace unit and tack through the lower end of the casing into the sheathing on the high side. Shim the other side of the window in the same manner, so the window sill is level. Tack the lower end of the casing on that side.

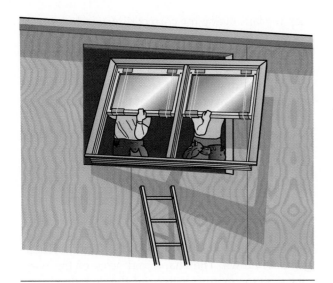

FIGURE 56–6 Windows can be installed from the inside, through the opening.

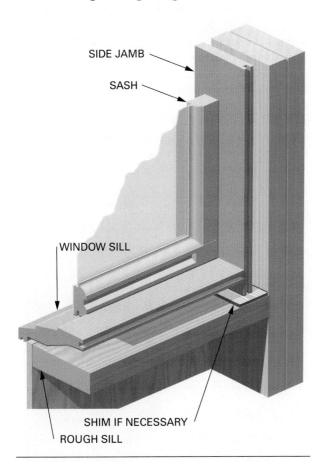

FIGURE 56–7 Use shims under the side jambs to level the window unit. (*Courtesy of Andersen Windows, Inc.*)

On wide windows with long sills, shim at intermediate points so that the sill is perfectly straight and level with no sag on wood-cased windows. Use either a long level or a shorter level in combination

with a straightedge. Also, sight the sill by eye from end to end to make sure it is straight.

Plumb the ends of the side jambs with a level. Tack the top ends of the side casings. Straighten the side jambs between sill and head jamb. Tack through the side casings at intermediate points. Straighten and tack the header casing.

Check the joint between sash and jamb. Make sure the sash operates properly. If not, make necessary adjustments. Then fasten the window permanently in place. Use galvanized casing or common nails spaced about 16 inches apart. Keep nails about 2 inches back from the ends of the casings to avoid splitting. Nail length depends on the thickness of the casing. Nails should be long enough to penetrate the sheathing and into the framing members. Set the nails so they can be puttied over later.

Vinyl-clad windows have a vinyl nailing flange. Large-head roofing nails are driven through the flange instead of through the casing **(Figure 56–8).**

Windows installed in masonry and brick veneer walls are usually attached to a treated wood buck. Adequate clearance should be left for caulking around the entire perimeter between the window and masonry **(Figure 56–9).**

Flashing the Window

The window perimeter is often flashed to the sheathing. The window flange is completely covered **(Figure 56–10).** Flashing material comes in rolls. It has a sticky side that is covered with a release paper. Begin with the bottom piece under the sill. Next do the sides and finish with the top. This will

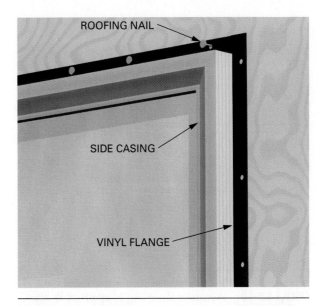

FIGURE 56–8 Roofing nails are used to fasten the flanges of vinyl-clad windows.

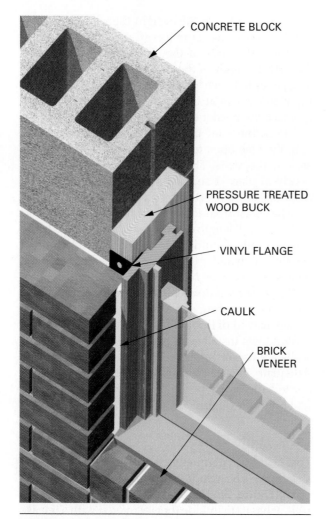

FIGURE 56–9 Windows are installed in masonry openings against wood bucks. *(Courtesy of Andersen Windows, Inc.)*

FIGURE 56–10 Window flashing is applied to the window perimeter.

keep the laps facing the correct direction in terms of any water runoff.

Additional head flashing may be applied over the window. This will help direct away any water from behind the siding **(Figure 56–11).** Its length should be equal to the overall width of the window. Do not let the ends project beyond the side casings. This will make the application of siding difficult. Place the flashing firmly on top of the head casing. Secure with fasteners along its top edge and into the wall sheathing.

METAL WINDOWS

Metal windows are available in the same styles as wood windows. The shape and sizes of the parts vary with the manufacturer and the intended use.

In frame construction, if metal windows are used, they may be set in a wood frame. The frame is then installed in the same manner as for wood windows. They may also be set in the opening with their flanges overlapping the siding or sheathing. Caulking is applied under the flanges. The unit is then screwed to the wall.

In masonry construction, wood **bucks** are fastened to the sides of the opening. Metal windows are installed against them. The flanges on the two sides and top are bedded in caulking **(Figure 56–12).** They are then screwed to the bucks.

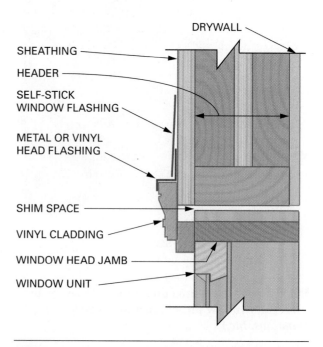

FIGURE 56–11 A drip cap is often installed as an extra layer of flashing over the top of the head casing.

FIGURE 56–12 Windows in masonry walls are fastened to window bucks.

Carefully follow the installation directions provided with the units, whether wood or metal. In areas prone to hurricanes, window installation must be approved by an inspector. This is done to ensure that every window installed is able to withstand the anticipated heavy wind loads.

CAULK

Caulk is a pliable material applied to fill gaps in building materials. It bonds to the surrounding material during cure and remains flexible after cure. This flexibility allows building materials to expand and contract with heat and moisture, maintaining the seal.

Caulks are made with a variety of materials, each with a different function. The materials include acrylic, butyl rubber, latex, polyurethane, and silicone. Many caulks are made with a mixture of these ingredients to blend the desirable characteristics

FIGURE 56–13 Caulk is made of a variety of materials.

(Figure 56–13). Read the manufacturer's recommendations on the tube to determine the best usage.

Generally speaking, acrylic and latex caulks are used when expected material movement is small. They perform best in interior applications. Silicone is used when resistance to mold and mildew are required such as in kitchens and bathrooms. Silicone can be used for interior and exterior applications and has very good flexibility. Polyurethane is designed for exterior applications where severe material movement is expected. It can fill large gaps and has superior bonding and flexibility characteristics.

Caulk performs best when it is installed with a backing material **(Figure 56–14).** Backing material

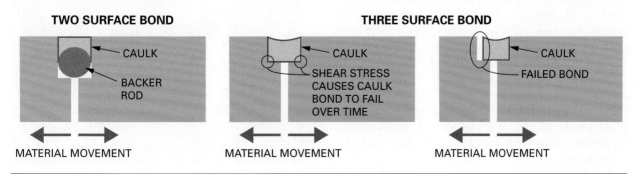

FIGURE 56–14 Large beads of caulk should be installed with a backer rod.

allows the caulk to bond to the materials on opposite sides of the joint. When the building material moves, the caulk is compressed and stretched. If the caulk is allowed to bond on three surfaces, as the material moves, the bond begins to tear in the corners and eventually the bond can be completely broken.

GLAZING

The art of cutting and installing lights of glass in sash and doors is called *glazing*. Those who do such work are called **glaziers.** Sometimes the carpenter may have to replace a light of glass.

Glazing Materials

Sashes are made so that the glass is usually held in place with **glazing points** and **glazing compound (Figure 56–15).** Glazing points are small triangular or diamond-shaped pieces of thin metal. They are driven into the sash parts to hold the glass in place. Glazing compound is commonly called **putty.** It is used to cover and seal around the edges of the light.

A light of glass is installed with its convex side, or crown of its bow, up in a thin bed of compound against the rabbet of the opening. Glass set in this manner is not as apt to break when installed.

Installing Glass

To replace or glaze a light of glass, first remove the broken glass.

CAUTION CAUTION CAUTION CAUTION

Use heavy gloves to handle the broken glass. Broken glass edges are sharp and can cut easily. ∎

Clean all compound and glazier points from the rabbeted section of the sash. Apply a thin bed of glazing compound to the surface of the rabbet on which the glass will lay. Lay the glass in the sash with its crowned side up. Carefully seat the glass in the bed by moving it back and forth slightly.

Fasten the glass in place with glazier points. Slide the driver along the glass. Do not lift the driver off the glass. Special glazier point driving tools prevent glass breakage. If a driving tool is not available, drive the points with the side of an old chisel or a putty knife.

Lay a bead of compound on the glass along the rabbet on top of the light of glass. Trim the compound at a bevel by drawing the putty knife along.

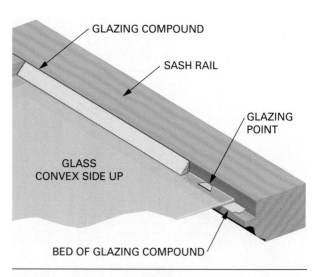

FIGURE 56–15   Panes of glass are secured with glazing points and glazing compound.

One edge should be flush with the inside edge of the glass opening. The other edge of the compound should feather to the outside edge of the opening. Prime the compound as soon as possible after glazing. Lap the paint about $\frac{1}{16}$″ onto the glass to make a seal against the weather.

Cutting Glass

Sometimes it may be necessary to cut a light of glass to size so it will fit in the opening. Lay the glass on a clean, smooth surface. Brush some mineral spirits along the line of cut. Hold a straightedge on the line. Draw a glass cutter with firm pressure along the straightedge to make a clean, uniformly scored line **(Figure 56–16).**

FIGURE 56–16   The glass must be scored with a single stroke of the glass cutter.

Do not score over the line more than once. This will dull the glass cutter. The line must be scored along the whole length the first time with no skips. Otherwise, the glass may not break where desired. Practice on scrap pieces to become proficient in making clean breaks.

Grab the glass with both hands close to the line using the thumbs and index fingers **(Figure 56–17)**. With outward pulling action and a twist of the wrist, jerk your hands up and down in a short, quick motion. The glass will break along the line. If it resists, move the glass so the scored line is even with the edge of the workbench. Apply downward pressure on the overhanging glass. If the glass is properly scored, it will break along the scored line.

FIGURE 56–17 After glass is scored completely, it will break cleanly along the scored line.

Key Terms

arch windows	extension jambs	lights	side jambs
argon gas	fixed windows	low emissivity	sill
awning window	glaziers	millwork	single-hung windows
bay window	glazing	mullion	single-strength glass
blind stops	glazing compound	muntins	skylight
bow window	glazing points	parting strip	sliding windows
bucks	grilles	putty	stiles
casement window	head casing	rails	tempered glass
caulk	head jambs	roof window	vinyl-clad
double-hung windows	hopper window	safety glass	window frame
double-strength glass	housewrap	sash	window schedule
drip cap	insulated glass	side casings	
	laminated glass		

Review Questions

Select the most appropriate answer.

1. A frame holding a pane of glass is called a
   a. light.   c. sash.
   b. mullion.   d. stile.

2. Small strips that divide the glass into smaller panes are called
   a. mantels.   c. mullions.
   b. margins.   d. muntins.

3. When windows are installed in multiple units, the joining of the side jambs forms a
   a. mantel.   c. mullion.
   b. margin.   d. muntin.

4. A window that consists of an upper and a lower sash, both of which slide vertically, is called a
   a. casement window.
   b. double-hung window.
   c. hopper window.
   d. sliding window.

5. A window that has a sash hinged on one side and swings outward is called
   a. an awning window.
   b. a casement window.
   c. a double-hung window.
   d. a hopper window.

6. The difference between a hopper and an awning window is that the hopper window
   a. swings inward instead of outward.
   b. swings outward instead of inward.
   c. is hinged at the top rather than at the bottom.
   d. is hinged on the side rather than on the bottom.

7. The term *fixed window* refers to a
   a. window with an unmovable sash.
   b. repaired window.
   c. window unit properly flashed.
   d. all of the above.

8. Single-hung windows
   a. have one sash.
   b. have one sash that moves.
   c. are installed with no other windows nearby.
   d. slide from side to side.

9. Multiple window units fastened together to form a large curved window are referred to as a
   a. bay window.
   b. bow window.
   c. double casement window.
   d. fixed double awning window.

10. Windows installed in areas prone to hurricanes should be
    a. flashed on all sides of the opening.
    b. caulked into place.
    c. inspected by an inspector.
    d. all of the above.

11. The best choice of caulk for interior bathroom applications is
    a. latex.
    b. silicone.
    c. urethane.
    d. all of the above.

12. The art of cutting and installing glass is called
    a. gauging.   c. gouging.
    b. glazing.   d. glassing.

UNIT 20

Exterior Doors

Exterior doors, like windows, are manufactured in millwork plants in a wide range of styles and sizes. Many entrance doors come prehung in frames, complete with exterior casings applied, and ready for installation. In other cases, the door is fitted and hinged to a door frame.

SAFETY REMINDER

Installing exterior doors may involve fastening large, heavy sections together. Be sure to lift and secure units properly to prevent injuries to people and damage to materials.

OBJECTIVES

After completing this unit, the student should be able to:

- name the parts of, build, and set an exterior door frame.
- describe the standard designs and sizes of exterior doors and name their parts.
- fit and hang an exterior door.
- install locksets in doors.

57 Door Frame Installation

Careful installation of a door frame and hanging of a door are essential. This contributes to smooth operation of the door and protection against the weather for many years.

PARTS OF AN EXTERIOR DOOR FRAME

Terms given to members of an exterior door frame are the same as similar members of a window frame. The bottom member is called a *sill* or **stool.** The vertical side members are called *side jambs.* The top horizontal part is a *head jamb.* The exterior door trim may consist of many parts to make an elaborate and eye-appealing entrance or a few parts for a more simple doorway. The **door casings,** if not too complex, are usually applied to the door frame before it is set **(Figure 57–1).** When more intricate trim is specified, it is usually applied after the frame is set **(Figure 57–2).**

Sills

In residential construction, door frames usually are designed and constructed for entrance doors that swing inward. Codes require that doors swing outward in buildings used by the general public. The shape of a wood door sill for an inswinging door is different from that for an outswinging door **(Figure 57–3).**

In addition to wood, extruded aluminum sills of many styles are manufactured for both inswinging and outswinging doors. They usually come with vinyl inserts to weatherstrip the bottom of the door. Some are adjustable for exact fitting between the sill and door **(Figure 57–4).**

Jambs

Side and header jambs are the same shape. Jambs may be square edge pieces of stock to which door stops are later applied. Or, they may be **rabbeted**

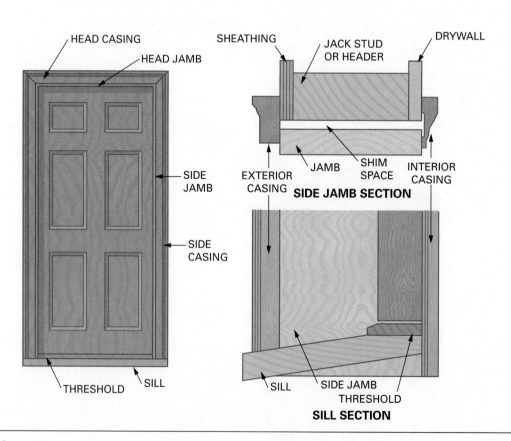

FIGURE 57–1   **Parts of an exterior door frame.**

jambs, with single or double rabbets. On double-rabbeted jambs, one rabbet is used as a stop for the main door. The other is used as a stop for storm and screen doors **(Figure 57–5).** Several jamb widths are available for different wall thicknesses. For walls of

FIGURE 57–2   Elaborate entrance trim is available. *(Courtesy of Morgan Manufacturing)*

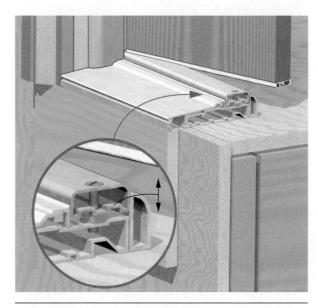

FIGURE 57–4   Some metal sills are adjustable for exact fitting at the bottom of the door.

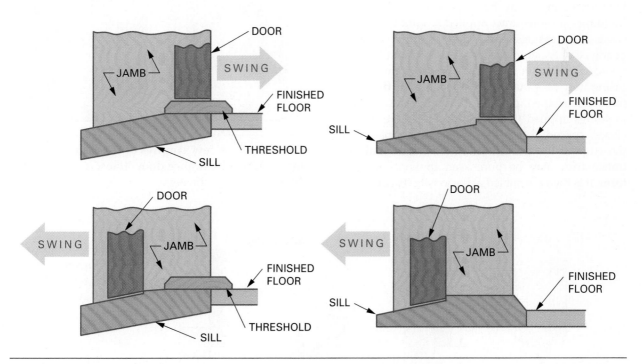

FIGURE 57–3   Wood sill shapes and styles vary according to the swing of the door.

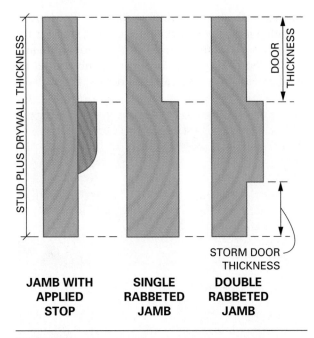

FIGURE 57–5 Door jamb cross-sections may be square edged, single rabbeted, or double rabbeted.

odd thicknesses, jambs, except double-rabbeted ones, may be ripped to any desired width.

Exterior Casings

Exterior casings may be plain square-edge stock. Moldings are sometimes applied around the outside edges of the casings. This is done to improve the appearance of the entrance. Because the main entrance is such a distinctive feature of a building, the exterior casing may be enhanced on the job with **fluted,** or otherwise shaped, pieces and appropriate caps and moldings applied **(Figure 57–6).** Flutes are narrow, closely spaced, concave grooves that run parallel to the edge of the trim. In addition, ornate main entrance trim may be purchased in knocked-down form. It is then assembled at the jobsite **(Figure 57–7).**

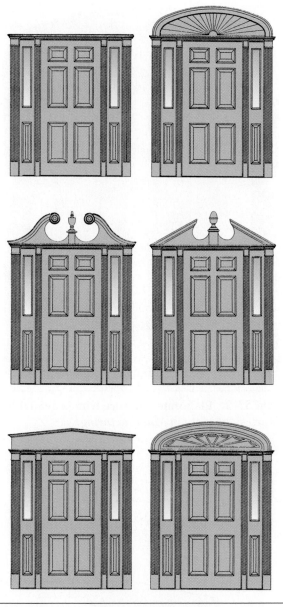

FIGURE 57–7 A few samples of the many manufactured entrance door styles. *(Courtesy of Morgan Manufacturing)*

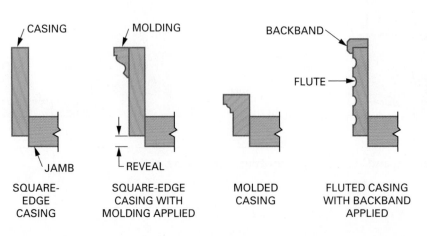

FIGURE 57–6 Exterior door casings may be enhanced by applying moldings and by shaping.

SETTING THE DOOR FRAME

When setting a door frame, cut off the horns that project beyond the sill and header jamb if necessary. Flash both sides of the rough door opening in the same manner as for windows. Run two rows of caulk on the subfloor where the door sill will sit. Set the door frame in the opening with the sill on the subfloor. Center the bottom of the frame in the opening. Level the sill by shimming under the side jambs, if necessary. Tack the lower end of both side casings to the wall.

The side jambs must be plumbed next. A 26- or 28-inch carpenter's level is not suitable when plumbing the sides because of any bow that may be in the jambs. Use either a 6-foot level or a straightedge in combination with a carpenter's level.

Block each end of the 6-foot level or straightedge with blocks of equal thickness. This will allow the level to read from the top and bottom of the frame only, not from the middle. Tack the top casing to the rough opening when plumb.

A plumb bob may also be used. To use a plumb bob, measure from the side jamb a short distance on the header jamb. Hang the plumb bob from the mark to the sill. Move the top of the frame sideways, one way or the other, until the same distance is obtained at the sill **(Procedure 57–A).** When the side jambs are plumb, tack the top of the side casings to the wall.

Straighten the jambs between the top and bottom using a long straightedge placed against their sides. Tack the side casings to the wall at intermediate points when they are straight. Make sure the head jamb is straight. Tack the header casing to the wall.

Flash under the sill before driving nails home. Flashing can then be installed under both of the side casings and the side flashings. Check all parts again for level and plumb. Then drive and set all nails. Apply any decorative molding or other trim, if specified. Install the door flashing and drip cap on the top edge of the head casing.

Setting Door Frames in Masonry Walls

In commercial construction, exterior wood or metal door frames are sometimes set in place before masonry walls are built. The frames must be set and firmly braced in a level and plumb position. The head jamb is checked for level. The bottom ends of the side jambs are secured in place. It may be necessary to

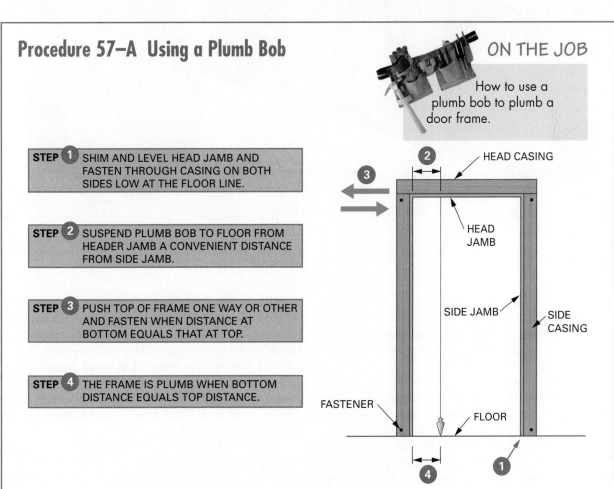

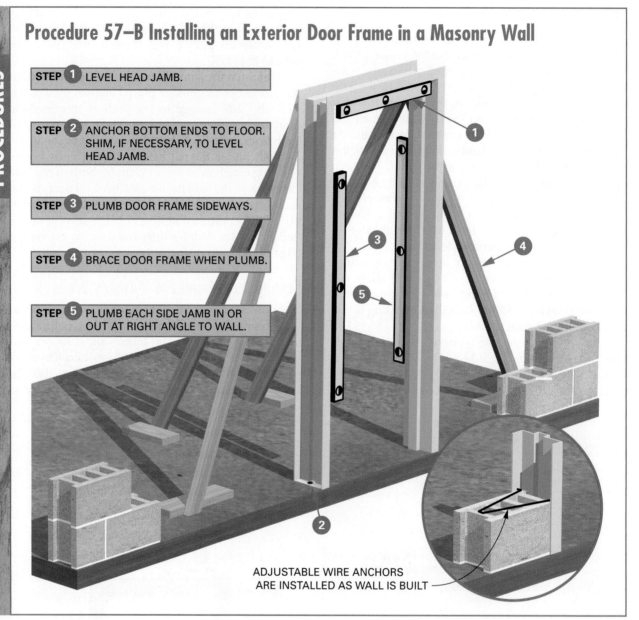

Procedure 57–B Installing an Exterior Door Frame in a Masonry Wall

STEP 1 LEVEL HEAD JAMB.

STEP 2 ANCHOR BOTTOM ENDS TO FLOOR. SHIM, IF NECESSARY, TO LEVEL HEAD JAMB.

STEP 3 PLUMB DOOR FRAME SIDEWAYS.

STEP 4 BRACE DOOR FRAME WHEN PLUMB.

STEP 5 PLUMB EACH SIDE JAMB IN OR OUT AT RIGHT ANGLE TO WALL.

ADJUSTABLE WIRE ANCHORS ARE INSTALLED AS WALL IS BUILT

shim one or the other side jamb in order to level the head jamb. The side jambs are then plumbed in a sideways direction. They are braced in position. Then, the frame is plumbed and braced at a right angle to the wall **(Procedure 57–B).**

Finally, the frame is checked to see if it has a **wind.** The term wind is pronounced the same way it's pronounced in the phrase "wind a clock." A wind is a twist in the door frame caused when the side jambs do not line up vertically with each other. No matter how carefully the side jambs of a door frame are plumbed, it is always best to check the frame to see if it has a wind.

One method of checking the door frame for a wind is to stand to one side. Sight through the door frame to see if the outer edge of one side jamb lines up with the inner edge of the other side jamb. If they do not line up, the frame is in wind. Make adjustments until they do. One way of making the adjustment is to plumb and brace one side at a right angle to the wall. Then sight, line up, and brace the other side jamb **(Figure 57–8).**

Some workers check for a wind by stretching two strings diagonally from the corners of the frame. If both strings meet accurately at their intersections, the frame does not have a wind **(Figure 57–9).**

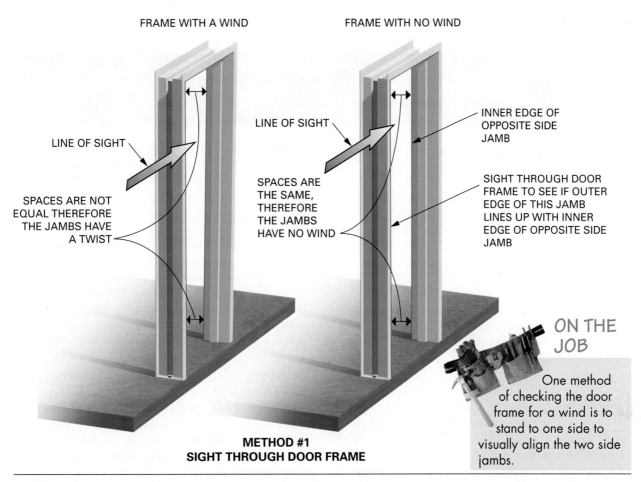

FRAME WITH A WIND

FRAME WITH NO WIND

LINE OF SIGHT

LINE OF SIGHT

INNER EDGE OF OPPOSITE SIDE JAMB

SPACES ARE NOT EQUAL THEREFORE THE JAMBS HAVE A TWIST

SPACES ARE THE SAME, THEREFORE THE JAMBS HAVE NO WIND

SIGHT THROUGH DOOR FRAME TO SEE IF OUTER EDGE OF THIS JAMB LINES UP WITH INNER EDGE OF OPPOSITE SIDE JAMB

METHOD #1
SIGHT THROUGH DOOR FRAME

ON THE JOB

One method of checking the door frame for a wind is to stand to one side to visually align the two side jambs.

FIGURE 57–8 Visual inspection method for checking for a wind or twist in a door frame.

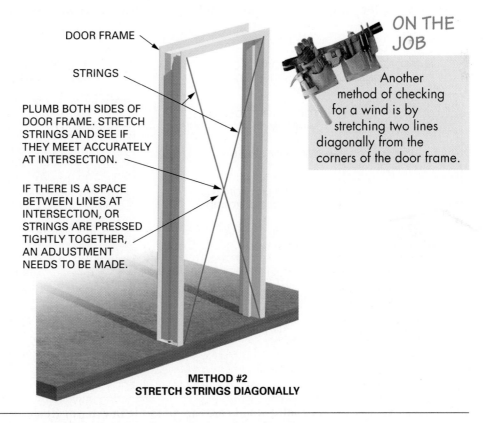

ON THE JOB

Another method of checking for a wind is by stretching two lines diagonally from the corners of the door frame.

DOOR FRAME

STRINGS

PLUMB BOTH SIDES OF DOOR FRAME. STRETCH STRINGS AND SEE IF THEY MEET ACCURATELY AT INTERSECTION.

IF THERE IS A SPACE BETWEEN LINES AT INTERSECTION, OR STRINGS ARE PRESSED TIGHTLY TOGETHER, AN ADJUSTMENT NEEDS TO BE MADE.

METHOD #2
STRETCH STRINGS DIAGONALLY

FIGURE 57–9 String method for checking for a wind or twist in a door frame.

58 Door Fitting and Hanging

Fitting and hanging of wood doors is still an important part of the carpentry trade despite the increasing use of prehung door units. There are many situations in new construction and in remodeling work that require fitting and hanging doors on the job. It is helpful to be able to identify door styles, know their sizes, and understand their construction.

DOOR STYLES AND SIZES

Exterior flush and panel doors are available in many styles. There are many choices when designing entrances.

Flush Doors

An exterior **flush door** has a smooth, flat surface of wood veneer or metal. It has a framed, *solid core* of staggered wood blocks or composition board. Wood *core blocks* are inserted in appropriate locations in composition cores. They serve as *backing* for door locks **(Figure 58–1)**. Openings may be cut in flush doors either in the factory or on the job. Lights of

various kinds and shapes are installed in them. Molding of various shapes may be applied in many designs to make the door more attractive.

Panel Doors

Panel doors are classified by one large manufacturer as *high-style, panel, sash, fire, insulated, French,* and *Dutch,* doors. **Sidelights,** although not actually doors, constitute part of some entrances. They are fixed in the door frame on one or both sides of the door **(Figure 58–2)**. **Transoms** are similar to sidelights. When used, they are installed above the door.

Panel Door Styles. High-style doors, as the name implies, are highly crafted designer doors. They may have a variety of cut-glass designs. **Panel doors** are made with raised panels of various shapes. Sash doors have panels of tempered or insulated glass that allow the passage of light **(Figure 58–3)**. Fire doors are used where required by codes. These doors prevent the spread of fire for a certain period of time. Insulated doors have thicker panels with Low E or argon-filled insulated glass. French doors may contain one,

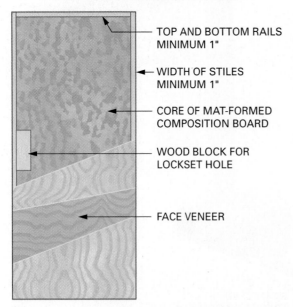

MAT-FORMED COMPOSITION BOARD CORE
7-PLY CONSTRUCTION ILLUSTRATED

- TOP AND BOTTOM RAILS MINIMUM 1"
- WIDTH OF STILES MINIMUM 1"
- CORE OF MAT-FORMED COMPOSITION BOARD
- WOOD BLOCK FOR LOCKSET HOLE
- FACE VENEER

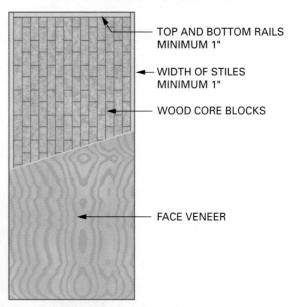

FRAMED BLOCK NON-GLUED CORE
5-PLY CONSTRUCTION ILLUSTRATED

- TOP AND BOTTOM RAILS MINIMUM 1"
- WIDTH OF STILES MINIMUM 1"
- WOOD CORE BLOCKS
- FACE VENEER

FIGURE 58–1 Composition or solid wood cores are possible in exterior flush doors. *(Courtesy of National Wood Window and Door Association)*

FIGURE 58–2 **Sidelights are installed on one or both sides of the main entrance door.** *(Courtesy of Morgan Manufacturing)*

five, ten, twelve, fifteen, or eighteen lights of glass for the total width and length inside the frame. A Dutch door consists of top and bottom units, hinged independently of each other.

Parts of a Panel Door. A panel door consists of a frame that surrounds panels of solid wood and glass, or louvers. Some door parts are given the same terms as a window sash.

The outside vertical members are called **stiles.** Horizontal frame members are called **rails.** The *top rail* is generally the same width as the stiles. The *bottom rail* is the widest of all rails. A rail situated at lockset height, usually 38 inches from the finish floor to its center, is called the **lock rail.** Almost all other rails are called *intermediate rails.* **Mullions** are vertical members between rails dividing panels in a door.

The molded shape on the edges of stiles, rails, mullions, and bars, adjacent to panels, is called the **sticking.** The name is derived from the molding machine, commonly called a sticker, used to shape the parts. Several shapes are used to *stick* frame members.

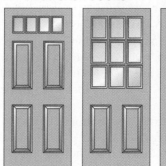

EXTERIOR PANEL DOORS

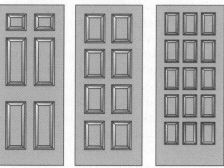

EXTERIOR SASH DOORS

SIDELIGHTS

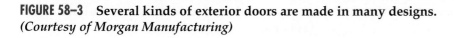

FIGURE 58–3 **Several kinds of exterior doors are made in many designs.** *(Courtesy of Morgan Manufacturing)*

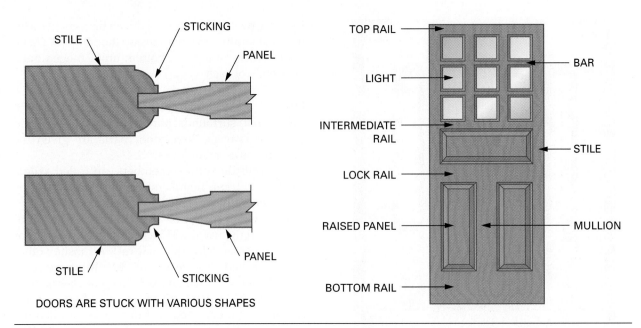

DOORS ARE STUCK WITH VARIOUS SHAPES

FIGURE 58–4   **The parts of an exterior paneled door.**

Bars are narrow horizontal or vertical rabbeted members. They extend the total length or width of a glass opening from rail to rail or from stile to stile. Door **muntins** are short members, similar to and extending from bars to a stile, rail, or another bar. Bars and muntins divide the overall length and width of the glass area into smaller lights.

Panels fit between and are usually thinner than the stiles, rails, and mullions. They may be raised on one side or on both sides for improved appearance **(Figure 58–4).**

EXTERIOR DOOR SIZES

Practically all exterior entrance doors are manufactured in a thickness of 1¾ inches, in widths of 2'–8" and 3'–0", and in heights of 6'–8" and 7'–0". Doors that are less than 6'–8" in height and less than 2'–8" in width are not used in entrances. Check local codes. A few styles are produced in widths of 2'–6" and 2'–10" and in heights of 6'–6", 6'–10", and 8'–0". Some styles, such as French, panel, and sash doors, are available in narrower widths, such as 1'–6", 2'–0", and 2'–4", when double doors are used, for instance. Sidelights are made in both 1⅜- and 1¾-inch thicknesses, and ½ inch greater in height than the entrance door. This allows for fitting the bottom against the sloping sill of the door frame.

There may be styles and sizes of stock doors other than those described above. It is not possible or desirable to describe all products here. Study door manufacturers' literature to become better informed about door styles and sizes.

PREHUNG DOORS

Most doors come already fitted, and hinged in the door frame with the **lockset** and outside casings installed. A **prehung door** unit is set in the rough opening similar to setting door frames.

Installing a Prehung Door Unit

Working from the outside, center the unit in the opening with the casings removed. The door should be in a closed position with factory-installed spacers between door and frame still in place. Level the sill. Shim between the side jambs and the jack stud at the bottom. Tack through the side jambs and shims into the stud at the bottom using finishing nails. Plumb the side jambs. Shim at the top and tack. Shim at intermediate points along the side jamb. Fasten through the shims **(Figure 58–5).**

Make sure the spacers between the door and the jamb are in place to maintain the proper joint between the door and jamb. Also make sure that the outside edge of the jamb is flush with the wall sheathing. Open the door to make sure it operates properly. Make any necessary adjustments. Drive additional nails as required. Set them all. Do not make any hammer marks on the finish. Drive the nail until it is almost flush. Then use a nail set to drive it the rest of the way **(Procedure 58–A).** Some installers prefer to leave the exterior casing on. They fasten the unit through the casing first. Then, they go inside to install the shims and additional fastening.

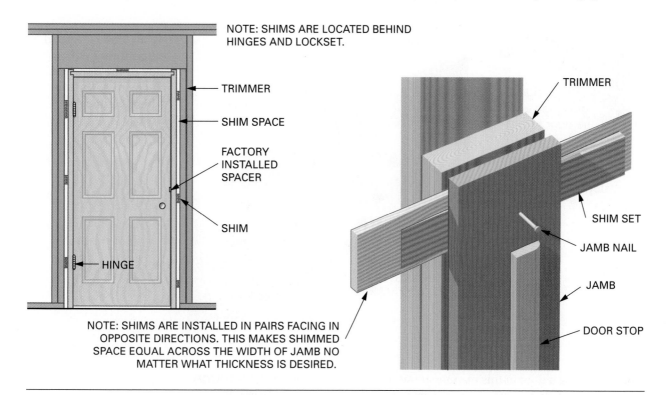

NOTE: SHIMS ARE LOCATED BEHIND HINGES AND LOCKSET.

TRIMMER

SHIM SPACE

FACTORY INSTALLED SPACER

SHIM

HINGE

NOTE: SHIMS ARE INSTALLED IN PAIRS FACING IN OPPOSITE DIRECTIONS. THIS MAKES SHIMMED SPACE EQUAL ACROSS THE WIDTH OF JAMB NO MATTER WHAT THICKNESS IS DESIRED.

TRIMMER

SHIM SET

JAMB NAIL

JAMB

DOOR STOP

FIGURE 58–5 Shimming techniques for setting a door frame into a rough opening.

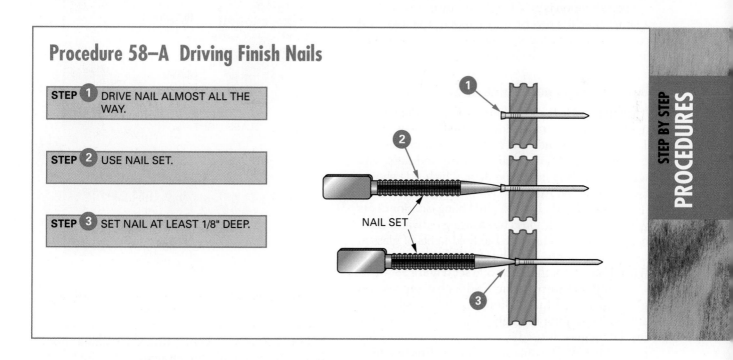

Procedure 58–A Driving Finish Nails

STEP 1 DRIVE NAIL ALMOST ALL THE WAY.

STEP 2 USE NAIL SET.

STEP 3 SET NAIL AT LEAST 1/8" DEEP.

NAIL SET

STEP BY STEP PROCEDURES

MAKING A DOOR UNIT

The first step in making a door unit is to determine the side that will close against the stops on the door frame. If the design of the door permits that either side may be used toward the stops, sight along the door stiles from top to bottom to see if the door is bowed. Hardly any doors are perfectly straight. Most are bowed to some extent.

Determining Stop Side of Door

The door should be fitted to the opening so that the hollow side of the bow will be against the door

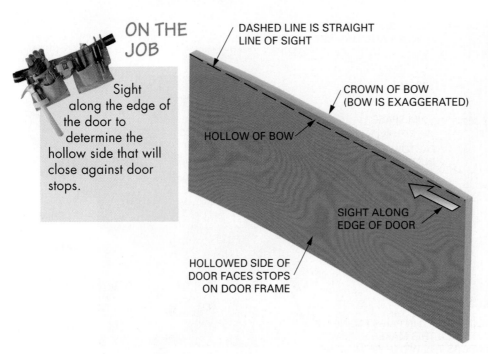

Sight along the edge of the door to determine the hollow side that will close against door stops.

DASHED LINE IS STRAIGHT LINE OF SIGHT

CROWN OF BOW (BOW IS EXAGGERATED)

HOLLOW OF BOW

SIGHT ALONG EDGE OF DOOR

HOLLOWED SIDE OF DOOR FACES STOPS ON DOOR FRAME

FIGURE 58–6 Method of determining which side of the door will rest against the top when the door is closed.

stops. Hanging a door in this manner allows the top and bottom of the closed door to come up tight against stops. The center comes up tight when the door is latched. Also, the door will not rattle.

If no attention is paid to which side the door stops against, then the reverse may happen. The door will come up against the stop at the center and away from the stop at the bottom and top **(Figure 58–6)**.

Determining Exposed Side of Sash Doors

It is important to hang exterior doors containing lights of glass with the proper side exposed to the weather. This prevents wind-driven rainwater from seeping through joints. Manufacturers clearly indicate, with warning labels glued to the door, which side should face the exterior. Any door warranty is voided if the door is improperly hung. Do not hang exterior doors with the removable **glass bead** facing outward. Glass bead is small molding used to hold lights of glass in the opening. The bead can be identified by holes made when fasteners of the bead were set **(Figure 58–7)**.

Some doors are manufactured with **compression glazing.** This virtually eliminates the possibility of any water seeping through the joints. When Low E insulating glass is used in a door, it is especially important to have the door facing in the direction indicated by the manufacturer.

Determining the Swing of the Door

Determine the swing of the door from the plans. A door is designated as being *right-hand* or *left-hand,* depending on the direction it swings. The designation is determined by standing on the side of the

SECTION THROUGH MUNTIN

MUNTIN

BEDDING COMPOUND

GLASS BEAD

GLASS LIGHT

EXTERIOR SIDE OF DOOR

INTERIOR SIDE OF DOOR

GLASS BEAD IS THE MOLDING APPLIED AROUND LIGHTS OF GLASS

FIGURE 58–7 Doors containing glass lights should face in the direction recommended by the manufacturer.

door that swings away from the viewer. It is a right-hand door if the hinges are on the viewer's right. It is a left-hand door if the hinges are on the left **(Figure 58–8)**. Lightly mark the door to identify the hinge edge and the hinge pin side of the door.

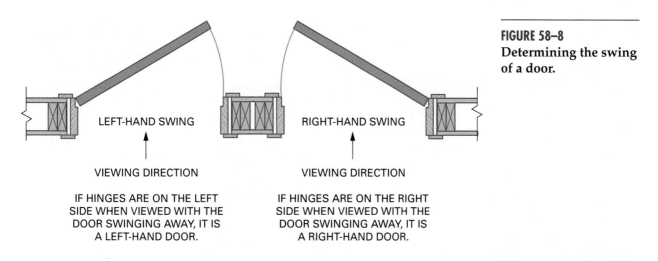

FIGURE 58–8
Determining the swing of a door.

LEFT-HAND SWING   RIGHT-HAND SWING

VIEWING DIRECTION   VIEWING DIRECTION

IF HINGES ARE ON THE LEFT
SIDE WHEN VIEWED WITH THE
DOOR SWINGING AWAY, IT IS
A LEFT-HAND DOOR.

IF HINGES ARE ON THE RIGHT
SIDE WHEN VIEWED WITH THE
DOOR SWINGING AWAY, IT IS
A RIGHT-HAND DOOR.

Fitting the Door

Place the door on sawhorses, with its face side up. Measure carefully the width and height of the door frame. The frame should be level and plumb, but this may not be the case and should not be taken for granted.

The process of fitting a door into a frame is called **jointing.** The door must be carefully jointed. An even joint of approximately ³⁄₃₂ inch must be made between the door and the frame on all sides. A wider joint of approximately ⅛ inch must be made to allow for swelling of the door and frame in extremely damp weather.

Use a **door jack** to hold the door steady. A manufactured or job-made jack may be used **(Figure 58–9).** Use a **jointer plane,** either hand or power. Plane the edges and ends of the door so that it fits snugly in the door frame.

Fit the top end against the head jamb. Then fit the bottom against the sill so that a proper joint is obtained. Careful jointing may require moving the door in and out of the frame several times.

Next, joint the hinge edge against the side jamb. Finally, plane the lock edge so the desired joint is obtained on both sides. The lock edge must also be planed on a *bevel.* This is so the back edge does not strike the jamb when the door is opened. The bevel is determined by making the same joint between the back edge of the door and the side jamb when the door is slightly open, as when the door is closed **(Figure 58–10).**

Extreme care must be taken when fitting doors not to get them undersize. The door can always be jointed a little more. However, it cannot be made wider or longer **(Figure 58–11).** Do not cut more than ½ inch total from the width of a door. Cut no more than 2 inches from its height. Cut equal amounts from ends and edges when approaching maximum amounts. Check the fit frequently by placing the

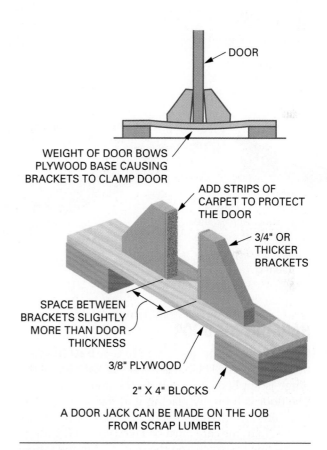

DOOR

WEIGHT OF DOOR BOWS
PLYWOOD BASE CAUSING
BRACKETS TO CLAMP DOOR

ADD STRIPS OF
CARPET TO PROTECT
THE DOOR

3/4" OR
THICKER
BRACKETS

SPACE BETWEEN
BRACKETS SLIGHTLY
MORE THAN DOOR
THICKNESS

3/8" PLYWOOD

2" X 4" BLOCKS

A DOOR JACK CAN BE MADE ON THE JOB
FROM SCRAP LUMBER

FIGURE 58–9 **Door jacks make working on a door edge easier.**

door in the opening, even if this takes a little extra time. Most entrance doors are very expensive. Care should be taken not to ruin one. Speed will come with practice. Handle the door carefully to avoid marring it or other finish. After the door is fitted, *ease* all sharp corners slightly with a block plane and sandpaper. To ease sharp corners means to round them over slightly.

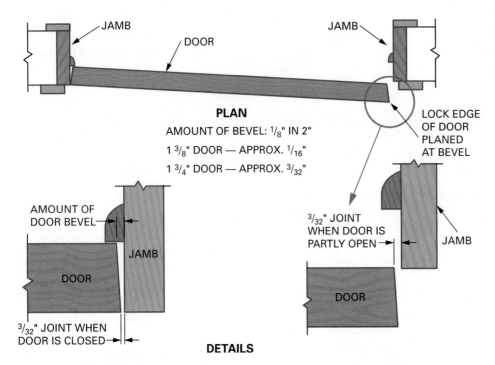

JAMB

DOOR

JAMB

PLAN

AMOUNT OF BEVEL: 1/8" IN 2"

1 3/8" DOOR — APPROX. 1/16"

1 3/4" DOOR — APPROX. 3/32"

LOCK EDGE
OF DOOR
PLANED
AT BEVEL

AMOUNT OF
DOOR BEVEL

JAMB

DOOR

3/32" JOINT
WHEN DOOR IS
PARTLY OPEN

JAMB

DOOR

3/32" JOINT WHEN
DOOR IS CLOSED

DETAILS

FIGURE 58–10 The lock edge of a door must be planed at a bevel.

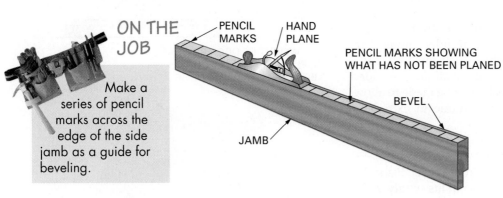

ON THE JOB

Make a series of pencil marks across the edge of the side jamb as a guide for beveling.

PENCIL
MARKS

HAND
PLANE

PENCIL MARKS SHOWING
WHAT HAS NOT BEEN PLANED

JAMB

BEVEL

FIGURE 58–11 Technique for seeing the amount of material removed while planing.

HANGING THE DOOR

On swinging doors, the loose-pin type **butt hinge** is ordinarily used. The pin is removed. Each leaf of the hinge is applied separately to the door and frame. The door is hung by placing it in the opening. The pins are inserted to rejoin the separated hinge leaves. Extreme care must be taken so that the hinge leaves line up exactly on the door and frame. Three 4 × 4 hinges on 1¾-inch doors 7′–0″, or less, in height are often used. Use four hinges on doors over 7′–0″ in height.

The hinge leaves are recessed flush with the door edge and only partway across. The recess for the hinge is called a **hinge gain** or **hinge mortise.** Hinge gains are only made partway across the edge of the door. This is so that the edge of the hinge is not exposed when the door is opened. The remaining distance, from the edge of the hinge to the side of the door, is called the **backset** of the hinge. Butt hinges must be wide enough so that the pin is located far enough be-

yond the door face to allow the door to clear the door trim when fully opened **(Figure 58–12).**

Location and Size of Door Hinges

On paneled doors, the top hinge is usually placed with its upper end in line with the bottom edge of the top rail. The bottom hinge is placed with its lower end in line with the top edge of the bottom rail. The middle hinge is centered between them.

On flush doors, the usual placement of the hinge is approximately 9 inches down from the top and 13 inches up from the bottom, as measured to the center of the hinge. A middle hinge is centered between the two **(Figure 58–13).**

Laying Out Hinge Locations

Place the door in the frame. Shim the top and bottom so the proper joint is obtained. Place shims between the lock edge of the door and side jamb of the frame.

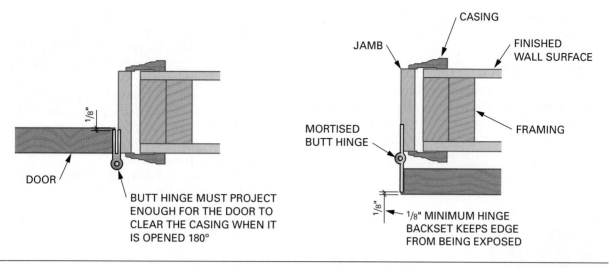

FIGURE 58–12   Hinges are set with the hinge barrel projecting out from the side of the door and jamb.

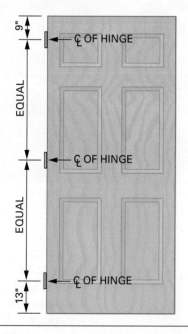

FIGURE 58–13   Recommended placement of hinges on doors.

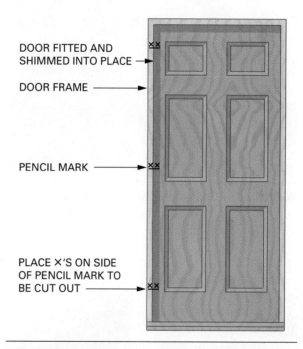

FIGURE 58–14   Mark the location of hinges on the door and frame with a sharp knife.

The hinge edge should be tightly against the door jamb.

Use a sharp knife. Mark across the door and jamb at the desired location for one end of each hinge. A knife is used because it makes a finer line than a pencil. The marks on the door and jamb should not be any longer than the hinge thickness. Place a small X, with a pencil, on both the door and the jamb. This indicates on which side of the knife mark to cut the gain for the hinge **(Figure 58–14)**. Care must be taken to cut hinge gains on the same side of the layout line on both the door and the door frame. Re-

move the door from the frame. Place it in the door jack with its hinge edge up in order to lay out and cut the hinge gains.

Laying Out the Hinge Gain

The first step in laying out a hinge gain is to mark its ends. Place a hinge leaf on the door edge with its end on the knife mark previously made. With the *barrel* of the leaf against the side of the door, hold the leaf firmly. Score a line along one end with a sharp knife. Then, tap the other end, until the leaf just covers the line. Score a line along the other end **(Figure 58–15).** Score only partway across the door edge.

Cutting Hinge Gains

Use a sharp knife to deepen the scored backset line. It may be necessary to draw the knife along the line several times to score to the bottom of the gain. Take care if using a chisel for scoring. Using a chisel will easily split the edge of the door **(Figure 58–16).**

With the bevel of the chisel down, cut a small chip from each end of the gain. The chips will break off at the scored end marks (see Procedure 12–B). With the flat of the chisel against the shoulders of the gain, cut down to the bottom of the gain.

Make a series of small chisel cuts along the length of the gain. Brush off the chips. Then, with the flat of the chisel down, pare and smooth the excess down to the depth of the gain. Be careful not to slip and cut off the backset **(Figure 58–17).**

After the gain is made, the hinge leaf should press-fit into it. It should be flush with the door edge. If the hinge leaf is above the surface, deepen the gain until the leaf lies flush. If the leaf is below

FIGURE 58–15   The ends of hinge gains may be laid out by scoring with a knife along a hinge leaf.

FIGURE 58–17   Chiseling out the hinge gain.

ON THE JOB

Use a knife to deepen the scored backset line when mortising for hinges.

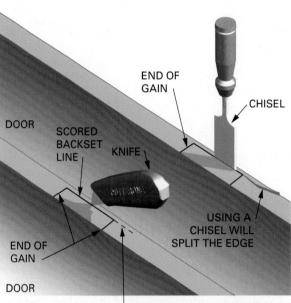

FIGURE 58–16   A utility knife is useful in deepening the cut that is parallel to the grain.

END OF GAIN

CHISEL

DOOR

SCORED BACKSET LINE

KNIFE

END OF GAIN

DOOR

USING A CHISEL WILL SPLIT THE EDGE

DO NOT SCORE OR DEEPEN BACKSET LINE BEYOND ENDS OF GAIN

the surface, it may be shimmed flush with thin pieces of cardboard from the hinge carton.

Applying Hinges

Press the hinge leaf in the gain. Drill pilot holes for the screws. Center the pilot holes carefully on the countersunk holes of the hinge leaf. A centering punch is often used for this purpose. Drilling off-center will cause the hinge to move from its position when the screw is driven. Fasten the hinge leaf with screws provided with the hinges. Cut gains and apply all hinge leafs on the door and frame in the same manner.

Shimming Hinges

Hang the door in the frame by inserting a pin in the barrels of the top hinge leaves first. Insert the other pins. Try the swing of the door. If the door binds against or is too close to the jamb on the hinged side, shim between the hinge leaves and the gain. Use a narrow strip of cardboard on the side of the screws nearest the pin. This will move the door toward the lock side of the door frame. If the door binds against or is too close to the jamb on the lock side, apply shims in the same manner. However, apply them on the opposite side of the hinge screws **(Figure 58–18).** Check the bevel on the lock edge of the door. Plane to the proper bevel, if necessary. Ease all sharp, exposed corners.

OTHER METHODS OF HANGING DOORS

Other methods are used to lay out and cut hinge gains. They may be faster and more efficient if the necessary tools and equipment are available.

Using Butt Hinge Markers

Instead of laying out hinge gains with a knife, butt hinge markers of several sizes are often used (see Figure 11–25). With markers, the location of the hinge leaves must still be marked on the door and jamb, as described. The width and length of the hinge gain are outlined by simply placing the marker in the proper location on the door edge. Tap it with a hammer **(Figure 58–19).** However, the depth of the gain must be scored with a butt gauge or some other gauging method. The gain is then chiseled out in the manner previously described.

Using a Butt Hinge Template and Router

A **butt hinge template** fixture or jig and portable electric router are usually used when many doors need to be hung. Most hinge routing jigs contain three adjustable templates secured on a long rod. The templates are positioned and tacked in place with pins. The jig is used on both the door and frame as a guide to rout hinge gains. The templates are adjustable for different size hinges, hinge locations, and door thicknesses. An attachment positions the hinge template jig to provide the required joint between the top of the door and the frame.

The gains are cut using a router with a special hinge mortising bit. The bit is set to cut the gain to the required depth. A template guide is attached to the base of the router. It rides against the jig template when routing the hinge gain **(Figure 58–20).**

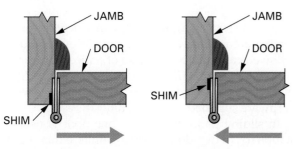

PLACING THE SHIM TOWARD THE OUTSIDE OF THE HINGE MOVES THE DOOR TOWARD THE LOCK EDGE

PLACING THE SHIM TOWARD THE INSIDE OF THE HINGE MOVES THE DOOR AWAY FROM THE LOCK EDGE

FIGURE 58–18 Shimming the hinge edges will move the door toward or away from the lock edge.

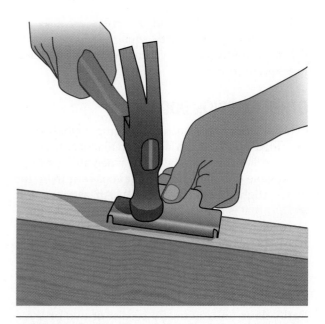

FIGURE 58–19 Sometimes butt hinge markers are used to lay out hinge gains.

FIGURE 58–20 A butt hinge template can be used on both door and frame when routing hinge gains. (*In memory of esteemed author Gaspar Lewis.*)

 CAUTION   CAUTION   CAUTION   CAUTION

Care must be taken not to clip the template with a rotating router bit when lifting the router from the template. It is best to let the router come to a stop before removing it. Both the template and bit could be damaged beyond repair. ■

Butt hinges with rounded corners are used in routed gains. Hinges with square corners require that the rounded corner of a routed gain be chiseled to a square corner.

OTHER EXTERIOR DOORS

Other exterior doors, such as swinging double doors, are fitted and hung in a similar manner to single doors. Allowance must be made, when fitting swinging double doors, for an **astragal** between them for weathertightness **(Figure 58–21)**. An astragal is a *molding* that is rabbeted on both edges. It is designed to cover the joint between double doors. One edge has a square rabbet. The other has a beveled rabbet to allow for the swing of one of the doors.

Patio Doors

Patio door units normally consist of two or three sliding or swinging glass doors completely assem-

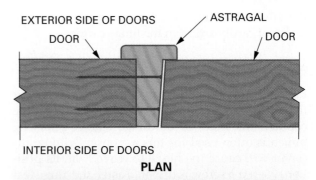

EXTERIOR SIDE OF DOORS   ASTRAGAL
DOOR   DOOR

INTERIOR SIDE OF DOORS
PLAN

FIGURE 58–21 An astragal is required between double doors for weathertightness. (*Courtesy of Andersen Corporation*)

bled in the frame **(Figure 58–22).** Instructions for assembly are included with the unit. They should be followed carefully. Installation of patio door frames is similar to setting frames for swinging doors. After the frame is set, the doors are installed using special hardware supplied with the unit.

In sliding patio door units, one door is usually stationary. In two-door swinging units, either one or the other door is the swinging door. In three-door units, the center door is usually the swinging door.

Garage Doors

Overhead garage doors come in many styles, kinds, and sizes. Two popular kinds used in residential construction are the *one-piece* and the *sectional* door.

FIGURE 58–22   Two or three doors usually are used in sliding- or swinging-type patio door units. *(Courtesy of Andersen Corporation)*

The rigid one-piece unit swings out at the bottom. It then slides up and overhead. The sectional type has hinged sections. These sections move upward on rollers and turn to a horizontal position overhead. A *rolling steel door,* used mostly in commercial construction, consists of narrow metal slats that roll up on a drum installed above the opening.

Special hardware, required for all types, is supplied with the door. Equipment is available for power operation of garage doors, including remote control. Also supplied are the manufacturers' directions for installation. These should be followed carefully. There are differences in the door design and hardware of many manufacturers.

59   Installing Exterior Door Locksets

After the door has been fitted and hung in the frame, the *lockset* and other door hardware are installed. A large variety of locks are available from several large manufacturers in numerous styles and qualities, providing a wide range of choices. Doors must be prepared to accept the lockset **(Figure 59–1).**

CYLINDRICAL LOCKSET

Cylindrical locksets are often called *key-in-the-knob locksets* **(Figure 59–2).** They are the most commonly used type in both residential and commercial con-struction. This is primarily because of the ease and speed of installing them. They may be obtained in several groups, from light-duty residential to extra-heavy-duty commercial and industrial applications.

In place of knobs, *lever handles* are provided on locksets likely to be used by handicapped persons or in other situations where a lever is more suited than a knob **(Figure 59–3). Deadbolt** locks are used for both primary and auxiliary locking of doors in residential and commercial buildings **(Figure 59–4).** They provide additional security. They also make an attractive design in combination with *grip-handle* locksets or latches **(Figure 59–5).**

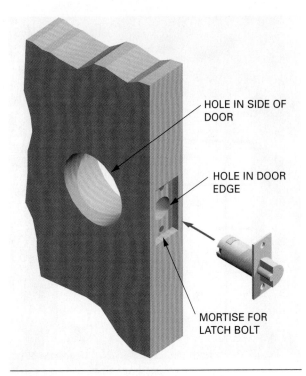

FIGURE 59–3   Locksets with lever handles are used when difficulty turning knobs is expected. *(Courtesy of Schlage Lock Co.)*

FIGURE 59–4   Deadbolt locks are used primarily as auxiliary locks for added security. *(Courtesy of Schlage Lock Co.)*

FIGURE 59–1   The method of preparing the door for cylindrical locksets.

FIGURE 59–2   Cylindrical locksets are the most commonly used type in both residential and commercial construction. *(Courtesy of Schlage Lock Co.)*

LOCK TRIM AND FINISH

The kind of trim and finish are factors, in addition to the kind and style, which determine the quality of a lockset. **Escutcheons** are decorative plates of various shapes that are installed between the door and the lock handle or knob. Locksets and escutcheons are available in various metals and finishes. More expensive locksets and trim are made of brass, bronze, or stainless steel. Less expensive ones may be of steel that is plated or coated with a finish. It is important that conditions and usage be taken into consideration when selecting locksets. This is especially the case in areas that have a humid climate or are near salt water.

FIGURE 59–5   A grip-handle lockset combines well with a deadbolt lock. *(Courtesy of Schlage Lock Co.)*

INSTALLING CYLINDRICAL LOCKSETS

To install a cylindrical lockset, first check the contents and read the manufacturer's directions carefully. There are so many kinds of locks manufactured that

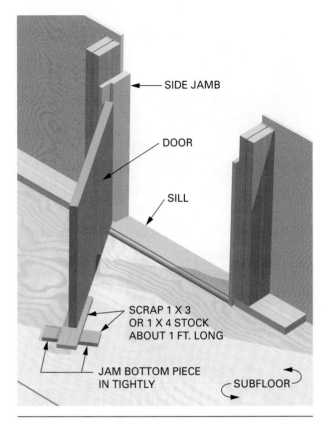

FIGURE 59–6 A door may be shimmed from the floor to hold it plumb during installation.

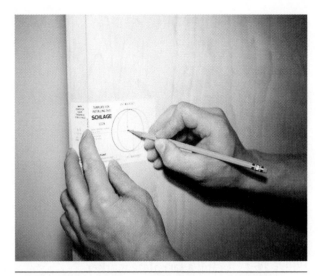

FIGURE 59–7 Using a template to lay out the centers of the holes for a lockset.

the mechanisms vary greatly. The directions included with the lockset must be followed carefully.

However, there are certain basic procedures. Open the door to a convenient position. Wedge the bottom to hold it in place **(Figure 59–6).** Measure up, from the floor, the recommended distance to the centerline of the lock. This is usually 36 to 40 inches. At this height, square a light line across the edge and stile of the door.

Marking and Boring Holes

Position the center of the paper template supplied with the lock on the squared lines. Lay out the centers of the holes that need to be bored **(Figure 59–7).** It is important that the template be folded over the high corner of the beveled door edge. The distance from the door edge to the center of the hole through the side of the door is called the *backset* of the lock. Usual backsets are 2⅜ inches for residential and 2¾ inches for commercial. Make sure the backset is marked correctly before boring the hole. One hole must be bored through the side and one into the edge of the door. The manufacturers' directions specify the hole sizes where a 1-inch hole for bolts and 2⅛-inch hole for locksets are common.

The hole through the side of the door should be bored first. Stock for the center of the boring bit is lost if the hole in the edge of the door is bored first. It can be bored with hand tools, using an expansion bit in a bit brace. However, it is a difficult job. If using hand tools, bore from one side until only the point of the bit comes through. Then bore from the other side to avoid splintering the door.

Using a Boring Jig. A **boring jig** is frequently used. It is clamped to the door to guide power-driven **multispur bits.** With a boring jig, holes can be bored completely through the door from one side. The clamping action of the jig prevents splintering **(Figure 59–8).**

After the holes are bored, insert the latchbolt in the hole bored in the door edge. Hold it firmly and

FIGURE 59–8 Boring jigs are frequently used to guide bits when boring holes for locksets.

FIGURE 59–9   **Using a faceplate marker.**

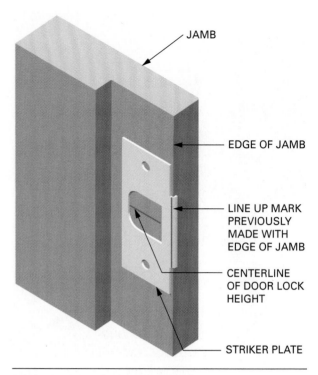

FIGURE 59–10   **Installing the striker plate.**

score around its faceplate with a sharp knife. Remove the latch unit. Deepen the vertical lines with the knife in the same manner as with hinges. Take great care when using a chisel along these lines. This may split out the edge of the door. Then, chisel out the recess so that the faceplate of the latch lays flush with the door edge.

Faceplate markers, if available, may be used to lay out the mortise for the latch faceplate. A marker of the appropriate size is held in the bored latch hole and tapped with a hammer **(Figure 59–9)**. Complete the installation of the lockset by following specific manufacturers' directions.

Installing the Striker Plate

The **striker plate** is installed on the door jamb so when the door is closed it latches tightly with no play. If the plate is installed too far out, the door will not close tightly against the stop. It will then rattle. If the plate is installed too far in, the door will not latch.

To locate the striker plate in the correct position, place it over the latch in the door. Close the door snugly against the stops. Push the striker plate in against the latch. Draw a vertical line on the face of the plate flush with the outside face of the door **(Figure 59–10)**.

Open the door. Place the striker plate on the jamb. The vertical line, previously drawn on it, should be in line with the edge of the jamb. Center the plate on the latch. Hold it firmly while scoring a line around the plate with a sharp knife. Chisel out the mortise so the plate lies flush with the jamb. Screw the plate in place. Chisel out the center to receive the latch.

Key Terms

astragal	deadbolt	glass bead	lockset
backset	door casings	hinge gain	mullions
boring jig	door jack	hinge mortise	multispur bit
butt hinge	escutcheons	jointer plane	muntins
butt hinge template	faceplate marker	jointing	panel door
compression glazing	flush door	lever handle	patio door
cylindrical lockset	fluted	lock rail	prehung door

rabbeted jamb	**sticking**	**stool**	**transom**
rail	**stile**	**striker plate**	**wind**
sidelights			

Review Questions

Select the most appropriate answer.

1. The standard thickness of exterior wood doors in residential construction is
   a. 1⅜ inches.   c. 1¾ inches.
   b. 1½ inches.   d. 2¼ inches.

2. The typical minimum width of exterior entrance doors is
   a. 3′–0″.   c. 2′–6″.
   b. 2′–8″.   d. 2′–4″.

3. The height of exterior entrance doors in residential construction is generally not less than
   a. 7′–0″.   c. 6′–8″.
   b. 6′–10″.   d. 6′–6″.

4. The term used to describe a door frame that twisted is
   a. twist.   c. bowed.
   b. warp.   d. wind.

5. A narrow member dividing the glass in a door into smaller lights is called a
   a. bar.   c. rail.
   b. mullion.   d. stile.

6. Shims should be installed
   a. in pairs.
   b. with nails through them.
   c. behind hinges.
   d. all of the above.

7. A left-hand door is one that
   a. is installed on the left side of the room.
   b. has hinges on the left when the door swings away.
   c. has the lockset on the left when the door swings away.
   d. requires a left hand to open it.

8. Before hanging a door, sight along its length for a bow. The hollow side of the bow should
   a. face the outside.
   b. face the door stops.
   c. face the inside.
   d. be straightened with a plane.

9. The joint between the door and door frame should be close to
   a. 3⁄32 inch.   c. ¼ inch.
   b. 3⁄64 inch.   d. 3⁄16 inch.

10. The top hinges on a paneled door has its top end placed
    a. in line with the bottom of the top rail.
    b. in line with the top edge of the intermediate rail.
    c. with the hinge center 13 inches down from the top.
    d. with the hinge center 7 inches down from the top.

11. The center of the hole for a lockset in a residential door is typically set back from the door edge
    a. 2 inches.
    b. 2⅛ inches.
    c. 2⅜ inches.
    d. 2¾ inches.

12. The lockset hole diameter in residential doors is typically
    a. 2 inches.
    b. 2⅛ inches.
    c. 2⅜ inches.
    d. 2¾ inches.

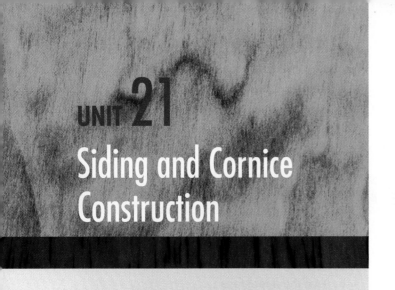

UNIT 21

Siding and Cornice Construction

The exterior finish work is the major visible part of the architectural design of a building. Because the exterior is so prominent, it is important that all finish parts be installed straight and true with well-fitted joints.

The portion of the finish that covers the vertical area of a building is the siding. Siding does not include masonry covering, such as stucco or brick veneer. Siding is used extensively in both residential and commercial construction.

That area where the lower portion of the roof, or eaves, overhangs the walls is called the cornice. Variations in cornice design and detail can set the appearance of one building apart from another.

Protecting a building from weather is the major function of exterior finish. Use care to install siding with the nature of rain and wind-blown water in mind. This will maximize the life of the finish and the building.

OBJECTIVES

After completing this unit, the student should be able to:

- describe the shapes, sizes, and grades of various siding products.
- install corner boards and prepare sidewalls for siding.
- apply horizontal and vertical lumber siding.
- apply plywood and hardboard panel and lap siding.
- apply wood shingles and shakes to sidewalls.
- apply aluminum and vinyl siding.
- estimate required amounts of siding.
- describe various types of cornices and name their parts.
- install gutters and downspouts.

60 Siding Types and Sizes

S iding is manufactured from solid lumber, plywood, hardboard, aluminum, concrete, and vinyl. It comes in many different patterns. Prefinished types eliminate the need to refinish for many years, if at all. Siding may be applied horizontally, vertically, or in other directions, to make many interesting designs **(Figure 60–1).**

WOOD SIDING

The natural beauty and durability of solid wood have long made it an ideal material for siding. The *Western Wood Products Association (WWPA)* and the *California Redwood Association (CRA)* are two major organizations whose member mills manufacture siding and other wood products. They have to meet standards supervised by their associations. Grade stamps of the WWPA and CRA and other associations of lumber manufacturers are placed on siding produced by member mills. Grade stamps ensure the consumer of a quality product that complies with established standards **(Figure 60–2).** WWPA member mills produce wood siding from species such as fir, larch, hemlock, pine, spruce, and cedar. Most redwood siding is produced by mills that belong to the CRA.

Wood Siding Grades

Grain. Some siding is available in *vertical grain, flat grain,* or *mixed grain.* In **vertical grain** siding the an-

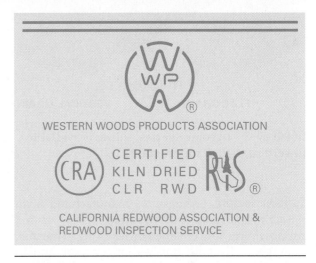

FIGURE 60–2   Association trademarks ensure that the product on which they appear has met established standards of quality. (Courtesy of Western Wood Products Association; Courtesy of California Redwood Association)

nual growth rings, viewed in cross-section, must form an angle of 45 degrees or more with the surface of the piece. All other lumber is classified as **flat grain (Figure 60–3).** Vertical grain siding is the highest quality. It warps less, takes and holds finishes better, has fewer defects, and is easier to work.

FIGURE 60–1   Wood siding is used in both residential and commercial construction. (Courtesy of California Redwood Association)

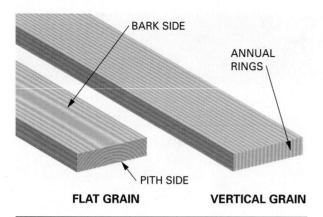

BARK SIDE

ANNUAL RINGS

PITH SIDE

FLAT GRAIN   **VERTICAL GRAIN**

FIGURE 60–3   **In some species, siding is available in vertical, flat, or mixed grain.**

Surface Texture. Sidings are manufactured with *smooth, rough,* or *saw-textured* surfaces. Saw-textured surface finishes are obtained by resawing in the mill. They generally hold finishes longer than smooth surfaces.

WWPA Grades. Grades published by the WWPA for siding products are shown in **Figure 60–4.** Siding

graded as *premium* has fewer defects such as knots and pitch pockets. The highest premium grade is produced from clear, all-heart lumber. *Knotty* grade siding is divided into #1, #2, and #3 *common*. The grade depends on the type and number of knots and other defects.

CRA Grades. There are over thirty grades of redwood lumber. The best grades are grouped in a category called *architectural.* They are used for high-quality exterior and interior uses, including siding **(Figure 60–5).**

Siding Patterns and Sizes

The names, descriptions, and sizes of siding patterns are shown in **Figure 60–6.** Some patterns can only be used for either horizontal or vertical applications. Others can be used for both. *Drop* and *tongue-and-grooved* sidings are manufactured in a variety of patterns other than shown. *Bevel* siding, more commonly known as *clapboards,* is a widely used kind **(Figure 60–7).**

PANEL AND LAP SIDING

Most *panel* and *lap* siding is manufactured from plywood, hardboard, and fiber-cement boards. They come in a variety of sizes, patterns, and surface textures. Plywood siding manufactured by American Plywood Association (APA) member mills is known as APA303 siding. It is produced in a variety of surface textures and patterns **(Figure 60–8).**

General Categories (Note that there are additional grades for bevel pattern)		Grades		
		Western Species		**Cedar**
		Selects	**Finish**	**Western & Canadian**
All Patterns	**Premium Grades**	C Select	Superior	Clear Heart A Grade
		D Select	Prime	
				B Grade
Additional Grades for Bevel Patterns	**Premium**		Superior Bevel	Clear VG Heart A Bevel B Bevel Rustic C Bevel
	Knotty		Prime Bevel	Select Knotty Quality Knotty
All Patterns	**Knotty Grades**	Commons	Alternate Boards	
		#2 Common	Select Merch.	
		#3 Common	Construction	Select Knotty Quality Knotty
			Standard	

FIGURE 60–4   **WWPA grade rules for siding products.** (*Courtesy of Western Wood Products Association*)

Grade	Description
Clear All Heart	A superior grade for fine sidings and architectural uses. It is all heartwood and the graded face of each piece is free of knots.
Clear	Similar in quality to clear all heart, except that it includes sapwood in varying amounts. Some boards may have one or two small, tight knots on the graded face.
Heart B	An economical all-heartwood grade containing a limited number of tight knots and characteristics not permitted in clear or clear all heart. It is graded on one face and one edge.
B Grade	An economical grade containing a limited number of tight knots with sapwood accenting the heartwood.

Redwood grades are established by the redwood inspection service

FIGURE 60–5   **Some redwood siding grades.** (*Courtesy of California Redwood Association*)

		Nominal Sizes
Siding Patterns		**Thickness & Width**
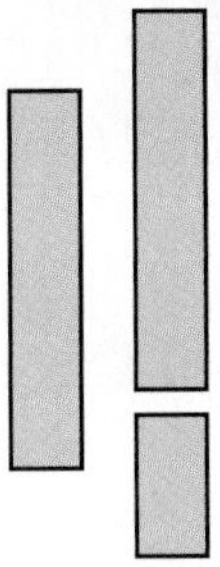	**TRIM** **BOARD-ON-BOARD** **BOARD-AND-BATTEN** Boards are surfaced smooth, rough or saw-textured. Rustic ranch-style appearance. Provide horizontal nailing members. Do not nail through overlapping pieces. Vertical applications only.	1 × 2 1 × 4 1 × 6 1 × 8 1 × 10 1 × 12 1¼ × 6 1¼ × 8 1¼ × 10 1¼ × 12
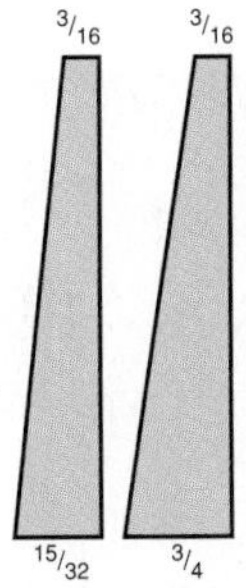	**BEVEL OR BUNGALOW** Bungalow ("Colonial") is slightly thicker than Bevel. Either can be used with the smooth or saw-faced surface exposed. Patterns provide a traditional-style appearance. Recommend a 1" overlap. Do not nail through overlapping pieces. Horizontal applications only. Cedar Bevel is also available in ⅞ × 10, 12.	½ × 4 ½ × 5 ½ × 6 ⅝ × 8 ⅝ × 10 ¾ × 6 ¾ × 8 ¾ × 10
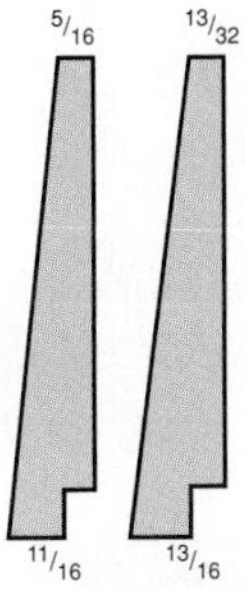	**DOLLY VARDEN** Dolly Varden is thicker than bevel and has a rabbeted edge. Surfaced smooth or saw-textured. Provides traditional-style appearance. Allows for ½" overlap, including an approximate ⅛" gap. Do not nail through overlapping pieces. Horizontal applications only. Cedar Dolly Varden is also available ⅞" × 10, 12.	Standard Dolly Varden ¾ × 6 ¾ × 8 ¾ × 10 Thick Dolly Varden 1 × 6 1 × 8 1 × 10 1 × 12
	**DROP** Drop siding is available in 13 patterns, in smooth, rough and saw-textured surfaces. Some are T&G (as shown), others are shiplapped. A variety of looks can be achieved with the different patterns. Do not nail through overlapping pieces. Horizontal or vertical applications.	¾ × 6 ¾ × 8 ¾ × 10

		Nominal Sizes
Siding Patterns		**Thickness & Width**
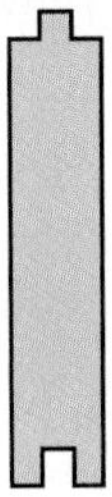	**TONGUE AND GROOVE** Tongue & groove siding is available in a variety of patterns. T&G lends itself to different effects aesthetically. Sizes given here are for Plain Tongue & Groove. Do not nail through overlapping pieces. Vertical or horizontal applications.	1 × 4 1 × 6 1 × 8 1 × 10 Note: T&G patterns may be ordered with ¼, ⅜ or ⁷⁄₁₆" tongues. For wider widths, specify the longer tongue and pattern.
	**CHANNEL RUSTIC** Channel Rustic has ½" overlap (including an approximate ⅛" gap) and a 1" to 1¼" channel when installed. The profile allows for maximum dimensional change without adversely affecting appearance in climates of highly variable moisture levels between seasons. Available smooth, rough or saw-textured. Do not nail through overlapping pieces. Horizontal or vertical applications.	¾ × 6 ¾ × 8 ¾ × 10
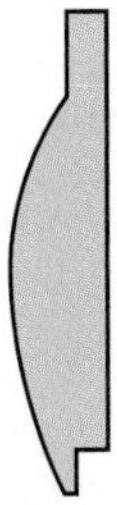	**LOG CABIN** Log Cabin siding is 1½" thick at the thickest point. Ideally suited to informal buildings in rustic settings. The pattern may be milled from appearance grades (Commons) or dimension grades (2× material). Allows for ½" overlap, including an approximately ⅛" gap. Do not nail through overlapping pieces. Horizontal or vertical applications.	1½ × 6 1½ × 8 1½ × 10 1½ × 12

FIGURE 60–6 Names, descriptions, and sizes of natural wood siding patterns. *(Courtesy of Western Wood Products Association)*

FIGURE 60–7 Bevel siding is commonly known as clapboards. *(Courtesy of California Redwood Association)*

BRUSHED

Brushed or relief-grain surfaces accent the natural grain pattern to create striking textured surfaces. Generally available in 11/32", 3/8", 1/2", 19/32" and 5/8" thicknesses. Available in redwood, Douglas fir, cedar, and other species.

KERFED ROUGH-SAWN

Rough-sawn surface with narrow grooves providing a distinctive effect. Long edges shiplapped for continuous pattern. Grooves are typically 4" OC. Also available with grooves in multiples of 2" OC Generally available in 11/32", 3/8", 1/2", 19/32" and 5/8" thicknesses. Depth of kerfgroove varies with panel thickness.

APA TEXTURE 1-11

Special 303 Siding panel with shiplapped edges and parallel grooves 1/4" deep, 3/8" wide; grooves 4" or 8" OC are standard. Other spacings sometimes available are 2", 6" and 12" OC, check local availability. T 1-11 is generally available in 19/32" and 5/8" thicknesses. Also available with scratch-sanded, overlaid, rough-sawn, brushed and other surfaces. Available in Douglas fir, cedar, redwood, southern pine, other species.

ROUGH-SAWN

Manufactured with a slight, rough-sawn texture running across panel. Available without grooves, or with grooves of various styles; in lap sidings, as well as in panel form. Generally available in 11/32", 3/8", 1/2", 19/32" and 5/8" thicknesses. Rough-sawn also available in Texture 1-11, reverse board-and-batten (5/8" thick), channel groove (3/8" thick), and V-groove (1/2" or 5/8" thick). Available in Douglas fir, redwood, cedar, southern pine, other species.

CHANNEL GROOVE

Shallow grooves typically 1/16" deep, 3/8" wide, cut into faces of 3/8" thick panels, 4" or 8" OC. Other groove spacings available. Shiplapped for continuous patterns. Generally available in surface patterns and textures similar to Texture 1-11 and in 11/32", 3/8" and 1/2" thicknesses. Available in redwood, Douglas fir, cedar, southern pine and other species.

REVERSE BOARD-AND-BATTEN

Deep, wide grooves cut into brushed, roughsawn, coarse sanded or other textured surfaces. Grooves about 1/4" deep, 1" to 1-1/2" wide, spaced 8", 12" or 16" OC with panel thickness of 19/32" and 5/8". Provides deep, sharp shadow lines. Long edges shiplapped for continuous pattern. Available in redwood, cedar, Douglas fir, southern pine and other species.

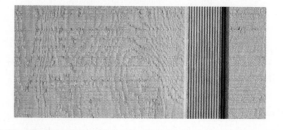

FIGURE 60–8 APA303 plywood panel siding is produced in a wide variety of sizes, surface textures, and patterns. *(Courtesy of American Plywood Association)*

Hardboard siding is made of high-density fiberboard with a hard tempered surface. It typically comes primed or prefinished in a variety of colors. It also comes in a variety of surface styles.

Fiber-cement siding is made of a fiber-reinforced cementitious material. It also comes in a variety of finishes and has excellent decay and termite resistance properties. Special considerations must be made when cutting this type of siding. Cutting of fiber-cement boards produces silica dust, which is known to cause cancer. All cutting should be done with a special dust-reducing circular saw blade or a set of shears. Cutting should be done outside and downwind of any other workers.

Panel Siding

Panel siding comes in 4-foot widths and lengths of 8 and 10 feet, and sometimes 9 feet. It is usually applied vertically, but may also be applied horizontally. Most panel siding is shaped with shiplapped edges for weathertight joints.

Lap Siding

Lap siding is applied horizontally and is manufactured in styles with rough-sawn, weathered wood grain, or other embossed surface textures so that they look like wood. Some surfaces are grooved or beaded with square or beveled edges **(Figure 60–9)**. They come in widths from 6 to 12 inches, and lengths of 12 or 16 feet.

Lap siding should overlap by at least 1¼ inches. It may be face nailed or blind nailed **(Figure 60–10)**. Check local codes for feasibility. Blind nails are placed 1 inch down from the top. Face nails are placed 1 inch up from the bottom. Nails on hardboard siding should penetrate only one layer of siding. This will allow the material to expand and contract with moisture. Fiber-cement siding may have two nails per board in vertical alignment.

Nails are driven just snug or flush with the surface. If nails are driven too deep, they lose holding

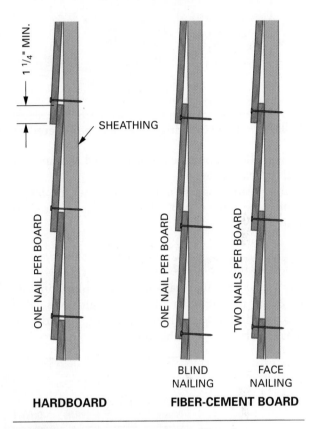

FIGURE 60–10   Recommended nailing for hardboard and fiber-cement board lap siding.

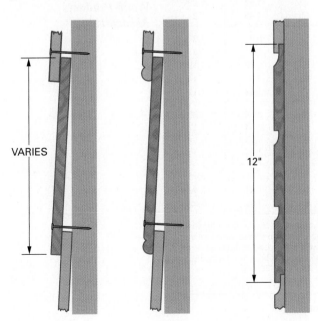

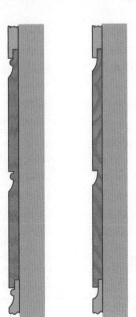

FIGURE 60–9   Styles of horizontal lap siding.

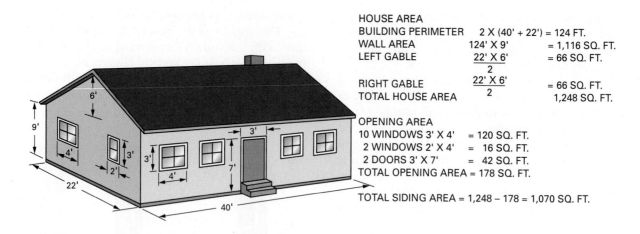

HOUSE AREA
BUILDING PERIMETER	2 X (40' + 22')	= 124 FT.
WALL AREA	124' X 9'	= 1,116 SQ. FT.
LEFT GABLE	$\frac{22' \times 6'}{2}$	= 66 SQ. FT.
RIGHT GABLE	$\frac{22' \times 6'}{2}$	= 66 SQ. FT.
TOTAL HOUSE AREA		1,248 SQ. FT.

OPENING AREA
10 WINDOWS 3' X 4'	= 120 SQ. FT.
2 WINDOWS 2' X 4'	= 16 SQ. FT.
2 DOORS 3' X 7'	= 42 SQ. FT.
TOTAL OPENING AREA	= 178 SQ. FT.

TOTAL SIDING AREA = 1,248 − 178 = 1,070 SQ. FT.

FIGURE 60–11 **Estimating the area to be covered by siding.**

power. These should be caulked closed and another nail driven nearby.

Butt seams may be made with either moderate contact or with a gap. The gap is later caulked shut.

ESTIMATING SIDING

First, calculate the wall area by multiplying its length by its height. To calculate the area of a gable end, multiply the length by the height. Then divide the result by two. Add the square foot areas together with the areas of other parts that will be sided, such as dormers, bays, and porches.

Because windows and doors will not be covered, their total surface areas must be deducted. Multiply the width by the height (large openings only). Subtract the results from the total area to be covered by siding **(Figure 60–11).**

The amount of siding to order depends on the kind of siding used. When calculating, use the *factor* for either linear feet or board feet, according to the way it is sold **(Figure 60–12).** Multiply square feet by factor to convert to lineal feet or board feet.

FIGURE 60–12
Estimating information for converting area to dimensions required for material purchase. *(Courtesy of Western Wood Products Association)*

Pattern	Nominal Width	Width		Factor for Converting SF to Lineal Feet	Factor for Converting SF to Board Feet
		Dressed	Exposed Face		
Bevel & Bungalow	4	3½	2½	4.8	1.60
	6	5½	4½	2.67	1.33
	8	7¼	6¼	1.92	1.28
	10	9¼	8¼	1.45	1.21
Dolly Varden	4	3½	3	4.0	1.33
	6	5½	5	2.4	1.2
	8	7¼	6¾	1.78	1.19
	10	9¼	8¾	1.37	1.14
	12	11¼	10¾	1.12	1.12
Drop T&G & Channel Rustic	4	3⅜	3⅛	3.84	1.28
	6	5⅜	5⅛	2.34	1.17
	8	7⅛	6⅞	1.75	1.16
	10	9⅛	8⅞	1.35	1.13
Log Cabin	6	5⁷⁄₁₆	4¹⁵⁄₁₆	2.43	2.43
	8	7⅛	6⅝	1.81	2.42
	10	9⅛	8⅝	1.39	2.32
Boards	2	1½	The exposed face width will vary depending on size selected and on how the boards-and-battens or boards-on-boards are applied.		
	4	3½			
	6	5½			
	8	7¼			
	10	9¼			

Coverage Estimator

61 Applying Vertical and Horizontal Siding

The method of siding application varies with the type. This chapter describes the application procedures for the most commonly used kinds of solid and engineered wood siding.

PREPARATION FOR SIDING APPLICATION

To maximize the sizing and paint longevity, drying potential must be built into the siding. This can be achieved by furring the siding off the sheathing and allowing air to circulate behind the siding **(Figure 61–1)**. As air circulates behind the siding, it mixes with soffit air and eventually leaves at the ridge.

Furring may be 1×3, 1×4, or 4-inch strips of plywood nailed to the sheathing over each stud. Screen is used to protect the airspace from insects. Furring should be installed before windows are installed. This maintains normal exterior finish details.

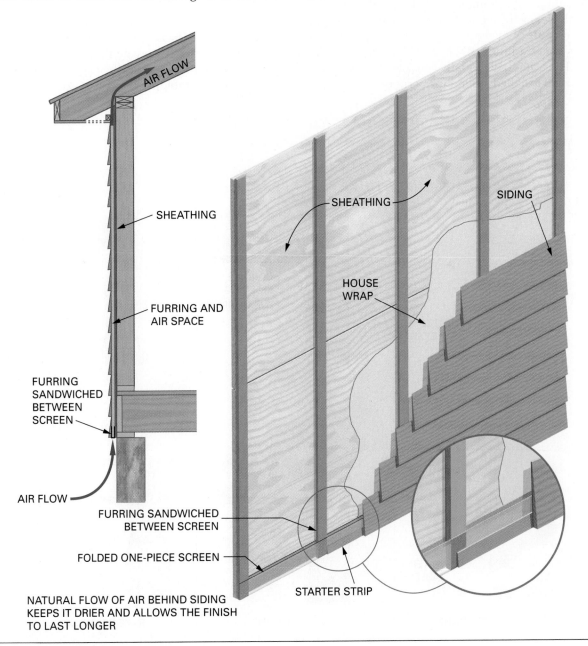

FIGURE 61–1 Siding lasts longer when it is furred off the sheathing.

A 6- to 8-inch-wide screen is smoothly stapled to the sheathing about 4 inches above the bottom edge of the siding. The extra screen is folded up and over after the furring is applied. Caulking or adhesive applied to the screen before the siding is applied seals the screen to the backside of the siding.

Next, it must be determined how the siding will be ended or treated at the foundation, eaves, and corners. The installation of various kinds of exterior wall trim may first be required.

Foundation Trim

In most cases, no additional trim is applied at the foundation. The siding is started so that it extends slightly below the top of the foundation. However, a **water table** may be installed for appearance. It sheds water a little farther away from the foundation. The water table usually consists of a board and a drip cap installed around the perimeter. Its bottom edge is slightly below the top of the foundation. The siding is started on top of the water table.

Eaves Treatment

At the eaves, the siding may end against the bottom edge of the frieze. The width of the frieze may vary. It is necessary to know its width when laying out horizontal siding courses. The siding may also terminate against the soffit, if no frieze is used. The joint between is then covered by a cornice molding. The size of the molding must be known to plan exposures on courses of horizontal siding.

Rake Trim

At the rakes, the siding may be applied under a furred-out rake fascia. When the **rake** overhangs the sidewall, the siding may be fitted against the **rake frieze.** When fitted against the rake soffit the joint is covered with a molding **(Figure 61–2).**

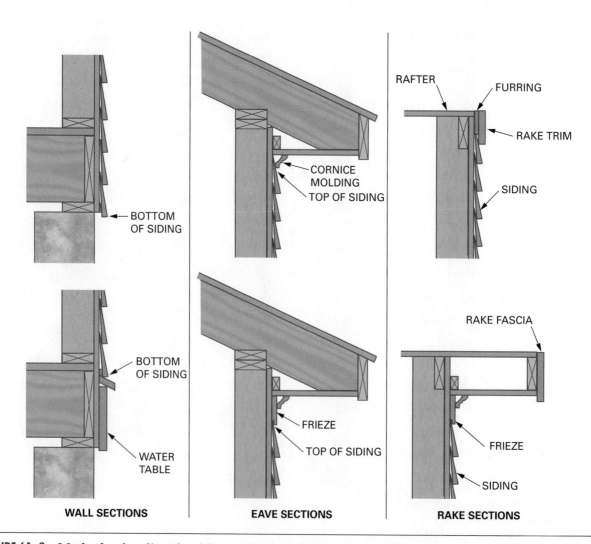

FIGURE 61-2 Methods of ending the siding at the foundation, eaves, and rakes.

Gable End Treatment

Sometimes a different kind of siding is used on the gable ends than on the sidewalls below the plate. The joint between the two types must be weathertight. One method of making the joint is to use a drip cap and flashing between the two types of material.

In another method, the plate and studs of the gable end are extended out from the wall a short distance. This allows the gable end siding to overlap the siding below **(Figure 61–3).** Furring strips may also be used on the gable end framing in place of extending the gable plate and studs.

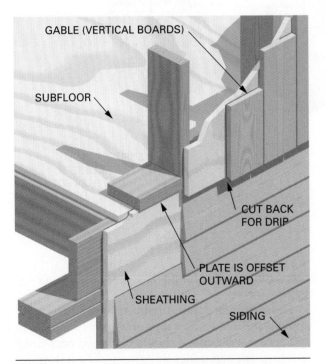

FIGURE 61–3 The upper gable end may be constructed to overhang the wall siding.

Treating Corners

One method of treating corners is with the use of **corner boards.** Horizontal siding may be mitered around exterior corners. Or, metal corners may be used on each course of siding. In interior corners, siding courses may butt against a square corner board or against each other **(Figure 61–4).** The thickness of corner boards depends on the type of siding used. The corner boards should be thick enough so that the siding does not project beyond the face of the corner board.

The width of the corner boards depends on the effect desired. However, one of the two pieces, making up an outside corner, should be narrower than the other by the thickness of the stock. Then, after the wider piece is fastened to the narrower one, the same width is exposed on both sides of the corner. The joint between the two pieces should be on the side of the building that is least viewed.

INSTALLING CORNER BOARDS

Before installing corner boards, flash both sides of the corner. Install a strip of #15 felt vertically on each side. One edge should extend beyond the edge of the corner board at least 2 inches. Tuck the top ends under any previously applied felt.

With a sharp plane, slightly back-bevel one edge of the narrower of the two pieces that make up the outside corner board. This ensures a tight fit between the two boards. Cut, fit, and fasten the narrow piece. Start at one end and work toward the bottom. Keep the beveled edge flush with the corner. Fasten with galvanized or other noncorroding nails spaced about 16 inches apart along its inside edge.

Cut, fit, and fasten the wider piece to the corner in a similar manner. Make sure its outside edge is flush with the surface of the narrower piece. The

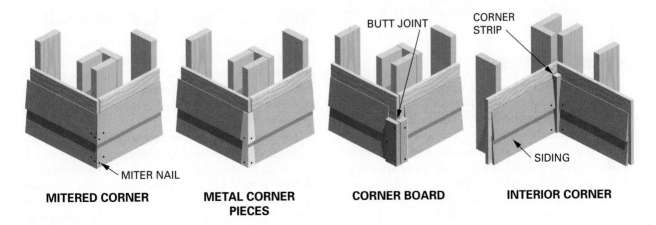

FIGURE 61–4 Methods for returning and ending courses of horizontal siding at corners.

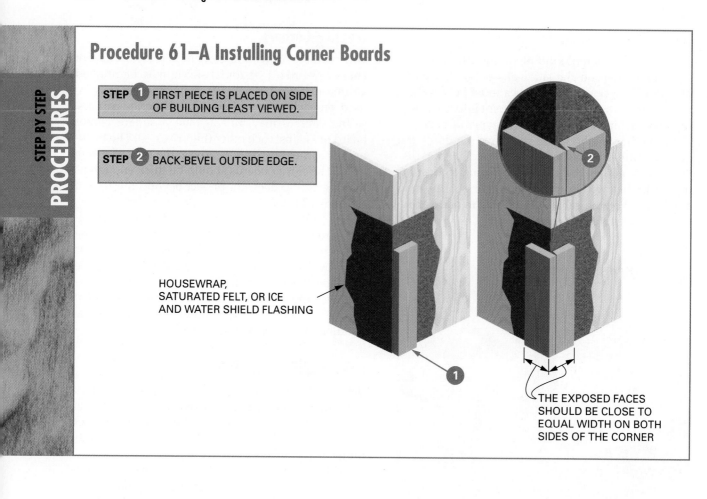

Procedure 61–A Installing Corner Boards

STEP 1 FIRST PIECE IS PLACED ON SIDE OF BUILDING LEAST VIEWED.

STEP 2 BACK-BEVEL OUTSIDE EDGE.

HOUSEWRAP, SATURATED FELT, OR ICE AND WATER SHIELD FLASHING

THE EXPOSED FACES SHOULD BE CLOSE TO EQUAL WIDTH ON BOTH SIDES OF THE CORNER

outside row of nails is driven into the edge of the narrower piece. Plane the outside edge of the wide piece wherever necessary to make it come flush. Slightly round over all sharp exposed corners by planing a small chamfer and sanding. Set all nails so they can be filled over later. Make sure a tight joint is obtained between the two pieces **(Procedure 61–A).**

Corner boards may also be applied by fastening the two pieces together first. Then install the assembly on the corner.

APPLYING HOUSEWRAP

A description of the kinds and purposes of housewrap has been previously given in Chapter 56. A *breathable* type material should be applied to sidewalls. It is typically applied to the whole building all at once. This serves as an immediate and temporary weather barrier as well as a backup water shed if a future leak develops in the siding.

Apply the paper horizontally. Start at the bottom of the wall. Make sure the paper lies flat with no wrinkles. Fasten it in position. If nailing, use large-head roofing nails. Fasten in rows near the bottom, the center, and the top about 16 inches apart. Each

succeeding layer should lap the lower layer by about 4 inches.

The sheathing paper should lap over any flashing applied at the sides and tops of windows and doors and at corner boards. It should be tucked under any flashings applied under the bottoms of windows or frieze boards. In any case, all laps should be over the paper below.

INSTALLING HORIZONTAL WOOD SIDING

One of the important differences between bevel siding and other types with tongue-and-groove, shiplap, or rabbeted edges is that exposure of courses of bevel siding can be varied somewhat. With other types, the amount exposed to the weather is constant with every course and cannot vary.

The ability to vary the exposure is a decided advantage. It is desirable from the standpoint of appearance, weathertightness, and ease of application to have a full course of horizontal siding exposed above and below windows and over the tops of doors **(Figure 61–5).** The exposure of the siding may vary gradually up to a total of ½ inch over the entire wall, but the width of each exposure should not vary more than ¼ inch from its neighbor.

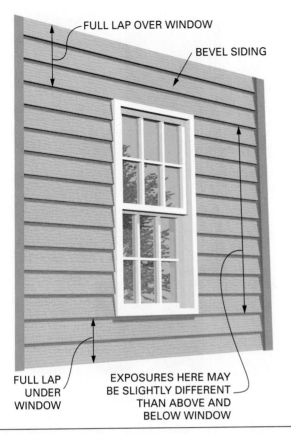

FULL LAP OVER WINDOW

BEVEL SIDING

FULL LAP UNDER WINDOW

EXPOSURES HERE MAY BE SLIGHTLY DIFFERENT THAN ABOVE AND BELOW WINDOW

FIGURE 61–5 **Bevel siding exposure may be varied from top to bottom.**

Determining Siding Exposure

To determine the siding exposure so that it is about equal both above and below the window sill, divide the overall height of the window frame by the amount of exposure. For example, consider a window that is 52 inches in height with the coverage required above and below the window being 12½ inches and 40½ inches, respectively **(Figure 61–6).** The exposure of each section may be adjusted to allow full laps above and below the window.

First, the number of courses for each section must be determined. This is done by dividing the coverage distance by the maximum exposure of the siding. Then round that number up to the next whole number of courses.

Next the number of courses is divided into the coverage distances to find the exposure for that area. Note that because windows vary in height around the house, this process does not always work out neatly. Sometimes all that can be done is to adjust the last exposure to the largest that it can be.

Layout lines are then transferred to a story pole. The story pole is used to lay out the courses all around the building **(Figure 61–7).**

Instead of calculating the number and height of siding courses mathematically, *dividers* may be used to space off the distances.

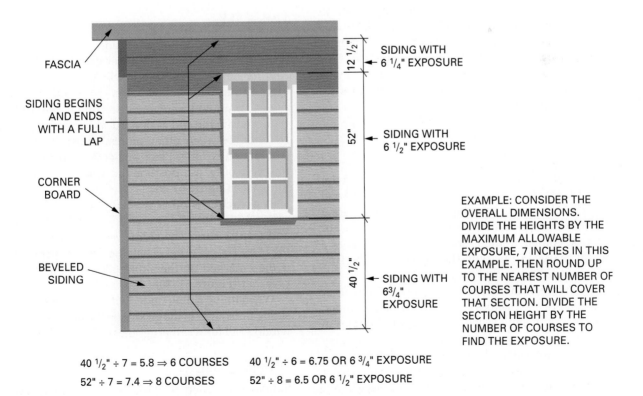

FASCIA

SIDING BEGINS AND ENDS WITH A FULL LAP

CORNER BOARD

BEVELED SIDING

12 ½" SIDING WITH 6 ¼" EXPOSURE

52" SIDING WITH 6 ½" EXPOSURE

40 ½" SIDING WITH 6¾" EXPOSURE

EXAMPLE: CONSIDER THE OVERALL DIMENSIONS. DIVIDE THE HEIGHTS BY THE MAXIMUM ALLOWABLE EXPOSURE, 7 INCHES IN THIS EXAMPLE. THEN ROUND UP TO THE NEAREST NUMBER OF COURSES THAT WILL COVER THAT SECTION. DIVIDE THE SECTION HEIGHT BY THE NUMBER OF COURSES TO FIND THE EXPOSURE.

40 ½" ÷ 7 = 5.8 ⇒ 6 COURSES   40 ½" ÷ 6 = 6.75 OR 6 ¾" EXPOSURE

52" ÷ 7 = 7.4 ⇒ 8 COURSES   52" ÷ 8 = 6.5 OR 6 ½" EXPOSURE

FIGURE 61–6 **Example of determining siding exposures around a window.**

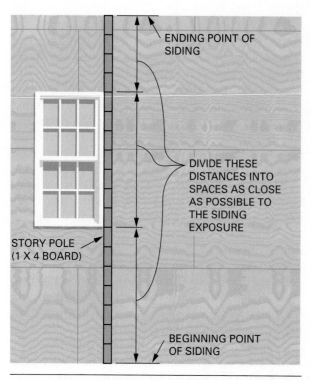

FIGURE 61–7   Laying out a story pole for courses of horizontal siding.

Starting Bevel Siding from the Top

Another advantage of using bevel siding is that application may be made starting at the top and working toward the bottom if more convenient. With this method, a number of chalk lines may be snapped without being covered by a previous course. This saves time. Any scaffolding already erected may be used and then dismantled as work progresses toward the bottom.

Starting Siding from the Bottom

Most horizontal siding, however, is usually started at the bottom. For bevel siding, a **furring strip,** called a **starter strip,** of the same thickness and width of the siding **headlap** is first fastened along the bottom edge of the sheathing **(Figure 61–8).** For the first course, a line is snapped on the wall at a height that is in line with the top edge of the first course of siding.

For siding with constant exposure, the only other lines snapped are across the tops of wide entrances, windows, and similar objects to keep the courses in alignment. For lap siding, with exposures that may vary, lines are snapped for each successive course in line with their top edge. Stagger joints in adjacent courses as far apart as possible. A small piece of felt paper often is used behind the butt seams to ensure the weathertightness of the siding.

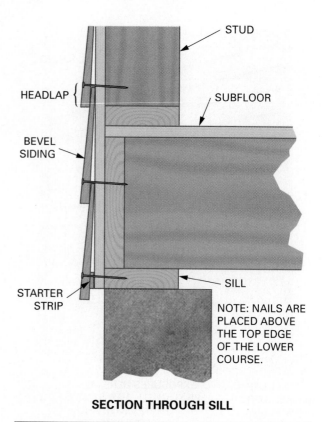

SECTION THROUGH SILL

FIGURE 61–8   For bevel siding, a strip of wood the same thickness and width of the headlap is used as a starter strip.

Fitting Siding

When applying a course of siding, start from one end and work toward the other end. With this procedure, only the last piece will have to be fitted. Tight-fitting butt joints must be made between pieces. If an end joint must be fitted, use a block plane to trim the end as needed. When a piece has to be fitted between other pieces, measure carefully. Cut it slightly long. Place one end in position. Bow the piece outward, position the other end, and snap into place **(Figure 61–9).**

A **preacher** is often used for accurate layout of siding where it butts against corner boards, casings, and similar trim. The siding is allowed to overlap the trim. The preacher is held against the trim. A line is then marked on the siding along the face of the preacher **(Figure 61–10).**

When fitting siding under windows, make sure the siding fits snugly in the groove on the underside of the window sill for weathertightness **(Figure 61–11).**

Fastening Siding

Siding is fastened to each bearing stud or about every 16 inches. On bevel siding, fasten through the butt edge just above the top edge of the course be-

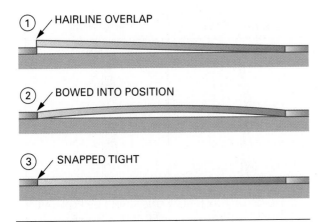

FIGURE 61–9 Method of cutting horizontal siding to fit snugly.

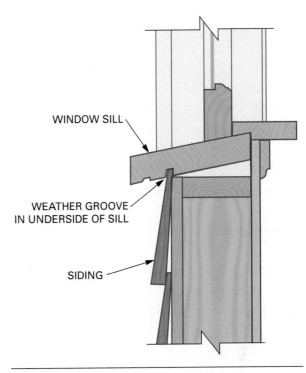

FIGURE 61–11 Fit siding into weather groove on the underside of the window sill.

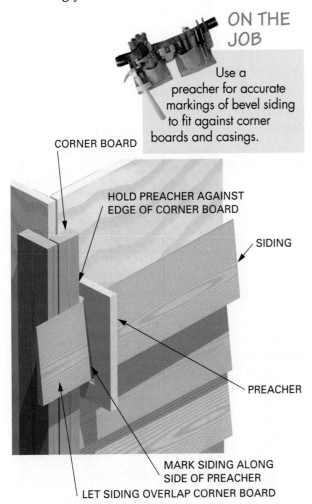

ON THE JOB

Use a preacher for accurate markings of bevel siding to fit against corner boards and casings.

FIGURE 61–10 A preacher may be used for accurately marking a siding piece for length.

low. Do not fasten through the lap. This prevents splitting of the siding that might be caused by slight swelling or shrinking due to moisture changes. Care must be taken to fasten as low as possible to avoid splitting the siding in the center. The location and

number of fasteners recommended for siding are shown in **Figure 61–12**.

VERTICAL APPLICATION OF WOOD SIDING

Bevel sidings are designed for horizontal applications only. **Board on board** and **board and batten** are applied only vertically. Almost all other patterns may be applied in either direction.

INSTALLING VERTICAL TONGUE-AND-GROOVE SIDING

Corner boards usually are not used when wood siding is applied vertically. The siding boards are fitted around the corner **(Figure 61–13).** Rip the grooved edge from the starting piece. Slightly back-bevel the ripped edge. Place it vertically on the wall with the beveled edge flush with the corner similar to making a corner board.

The tongue edge should be plumb, the bottom end should be about 1 inch below the sheathing **(Figure 61–14).** The top end should butt or be tucked under any trim above. **Face nail** the edge nearest the corner. **Blind nail** into the tongue edge. Nails should be placed from 16 to 24 inches apart. Blocking must be provided between studs if siding is applied directly to the frame.

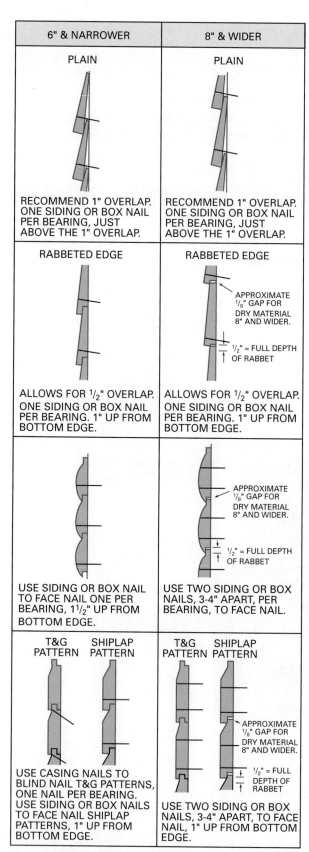

6" & NARROWER	8" & WIDER
PLAIN	PLAIN
USE ONE CASING NAIL PER BEARING TO BLIND NAIL.	USE TWO SIDING OR BOX NAILS, 3-4" APART TO FACE NAIL.
USE ONE SIDING OR BOX NAIL TO FACE NAIL ONCE PER BEARING. 1" UP FROM BOTTOM.	← APPROXIMATE 1/8" GAP FOR DRY MATERIAL 8" AND WIDER. 1/2" = FULL DEPTH OF RABBET USE TWO SIDING OR BOX NAILS, 3-4" APART, PER BEARING.
BOARD AND BATTEN RECOMMEND 1/2" OVERLAP. ONE SIDING OR BOX NAIL PER BEARING.	BOARD AND BATTEN / BOARD ON BOARD 1/2" INCREASE OVERLAP PROPORTIONATELY. USE TWO SIDING OR BOX NAILS, 3-4" APART.

SIDING USUALLY APPLIED HORIZONTALLY

6" & NARROWER	8" & WIDER
PLAIN	PLAIN
RECOMMEND 1" OVERLAP. ONE SIDING OR BOX NAIL PER BEARING, JUST ABOVE THE 1" OVERLAP.	RECOMMEND 1" OVERLAP. ONE SIDING OR BOX NAIL PER BEARING, JUST ABOVE THE 1" OVERLAP.
RABBETED EDGE ALLOWS FOR 1/2" OVERLAP. ONE SIDING OR BOX NAIL PER BEARING. 1" UP FROM BOTTOM EDGE.	RABBETED EDGE → APPROXIMATE 1/8" GAP FOR DRY MATERIAL 8" AND WIDER. 1/2" = FULL DEPTH OF RABBET ALLOWS FOR 1/2" OVERLAP. ONE SIDING OR BOX NAIL PER BEARING. 1" UP FROM BOTTOM EDGE.
USE SIDING OR BOX NAIL TO FACE NAIL ONE PER BEARING, 1 1/2" UP FROM BOTTOM EDGE.	← APPROXIMATE 1/8" GAP FOR DRY MATERIAL 8" AND WIDER. 1/2" = FULL DEPTH OF RABBET USE TWO SIDING OR BOX NAILS, 3-4" APART, PER BEARING, TO FACE NAIL.
T&G PATTERN / SHIPLAP PATTERN USE CASING NAILS TO BLIND NAIL T&G PATTERNS, ONE NAIL PER BEARING. USE SIDING OR BOX NAILS TO FACE NAIL SHIPLAP PATTERNS, 1" UP FROM BOTTOM EDGE.	T&G PATTERN / SHIPLAP PATTERN → APPROXIMATE 1/8" GAP FOR DRY MATERIAL 8" AND WIDER. 1/2" = FULL DEPTH OF RABBET USE TWO SIDING OR BOX NAILS, 3-4" APART, TO FACE NAIL, 1" UP FROM BOTTOM EDGE.

SIDING USUALLY APPLIED HORIZONTALLY

FIGURE 61–12 Location and number of fasteners recommended for wood siding. (*Courtesy of Western Wood Products Association*)

FIGURE 61–13 Vertical tongue-and-groove siding needs little accessory trim, such as corner boards. *(Courtesy of California Redwood Association)*

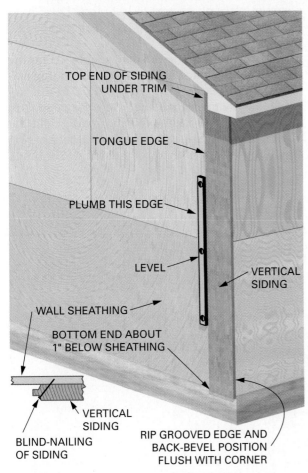

TOP END OF SIDING UNDER TRIM

TONGUE EDGE

PLUMB THIS EDGE

LEVEL

VERTICAL SIDING

WALL SHEATHING

BOTTOM END ABOUT 1" BELOW SHEATHING

VERTICAL SIDING

BLIND-NAILING OF SIDING

RIP GROOVED EDGE AND BACK-BEVEL POSITION FLUSH WITH CORNER

FIGURE 61–14 Starting the application of vertical board siding.

Fasten a temporary piece on the other end of the wall projecting below the sheathing by the same amount. Stretch a line to keep the bottom ends of other pieces in a straight line. Apply succeeding pieces by toenailing into the tongue edge of each piece.

Make sure the edges between boards come up tight. If they do not come up tight by nailing alone, drive a chisel, with its beveled edge side against the tongue, into the sheathing. Use it as a pry to force the board up tight. When the edge comes up tight, fasten it close to the chisel. If this method is not successful, toenail a short block of the siding with its grooved edge into the tongue of the board **(Figure 61–15)**. Drive the nail home until it forces the board up tight. Drive nails into the siding on both sides of the scrap block. Remove the scrap block.

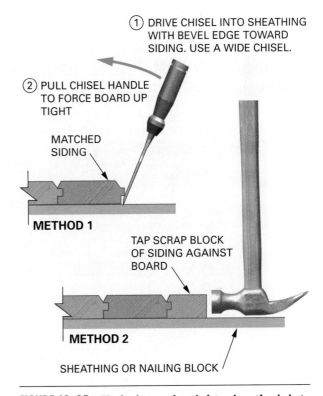

① DRIVE CHISEL INTO SHEATHING WITH BEVEL EDGE TOWARD SIDING. USE A WIDE CHISEL.

② PULL CHISEL HANDLE TO FORCE BOARD UP TIGHT

MATCHED SIDING

METHOD 1

TAP SCRAP BLOCK OF SIDING AGAINST BOARD

METHOD 2

SHEATHING OR NAILING BLOCK

FIGURE 61–15 Techniques for tightening the joints of boards during installation.

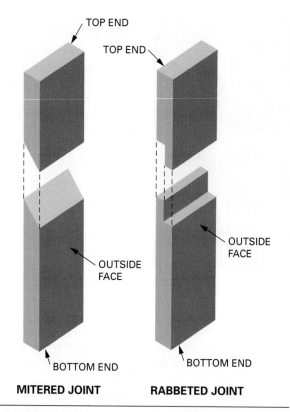

FIGURE 61–16   Use mitered or rabbeted end joints between lengths of vertical board siding.

Continue applying pieces in the same manner. Make sure to keep the bottom ends in a straight line. Avoid making horizontal joints between lengths. If joints are necessary, use a mitered or rabbeted joint for weathertightness **(Figure 61–16).**

Fitting around Doors and Windows

Vertical siding is fitted tightly around window and door casings with different methods than those used for horizontal siding.

Approaching a Wall Opening. When approaching a door or window, cut and fit the piece just before the one to be fitted against the casing. Then remove it. Set it aside for the time being.

Cut, fit, and tack the piece to be fitted against the casing in place of the next to last piece. Level from the top of the window casing and the bottom of the sill. Mark the piece.

To lay out the piece so it will fit snugly against the side casing, first cut a scrap block of the siding material, about 6 inches long **(Procedure 61–B).** Remove the tongue from one edge. Be careful to remove all of the tongue, but no more. Hold the block so its grooved edge is against the side casing and the other edge is on top of the siding to be fitted.

STEP BY STEP PROCEDURES

Procedure 61–B Method for Fitting Vertical Board Siding When Approaching a Window Casing

STEP ❶ TACKED PIECE THAT WILL LATER BE CUT AROUND OPENING.

STEP ❷ MOVE SCRAP BLOCK OF SIDING ALONG SIDE CASING OF WINDOW. HOLD PENCIL AGAINST THIS EDGE AND MARK SIDING TO BE FITTED.

STEP ❸ LEVEL FROM TOP OF WINDOW CASING.

PREVIOUSLY APPLIED SIDING

LEVEL FROM BOTTOM OF WINDOW SILL. MAKE ALLOWANCE TO FIT IN GROOVE ON UNDERSIDE.

Procedure 61–C Method of Fitting Vertical Board Siding When Leaving a Window Casing

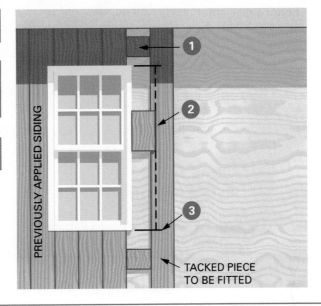

STEP 1 INSTALL TEMPORARY BLOCKS OF SIDING.

STEP 2 HOLD PENCIL AGAINST THIS EDGE. MOVE BLOCK AGAINST CASING WHILE MARKING SIDING.

STEP 3 LEVEL FROM TOP AND BOTTOM OF CASING.

PREVIOUSLY APPLIED SIDING

TACKED PIECE TO BE FITTED

STEP BY STEP PROCEDURES

Mark the piece by holding a pencil against the outer edge of the block while moving the block along the length of the side casing.

Cut the piece, following the layout lines carefully. When laying out to fit against the bottom of the sill, make allowance to rabbet the siding to fit in the weathergroove on the bottom side of the window sill. Place and fasten the pieces in position.

Continue to apply the siding with short lengths across the top and bottom of the window. Each length under a window must be rabbeted at the top end to fit in the weather groove at the sill.

Leaving a Wall Opening. A full length must also be fitted to the casing on the other side of the window. To mark the piece, first tack a short length of scrap siding above and below the window and against the last pieces of siding installed. Tack the length of siding to be fitted against these blocks. Level from the top and bottom of the window. Mark the piece for the horizontal cuts.

To lay out the piece for the vertical cut that fits against the side casing, use the same block with the tongue removed that was used previously. Hold the grooved edge against the side casing. With a pencil against the other edge, ride the block along the side casing while marking the piece to be fitted **(Procedure 61–C).**

Remove the piece and the scrap blocks from the wall. Carefully cut the piece to the layout lines. Then fasten in position. Continue applying the rest of the siding until you are almost to the other end of the wall.

Method of Ending Vertical Siding

The last piece of vertical siding should be close to the same width of previously installed pieces. If siding is installed in random widths, plan the application. The width of the last piece should be equal, at least, to the width of the narrowest piece. It is not good practice to allow vertical siding to end with a narrow sliver.

Stop several feet from the end. Space off, and determine the width of the last piece. If it will not be a satisfactory width, install narrower or wider pieces for the remainder, as required. It may be necessary to rip available siding to narrower widths and reshape the grooves **(Figure 61–17).**

When the corner is reached, the board is ripped to width along its tongue edge. It is slightly backbeveled for the first piece on the next wall to butt against. When the last corner is reached, the board is ripped in a similar manner. However, it is smoothed to a square edge to fit flush with the surface of the first piece installed. All exposed sharp corners should be eased or slightly rounded.

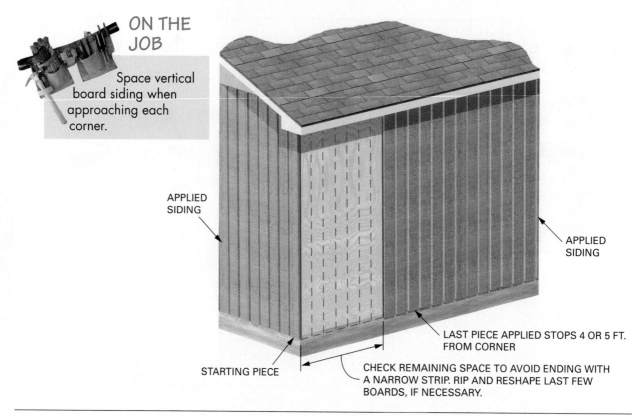

ON THE JOB

Space vertical board siding when approaching each corner.

APPLIED SIDING

APPLIED SIDING

LAST PIECE APPLIED STOPS 4 OR 5 FT. FROM CORNER

STARTING PIECE

CHECK REMAINING SPACE TO AVOID ENDING WITH A NARROW STRIP. RIP AND RESHAPE LAST FEW BOARDS, IF NECESSARY.

FIGURE 61–17 Joints in vertical siding may be adjusted slightly to ensure that the corner piece is nearly as wide as the others.

INSTALLING PANEL SIDING

Plywood, hardboard, and other panel siding is usually installed vertically. It can be installed horizontally, if desired. Lap siding panels are ordinarily applied horizontally.

Installing Vertical Panel Siding

Start a vertical panel so it is plumb, with one edge squared and flush with the starting corner. The inner edge should fall on the center of a stud. Fasten panels, of ½-inch thickness or less, with 6d siding nails. Use 8d siding nails for thicker panels. Fasten panel edges about every 6 inches and about every 12 inches along intermediate studs.

Apply successive sheets. Leave a ⅛-inch space between panels. Panels must be installed with their bottom ends in a straight line. There should be a minimum of 6 inches above the finished grade line.

Installing Horizontal Panel Siding

Mark the height of the first course of horizontal panel siding on both ends of the wall. Snap a chalk line between marks. Fasten a full-length panel with its top edge to the line, its inner end on the center of a stud, and its outer end flush with the corner. Fasten in the same way as for vertical panels.

Apply the remaining sheets in the first course in like manner. Trim the end of the last sheet flush with the corner. Start the next course so joints will line up with those in the course below.

Both vertical and horizontal panels may be applied to sheathing or directly to studs if backing is provided for all joints **(Figure 61–18).**

Carefully fit and caulk around doors and windows. It is important that horizontal butt joints be either offset and lapped, rabbeted, or flashed **(Figure 61–19).** Vertical joints are either shiplapped or covered with **battens.**

Applying Lap Siding Panels

Panels of lap siding are applied in much the same manner as wood bevel siding with some exceptions. First, install a strip, of the same thickness and width of the siding headlap, along the bottom of the wall. Determine the height of the top edge of the first course. Snap a chalk line across the wall. Apply the first course with its top edge to the snapped line.

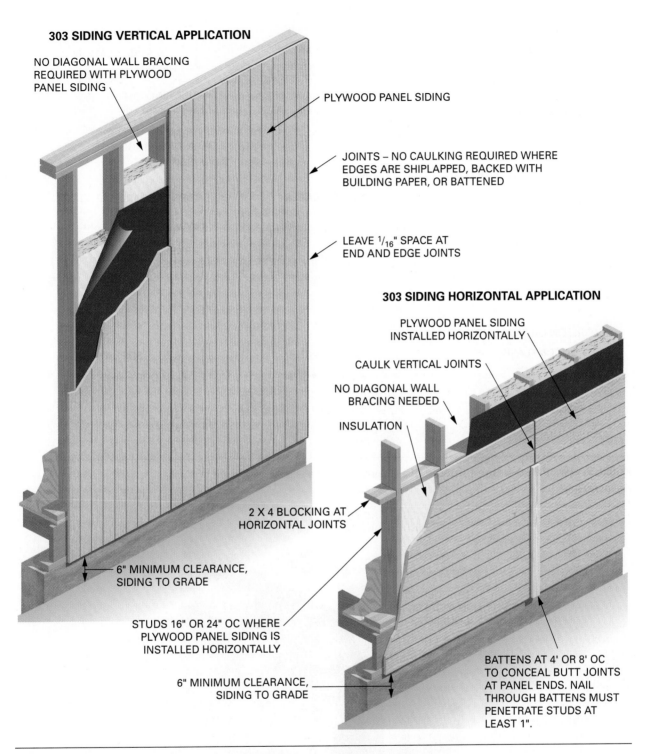

303 SIDING VERTICAL APPLICATION

NO DIAGONAL WALL BRACING
REQUIRED WITH PLYWOOD
PANEL SIDING

PLYWOOD PANEL SIDING

JOINTS – NO CAULKING REQUIRED WHERE
EDGES ARE SHIPLAPPED, BACKED WITH
BUILDING PAPER, OR BATTENED

LEAVE $\frac{1}{16}$" SPACE AT
END AND EDGE JOINTS

303 SIDING HORIZONTAL APPLICATION

PLYWOOD PANEL SIDING
INSTALLED HORIZONTALLY

CAULK VERTICAL JOINTS

NO DIAGONAL WALL
BRACING NEEDED

INSULATION

2 X 4 BLOCKING AT
HORIZONTAL JOINTS

6" MINIMUM CLEARANCE,
SIDING TO GRADE

STUDS 16" OR 24" OC WHERE
PLYWOOD PANEL SIDING IS
INSTALLED HORIZONTALLY

6" MINIMUM CLEARANCE,
SIDING TO GRADE

BATTENS AT 4' OR 8' OC
TO CONCEAL BUTT JOINTS
AT PANEL ENDS. NAIL
THROUGH BATTENS MUST
PENETRATE STUDS AT
LEAST 1".

**FIGURE 61–18   Panel siding may be applied vertically or horizontally to
sheathing or directly to studs.** *(Courtesy of American Plywood Association)*

When applied over nailable sheathing, space nails
8 inches apart in a line about ¾ inch above the bottom
edge of the siding. When applied directly to framing,
fasten at each stud location. A ⅛-inch, caulked joint between the ends of siding and trim is recommended.
Joints between ends of siding may also be flashed with
a narrow strip of #15 felt centered behind the joint and
backed with a wood shingle wedge **(Figure 61–20).**

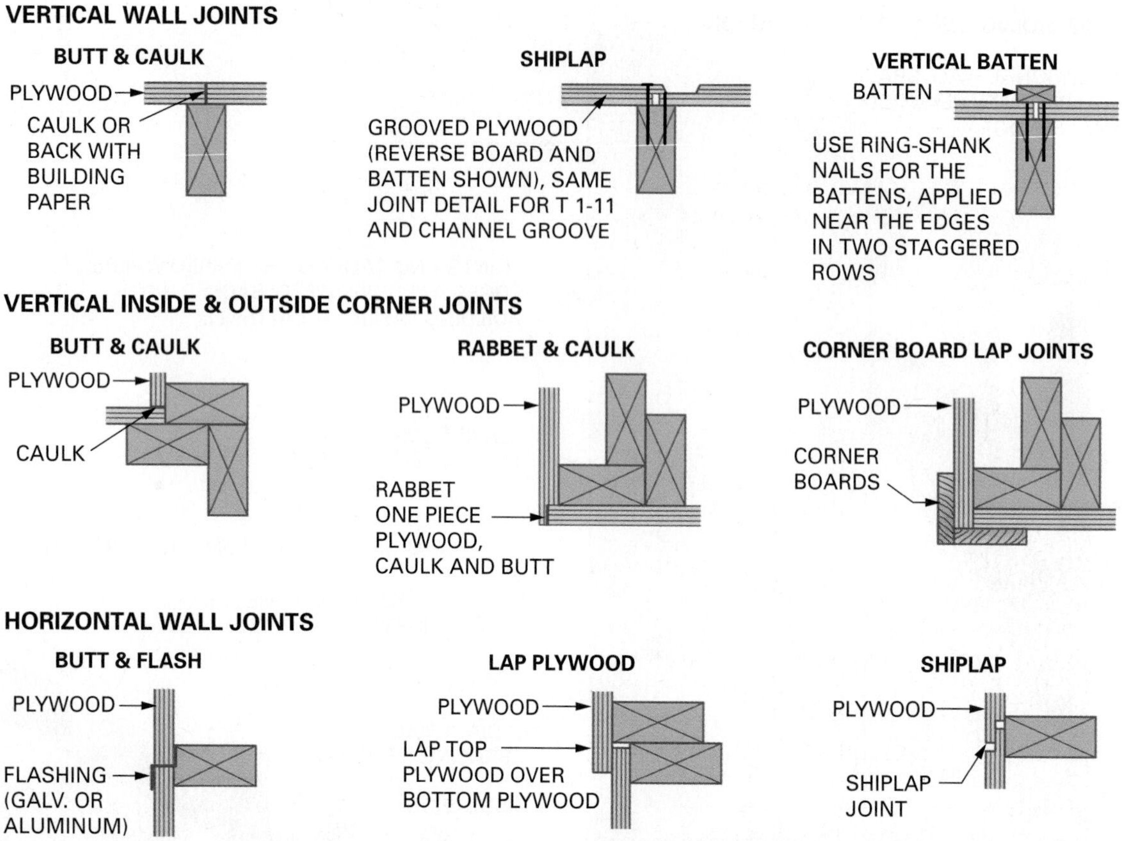

VERTICAL WALL JOINTS

BUTT & CAULK

PLYWOOD →

CAULK OR
BACK WITH
BUILDING
PAPER

SHIPLAP

GROOVED PLYWOOD
(REVERSE BOARD AND
BATTEN SHOWN), SAME
JOINT DETAIL FOR T 1-11
AND CHANNEL GROOVE

VERTICAL BATTEN

BATTEN →

USE RING-SHANK
NAILS FOR THE
BATTENS, APPLIED
NEAR THE EDGES
IN TWO STAGGERED
ROWS

VERTICAL INSIDE & OUTSIDE CORNER JOINTS

BUTT & CAULK

PLYWOOD →

CAULK

RABBET & CAULK

PLYWOOD →

RABBET
ONE PIECE
PLYWOOD,
CAULK AND BUTT

CORNER BOARD LAP JOINTS

PLYWOOD →

CORNER
BOARDS

HORIZONTAL WALL JOINTS

BUTT & FLASH

PLYWOOD →

FLASHING →
(GALV. OR
ALUMINUM)

LAP PLYWOOD

PLYWOOD →

LAP TOP
PLYWOOD OVER
BOTTOM PLYWOOD

SHIPLAP

PLYWOOD →

SHIPLAP
JOINT

FIGURE 61–19 Panel siding joint details. *(Courtesy of American Plywood Association)*

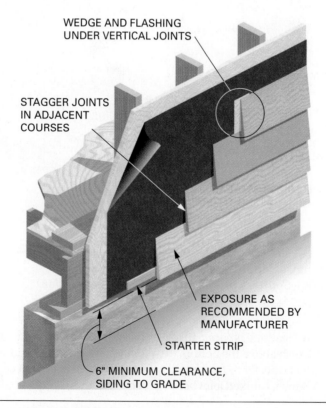

WEDGE AND FLASHING
UNDER VERTICAL JOINTS

STAGGER JOINTS
IN ADJACENT
COURSES

EXPOSURE AS
RECOMMENDED BY
MANUFACTURER

STARTER STRIP

6" MINIMUM CLEARANCE,
SIDING TO GRADE

FIGURE 61–20 Lap siding application details.

62 Wood Shingle and Shake Siding

Wood shingles and shakes may be used for siding, as well as roofing **(Figure 62–1).** Those previously described in Chapter 53 for roofing may also be applied to sidewalls.

FIGURE 62–1   Wood shingles and shakes may be used as siding. *(Courtesy of Cedar Shake and Shingle Bureau)*

SIDEWALL SHINGLES AND SHAKES

Some kinds of shingles and shakes are designed for sidewall use only **(Figure 62–2).** Rebutted and rejointed ones are machine trimmed with parallel edges and square butts for sidewall application. Rebutted and rejointed machine-grooved, sidewall shakes have **striated** faces.

Special **fancy butt** shingles are available in a variety of designs. They provide interesting patterns, in combination with square butts or other types of siding **(Figure 62–3).**

APPLYING WOOD SHINGLES AND SHAKES

Wood shingles and shakes may be applied to sidewalls in either single-layer or double-layer courses. In **single coursing,** shingles are applied to walls with a single layer in each course, in a way similar to roof application. However, greater exposures are allowed on sidewalls than on roofs **(Figure 62–4).**

In **double coursing,** two layers are applied in one course. Consequently, even greater weather exposures are allowed. Double coursing is used when

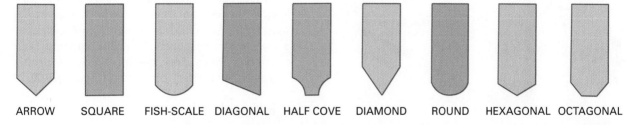

| ARROW | SQUARE | FISH-SCALE | DIAGONAL | HALF COVE | DIAMOND | ROUND | HEXAGONAL | OCTAGONAL |

FANCY BUTT RED CEDAR SHINGLES. NINE OF THE MOST POPULAR DESIGNS ARE SHOWN. FANCY BUTT SHINGLES CAN BE CUSTOM PRODUCED TO INDIVIDUAL ORDERS.

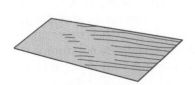

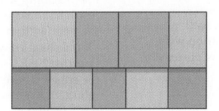

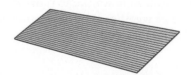

REBUTTED AND REJOINTED. MACHINE TRIMMED FOR PARALLEL EDGES WITH BUTTS SAWN AT RIGHT ANGLES. FOR SIDEWALL APPLICATION WHERE TIGHTLY FITTING JOINTS ARE DESIRED.

PANELS. WESTERN RED CEDAR SHINGLES ARE AVAILABLE IN 4- AND 8-FOOT PANELIZED FORM.

MACHINE GROOVED. MACHINE-GROOVED SHAKES ARE MANUFACTURED FROM SHINGLES AND HAVE STRIATED FACES AND PARALLEL EDGES. USED DOUBLE-COURSED ON EXTERIOR SIDEWALLS.

FIGURE 62–2   Some wood shingles and shakes are made for sidewall applications only. *(Courtesy of Cedar Shake and Shingle Bureau)*

ROUND FANCY BUTT SHINGLES

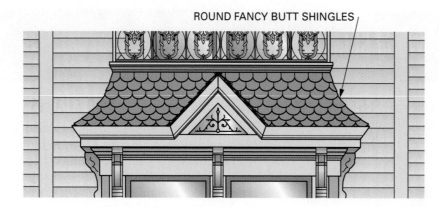

FIGURE 62–3 **Fancy butt shingles are still used to accent sidewalls with distinctive designs.**

SINGLE COURSING

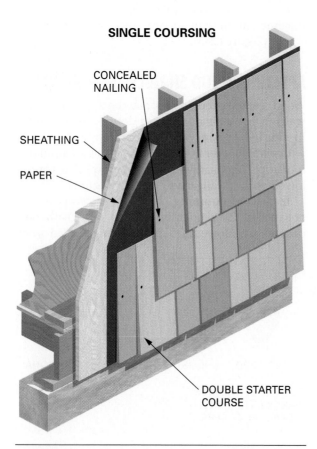

CONCEALED NAILING

SHEATHING

PAPER

DOUBLE STARTER COURSE

FIGURE 62–4 **Single-coursed shingle wall application is similar to roof application with greater weather exposures allowed.** *(Courtesy of Cedar Shake and Shingle Bureau)*

wide courses with deep, bold shadow lines are desired **(Figure 62–5).**

Applying the Starter Course

The starter course of sidewall shingles and shakes is applied in much the same way as the starter

course on roofs. A double layer is used for single-course applications. A triple layer is used for triple coursing. Less expensive **undercourse** shingles are used for underlayers.

Fasten a shingle on both ends of the wall with its butt about 1 inch below the top of the foundation. Stretch a line between them at the butts. Sight the line for straightness. Fasten additional shingles at necessary intervals. Attach the line to their butts to straighten it **(Figure 62–6).** Even a tightly stretched line will sag in the center over a long distance.

Apply a single course of shingles so the butts are as close to the chalk line as possible without touching it. Remove the line. Apply another course on top of the first course. Offset the joints in the outer layer at least 1½ inches from those in the bottom layer. Untreated shingles should be spaced ⅛ to ¼ inch apart to allow for swelling and to prevent buckling. Shingles can be applied close together if factory primed or if treated soon after application **(Figure 62–7).**

Single Coursing

A story pole may be used to lay out shingle courses in the same manner as with horizontal wood siding. Snap a chalk line across the wall, at the shingle butt line, to apply the first course. Using only as many finish nails as necessary, tack 1 × 3 straightedges to the wall with their top edges to the line. Lay individual shingles with their butts on the straightedge **(Figure 62–8).** Use a shingling hatchet to trim and fit the edges, if necessary. Butt ends are not trimmed. If rebutted and rejointed shingles are used, no trimming should be necessary.

At times it may be necessary to fit a shingle between others in the same course. Tack the next to last shingle in place with one nail. Slip the last shingle under it. Score along the overlapping edge with the hatchet. Cut along the scored line. Fasten both shingles in place **(Figure 62–9).**

DOUBLE COURSING

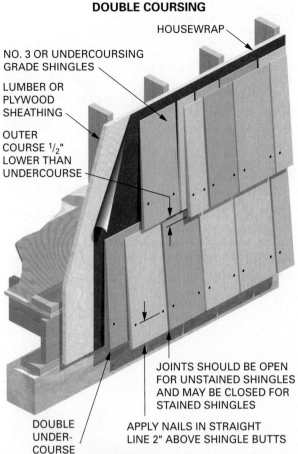

NO. 3 OR UNDERCOURSING GRADE SHINGLES

LUMBER OR PLYWOOD SHEATHING

OUTER COURSE ½" LOWER THAN UNDERCOURSE

HOUSEWRAP

JOINTS SHOULD BE OPEN FOR UNSTAINED SHINGLES AND MAY BE CLOSED FOR STAINED SHINGLES

DOUBLE UNDER- COURSE

APPLY NAILS IN STRAIGHT LINE 2" ABOVE SHINGLE BUTTS

Max. Weather Exposure		
Shingle Length	**Single Course**	**Double Course**
16"	7"	12"
18"	8"	14"
24"	10½"	16"
Shake Length & Type	**Single Course**	**Double Course**
16" Centigroove	7"	12"
18" Centigroove	8"	14"
18" resawn	8"	14"
24" resawn	10½"	16"
18" taper-sawn	8"	14"
24" tapersplit	10½"	18"
24" taper-sawn1	0½"	18"
18" straight-split	8"	16"

FIGURE 62–5 Double-coursed shingles, with two layers in each row, permit even greater weather exposures. (*Courtesy of Cedar Shake and Shingle Bureau*)

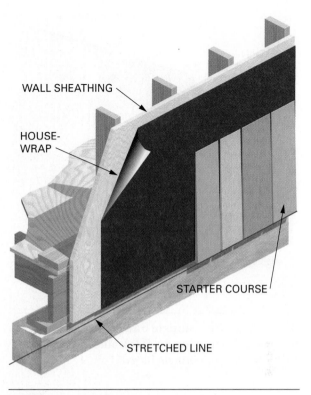

WALL SHEATHING

HOUSE- WRAP

STARTER COURSE

STRETCHED LINE

FIGURE 62–6 Stretch a straight line as a guide for the butts of the starter course.

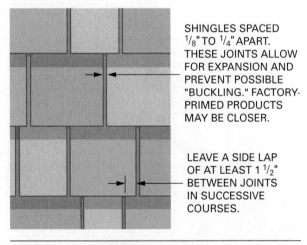

SHINGLES SPACED ⅛" TO ¼" APART. THESE JOINTS ALLOW FOR EXPANSION AND PREVENT POSSIBLE "BUCKLING." FACTORY-PRIMED PRODUCTS MAY BE CLOSER.

LEAVE A SIDE LAP OF AT LEAST 1 ½" BETWEEN JOINTS IN SUCCESSIVE COURSES.

FIGURE 62–7 Stagger joints between shingle courses. (*Courtesy of Cedar Shake and Shingle Bureau*)

Fasteners. Each shingle, up to 8 inches wide, is fastened with two nails or staples about ¾ inch in from each edge. On shingles wider than 8 inches, drive two additional nails about 1 inch apart near the center. Fasteners should be hot-dipped galvanized, stainless steel, or aluminum. They should be driven

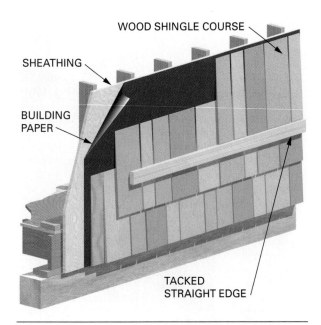

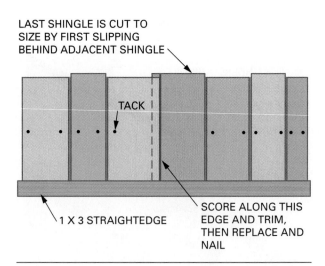

FIGURE 62–8 Rest shingle butts on a straightedge when single coursing on sidewalls. *(Courtesy of Cedar Shake and Shingle Bureau)*

FIGURE 62–9 Technique for cutting last shingle to the proper size.

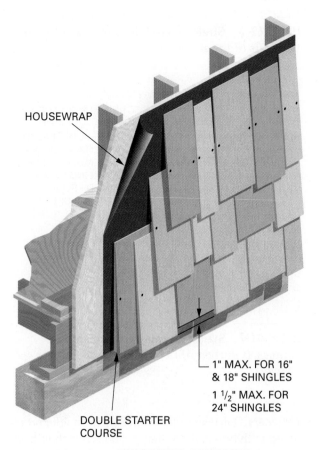

STAGGERED COURSING

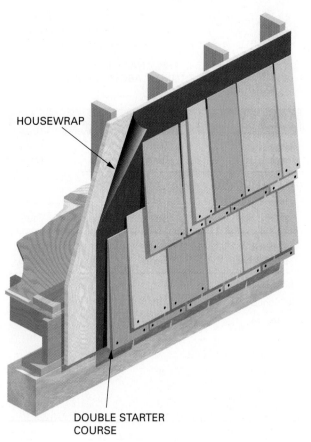

RIBBON COURSING

FIGURE 62–10 Staggered and ribbon courses are alternatives to straight-line courses. *(Courtesy of Cedar Shake and Shingle Bureau)*

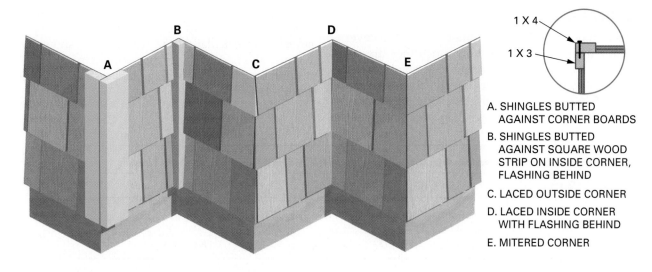

A. SHINGLES BUTTED AGAINST CORNER BOARDS

B. SHINGLES BUTTED AGAINST SQUARE WOOD STRIP ON INSIDE CORNER, FLASHING BEHIND

C. LACED OUTSIDE CORNER

D. LACED INSIDE CORNER WITH FLASHING BEHIND

E. MITERED CORNER

FIGURE 62–11 **Corner details for wood shingle siding.**

about 1 inch above the butt line of the next course. Fasteners must be long enough to penetrate the sheathing by at least ¾ inch.

Staggered and Ribbon Coursing. An alternative to straight-line courses are staggered and ribbon coursing. In **staggered coursing,** the butt lines of the shingles are alternately offset below, but not above, the horizontal line. Maximum offsets are one inch for 16- and 18-inch shingles and 1½ inches for 24-inch shingles.

In **ribbon coursing,** both layers are applied in straight lines. The outer course is raised about 1 inch above the inner course **(Figure 62–10).**

Corners

Shingles may be butted to corner boards like any horizontal wood siding. On outside corners, they may also be applied by alternately overlapping each course in the same manner as in applying a wood shingle ridge. Inside corners may be woven by alternating the corner shingle first on one wall and then the other **(Figure 62–11).**

Double Coursing

When **double coursing,** the first course is tripled. The outer layer of the course is applied ½ inch lower than the inner layer. For ease in application, use a rabbeted straightedge or one composed of two pieces with offset edges **(Figure 62–12).**

Fastening. Each inner layer shingle is applied with one fastener at the top center. Each outer course shingle is face-nailed with two 5d galvanized box or

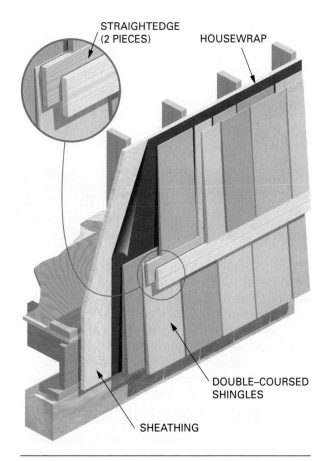

FIGURE 62–12 **Use a straightedge made of two pieces with offset edges for double-coursed application.**

special 14-gauge shingle nails. The fasteners are driven about 2 inches above the butts, and about ¾ inch in from each edge.

ESTIMATING SHINGLE SIDING

The number of squares of shingles needed to cover a certain area depends on how much they are exposed to the weather. One square of shingles will cover 100 square feet when 16-inch shingles are exposed 5 inches, 18-inch shingles exposed 5½ inches, and 24-inch shingles exposed 7½ inches.

A square of shingles will cover more area with greater exposures and less area with smaller exposures. After calculating the wall area as shown in Figure 60–11, determine the amount of shingles needed by using the coverage tables shown in **Figure 62–13.**

Length	Approximate Coverage of One Square (4-bundle roof-pack) of Shingles at Indicated Weather Exposures:												
	3½″	4″	4½″	5″	5½″	6″	6½″	7″	7½″	8″	8½″	9″	9½″
16″	70	80	90	100	110	120	130	140	150	160	170	180	190
18″	—	72½	81½	90½	100	109	118	127	136	145½	154½	163½	172½
24″	—	—	—	—	73½	80	86½	93	100	106½	113	120	126½

Length	10″	10½″	11″	11½″	12″	12½″	13″	13½″	14″	14½″	15″	15½″	16″
16″	200	210	220	230	240	—	—	—	—	—	—	—	—
18″	181½	191	200	209	218	227	236	245½	254½	—	—	—	—
24″	133	140	146½	153	160	166½	173	180	186½	193	200	206½	213

FIGURE 62–13   Sidewall shingle coverage at various exposures. *(Courtesy of Cedar Shake and Shingle Bureau)*

63 Aluminum and Vinyl Siding

Except for the material, aluminum and vinyl siding systems are similar. *Aluminum siding* is finished with a baked-on enamel. In *vinyl siding*, the color is embedded in the material itself. Both kinds are manufactured with interlocking edges, for horizontal and vertical applications. Descriptions and instructions are given here for vinyl siding systems, much of which can be applied to aluminum systems.

SIDING PANELS AND ACCESSORIES

Siding systems are composed of siding panels and several specially shaped moldings. Moldings are used on various parts of the building to trim the installation. In addition, the system includes shapes for use on soffits.

Siding Panels

Siding panels, for horizontal application, are made in 8- and 12-inch widths. They come in configura-tions to simulate one, two, or three courses of bevel or drop siding. Panels designed for vertical application come in 12-inch widths. They are shaped to resemble boards. They can be used in combination with horizontal siding. Vertical siding panels with solid surfaces may also be used for soffits. For ventilation, perforated soffit panels of the same configuration are used **(Figure 63–1).**

Siding System Accessories

Siding systems require the use of several specially shaped accessories. *Inside* and *outside corner posts* are used to provide a weather-resistant joint to corners. Corner posts are available with channels of appropriate widths to accommodate various configurations of siding.

Some other accessories include **J-channels,** *starter strips,* and **undersill trim,** also known as *finish trim.* J-channels are made with several opening sizes. They are used in a number of places such as

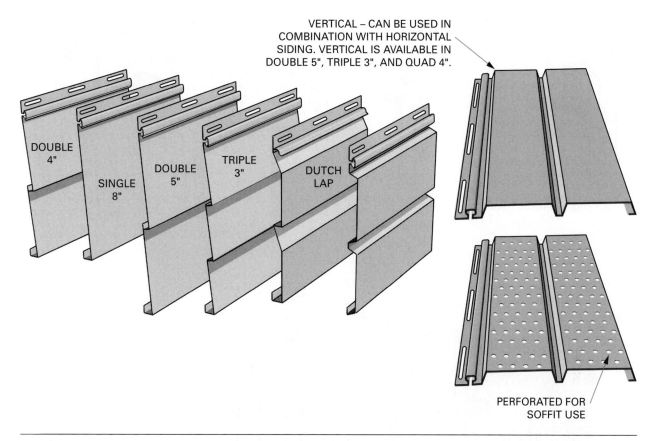

VERTICAL – CAN BE USED IN COMBINATION WITH HORIZONTAL SIDING. VERTICAL IS AVAILABLE IN DOUBLE 5", TRIPLE 3", AND QUAD 4".

DOUBLE 4"

SINGLE 8"

DOUBLE 5"

TRIPLE 3"

DUTCH LAP

PERFORATED FOR SOFFIT USE

FIGURE 63–1 Commonly used configurations of horizontal and vertical vinyl siding. *(Courtesy of Vinyl Siding Institute)*

around doors and windows, at transition of materials, against soffits, and in many other places **(Figure 63–2).** The majority of vinyl siding panels and accessories are manufactured in 12′–6″ lengths. Trim accessories and molding are typically 12 feet long.

APPLYING HORIZONTAL SIDING

The siding may expand and contract as much as ¼ inch over a 12′–6″ length with changes in temperature. For this reason, it is important to center fasteners in the slots. Do not drive them too tightly. There should be about 1⁄32 inch between the head of the fastener, when driven, and the siding **(Figure 63–3).** After the panel is fastened, it should be easily moved from side to side in the nail slots. Space fasteners 16 inches apart for horizontal siding and 6 to 12 inches apart for accessories unless otherwise specified by the manufacturer.

Applying the Starter Strip

Snap a level line to the height of the starter strip all around the bottom of the building. Fasten the strips to the wall with their top edges to the chalk line. Leave a ¼-inch space between them and other accessories to allow for expansion **(Figure 63–4).** Make sure the starter strip is applied as straight as possible. It controls the straightness of entire installation.

Installing Corner Posts

Corner posts are installed in corners ¼ inch below the starting strip and ¼ inch from the top. Attach the posts by fastening in the top of the upper slot on each side. The posts will hang on these fasteners. The rest of the fasteners should be centered on the slots. Make sure the posts are straight and true from top to bottom **(Figure 63–5).**

Installing J-Channel

Install J-channel pieces across the top and along the sides of window and door casings. They may also be installed under windows or doors with the undersill trim nailed inside of the channel. To miter the corners, cut all pieces to extend, on both ends, beyond the casings and sills a distance equal to the width of the channel face. On both ends of the side J-channels, cut a ¾-inch notch out of the bottom of the J-channel. Fasten in place. On both ends of the top and bottom channels, make ¾-inch cuts. Bend

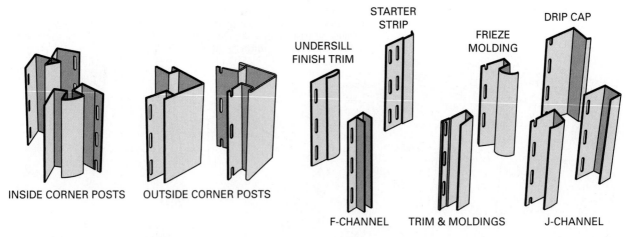

INSIDE CORNER POSTS   OUTSIDE CORNER POSTS

UNDERSILL FINISH TRIM

STARTER STRIP

FRIEZE MOLDING

DRIP CAP

F-CHANNEL   TRIM & MOLDINGS   J-CHANNEL

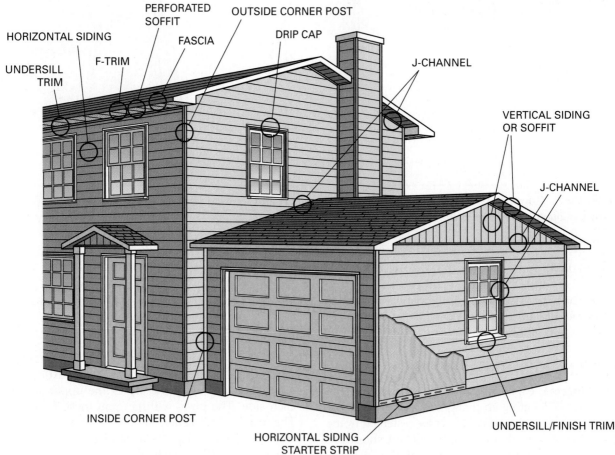

HORIZONTAL SIDING

PERFORATED SOFFIT

OUTSIDE CORNER POST

UNDERSILL TRIM

F-TRIM

FASCIA

DRIP CAP

J-CHANNEL

VERTICAL SIDING OR SOFFIT

J-CHANNEL

INSIDE CORNER POST

HORIZONTAL SIDING STARTER STRIP

UNDERSILL/FINISH TRIM

FIGURE 63–2   **Various accessories are used to trim a vinyl siding installation.**
(Courtesy of Georgia-Pacific Corporation)

down the tabs and miter the faces. Install them so the mitered faces are in front of the faces of the side channels **(Figure 63–6)**.

Installing Siding Panels

Snap the bottom of the first panel into the starter strip. Fasten it to the wall. Start from a back corner,

leaving a ¼-inch space in the corner post channel. Work toward the front with other panels. Overlap each panel about 1 inch **(Figure 63–7)**.

The seams of overlapped panels are more visible from one direction and less so from the other. Lap the panels so they are visible from the direction least traveled. This will put the best side of the siding toward the most often viewed side. If either direction

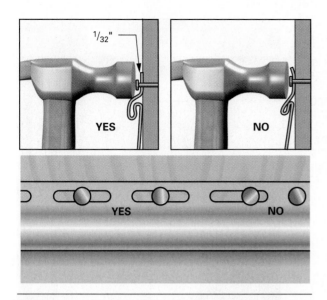

FIGURE 63–3 Fasten siding so as to allow for expansion and contraction.

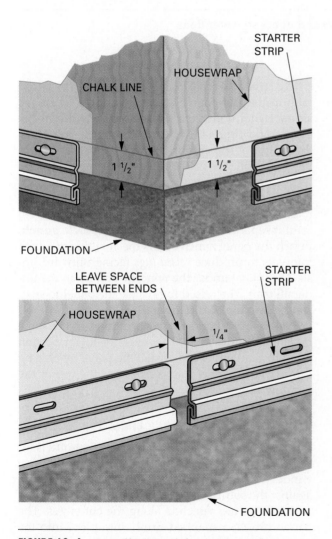

FIGURE 63–4 Installation of the starter strip.

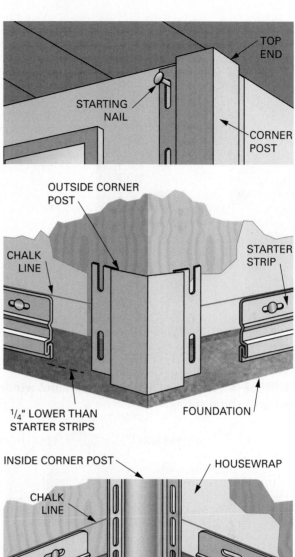

FIGURE 63–5 Inside and outside corner posts are installed in a similar way.

is equally traveled, alternate the direction of the lap on each course. This will reduce the number of seams visible to almost a half.

Also keep vertical seams as far away from each other as possible. Do not let them align vertically. A random pattern in the seams will be less noticeable to the eye than vertically aligned seams.

Install successive courses by interlocking them with the course below. Use tin snips, hacksaw, utility knife, or circular saw. Reverse the blade if a circular saw is used, for smooth cutting through the vinyl.

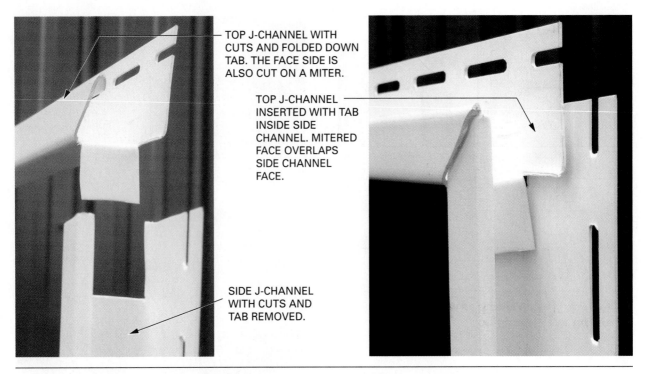

TOP J-CHANNEL WITH CUTS AND FOLDED DOWN TAB. THE FACE SIDE IS ALSO CUT ON A MITER.

TOP J-CHANNEL INSERTED WITH TAB INSIDE SIDE CHANNEL. MITERED FACE OVERLAPS SIDE CHANNEL FACE.

SIDE J-CHANNEL WITH CUTS AND TAB REMOVED.

FIGURE 63–6   Cutting J-channel to fit around windows and doors so water does not get behind siding.

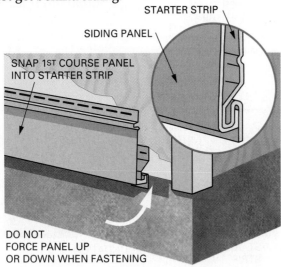

STARTER STRIP

SIDING PANEL

SNAP 1ST COURSE PANEL INTO STARTER STRIP

DO NOT FORCE PANEL UP OR DOWN WHEN FASTENING

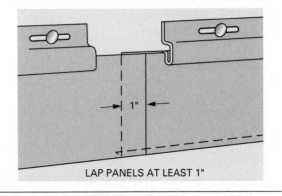

1"

LAP PANELS AT LEAST 1"

FIGURE 63–7   Installing the first course of horizontal siding panels.

Fitting around Windows. Plan so there will be no joint in the last course under a window. Hold the siding panel under the sill. Mark the width of the cutout, allowing ¼-inch clearance on each side. Mark the height of the cutout, allowing ¼-inch clearance below the sill.

Make vertical cuts with tin snips. Score the horizontal layout lines with a utility knife or scoring tool. Bend the section to be removed back and forth until it separates. Using a special **snaplock punch,** punch the panel ¼ inch below the cut edge at 6-inch intervals to produce *raised lugs* facing outward. Install the panel under the window and up in the undersill trim. The raised lugs cause the panel to snap snugly into the trim **(Figure 63–8).**

Panels are cut and fit over windows in the same manner as under them. However, the lower portion is cut instead of the top. Install the panel by placing it into the J-channel that runs across the top of the window **(Figure 63–9).**

Installing the Last Course under the Soffit

The last course of siding panels under the soffit is installed in a manner similar to that for fitting under a window. An undersill trim is applied on the wall, up against the soffit. Panels in the last course are cut to width. Lugs are punched along the cut edges. The panels are then snapped firmly into place into the undersill trim **(Procedure 63–A).**

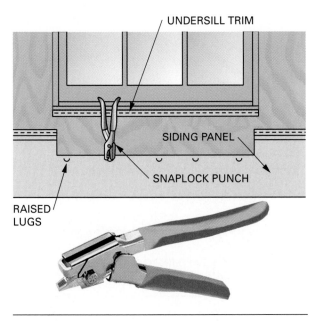

FIGURE 63–8 Method of fitting a panel under a window.

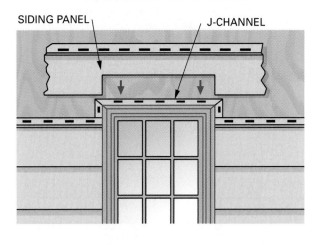

FIGURE 63–9 Fitting a panel over a window.

Procedure 63–A Fitting the Last Course of Horizontal Siding under the Soffit

STEP 1 MEASURE FOR LAST COURSE OF SIDING.

STEP 2 CUT PIECE AND CREATE RAISED LUGS WITH SNAPLOCK PUNCH.

STEP 3 SNAP CUT EDGE INTO UNDERSILL TRIM AND BOTTOM EDGE INTO COURSE BELOW.

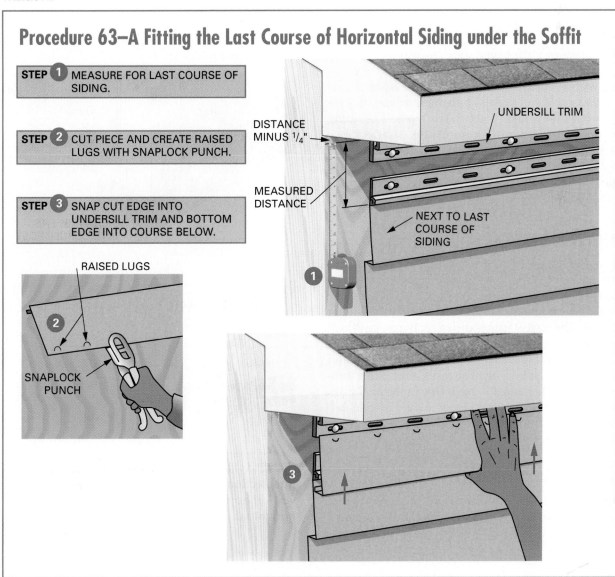

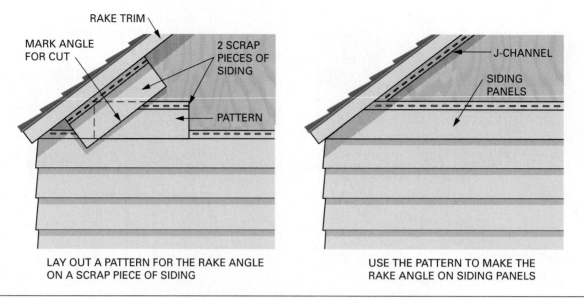

LAY OUT A PATTERN FOR THE RAKE ANGLE
ON A SCRAP PIECE OF SIDING

USE THE PATTERN TO MAKE THE
RAKE ANGLE ON SIDING PANELS

FIGURE 63–10   Fitting horizontal siding panels to the rakes.

Gable End Installation

The rakes of a gable end are first trimmed with J-channels. The panel ends are inserted into the channel with a ¼-inch expansion gap. Make a pattern for cutting gable end panels at an angle where they intersect with the rake. Use two scrap pieces of siding to make the pattern **(Figure 63–10)**. Interlock one piece with an installed siding panel below. Hold the other piece on top of it and against the rake. Mark along the bottom edge of the slanted piece on the face of the level piece.

APPLYING VERTICAL SIDING

The installation of vertical siding is similar to that for horizontal siding with a few exceptions. The method of fastening is the same. However, space fasteners about 12 inches apart for vertical siding panels. The starter strip is different. It may be ½-inch J-channel or drip cap flush with and fitted into the corner posts **(Figure 63–11)**. Around windows and doors, under soffits, against rakes, and other locations, ½-inch J-channel is used. One of the major differences is that a vertical layout should be planned so that the same and widest possible piece is exposed at both ends of the wall.

Installing the First Panel

To install the first panel, start by determining the widest possible width of the first and last panel.

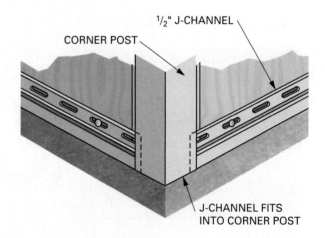

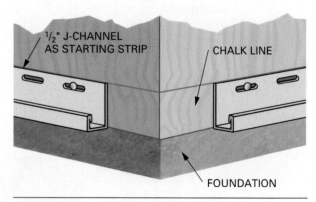

FIGURE 63–11   The starter strip shape and its intersection with corner posts is different for vertical application of vinyl siding compared to horizontal.

This is done by measuring, between the corner posts, the width of the face to be sided. Divide this number by the exposure of one panel. Take the decimal remainder and add one to it. Divide this number by two. This will be the size of the first and last panel **(Figure 63–12)**.

Install the first vertical panel plumb on the starter strip with one edge into the corner post. Allow ¼ inch at top and bottom. Place the first nails in the uppermost end of the top nail slots to hold it in position.

EXAMPLE What is the starting and finishing widths for a wall section that measures 18'–9" for siding that is 12" wide?

Convert this measurement to a decimal by first dividing the inches portion by 12 and then adding it to the feet to get 18.75'.

Divide this by the siding exposure, in feet: 18.75 ÷ 1 foot = 18.75 pieces.

Subtract the decimal portion along with one full piece giving 1.75 pieces. Next 1.75 ÷ 2 = 0.875, multiplied by 12 gives 10½".

This is the size of the starting and finishing piece. Thus there are 17 full-width pieces and two 10½" wide pieces.

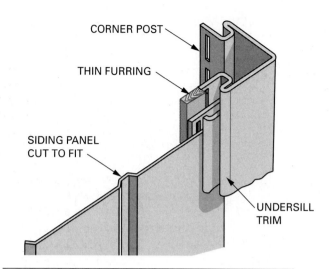

FIGURE 63–13 **Undersill trim and furring are required when vertical siding is cut to fit into corner posts and J-channels.** *(Courtesy of Vinyl Siding Institute)*

The edge of the panel may need to be cut in order for equal widths to be exposed on both ends of the wall. If the panel is cut on the flat surface, place a piece of undersill trim backed by furring into the channel of the corner post. Punch lugs along the cut edge of the panel at 6-inch intervals. Snap the panel into the undersill trim. Edges of vertical panels cut to fit in J-channels around windows and doors are treated in the same way **(Figure 63–13)**.

ALUMINUM ACCESSORIES

Fascia is completed by covering the wood subfascia with an L-shaped piece of siding. This piece may be made of either metal or vinyl. The top edge is installed by slipping it under the metal drip edge. The bottom is held in place with nails **(Figure 63–14)**. Nails are aluminum or stainless steel painted to the same color as the trim.

Aluminum trim pieces are often fabricated on the job. This is done by using a tool, called a **brake,** to bend sheet metal **(Figure 63–15)**. Light-duty brakes are designed for aluminum only, whereas others will bend heavy sheet metal.

Aluminum stock is sold in 50-foot-long rolls of various widths ranging from 12 to 24 inches. These rolls are referred to as **coil stock.** Each side of the sheet is colored with a baked-on enamel finish. One side is usually white and the other one of a variety of colors produced by the manufacturer.

Coil stock can be cut with a utility knife, but using a straightedge makes the cut look professional. Stock is cut with one score of the knife. The cut is

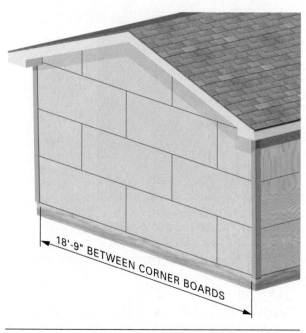

FIGURE 63–12 **Example for finding first and last panel width of vertical siding.**

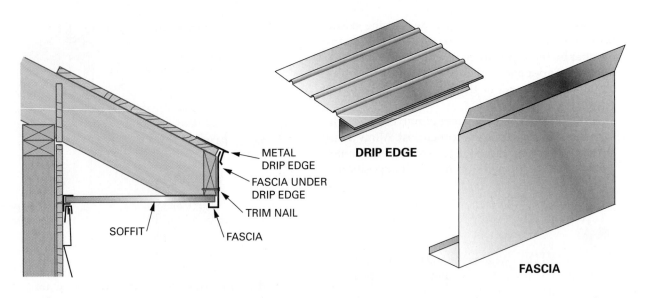

METAL
DRIP EDGE

FASCIA UNDER
DRIP EDGE

TRIM NAIL

SOFFIT

FASCIA

DRIP EDGE

FASCIA

FIGURE 63–14   Metal fascia material is fitted under metal drip edge.

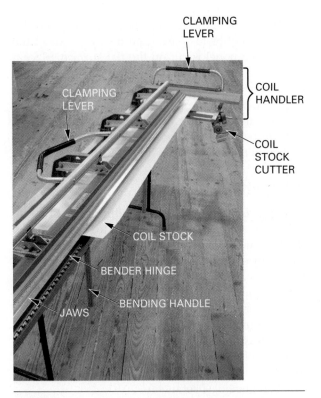

CLAMPING
LEVER

CLAMPING
LEVER

COIL
HANDLER

COIL
STOCK
CUTTER

COIL STOCK

BENDER HINGE

BENDING HANDLE

JAWS

FIGURE 63–15   An aluminum brake is used to shape coil stock.

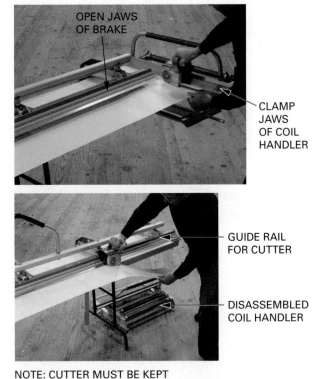

OPEN JAWS
OF BRAKE

CLAMP
JAWS
OF COIL
HANDLER

GUIDE RAIL
FOR CUTTER

DISASSEMBLED
COIL HANDLER

NOTE: CUTTER MUST BE KEPT
TIGHT TO GUIDE RAIL AS CUT IS MADE.

FIGURE 63–16   A special cutter can be used to cut stock to length and width.

then completed by bending the piece back and forth through the cut.

Stock can also be cut with a cutter designed to work with the brake **(Figure 63–16).** A coil handler makes unrolling the coil easier and the jaws provide

the rail for the cutter to ride. Coil stock is ripped to the desired width using the same cutter and the jaws of the brake. Brake jaws are clamped and unlocked using a clamping lever.

Once the pieces are cut to width, they are then bent to the desired configuration. Care must be taken to visualize the piece as it is bent. Mistakes are easy to make. For example, the piece may have the correct shape but the wrong color facing outward.

Bending the stock begins with careful positioning of the piece in the jaws. The same amount of stock should be revealed from the jaws on both ends. This will ensure that the piece is not tapered. Locking the jaws tight secures the piece during the bend. The bend is made by raising the handles of the brake **(Figure 63–17).** The bend may be any angle from 0 to 180 degrees. Care must be taken to bend the stock to the desired angle. Making repetitive stops at the same angle takes some practice.

Making an edge return uses a full 180-degree bend, which takes two steps. First align the piece to as

small a bend as possible, usually about ⅝ inch. The piece should be flush with the brake bender. Then lock and bend as far as the brake will allow. This is about 150 degrees. The jaws are then unlocked, the piece is removed, and the jaws relocked. The piece is then placed on top of the jaws **(Figure 63–18).** The final bend to 180 degrees is made while the piece is placed on top of the jaws.

ESTIMATING ALUMINUM AND VINYL SIDING

Aluminum and vinyl siding panels are sold by the square. Determine the wall area to be covered. Add 10 percent of the area for waste. Divide by 100. This gives you the number of squares needed.

Become familiar with accessories and how they are used. Measure the total linear feet required for each item.

FIGURE 63–17 Bending metal is done by clamping jaws and lifting handles.

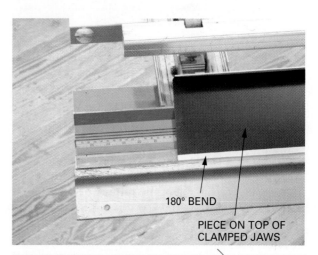

180° BEND

PIECE ON TOP OF CLAMPED JAWS

FIGURE 63–18 Full 180-degree bends are completed on top of the jaws.

64 Cornice Terms and Design

Cornice terms from earlier times still remain in use. However, cornice design has changed considerably. In earlier times, cornices were very elaborate and required much time to build. Now, cornice design, in most cases, is much more simplified. Only occasionally is a building designed with an ornate cornice similar to those built in years gone by.

PARTS OF A CORNICE

Several finish parts are used to build the cornice. In some cases, additional framing members are required.

Subfascia

The **subfascia** is sometimes called the false fascia or rough fascia **(Figure 64–1).** It is a horizontal

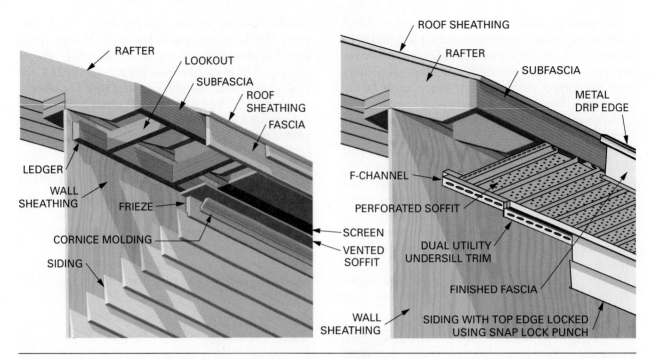

FIGURE 64–1 Cornices may be constructed of different materials.

framing member fastened to the rafter tails. It provides an even, solid, and continuous surface for the attachment of other cornice members. When used, the subfascia is generally a nominal 1- or 2-inch-thick piece. Its width depends on the slope of the roof, the tail cut of the rafters, or the type of cornice construction.

Soffit

The finished member on the underside of the cornice is called a **plancier** and is often referred to as a soffit **(Figure 64–2).** Soffit material may include solid lumber, plywood, strand board, fiberboard, or corrugated aluminum and vinyl panels. Soffits may be perforated or constructed with screen openings to allow for ventilation of the rafter cavities. Soffits may be fastened to the bottom edge of the rafter tails to the slope of the roof. The soffit is an ideal location for the placement of attic ventilation.

Fascia

The fascia is fastened to the subfascia or to the ends of the rafter tails. It may be a piece of lumber grooved to receive the soffit. It also may be made from bent aluminum and vinyl material used to wrap the subfascia. Fascia provides a surface for the attachment of a gutter. The fascia may be built up from one or more members to enhance the beauty of the cornice.

The bottom edge of the fascia usually extends below the soffit by ¼ to ⅜ inch. The portion of the fascia that extends below the soffit is called the **drip.** The drip is necessary to prevent rainwater from being swept back against the walls of the building. In addition, a drip makes the cornice more attractive.

Frieze

The **frieze** is fastened to the sidewall with its top edge against the soffit. Its bottom edge is sometimes rabbeted to receive the sidewall finish. In other cases, the frieze may be furred away from the sidewall to allow the siding to extend above and behind its bottom edge. However, the frieze is not always used. The sidewall finish may be allowed to come up to the soffit. The joint between the siding and the soffit is then covered by a molding.

Cornice Molding

The **cornice molding** is used to cover the joint between the frieze and the soffit. If the frieze is not used, the cornice molding covers the joint between the siding and the soffit.

Lookouts

Lookouts are framing members, usually 2 × 4 stock, that are used to provide a fastening surface for the soffit. They run horizontally from the end of the rafter to the wall, adding extra strength to larger overhangs. Lookouts may be installed at every rafter or spaced 48 inches on center (OC), depending on the material being used for the soffit (Figure 64–2).

CORNICE DESIGN

Cornices are generally classified into three main types: box, snub, and open **(Figure 64–3).**

The Box Cornice

The **box cornice** is probably most common. It gives a finished appearance to this section of the exterior.

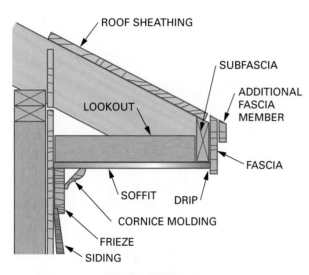

ROOF SHEATHING
SUBFASCIA
ADDITIONAL FASCIA MEMBER
LOOKOUT
FASCIA
SOFFIT   DRIP
CORNICE MOLDING
FRIEZE
SIDING

WOOD CORNICE

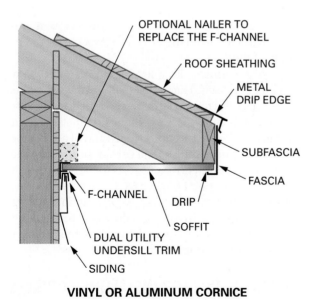

OPTIONAL NAILER TO REPLACE THE F-CHANNEL
ROOF SHEATHING
METAL DRIP EDGE
SUBFASCIA
FASCIA
F-CHANNEL   DRIP
SOFFIT
DUAL UTILITY UNDERSILL TRIM
SIDING

VINYL OR ALUMINUM CORNICE

FIGURE 64–2 **The soffit is the bottom finish member of the cornice.**

Because of its overhang, it helps protect the sidewalls from the weather. It also provides shade for windows.

Box cornices may be designated as narrow or wide. They may be constructed with level or sloping soffits. A *narrow box cornice* is one in which the cuts on the rafters serve as nailing surfaces for the cornice members. A *wide box cornice* may be constructed with a level or sloping soffit. A wide, level soffit requires the installation of lookouts.

The Snub Cornice

The **snub cornice** is not as attractive as some of the other designs, nor does it give as much protection to the sidewalls of a building because of its small overhang. There is no rafter projection beyond the wall. The snub cornice is chosen primarily to cut down the cost of material and labor. A snub cornice is frequently used on the rakes of a gable end in combination with a boxed cornice on the sides of a building.

The Open Cornice

The **open cornice** has no soffit. It is used when it is desirable to expose the rafter tails. It is often used when the rafters are large, laminated or solid beams with a wide overhang that exposes the roof decking on the bottom side. Open cornices give a contemporary or rustic design look to post-and-beam framing. They provide protection to sidewalls at low cost. This cornice might also be used for conventionally framed buildings for reasons of design and to reduce costs.

By adding a soffit, a *sloped cornice* is created. The soffit is installed directly to the underside of the rafter tails. This is sometimes done to simplify the cornice detail when there is also an overhang over the gable end of the building.

Rake Cornices

The main cornice is constructed on the rafter tails where they meet the walls of a building. On buildings with hip or mansard roofs, the main cornice, regardless of the type, extends around the entire building.

On buildings with gable roofs, a boxed main cornice with a sloping soffit, attached to the bottom edge of the rafter tails, may be returned up the rakes to the ridge. The cornice that runs up the rakes is called a **rake cornice (Figure 64–4).**

Cornice Returns. A main cornice with a horizontal soffit attached to level lookouts may, at times, be terminated at each end wall against a snub rake cornice

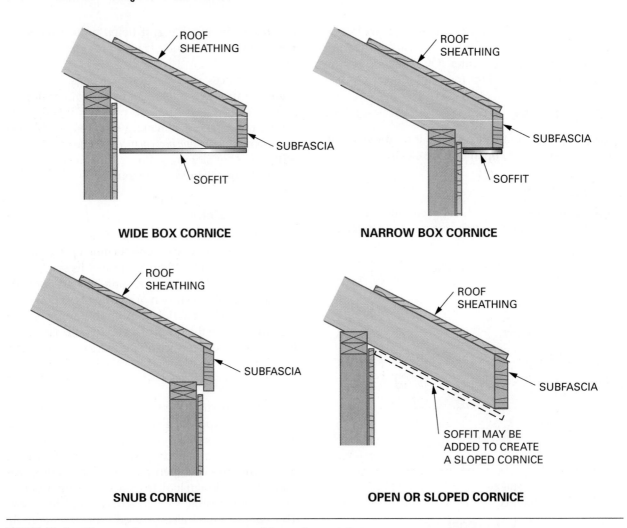

WIDE BOX CORNICE

NARROW BOX CORNICE

SNUB CORNICE

OPEN OR SLOPED CORNICE

FIGURE 64–3   The cornice may be constructed in various styles.

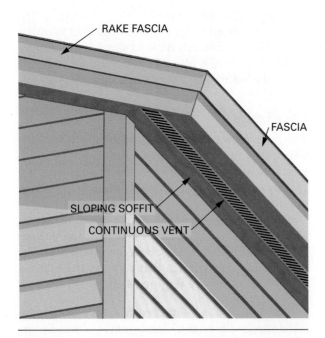

FIGURE 64–4   **A boxed cornice with a sloping soffit may be returned up the rakes of a gable roof.**

(Figure 64–5). At other times, a **cornice return** must be constructed to change the level box cornice to the angle of the roof.

A main cornice with a level soffit may also be returned upon itself. That is, the main cornice is mitered at each end. It is turned 90 degrees toward and beyond the corner as much as it overhangs on the side of the building. This short section on each end of a gable roof is called a cornice return. The cornice return provides a stop for the rake cornice. It adds to the design of the building at this point **(Figure 64–6).** However, cornice returns of this type are rarely built today. A large amount of labor is involved in their construction. The main cornice is returned on the rakes of the gable end in a much more simple fashion as described.

PRACTICAL TIPS FOR INSTALLING CORNICES

Install cornice members in a straight and true line with well-fitting and tight joints. Do not dismiss the use of hand tools for cutting, fitting, and fastening exterior trim.

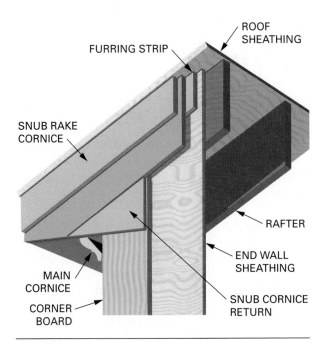

FIGURE 64–5 A boxed main cornice with a level soffit may be terminated at the gable ends against the rake cornice.

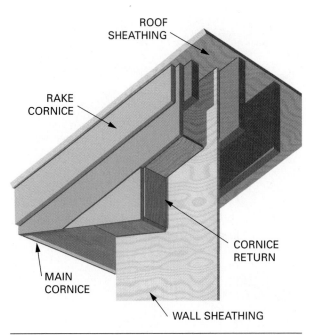

FIGURE 64–6 A boxed main cornice with a level soffit may be changed to the angle of the roof with a cornice return.

All fasteners should be noncorrosive. Stainless steel fasteners offer the best protection against corrosion. Hot-dipped galvanized fasteners provide better protection than plated ones. Fasteners used in wood should be well set. They should be puttied to conceal the fastener. This also prevents the heads

from corroding. Setting and concealing fasteners in exterior trim is a mark of quality.

Paint or otherwise seal and protect all exterior trim as soon as possible after installation. Properly installed and protected wood exterior trim will last indefinitely.

65 Gutters and Downspouts

A **gutter** is a shallow trough or conduit set below the edge of the roof along the fascia. It catches and carries off rainwater from the roof. A **downspout,** also called a *conductor,* is a rectangular or round pipe. It carries water from the gutter downward and away from the foundation **(Figure 65–1).**

GUTTERS

Gutters or *eavestroughs* may be made of wood, galvanized iron, aluminum, copper, or vinyl. Copper gutters require no finishing. Vinyl and aluminum gutters are prefinished and ready to install. Wood and galvanized metal gutters need several protective coats of finish after installation.

The size of a gutter is determined by area of the roof for which it handles the water runoff. Under ordinary conditions, 1 square inch of gutter cross-section is required for every 100 square feet of roof

area. For instance, a 4 × 5 inch gutter has a cross-section area of 20 square inches. It is, therefore, capable of handling the runoff from 2,000 square feet of roof surface.

Wood Gutters

Wood gutters usually are made of Douglas fir, western red cedar, or California redwood. They come in sizes of 3 × 4, 4 × 5, 4 × 6, and 5 × 7, and lengths of over 30 feet. Although wood gutters are not used as extensively as in the past, they enhance the cornice design. When properly installed and maintained, they will last as long as the building itself.

Metal Gutters

Metal gutters are made in rectangular, beveled, ogee, or semicircular shapes **(Figure 65–2).** They come in a variety of sizes, from 2½ inches to 6 inches in height

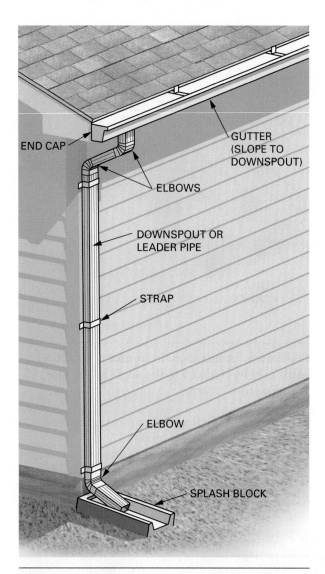

FIGURE 65–1 Gutters and downspouts form an important system for conducting water away from the building.

and from 3 inches to 8 inches in width. Stock lengths run from 10 to 40 feet. Forming machines are often brought to the jobsite to form aluminum into gutters of practically any desired length.

Besides straight lengths, metal gutter systems have components comprised of *inside* and *outside corners, joint connectors, outlets, end caps,* and others. *Metal brackets* or *spikes and ferrules* are used to support the gutter sections on the fascia **(Figure 65–3).**

Laying Out the Gutter Position

Gutters should be installed with a slight pitch to allow water to drain toward the downspout. A gutter of 20 feet or less in length may be installed to slope in one direction with a downspout on one end. On longer buildings, the gutter is usually crowned in the center. This allows water to drain to both ends.

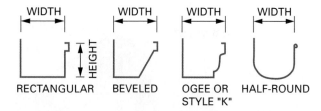

OTHER SHAPES ARE AVAILABLE

FIGURE 65–2 Metal gutters are available in several shapes.

On both ends of the fascia, mark the location of the bottom side of the gutter. The top outside edge of the gutter should be in relation to a straight line projected from the top surface of the roof. The height of the gutter depends on the pitch of the roof **(Figure 65–4).**

Stretch a chalk line between the two marks. Move the center of the chalk line up enough to give the gutter the proper pitch. Thumb the line down. Snap it on both sides of the center. For a slope in one direction only, snap a straight line lower on one end than the other to obtain the proper pitch. It is important to install gutters on the chalk line. This avoids any dips that may prevent complete draining of the gutter.

INSTALLING METAL AND VINYL GUTTERS

Fasten the gutter brackets to the chalk line on the fascia with screws. All screws should be made of stainless steel or other material that is corrosion resistant. Aluminum brackets may be spaced up to 30 inches on center (OC). Steel brackets may be spaced up to 48 inches OC. Install the gutter sections in the brackets. Use slip-joint connectors to join the sections. Apply the recommended gutter sealant to connectors before joining.

Locate the outlet tubes as required, keeping in mind that the downspout should be positioned plumb with the building corner and square with the building. Join with a connector. Add the end cap. Use either inside or outside corners where gutters make a turn.

Vinyl gutters and components are installed in a manner similar to metal ones.

INSTALLING DOWNSPOUTS

Metal or vinyl downspouts are fastened to the wall of the building in specified locations with aluminum straps. Downspouts should be fastened at the top and bottom and every 6 feet in between for long lengths.

The connection between the downspout and the gutter is made with 45-degree elbows and short

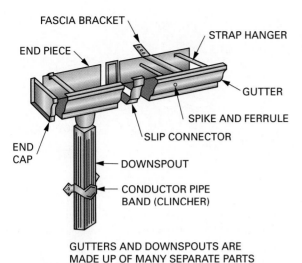

FASCIA BRACKET
STRAP HANGER
END PIECE
GUTTER
SPIKE AND FERRULE
END CAP
SLIP CONNECTOR
DOWNSPOUT
CONDUCTOR PIPE BAND (CLINCHER)

GUTTERS AND DOWNSPOUTS ARE MADE UP OF MANY SEPARATE PARTS

PIECE NEEDED	DESCRIPTION
	GUTTER COMES IN VARIOUS LENGTHS
	SLIP JOINT CONNECTOR USED TO CONNECT JOINTS OF GUTTER
	END CAPS – WITH OUTLET USED AT ENDS OF GUTTER RUNS
	END PIECE – WITH OUTLET USED WHERE DOWNSPOUT CONNECTS
	OUTSIDE MITER USED FOR OUTSIDE TURN IN GUTTER
	INSIDE MITER USED FOR INSIDE TURN IN GUTTER

PIECE NEEDED	DESCRIPTION	PIECE NEEDED	DESCRIPTION
	FASCIA BRACKET USED TO HOLD GUTTER TO FASCIA ON WALL		ELBOW – STYLE B FOR DIVERTING DOWNSPOUT TO LEFT OR RIGHT
	STRAP HANGER CONNECTS TO EAVE OF ROOF TO HOLD GUTTER		CONNECTOR PIPE BAND OR CLINCHER USED TO HOLD DOWNSPOUT TO LEFT OR RIGHT
	STRAINER CAP SLIPS OVER OUTLET IN END PIECE AS A STRAINER		SHOE USED TO LEAD WATER TO SPLASHER BLOCK
	DOWNSPOUT COMES IN 10' LENGTHS		MASTIC USED TO SEAL ALUMINUM GUTTERS AT JOINTS
	ELBOW – STYLE A FOR DIVERTING DOWNSPOUT IN OR OUT FROM WALL		SPIKE AND FERRULE USED TO HOLD GUTTER TO EAVES OF ROOF

FIGURE 65–3 **Components of a metal gutter system.**

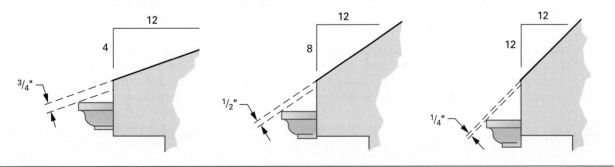

FIGURE 65–4 **The height of the gutter on the fascia is in relation to the slope of the roof.**

straight lengths of downspout **(Figure 65–5).** The connection will depend on the offset of the gutter from the downspout. Because water runs downhill, care should be taken when putting the downspout pieces together. The downspout components are assembled where the upper piece is inserted into the lower one **(Figure 65–6).** This makes the joint lap

such that the water cannot escape until it reaches the bottom-most piece. An elbow, called a *shoe*, should be used with a splash block at the bottom of the downspout. This leads water away from the foundation. An alternate method is to connect the downspout with underground piping that carries the water away from the foundation. The piping can be

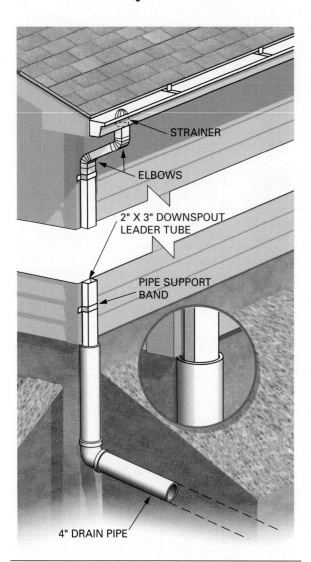

FIGURE 65–5 Typical downspout leader tubes fastened in place with support bands.

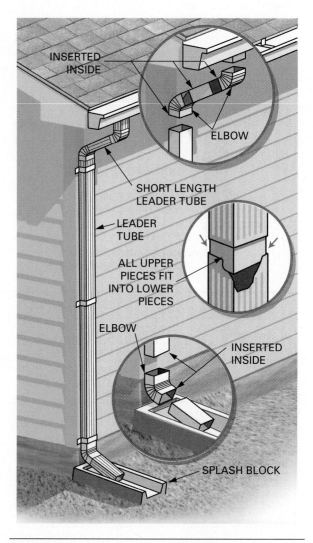

FIGURE 65–6 Upper gutter components are inserted into lower ones to ensure that the downspout does not leak.

connected to storm drains and drywells or piped to the surface elsewhere. Storm water as found in gutter downspouts should never be connected to footing or foundation drains. Strainer caps should be placed over gutter outlets if water is conducted by this alternate method. Leaves and other debris that fall into gutters flow into the drainage system and can cause clogging problems.

Key Terms

battens	cornice return	J-channel	snub cornice
blind nail	double coursing	lookouts	staggered coursing
board and batten	downspout	open cornice	starter strip
board on board	drip	plancier	striated
box cornice	face nail	preacher	subfascia
brake	fancy butt	rake cornice	undercourse
coil stock	frieze	ribbon coursing	undersill trim
corner board	gutter	single coursing	water table
cornice molding	headlap	snaplock punch	

Review Questions

Select the most appropriate answer.

1. Bevel siding is applied
   a. horizontally.
   b. vertically.
   c. horizontally or vertically.
   d. horizontally, vertically, or diagonally.

2. Fiber-cement siding products are
   a. made to look like concrete blocks.
   b. installed similar to wood siding.
   c. usually cut to size indoors.
   d. only installed horizontally.

3. A particular advantage of bevel siding over other types of horizontal siding is that
   a. it comes in a variety of prefinished colors.
   b. it has a constant weather exposure.
   c. the weather exposure can be varied slightly.
   d. application can be made in any direction.

4. When applying horizontal siding, it is desirable to
   a. maintain exactly the same exposure with every course.
   b. apply full courses above and below windows.
   c. work from the top down.
   d. use a water table.

5. In order to lay out a story pole for courses of horizontal siding, which of the following must be known?
   a. the width of windows and doors
   b. the kind and size of finish at the eaves and foundation
   c. the location of windows, doors, and other openings
   d. the length of the wall to which siding is to be applied

6. With wood shingle siding
   a. the exposure may be varied.
   b. butt seams should line up vertically.
   c. each piece should be fastened with four nails.
   d. all of the above.

7. When installing aluminum or vinyl siding, drive nails
   a. tightly against the flange.
   b. loosely against the flange.
   c. with small heads.
   d. colored the same as the siding.

8. Vinyl siding is installed with
   a. vertical butt seams randomly placed.
   b. butt seam overlapping away from view.
   c. a loose fit to trim pieces.
   d. all of the above.

9. To allow for expansion when installing solid vinyl starter strips, leave a space between the ends of at least
   a. ⅛ inch.   c. ⅜ inch.
   b. ¼ inch.   d. ½ inch.

10. The exterior trim that extends up the slope of the roof on a gable end is called the
   a. box finish.   c. return finish.
   b. rake finish.   d. snub finish.

11. A member of the cornice fastened in the vertical position to the rafter tails is called the
   a. drip.   c. soffit.
   b. fascia.   d. frieze.

12. A soffit is the part of a cornice that
   a. may be horizontal under the rafter tails.
   b. is vertical and attached to the ends of rafters.
   c. serves as a drip cap.
   d. is often made of 2 × 4s.

13. Care should be taken when installing gutters so that
   a. they are level with the fascia.
   b. downspouts are in the center.
   c. downspout leader tubes are connected to the foundation drains.
   d. parts are installed with the idea that water runs downhill.

14. The square feet (SF) area of siding needed for a rectangular house with a hip roof measuring 30 × 40 feet where the wall height is 8 feet is
   a. unable to be determined because roof height is not known.
   b. 78 SF minus window and door areas.
   c. 1,120 SF minus window and door areas.
   d. 9,600 SF minus window and door areas.

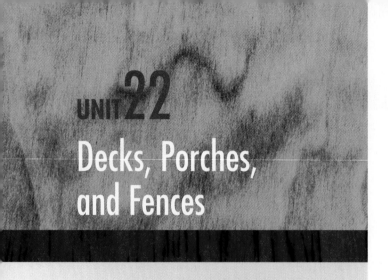

Among the final steps in finishing the exterior is the building of porches, decks, fences, and other accessory structures. Plans may not always show specific construction details. Therefore, it is important to know some of the techniques used to build these structures.

Accents to a building may come in the form of decks and porches. They serve as reminders to the overall workmanship of the house. Take care to install material in a neat and professional manner.

OBJECTIVES

After completing this unit, the student should be able to:

- describe the construction of and kinds of materials used in decks and porches.
- lay out and construct footings for decks and porches.
- install supporting posts, girders, and joists.
- apply decking in the recommended manner and install flashing, for an exposed deck, against a wall.
- construct deck stairs and railings.
- describe several basic fence styles.
- design and build a straight and sturdy fence.

66 Deck and Porch Construction

ood porches and decks are built to provide outdoor living areas for various reasons, in both residential and commercial construction **(Figure 66–1)**. The construction of both is similar. However, a porch is covered by a roof. Its walls may be enclosed with wire mesh screens for protection against insects. With screen and storm window combinations, glass replaces screens to keep the porch comfortably warm in the cold months.

DECK MATERIALS

Decking materials must be chosen for strength and durability, as well as appearance and resistance to decay. Redwood, cedar, and pressure-treated southern yellow pine are often used as decking boards. Other decking materials available include Timber Tech® and Trex®. These decking products are made from a mixture of plastic and sawdust. They are cut, fit, and fastened in the same manner as wood and have the added benefit of being made mostly from recycled material.

If not specified, the kind, grade, and sizes of material must be selected before building a deck. Also, the size and kind of fasteners, connectors, anchors, and other hardware must be determined.

Lumber

All lumber used to build decks should be **pressure-treated** with preservatives or be from a decay-resistant species, such as redwood or cedar. Remember, it is the heartwood of these species that is resistant to decay, not the sapwood. Either *all-heart* or pressure-treated lumber should be used wherever there is a potential for decay. This is essential for posts that are close to the ground and other parts subject to constant moisture. (A description of pressure-treated lumber and its uses can be found in Chapter 33.)

Lumber Grades. For pressure-treated southern pine and western cedar, #2 grade is structurally adequate for most applications. Appearance can be a deciding factor when choosing a grade. If a better appearance is desired, higher grades should be considered.

A grade called **construction heart** is the most suitable and most economical grade of California redwood for deck posts, beams, and joists. For decking and rails, a grade called construction common is recommended. Better appearing grades are available. However, they are more expensive.

Two grades of redwood, called redwood deck heart and redwood deck common, are manufactured especially for exterior walking surfaces. Two grades of decking, standard and premium, are also available in pressure-treated southern pine. Special decking grades of western cedar may also be obtained. The lumber grade or special purpose is shown in the grade stamp **(Figure 66–2)**.

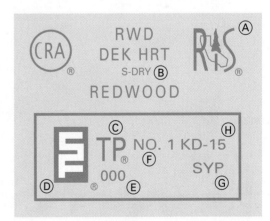

A) REDWOOD INSPECTION SERVICE
B) NOT MORE THAN 19% M.C.
C) INSPECTION SERVICE: TIMBER PRODUCTS INSPECTION
D) SYMBOL FOR MEMBERS OF SFPA (OPTIONAL)
E) MILL IDENTIFICATION NUMBER
F) LUMBER GRADE
G) LUMBER SPECIES
H) KILN DRIED (MAX. 15% M.C.)

FIGURE 66–2 Grade stamps show the grade and special purpose of lumber used in deck construction. *(Courtesy of California Redwood Association; Courtesy of Southern Forest Products Association)*

FIGURE 66–1 Wood decks are built in many styles. This multilevel deck blends well with the landscape. *(Courtesy of California Redwood Association)*

Lumber Sizes. Specific sizes of supporting posts, girders, and joists depend on the spacing and height of supporting posts and the spacing of girders and joists. In addition, the sizes of structural members depend on the type of wood used and the weight imposed on the members. Too many factors prohibit generalization about sizes of structural members. Check with local building officials or with a professional to determine the sizes of structural members for specific deck construction. Determining sizes with incomplete information may result in failure of undersized members. Unnecessary expense is incurred with the use of oversized members.

Fasteners

All nails, fasteners, and hardware should be stainless steel, aluminum, or hot-dipped galvanized. Electroplated galvanizing is not acceptable because the coating is too thin. In addition to corroding and eventual failure, poor quality fasteners will react with substances in decay-resistant woods and cause unsightly stains.

BUILDING A DECK

Most wood decks consist of posts, set on footings, supporting a platform of *girders* and *joists* covered with **deck boards.** Posts, **rails, balusters,** and other special parts make up the railing **(Figure 66–3).** Other parts, such as shading devices, privacy screens, benches, and planters, lend finishing touches to the area.

Installing a Ledger

If the deck is to be built against a building, a **ledger,** usually the same size as the joists, is bolted to the wall for the entire length of the deck **(Figure 66–4).** The ledger acts as a beam to support joists that run at right angles to the wall. It is installed to a level line. Its top edge is located to provide a comfortable step down from the building after the decking is applied. The ledger height may be used as a benchmark for establishing the elevations of supporting posts and girders.

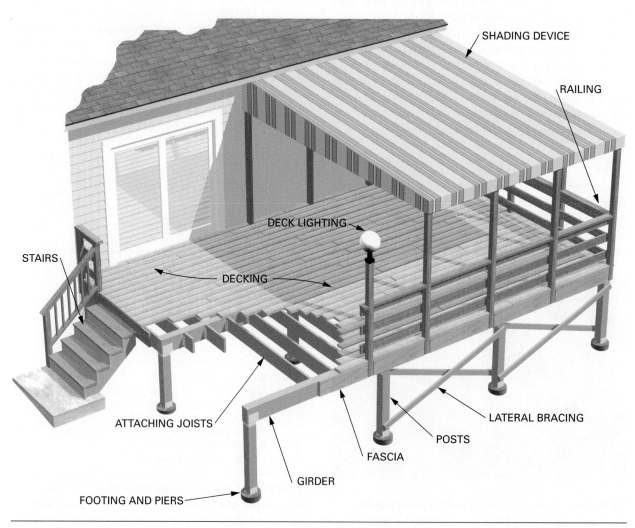

FIGURE 66–3   The components of a deck.

After the deck is applied, a flashing is installed under the siding and on top of the deck board. Caulking is applied between the deck and the flashing. The flashing is then fastened, close to and along its outside edge, with nails spaced closely together. The outside edge of the flashing should extend beyond the ledger.

Footing Layout and Construction

Footings for the supporting posts must be accurately located. To determine their location, erect batter boards and stretch lines in a manner previously described in Chapter 26. All footings require digging a hole and filling it with concrete **(Figure 66–5)**.

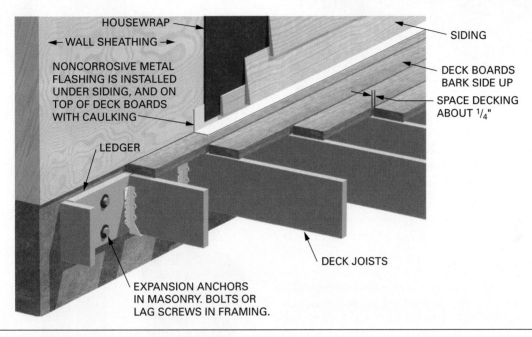

FIGURE 66–4   **A ledger is made weathertight with flashing.**

FOOTING AND POST LAYOUT AND EXCAVATION

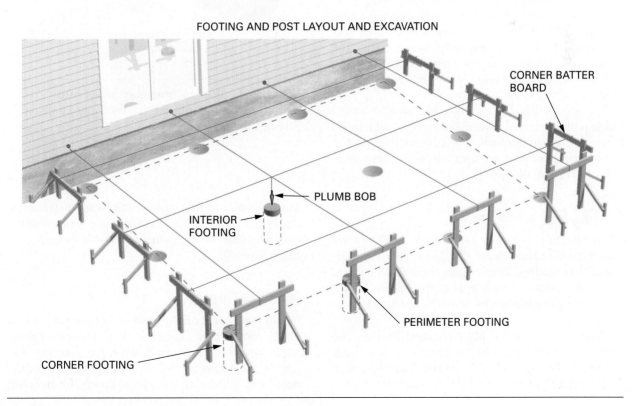

FIGURE 66–5   **Footing and post layout and excavation.**

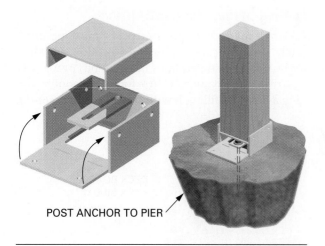

POST ANCHOR TO PIER

FIGURE 66–6   **Post anchors may be fastened to concrete using an expansion bolt.** *(Courtesy of Simpson Strong-Tie Company)*

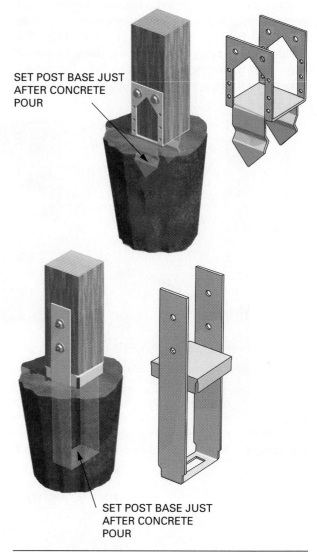

SET POST BASE JUST AFTER CONCRETE POUR

SET POST BASE JUST AFTER CONCRETE POUR

FIGURE 66–7   **Post anchors may be set into concrete after it is placed.** *(Courtesy of Simpson Strong-Tie Company)*

Footing Size and Style

In stable soil and temperate climate, the footing width is usually made twice the width of the post it is supporting. The footing depth reaches undisturbed soil, at least 12 inches below grade. In cold climates the footing should extend below the frost line.

Several footing styles are commonly used. One method is to partially fill the footing hole with concrete to within a few inches from the top. Set a precast pier 2 inches into the wet concrete. After the concrete has set, attach a post anchor to the top of the precast pier **(Figure 66–6)**. Align piers and anchors with the layout lines.

The top of the footing may be brought above grade with the use of a wood box or fiber-tube form. Place concrete in the footing hole. Bring it to the top of the form. Set post anchors while the concrete is still wet **(Figure 66–7)**.

Erecting Supporting Posts

All supporting posts are set on footings. They are then plumbed and braced. Cut posts, for each footing, a few inches longer than their final length. Tack the bottom of each post to the anchor. Brace them in a plumb position in both directions **(Procedure 66–A)**.

When all posts are plumbed and braced, the tops must be cut level to the proper height. From the height of the deck, deduct the deck thickness and the depth of the girder. Mark on a corner post.

Mark the other posts by leveling from the first post marked. Mark each post completely around using a square. Cut the tops with a portable circular saw **(Figure 66–8)**.

Installing Girders

Install the girders on the posts using post-and-beam metal connectors. The deck should slope slightly, about ⅛ inch per foot, away from the building. The size of the connector will depend on the size of the posts and girders. Install girders with the crowned

Procedure 66–A Erecting and Bracing Supporting Posts

(Courtesy of Simpson Strong-Tie Company)

STEP **1** FASTEN POST TO ANCHOR WITH NAILS OR BOLTS.

STEP **2** NAIL BRACE TOPS TO POSTS.

STEP **3** DRIVE STAKE AT THE END OF EACH BRACE.

STEP **4** PLUMB POSTS EACH WAY.

STEP **5** FASTEN BRACES TO STAKES.

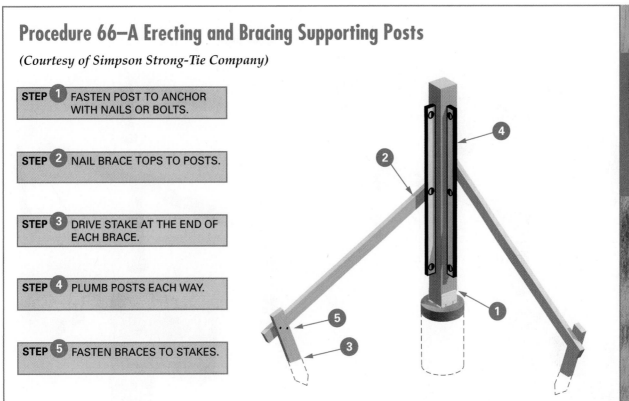

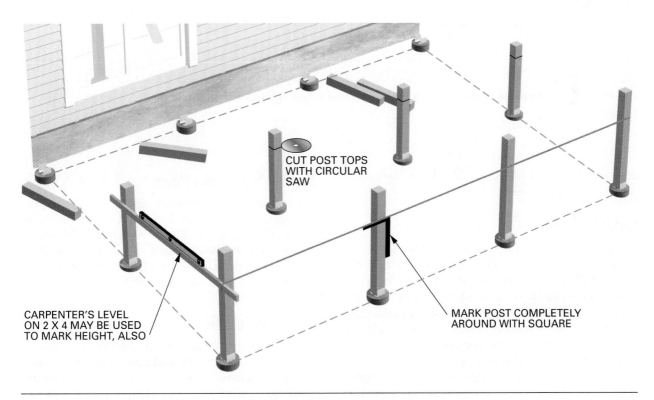

CUT POST TOPS WITH CIRCULAR SAW

CARPENTER'S LEVEL ON 2 X 4 MAY BE USED TO MARK HEIGHT, ALSO

MARK POST COMPLETELY AROUND WITH SQUARE

FIGURE 66–8 The post height is determined and the tops cut level with each other.

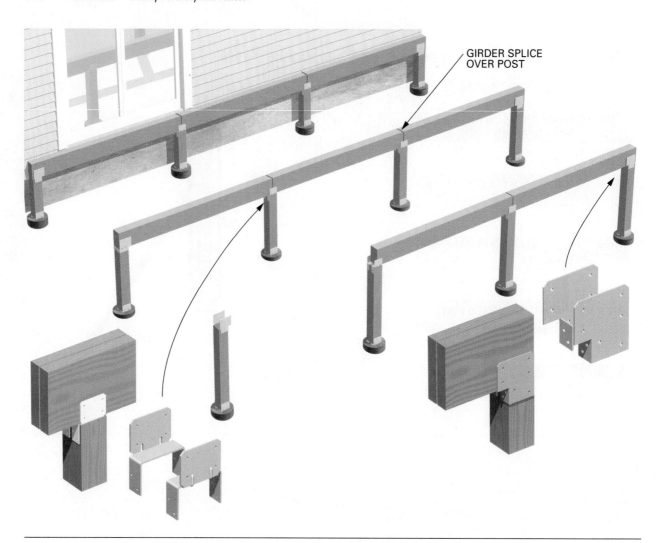

GIRDER SPLICE
OVER POST

FIGURE 66–9 Girders are installed with their crowned edges up and anchored to supporting posts.

edge up. Any splice joints should fall over the center of the post **(Figure 66–9).**

Installing Joists

Joists may be placed over the top or between the girders. When joists are hung between the girders, the overall depth of the deck is reduced. This provides more clearance between the frame and the ground. For decking run at right angles, joists may be spaced 24 inches on center. Joists should be spaced 16 inches on center for diagonal decking.

Lay out and install the joists in the same manner as described in earlier chapters. Use appropriate hangers if joists are installed between girders **(Figure 66–10).** When joists are installed over girders, use recommended framing anchors. Make sure all joists are installed with their crowned edges facing upward.

Bracing Supporting Posts

If the deck is 4 feet or more above the ground, the supporting posts should be braced in a manner similar to that shown in **Figure 66–11.** Use minimum 1 × 6 braces for heights up to 8 feet. Use minimum 1 × 8 braces applied continuously around the perimeter for higher decks.

Applying Deck Boards

Specially shaped **radius edge decking** is available in both pressure-treated and natural decay-resistant lumber. It is usually used to provide the surface and walking area of the deck. Dimension lumber of 2-inch thickness and widths of 4 and 6 inches is also widely used. Lay boards with the bark side up to minimize cupping **(Figure 66–12).**

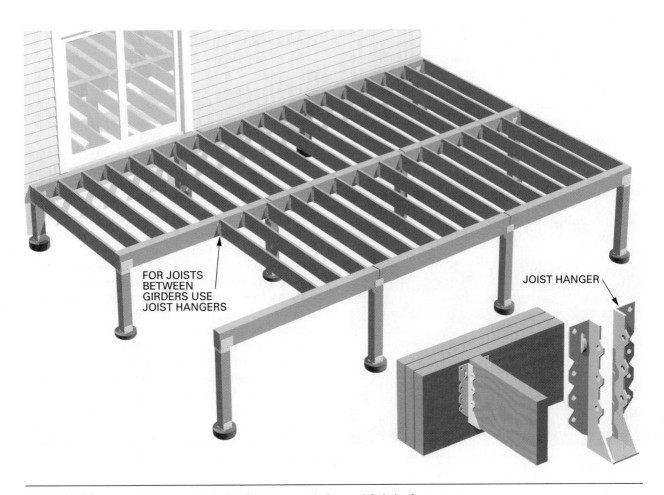

FIGURE 66-10 Deck joists are installed between girders with joist hangers.

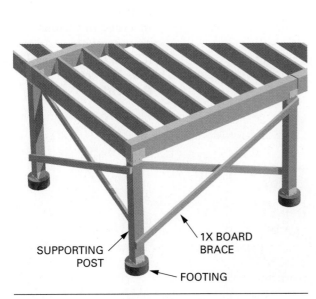

FIGURE 66-11 Supporting posts must be braced if the deck is 4 feet or more above the ground.

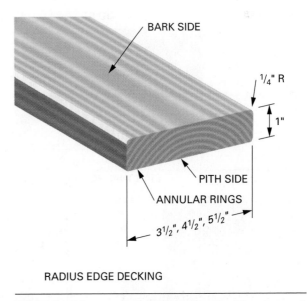

RADIUS EDGE DECKING

FIGURE 66-12 Special radius edge decking is used and applied with the bark side up to resist cupping. (*Courtesy of Southern Pine Marketing Council*)

Boards are usually laid parallel with the long dimension of the deck **(Figure 66–13)**. However, because **deck boards** usually do not come longer than 16 feet, it may be desirable to lay the boards parallel to the short dimension, if their length will span it, to eliminate end joints in the decking. Boards may also be laid in a variety of patterns including diagonal, herringbone, and parquet. Make sure the support-

ing structure has been designed and built to accommodate the design **(Figure 66–14)**.

Much care should be taken with the application of deck boards. Snap a straight line as a guide to apply the starting row. Start at the outside edge if the deck is built against a building. A ripped and narrower ending row of decking is not as noticeable against the building as it is on the outside edge of the deck **(Figure 66–15)**.

Straighten boards as they are installed. Maintain about a ¼-inch space between dry boards. If the decking boards are wet, as with most pressure-treated boards, they will shrink as they dry. Nailing them tight together is the preferred method as the ¼" space will appear when the lumber reaches equilibrium moisture content.

If the deck boards do not span the entire width of the deck, cut the boards so their ends are centered over joists. Make tight-fitting end joints. Stagger them between adjacent rows. Predrill holes for fasteners to prevent splitting the ends. Let the end of each row overhang the deck. Use two screws or nails in each joist. Drive nails at an angle. Set the heads below the surface. This will keep the nails from working loose and the heads from staining the surface.

When approaching the end, it is better to increase or decrease the spacing of the last six or seven rows of decking. This way you will end with a row that is nearly equal in width to all the rest, rather than with a narrow strip. When all the deck boards are laid, snap lines and cut the overhanging ends. Apply a preservative to them.

FIGURE 66–13 **Deck boards are often installed parallel to the length of the deck.** *(Courtesy of California Redwood Association)*

FIGURE 66–14 **Deck boards may be installed in various arrangements.** *(Courtesy of Southern Pine Marketing Council)*

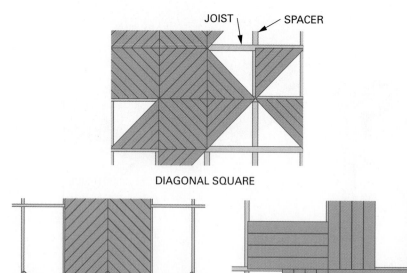

DIAGONAL SQUARE

HERRINGBONE

PARQUET

DECK BOARD PATTERNS

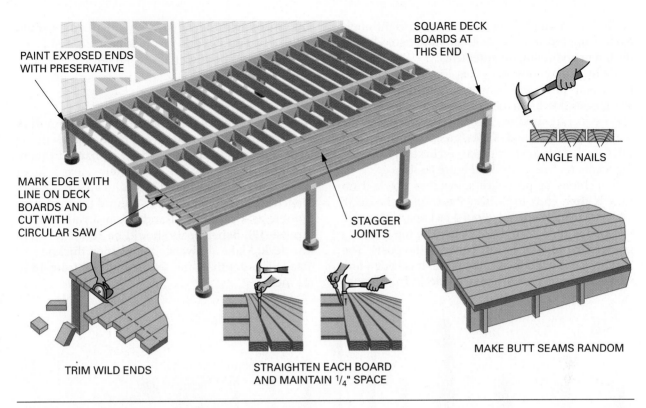

FIGURE 66–15 Techniques for installing deck boards.

Applying Trim

A fascia board may be fastened around the perimeter of the deck. Its top edge should be flush with the deck surface. The fascia board conceals the cut ends of the decking and the supporting members below. The fascia board is optional.

Stairs and Railings

Stairs. Most decks require at least one or two steps leading to the ground. To protect the bottom ends of the stair carriage, they should be treated with preservative and supported by an above-grade concrete pad **(Figure 66–16)**. Stair layout and construction are described in Chapter 48. Stairs with more than two risers are generally required to have at least one handrail. The design and construction of the stair handrail should conform to that of the deck railing.

Rails. There are numerous designs for deck railings. All designs must conform to certain code requirements. Most codes require at least a 36-inch-high railing around the exposed sides, if the deck is more than 30 inches above the ground. In addition, some codes specify that no openings in the railing should allow a 4-inch sphere to pass through it. Each linear foot of railing must be strong enough to resist a pressure of

FIGURE 66–16 Stairs for decks are usually constructed with a simple basement or utility design.

20 lbs./sq. ft. applied horizontally at a right angle against the top rail. Check local building codes for deck stair and railing requirements.

Railings may consist of posts, top, bottom, and intermediate rails, and balusters. Sometimes **lattice work** is used to fill in the rail spaces above the deck. It is frequently used to close the space between the deck and the ground. Posts, rails, balusters, and other deck parts are manufactured in several shapes especially for use on decks **(Figure 66–17)**.

Stanchions or posts are sometimes notched on their bottom ends to fit over the edge of the deck. They are usually spaced about 4 feet apart. They are fastened with lag screws or bolts. The top rail may go over the tops or be cut between the posts. The bottom rail is cut between the posts. It is kept a few inches (no more than 8) above the deck. The remain-ing space may be filled with intermediate rails, balusters, lattice work, or other parts in designs as desired or as specified **(Figure 66–18)**.

Deck Accessories

There are many details that can turn a plain deck into an attractive and more comfortable living area. *Shading structures* are built in many different designs. They may be completely closed in or spaced to provide filtered light and air circulation. Benches partially or entirely around the deck may double as a place to sit and act as a railing **(Figure 66–19)**. Bench seats should be 18 inches from the deck. Make allowance for cushion thickness, if used. The depth of the seat should be from 18 to 24 inches.

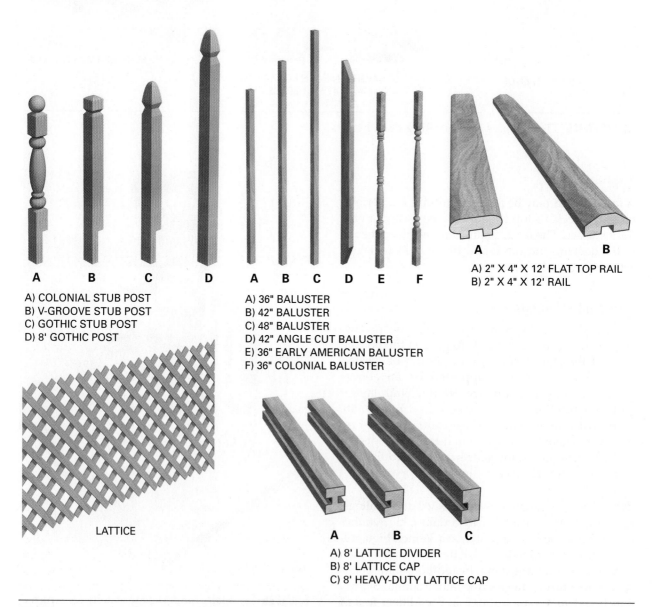

A) COLONIAL STUB POST
B) V-GROOVE STUB POST
C) GOTHIC STUB POST
D) 8' GOTHIC POST

A) 36" BALUSTER
B) 42" BALUSTER
C) 48" BALUSTER
D) 42" ANGLE CUT BALUSTER
E) 36" EARLY AMERICAN BALUSTER
F) 36" COLONIAL BALUSTER

A) 2" X 4" X 12' FLAT TOP RAIL
B) 2" X 4" X 12' RAIL

LATTICE

A) 8' LATTICE DIVIDER
B) 8' LATTICE CAP
C) 8' HEAVY-DUTY LATTICE CAP

FIGURE 66–17 Railing parts are manufactured in many shapes.

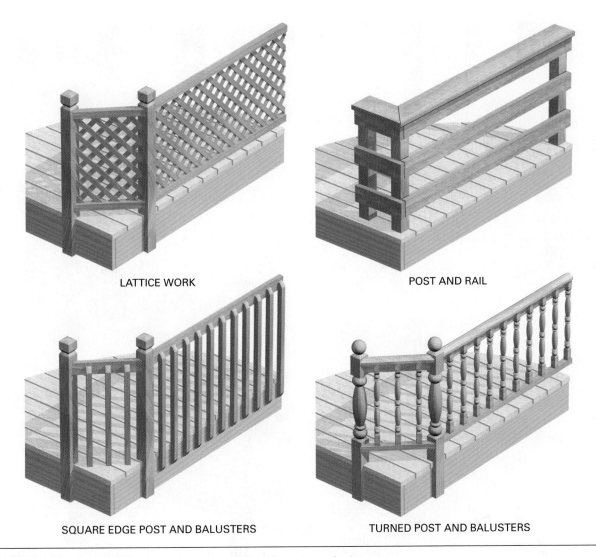

LATTICE WORK

POST AND RAIL

SQUARE EDGE POST AND BALUSTERS

TURNED POST AND BALUSTERS

FIGURE 66–18 Deck railings are constructed with various designs.

FIGURE 66–19 **A plain deck can be made more attractive and comfortable with benches and shading devices.** *(Courtesy of California Redwood Association)*

PORCHES

The porch deck is constructed in a similar manner to an open deck. Members of the supporting structure may need to be increased in size and the spans decreased to support the weight of the walls and roof above. Work from plans drawn by professionals or check with building officials before starting. Porch walls and roofs are framed and finished as described in previous chapters **(Figure 66–20).**

SUMMARY

Decks and porches are designed in many different ways. There are other ways to construct them besides the procedures described in this chapter. However, making the layout, building the supporting structure, applying the deck, constructing the railing and stairs, and, in the case of a porch, building the walls and roof are basic steps that can be applied to the construction of practically any deck or porch.

FIGURE 66–20 A porch is composed of a deck enclosed by walls and a roof.

67 Fence Design and Erection

A superior fence combines utility and beauty **(Figure 67–1).** It may define space, create privacy, provide shade or shelter, screen areas from view, or form required barriers around swimming pools and other areas for the protection of small children.

The design of a fence is often the responsibility of the builder. With creativity and imagination, he or she can construct an object of beauty and elegance that also fulfills its function. The objective of this chapter is to create an awareness of the importance

FIGURE 67–1 A fence can serve its intended purpose and also enhance the surroundings. *(Courtesy of California Redwood Association)*

FIGURE 67–2 Sometimes the site, such as a steep slope, may affect the design of a fence.

of design by showing several styles of fences and the methods used to erect them.

FENCE MATERIAL

Fences are not supporting structures. Lower, knotty grades of lumber may be used to build them. This is not the case if appearance is a factor and you want to show the natural grain of the wood. Many kinds of softwood may be used as long as they are shielded from the weather by protective coatings of paint, stains, or preservatives. However, for wood posts that are set into the ground and other parts that may be subjected to constant moisture, pressure-treated or all-heart, decay-resistant lumber must be used.

The same type of fasteners that are used on exposed decks should be used to build fences. Inferior hardware and fasteners will corrode and cause unsightly stains when in contact with moisture.

FENCE DESIGN

Fences consist of posts, rails, and fence boards. Fences may be constructed in almost limitless designs. Zoning regulations sometimes restrict their height or placement. Often the site will affect the design. For example, the fence may have to be stepped like a staircase on a steep slope **(Figure 67–2).** Fences on property lines can be designed to look attractive from either side. Fence designs may block wind and sunlight. Fence boards may be spaced in many attractive patterns. Most fences are constructed in several basic styles, each of which can be designed in numerous ways. Provisions should be made in the fence design to drain water from any area where it may otherwise be trapped.

Picket Fence

The **picket fence** is commonly used on boundary lines or as barriers for pets and small children. Usually not more than 4 feet high, the pickets are spaced to provide plenty of air and also to conserve material. The tops of the pickets may be shaped in various styles. Or, they may be cut square, with ends exposed or capped with a molding. The pickets may be applied with their tops in a straight line or in curves between posts **(Figure 67–3).** When pickets are applied with their edges tightly together, the assembly is called a **stockade fence.** Stockade fences are usually higher and are used when privacy is desired.

Board-on-Board Fence

The **board-on-board** fence is similar to a picket fence. However, the boards are alternated from side to side so that the fence looks the same from both sides **(Figure 67–4).** The boards may vary in height and spacing according to the degree of privacy and protection from wind and sun desired. The tops or edges of the boards may be shaped in many different designs.

Lattice Fence

This fence gets its name from narrow strips of wood called **lattice.** Strips are spaced by their own width. Two layers are applied at right angles to each other,

FIGURE 67–3   The picket fence is constructed in many styles. *(Courtesy of California Redwood Association)*

FIGURE 67–4   A board-on-board fence is similar to a picket fence. However, it looks the same on both sides.

either diagonally or in a horizontal and vertical fashion, to form a lattice work panel. Panels of various sizes can be prefabricated and installed between posts and rails, similar to a lattice work deck railing as shown in Figure 66–17.

Panel Fence

The **panel** fence creates a solid barrier with boards or panels fitted between top and bottom rails **(Figure 67–5).** Fence boards may be installed diagonally or in other appealing designs. Alternating the panel de-

sign provides variety and adds to the visual appeal of the fence. A small space should be left between panel boards to allow for swelling in periods of high humidity. In many cases, two or more basic styles may be combined to enhance the design. **Figure 67–6** shows lattice panels combined with board panels.

FIGURE 67–5   The panel fence provides shade and privacy, and it also restricts air movement. *(Courtesy of California Redwood Association)*

FIGURE 67–6 Solid panels are combined with lattice work panels for an attractive design. (*Courtesy of California Redwood Association*)

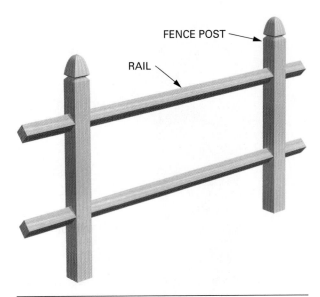

FIGURE 67–8 The post-and-rail fence is an inexpensive design for long boundary lines.

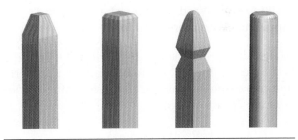

FIGURE 67–9 Fence post tops may be shaped in various ways to enhance the design of the fence.

FIGURE 67–7 The louvered fence provides privacy and lets air through. (*Courtesy of California Redwood Association*)

Louvered Fence

The **louvered fence** is a panel fence with vertical boards set at an angle **(Figure 67–7)**. The fence permits the flow of air through it and yet provides privacy. This fence is usually used around patios and pools.

Post-and-Rail Fence

The **post-and-rail fence** is a basic and inexpensive style normally used for long boundaries **(Figure 67–8)**. Designs include two or more square edge or round rails, of various thicknesses, widths, and diameters, cut between the posts or fastened to their edges. Most post-and-rail fence designs have large openings. They are not intended as barriers.

FENCE POST DESIGN

Fence posts are usually wood or iron. Wood posts are usually 4 × 4. Larger sizes may be used, depend-

ing on the design. The post tops may be shaped in various ways to enhance the design of the fence **(Figure 67–9)**. To conserve material and reduce expenses, a 4 × 4 post may be made to appear larger by applying furring and then boxing it in with 1-inch boards. The top may then be capped with shaped members and molding in various designs to make an attractive fence post **(Figure 67–10)**.

Iron posts may be pipe or solid rod ranging in size from 1 inch to 1¼ inches in diameter for fences 3 to 4 feet tall. Larger diameters should be used for higher fences. Iron fence posts may be boxed in to simulate large wood posts. The tops are then usually capped with various shaped members similar to boxed wood posts **(Figure 67–11)**. Iron posts should be galvanized or otherwise coated to resist corrosion.

BUILDING A FENCE

The first step in building a fence is to set the fence posts. Locate and stretch a line between the end posts. If it is not possible to stretch a line because of steep sloping land, set up a transit-level to lay out a

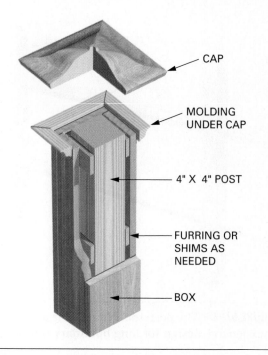

FIGURE 67–10 Wood posts may be boxed and capped to give a more solid look.

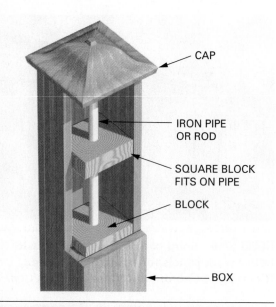

FIGURE 67–11 An iron fence post may be boxed and capped.

straight line. (See Chapter 26, Laying Out Foundation Lines.) If the fence is to be built on a property line, make sure that the exact locations of the boundary markers are known.

Setting Posts

Posts are generally placed about 8 feet apart, to their center lines. Mark the post locations with stakes along the fence line. Dig holes about 10 inches in diameter with a post hole digger. The depth of the hole depends on the height of the fence. Higher fences require that posts be set deeper **(Figure 67–12).** The bottom of the hole should be filled with gravel or stone. This provides drainage and helps eliminate moisture to extend the life of the post.

Posts may be set in the earth. For the strongest fences, however, set the posts in concrete. Set the end

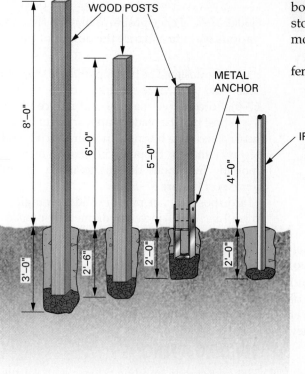

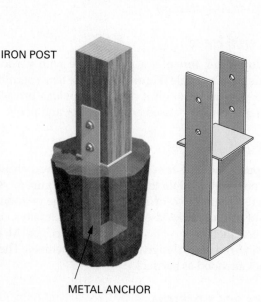

FIGURE 67–12 Design requirements for fence posts.

Procedure 67–A Setting Fence Posts

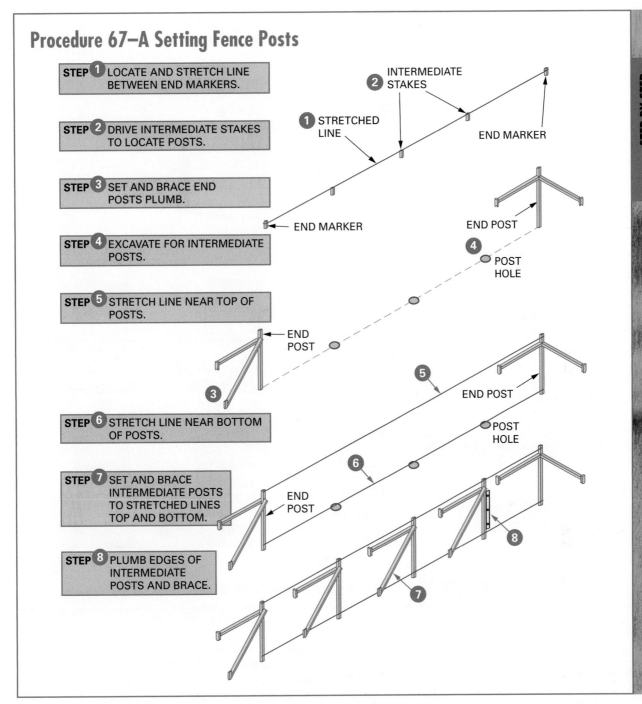

STEP ❶ LOCATE AND STRETCH LINE BETWEEN END MARKERS.

STEP ❷ DRIVE INTERMEDIATE STAKES TO LOCATE POSTS.

STEP ❸ SET AND BRACE END POSTS PLUMB.

STEP ❹ EXCAVATE FOR INTERMEDIATE POSTS.

STEP ❺ STRETCH LINE NEAR TOP OF POSTS.

STEP ❻ STRETCH LINE NEAR BOTTOM OF POSTS.

STEP ❼ SET AND BRACE INTERMEDIATE POSTS TO STRETCHED LINES TOP AND BOTTOM.

STEP ❽ PLUMB EDGES OF INTERMEDIATE POSTS AND BRACE.

posts first. Use a level to ensure that the posts are plumb in both directions. Brace them securely. The height of the post is determined by measuring from the ground, especially when the ground slopes.

If top fence rails run across the tops of the posts, then the posts are left long. The tops are cut later in a straight or contoured line, as necessary. Make sure the face edges of the posts are aligned with the length of the fence. After the posts are braced, stretch lines between the end posts at the top and at the bottom.

Set and brace intermediate posts. The face edges of the posts are kept to the stretched line at the top

and bottom. They are plumbed in the other direction with a level. Lag screws or spikes partially driven into the bottom of the post strengthen the set of fence posts in concrete.

Place concrete around the posts. The concrete may be placed in the hole dry as the moisture from the ground will provide enough water to hydrate the cement. Tamp it well into the holes. Form the top so the surface pitches down from the posts. Instead of setting the posts in concrete, metal anchors may be embedded. The posts are installed on them after the concrete has hardened **(Procedure 67–A).**

Procedure 67–B Cutting Railing to Meet a Twisted Post

FENCE POST

RAILS PREVIOUSLY INSTALLED

RAIL

SUPPORT RAILS WITH NAILS ON CHALK LINE

RAILS TO BE INSTALLED

FENCE POST

NAIL

RAILS BEING LAID OUT

CHALK LINE

STEP 1 DRAW LINE ON RAIL IN LINE WITH EDGE OF FENCE POST.

STEP 2 SQUARE LINE ACROSS FENCE RAIL.

STEP 3 CUTTING LINE IS DRAWN PARALLEL TO FIRST LINE.

STEP 4 DRAW LINE ON RAIL IN LINE WITH EDGE OF FENCE POST.

STEP 5 SQUARE LINE TO EDGE OF FENCE RAIL.

STEP 6 DRAW CUTTING LINE PARALLEL TO FIRST LINE.

TWIST OF FENCE POSTS IS EXAGGERATED FOR CLARITY

TWISTED FENCE POSTS

RAIL

90°

POSTS TWISTED IN OTHER DIRECTION

RAIL

90°

NOTE: BOTH ENDS MUST BE MARKED WITHOUT MOVING RAIL

Installing Fence Rails

Usually two or three horizontal rails are used on most fences. However, the height or the design may require more. Rails may run across the face or be cut in between the fence posts. If rails are cut between posts, they are installed to a snapped chalk line across the edges of the posts. They are secured by toenailing or with metal framing connectors. The bottom rail should be kept at least 6 inches above the ground.

Very often, the faces of wood posts are not in the same line as the rails. Therefore, rail ends must be cut with other than square ends to fit between posts. **Procedure 67–B** shows a method of laying out rail ends to fit between twisted posts.

When iron fence posts are used, holes are bored in the rails. The rails are then installed over the previ-

ously set iron posts. Splices are made on the ends to continue the rail in a straight, unbroken line. Special metal pipe grips are made to fasten fence rails to iron posts **(Figure 67–13).** If iron posts are boxed with wood, then the rails are installed in the same manner as for wood posts.

On some rustic-style post-and-rail fences, the rails are doweled into the posts. In this case, post and rails must be installed together, one section at a time.

Applying Spaced Fenceboards and Pickets

Fasten pickets in plumb positions with their tops to the correct height at the starting and ending points. Stretch a line tightly between the tops of the two pickets. If the fence is long, temporarily install intermediate pickets to support the line from sagging.

METAL GRIP TIES ARE MANUFACTURED TO
ATTACH FENCE RAILS TO PIPE FENCE POSTS

PIPE GRIP TIE

FIGURE 67–13 **Fence rails may be attached to iron fence posts.**

Sight the line by eye to see if it is straight. If not, make adjustments and add more support pickets if necessary. Use a picket for a spacer, and fasten pickets to the rails. If the spacing is different, rip a piece of lumber for use as a spacer.

Cut only the bottom end of the pickets when trimming their height. The bottom of pickets should not touch the ground. Place a 2-inch block on the ground. Turn the picket upside down with its top end on the block. Mark it at the chalk line. Fasten the picket with its top end to the stretched line (**Procedure 67–C**).

Continue cutting and fastening pickets using the spacer. Keep their tops to the line. Check the pickets for plumb frequently. If not plumb, bring back into plumb gradually with the installation of three or four pickets.

Stop 3 or 4 feet from the end. Check to see if the spacing will come out even. Usually the spacing has to be either increased or decreased slightly. Set the dividers for the width of a picket plus a space, increased or decreased slightly, whichever is appropriate. Space off the remaining distance. Adjust the dividers until the spacing comes out even. Any slight difference in a few spaces is not noticeable. This is much better than ending up with one narrow, conspicuous space (**Figure 67–14**).

Procedure 67–C Installing Spaced Pickets

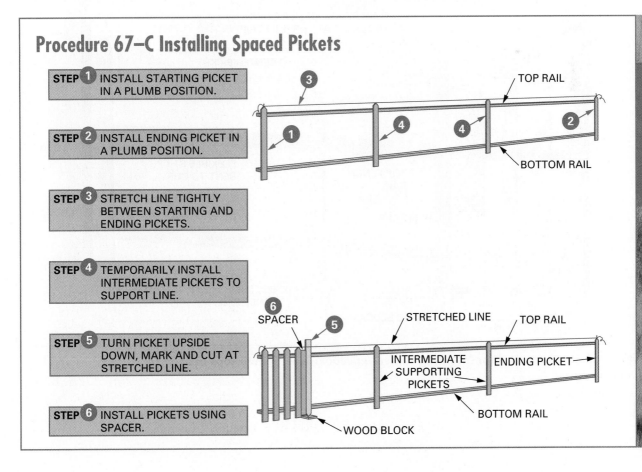

STEP 1 INSTALL STARTING PICKET IN A PLUMB POSITION.

STEP 2 INSTALL ENDING PICKET IN A PLUMB POSITION.

STEP 3 STRETCH LINE TIGHTLY BETWEEN STARTING AND ENDING PICKETS.

STEP 4 TEMPORARILY INSTALL INTERMEDIATE PICKETS TO SUPPORT LINE.

STEP 5 TURN PICKET UPSIDE DOWN, MARK AND CUT AT STRETCHED LINE.

STEP 6 INSTALL PICKETS USING SPACER.

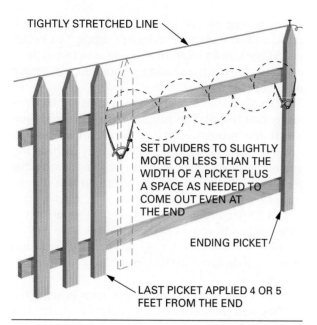

TIGHTLY STRETCHED LINE

SET DIVIDERS TO SLIGHTLY
MORE OR LESS THAN THE
WIDTH OF A PICKET PLUS
A SPACE AS NEEDED TO
COME OUT EVEN AT
THE END

ENDING PICKET

LAST PICKET APPLIED 4 OR 5
FEET FROM THE END

FIGURE 67–14 **Technique of dividing the remaining distance into equal spaces when ending a picket fence installation.**

Installing Pickets in Concave Curves

In some cases, the pickets or other fenceboards are installed with their tops in concave curved lines between fence posts. If the fenceboards have shaped tops that cannot be cut, erect the fence using the following procedure.

Install a picket on each end at the high point of the curve. In the center, temporarily install a picket with its top to the low point of the curve. Fasten a flexible strip of wood in a curve to the top of the three pickets. Start from the center. Work both ways to install the remainder of the pickets with their tops to the curved strip. Space the pickets to each end **(Figure 67–15)**. Other fenceboards, such as board-on-board and louvers, are installed in a similar manner.

If fenceboard tops are to be cut in the shape of the curve, tack them in place with their tops above the curve. Bend the flexible strip to the curve. Mark all the fenceboard tops. Remove the fenceboards, if necessary. Cut the tops and replace them.

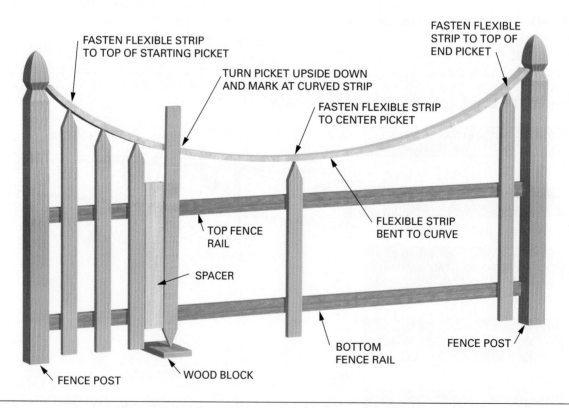

FASTEN FLEXIBLE STRIP
TO TOP OF STARTING PICKET

TURN PICKET UPSIDE DOWN
AND MARK AT CURVED STRIP

FASTEN FLEXIBLE
STRIP TO TOP OF
END PICKET

FASTEN FLEXIBLE STRIP
TO CENTER PICKET

FLEXIBLE STRIP
BENT TO CURVE

TOP FENCE
RAIL

SPACER

FENCE POST

BOTTOM
FENCE RAIL

FENCE POST

WOOD BLOCK

FENCE POST

FIGURE 67–15 **A method for installing spaced fenceboards with their tops in a concave curve.**

Key Terms

balusters	lattice work	radius edge decking	stockade fence
deck board	pressure treated	rail	

Review Questions

Select the most appropriate answer.

1. Fasteners and hardware used on exposed decks and fences should not be
   a. aluminum.
   b. hot-dipped galvanized.
   c. electroplated.
   d. stainless steel.

2. A ledger is a beam
   a. attached to the side of a building.
   b. supported by a girder.
   c. used to support girders.
   d. installed on supporting posts.

3. A footing for supporting fence posts must extend
   a. at least 12 inches below grade.
   b. below the frost line.
   c. to stable soil.
   d. all of the above.

4. Deck joists must be installed
   a. between girders.
   b. over girders.
   c. crowned edge up.
   d. using adhesive or glue.

5. A railing is required on deck stairs with more than
   a. 30 inches total rise.
   b. 2 risers.
   c. 3 feet total rise.
   d. 4 risers.

6. A railing is required on decks
   a. more than 3 stair risers above the ground.
   b. 30 or more inches above the ground.
   c. 4 feet or more above the ground.
   d. Check local building codes.

7. The usual height of a bench seat without a cushion is
   a. 14 inches.   c. 18 inches.
   b. 16 inches.   d. 20 inches.

8. A high fence with picket edges applied tightly together is called a
   a. board-on-board fence.
   b. panel fence.
   c. post-and-rail fence.
   d. stockade fence.

9. The bottom rail of any type fence is installed above the ground by, at least,
   a. the thickness of a 2 × 4 block.
   b. the width of a 2 × 4 block.
   c. 6 inches.
   d. 8 inches.

10. When applying fence pickets to rails
    a. plumb each one before fastening.
    b. plumb them frequently.
    c. cut the top ends to the line.
    d. fasten bottom ends flush with bottom rail.

SECTION FOUR

INTERIOR FINISH

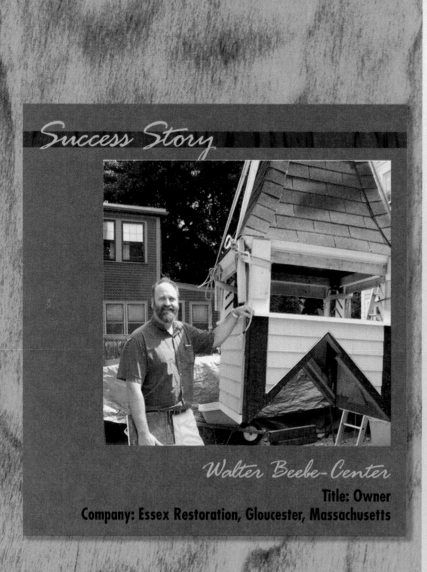

Success Story

Walter Beebe-Center

Title: Owner
Company: Essex Restoration, Gloucester, Massachusetts

EDUCATION

After graduating from Columbia University in New York City with a degree in economics, Walter took a temporary project to kill time: He agreed to fix up a dilapidated old family home, in exchange for living there for a year. Supporting himself by waiting tables, Walter ended up remodeling the entire house, learning solely by making mistakes and correcting them. "There was a moment partway through the project when I thought to myself, 'If I could do this every day for the rest of my life, that would be gratifying.'" It proved more than a distraction: He finished the home, and enjoyed the experience so much that he spent the next three years working as a carpenter. Finding himself in need of more formal training (and being in the midst of a recession), he went on to a program at the North Bennet Street School in Boston, where he would specialize in preservation carpentry.

HISTORY

As soon as he graduated in 1995, he started his own company, Essex Restoration, with the purpose of remodeling

644

homes in a way that keeps their original design sensibilities. "Our goal is to have every part of the house look almost as it did when the original builder walked off the job on that last day," he says.

ON THE JOB

Today, Walter works with ten employees on jobs of all sizes. "Often we're taking an older home that either hasn't been updated or has been, but without respect to architectural style," His team then works with the architect to undo things done in ways the new owner doesn't like and upgrade other areas while preserving the style of the structure. "We're often expanding kitchens," says Walter. "When colonial and Victorian homes were built, much of the cooking was done outside. Now kitchens are gathering places."

BEST ASPECTS

Walter's favorite part of the job is that it's always exciting. "Every house is different, so nothing is ever routine and you never get bored," he says. "You commute to one house for a few months then move on to a new town, new client, new architect." Besides, there's the inherent creativity of working with a project. "Shelter is our most basic need and yet we don't just build the simplest hut. We choose colors, shapes of molding, et cetera . . . and in the end it becomes aesthetic, like sculpture."

CHALLENGES

A restoration project requires many people to work together to get something done; Walter says it's a sociological experiment having to coordinate the architect, carpenters, plumbers, etc., and get them all to finish on time and on budget with a beautiful end product. "It's difficult, but it's also more rewarding in the end because it was so challenging along the way."

IMPORTANCE OF EDUCATION

"Formal training accelerates your career path," says Walter. "It can get you more quickly to a position of higher responsibility and higher pay. Not to mention that you'll get to do more interesting work."

FUTURE OPPORTUNITIES

Having only run his own business for a decade, Walter says his real goal is to keep his company profitable so that "we all continue to learn new things and how to do what we do even better, as well as to be rewarded with more interesting, more challenging, and more profitable jobs."

WORDS OF ADVICE

Get an education. "Yes, often people get into the building trades by learning on the job," says Walter. "But by taking formal vocational training, a person can benefit in three important ways: A trained carpenter gets a higher pay rate on his first job; he will get more interesting work faster; and he learns the best (rather than most expedient) techniques to get the job done."

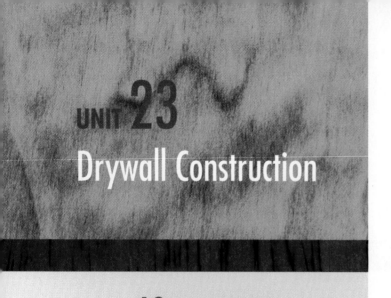

UNIT 23

Drywall Construction

The term *drywall construction* generally means the application of gypsum board. Drywall is used extensively as interior finish and is produced from gypsum, which is mined from the earth.

Installing the interior finish covers up other work performed. Check that all the work; framing, insulating and mechanicals, is truly complete before proceeding.

OBJECTIVES

After completing this unit, the student should be able to:

- describe various kinds, sizes, and uses of gypsum panels.
- describe the kinds and sizes of nails, screws, and adhesives used to attach gypsum panels.
- make single-ply and multi-ply gypsum board applications to interior walls and ceilings.
- conceal gypsum board fasteners and corner beads.
- reinforce and conceal joints with tape and compound.

68 Gypsum Board

Gypsum board is sometimes called *wallboard, plasterboard*, **drywall,** or *Sheetrock®*. It is used extensively in construction **(Figure 68–1).** The term Sheetrock is a brand name for gypsum panels made by the U.S. Gypsum Company. However, the brand name is in such popular use, it has become a generic name for gypsum panels. Gypsum board makes a strong, high-quality, fire-resistant wall and ceiling covering. It is readily available, easy to apply, decorate, or repair, and relatively inexpensive.

GYPSUM BOARD

Many types of gypsum board are available for a variety of applications. The board or panel is composed of a gypsum core encased in a strong, smooth-finish paper on the **face** side and a natural-finish paper on the back side. The face paper is folded around the long edges. This reinforces and protects the core. The long edges are usually tapered. This allows the joints to be concealed with compound without any noticeable crown joint **(Figure 68–2).** A crowned joint is a buildup of the compound above the surface.

Types of Gypsum Panels

Most gypsum panels can be purchased, if desired, with an aluminum foil backing. The backing functions as a vapor retarder. It helps prevent the passage of interior water vapor into wall and ceiling spaces.

FIGURE 68–1 The application of gypsum board to interior walls and ceilings is called drywall construction. *(Courtesy of U.S. Gypsum Company)*

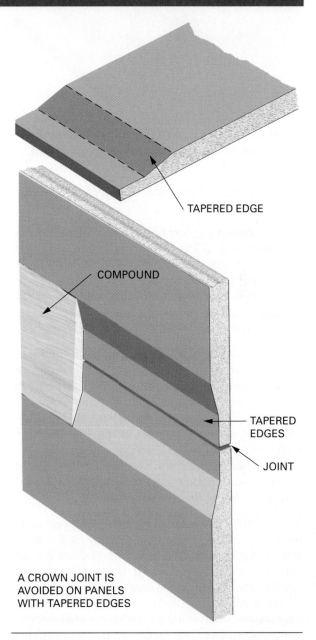

TAPERED EDGE

COMPOUND

TAPERED EDGES

JOINT

A CROWN JOINT IS AVOIDED ON PANELS WITH TAPERED EDGES

FIGURE 68–2 The long edges of gypsum panels usually are tapered for effective joint concealment.

Regular. Regular gypsum panels are most commonly used for single- or multilayer application. They are applied to interior walls and ceilings in new construction and remodeling.

Eased Edge. An **eased edge** gypsum board has a special tapered, rounded edge. This produces a much

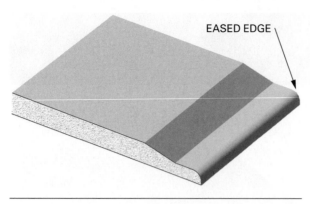

EASED EDGE

FIGURE 68–3 **An eased edge panel has a rounded corner that produces a stronger concealed joint.**

stronger concealed joint than a tapered, square edge **(Figure 68–3).**

Type X. **Type X** gypsum board is typically known as firecode board. It has greater resistance to fire because of special additives in the core. Type X gypsum board is manufactured in several degrees of resistance to fire. Type X looks the same as regular gypsum board. However, it is labeled Type X on the edge or on the back.

Water-Resistant. Water-resistant or *moisture resistant* (MR) gypsum board consists of a special moisture-resistant core and paper cover that is chemically treated to repel moisture. It is used frequently as a base for application of wall tile in bath, showers, and other areas subjected to considerable moisture. It is easily recognized by its distinctive green face. It is frequently called *green board* by workers in the field. Water-resistant panels are available with a Type X core for increased fire resistance.

Special Purpose. **Backing board** is designed to be used as a base layer in multilayer systems. It is available with regular or Type X cores. *Core-board* is available in 1-inch thicknesses. It is used for various applications, including the core of solid gypsum partitions. It comes in 24-inch widths with a variety of edge shapes. *Predecorated* panels have coated, printed, or overlay surfaces that require no further treatment. *Liner board* has a special fire-resistant core encased in a moisture-repellent paper. It is used to cover shaft walls, stairwells, chaseways, and similar areas.

Veneer Plaster Base. Veneer plaster bases are commonly called blue board. They are large, 4-foot-wide gypsum board panels faced with a specially treated blue paper. This paper is designed to receive applications of veneer plaster. Conventional plaster is applied about ⅜ inch thick and takes considerable time to dry. In contrast, specially formulated veneer plaster is applied in one coat of about ¹⁄₁₆ inch, or two

coats totaling about ⅛ inch. It takes only about forty-eight hours to dry.

Gypsum lath is used as a base to receive conventional plaster. Other gypsum panels, such as soffit board and sheathing, are manufactured for exterior use.

Sizes of Gypsum Panels

Widths and Lengths. Coreboards and liner boards come in 2-foot widths and from 8 to 12 feet long. Other gypsum panels are manufactured 4 feet wide and in lengths of 8, 9, 10, 12, 14, or 16 feet. Gypsum board is made in a number of thicknesses. Not all lengths are available in every thickness.

Thicknesses.

- A ¼-inch thickness is used as a base layer in multilayer applications. It is also used to cover existing walls and ceilings in remodeling work. It can be applied in several layers for forming curved surfaces with short radii.

- A ⅜-inch thickness is usually applied as a face layer in repair and remodeling work over existing surfaces. It is also used in multilayer applications in new construction.

- Both ½ inch and ⅝ inch are commonly used thicknesses of gypsum panels for single-layer wall and ceiling application in residential and commercial construction. The ⅝ inch-thick panel is more rigid and has greater resistance to impacts and fire than does the ½-inch panel.

- Coreboards and liner boards come in thicknesses of ¾ and 1 inch.

CEMENT BOARD

Like gypsum board, **cement board** and *wonder board* are panel products. However, they have a core of portland cement reinforced with a glass fiber mesh embedded in both sides **(Figure 68–4).** The core resists water penetration and will not deteriorate when wet. It is designed for use in areas that may be subjected to high-moisture conditions. It is used extensively in bathtub, shower, kitchen, and laundry areas as a base for ceramic tile. In fact, some building codes require its use in these areas.

Panels are manufactured in sizes designed for easy installation in tub and shower areas with a minimum of cutting. Standard cement board panels come in a thickness of ½ inch, in widths of 32 or 36 inches, and in 5-foot lengths. Custom panels are available in a thickness of ⅝ inch, widths of 32 or 48 inches, and lengths from 32 to 96 inches.

Cement board is also manufactured in a ⁵⁄₁₆-inch thickness. It is used as an underlayment for floors and

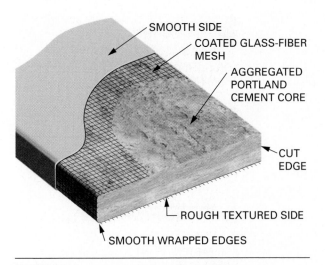

FIGURE 68–4 Composition of cement board.

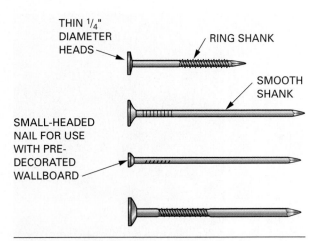

FIGURE 68–5 Special nails are required to fasten gypsum board.

countertops. Exterior cement board is used primarily as a base for various finishes on building exteriors.

DRYWALL FASTENERS

Specially designed nails and screws are used to fasten drywall panels. Ordinary nails or screws are not recommended. The heads of common nails are too small in relation to the shank. They are likely to cut the paper surface when driven. Staples may only be used to fasten the base layer in multilayer applications. They must penetrate at least ⅝ inch into supports. Using the correct fastener is extremely important for proper performance of the application. Fasteners with corrosion-resistant coatings must be used when applying water-resistant gypsum board or cement board. Care should be taken to drive the fasteners straight and at right angles to the wallboard to prevent the fastener head from breaking the face paper.

Nails

Gypsum board nails should have flat or concave heads that taper to thin edges at the rim. Nails should have relatively small-diameter shanks with heads at least ¼ inch, but no more than ⁵⁄₁₆ inch in diameter. For greater holding power, nails with annular ring shanks are used **(Figure 68–5).**

Smooth shank nails should penetrate at least ⅞ inch into framing members. Only ¾-inch penetration is required when ring shank nails are used. Greater nail penetrations are required for fire-rated applications.

Nails should be driven with a drywall hammer that has a convex face. This hammer is designed to compress the gypsum panel face to form a dimple of not more than ¹⁄₁₆ inch when the nail is driven home

(Figure 68–6). The dimple is made so the nail head can later be covered with compound.

Screws

Special drywall screws are used to fasten gypsum panels to steel or wood framing or to other panels. They are made with Phillips heads designed to be driven with a drywall screwgun **(Figure 68–7).** A proper setting of the nosepiece on the power screwdriver ensures correct countersinking of the screwhead. When driven correctly, the specially contoured bugle head makes a uniform depression in the panel surface without breaking the paper.

Different kinds of drywall screws are used for fastening into wood, metal, and gypsum panels. *Type W* screws are used for wood, *Type S* and *Type S-12* for metal framing, and *Type G* for fastening to gypsum backing boards **(Figure 68–8).**

Type W screws have sharp points. They have specially designed threads for easy penetration and

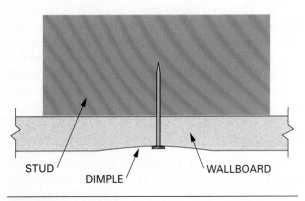

FIGURE 68–6 Fasteners are set with a dimple in the board for easier concealing with compound.

FIGURE 68–7 Drywall screws are driven with a screwgun to the desired depth. *(Courtesy of U.S. Gypsum Company)*

TYPE S

FOR LIGHT-GAUGE METAL FRAMING

TYPE S-12

FOR 20-GAUGE OR HEAVIER METAL FRAMING

TYPE G

FOR FASTENING INTO BASE LAYERS OF GYPSUM BOARD

TYPE W

FOR WOOD FRAMING

FIGURE 68–8 Several types of screws are used to fasten gypsum panels. Selection of the proper type is important.

excellent holding power. The screw should penetrate the supporting wood frame by at least ⅝ inch.

Type S screws are self-drilling and self-tapping. The point is designed to penetrate sheet metal with little pressure. This is an important feature because thin metal studs have a tendency to bend away when driving screws. Type S-12 screws have a different drill point designed for heavier gauge metal framing.

Type G screws have a deep, special thread design for effectively fastening gypsum panels together. These screws must penetrate into the supporting board by at least ½ inch. If the supporting board is not thick enough, longer fasteners should be used. Make sure there is sufficient penetration into framing members.

Adhesives

Drywall adhesives are used to bond single layers directly to supports or to laminate gypsum board to base layers. Adhesives used to apply gypsum board are classified as stud adhesives and laminating adhesives.

For bonding gypsum board directly to supports, special drywall stud adhesive or approved *construction adhesive* is used. Supplemental fasteners must be used with stud adhesives. Stud adhesives are available in large cartridges. They are applied to framing members with hand or powered adhesive guns **(Figure 68–9)**.

FIGURE 68–9 Applying drywall adhesive to studs.

CAUTION CAUTION CAUTION CAUTION

Some types of drywall adhesives may contain a flammable solvent. Do not use these types near an open flame or in poorly ventilated areas. ■

For laminating gypsum boards to each other, joint compound adhesives and **contact adhesives** are used. Joint compound adhesives are applied over the entire board with a suitable spreader prior to lamination. Boards laminated with joint compound adhesive require supplemental fasteners.

When contact adhesives are used, no supplemental fasteners are necessary. However, the board cannot be moved after contact has been made. The adhesive is applied to both surfaces by brush, roller, or spray gun. It is allowed to dry before laminating. A modified contact adhesive is also used. It permits an open time of up to thirty minutes during which the board can be repositioned, if necessary.

69 Single-Layer and Multilayer Drywall Application

Single-layer gypsum board applications are widely used in light commercial and residential construction. They adequately meet building code requirements. Multilayer applications are more often used in commercial construction. They have increased resistance to fire and sound transmission. Both systems provide a smooth, unbroken, quality surface if recommended application procedures are followed.

SINGLE-LAYER APPLICATION

Drywall should be not be delivered to the jobsite until shortly before installation begins. Boards stored on the job for long periods are subject to damage. The boards must be stored under cover and stacked flat on supports. Supports should be at least 4 inches wide and placed fairly close together **(Figure 69–1).** Leaning boards against framing for long periods may cause the boards to warp. This makes application more difficult. To avoid damaging the edges, carry the boards. Do not drag them. Then, set the boards down gently. Be careful not to drop them.

GYPSUM PANELS
STICKERS SHOULD BE AT LEAST 4" WIDE

FIGURE 69–1 Correct method of stacking gypsum board.

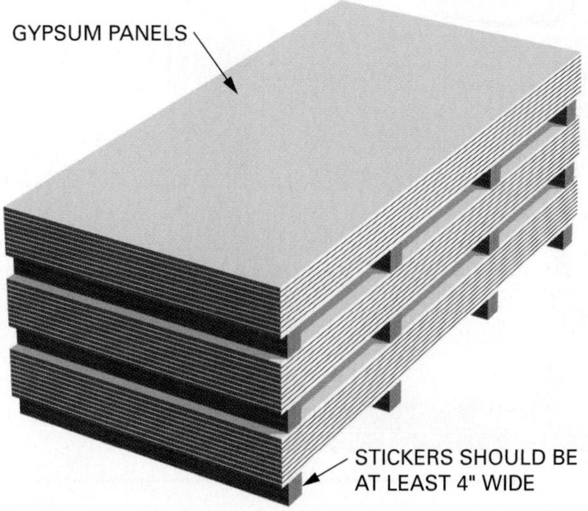

Cutting and Fitting Gypsum Board

Take measurements accurately. Cut the board by first scoring the face side through the paper to the core. Use a utility knife. Guide it with a *drywall T-square,* if cutting a square end **(Figure 69–2).** The board is then broken along the scored face. The back

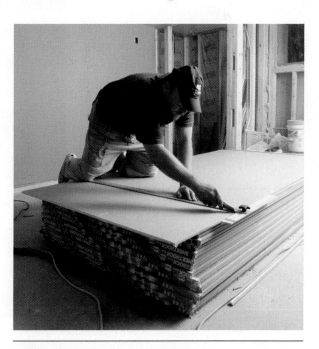

FIGURE 69–2 Using a drywall T-square as a guide when scoring across the width of a board. ***(Courtesy of U.S. Gypsum Company)***

Procedure 69–A Scoring and Breaking Drywall

STEP **1** BREAK BY LIFTING SHEET AND STEPPING BACK WITH THE CUT.

STEP **2** SCORE BACK SIDE OF SHEET LEAVING TOP AND BOTTOM PAPER TO ACT AS A HINGE.

STEP **3** RAISE SHEET ONTO TOE, THEN SNAP SHEET BACK BY SWINGING IT TOWARD YOU.

FIGURE 69–3 Technique for cutting gypsum board parallel to the edge.

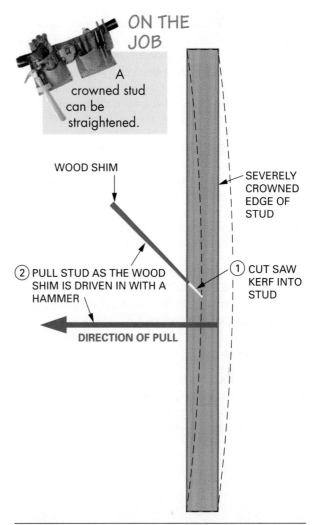

FIGURE 69–4 Technique for straightening a severely crowned stud.

paper is scored along the fold. The sheet is then broken by snapping the board in the reverse direction **(Procedure 69–A).**

To make cuts parallel to the long edges, the board is often gauged with a tape and scored with a utility knife. A *tape guide* and *tape tip* are sometimes used to aid the procedure. The tape guide permits more accurate gauging and protects the fingers. The tape tip contains a slot into which the knife is inserted. This prevents slipping off the end of the tape **(Figure 69–3).**

When making parallel cuts close to long edges, it is usually necessary to score both sides of the board to obtain a clean break.

Smooth ragged edges with a drywall rasp, coarse sanding block, or knife. A job-made drywall rasp can be made by fastening a piece of metal lath to a wood block. Cut panels should fit easily into place without being forced. Forcing the panel may cause it to break.

Aligning Framing Members

Before applying the gypsum board, check the framing members for alignment. Stud edges should not be out of alignment more than ⅛ inch with adjacent studs. A wood stud that is out of alignment can be straightened by the procedure shown in **Figure 69-4.** Ceiling joists are sometimes brought into alignment with the installation of a strongback across the tops of the joists at about the center of the span **(Figure 69–5).**

Fastening Gypsum Panels

Drywall is fastened to framing members with nails or screws. Hand pressure should be applied on the

panel next to the fastener being driven. This ensures that the panel is in tight contact with the framing member. The use of adhesives reduces the number of nails or screws required. A single or double method of nailing may be used.

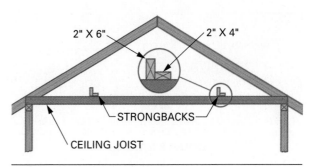

FIGURE 69–5 A strongback is sometimes used to align ceiling joists or the bottom chord of roof trusses.

Single Nailing Method. With this method, nails are spaced a maximum of 7 inches on center (OC) on ceilings and 8 inches OC on walls into frame members. Nails should be first driven in the center of the board and then outward toward the edges. Perimeter fasteners should be at least ⅜ inch, but not more than 1 inch from the edge.

Double Nailing Method. In double nailing, the perimeter fasteners are spaced as for single nailing. In the field of the panel, space a first set of nails 12 inches OC. Space a second set 2 to 2½ inches from the first set. The first nail driven is reseated after driving the second nail of each set. This assures solid contact with framing members **(Figure 69–6).**

Screw Attachment. Screws are spaced 12 inches OC on ceilings and 16 inches OC on walls when framing members are spaced 16 inches OC. If framing members are spaced 24 inches OC, then screws are spaced a maximum of 12 inches OC on both walls and ceilings.

Using Adhesives. Apply a straight bead about ¼ inch in diameter to the centerline of the stud edge. On studs where panels are joined, two parallel beads of adhesive are applied, one on each side of the center-line. Zigzag beads should be avoided to prevent the adhesive from squeezing out at the joint.

On wall applications, supplemental fasteners are used around the perimeter. Space them about 16 inches apart. On ceilings, in addition to perimeter fastening, the field is fastened at about 24-inch intervals **(Figure 69–7).**

Gypsum panels may be prebowed. This reduces the number of supplemental fasteners required. Prebow the panels by one of the methods shown in **Figure 69–8.** Make sure the finish side of the panel faces in the correct direction. Allow them to remain overnight or until the boards have a 2-inch permanent bow. Apply adhesive to the studs. Fasten the panel at top and bottom plates. The bow keeps the center of the board in tight contact with the adhesive until bonded.

Ceiling Application

Gypsum panels are applied first to ceilings and then to the walls. Panels may be applied parallel, or at right angles, to joists or furring. If applied parallel, edges and ends must bear completely on framing. If applied at right angles, the edges are fastened where they cross over each framing member. Ends must be fastened completely to joists or furring strips.

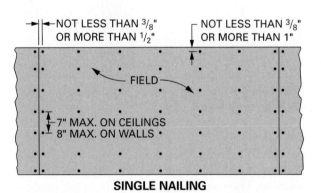

NOT LESS THAN ³/₈"
OR MORE THAN ¹/₂"

NOT LESS THAN ³/₈"
OR MORE THAN 1"

FIELD

7" MAX. ON CEILINGS
8" MAX. ON WALLS

SINGLE NAILING

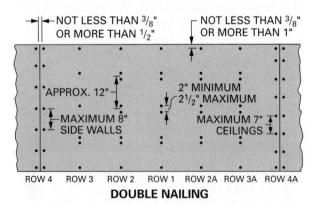

NOT LESS THAN ³/₈"
OR MORE THAN ¹/₂"

NOT LESS THAN ³/₈"
OR MORE THAN 1"

APPROX. 12"

2" MINIMUM
2¹/₂" MAXIMUM

MAXIMUM 8"
SIDE WALLS

MAXIMUM 7"
CEILINGS

ROW 4   ROW 3   ROW 2   ROW 1   ROW 2A   ROW 3A   ROW 4A

DOUBLE NAILING

FIGURE 69–6 Spacing of single nailed or double nailed panels. Greater fastener spacing is used when panels are screwed.

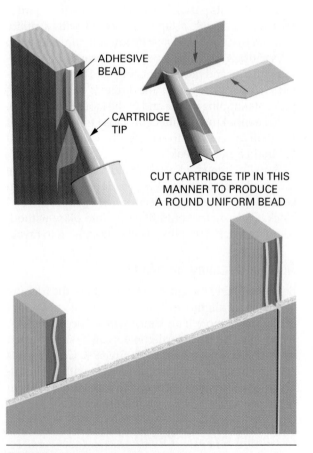

ADHESIVE BEAD

CARTRIDGE TIP

CUT CARTRIDGE TIP IN THIS MANNER TO PRODUCE A ROUND UNIFORM BEAD

FIGURE 69–7 Two beads of adhesive are applied under joints in the board.

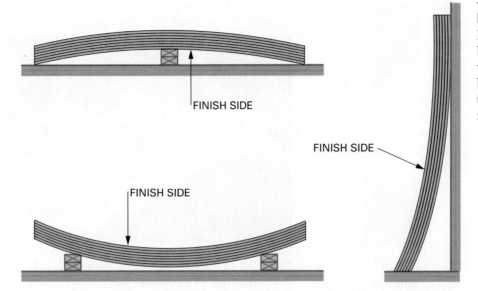

FINISH SIDE

FINISH SIDE

FINISH SIDE

FIGURE 69–8
Prebowing keeps the board in tight contact with the adhesive until bonded and reduces the number of fasteners required.

Carefully measure and cut the first board to width and length. Cut edges should be against the wall. Lay out lines on the panel face indicating the location of the framing in order to place fasteners accurately.

Gypsum board panels are heavy. At least two or more people are needed for ceiling application un-

less a drywall jack is available. Lift the panel overhead and place it in position **(Figure 69–9).** Install two **deadmen** under the panel to hold it in position.

Deadmen are supports made in the form of a "T." They are easily made on the job using 1×3 lumber with short braces from the vertical member to the horizontal member. The leg of the support is made about ¼ inch longer than the floor-to-ceiling height. The deadmen are wedged between the floor and the panel. They hold the panel in position while it is being fastened **(Figure 69–10).** Using deadmen is much easier than trying to hold the sheet in position and fasten it at the same time.

FIGURE 69–9 Drywall jacks are often used to raise drywall panels to the ceiling while fastening.

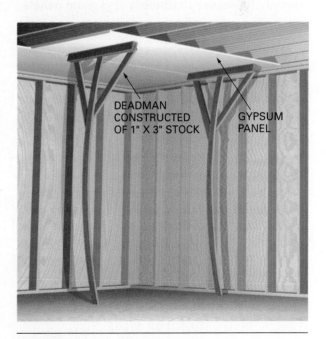

DEADMAN CONSTRUCTED OF 1" X 3" STOCK

GYPSUM PANEL

FIGURE 69–10 A deadman may be used to hold a board in place while fastening.

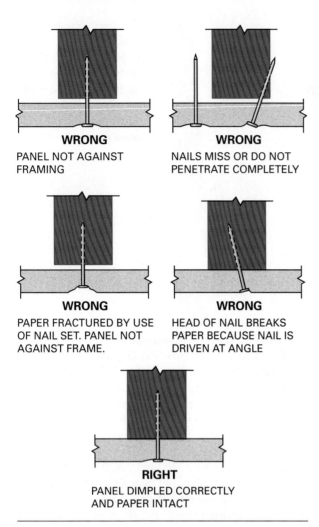

WRONG
PANEL NOT AGAINST
FRAMING

WRONG
NAILS MISS OR DO NOT
PENETRATE COMPLETELY

WRONG
PAPER FRACTURED BY USE
OF NAIL SET. PANEL NOT
AGAINST FRAME.

WRONG
HEAD OF NAIL BREAKS
PAPER BECAUSE NAIL IS
DRIVEN AT ANGLE

RIGHT
PANEL DIMPLED CORRECTLY
AND PAPER INTACT

FIGURE 69–11 It is important to drive fasteners correctly for secure attachment of gypsum panels.

Fasten the sheet in one of the recommended manners. Hold the board firmly against framing to avoid nail pops or protrusions. Drive fasteners straight into the member. Fasteners that miss supports should be removed. The nail hole should be dimpled so that later it can be covered with joint compound **(Figure 69–11).**

Continue applying sheets in this manner, staggering end joints, until the ceiling is covered.

To cut a corner out of a panel to accommodate a protrusion in the wall, make the shortest cut with a drywall saw. Then, score and snap the sheet in the other direction **(Figure 69–12).** To cut a circular hole, mark the circle with pencil dividers, twist and push the drywall saw through the board. Cut out the hole, following the circular line.

Horizontal Application on Walls

When walls are less than 8'–1" high, wallboard is usually installed horizontally, at right angles to the

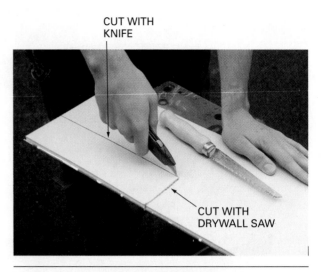

CUT WITH
KNIFE

CUT WITH
DRYWALL SAW

FIGURE 69–12 Use a knife and drywall saw to cut the corner out of a gypsum panel.

studs. If possible, use a board of sufficient length to go from corner to corner. Otherwise, use as long a board as possible to minimize end joints because they are difficult to conceal. Stagger end joints or center them over and below window and door openings if possible. That way not so much of the joint is visible. End joints should not fall on the same stud as those on the opposite side of the partition.

The top panel is installed first. Cut the board to length to fit easily into place without forcing it. Stand the board on edge against the wall. Start fasteners along the top edge opposite each stud. Raise the sheet so the top edge is firmly against the ceiling and drive nails. Fasten the rest of the sheet in the recommended manner **(Figure 69–13).**

Measure and cut the bottom panel to width and length. Cut the width about ¼ inch narrower than the distance measured. Lay the panel against the wall. Raise it with a **drywall foot lifter** against the bottom edge of the previously installed top panel. A drywall foot lifter is a tool especially designed for this purpose. However, one can be made on the job by tapering the end of a short piece of 1 × 3 or 1 × 4 lumber **(Figure 69–14).** Fasten the sheet as recommended. Install all others in a similar manner. Stagger any necessary end joints. Locate them as far from the center of the wall as possible so they will be less conspicuous. Avoid placing end joints over the ends of window and door headers. This will reduce the potential for wallboard cracks.

Where end joints occur on steel studs, attach the end of the first panel to the open or unsupported edge of the stud. This holds the stud flange in a rigid position for the attachment of the end of the adjoining panel. Making end joints in the opposite manner usually causes the stud edge to deflect. This results in an uneven surface at the joint **(Figure 69–15).**

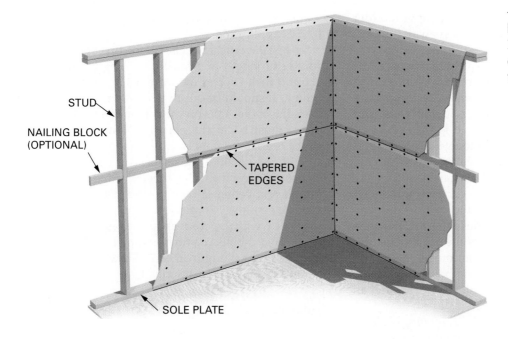

FIGURE 69–13
Horizontal application of gypsum panels to walls.

STUD

NAILING BLOCK (OPTIONAL)

TAPERED EDGES

SOLE PLATE

FIGURE 69–14 A drywall foot lifter is used to lift gypsum panels in the bottom course up against the top panels.

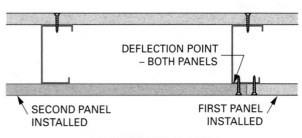

DEFLECTION POINT – BOTH PANELS

SECOND PANEL INSTALLED

FIRST PANEL INSTALLED

CORRECT APPLICATION

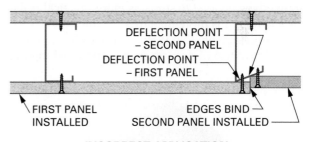

DEFLECTION POINT – SECOND PANEL

DEFLECTION POINT – FIRST PANEL

FIRST PANEL INSTALLED

EDGES BIND

SECOND PANEL INSTALLED

INCORRECT APPLICATION

FIGURE 69–15 Sequence of making end joints when attaching gypsum panels to steel studs. *(Courtesy of U.S. Gypsum Company)*

Making Cutouts in Wall Panels. There are several ways of making cutouts in wall panels for electrical outlet boxes, ducts, and similar objects. Care must be taken not to make the cutout much larger than the outlet. Most cover plates do not cover by much. If cut too large, much extra time has to be taken to patch up around the outlet, replace the panel, or install oversize outlet cover plates.

Plumb the sides of the outlet box down to the floor, or up to the previously installed top panel, whichever is more convenient. The panel is placed in position. Lines for the sides of the box are plumbed on it from the marks on the floor or on the panel. The top and bottom of the box are laid out by measuring down from the bottom edge of the top panel. With a saw or utility knife, cut the outline of the box. Take care not to damage the vapor retarder by pulling the lower end of the sheet away from the wall as you cut.

A fast, easy, and accurate way of making cutouts is with the use of a portable electric **drywall cutout tool (Figure 69–16).** The approximate location of the center of the outlet box is determined and marked on the panel. The panel is then installed over the box. Using the cutout tool, a hole is plunged through the panel in the approximate center of the outlet box. Care must be taken not to make contact with wiring. The tool is not recommended for use around live wires. The tool is moved in any direction until the bit hits a side of the box. It is then withdrawn slightly to ride over the edge to the outside of the box. The tool is then moved so the bit rides around the outside of the box to make the cutout. Usually cutouts are made for outlets after all the panels in a room have been installed.

To make cutouts around door openings, either mark and cut out the panel before it is applied, or make the cutout after it is applied. To make the cutout after the panel is applied, use a saw to cut in one direction. Then score it flush with the opening on the back side in the other direction. Bend and score it on the face to make the cutout. Another method uses the drywall cutout tool around framing.

Vertical Application on Walls

Vertical application of gypsum panels on walls, with long edges parallel to studs, is more practical if the ceiling height is more than 8'–1" or the wall is 4'–0" wide or less. Note that vertical application requires more lineal feet of drywall seams, and finishing the

FIGURE 69–16 A portable electric drywall cutout tool often is used to make cutouts for outlet boxes and similar objects. (*Courtesy of Porter-Cable*)

seams is more physically demanding than with horizontal application.

To install vertical panels, cut the first board, in the corner, to length and width. Its length should be about ¼ inch shorter than the height from floor to ceiling. It should be cut to width so the edge away from the corner falls on the center of a stud. All cut edges must be in the corners. None should butt edges of adjacent panels.

With a foot lifter, raise the sheet so it is snug against the ceiling. The tapered edge should be plumb and centered on the stud. Fasten it in the specified manner. Continue applying sheets around the room with tapered uncut edges against each other. There should be no horizontal joints between floor and ceiling **(Figure 69–17).** Make any necessary cutouts as previously described.

Floating Interior Angle Construction

To help prevent nail popping and cracking due to structural stresses where walls and ceilings meet, the *floating angle* method of drywall application may be used. When joists or furring strips are at right angles to walls, fasteners in ceiling panels are located 7 inches from the wall for single nailing, and 11 to 12 inches for double nailing or screw attachment. When joists or furring are parallel to the wall, nailing is started at the intersection.

Gypsum panels applied on walls are fitted tightly against ceiling panels. This provides a firm support for the floating edges of the ceiling panels. The top fastener into each stud is located 8 inches down from the ceiling for single nailing and 11 to 12 inches down for double nailing or screw attachment.

At interior wall corners, the underlying wallboard is not fastened. The overlapping board is fitted snugly against the underlying board. This brings it in firm contact with the face of the stud. The overlapping panel is nailed or screwed into the interior corner stud **(Figure 69–18).**

APPLYING GYPSUM PANELS TO CURVED SURFACES

Gypsum panels may be applied to curved surfaces. However, closer spacing of the frame members may be required to prevent flat areas from occurring on the face of the panel. If the paper and core of gypsum panels are moistened, they may be bent to curves with shorter radii than when dry. After the boards are thoroughly moistened, they should be stacked on a flat surface. They should be allowed to stand for at least one hour before bending. Moistened boards must be handled very carefully. They will regain their original hardness after drying. Wallboard mar-

GYPSUM BOARD

TAPERED EDGES

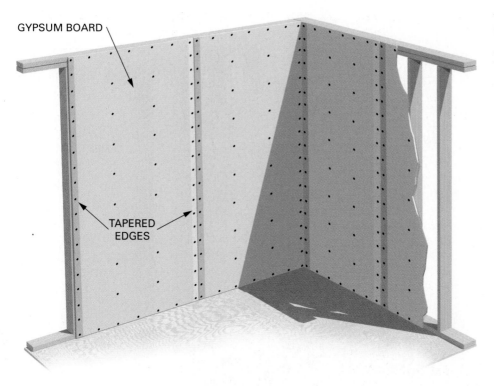

FIGURE 69–17 Applying gypsum panels vertically.

7" TO FIRST FASTENER ON CEILING

8"–11"

7"

8" TO FIRST FASTENER ON WALL

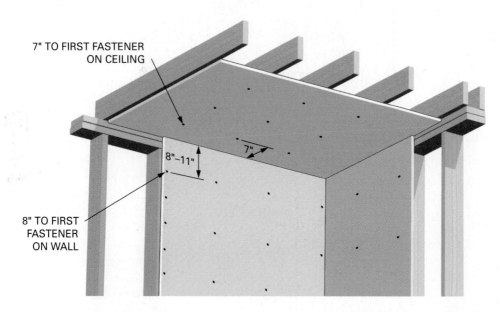

FIGURE 69–18 Floating angle method of applying drywall has no fasteners at the corners of sheets.

keted as *bendable* does not need to be wet before it is shaped. The minimum bending radii for dry and wet gypsum panels are shown in **Figure 69–19**.

To apply panels to a convex surface, fasten one end to the framing. Gradually work to the other end by gently pushing and fastening the panel progressively to each framing member. When applying panels to a concave curve, fasten a stop at one end. Carefully push on the other end to force the center of the panel against the framing. Work from the end against the stop. Fasten the panel successively to each framing member.

Board Thickness in Inches	Dry	Wet
¼	5 ft.	2 to 2½ ft.
⅜	7½ ft.	3 to 3½ ft.
½	20 ft.	4 to 4½ ft.

FIGURE 69–19 Minimum bending radii of gypsum panels.

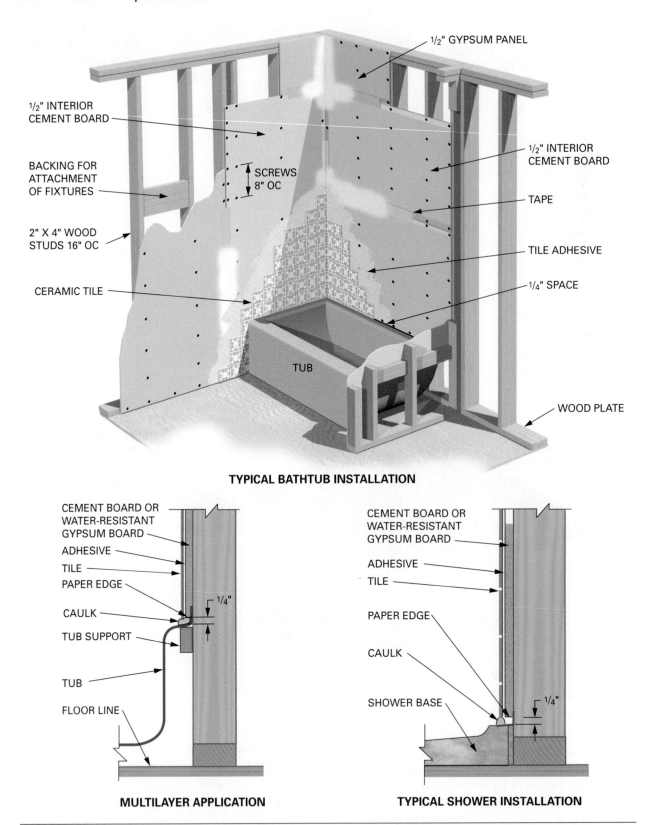

½" GYPSUM PANEL

½" INTERIOR CEMENT BOARD

BACKING FOR ATTACHMENT OF FIXTURES

2" X 4" WOOD STUDS 16" OC

CERAMIC TILE

SCREWS 8" OC

½" INTERIOR CEMENT BOARD

TAPE

TILE ADHESIVE

¼" SPACE

TUB

WOOD PLATE

TYPICAL BATHTUB INSTALLATION

CEMENT BOARD OR WATER-RESISTANT GYPSUM BOARD

ADHESIVE

TILE

PAPER EDGE

CAULK

TUB SUPPORT

TUB

FLOOR LINE

¼"

MULTILAYER APPLICATION

CEMENT BOARD OR WATER-RESISTANT GYPSUM BOARD

ADHESIVE

TILE

PAPER EDGE

CAULK

SHOWER BASE

¼"

TYPICAL SHOWER INSTALLATION

FIGURE 69–20   Installation details around bathtubs and showers.

Gypsum board may be applied to the curved inner surfaces of arched openings. If the dry board cannot be bent to the desired curve, it may be moistened or parallel knife scores made about 1 inch apart across its width.

DRYWALL APPLICATION TO BATH AND SHOWER AREAS

Water-resistant gypsum board and cement board panels are used in bath and shower areas as bases for the application of ceramic tile. (Some areas only cement board; check your local codes.) Framing should be 16 inches OC. Steel framing should be at least 20-gauge thickness.

Apply panels horizontally with the bottom edge not less than ¼ inch above the lip of the shower pan or tub. The bottom edges of gypsum panels should be uncut and paper covered.

Check the alignment of the framing. If necessary, apply furring strips to bring the face of the board flush with the lip of the tub or shower pan **(Figure 69–20)**.

Provide blocking between studs about 1 inch above the top of the tub or shower pan. Install additional blocking between studs behind the horizontal joint of the panels above the tub or shower pan.

Cement board panels are cut using a masonry blade in a circular saw. Care must be taken to reduce exposure to dust. Before attaching panels, apply thinned ceramic-tile adhesive to all cut edges around holes and other locations.

Attach panels with corrosion-resistant screws or nails spaced not more than 8 inches apart. When ce-

ramic tile more than ⅜ inch thick is to be applied, the nail or screw spacing should not exceed 4 inches OC.

MULTILAYER APPLICATION

A multilayer application has one or more layers of gypsum board applied over a base layer. This layering provides greater strength, higher fire resistance, and better sound control. The base layer may be gypsum backing board, regular gypsum board, or other gypsum base material.

Base Layer

The base layer is fastened in the same manner as single-layer panels. However, double nailing is not necessary and staples may be used in wood framing. On ceilings, panels are applied with the long edges either at right angles or parallel to framing members. On walls, the panels are applied with the long edges parallel to the studs.

Face Layer

Joints in the face layer are offset at least 10 inches from joints in the base layer. The face layer is applied either parallel to or at right angles to framing, whichever minimizes end joints and results in the least amount of waste.

The face layer may be attached with nails, screws, or adhesives. If nails or screws are used without adhesive, the maximum spacing and minimum penetration into framing should be the same as for single-layer application.

70   Concealing Fasteners and Joints

After the gypsum board is installed, it is necessary to conceal the fasteners and to reinforce and conceal the joints. One of several levels of finish may be specified for a gypsum board surface. The lowest level of finish may simply require the taping of wallboard joints and *spotting* of fastener heads on surfaces. This is done in warehouses and other areas where appearance is normally not critical. The level of finish depends, among other things, on the number of coats of compound applied to joints and fasteners **(Figure 70–1)**.

DESCRIPTION OF MATERIALS

Fasteners are concealed with *joint compound*. Joints are reinforced with *joint tape* and covered with joint

compound. Exterior corners are reinforced with *corner bead*. Other kinds of drywall trim may be used around doors, windows, and other openings.

Joint Compounds

Drying type **joint compounds** for joint finishing and fastener spotting are made in both a dry powder form and a ready-mixed form in three general types. Drying type compounds provide smooth application and ample working time. A *taping compound* is used to embed and adhere tape to the board over the joint. A *topping compound* is used for second and third coats over taped joints. An *all-purpose compound* is used for both bedding the tape and finishing the joint. All-purpose compounds do not possess

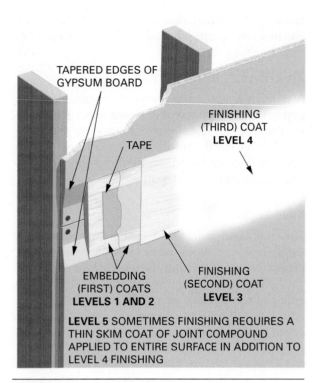

FIGURE 70–1 labels:

TAPERED EDGES OF GYPSUM BOARD

TAPE

FINISHING (THIRD) COAT
LEVEL 4

EMBEDDING (FIRST) COATS
LEVELS 1 AND 2

FINISHING (SECOND) COAT
LEVEL 3

LEVEL 5 SOMETIMES FINISHING REQUIRES A THIN SKIM COAT OF JOINT COMPOUND APPLIED TO ENTIRE SURFACE IN ADDITION TO LEVEL 4 FINISHING

FIGURE 70–1 **The level of finish varies with the type of final decoration to be applied to drywall panels.**

the strength or workability of two-step taping and topping compound systems.

Setting type joint compounds are used when a faster setting time than that of drying types is desired. Drying type compounds harden through the loss of water by evaporation. They usually cannot be recoated until the next day. Setting type compounds harden through a chemical reaction when water is added to the dry powder. Therefore, they come only in a dry powder form and not ready-mixed. They are formulated in several different setting times. The fastest setting type will harden in as little as twenty to thirty minutes. The slowest type takes four to six hours to set up. Setting type joint compounds permit finishing of drywall interiors in a single day.

Joint Reinforcing Tape

Joint reinforcing tape is used to cover, strengthen, and provide crack resistance to drywall joints. One type is made of *high-strength fiber paper.* It is designed for use with joint compounds on gypsum panels. It is creased along its center to simplify folding for application in corners **(Figure 70–2).**

Another type is made of *glass fiber mesh.* It is designed to reinforce joints on veneer plaster gypsum panels. It is not recommended for use with conventional compounds for general drywall joint finishing. It may be used with special high-strength setting com-

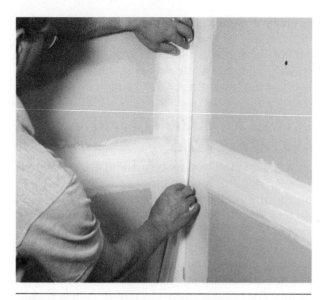

FIGURE 70–2 **Applying joint tape to an interior corner.**

pounds. Glass fiber mesh tape is available with a plain back or with an adhesive backing for quick application **(Figure 70–3).** Joint tape is normally available 2 and 2½ inches wide in 300-foot rolls.

Corner Bead and Other Drywall Trim

Corner beads are applied to protect exterior corners of drywall construction from damage by impact. One type with solid metal flanges is widely used.

FIGURE 70–3 **An adhesive-backed glass fiber mesh tape is quickly applied to drywall joints.** *(Courtesy of U.S. Gypsum Company)*

Another type has flanges of expanded metal with a fine mesh. This provides excellent keying of the compound **(Figure 70–4)**.

Corner bead is fastened through the drywall panel into the framing with nails or staples. Instead of using fasteners, a *clinching tool* is sometimes used. It crimps the solid flanges and locks the bead to the corner **(Figure 70–5)**.

Metal corner tape is applied by embedding it in joint compound. It is used for corner protection on arches, windows with no trim, and other locations **(Figure 70–6)**.

A variety of *metal trim* is used to provide protection and finished edges to drywall panels. Metal trim is used at windows, doors, inside corners, and intersections. Such trim is fastened through their flanges into the framing **(Figure 70–7)**.

Control joints are metal strips with flanges on both sides of a ¼-inch, V-shaped slot. Control joints are placed in large drywall areas. They relieve stresses induced by expansion or contraction. They are used from floor to ceiling in long partitions and from wall to wall in large ceiling areas **(Figure 70–8)**. The flanges are concealed with compound in a manner similar to corner beads and other trim.

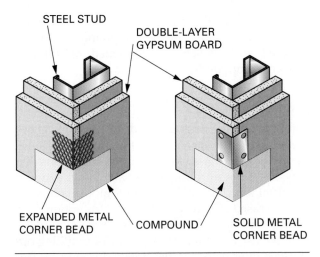

FIGURE 70–4 Corner beads are used to finish and protect exterior corners of drywall panels.

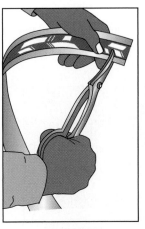

CUT TAPE WITH SNIPS

EMBED IN JOINT COMPOUND

FIGURE 70–5 A clinching tool is sometimes used to fasten corner beads to exterior corners.

FIGURE 70–6 Flexible metal corner tape is applied to exterior corners by embedding in compound. *(Courtesy of U.S. Gypsum Company)*

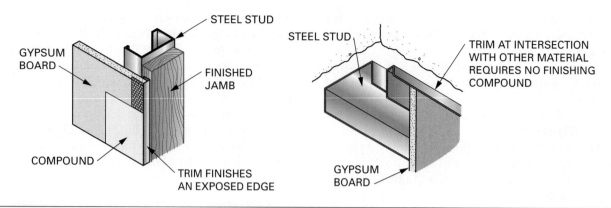

FIGURE 70–7 Various types of metal trim are used to provide finished edges to gypsum panels.

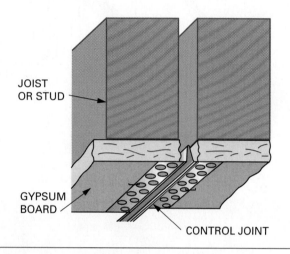

FIGURE 70–8 Control joints are used in large wall and ceiling areas subject to movement by expansion and contraction.

A wide assortment of rigid vinyl drywall accessories is available **(Figure 70–9)** including the metal trim previously discussed. They are designed for easy installation and workability to reduce installation time. Most have edges to guide the drywall knife, which allows for an even application of joint compound. Some have edges that are later torn away when the painting is done. This allows the finish to be applied more quickly and at the same time more uniformly. Vinyl accessories make it possible to create smooth joints easily, whether they are curved or straight.

APPLYING JOINT COMPOUND AND TAPE

In cold weather, care should be taken to maintain the interior temperature at a minimum of 50°F for twenty-four hours before and during application of joint compound, and for at least four days after application has been completed.

Care should also be taken to use clean tools. Avoid contamination of the compound by foreign material, such as sawdust, hardened, or different types of compounds.

Prefilling Joints

Before applying compound to drywall panels, check the surface for fasteners that have not been sufficiently recessed. Also look for other conditions that may affect the finishing. Prefill any joints between panels of ¼ inch or more and all V-groove joints between eased-edged panels with compound. A twenty-four-hour drying period can be eliminated with the use of setting compounds for prefilling operations. Normally the flat, tapered seams are finished first before the corners.

Embedding Joint Tape

Fill the recess formed by the tapered edges of the sheets with the specified type of joint compound. Use a joint knife **(Figure 70–10)**. Center the tape on the joint. Lightly press it into the compound. Draw the knife along the joint with sufficient pressure to *embed* the tape and remove excess compound **(Figure 70–11)**.

There should be enough compound under the tape for a proper bond, but not over ½₂ inch under the edges. Make sure there are no air bubbles under the tape. The tape edges should be well adhered to the compound. If not satisfactory, lift the portion. Add compound and embed the tape again. A *taping tool* sometimes is used. It applies the compound and embeds the tape at the same time **(Figure 70–12)**.

Immediately after embedding, apply a thin coat of joint compound over the tape. This helps prevent the edges from wrinkling. It also makes easier concealment of the tape with following coats. Draw the knife to bring the coat to a feather edge on both sides of the joint. Make sure there is no excess compound

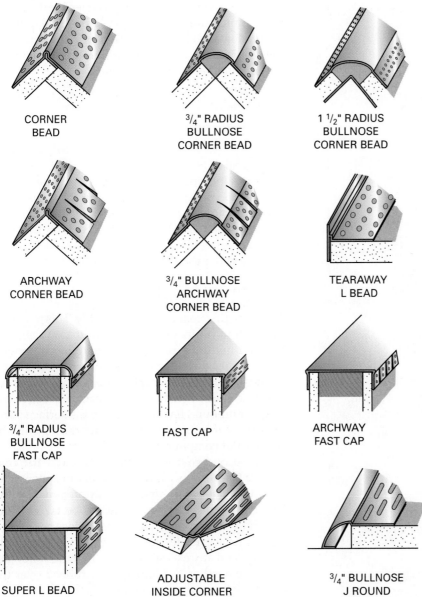

FIGURE 70–9   Many rigid vinyl drywall accessories are available.

CORNER BEAD

3/4" RADIUS BULLNOSE CORNER BEAD

1 1/2" RADIUS BULLNOSE CORNER BEAD

ARCHWAY CORNER BEAD

3/4" BULLNOSE ARCHWAY CORNER BEAD

TEARAWAY L BEAD

3/4" RADIUS BULLNOSE FAST CAP

FAST CAP

ARCHWAY FAST CAP

SUPER L BEAD

ADJUSTABLE INSIDE CORNER

3/4" BULLNOSE J ROUND

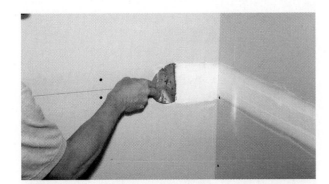

FIGURE 70–10   Taping compound is first applied to the channel formed by the tapered edges between panels.

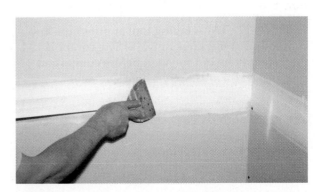

FIGURE 70–11   The tape is embedded into the compound.

A

B

FIGURE 70–12 **(A) A taping tool applies tape and compound at the same time. (B) The corner is then smoothed and finished before compound sets.** *(Courtesy of U.S. Gypsum Company)*

left on the surface. After the compound has set up, but not completely hardened, wipe the surface with a damp sponge. This eliminates the need for sanding any excess after the compound has hardened.

Spotting Fasteners

Fasteners should be *spotted* immediately before or after embedding joint tape. Spotting is the applica-

tion of compound to conceal fastener heads. Apply enough pressure on the taping knife to fill only the depression. Level the compound with the panel surface. Spotting is repeated each time additional coats of compound are applied to joints.

Applying Compound to Corner Beads and Other Trim

The first coat of compound is applied to corner beads and other metal trim when first coats are given to joints and fasteners. The nose of the bead or trim serves as a guide for applying the compound. The compound is applied about 6 inches wide from the nose of the bead to a feather edge on the wall. Each subsequent finishing coat is applied about 2 inches wider than the previous one.

Fill and Finishing Coats

Allow the first coat to dry thoroughly. This may take twenty-four hours or more depending on temperature and humidity unless a setting type compound has been used. It is common to use setting compounds for first coats and slower setting types for finishing coats. Feel the entire surface to see if any excess compound has hardened on the surface. Sand any excess, if necessary, to avoid interfering with the next coat of compound.

The second coat is sometimes called a **fill coat** **(Figure 70–13)**. It is feathered out about 2 inches beyond the edges of all first coats, approximately 7 to 10 inches wide. Care must always be taken to remove all excess compound so that it does not harden on the surface. Professional drywall finishers rarely have to sand any excess in preparation for following coats. Remember, a damp sponge rubbed over the joint after the compound starts to set up will remove

FIGURE 70–13   **Applying a coat of compound to a drywall joint.**

any small particles of excess. It will also help bring edges to a feather edge.

If the level of finish requires it, apply a third and **finishing coat** of compound over all fill coats. The edges of the finishing coat should be feathered out about 2 inches beyond the edges of the second coat. Some drywall finishers apply a skim coat of compound over the entire wall surface **(Figure 70–14)**.

FIGURE 70–14 **Some walls are finished with a skim coat of compound.** *(Courtesy of U.S. Gypsum Company)*

This provides a more uniform surface where reflected light shows variations between the panels and compound.

Interior Corners

Interior corners are finished in a similar way. However, the tape is folded in the center to fit in the corner. After the tape is embedded, drywall finishers usually apply a setting compound to one side only of each interior corner. By the time they have finished all interior corners in a room, the compound has set enough to finish the other side of the corners.

ESTIMATING DRYWALL

To estimate the amount of drywall material needed, determine the area of the walls and ceilings to be covered. To find the ceiling area, multiply the length of the room by its width. To find the wall area, multiply the perimeter of the room by the height. Subtract all large wall openings. Combine all areas to find the total number of square feet of drywall required. Add about 5 percent of the total for waste. The number of drywall panels can then be determined by dividing the total area to be covered by the area of one panel.

About 1,000 screws are needed for each 1,000 square feet of drywall when applied to framing 16 inches OC, and 850 screws are needed for 24 inch OC framing. About 5 pounds of nails are required to fasten each 1,000 square feet of drywall.

Approximately 7½ gallons or 1½ five-gallon pails of joint compound and 370 feet of joint tape will be needed to finish every 1,000 square feet of drywall.

Key Terms

backing board	drywall	fill coat	joint reinforcing tape
cement board	drywall cutout tool	finishing coat	
contact adhesives	eased edge	gypsum lath	metal corner tape
corner beads	face	joint compounds	type X
deadman			

Review Questions

Select the most appropriate answer.

1. Standard gypsum board width is
   - **a.** 36 inches.
   - **b.** 48 inches.
   - **c.** 54 inches.
   - **d.** 60 inches.

2. Standard gypsum board lengths are
   - **a.** 8, 10, and 12 feet.
   - **b.** 8, 10, 12, and 14 feet.
   - **c.** 8, 9, 10, 12, and 14 feet.
   - **d.** 8, 9, 10, 12, 14, and 16 feet.

3. When fastening drywall, minimum penetration of ring-shanked nails into the framing member is
   a. ½ inch.
   c. ⅞ inch.
   b. ¾ inch.
   d. 1 inch.

4. Gypsum board is usually installed vertically on walls when the wall height is greater than
   a. 8'–0".
   c. 8'–4".
   b. 8'–1".
   d. 8'–6".

5. Ceiling joists are sometimes aligned by the use of a
   a. deadman.
   c. strongback.
   b. dutchman.
   d. straightedge.

6. In the single nailing method, nails are spaced a maximum of
   a. 8 inches OC on walls; 7 inches OC on ceilings.
   b. 10 inches OC on walls; 8 inches OC on ceilings.
   c. 12 inches OC on walls; 10 inches OC on ceilings.
   d. 12 inches OC on walls and ceilings.

7. Screws are spaced
   a. 12 inches OC on walls; 10 inches OC on ceilings.
   b. 12 inches OC on walls and ceilings.
   c. 16 inches OC on walls and ceilings.
   d. 16 inches OC on walls; 12 inches OC on ceilings.

8. Joints in the face layer of a multilayer application are offset from joints in the base layer by at least
   a. 6 inches.
   c. 10 inches.
   b. 8 inches.
   d. 12 inches.

9. The paper-covered edge of water-resistant gypsum board is applied above the lip of the tub or shower pan not less than
   a. ¼ inch.
   c. ½ inch.
   b. ⅜ inch.
   d. ¾ inch.

10. When ceramic tile more than ⅜-inch thick is to be applied over water-resistant gypsum board, fasten the board with screws or nails spaced not more than
   a. 4 inches OC.
   c. 8 inches OC.
   b. 6 inches OC.
   d. 10 inches OC.

UNIT 24

Wall Paneling and Ceramic Tile

Plans and specifications often call for the installation of *wall paneling* in certain rooms of both residential and commercial construction. *Ceramic tile* is widely used in rest rooms, baths, showers, kitchens, and similar areas.

Interior finishes serve as accents to design and protection of the structure from water. Long-term use and function relies on proper workmanship.

OBJECTIVES

After completing this unit, the student should be able to:

- describe and apply several kinds of sheet wall paneling.
- describe and apply various patterns of solid lumber wall paneling.
- describe and install ceramic wall tile to bathroom walls.
- estimate quantities of wall paneling and ceramic wall tile.

71 Types of Wall Paneling

Two basic kinds of wall paneling are sheets of various prefinished material and solid *wood boards.* Many compositions, colors, textures, and patterns are available in sheet form. Solid wood boards of many species and shapes are used for both rustic and elegant interiors **(Figure 71–1).**

DESCRIPTION OF SHEET PANELING

Sheets of prefinished plywood, **hardboard, particleboard, plastic laminate,** and other material are used to panel walls.

Plywood

Prefinished plywood is probably the most widely used sheet paneling. A tremendous variety is available in both **hardwoods** and **softwoods.** The more expensive types have a face veneer of real wood. The less expensive kinds of plywood paneling are prefinished with a printed wood grain or other design on a thin vinyl covering. Care must be taken not to scratch or scrape the surface when handling these types. Unfinished plywood panels are also available.

Some sheets are scored lengthwise at random intervals to imitate solid wood paneling. There is always a score 16, 24, and 32 inches from the edge. This facilitates fastening of the sheets and in case the sheet has to be ripped lengthwise to fit stud spacing.

Most commonly used panel thicknesses are ³⁄₁₆ and ¼ inch. Sheets are normally 4 feet wide and 7 to 10 feet long. An 8-foot length is most commonly used. Panels may be shaped with square, beveled, or shiplapped edges **(Figure 71–2).** Matching molding is available to cover edges, corners, and joints. Thin

FIGURE 71–1   Solid wood board paneling provides warmth and beauty to interiors of buildings. *(Courtesy of California Redwood Association)*

ring-shanked nails, called *color pins,* are available in colors to match panels. They are used when exposed fastening is necessary.

Hardboard

Hardboard is available in many man-made surface colors, textures, and designs. Some of these designs simulate stone, brick, stucco, leather, weathered or smooth wood, and other materials. Unfinished hardboard is also used that has a smooth, dark brown surface suitable for painting and other decorating. Unfinished hardboard may be solid or perforated in a number of designs.

Tileboard is a hardboard panel with a baked-on plastic finish. It is embossed to simulate ceramic wall tile. The sheets come in a variety of solid colors, marble, floral, and other patterns. Tileboard is designed for use in bathrooms, kitchens, and similar areas.

Hardboard paneling comes in widths of 4 feet and in lengths of from 8 to 12 feet. Commonly used thicknesses are from ⅛ to ¼ inch. Color-coordinated molding and trim are available for use with hardboard paneling.

Particleboard

Panels of particleboard come with wood grain or other designs applied to the surface, similar to plywood and hardboard. Sheets are usually ¼ inch thick, 4 feet wide, and 8 feet long. Prefinished particleboard is used when an inexpensive wall covering is desired. Because the sheets are brittle and break easily, care must be taken when handling them. They must be applied only on a solid wall backing.

Unfinished particleboard is not usually used as an interior wall finish. One exception, made from aromatic cedar chips, is used to cover walls in closets to repel moths.

Plastic Laminates

Plastic laminates are widely used for surfacing kitchen cabinets and countertops. They are also used to cover walls or parts of walls in kitchens, bathrooms, restrooms, and similar areas where a durable, easy-to-clean surface is desired. Laminates can be scorched by an open flame. However, they resist mild heat, alcohol, acids, and stains. They clean easily with a mild detergent.

Laminates are manufactured in many colors and designs, including wood grain patterns. Surfaces are available in gloss, satin, and textured finishes, among others.

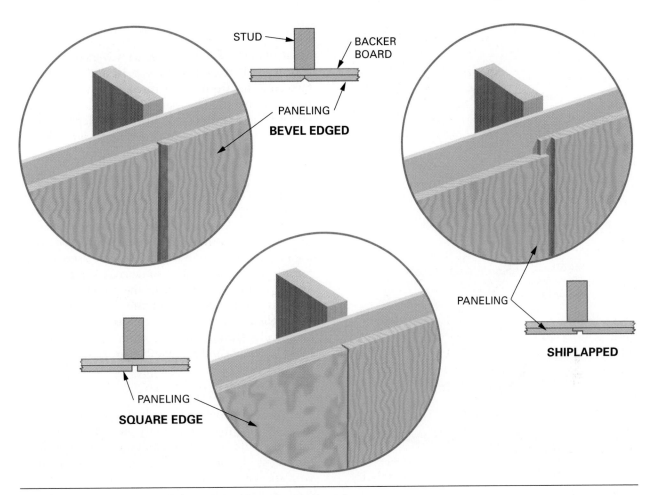

FIGURE 71–2 Sheet paneling comes with various edge shapes.

Laminates are ordinarily used in two thicknesses. Vertical-type laminate is relatively thin (about ½ inch). It is used for vertical surfaces, such as walls and cabinet sides. Vertical-type laminate is available only in widths of 4 feet or 8 feet. Regular or standard laminate is about ⅟₁₆ inch thick. It comes in widths of 24, 36, 48, and 60 inches and in lengths of 5, 6, 8, 10, and 12 feet. It is generally used on horizontal surfaces, such as countertops. It can be used on walls, if desired, or if the size required is not available in vertical type. Sheets are usually manufactured 1 inch wider and longer than the nominal size.

Laminates are difficult to apply to wall surfaces because they are so thin and brittle. Also, because a *contact cement* is used, the sheet cannot be moved once it makes contact with the surface. Thus, prefabricated panels, with sheets of laminate already bonded to a backer, are normally used to panel walls. See Chapter 81 for installation techniques for laminates.

DESCRIPTION OF BOARD PANELING

Board paneling is used on interior walls when the warmth and beauty of solid wood is desired. Wood paneling is available in softwoods and hardwoods of many species. Each has its own distinctive appearance, unique grain, and knot pattern.

Wood Species

Woods may be described as light, medium, and dark toned. Light tones include birch, maple, spruce, and white pine. Some medium tones are cherry, cypress, hemlock, oak, ponderosa pine, and fir. Among the darker-toned woods are cedar, mahogany, redwood, and walnut. For special effects, knotty pine, wormy chestnut, pecky cypress, and white-pocketed Douglas fir board paneling may be used.

Surface Textures and Patterns

Wood paneling is available in many shapes. It is either planed for smooth finishing, or rough-sawn for a rustic, informal effect. Square-edge boards may be joined edge to edge, spaced on a dark background, or applied in *board-and-batten* or *board-on-board* patterns. *Tongue-and-grooved* or *shiplapped* paneling comes in patterns, a few of which are illustrated in **Figure 71–3.**

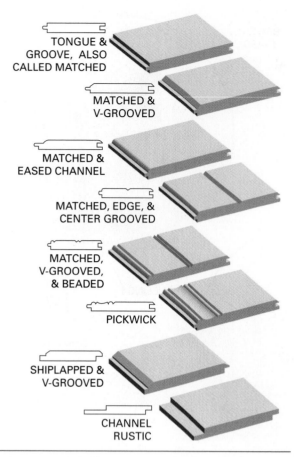

FIGURE 71-3   **Solid wood paneling is available in a number of patterns.**

Sizes

Most wood paneling comes in a ¾-inch thickness and in nominal widths of 4, 6, 8, 10, and 12 inches. A few patterns are manufactured in a ⁹⁄₁₆-inch thickness. Aromatic cedar paneling is used in clothes closets. It runs from ⅜ to ⁵⁄₁₆ inch thick, depending on the mill. It is usually *edge-* and *end-matched* (tongue-and-grooved) for application to a backing surface.

Moisture Content

To avoid shrinkage, wood paneling, like all interior finish, should be dried to a *moisture equilibrium* content. That is, its moisture content should be about the same as the area in which it is to be used. Except for arid, desert areas and some coastal regions, the average moisture content of the air in the United States is about 8 percent **(Figure 71–4).** Interior finish applied with an excessive moisture content will eventually shrink, causing open joints, warping, loose fasteners, and many other problems.

OTHER PANELING

The types of paneling described in this chapter are those that are most commonly used. To become acquainted with other types and methods of application, a study of manufacturers' catalogs dealing with sheet and board paneling, found in *Sweet's Architectural File,* is suggested.

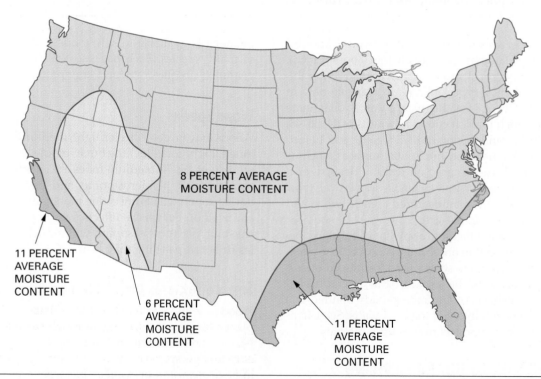

FIGURE 71-4   **Recommended average moisture content for interior finish woodwork in different parts of the United States.**

72 Application of Wall Paneling

Sheet paneling, such as prefinished plywood and hardboard, is usually applied to walls with long edges vertical. *Board paneling* may be installed vertically, horizontally, diagonally, or in many interesting patterns **(Figure 72–1).**

INSTALLATION OF SHEET PANELING

A backer board layer, at least ⅜ inch thick, should be installed on walls prior to the application of sheet paneling. The backing makes a stronger and more fire-resistant wall, helps block sound transmission, tends to bring studs in alignment, and provides a rigid finished surface for application of paneling **(Figure 72–2).**

Furring strips must be applied to masonry walls before paneling is installed. The strips are usually applied by driving hardened nails. Instead of using furring strips, a freestanding wood wall close to the masonry wall can be built, if enough space is available.

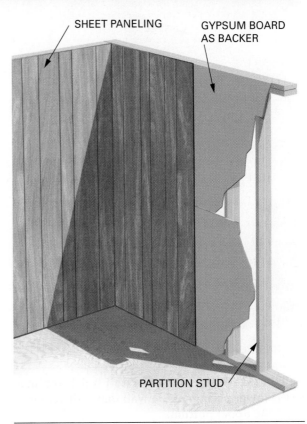

SHEET PANELING

GYPSUM BOARD AS BACKER

PARTITION STUD

FIGURE 72–2   Apply sheet paneling over a gypsum wallboard base.

FIGURE 72–1   Wood paneling can be applied in many directions. *(Courtesy of California Redwood Association)*

Preparation for Application

Mark the location of each stud in the wall on the floor and ceiling. Paneling edges must fall on stud centers, even if applied with adhesive over a backer board, in case supplemental nailing of the edges is necessary. Panels are usually fastened with a combination of color pins and adhesive.

Apply narrow strips of color on the wall from floor to ceiling where joints between paneling will occur. Use paint or magic marker colored the same as the joint of the panel. If joints between sheets open slightly because of shrinkage during extended dry periods or heating seasons, it is not so noticeable with a similar color behind the joint.

Sometimes paneling does not extend to the ceiling but covers only the lower portion of the wall. This partial paneling is called **wainscoting.** It is usually installed about 3 feet above the floor **(Figure 72–3).** If the wall is to be wainscoted, snap a horizontal line across the wall to indicate its height.

Stand panels on their long edge against the wall for at least forty-eight hours before installation. This allows them to adjust to room temperature and humidity. Otherwise sheets may buckle after installation, especially if they were very dry when installed.

Just prior to application, stand panels on end, side by side, around the room. Arrange them by matching the grain and color to obtain the most pleasing appearance.

Starting the Application

Start in a corner and continue installing consecutive sheets around the room. Select the starting corner, remembering that it will also be the ending point. This corner should be the least visible, such as behind an often-open door. Cut the first sheet to a length about ¼ inch less than the wall height. Place the sheet in the corner. Plumb the outside edge and tack it in the plumb position. Set the distance between the points of the dividers the same as the amount the sheet overlaps the center of the stud **(Figure 72–4).** Scribe this amount on the edge of the sheet butting the corner **(Figure 72–5).**

Remove the sheet from the wall. Place it on sawhorses on which a sheet of plywood, or two 8-foot lengths of 2 × 4s, have been placed for support. Cut to the scribed lines with a sharp, fine-toothed, hand crosscut saw. Handsaws may be used on the face side because they cut on the downstroke. This action will keep the splintering of the cut on the backside of the panel. Power saws should be used to cut from the backside.

Replace the sheet with the cut edge fitting snugly in the corner. The joint at the ceiling need not be fit-

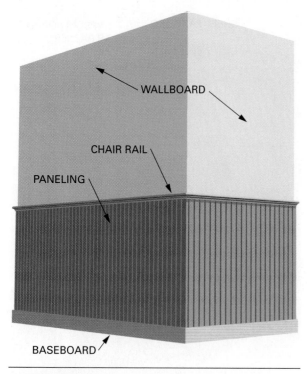

FIGURE 72–3 Wainscoting is a wall finish applied to the lower portion of the wall that is different from the upper portion.

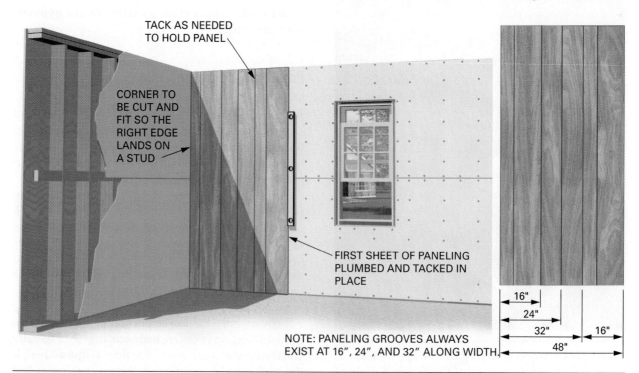

FIGURE 72–4 The first sheet must be set plumb in the corner.

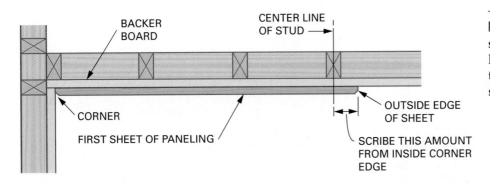

BACKER BOARD

CENTER LINE OF STUD

CORNER

FIRST SHEET OF PANELING

OUTSIDE EDGE OF SHEET

SCRIBE THIS AMOUNT FROM INSIDE CORNER EDGE

FIGURE 72–5 Set scribers equal to the largest space and scribe the edge of the first sheet to the corner.

ted if a molding is to be used. If a tight fit between the panel and ceiling is desired, set the dividers and scribe a small amount at the ceiling line. Remove the sheet again. Cut to the scribed line. Replace the sheet, and raise it snugly against the ceiling. The space at the bottom will be covered later by a baseboard.

Fastening

If only nails are used, fasten about 6 inches apart along edges and about 12 inches apart on intermediate studs for ¼-inch thick paneling. Nails may be spaced farther apart on thicker paneling. Drive nails at a slight angle for better holding power **(Figure 72–6)**.

If adhesives are used, apply a ⅛-inch continuous bead where panel edges and ends make contact. Apply beads 3 inches long and about 6 inches apart on all intermediate studs. Put the panel in place. Tack it at the top. Be sure the panel is properly placed in position. Press on the panel surface to make contact with the adhesive. Use firm, uniform pressure to spread the adhesive beads evenly between the wall and the panel. Then, grasp the panel and slowly pull the bottom of the sheet a few inches away from the wall **(Figure 72–7)**. Press the sheet back into position after about two minutes. After about twenty minutes, recheck the panel. Apply pressure to ensure

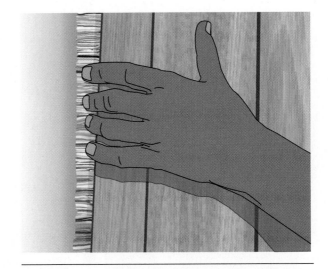

FIGURE 72–7 Pull the sheet a short distance away from the wall to allow the adhesive to dry slightly.

thorough adhesion and to smooth the panel surface. Apply successive sheets in the same manner. Do not force panels in position. Panels should touch very lightly at joints.

Wall Outlets

To lay out for wall outlets, plumb and mark both sides of the outlet to the floor. If the opening is close to the ceiling, plumb upward and mark lightly on the ceiling. Level the top and bottom of the outlet on the wall beyond the edge of the sheet to be installed. Or, level on the adjacent sheet, if closer. Cut, fit, and tack the sheet in position. Level and plumb marks from the wall and floor onto the sheet for the location of the opening **(Procedure 72–A)**.

Another method is to rub a cake of carpenter's chalk on the edges of the outlet box. Fit and tack the sheet in position. Tap on the sheet directly over the outlet to transfer the chalked edges to the back of the sheet. Remove the sheet. Cut the opening for the outlet. Openings for wall outlets, such as electrical boxes must be cut fairly close to the location

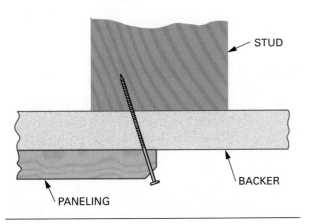

STUD

PANELING

BACKER

FIGURE 72–6 Drive nails or color pins at a slight angle for better holding power.

STEP BY STEP
PROCEDURES

Procedure 72–A Cutting Outlet Holes in Paneling

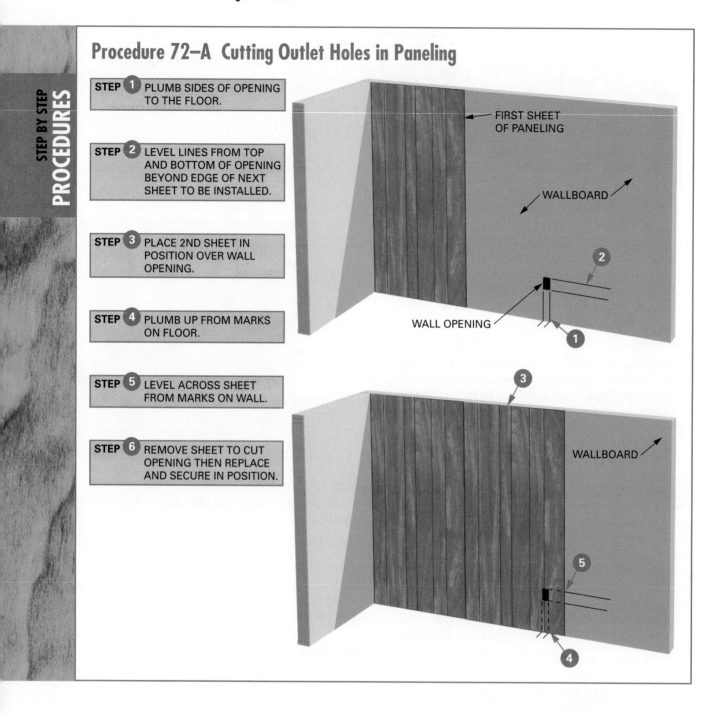

STEP ❶ PLUMB SIDES OF OPENING TO THE FLOOR.

STEP ❷ LEVEL LINES FROM TOP AND BOTTOM OF OPENING BEYOND EDGE OF NEXT SHEET TO BE INSTALLED.

STEP ❸ PLACE 2ND SHEET IN POSITION OVER WALL OPENING.

STEP ❹ PLUMB UP FROM MARKS ON FLOOR.

STEP ❺ LEVEL ACROSS SHEET FROM MARKS ON WALL.

STEP ❻ REMOVE SHEET TO CUT OPENING THEN REPLACE AND SECURE IN POSITION.

FIRST SHEET OF PANELING

WALLBOARD

WALL OPENING

WALLBOARD

and size. The cover plate may not cover if the cutout is not made accurately. This could require replacement of the sheet. A saber saw may be used to cut these openings. When using the saber saw, cut from the back of the panel to avoid splintering the face **(Figure 72–8).**

Ending the Application

The final sheet in the wall need not fit snugly in the corner if the adjacent wall is to be paneled or if interior corner molding is to be used. Take measure-

ments at the top, center, and bottom. Cut the sheet to width, and install.

If the last sheet butts against a finished wall and no corner molding is used, the sheet must be cut to fit snugly in the corner. To mark the sheet accurately, first measure the remaining space at the top, bottom, and about the center. Rip the panel about ½ inch wider than the greatest distance. Place the sheet with the cut edge in the corner and the other edge overlapping the last sheet installed. Tack the sheet in position. The amount of overlap should be exactly the same from top to bottom. Set the dividers for the

FIGURE 72–8 **A saber saw may be used to cut an opening for an electrical outlet box.**

amount of overlap. Scribe this amount on the edge in the corner **(Procedure 72–B).** Instead of dividers, it is sometimes more exact to use a small block of wood for scribing. The width is cut the same as the amount of overlap. Care must be taken to keep from turning the dividers while scribing along a surface **(Figure 72–9).**

Cut to the scribed line. If the line is followed carefully, the sheet should fit snugly between the last sheet installed and the corner, regardless of any irregularities. On exterior corners, a quarter-round molding is sometimes installed against the edges of the sheets. Or, the joint may be covered with a wood, metal, or vinyl corner molding **(Figure 72–10).**

STEP BY STEP PROCEDURES

Procedure 72–B Scribing the Last Piece of Paneling

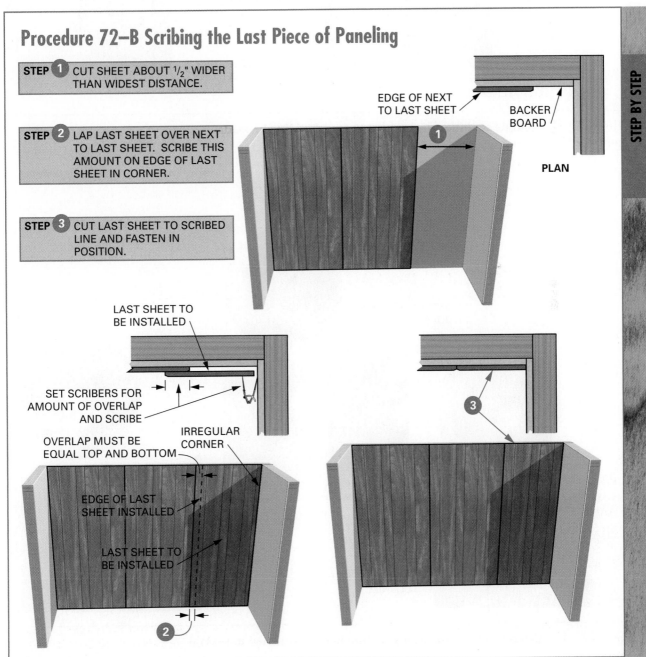

STEP 1 CUT SHEET ABOUT ½" WIDER THAN WIDEST DISTANCE.

STEP 2 LAP LAST SHEET OVER NEXT TO LAST SHEET. SCRIBE THIS AMOUNT ON EDGE OF LAST SHEET IN CORNER.

STEP 3 CUT LAST SHEET TO SCRIBED LINE AND FASTEN IN POSITION.

EDGE OF NEXT TO LAST SHEET

BACKER BOARD

PLAN

LAST SHEET TO BE INSTALLED

SET SCRIBERS FOR AMOUNT OF OVERLAP AND SCRIBE

OVERLAP MUST BE EQUAL TOP AND BOTTOM

IRREGULAR CORNER

EDGE OF LAST SHEET INSTALLED

LAST SHEET TO BE INSTALLED

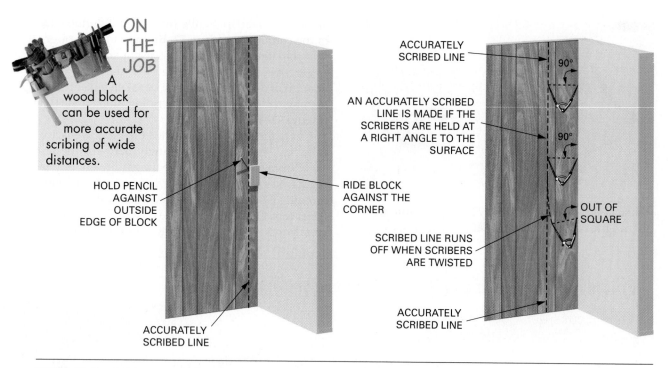

ON THE JOB

A wood block can be used for more accurate scribing of wide distances.

HOLD PENCIL AGAINST OUTSIDE EDGE OF BLOCK

RIDE BLOCK AGAINST THE CORNER

ACCURATELY SCRIBED LINE

ACCURATELY SCRIBED LINE

90°

AN ACCURATELY SCRIBED LINE IS MADE IF THE SCRIBERS ARE HELD AT A RIGHT ANGLE TO THE SURFACE

90°

OUT OF SQUARE

SCRIBED LINE RUNS OFF WHEN SCRIBERS ARE TWISTED

ACCURATELY SCRIBED LINE

FIGURE 72–9   Accurate scribing requires that the marked line be made perpendicular to the corner.

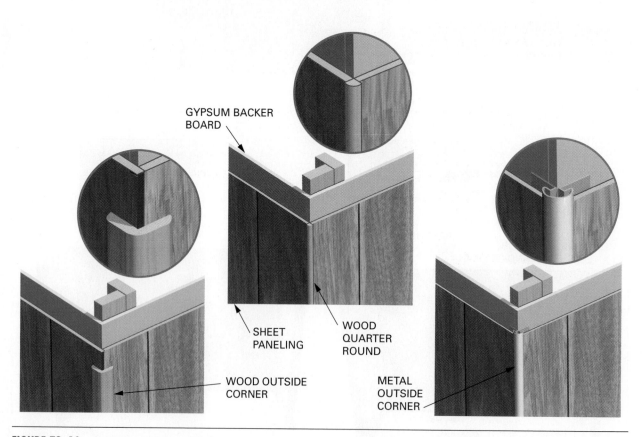

GYPSUM BACKER BOARD

SHEET PANELING

WOOD QUARTER ROUND

WOOD OUTSIDE CORNER

METAL OUTSIDE CORNER

FIGURE 72–10   Exterior corners of sheet paneling may be finished in several ways.

INSTALLING SOLID WOOD BOARD PANELING

Horizontal board paneling may be fastened to studs in new and existing walls **(Figure 72–11).** For vertical application of board paneling in a frame wall, blocking must be provided between studs **(Figure 72–12).** On existing and masonry walls, horizontal furring strips must be installed. Blocking or furring must be provided in appropriate locations for diagonal or pattern applications of board paneling.

Allow the boards to adjust to room temperature and humidity by standing them against the walls around the room. At the same time, put them in the order of application. Match them for grain and color. If tongue-and-grooved boards are to be eventually stained or painted, apply the same finish to the tongues so that an unfinished surface is not exposed if the boards shrink after installation.

Starting the Application

Select a straight board to start with. Cut it to length, about ¼ inch less than the height of the wall. If tongue-and-grooved stock is used, tack it in a plumb position with the grooved edge in the corner. If a tight fit is desired, adjust the dividers to scribe an amount a little more than the depth of the groove.

FIGURE 72–11   No wall blocking is required for horizontal application of board paneling. *(Courtesy of California Redwood Association)*

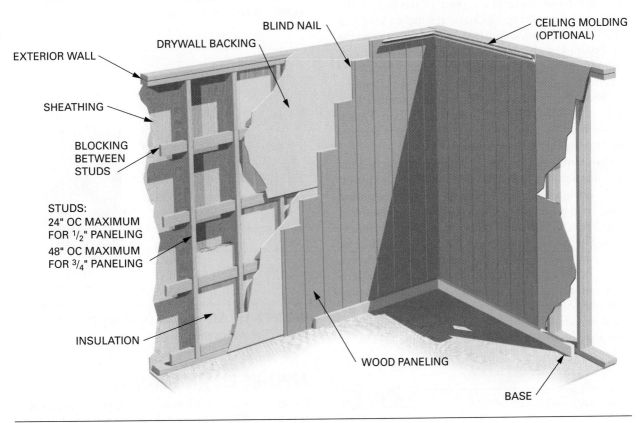

FIGURE 72–12   Blocking must be provided between studs for vertical board paneling.

Rip to the scribed line. Face nail along the cut edge into the corner with finish nails about 16 inches apart. Blind nail the other edge through the tongue.

Continuing the Application

Apply succeeding boards by blind nailing into the tongue only **(Figure 72–13)**. Make sure the joints between boards come up tightly. See Figure 61–15 for methods of bringing edge joints of matched boards up tightly if they are slightly crooked. Severely warped boards should not be used. As installation progresses, check the paneling for plumb. If out of plumb, gradually bring back by driving one end of several boards a little tighter than the other end. Cut out openings in the same manner as described for sheet paneling.

Applying the Last Board

If the last board in the installation must fit snugly in the corner without a molding, the layout should be planned so that the last board will be as wide as possible. If boards are a uniform width, the width of the starting board must be planned to avoid ending with a narrow strip. If random widths are used, they can be arranged when nearing the end.

Cut and fit the next to the last board. Then remove it. Tack the last board in the place of the next to the last board. Cut a scrap block about 6 inches long and equal in width to the finished face of the next to the last board. Use this block to scribe the last board by running one edge along the corner and holding a pencil against the other edge **(Figure 72–14)**. Remove the board from the wall. Cut it to the scribed line. Fasten the next to the last board in position. Fasten the last board in position with the cut edge in the corner. Face nail the edge nearest the corner.

Horizontal application of wood paneling is done in a similar manner. However, blocking between studs on open walls or furring strips on existing walls are not necessary. On existing walls, locate and snap lines to indicate the position of stud centerlines. The thickness of wood paneling should be at least ⅜ inch for 16-inch spacing of frame members and ⅝ inch for 24-inch spacing. Diagonal and pattern application of board paneling is similar to vertical and horizontal applications. If wainscoting is applied to a wall, the joint between the different materials may treated in several ways **(Figure 72–15)**.

FIGURE 72–13   **Tongue-and-grooved paneling is blind nailed.**

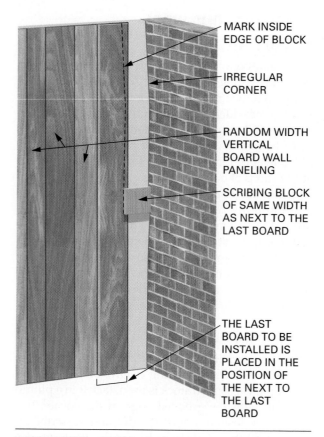

MARK INSIDE EDGE OF BLOCK

IRREGULAR CORNER

RANDOM WIDTH VERTICAL BOARD WALL PANELING

SCRIBING BLOCK OF SAME WIDTH AS NEXT TO THE LAST BOARD

THE LAST BOARD TO BE INSTALLED IS PLACED IN THE POSITION OF THE NEXT TO THE LAST BOARD

FIGURE 72–14   **Laying out the last board to fit against a finished corner.**

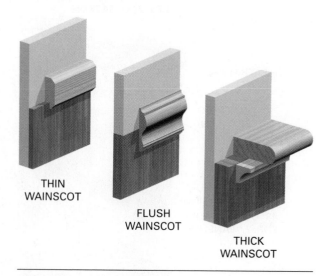

THIN WAINSCOT

FLUSH WAINSCOT

THICK WAINSCOT

FIGURE 72–15   **Methods of finishing the joint at the top of wainscoting.**

APPLYING PLASTIC LAMINATES

Plastic laminates are not usually applied to walls unless first prefabricated to sheets of plywood or similar material. They are then installed on walls in the same manner as sheet paneling. Special matching molding is used between sheets and on interior and exterior corners. (The application of plastic laminates is described in greater detail in Unit 30 on kitchen cabinets and countertops.)

ESTIMATING PANELING

Sheet Paneling

To determine the number of sheets of paneling needed, measure the perimeter of the room. Divide the perimeter by the width of the panels to be used. Deduct from this number any large openings such as doors, windows, or fireplaces. Deduct ⅔ of a panel for a door and ½ for a window or fireplace. Round off any remainder to the next highest number.

Board Paneling

Determine the square foot area to be covered by multiplying the perimeter by the height of each room. Deduct the area of any large openings. An additional percentage of the total area to be covered is needed because of the difference in the nominal size of lumber and its actual size.

Multiply the area to be covered by the area factor shown in **Figure 72–16.** Add 5 percent for waste in cutting. For example, the total area to be covered is 850 square feet, and 1 × 8 tongue-and-groove board paneling is to be used. Multiply 850 by 1.21, the sum of the coverage factor of 1.16 found in the table and 0.05 for waste in cutting. Round the answer of 1028.50 to 1029 for the number of board feet of paneling needed. To reduce waste in cutting, order suitable lengths.

Nominal Size	Width		Area Factor*
	Dress	**Face**	
Shiplap			
1 × 6	5½	5⅛	1.17
1 × 8	7¼	6⅞	1.16
1 × 10	9¼	8⅞	1.13
1 × 12	11¼	10⅞	1.10
Tongue-and-Groove			
1 × 4	3⅜	3⅛	1.28
1 × 6	5⅜	5⅛	1.17
1 × 8	7⅛	6⅞	1.16
1 × 10	9⅛	8⅞	1.13
1 × 12	11⅛	10⅞	1.10
S4S			
1 × 4	3½	3½	1.14
1 × 6	5½	5½	1.09
1 × 8	7¼	7¼	1.10
1 × 10	9¼	9¼	1.08
1 × 12	11¼	11¼	1.07
Paneling and Siding Patterns			
1 × 6	5⁷⁄₁₆	5¹⁄₁₆	1.19
1 × 8	7⅛	6¾	1.19
1 × 10	9⅛	8¾	1.14
1 × 12	11⅛	10¾	1.12

*Number multiplied by square feet to convert square feet to board feet.

FIGURE 72–16 **Factors used to estimate amounts of board paneling.**

73 Ceramic Tile

Ceramic tile is used to cover floors and walls in rest rooms, baths, showers, and other high-moisture areas that need to be cleaned easily and frequently **(Figure 73–1).** On large jobs, the tile is usually applied by specialists. On smaller jobs, it is sometimes more expedient for the general carpenter to install tile. Ceramic tile is usually set in place using **thin set,** a mortar-type adhesive, but is sometimes set using an organic adhesive. Thin set is used when tile is likely to get soaked with water such as when used in showers. Organic adhesive is used when tile will not get soaked such as on a kitchen counter backsplash.

DESCRIPTION OF CERAMIC TILE

Ceramic tiles are usually rectangular or square, but many geometric shapes, such as hexagons and octagons, are manufactured. Many solid colors, patterns, designs, and sizes give a wide choice to achieve the desired wall effect.

The most commonly used tiles are nominal 4- and 6-inch squares, in ¼-inch thickness. Many other sizes

FIGURE 73–1 Ceramic tile is used extensively in high-moisture areas.

are also available including 1- through 12-inch-square tiles. Special pieces such as *base, caps, inside corners,* and *outside corners* are used to trim the installation **(Figure 73–2).**

WALL PREPARATION FOR TILE APPLICATION

Cement board is the recommended backing for ceramic tile when tile will become wet daily. Otherwise, water-resistant gypsum board may be used. Installation instructions for these products is given in Chapter 68.

Minimum Backer and Tile Area in Baths and Showers

Tiles should overlap the lip and be applied down to the shower floor or top edge of the bathtub. On tubs without a showerhead, they should be installed to extend to a minimum of 6 inches above the rim. Around bathtubs with showerheads, tiles should extend a minimum of 5 feet above the rim or 6 inches above the showerhead, whichever is higher.

In shower stalls, tiles should be a minimum of 6 feet above the shower floor or 6 inches above the showerhead, whichever is higher. A 4-inch minimum extension of the full height is recommended beyond the outside face of the tub or shower **(Figure 73–3).**

Calculating Border Tiles

Before beginning the application of ceramic wall tile, the width of the *border tiles* must be calculated. Border tiles are those that fit to the corners and edges of the tiled area. The installation has a professional appearance if the border tiles are the same width and also as wide as possible **(Figure 73–4).**

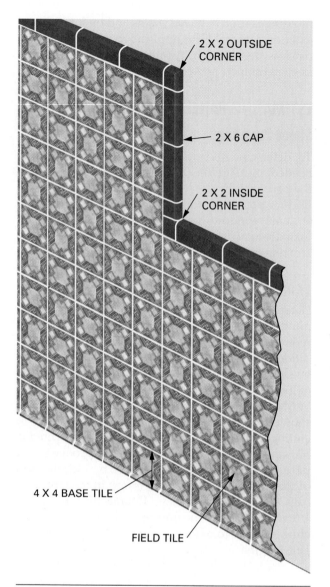

FIGURE 73–2 Special pieces are used to trim a ceramic tile installation.

Measure the width of the wall from corner to corner. Change the measurement to inches. Divide it by the width of a tile. Measure the tile accurately. Sometimes, the actual size of a tile is different than its nominal size. Add the width of a tile to the remainder. Divide by 2 to find the width of border tiles.

EXAMPLE A wall section measures 8′–4″, or 100 inches from wall to wall. If 4¼-inch, actual size, tiles are used, dividing 100 by 4¼ equals 23 full tiles with 0.53 of a tile remainder. Add 1 to the decimal remainder and multiply it by the width of a tile, or 1.53 × 4.25 = 6.5 inches. Divide this by 2 to get 3¼ inches. This will give 22 full tiles and two 3¼-inch border tiles across the tiled area.

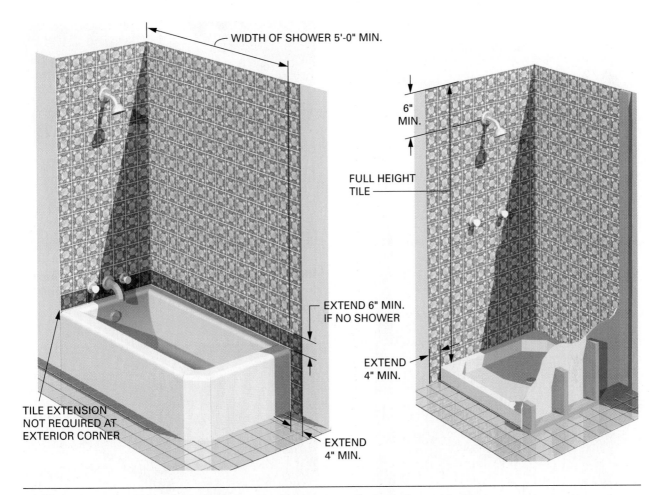

FIGURE 73–3 Minimum areas recommended for the installation of ceramic tile around bathtubs and shower stalls.

Tile Layout

A layout line needs to be placed as a guide for the individual tiles. Measure out from the corner and mark a point that will be the edge of a full tile. If the section to be tiled is a wall, plumb a line from top to bottom. If the section is a floor, measure and mark another point from the other corner. Snap a line through these points.

For wall applications, check the level of the floor. If the floor is level, tiles may be applied by placing the bottom edge of the first row on the floor using plain tile or base tile. If the floor is not level, the bottom edge of the tiles must be cut to fit the floor while keeping the top edges level. Place a level on the floor and find the low point of the tile installation. From this point, measure up and mark on the wall the height of a tile. Draw a level line on the wall through the mark. Tack a straightedge on the wall with its top edge to the line. Tiles are then laid to the straightedge. When tiling is completed above, the

straightedge is removed. The tiles in the bottom row are then cut and fitted to the floor.

For floor applications of tile, check that the line is parallel to the opposite wall. Adjust the line to split any differences. This line is best placed where the tile will be most often viewed from a distance. This will ensure the tiles are as straight as possible where they are most visible.

TILE APPLICATION

When tiling walls such as a shower, it usually is best to install tile on the back wall first. This way, the joint in the corner is the least visible and the most watertight. Apply the recommended adhesive to the wall or floor with a trowel. Use a flat trowel with grooved edges that allows the recommended amount of adhesive to remain on the surface **(Figure 73–5).** Too heavy a coat results in adhesive being squeezed out between the joints of applied tile, causing a mess. Too little adhesive results in failure of tiles to adhere

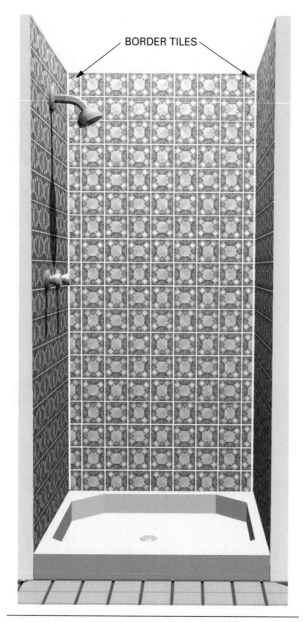

BORDER TILES

FIGURE 73–4 Border tiles should be as wide as possible and close to the same width on both sides of the wall.

FIGURE 73–5 Use a trowel that is properly grooved to allow the correct amount of adhesive to remain on the surface.

FIGURE 73–6 Rubber tile spacers may be used to maintain a uniform grout spacing.

to the wall. Follow the manufacturer's directions for the type of trowel to use and the amount of adhesive to be spread at any one time. Be careful not to spread adhesive beyond the area to be covered.

Apply the first tile to the guide line. Press the tile firmly into the adhesive. Apply other tiles in the same manner. Start from the center guide line and pyramid upward and outward. As tiles are applied, slight adjustments may need to be made to keep them lined up. Tile spacers may be used to help keep tile properly spaced **(Figure 73–6).** Tile spacers are rubber pieces with the same dimensions as the joints between tiles. Keep fingers clean

and adhesive off the face of the tiles. Clean tiles with a damp cloth.

Cutting Border Tiles

After all *field tiles* are applied, it is necessary to cut and apply border tiles. Field tiles are whole tiles that are applied in the center of the wall. Ceramic tile may be cut in any of several ways. Tile saws can be used to cut all types of tile **(Figure 73–7).** They are operated in a manner similar to that of a power miter box except that the material is eased into the blade on a sliding tray. Water is pumped from the reservoir below to the blade, keeping it cool during the cutting process. Thin ceramic tile is often cut using a hand cutter **(Figure 73–8).** First the tool scores the tile. The back edge of the cutter is then

FIGURE 73–7   Tile cutters are often used for cutting ceramic tile.

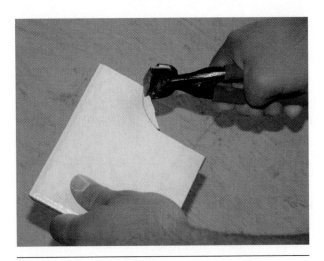

FIGURE 73–9   Nibblers are used to cut curves.

FIGURE 73–8   Hand cutter for thin ceramic tile.

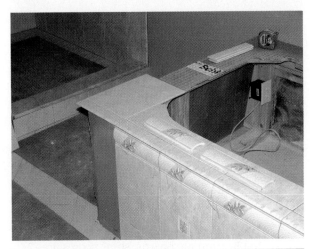

FIGURE 73–10   Special trim pieces are often used to add accents to a tiled surface.

pressed against the tile to break it. This tool cannot make small or narrow cuts. A nibbler is used to make small or irregular cuts **(Figure 73–9).** Nibblers chip small pieces off in sometimes random directions. Care should be taken to cut only small pieces at a time. This will allow a successful cut to be made gradually.

To finish the edges and ends of the installation, *caps* are sometimes used. Caps may be 2 × 6 or 4 × 4 pieces with one rounded finished edge. Special trim pieces are used to finish interior and exterior corners **(Figure 73–10).**

Grouting Tile Joints

After all tile has been applied, the joints are filled with **tile grout.** Grout comes in a powder form. It is

mixed with water to form a paste of the desired consistency. It is spread over the face of the tile with a *rubber trowel* to fill the joints. The grout is worked into the joints. Then, the surface is wiped as clean as possible with the trowel.

Wall tile grout is allowed to set up, but not harden. The joints are then *pointed*. Removing excess grout and smoothing joints is called *pointing*. A small hardwood stick with a rounded end can be used as a pointing tool. The entire surface is wiped clean with a dry cloth after the grout has dried.

After floor tile grout has set up slightly, the excess is removed. Wipe the tile across the grout lines at a 45-degree angle with a damp sponge. Wipe once, then turn the sponge over and wipe another area. Rinse and repeat. The key is to keep the sponge clean for one wiping at a time. Let it set up more and repeat. Finish the grouting by buffing with a dry clean rag.

After the grout has set and cured for several weeks, silicone grout sealer may be applied. This product seals tiny pores in the grout, making it more resistant to staining. Sealer is liberally brushed on the grout, allowing it to soak in. The excess is wiped clean before it dries.

ESTIMATING TILE

First, determine the square foot area to be covered for the amount of tile to order. To find the number of tiles needed, multiply the area covered by the number of tiles in one square foot.

EXAMPLE   4 × 4 tiles are being used to cover 120 square feet of surface. Each 4 × 4 tile = 16 square inches. Then one square foot or 144 square inches is divided by 16, equaling 9 tiles needed to cover 1 square foot. Thus, if 120 square feet is to be covered, then 120 times 9 equals 1,080 tiles.

The number of straight pieces of cap is found by determining the total linear feet to be covered and dividing by the length of the cap. The number of interior and exterior corners is determined by counting from a layout of the installation.

Key Terms

board paneling	hardwoods	softwoods	tile grout
ceramic tile	particleboard	thin set	wainscoting
hardboard	plastic laminate		

Review Questions

Select the most appropriate answer.

1. On prefinished plywood paneling that is scored to simulate boards, some scores are always placed in from the edge
   a. 12 and 16 inches.
   b. 16 and 20 inches.
   c. 12 and 24 inches.
   d. 16 and 32 inches.

2. A wainscoting is a wall finish
   a. applied diagonally.
   b. applied partway up the wall.
   c. used as a coating on prefinished wall panels.
   d. used around tubs and showers.

3. For most parts of the country, wood used for interior finish should be dried to a moisture content of about
   a. 8 percent.
   b. 12 percent.
   c. 15 percent.
   d. 20 percent.

4. The thickest plastic laminate described in this unit is called
   a. vertical type.
   b. regular type.
   c. backer type.
   d. all-purpose type.

5. If ¾-inch thick wood paneling is to be applied vertically over open studs, wood blocking must be provided between studs for nailing at intervals of
   a. 16 inches.
   b. 24 inches.
   c. 32 inches.
   d. 48 inches.

6. When applying board paneling vertically to an existing wall
   a. nail paneling to existing studs.
   b. apply horizontal furring strips.
   c. remove the wall covering and install blocking between studs.
   d. use adhesives.

7. The recommended backing for use behind ceramic tile for showers is
   a. waterproof drywall.
   b. water-resistant gypsum board.
   c. cement board.
   d. all of the above.

8. Cutting ceramic tile is done with a
   a. nibbler.
   b. ceramic tile saw.
   c. hand cutter.
   d. all of the above.

9. Bathroom tile should extend over the tops of showerheads a minimum of
   a. 4 inches.
   b. 6 inches.
   c. 8 inches.
   d. 12 inches.

10. The number of 4×4 ceramic tiles needed to cover 150 square feet of wall area is
    a. 1,152.
    b. 1,350.
    c. 1,500.
    d. 1,674.

UNIT 25
Ceiling Finish

CHAPTER **74** Suspended Ceilings

CHAPTER **75** Ceiling Tile

Inexpensive and highly attractive ceilings may be created by installing suspended ceilings or ceiling tiles. They may be installed in new construction beneath exposed joists or when remodeling below existing ceilings.

Suspended or tile ceiling finish is installed in sections and pieces. The layout of the pattern and border tiles is important for the ceiling to look professionally installed.

OBJECTIVES

After completing this unit, the student should be able to:

- describe the sizes, kinds, and shapes of ceiling tile and suspended ceiling panels.
- lay out and install suspended ceilings.
- lay out and install ceiling tile.
- estimate quantities of ceiling finish materials.

74 Suspended Ceilings

Suspended ceilings are widely used in commercial and residential construction as a ceiling finish. They also provide space for recessed lighting, ducts, pipes, and other necessary conduits **(Figure 74–1).** Besides improving the appearance of a room, a suspended ceiling conserves energy by increasing the insulating value of the ceiling. It also aids in controlling sound transmission. In remodeling work, a suspended system can be easily installed beneath an existing ceiling. In basements, where overhead pipes and ducts may make other types of ceiling application difficult, a suspended type is easily installed. In addition, removable panels make pipes, ducts, and wiring accessible.

SUSPENDED CEILING COMPONENTS

A suspended ceiling system consists of panels that are laid into a metal grid. The grid consists of **main runners, cross tees,** and **wall angles.** It is constructed in a 2 × 4 rectangular or 2 × 2 square pattern **(Figure 74–2).** Grid members come prefinished in white, black, brass, chrome, and wood grain patterns, among others.

Wall Angles

Wall angles are L-shaped pieces that are fastened to the wall to support the ends of main runners and cross tees. They come in 10- and 12-foot lengths.

FIGURE 74–1 **Installing panels in a suspended ceiling grid.** (*Courtesy of Armstrong World Industries*)

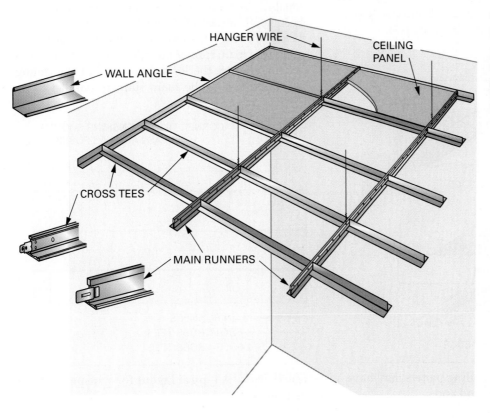

HANGER WIRE

CEILING PANEL

WALL ANGLE

CROSS TEES

MAIN RUNNERS

FIGURE 74–2 **A suspended ceiling consists of grid members and ceiling panels.**

They provide a continuous finished edge around the perimeter of the ceiling, where it meets the wall.

Main Runners

Main runners or tees are shaped in the form of an upside-down T. They come in 12-foot lengths. End splices make it possible to join lengths of main runners together. Slots are punched in the side of the runner at 12-inch intervals to receive cross tees. Along the top edge, punched holes are spaced at intervals for suspending main runners with *hanger* wire. Main runners extend from wall to wall. They are the primary support of a ceiling's weight.

Cross Tees

Cross tees come in 2- and 4-foot lengths. A slot, of similar shape and size as those in main runners, is centered on the 4-foot cross tees for use when turning 2 × 4 grid into a 2 × 2 grid. They come with connecting tabs on each end. These tabs are inserted and locked into main runners and other cross tees.

Ceiling Panels

Ceiling panels are manufactured of many different kinds of material, such as gypsum, glass fibers, mineral fibers, and wood fibers. Panel selection is based on considerations such as fire resistance, sound control, thermal insulation, light reflectance, moisture resistance, maintenance, appearance, and cost. Panels are given a variety of surface textures, designs, and finishes. They are available in 2 × 2 and 2 × 4 sizes with square or rabbeted edges **(Figure 74–3)**.

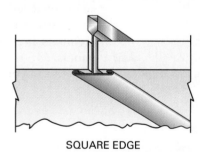

SQUARE EDGE

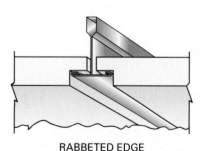

RABBETED EDGE

FIGURE 74–3   Suspended ceiling panels may have square or rabbeted edges and ends.

SUSPENDED CEILING LAYOUT

Before the actual installation of a suspended ceiling, a scaled sketch of the ceiling grid should be made. The sketch should indicate the direction and location of the main runners, cross tees, light panels, and border panels.

Main runners usually are spaced 4 feet apart. They usually run parallel with the long dimension of the room. For a standard 2 × 4 pattern, 4-foot cross tees are then spaced 2 feet apart between main runners. If a 2 × 2 pattern is used, 2-foot cross tees are installed between the midpoints of the 4-foot cross tees. Main runners and cross tees should be located in such a way that *border panels* on both sides of the room are equal and as large as possible **(Figure 74–4)**. Sketching the ceiling layout also helps when estimating materials.

Sketching the Layout

Sketch a grid plan by first drawing the overall size of the ceiling to a convenient scale. Use special care in measuring around irregular walls.

Locating Main Runners. To locate main runners, change the width of room to inches and divide by 48. Add 48 inches to any remainder. Divide the sum by 2 to find the distance from the wall to the first main runner. This distance is also the length of border panels.

For example, if the width of the room is 15'–8" changing to inches equals 188. Dividing 188 by 48 equals 3, with a remainder of 44 inches. Adding 48 to 44 equals 92 inches. Dividing 92 by 2 equals 46 inches, the distance from the wall to the first main runner.

Draw a main runner the calculated distance from, and parallel to, the long dimension of the ceiling. Draw the rest of the main runners parallel to the

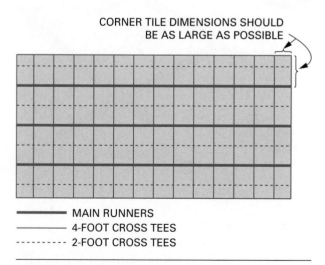

CORNER TILE DIMENSIONS SHOULD BE AS LARGE AS POSSIBLE

——— MAIN RUNNERS
——— 4-FOOT CROSS TEES
-------- 2-FOOT CROSS TEES

FIGURE 74–4   A typical layout for a suspended ceiling grid.

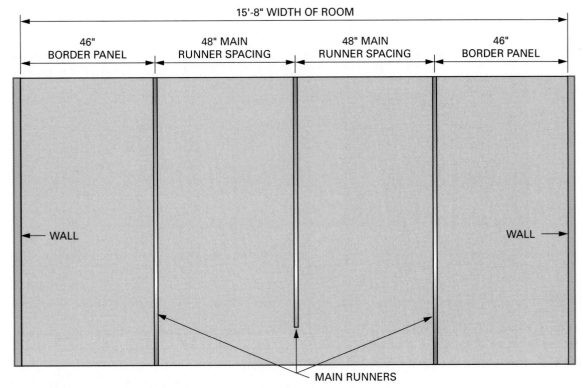

15'-8" WIDTH OF ROOM

| 46" BORDER PANEL | 48" MAIN RUNNER SPACING | 48" MAIN RUNNER SPACING | 46" BORDER PANEL |

WALL

WALL

MAIN RUNNERS

1. CHANGE ROOM WIDTH DIMENSION TO INCHES: 15 X 12 + 8 = 188"
2. DIVIDE 188 BY 48 = 3.91667 TILES
3. ADD 1 TO REMAINDER = 1.91667 TILES
4. DIVIDE THIS NUMBER BY 2: 1.91667 ÷ 2 = 0.95833
5. MULTIPLY FRACTION OF A TILE BY 48: 0.95833 X 48 = 46" BORDER TILE LENGTH

FIGURE 74–5 Method of determining the location of main runners.

first, and at 4-foot intervals. The distance between the last main runner and the wall should be the same as the distance between the first main runner and the opposite wall **(Figure 74–5).**

Locating Cross Tees. To locate 4-foot cross tees between main runners, first change the long dimension of the ceiling to inches. Divide by 24. Add 24 to the remainder. Divide the sum by 2 to find the width of the border panels on the other walls.

For example, if the long dimension of the room is 27'–10", changing it to inches equals 334. Dividing 334 by 24 equals 13, with a remainder of 22 inches. Adding 24 to 22 equals 46 inches. Dividing 46 by 2 equals 23 inches, the distance from the wall to the first row of cross tees.

Draw the first row of cross tees the calculated distance from, and parallel to, the short wall. Draw the remaining rows of cross tees parallel to each other at 2-foot intervals. The distance from the last row of cross tees to the wall should be the same as the distance from the first row of cross tees to the opposite wall **(Figure 74–6).**

CONSTRUCTING THE CEILING GRID

The ceiling grid is constructed by first installing *wall angles*, then installing *suspended ceiling lags* and *hanger wires*, suspending the *main runners*, inserting full-length *cross tees*, and, finally, cutting and inserting *border* cross tees.

Installing Wall Angles

A suspended ceiling must be installed with at least 3 inches for clearance below the lowest air duct, pipe, or beam. This clearance provides enough room to insert ceiling panels in the grid. If recessed lighting is to be used, allow a minimum of 6 inches for clearance. The height of the ceiling may be located by measuring up from the floor. If the floor is rough or out of level, the ceiling line may be located with various leveling devices previously described. A combination of a hand level and straightedge, a water level, builders' level, transit-level, or laser level can be used **(Figure 74–7).** Snap chalk lines on all walls around the room to the height of the top edge of the wall angle.

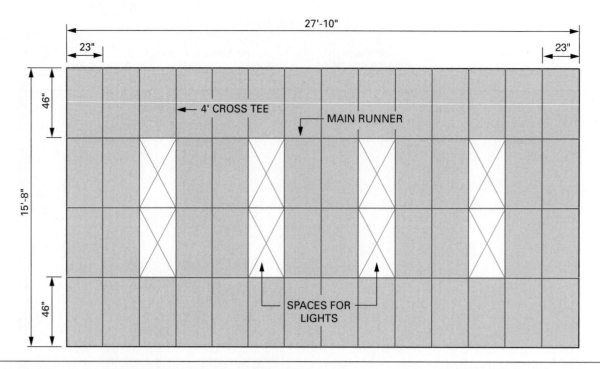

FIGURE 74–6 Completed sketch of a suspended ceiling layout.

FIGURE 74–7 A laser level may be used to install a suspended ceiling wall angle.

FIGURE 74–8 Installing wall angles.

Fasten wall angles around the room with their top edge lined up with the chalk line. It may be easier to fasten the wall angle by prepunching holes with a center punch or spike. Fasten into framing wherever possible, not more than 24 inches apart **(Figure 74–8).**

To fasten wall angles to concrete walls, short masonry nails sometimes are used. However, they are difficult to hold and drive. Use a small strip of cardboard to hold the nail while driving it with the hammer **(Figure 74–9).** Lead or plastic inserts and screws may also be used to fasten the wall angles. Their use does require more time. If available, power nailers can be used for efficient fastening of wall angles to masonry walls.

Make miter joints on outside corners. Make butt joints in interior corners and between straight lengths of wall angle **(Figure 74–10).** Use a combination square to layout and draw the square and angled lines. Cut carefully along the lines with snips.

To fasten wall angles to concrete walls, short masonry nails sometimes are used. However, they are difficult to hold and drive. Use a small strip of cardboard to hold the nail while driving it with the hammer.

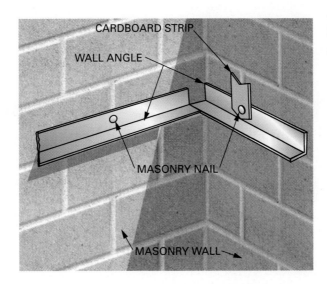

CARDBOARD STRIP

WALL ANGLE

MASONRY NAIL

MASONRY WALL

FIGURE 74–9 Technique for driving short nails.

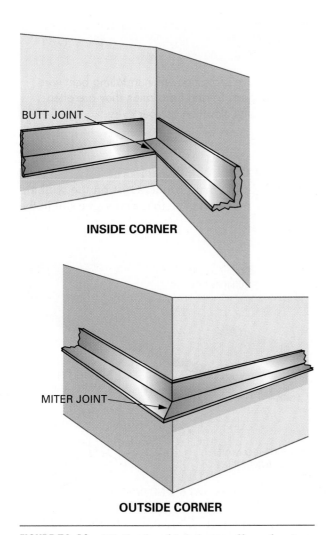

BUTT JOINT

INSIDE CORNER

MITER JOINT

OUTSIDE CORNER

FIGURE 74–10 Methods of joining wall angle at corners.

⚠️ **CAUTION CAUTION CAUTION CAUTION**

Use care in handling the cut ends of wall angle and other grid members. Cut metal ends are very sharp and can cause serious injury. ∎

Installing Hanger Lags

From the ceiling sketch, determine the position of the first main runner. Stretch a line at this location across the room from the top edges of the wall angle. Stretch the line tightly on nails inserted between the wall and wall angle **(Figure 74–11).** The line serves as a guide for installing *hanger lags* or *screw eyes* and *hanger wires* from which main runners are suspended.

Install hanger lags not over 4 feet apart and directly over the stretched line. Hanger lags should be of the type commonly used for suspended ceilings. They must be long enough to penetrate wood joists a minimum of 1 inch to provide strong support. *Eye pins* are driven into concrete with a *powder-actuated* fastening tool (Figure 16–13). Hanger wires may also be attached directly around the lower chord of bar joists or trusses.

Installing Hanger Wire

Cut a number of hanger wires using wire cutters. The wires should be about 12 inches longer than the distance between the overhead construction and the

Stretch lines for main runners on nails inserted between the wall and wall angle.

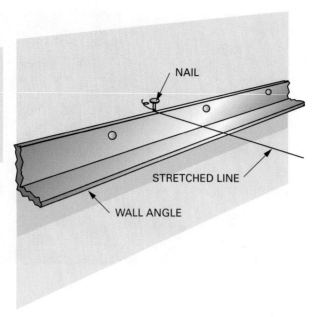

NAIL

STRETCHED LINE

WALL ANGLE

FIGURE 74–11
Technique for stretching a line between wall angles.

stretched line. For residential work, 16-gauge wire is usually used. For commercial work, 12-gauge and heavier wire is used.

Attach the hanger wires to the hanger lags. Insert about 6 inches of the wire through the screw eye. Securely wrap the wire around itself three times. Pull on each wire to remove any kinks. Then make a 90-degree bend where it crosses the stretched line **(Figure 74–12).** Stretch lines, install hanger lags, and attach and bend hanger wires in the same manner at each main runner location. Leave the last line stretched tightly in position.

> ⚠ **CAUTION   CAUTION   CAUTION   CAUTION**
>
> **W**ear eye protection when installing bent wire hangers. During installation they are often at the same elevation as your eyes. ∎

Installing Main Runners

The ends of the main runners rest on the wall angles. They must be cut so that a cross-tee slot in the

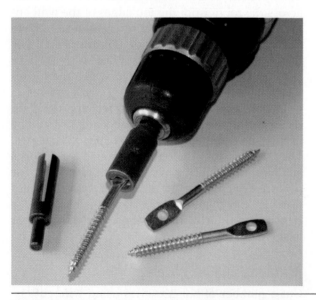

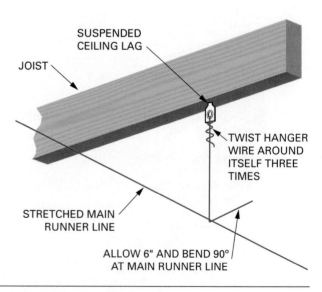

SUSPENDED CEILING LAG

JOIST

TWIST HANGER WIRE AROUND ITSELF THREE TIMES

STRETCHED MAIN RUNNER LINE

ALLOW 6" AND BEND 90° AT MAIN RUNNER LINE

FIGURE 74–12   Suspended ceiling lags are used to support hanger wire. Wire may be prebent to accept main runners.

web of the runner lines up with the first row of cross tees. A cross-tee line must be stretched, at wall angle height, across the short dimension of the room to line up the slots in the main runners. The line must run exactly at right angles to the main runner line and at a distance from the wall equal to the width of the border panels. If the walls are at right angles to each other, the location of the cross-

tee line can be determined by measuring out from both ends of the wall.

When the walls are not at right angles, the Pythagorean theorem (Figure 26–6) is used to square the grid system **(Figure 74–13)**. After the main runner line is installed, measure out from the short wall, along the stretched main runner line, a distance equal to the width of the border panel. Mark the line.

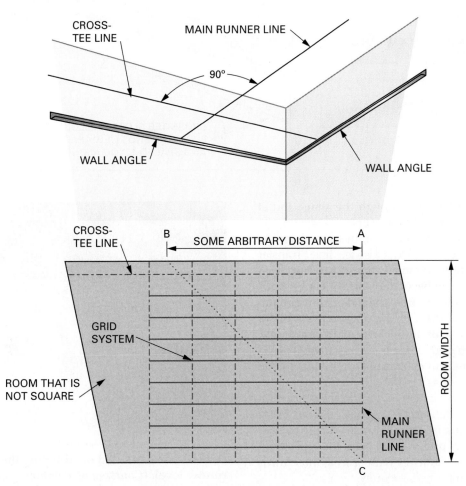

1) Starting at point A, measure the room width keeping your tape as square as possible with the wall.
2) From A measure some distance to B. The actual distance does not matter; it merely needs to be large enough to make a big triangle.
3) Use these numbers in the Pythagorean theorem to determine the distance from B to C.
4) Measure and mark the distance from B to C and mark C. C is now square with point A.
5) Connect A and C with a string and measure each successive row of main runner from it.
6) **EXAMPLE** If AC, the room width, is 18'–9", then measure AB to be, say, 16'–0". Convert these dimensions to inches. 16'–0" becomes 192" and 18'–9" becomes 225" (18 × 12 + 9 = 225"). Put these dimensions in the Pythagorean theorem. $C^2 = A^2 + B^2$

$C = \sqrt{192^2 + 225^2} = \sqrt{36864 + 50625} = \sqrt{87489} = 295.7853952"$

To convert the decimal to sixteenths

0.7853952 × 16 = 12.566 sixteenths ⇒ $^{13}/_{16}"$

Thus, the measurement from B to C is 295 $^{13}/_{16}$ inches.

FIGURE 74–13 Method of stretching two perpendicular lines using the Pythagorean theorem.

Procedure 74–A   Cutting Main Runners so Cross-Tee Slots Align

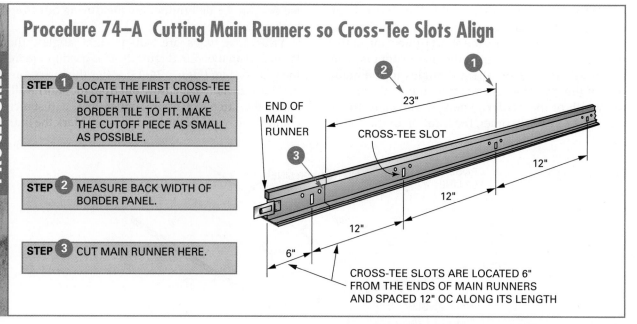

STEP 1 LOCATE THE FIRST CROSS-TEE SLOT THAT WILL ALLOW A BORDER TILE TO FIT. MAKE THE CUTOFF PIECE AS SMALL AS POSSIBLE.

STEP 2 MEASURE BACK WIDTH OF BORDER PANEL.

STEP 3 CUT MAIN RUNNER HERE.

END OF MAIN RUNNER

CROSS-TEE SLOT

23"

12"

12"

12"

6"

CROSS-TEE SLOTS ARE LOCATED 6" FROM THE ENDS OF MAIN RUNNERS AND SPACED 12" OC ALONG ITS LENGTH

Stretch the cross-tee line through this mark and at right angles to the main runner line.

At each main runner location, measure from the short wall to the stretched cross-tee line. Transfer the measurement to the main runner. Measure from the first cross-tee slot beyond the measurement, so as to cut as little as possible from the end of the main runner **(Procedure 74–A)**. Cut the main runners about ⅛-inch less to allow for the thickness of the wall angle. Backcut the web slightly for easier installation at the wall. Measure and cut main runners individually. Do not use the first one as a pattern to cut the rest.

Hang the main runners by resting the cut end on the wall angle and inserting suspension wires in the appropriate holes in the top of the main runner. Bring the runners up level and bend the wires **(Figure 74–14)**. Twist the wires with at least three turns to hold the main runners securely.

More than one length of main runner may be needed to reach the opposite wall. Connect lengths of main runners together by inserting tabs into matching ends. Make sure end joints come up tight. The length of the last section is measured from the end of the last one installed to the opposite wall, allowing about ⅛-inch less for the thickness of the wall angle.

Installing Cross Tees

Cross tees are installed by inserting the tabs on the ends into the slots in the main runners. These fit into position easily, although the method of attaching varies from one manufacturer to another.

FIGURE 74–14   Method of adjusting the main runner level. *(Courtesy of Trimble)*

Install all full-length cross tees between main runners first. Lay in a few full-size ceiling panels. This stabilizes the grid while installing cross tees for border panels. Cut and install cross tees along the border. Insert the connecting tab of one end in the main runner and rest the cut end on the wall angle **(Figure 74–15)**. If the walls are not straight or square, it is necessary to cut cross tees for border tiles individually. For 2 × 2 panels, install 2-foot cross tees at the midpoints of the 4-foot cross tees. After the grid is complete, sight sections by eye. Straighten where necessary by making adjustments to border cross tees or hanger wires.

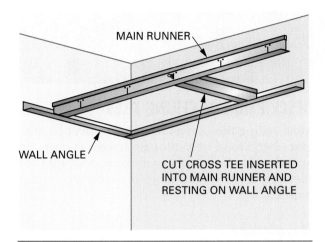

FIGURE 74–15 **Inserting a cross tee in the main runner.**

FIGURE 74–16 **Installing ceiling panels.**

Installing Ceiling Panels

Ceiling panels are placed in position by tilting them slightly, lifting them above the grid, and letting them fall into place **(Figure 74–16).** When handling panels, be careful to avoid marring the finished surface. Install all border panels first. Then install full-sized field panels. Measure each border panel individually, if necessary. Cut them slightly smaller than measured so they can drop into place easily. Cut the panels with a sharp utility knife using a straightedge as a guide. A scrap piece of cross-tee material can be used as a straightedge. Always cut with the finished side of the panel up.

Cutting Ceiling Panels around Columns

When a column is near the center of a ceiling panel, cut the panel at the midpoint of the column. Cut semicircles from the cut edge to the size required for the panel pieces to fit snugly around the column. After the two pieces are rejoined around the column, glue scrap pieces of panel material to the back of the installed panel.

If the column is close to the edge or end of a panel, cut the panel from the nearest edge or end to fit around the column. The small piece is also fitted around the column and joined to the panel by gluing scrap pieces to its back side **(Figure 74–17).**

ESTIMATING SUSPENDED CEILING MATERIALS

- Divide the perimeter of the room by 10, the length of a wall angle, to find the number needed.

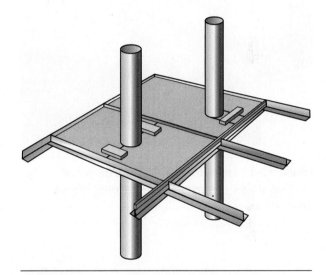

FIGURE 74–17 **Fitting ceiling panels around columns.**

- Find the number of main runners needed from the sketch. No more than two pieces can be cut from one 12-foot length.
- Count the number of 2- and 4-foot cross tees from the sketch. Border cross tees must be cut from full-length tees.
- From the sketch, count the number of hanger wires and screw eyes needed. Multiply the number needed by the length of each hanger wire to find the total linear feet of hanger wire needed.
- Count the number of ceiling panels from the sketch. Each border panel requires a full-size ceiling panel. There will be one less for each light fixture.

75 Ceiling Tile

Ceiling tile is usually stapled to furring strips that are fastened to exposed joists. They may also be cemented to existing ceilings provided the ceilings are solid and level. If the existing ceiling is not sound, furring strips should be installed and fastened through the ceiling into the joists above **(Figure 75–1).**

FIGURE 75–1   Installing ceiling tile on wood furring strips. *(Courtesy of Armstrong World Industries)*

DESCRIPTION OF CEILING TILES

Most ceiling tiles are made of wood fiber or mineral fiber. Wood fiber tiles are lowest in cost and are adequate for many installations. Mineral fiber tiles are used when a more fire-resistant type is required.

Manufacture

Wood fibers are pressed into large sheets that are ⁷⁄₁₆ to ¾ inch thick. Mineral fiber tiles are made of rock that is heated to a molten state. The fibers are then sprayed into a sheet form. The surfaces of some sheets are **fissured** or **perforated** for sound absorption. The surfaces of other sheets are embossed with different designs or left smooth. Then, they are given a factory finish and cut into individual tiles. Most tiles are cut with *chamfered, tongue-and-grooved* edges with two adjacent *stapling flanges* for concealed fastening **(Figure 75–2).**

Sizes

The most popular sizes of ceiling tile are a 12-inch square and a 12- × 24-inch rectangle in thicknesses of ½ inch. Tiles are also manufactured in squares of 16 inches and in rectangles of 16 × 32 inches.

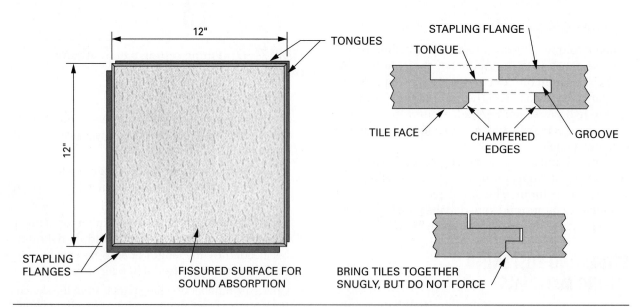

FIGURE 75–2   A typical ceiling tile has tongue-and-grooved edges with stapling flanges.

CEILING TILE LAYOUT

Before installation begins, it is necessary to calculate the size of *border tiles* that run along the walls. It is desirable for border tiles to be as wide as possible and of equal widths on opposite walls.

Calculating Border Tile Sizes

To find the width of the border tiles along the long walls of a room, first determine the dimension of the short wall. In most cases, the measurement will be a number of full feet, plus a few inches. Not counting the foot measurement, add 12 more inches to the remaining inch measurement. Divide the sum by 2 to find the width of border tiles for the long wall. The following example is applicable for a distance between the long walls of 10'–6".

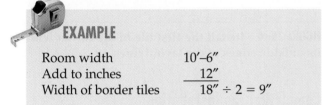

EXAMPLE

Room width	10'–6"
Add to inches	12"
Width of border tiles	18" ÷ 2 = 9"

The width of border tiles along the short walls is calculated in the same manner. The following example applies when the distance between the short walls is 19'–8".

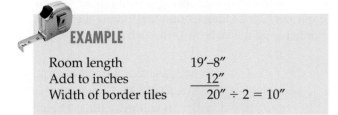

EXAMPLE

Room length	19'–8"
Add to inches	12"
Width of border tiles	20" ÷ 2 = 10"

PREPARATION FOR CEILING TILE APPLICATION

Unless an adhesive application to an existing ceiling is to be made, **furring strips** must be provided on which to fasten ceiling tiles. Furring strips are usually fastened directly to exposed joists. They are sometimes applied to an existing ceiling and fastened into the concealed joists above.

Locating Concealed Joists

If the joists are hidden by an existing ceiling, tap on the ceiling with a hammer. Drive a nail into the spot where a dull thud is heard, to find a concealed joist. Locate other joists by the same method or by measuring from the first location. Usually, ceiling joists are spaced 16 inches on center (OC) and run parallel to the short dimension of the room. When all joists are located, snap lines on the existing ceiling directly below and in line with the concealed joists.

LAYING OUT AND APPLYING FURRING STRIPS

For fastening 12-inch tiles, furring strips must be installed 12 inches OC. From the corner, measure out the width of the border tiles. This measurement is the center of the first furring strip away from the wall. To mark the edge, measure from the center, in either direction, half the width of the furring strip. Mark an *X* on the side of the mark toward the center of the furring strip. From the edge of the first furring strip, measure and mark, every 12 inches, across the room. Place Xs on the same side of the mark as the first one **(Figure 75–3)**.

Lay out the other end of the room in the same manner. Snap lines between the marks for the location of the furring strips. The strips are fastened by keeping one edge to the chalk line with the strip on the side of the line indicated by the *X*. Fasteners must penetrate at least 1 inch into the joist. Starting and ending furring strips are also installed against both walls.

Squaring the Room

First, snap a chalk line on a furring strip as a guide for installing border tiles against the long wall. The line is snapped parallel to, and the width of the border tiles, away from the wall.

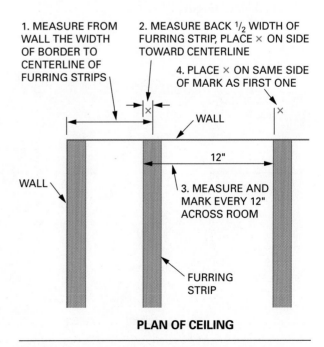

1. MEASURE FROM WALL THE WIDTH OF BORDER TO CENTERLINE OF FURRING STRIPS

2. MEASURE BACK 1/2 WIDTH OF FURRING STRIP, PLACE × ON SIDE TOWARD CENTERLINE

4. PLACE × ON SAME SIDE OF MARK AS FIRST ONE

WALL

12"

WALL

3. MEASURE AND MARK EVERY 12" ACROSS ROOM

FURRING STRIP

PLAN OF CEILING

FIGURE 75–3 Furring strip layout for ceiling tile.

A second chalk line must be snapped to guide the application of the short wall border tiles. The line must be snapped at exactly at 90 degrees to the first chalk line. Otherwise tiles will not line up properly. From the short wall, measure in along the first chalk line, the width of the short wall border tiles. From this point, use the Pythagorean theorem and snap another line at a right angle to the first line. This method of squaring lines has been previously described in Figure 74–13 and Figure 26–6.

INSTALLING CEILING TILE

Ceiling tiles should be allowed to adjust to normal interior conditions for twenty-four hours before installation. Some carpenters sprinkle talcum powder or corn starch on their hands to keep them dry. This prevents fingerprints and smudges on the finished ceiling. Cut tiles face up with a sharp utility knife guided by a straightedge. All cut edges should be against a wall.

Starting the Installation

To start the installation, cut a tile to fit in the corner. The outside edges of the tile should line up exactly with both chalk lines. Because this tile fits in the corner, it must be cut to the size of border tiles on both long and short sides of the room. For example, if the border tiles on the long wall are 9 inches and those on the short wall are 10 inches, the corner tile should be cut twice to make it 9 × 10. Staple the tile in position. Be careful to line up the outside edges with both chalk lines **(Figure 75–4).**

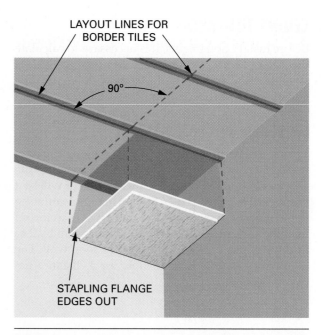

FIGURE 75–4   **Install the first tile in the corner with its outside edges to the layout lines.**

Completing the Installation

After the corner tile is in place, work across the ceiling. Install two or three border tiles at a time. Then fill in with full-sized field tiles **(Figure 75–5).** Tiles are applied so they are snug to each other, but not jammed tightly. Fasten each tile with four ½- or ⁹⁄₁₆-inch staples. Place two in each flange, using a hand stapler. Use six staples in 12 × 24 tiles. Continue applying tiles in this manner until the last row is reached.

FIGURE 75–5   **Install a few border tiles. Then fill in with full-sized field tiles.**

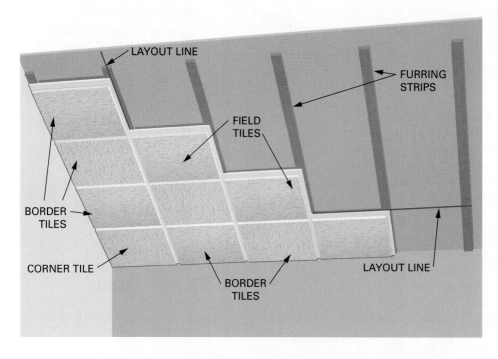

When the last row is reached, measure and cut each border tile individually. Cut the tiles slightly less than measured for easy installation. Do not force the tiles in place. Face nail the last tile in the corner near the wall where the nailhead will be covered by the wall molding. After all tiles are in place, the ceiling is finished by applying molding between the wall and ceiling. (The application of wall molding is discussed in a later unit.)

Adhesive Application of Ceiling Tile

Ceiling tile is sometimes cemented to existing plaster and drywall ceilings. These ceilings must be completely dry, solid, level, and free of dust and dirt. If the existing ceiling is in poor condition or has loose paint, adhesive application is not recommended.

Special tile adhesive is used to cement tiles to ceilings. Four daubs of cement are used on 12 × 12 tile. Six daubs are used on 12 × 24 tile. Before applying the daubs, prime each spot by using the putty knife to force a thin layer into the back surface of the tile. Apply the daubs, about the size of a walnut. Press

the tile into position. Keep the adhesive away from the edges to allow for spreading when the tile is pressed into position. No staples are required to hold the tile in place.

ESTIMATING MATERIALS FOR TILE INSTALLATION

To estimate ceiling tile, measure the width and length of the room to the next whole foot measurement. Multiply these figures together to find the area of the ceiling in square feet. Divide the ceiling area by the number of square feet contained in one of the ceiling tiles being used to find the number of ceiling tiles needed.

To estimate furring strips, measure to the next whole foot the length of the room in the direction that the furring strips are to run. Multiply this by the number of rows of furring strips to find the total number of linear feet of furring strip stock needed. To find the number of rows, divide the width of the room by the furring strip spacing and add one.

Key Terms

cross tee

furring strips

fissured

main runner

perforated

wall angle

Review Questions

Select the most appropriate answer.

1. The most common sizes in inches of suspended ceiling panels are
   a. 8 × 12 and 12 × 12.
   b. 12 × 12 and 16 × 16.
   c. 12 × 12 and 12 × 24.
   d. 24 × 24 and 24 × 48.

2. The dimension of a border panel for a 12'–6" room, when 24" × 24" suspended ceiling panels are used, is
   a. 3.   c. 15.
   b. 6.   d. 16.

3. The diagonal measurement of a 16" × 24" rectangle is
   a. 384".
   b. 832".
   c. 28⅞".
   d. 28'–10⅛".

4. The number of usable pieces that may be cut from one main runner is
   a. 2.
   b. 3.
   c. 4.
   d. more than 4.

5. In a suspended ceiling, hanger wire is used to suspend
   a. cross tees.
   b. main runners.
   c. wall angle.
   d. furring strips.

6. The first step in installing a suspended ceiling is to
   a. square the room.
   b. make a sketch of the planned ceiling.
   c. calculate border tiles.
   d. install the wall angle.

7. The number of 2 × 2 suspended ceiling tiles estimated for a room that measures 17′–9″ × 23′–9″ is
   a. 88.
   b. 108.
   c. 130.
   d. 422.

8. The most common sizes in inches of stapled-place ceiling tile are
   a. 8 × 12 and 12 × 12.
   b. 12 × 12 and 16 × 16.
   c. 12 × 12 and 12 × 24.
   d. 14 × 24 and 24 × 48.

9. Nails used to fasten furring to ceiling joists must penetrate into the joists at least
   a. ¾ inch.
   b. 1 inch.
   c. 1¼ inches.
   d. 1½ inches.

10. Fastening 12″ × 12″ ceiling tiles requires
    a. two staples.
    b. three staples.
    c. four staples.
    d. six staples.

UNIT 26

Interior Doors and Door Frames

nterior doors used in residential and light commercial buildings are less dense than exterior doors. They are not ordinarily subjected to as much use and are not exposed to the weather. In commercial buildings, such as hospitals and schools, heavier and larger interior doors are specified that meet special conditions.

Doors must be installed level and plumb to operate properly. Otherwise they may not remain in the position last placed, moving on their own.

OBJECTIVES

After completing this unit, the student should be able to:

- describe the sizes and kinds of interior doors.
- make and set interior door frames.
- hang an interior swinging door.
- install locksets on interior swinging doors.
- set a prehung door and frame.
- install bypass, bifold, pocket, and folding doors.

76 Description of Interior Doors

nterior doors are classified by style as *flush, panel, French, louver,* and *café* doors. Interior flush doors have a smooth surface, are usually less expensive, and are widely used when a plain appearance is desired. Some of the other styles have special uses. Doors are also classified by the way they operate, such as *swinging, sliding,* or *folding.*

INTERIOR DOOR SIZES AND STYLES

For residential and light commercial use, most interior doors are manufactured in 1⅜-inch thickness. Some, like café and bifold doors, are also made in 1¼- and 1⅛-inch thicknesses. Most doors are manufactured in 6'–8" heights. Some types may be obtained in heights of 6'–0" and 6'–6". Door widths range from 1'–0" to 3'–0" in increments of 2 inches. However, not all sizes are available in every style.

Flush Doors

Flush doors are made with *solid* or *hollow* cores. Solid core doors are generally used as entrance or fire-rated doors. (They have been previously described in Unit 20, Exterior Doors.) Hollow-core doors are commonly used in the interior except when fire resistance or sound transmission is critical.

A hollow-core door consists of a light perimeter frame. This frame encloses a mesh of thin wood or composition material supporting the faces of the door. Solid wood blocks are appropriately placed in the core for the installation of locksets. The frame and mesh are covered with a thin plywood called a *skin. Lauan* plywood is used extensively for flush door skins. Flush doors are also available with veneer faces of *birch, gum, oak,* and *mahogany,* among others **(Figure 76–1)**. When flush doors are to be painted, an overlay plywood or tempered hardboard may be used for the skin.

Panel Doors

Interior **panel doors** consist of a frame with usually one to eight wood panels in various designs **(Figure 76–2)**. They are similar in style to some exterior panel doors. (The construction of panel doors has been previously described in Unit 20, Exterior Doors.)

French Doors

French doors may contain from one to fifteen lights of glass. They are made in a 1¾-inch thickness for exterior doors and 1⅜-inch thickness for interior doors **(Figure 76–3)**.

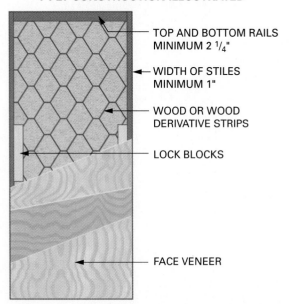

**MESH OR CELLULAR CORE
7 PLY CONSTRUCTION ILLUSTRATED**

- TOP AND BOTTOM RAILS MINIMUM 2 ¼"
- WIDTH OF STILES MINIMUM 1"
- WOOD OR WOOD DERIVATIVE STRIPS
- LOCK BLOCKS
- FACE VENEER

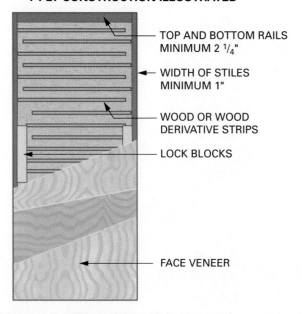

**LADDER CORE
7 PLY CONSTRUCTION ILLUSTRATED**

- TOP AND BOTTOM RAILS MINIMUM 2 ¼"
- WIDTH OF STILES MINIMUM 1"
- WOOD OR WOOD DERIVATIVE STRIPS
- LOCK BLOCKS
- FACE VENEER

FIGURE 76–1 The construction of hollow-core flush doors.

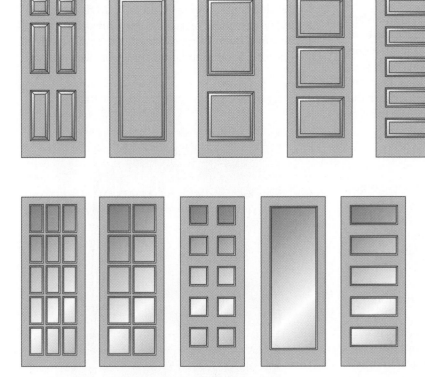

FIGURE 76–2   Styles of commonly used interior panel doors.

FIGURE 76–3   French doors are used in the interior as well as for entrances.

Louver Doors

Louver doors are made with spaced horizontal slats called louvers used in place of panels. The louvers are installed at an angle to obstruct vision but permit the flow of air through the door. Louvered doors are widely used on clothes closets **(Figure 76–4)**.

Café Doors

Café doors are short panel or louver doors. They are hung in pairs that swing in both directions. They are used to partially screen an area, yet allow easy and safe passage through the opening. The tops and bottoms of the doors are usually shaped in a pleasing design **(Figure 76–5)**.

METHODS OF INTERIOR DOOR OPERATION

Doors are also identified by their method of operation. Doors either swing on hinges or slide on tracks. The choice of door operation depends on such factors as convenience, cost, safety, and space.

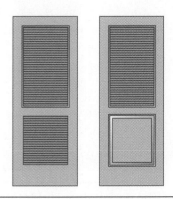

FIGURE 76–4   Louver doors obstruct vision, but permit the circulation of air.

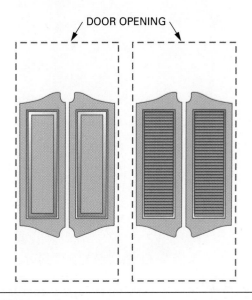

FIGURE 76–5   Café doors usually are used between kitchens and dining areas.

Swinging Doors

Swinging doors are hinged on one edge. They swing out of the opening. When closed, they cover the total opening. Swinging doors that swing in one direction are called *single-acting doors* **(Figure 76–6).** With special hinges, they can swing in both directions. They are then called *double-acting doors.* Swinging doors are the most commonly used type of door. They have the disadvantage of requiring space for the swing.

Bypass Doors

Bypass doors are commonly used on wide clothes closet openings. A double track is mounted on the header jamb of the door frame. Rollers that ride in the track are attached to the doors so that they slide by each other. A floor guide keeps the doors in alignment at the bottom **(Figure 76–7).** Usually two doors are used in a single opening. Three or more doors may be used, depending on the situation.

The disadvantage of bypass doors is that although they do not project out into the room, access to the complete width of the opening is not possible. They are easy to install, but are not practical in openings less than 6 feet wide.

Pocket Doors

The **pocket door** is opened by sliding it sideways into the interior of the partition. When opened, only the lockedge of the door is visible **(Figure 76–8).** Pocket doors may be installed as a single unit, sliding in one direction, or as a double unit sliding in opposite directions. When opened, the total width of the opening is obtained, and the door does not pro-

FIGURE 76–7 Bypass doors are used on wide closet openings.

FIGURE 76–6 A single-acting swinging door is the most widely used type of interior door.

FIGURE 76–8 The pocket door slides into the interior of the partition.

ject out into the room. Pocket doors are used when these advantages are desired.

The installation of pocket doors requires more time and material than other methods of door operation. A special pocket door frame unit and track must be installed during the rough framing stage **(Figure 76–9)**. The rough opening in the partition must be large enough for the door opening and the pocket.

Bifold Doors

Bifold doors are made in flush, panel, louver, or combination panel and louver styles. They are made in narrower widths than other doors. This allows them to operate in a folding fashion on closet and similar type openings **(Figure 76–10)**.

Bifold doors consist of pairs of doors hinged at their edges. The doors on the jamb side swing on pivots installed at the top and bottom. Other doors fold up against the jamb door as it is swung open. The end door has a guide pin installed at the top. The pin rides in a track to guide the set when opening or closing **(Figure 76–11)**. On very wide openings the guide pin is replaced by a combination guide and support to keep the doors from sagging.

Bifold doors may be installed in double sets, opening and closing from each side of the opening. They have the advantage of providing access to almost the total width of the opening, yet they do not project out much into the room.

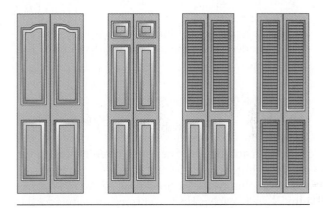

FIGURE 76–10 Bifold doors are manufactured in many styles.

FIGURE 76–9 A pocket door frame comes preassembled from the factory. It is installed when the interior partitions are framed.

FIGURE 76–11 Bifold doors provide access to almost the total width of the opening.

77 Installation of Interior Doors and Door Frames

Many interior doors come **prehung** in their frames for easier and faster installation on the job. However, it is often necessary to build and set the door frame, hang the door, and install the locksets.

INTERIOR DOOR FRAMES

Special rabbeted jamb stock or nominal 1-inch square-edge lumber is used to make interior door frames. If square-edge lumber is used, a separate *stop*, if needed, is applied to the inside faces of the door frame.

Checking Rough Openings

The first step in making an interior door frame is to measure the door opening to make sure it is the correct width and height. The rough opening width for single-acting swinging doors should be the width of the door plus twice the thickness of the side jamb, plus ½ inch each side for shimming between the door frame and the opening. For example, if the thickness of the side jamb beyond the door is ¾ inch, the rough opening width is 2½ inches more than the door width.

The rough opening height should be the height of the door, plus the thickness of the header jamb, plus ½ inch for clearance at the top, plus the thickness of the finished floor, plus a desired clearance under the door. An allowance of ½ to 1 inch is usually made for clearance between the finished floor and the door.

For example, if the header jamb and finished floor thickness are both ¾ inch, the rough opening height should be 2½ inches over the door height, if ½ inch is allowed for clearance under the door **(Figure 77–1)**.

The rough opening size for other than single-acting swinging doors, such as bypass and bifold doors, should be checked against the manufacturer's directions. The sizes of the doors and allowances for hardware may differ with the manufacturer.

Making an Interior Door Frame

Interior door frames are constructed like exterior door frames except they have no sill. Interior door frames usually are installed after the interior wall covering has been applied. Measure the total thickness of the wall, including the wall covering, to find the jamb width.

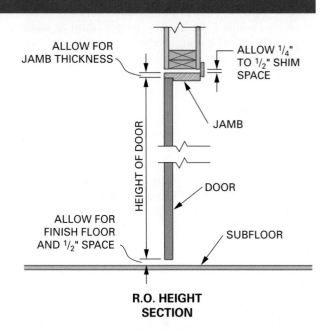

R.O. HEIGHT SECTION

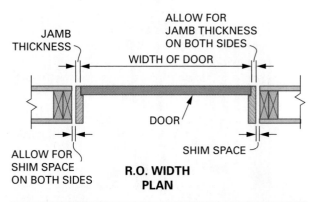

R.O. WIDTH PLAN

FIGURE 77–1   The size of rough openings for doors needs to take into account space for jambs and shimming.

Cutting Jambs to Width

If necessary, rip the door jamb stock so its width is the same as the wall thickness. If rabbeted jamb stock is used, cut the edge opposite the rabbet. Plane and smooth both edges to a slight *back bevel*. The back bevel permits the door casings, when later applied, to fit tightly against the edges of the door frame in case there are irregularities in the wall **(Figure 77–2)**. *Ease* all sharp exposed corners.

Cutting Jambs to Length

On interior door frames, the *head jamb* is usually cut to fit between, and is dadoed into, the *side jambs*. The

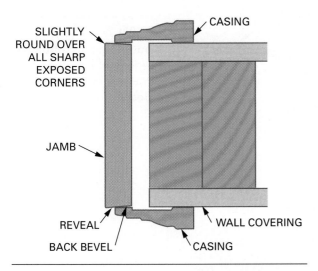

FIGURE 77–2   Back bevel jamb edges slightly to permit casings to fit snugly against them.

side jambs run the total height of the rough opening. Cut both side jambs to a length equal to the height of the opening.

Head Jambs of Door Frames for Hinged Doors. If rabbeted jambs are used, the length of the head jamb is the width of the door plus ³⁄₁₆ inch. The extra ³⁄₁₆ inch is for joints of ³⁄₃₂ inch on each side, between the edges of the door and the side jambs. If square-edge lumber is used, its length is the same as a rabbeted head jamb. However, ½ inch is added for dadoing ¼ inch deep into each side jamb **(Figure 77–3).** Cut the head jamb to length with both ends square.

Side Jambs of Door Frames for Hinged Doors. Measure up from the bottom ends. Square lines across the side jambs to mark the location of the bottom side of the head jamb. This dimension is the sum of:

- the thickness of the finish floor, if the door frame rests on the subfloor,
- an allowance of ½ inch minimum between the door and the finish floor,
- the height of the door, and
- ³⁄₃₂ inch for a joint between the door and the head jamb;
- on rabbeted jambs, subtract ½ inch for the depth of the rabbet.

Hold a scrap piece of jamb stock to the squared lines. Mark its other side to lay out the width of the dado. Mark the depth of the dadoes on both edges of the side jambs. Cut the dadoes to receive the head jamb. On rabbeted jambs, dado depth is to the face of the rabbet. A dado depth of ¼ inch is sufficient on plain jambs **(Figure 77–4).**

Jamb Lengths for Other Types of Doors. The length of head and side jambs for other types of doors, such as bypass and bifold, must be determined from instructions provided by the manufacturer of the hardware and the door. Door hardware and door sizes differ with the manufacturer. This affects the length of the door jambs.

Plain, square-edge lumber jambs are used to make door frames for doors other than single-acting swinging doors. The rest of the procedure, such as checking

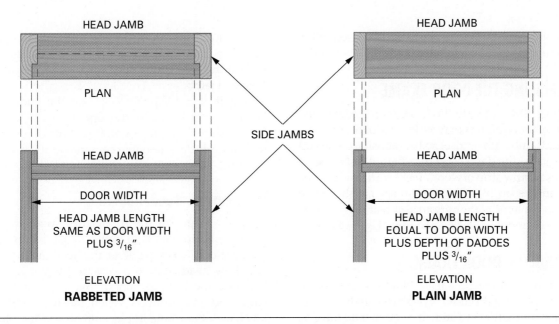

FIGURE 77–3   Length of plain and rabbeted head jambs of door frames for swinging doors.

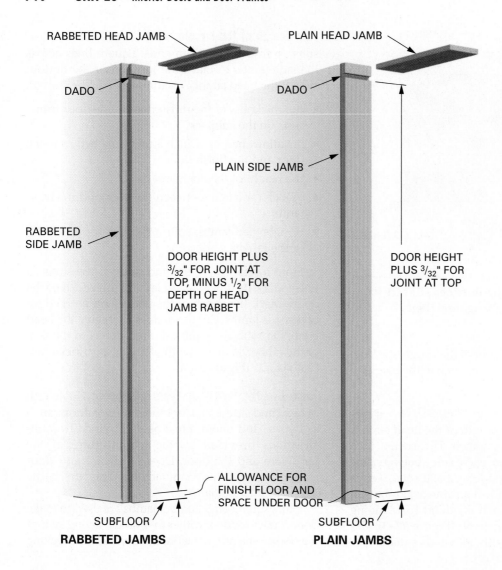

RABBETED HEAD JAMB

PLAIN HEAD JAMB

DADO

DADO

PLAIN SIDE JAMB

RABBETED SIDE JAMB

DOOR HEIGHT PLUS 3/32" FOR JOINT AT TOP, MINUS 1/2" FOR DEPTH OF HEAD JAMB RABBET

DOOR HEIGHT PLUS 3/32" FOR JOINT AT TOP

ALLOWANCE FOR FINISH FLOOR AND SPACE UNDER DOOR

SUBFLOOR

RABBETED JAMBS

SUBFLOOR

PLAIN JAMBS

FIGURE 77–4 Laying out plain and rabbeted side jambs of door frames for swinging doors.

rough openings, cutting jambs to width, assembling, and setting, is the same for all door frames.

ASSEMBLING THE DOOR FRAME

Fasten the side jambs to the head jamb, keeping the edges flush. If there is play in the dado, first wedge the head jamb with a chisel so the face side comes up tight against the dado shoulders before fastening.

Cut a narrow strip of wood. Tack it to the side jambs a few inches up from the bottom so that the frame width is the same at the bottom as it is at the top. This strip is commonly called a **spreader (Figure 77–5).**

SETTING THE DOOR FRAME

Door frames must be set so that the jambs are straight, level, and plumb. They are usually set before the finish floor is laid. If a rabbeted frame is used, determine the swing of the door so that the rabbet is facing toward the correct side.

Cut any *horns* from the top ends of side jambs. Place the frame in the opening. The horns are cut off in case the side jambs need to be shimmed to level the head jamb.

Install shims directly opposite the ends of the header jamb between the opening and the side jambs. Shim an equal amount on both sides so that the frame is close to being centered at the top of the opening. Drive shims up snugly not tightly.

Leveling the Header Jamb

Keep the edges of the frame flush with the wall. With the bottom ends of both side jambs resting on the subfloor, check whether the head jamb is level. Level the head jamb, if necessary, by placing shims between the bottom end of the appropriate side jamb and the subfloor. When the header jamb is level, tack the frame in place on both sides, close to the top. Drive fasteners through side jambs and shims into the studs.

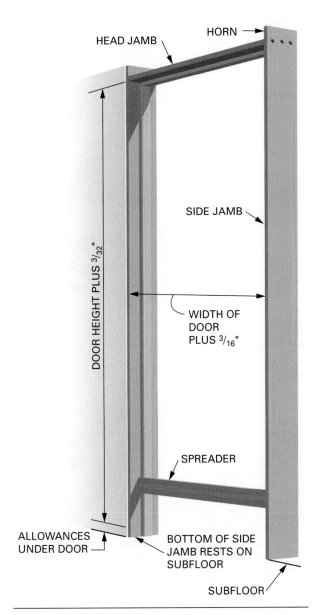

FIGURE 77–5 An assembled interior rabbeted door frame.

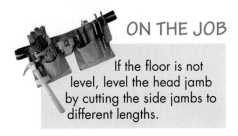

ON THE JOB

If the floor is not level, level the head jamb by cutting the side jambs to different lengths.

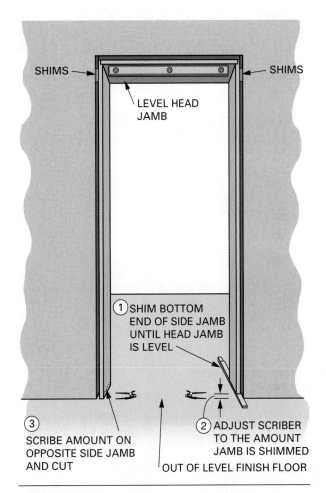

FIGURE 77–6 Technique for cutting side jambs to make the head jamb level.

If the door frame rests on a finish floor, then a tight joint must be made between the bottom ends of the side jambs and the finish floor. If the floor is level, side jambs will fit the floor if their lengths are exactly the same and their bottom ends have been cut square. Once the frame is set, the head jamb should be level when the ends of the side jamb are resting on the floor.

If the floor is not quite level, or if the side jambs are of unequal length, level the head jamb by shimming under the bottom end of the side jamb on the low side. Set the dividers for the amount the jamb has been shimmed. Scribe that amount on the bottom of the opposite side jamb. Remove the frame from the opening. Cut to the scribed line. Replace the frame in the opening. The head jamb should be

level. The bottom ends of both side jambs should fit snugly against the finish floor **(Figure 77–6)**.

Plumbing the Side Jambs

Several ways of plumbing door frames have been previously described. An accurate and fast method is with the use of a 6-foot level. When one side jamb is plumb, shim and tack its bottom end in place. Only one side needs to be plumbed. Locate the bottom end of the other side jamb by measuring across. The door frame width at the bottom should be the same as on the top. Shim and tack the bottom end of the other side jamb in place.

STEP BY STEP PROCEDURES

Procedure 77–A Installing a Door Jamb

STEP 1 CUT HORNS FROM TOP OF SIDE JAMBS IF NECESSARY.

STEP 2 SET FRAME IN OPENING. SHIM ON BOTH SIDES OPPOSITE HEAD JAMB. LEVEL HEAD JAMB AND FASTEN AT TOP THROUGH SIDE JAMB SHIMS.

STEP 3 PLUMB SIDE JAMBS SHIM AND TACK AT BOTTOM.

STEP 4 STRAIGHTEN SIDE JAMBS, INSTALL INTERMEDIATE SHIMS AND TACK IN PLACE.

STEP 5 SIGHT THROUGH THE FRAME TO CHECK FOR A WIND. DRIVE AND SET ALL NAILS WHEN EDGES LINE UP.

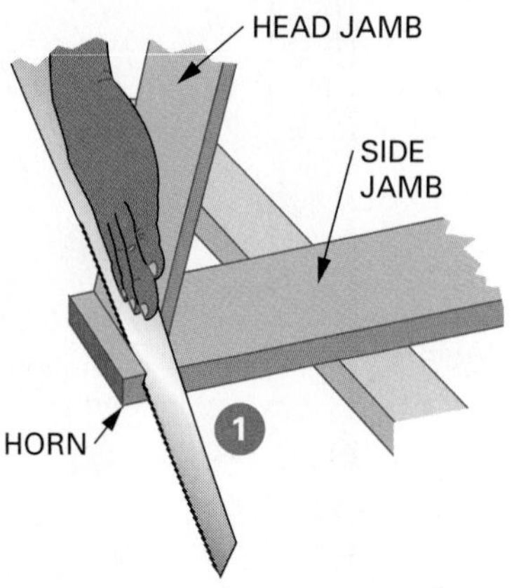

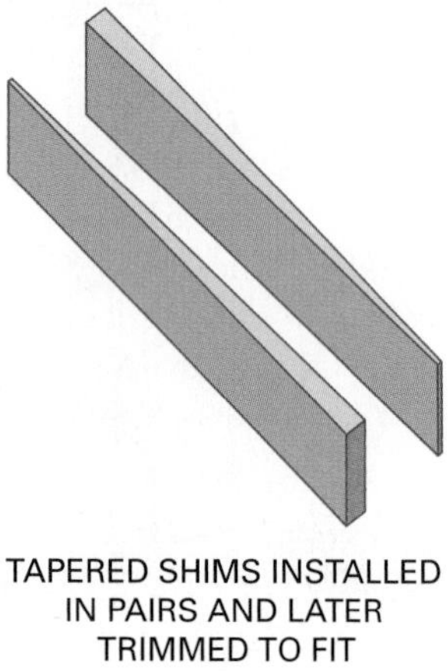

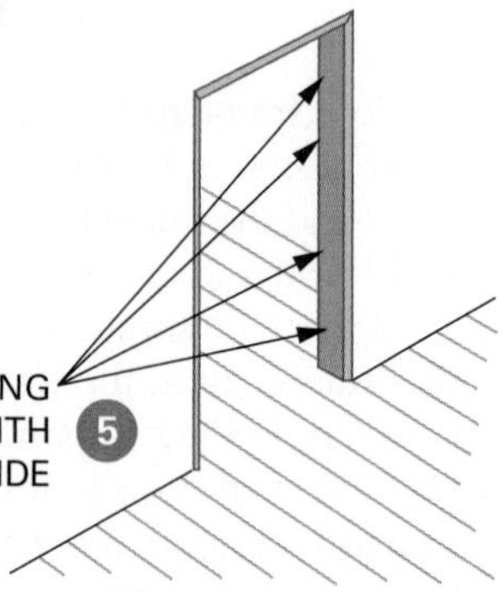

HEAD JAMB

SIDE JAMB

HORN

①

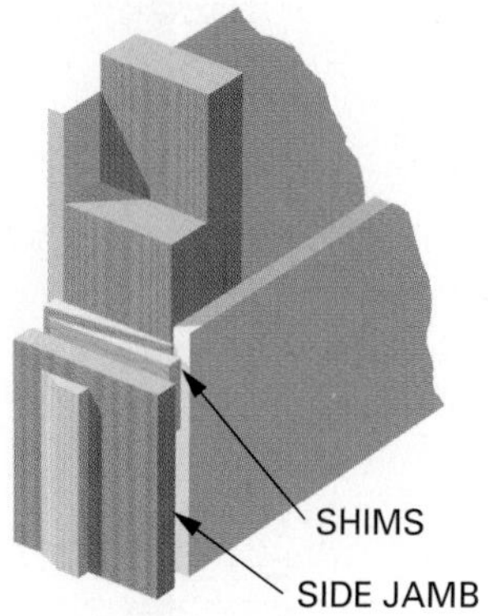

SHIMS

SIDE JAMB

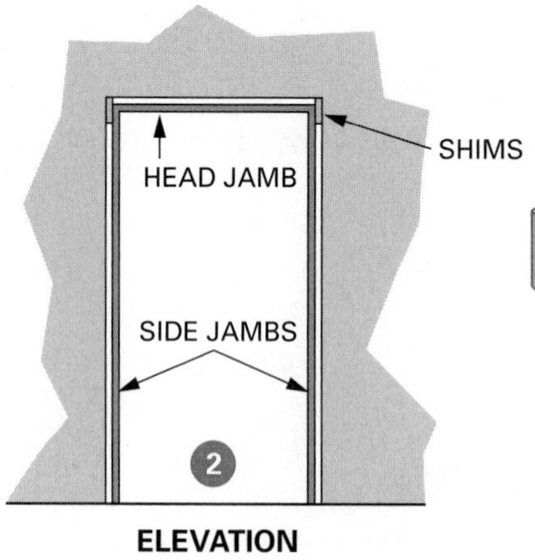

HEAD JAMB

SHIMS

SIDE JAMBS

②

ELEVATION

TAPERED SHIMS INSTALLED IN PAIRS AND LATER TRIMMED TO FIT

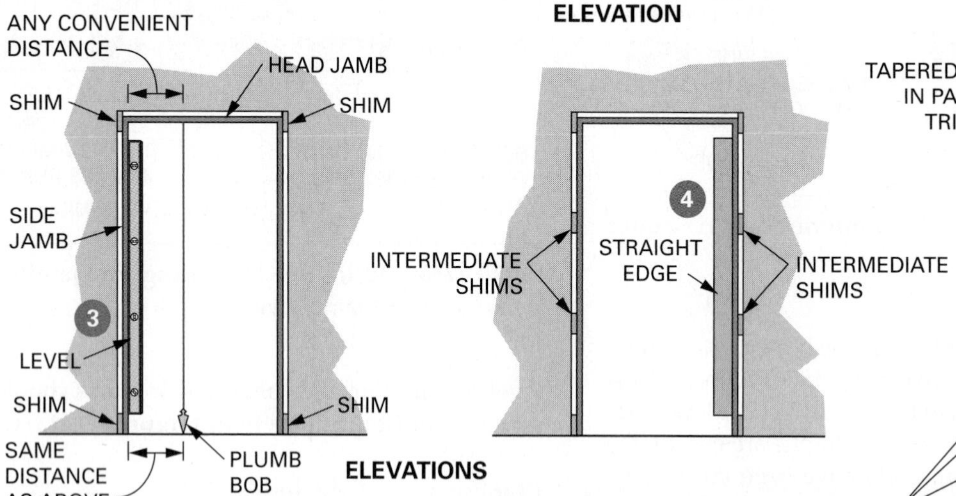

ANY CONVENIENT DISTANCE

HEAD JAMB

SHIM

SHIM

SIDE JAMB

③

LEVEL

SHIM

SHIM

SAME DISTANCE AS ABOVE

PLUMB BOB

ELEVATIONS

④

INTERMEDIATE SHIMS

STRAIGHT EDGE

INTERMEDIATE SHIMS

SIGHT FROM ONE SIDE OF THE OPENING TO LINE UP THE EDGE ON ONE SIDE WITH THE OPPOSITE EDGE ON THE OTHER SIDE

⑤

Straightening Jambs

Use a 6-foot straightedge against the side jambs. Straighten them by shimming at intermediate points. Besides other points, shims should be placed opposite hinge and lockset locations. Fasten the jambs by nailing through the shims. Header jambs on wide door frames are straightened, shimmed, and fastened in a similar way.

Sighting the Door Frame for a Wind

Before any nails are set, sight the door frame to see if it has a *wind*. The frame must be sighted by eye to make sure that side jambs line up vertically with each other and that the frame is not twisted. This is important when installing rabbeted jambs. The method of checking for a wind in door frames has been previously described (see Figure 57–8).

If the frame has a wind, move the top or bottom ends of the side jambs slightly until they line up with each other. Fasten top and bottom ends of the side jambs securely. Set the nails **(Procedure 77–A).**

Applying Door Stops

At this time, *door stops* may be applied to plain jambs. The stops are not permanently fastened, in case they have to be adjusted when locksets are installed. A **back miter** joint is usually made between molded side and header stops. A butt joint is made between square-edge stops **(Figure 77–7).**

HANGING INTERIOR DOORS

The method of fitting and hanging single-acting, hinged, interior doors is similar to that for exterior doors.

Double-Acting Doors

Double-acting doors are installed with either special pivoting hardware installed on the floor and the head jamb or with spring-loaded double-acting hinges. Both types return the door to a closed position after being opened. When opened wide, the doors can be held in the open position. A different type of light-duty, double-acting hardware is used on café doors. To install double-acting door hardware, follow the manufacturer's directions.

Installing Bypass Doors

Bypass doors are installed so they overlap each other by about 1 inch when closed. Cut the track to length. Install it on the header jamb according to the manufacturer's directions.

Installing Rollers. Install pairs of *roller hangers* on each door. The roller hangers may be offset a differ-

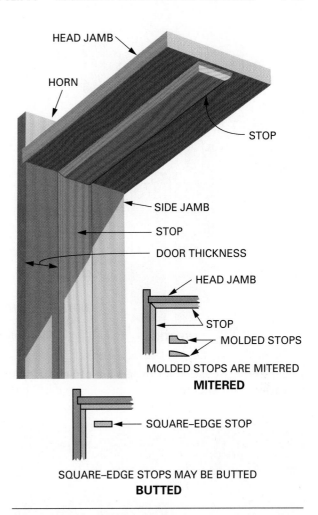

FIGURE 77–7 A back miter joint is used for molded stops, and a butt joint is usually used on square-edge stops.

ent amount for the door on the outside than the door on the inside. They are also offset differently for doors of various thicknesses. Make sure that rollers with the same and correct offset are used on each door **(Figure 77–8).** The location of the rollers from the edge of the door is usually specified in the manufacturer's instruction sheet.

Installing Door Pulls. Mark the location and bore holes for door pulls. Flush pulls must be used so that bypassing is not obstructed **(Figure 77–9).** The proper size hole is bored partway into the door. The pull is tapped into place with a hammer and wood block. The press fit holds the pull in place. Rectangular flush pulls, also used on bypass doors, are held in place with small recessed screws.

Hanging Doors. Hang the doors by holding the bottom outward. Insert the rollers in the overhead track. Then gently let the door come to a vertical position. Install the inside door first, then the outside door **(Figure 77–10).**

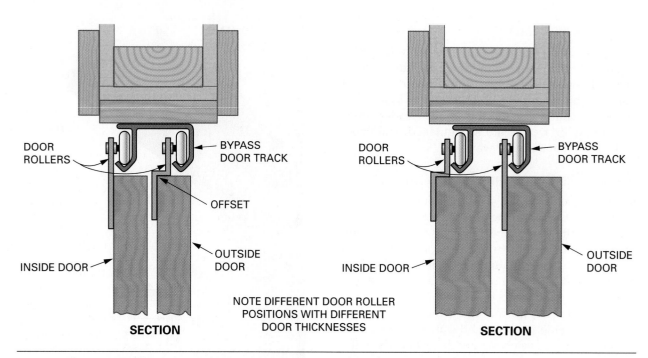

NOTE DIFFERENT DOOR ROLLER
POSITIONS WITH DIFFERENT
DOOR THICKNESSES

FIGURE 77–8   Bypass door rollers are offset different distances for use on doors of various thicknesses.

FIGURE 77–9   Bypass doors must have flush pulls.

Fitting Doors. Test the door operation and the fit against side jambs. Door edges must fit against side jambs evenly from top to bottom. If the top or bottom portion of the edge strikes the side jamb first, it may cause the door to jump from the track. The door rollers have adjustments for raising and lowering. Adjust one or the other to make the door edges fit against side jambs.

Installing Floor Guides. A **floor guide** is included with bypass door hardware to keep the doors in

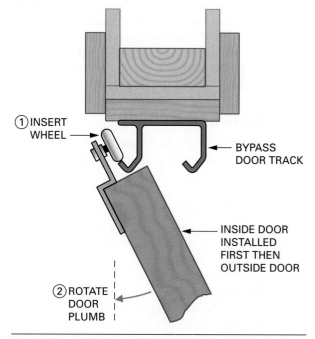

FIGURE 77–10   Bypass doors are hung on the overhead track by holding the bottom of the door outward.

alignment. The guide is centered on the lap to steady the doors at the bottom. Mark the location of the guide. Remove the doors. Install the inside section of the guide. Replace the inside door. Replace the outside door. Install the rest of the guide (**Figure 77–11**).

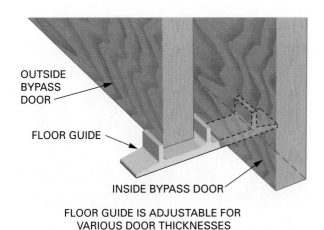

OUTSIDE
BYPASS
DOOR

FLOOR GUIDE

INSIDE BYPASS DOOR

FLOOR GUIDE IS ADJUSTABLE FOR
VARIOUS DOOR THICKNESSES

FIGURE 77–11   **A floor guide is installed to keep bypass doors aligned.**

Installing Bifold Doors

Before installing bifold doors, make sure the opening size is as specified by the hardware or door manufacturer. Usually bifold doors come hinged together in pairs. The hardware consists of the track, pivot sockets, pivot pins and guides, door aligners, door pulls, and necessary fasteners **(Figure 77–12)**.

Installing the Track.   Cut the track to length. Fasten it to the header jamb with screws provided in the kit. The track contains adjustable sockets for the door *pivot pins.* Make sure these are inserted before fastening the track in position. The position of the track on the header jamb is not critical. It may be positioned as desired **(Figure 77–13)**.

Installing Bottom Pivot Sockets.   Locate the bottom pivot sockets. Fasten one on each side, at the bottom of the opening. The pivot socket bracket is L-shaped. It rests on the floor against the side jamb. It is centered on a plumb line from the center of the pivot sockets in the track on the header jamb above.

Installing Pivot and Guide Pins.   In most cases, bifold doors come with prebored holes for *pivot* and *guide pins.* If not, it is necessary to bore them. Follow the manufacturer's directions as to size and location. Install pivot pins at the top and bottom ends of the door in the prebored holes closest to the jamb. Sometimes the top pivot pin is spring loaded. It can then be depressed for easier installation of the door. The bottom pivot pin is threaded and can be adjusted for height. The guide pin rides in the track. It is installed in the hole provided at the top end of the door farthest away from the jamb.

FIGURE 77–12   **Installation of the bifold door requires several kinds of special hardware.**

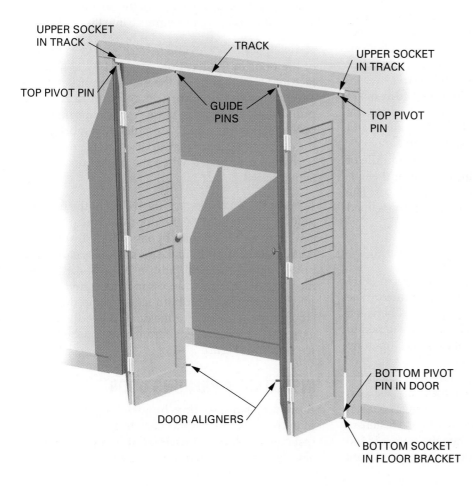

UPPER SOCKET
IN TRACK

TRACK

UPPER SOCKET
IN TRACK

TOP PIVOT PIN

GUIDE
PINS

TOP PIVOT
PIN

BOTTOM PIVOT
PIN IN DOOR

DOOR ALIGNERS

BOTTOM SOCKET
IN FLOOR BRACKET

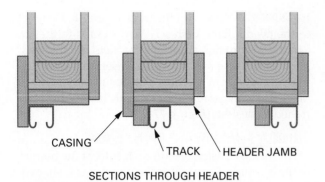

CASING

TRACK   HEADER JAMB

SECTIONS THROUGH HEADER

FIGURE 77–13 The bifold door track may be located in any position on the header jamb in several ways. Trim conceals the track from view.

Hanging the Doors. After all the necessary hardware has been applied, the doors are ready for installation. Loosen the set screw in the top pivot socket. Slide it along the track toward the center of the opening about one foot away from the side jamb. Place the door in position by inserting the bottom pivot pin in the bottom pivot socket. Tilt the doors to an upright position. At the same time insert the top pivot pin in the top socket, and the guide pin in the track, while sliding the socket toward the jamb.

Adjusting the Doors. Adjust top and bottom pivot sockets in or out so the desired joint is obtained between the door and the jamb. Lock top and bottom pivot sockets in position. Adjust the bottom pivot pin to raise or lower the doors, if necessary.

If more than one set of bifold doors is to be installed in an opening, install the other set on the opposite side in the same manner. Install knobs in the manner and location recommended by the manufacturer.

Where sets of bifold doors meet at the middle of an opening, door aligners are installed, near the bottom, on the inside of each of the meeting doors. The door aligners keep the faces of the center doors lined up when closed **(Figure 77–14).**

Installing Pocket Doors

The pocket door frame, complete with track, is installed when interior partitions are framed. The pocket consists of two ladder-like frames between which the door slides. A steel channel is fastened to the floor. The channel keeps the pocket opening spread the proper distance apart.

The frame, which is usually preassembled at the factory, is made of nominal 1-inch stock. The pocket is covered by the interior wall finish. Care must be taken when covering the pocket frame not to use fasteners that are so long that they penetrate the frame. If fasteners penetrate through the pocket door frame, they

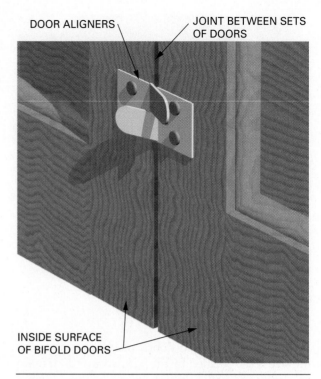

DOOR ALIGNERS   JOINT BETWEEN SETS OF DOORS

INSIDE SURFACE OF BIFOLD DOORS

FIGURE 77–14 Door aligners are used near the bottom where sets of bifold doors meet.

will probably scratch the side of the door as it is operated or stop its complete entrance into the pocket.

Installing Door Hardware. Attach rollers to the top of the door in the location specified by the manufacturer. Install pulls on the door. On pocket doors an edge pull is necessary, in addition to recessed pulls on the sides of the door. A special pocket door pull contains edge and side pulls. It is mortised in the edge of the door. In most cases, all the necessary hardware is supplied when the pocket door frame is purchased.

Hanging the Door. Engage the rollers in the track by holding the bottom of the door outward in a way similar to that used with bypass doors. Test the operation of the door to make sure it slides easily and butts against the side jamb evenly. Make adjustments to the rollers, if necessary. Stops are later applied to the jambs on both sides of the door. The stops serve as guides for the door. When the door is closed, the stops prevent it from being pushed out of the opening **(Figure 77–15).**

INSTALLING A PREHUNG DOOR

A prehung single-acting, hinged door unit consists of a door frame with the door hinged and casings installed. Holes are provided, if locksets have not already been installed. Small cardboard shims are

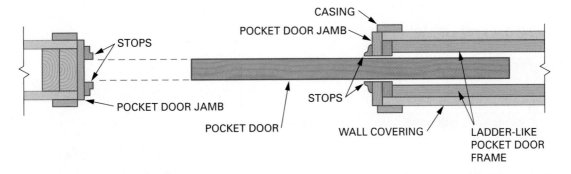

FIGURE 77–15 **Plan view of a pocket door.**

stapled to the lock edge and top end of the door to maintain proper clearance between the door and frame.

Prehung units are available in several jamb widths to accommodate various wall thicknesses. Some prehung units have split jambs that are adjustable for varying wall thicknesses **(Figure 77–16)**.

A prehung door unit can be set in a matter of minutes. Many prehung units come without the casing attached, but if it is attached, remove the casings carefully from one side of the solid jamb units. Center the unit in the opening, so the door will swing in the desired direction. Be sure the door is closed and spacer shims are in place between the jamb and door. Plumb the door unit. Tack it to the wall through the casing.

Open the door and move to the other side. Install shims between the side jambs and the rough opening at intermediate points, keeping side jambs straight. Nail through the side jambs and shims. Remove spacers. Check the operation of the door. Make any necessary adjustments. Replace the previously removed casings. Drive and set all nails **(Procedure 77–B).**

Prehung door units with split jambs are set in a similar manner. However, there is no need to remove the casings. One section is installed as described earlier. The remaining section is inserted into the one already in place.

INSTALLING LOCKSETS

Locksets are installed on interior doors in the same manner as for exterior doors and as described in Chapter 59, Installing Exterior Door Locksets. Although their installation is basically the same, some locks are used exclusively on interior doors.

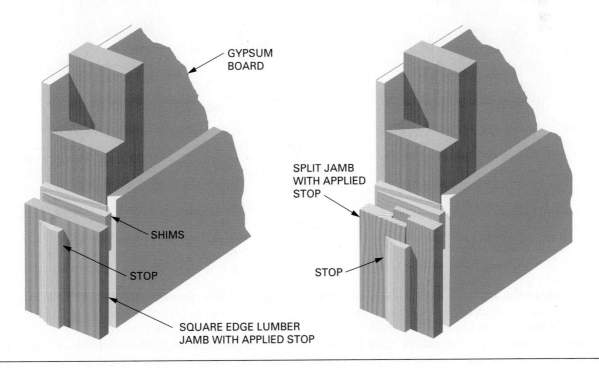

FIGURE 77–16 **Prehung door units come with solid or split jambs.**

Procedure 77–B Installing a Prehung Door

STEP ❶ REMOVE CASINGS FROM ONE SIDE OF UNIT.

STEP ❷ PLACE UNIT IN OPENING AND PLUMB SIDE CASING.

STEP ❸ FASTEN THROUGH CASINGS INTO WALL.

STEP ❹ MOVE TO OTHER SIDE OF DOOR.

STEP ❺ INSTALL SHIMS BETWEEN JAMB AND WALL.

STEP ❻ FASTEN THROUGH JAMB AND SHIMS. CHECK THAT JAMBS ARE PLUMB AND STRAIGHT.

STEP ❼ REPLACE CASINGS THAT WERE PREVIOUSLY REMOVED.

STEP ❽ FASTEN THROUGH CASINGS INTO WALL.

DOOR

CASING

HINGE

LEVEL

A   A

CASINGS REMOVED FROM THIS SIDE

JAMB

DOOR

CASING

SECTION A-A

DOOR

SHIM

B   B

SHIM

JAMB

❹

CASING

DOOR

FASTEN THROUGH JAMB AND SHIMS

❺ SHIMS   ❻

SECTION B-B

DOOR

❼   CASING

STOP

JAMB

❽ FASTEN CASING

SECTION C-C

C   C

The **privacy lock** is often used on bathroom and bedroom doors. It is locked by pushing or turning a button on the room side. On most privacy locks, a turn of the knob on the room side unlocks the door. On the opposite side, the door can be unlocked by a pin or key inserted into a hole in the knob. The un-

locking device should be kept close by, in a prominent location, in case the door needs to be opened quickly in an emergency.

The **passage lockset** has knobs on both sides that are turned to unlatch the door. This lockset is used when it is not desirable to lock the door.

Key Terms

back miter	double-acting doors	louver doors	prehung
bifold doors	floor guide	panel doors	privacy lock
bypass doors	flush doors	passage lockset	spreader
café doors	french doors	pocket doors	

Review Questions

Select the most appropriate answer.

1. Most interior doors are manufactured in a thickness of
   a. 1 inch.
   b. 1⅜ inches.
   c. 1½ inches.
   d. 1¾ inches.

2. The height of most interior doors is
   a. 6′–0″.
   b. 6′–6″.
   c. 6′–8″.
   d. 7′–0″.

3. Interior door widths usually range from
   a. 1′–6″ to 2′–6″.
   b. 2′–2″ to 2′–8″.
   c. 2′–6″ to 2′–8″.
   d. 1′–0″ to 3′–0″.

4. Used extensively for flush door skins is
   a. fir plywood.
   b. lauan plywood.
   c. metal.
   d. plastic laminate.

5. The usual distance between the finish floor and the bottom of swinging doors for clearance is
   a. ¼ to ½ inch.
   b. ½ to ¾ inch.
   c. ¾ to 1 inch.
   d. ½ to 1 inch.

6. A disadvantage of bypass doors is that they
   a. project out into the room.
   b. cost more and require more time to install.
   c. are difficult to operate.
   d. do not provide total access to the opening.

7. If the jamb stock is ¾ inch thick, the rough opening width for a swinging door should be the door width plus
   a. ¾ inch.
   b. 1½ inches.
   c. 2 inches.
   d. 2½ inches.

8. If the jamb stock and the finished floor are both ¾ inch thick and the space under the door is ½″, the rough opening height for a 6′–8″ swinging door should be
   a. 7′–0″.
   b. 6′–11½″.
   c. 6′–10½″.
   d. 6′–9½″.

9. When a plain door frame is made for a swinging door, the header jamb length is the door width plus
   a. the dado depth on both ends.
   b. the dado depth plus ³⁄₃₂″ space on both ends.
   c. the dado depth plus ³⁄₃₂″ space plus ½″ shim space both ends.
   d. none of the above.

10. An accurate and fast method of plumbing side jambs of a door frame is by the use of a
    a. builder's level.
    b. carpenter's 26-inch hand level.
    c. plumb bob.
    d. straightedge.

UNIT 27

Interior Trim

*I*nterior trim, also called *interior finish*, involves the application of molding around windows and doors; at the intersection of walls, floor, and ceilings; and to other inside surfaces. Moldings are strips of material, shaped in numerous patterns, for use in a specific location. Wood is used to make most moldings, but some are made of plastic or metal.

Interior trim is among the final materials installed and requires the installer to take care not to mar the finish. Blemishes and dings in the finish may be visible for the life of the building.

OBJECTIVES

After completing this unit, the student should be able to:

- identify standard moldings and describe their use.
- apply ceiling and wall molding.
- apply interior door casings, baseboard, base cap, and base shoe.
- install window trim, including stools, aprons, jamb extensions, casings, and stop beads.
- install closet shelves and closet pole.
- install mantels.

78 Description and Application of Molding

oldings are available in many *standard* types. Each type is manufactured in several sizes and patterns. Standard patterns are usually made only from softwood. When other kinds of wood, or special patterns, are desired, mills make *custom* moldings to order. All moldings must be applied with tight-fitting joints to present a suitable appearance.

STANDARD MOLDING PATTERNS

Standard moldings are designated as bed, crown, cove, full round, half round, quarter round, base, base shoe, base cap, casing, chair rail, back band, apron, stool, stop, and others **(Figure 78–1).**

Molding usually comes in lengths of 8, 10, 12, 14, and 16 feet. Some moldings are available in odd lengths. Door casings, in particular, are available in lengths of 7 feet to reduce waste.

Finger-jointed lengths are made of short pieces joined together. These are used only when a paint finish is to be applied. The joints show through a stained or natural finish.

MOLDING SHAPE AND USE

Some moldings are classified by the way they are shaped. Others are designated by location. For example, *beds, crowns,* and *coves* are terms related to shape. Although they may be placed in other locations, they are usually used at the intersections of walls and ceilings **(Figure 78–2).** Also classified by their shape are **full rounds, half rounds,** and **quarter rounds.** They are used in many locations. Full rounds are used for such things as closet poles. Half rounds may be used to conceal joints between panels or to trim shelf edges. Quarter rounds may be

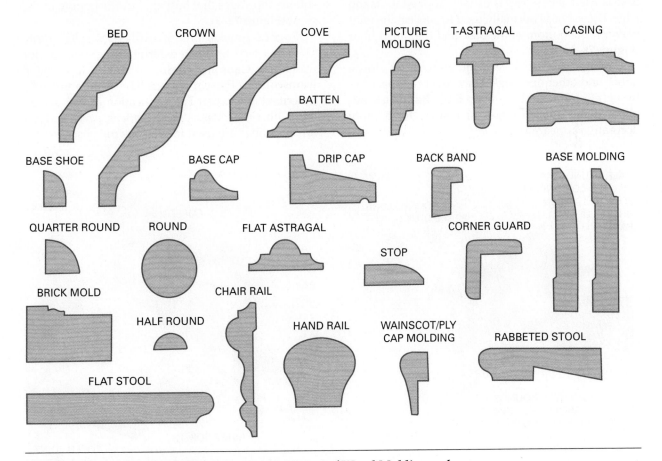

FIGURE 78–1   Standard molding patterns. *(Courtesy of Wood Molding and Millwork Producers, Inc., P.O. Box 25278, Portland, Oregon 97225)*

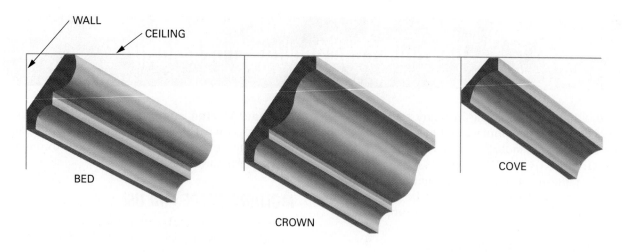

FIGURE 78–2 Bed, crown, and cove moldings are often used at the intersections of walls and ceilings.

used to trim outside corners of wall paneling and for many other purposes **(Figure 78–3).**

Designated by location, **base, base shoe,** and **base cap** are moldings applied at the bottom of walls where they meet the floor. When square-edge base is used, a base cap is usually applied to its top edge. Base shoe is normally used to conceal the joint between the bottom of the base and the finish floor **(Figure 78–4).**

Casings are used to trim around windows, doors, and other openings. They cover the space between the frame and the wall. **Back bands** are applied to the outside edges of casings for a more decorative appearance.

Aprons, stools, and **stops** are parts of window trim. Stops are also applied to door frames. On the same window, aprons should have the same molded shape as casings. Aprons, however, are not *backed out.* They have straight, smooth backs and sharp, square, top edges that butt against the bottom of the stool **(Figure 78–5).**

Corner guards are also called *outside corners.* They are used to finish exterior corners of interior wall finish. **Caps** and **chair rail** trim the top edge of wainscoting. (These moldings have been previously described in Chapter 72, Application of Wall Paneling.) Others, such as *astragals, battens, panel,* and *picture* moldings are used for various purposes.

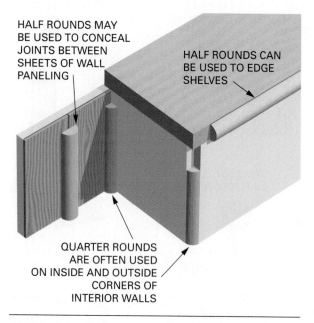

FIGURE 78–3 Half round and quarter round moldings are used for many purposes.

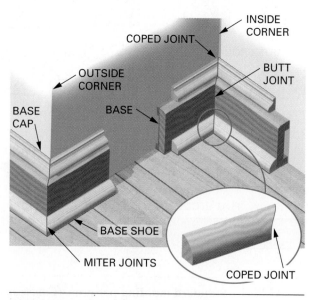

FIGURE 78–4 Base, base shoe, and base cap are used to trim the bottom of the wall.

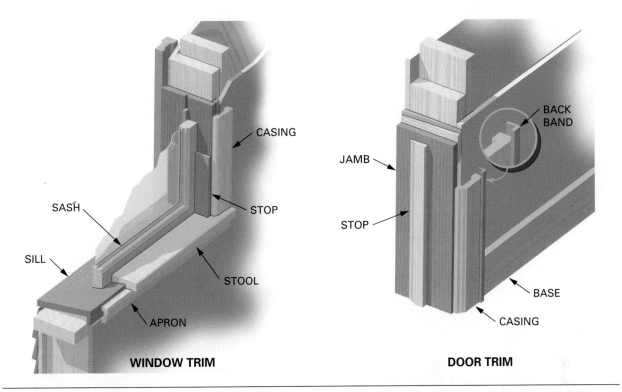

CASING

SASH

STOP

SILL

STOOL

APRON

WINDOW TRIM

BACK
BAND

JAMB

STOP

BASE

CASING

DOOR TRIM

FIGURE 78–5   Casing, back bands, and stops are used for window and door trim. Stools and aprons are part of window trim.

MAKING JOINTS ON MOLDING

End joints between lengths of ceiling molding may be made square or at a 45-degree angle. Many carpenters prefer to make square joints between moldings because less joint line is shown. Also, the square end acts as a stop when bowing and snapping the last length of molding into place at a corner.

Usually, the last piece of molding along a wall is cut slightly longer. It is bowed outward in the center, then pressed into place when the ends are in position. This makes the joints come up tight. After the molding has been fastened, joints between lengths should be sanded flush, except on prefinished moldings. Failure to sand butted ends flush with each other results in a shadow being cast at the joint line. This gives the appearance of an open joint.

Joints on exterior corners are **mitered.** (Miter joints are defined in Chapter 12, Boring and Cutting Tools.) Joints on interior corners are usually **coped,** especially on large moldings. A coped joint is made by fitting the piece on one wall with a square end into the corner. The end of the molding on the other wall is cut to fit against the shaped face of the molding on the first wall **(Figure 78–6).**

Methods of Mitering Using Miter Boxes

Moldings of all types may be mitered by using either hand or power **miter boxes.** A miter box is a tool

FIGURE 78–6   A coped joint is made by fitting the end of one piece of molding against the shaped face of the other piece.

that cuts a piece of material at an angle. The most common angle is 45 degrees. A job built version of a miter box used to guide a handsaw is made from wood scraps. Another style of box is metal with a backsaw attached and easily adjustable to cut different angles **(Figure 78–7).** These are effective tools in cutting a miter.

The most popular way to cut miters and other end cuts on trim is with a power miter box **(Figure 78–8).** With this tool, a carpenter is able to cut virtually any angle with ease, whether it is a simple or a compound

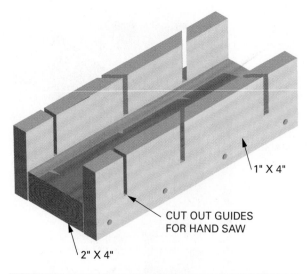

CUT OUT GUIDES
FOR HAND SAW

1" X 4"

2" X 4"

FIGURE 78–7 Miter boxes for a handsaw.

BOSCH

FIGURE 78–8 A power miter box makes easy work
of cutting molding.

miter (one with two angles). Fine adjustments to a
piece of trim, ±⅟₆₄ inch, can be made with great speed
and accuracy. The power miter box is discussed in
more depth in Chapter 18.

Positioning Molding in the Miter Box

Placing molding in the correct position in the miter
box is essential for accurate mitering. Cut all moldings with their face sides or edges up or toward the
operator so that the saw splinters out the back side,
not the face side. Position the molding with one back
side or edge against the bottom of the miter box and
the other against the side of the miter box. On a
wood miter box, which ordinarily has two sides,
hold the molding against the side farthest away
from the worker.

Flat miters are cut by holding the molding with
its face side up and its thicker edge against the side
of the miter box. Some moldings, such as base, base
cap and shoe, and chair rail, are held right side up.
Their bottom edge should be against the bottom of
the miter box and their back against the side of the
miter box **(Figure 78–9)**.

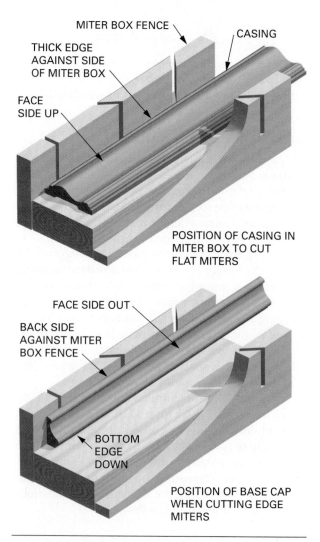

MITER BOX FENCE

CASING

THICK EDGE
AGAINST SIDE
OF MITER BOX

FACE
SIDE UP

POSITION OF CASING IN
MITER BOX TO CUT
FLAT MITERS

FACE SIDE OUT

BACK SIDE
AGAINST MITER
BOX FENCE

BOTTOM
EDGE
DOWN

POSITION OF BASE CAP
WHEN CUTTING EDGE
MITERS

FIGURE 78–9 Position of molding in a miter box to
cut square-edged miters.

BACK
AGAINST
MITER
FENCE

FACE SIDE
OUT

TOP EDGE AGAINST
MITER BASE

FIGURE 78–10 To miter ceiling moldings, they must be positioned upside down in the miter box. It may be helpful to think of the fence as the wall and the base as the ceiling.

FIGURE 78–11 Large ceiling moldings may be mitered while held flat to miter base and cutting at a compound angle.

Mitering Bed, Crown, and Cove Molding. **Bed, crown,** and **cove** molding must be positioned upside down in the miter box. Their top edge is placed against the bottom of the miter box and their bottom edge against the far side **(Figure 78–10)**. This position is convenient because the bottom edge of this type of molding is the edge that is usually marked for cutting to length. With the bottom edge up, the mark, on which to start the cut, can be easily seen.

Large moldings may also be cut while they are laying flat on the miter saw base. The saw is set to make a compound miter by adjusting to two angles **(Figure 78–11)**. High-quality miter saw manufacturers include an instruction booklet describing the various setting angles.

A thin, narrow strip of wood fastened to the bottom of the miter box and against the molding helps prevent the molding from moving when being mitered. The strip also ensures that subsequent pieces of the same type of molding will be positioned at the same angle. Therefore, they will be mitered the same as the first piece **(Figure 78–12)**.

To position corner guards and other rabbeted molding, such as back bands and caps, for mitering, temporarily fasten a small strip of wood into the far corner of the miter box. The width and thickness of the strip should be slightly more than the rabbet of the molding. The molding is positioned on the strip and held steady while being mitered **(Figure 78–13)**.

ON
THE
JOB
Fasten a thin, narrow strip of wood to the bottom of the miter box. This ensures accurate positioning of moldings.

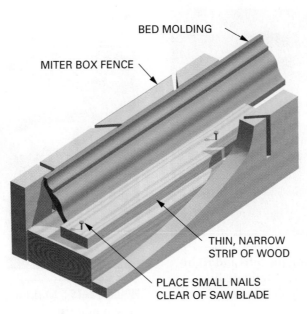

BED MOLDING

MITER BOX FENCE

THIN, NARROW
STRIP OF WOOD

PLACE SMALL NAILS
CLEAR OF SAW BLADE

FIGURE 78–12
Technique for holding a wide piece of molding in the proper position for cutting.

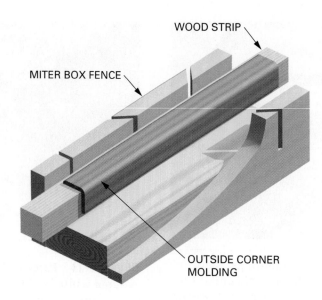

ON THE JOB

Fasten a strip of wood in the corner of the miter box on which to position rabbeted-type moldings, such as outside corners, back bands, and caps.

WOOD STRIP

MITER BOX FENCE

OUTSIDE CORNER MOLDING

FIGURE 78–13
Technique for holding an outside corner for cutting.

Mitering with a Table Saw

Miters may also be made by using the table saw or the radial arm saw. The use of mitering jigs is helpful when making flat miters on window and door casings. The jigs allow both right- and left-hand miters to be cut quickly and easily without any changes in the setup **(Figure 78–14).** (The construction and use of mitering jigs for saws are more fully described in Unit 7.)

Making a Coped Joint

To cope the end of molding, first make a **back miter** on the end. A back miter starts from the end and is cut back on the face of the molding **(Figure 78–15).** The edge of the cut along the face forms the profile

of the cope. Rub the side of a pencil point lightly along the profile to outline it more clearly.

Use a coping saw with a fine-tooth blade. Cut along the outlined profile with a slight undercut. Cut with the handle of the coping saw above the work and the teeth of the blade pointing away from the handle **(Figure 78–16).** Hold the molding so it is over the end of a sawhorse. The side of the molding that will butt the wall should be lying flat on the top of the sawhorse. It is important that the molding be held in this position. Holding it any other way makes it difficult to cut the cope with an undercut so that it will fit properly. It may be necessary to touch up the cut with a wood file or sharp utility knife.

Coped joints are used on interior wall corners. They will not open up when the molding is nailed in place, especially if the backing is not solid. Miter joints may open up in interior corners when the ends are fastened.

FIGURE 78–14 With a mitering jig, left- and right-hand miters can be made quickly and easily without changing the setup *(guard has been removed for clarity).*

FIGURE 78–15 Making a back miter on a piece of crown molding.

FIGURE 78–16 Cutting along the face edge of a cove molding to create a coped joint.

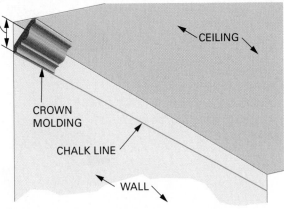

MARK THIS DISTANCE FROM THE CEILING ON BOTH ENDS OF WALL AND SNAP A LINE BETWEEN MARKS

CEILING

CROWN MOLDING

CHALK LINE

WALL

FIGURE 78–17 Hold a scrap piece of molding against the ceiling and wall to determine the distance from the ceiling to its bottom edge.

APPLYING MOLDING

To apply chair rail, caps, or some other type of molding located on the wall, chalk lines should be snapped. This ensures that molding is applied in a straight line. No lines need to be snapped for base moldings or for small-size moldings applied at the intersection of walls and ceiling.

For large-size ceiling moldings, such as beds, crowns, and coves, a chalk line should be snapped. This ensures straight application of the molding and easier joining of the pieces. Without a straight line to guide application, the molding may be forced at different angles along its length. This results in a noticeably crooked bottom edge and difficulty making tight-fitting miters and copes.

Hold a short scrap piece of the large-size molding at the proper angle at the wall and ceiling intersection **(Figure 78–17)**. Lightly mark the wall along the bottom edge of the molding. Measure the distance from the ceiling down to the mark. Measure and mark this same distance down from the ceiling on each end of each wall to which the molding is to be applied. Snap lines between the marks. Apply the molding so its bottom edge is to the chalk line.

Apply the molding to the first wall with square ends in both corners. If more than one piece is required to go from corner to corner, install the first piece with both ends square. If mitered joints between lengths are desired, cut a square end into the corner and a back miter on the other end. On subsequent lengths, make matching end joints until the other corner is reached.

On some moldings, such as quarter rounds and small cove moldings, the straight, back surfaces should, but may not always, be of equal width. One of the back surfaces of these moldings should be marked with a pencil to ensure positioning them in the miter box the same way each time. Mitering the molding with the same side down each time helps make fitting more accurate, faster, and easier **(Figure 78–18)**.

If a small-size molding is used, fasten it with finish nails in the center. Use nails of sufficient length to penetrate into solid wood at least 1 inch. If large-size molding is used, fastening is required along both edges **(Figure 78–19)**.

Press the molding in against the wall or intersection with one hand while driving the nail almost home. Then set the nail below the surface. Nail at about 16-inch intervals and in other locations as necessary to bring the molding tight against the surface. End nails should be placed 2 to 3 inches from the end to keep from splitting the molding. If it is likely that the molding may split, blunt the pointed end of the nail. Or, drill a hole slightly smaller than the nail diameter.

Install the last piece on the first wall by first squaring one end. Place the square end in the corner. Let the other end overlap the first piece. Mark and cut it at the overlap. This method is more accurate than measuring and then transferring the measurement to the piece. Mark all pieces of interior trim for length in this manner whenever possible. Cut and fasten the last piece with its square end into the corner.

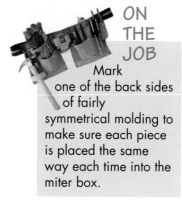

ON THE JOB

Mark one of the back sides of fairly symmetrical molding to make sure each piece is placed the same way each time into the miter box.

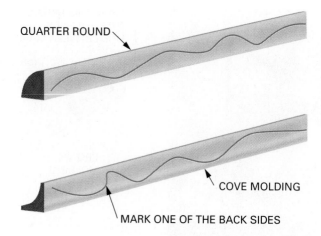

QUARTER ROUND

COVE MOLDING

MARK ONE OF THE BACK SIDES

FIGURE 78–18 **Technique for reducing the confusion of working with molding that has a fairly symmetrical cross section.**

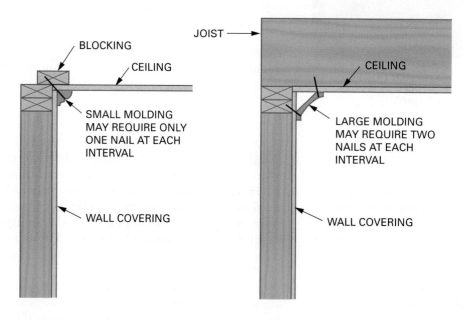

BLOCKING

CEILING

SMALL MOLDING MAY REQUIRE ONLY ONE NAIL AT EACH INTERVAL

WALL COVERING

JOIST

CEILING

LARGE MOLDING MAY REQUIRE TWO NAILS AT EACH INTERVAL

WALL COVERING

FIGURE 78–19 **Methods of fastening molding.**

Cope the starting end of the first piece on each succeeding wall against the face of the last piece installed on the previous wall. Work around the room in one direction, either clockwise or counterclockwise. The end of the last piece installed must be coped to fit against the face of the first piece.

79 Application of Door Casings, Base, and Window Trim

In addition to wall and ceiling molding, the application of door casings, base, base cap, base shoe, and window trim is a major part of interior finish work. Care must be taken to avoid marring the work and to make neat, tight-fitting joints.

DOOR CASINGS

Door casings are moldings applied around the door opening. They trim and cover the space between the door frame and the wall. Casings must be applied before any base moldings because the base butts against the edge of door casing **(Figure 79–1).** Door casings extend to the floor.

Design of Door Casings

Moldings or *S4S* stock may be used for door casings. S4S is the abbreviation for **surfaced four sides.** It is used to describe smooth, square-edge lumber. When molded casings are used, the joint at the head must be mitered unless butted against **plinth blocks.**

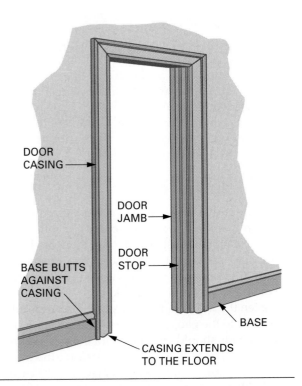

FIGURE 79–1 Door casings are applied before the base is installed.

Plinth blocks are small decorative blocks. They are thicker and wider than the door casing. They are used as part of the door trim at the base and at the head **(Figure 79–2)**.

When using S4S lumber, the joint may be mitered or butted. If a butt joint is used, the head casing overlaps the side casing. The appearance of S4S casings and some molded casings may be enhanced with the application of back bands **(Figure 79–3)**.

Molded casings usually have their back sides backed out. In cases where the jamb edges and the wall surfaces may not be exactly flush with each other, the backed out surfaces allow the casing to come up tight on both wall and jamb **(Figure 79–4)**. If S4S casings are used, they must be backed out on the job. (A method of backing out S4S lumber is described in Chapter 19, Table Saws, and is illustrated in Figure 19–6.)

Applying Door Casings

Door casings are set back from the inside face of the door frame a distance of about 5⁄16 inch. This allows room for the door hinges and the striker plate of the door lock. This setback is called a **reveal** **(Figure 79–5)**. The reveal also improves the appearance of the door trim.

Set the blade of the combination square so that it extends 5⁄16 inch beyond the body of the square. Gauge lines at intervals along the side and head jamb edges by riding the square against the inside face of the jamb. Let the lines intersect where side and head jambs meet. Mark lightly with a sharp pencil or mark with a utility knife. The knife leaves no pencil lines to erase later.

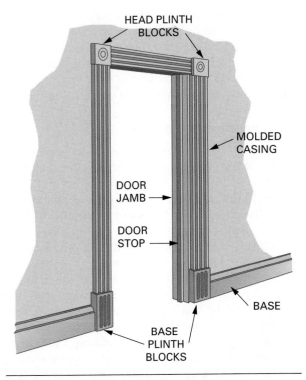

FIGURE 79–2 Molded casings are mitered at the head unless plinth blocks are used.

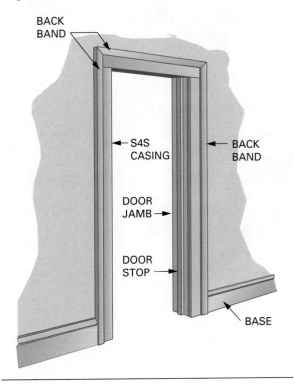

FIGURE 79–3 Back bands may be applied to improve the appearance of door casings.

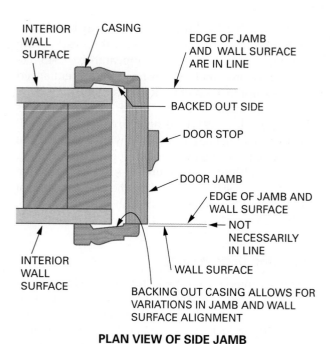

INTERIOR WALL SURFACE

CASING

EDGE OF JAMB AND WALL SURFACE ARE IN LINE

BACKED OUT SIDE

DOOR STOP

DOOR JAMB

EDGE OF JAMB AND WALL SURFACE

NOT NECESSARILY IN LINE

INTERIOR WALL SURFACE

WALL SURFACE

BACKING OUT CASING ALLOWS FOR VARIATIONS IN JAMB AND WALL SURFACE ALIGNMENT

PLAN VIEW OF SIDE JAMB

FIGURE 79–4   Backing out door casings allows for a tight fit on wall and jamb.

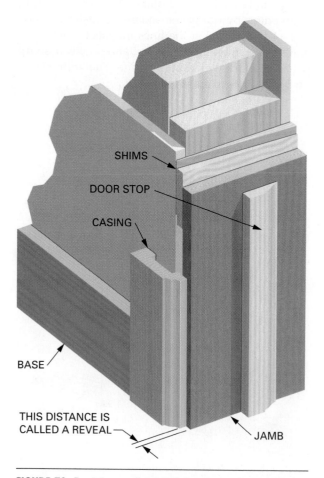

SHIMS

DOOR STOP

CASING

BASE

THIS DISTANCE IS CALLED A REVEAL

JAMB

FIGURE 79–5   The setback of the door casing on the jamb is called a reveal.

The following procedure applies to molded door casings mitered at the head. If several door openings are to be cased, cut the necessary number of casings to rough lengths with a miter cut on one end of each piece. Rough lengths are a few inches longer than actually needed. For each interior door opening, four side casings and two head casings are required. Cut side casings in pairs with right- and left-hand miters for use on both sides of the opening.

Applying the Head Casing

Miter one end of the head casing. Hold it against the head jamb of the door frame so that the miter is on the intersection of the gauged lines. Mark the length of the head casing at the intersection of the gauged lines on the opposite side of the door frame. Miter the casing to length at the mark.

Fasten the head casing in position. Its inside edge should be to the gauged lines on the head jamb. The mitered ends should be in line with the gauged lines on the side jambs. Use finish nails along the inside edge of the casing into the header jamb. If the casing edge is thin, use 3d or 4d finish nails spaced about 12 inches apart. Keep the edge of the casing to the gauged lines on the jamb. Straighten the casing as necessary as nailing progresses. Drive nails at the proper angle to keep them from coming through the face or back side of the jamb. Pneumatic finish nailers speed up the job of fastening interior trim.

Fasten the top edge of the casing into the framing. The outside edge is thicker, so longer nails are used, usually 6d or 8d finish nails. They are spaced farther apart, about 16 inches on center (OC) **(Procedure 79–A)**. Do not drive end nails at this time. It may be necessary to move the ends slightly to fit the mitered joint between head and side casings.

Applying the Side Casings

Mark one of the previously mitered side casings by turning it upside down with the point of the miter touching the floor. If the finish floor has not been laid, hold the point of the miter on a scrap block of wood that is equal in thickness to the finish floor. Mark the side casing in line with the top edge of the head casing **(Procedure 79–B)**. Make a square cut on the casing at the mark.

Place the side casing in position. Try the fit at the mitered joint. If the joint needs fitting, trim the mitered end of the side casing by planing thin shavings with a sharp block plane. The joint may also be fitted by shimming the casing away from the side of the chop saw and making a thin corrective cut. Shim either near or far from the saw blade as needed to hold the casing at the desired angle **(Figure 79–6)**. When fitted, apply a little glue to the

Procedure 79–A Cutting a Head Casing to Fit

STEP 1 MARK LIGHT GAUGE LINES ON THE EDGE OF JAMB TO INDICATE DESIRED REVEAL.

STEP 2 CUT A HEAD CASING PIECE SLIGHTLY LONG WITH ONLY ONE MITER. POSITION IT INTO PLACE.

STEP 3 ALIGN MITER TO THE REVEAL GAUGE LINES.

STEP 4 MARK THE SECOND MITER AT THE REVEAL LINE.

STEP 5 TACK HEAD CASING IN PLACE AND DO NOT SET THE NAILS YET.

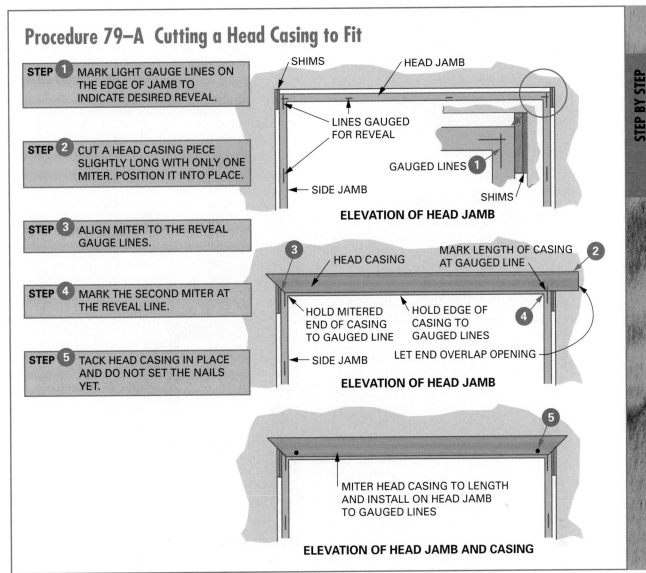

SHIMS   HEAD JAMB

LINES GAUGED FOR REVEAL

GAUGED LINES 1

SIDE JAMB

SHIMS

ELEVATION OF HEAD JAMB

3   HEAD CASING   MARK LENGTH OF CASING AT GAUGED LINE   2

HOLD MITERED END OF CASING TO GAUGED LINE

HOLD EDGE OF CASING TO GAUGED LINES

4

LET END OVERLAP OPENING

SIDE JAMB

ELEVATION OF HEAD JAMB

5

MITER HEAD CASING TO LENGTH AND INSTALL ON HEAD JAMB TO GAUGED LINES

ELEVATION OF HEAD JAMB AND CASING

ON THE JOB

Slight adjustments may be made in miter box cuts by shimming molding against fence.

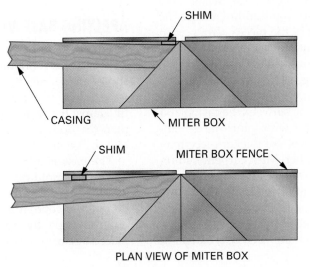

SHIM

CASING   MITER BOX

SHIM   MITER BOX FENCE

PLAN VIEW OF MITER BOX

FIGURE 79–6   Technique for making small adjustments to the angle of a miter.

Procedure 79–B Cutting Side Casings to Fit

**STEP ① ** CUT SIDE CASINGS SLIGHTLY LONG WITH A MITER. BE SURE THEY ARE MIRROR IMAGES OF EACH OTHER. MARK THE CASING LENGTH AT THE TOP OF HEAD CASING.

**STEP ② ** INSTALL CASING TO REVEAL LINES AND SET NAILS.

MARK SIDE CASING IN LINE WITH TOP EDGE OF HEAD CASING

HEAD CASING INSTALLED

DOOR JAMB

MITERED END OF SIDE CASING AGAINST FLOOR

CUT SQUARE END AND PLACE ON FLOOR TO INSTALL SIDE CASING

MARK AND INSTALL OTHER SIDE IN LIKE MANNER

EDGE OF OPENING

DOOR JAMB

ELEVATIONS OF DOOR OPENING

joint. Nail the side casing in the same manner as the head casing.

Avoid sanding the joint to bring the casing faces flush. It is difficult to keep from sanding across the grain on one or the other of the pieces. Cross-grain scratches will be very noticeable, especially if the trim is to have a stained finish. Bring the faces flush, if necessary, by shimming between the back of the casing and the wall. Usually, only very thin shims are needed. Any small space between the casing and the wall can be filled later with joint filling compound. Also, the backside of the thicker piece may be planed or chiseled thinner. Most carpenters prefer to do these rather than try to sand the joint.

Drive a 4d finish nail into the edge of the casing and through the mitered joint. Drive end nails. Then set all fasteners. Keep nails 2 or 3 inches from the end to avoid splitting the casing. If there is danger of

splitting, blunt the pointed end or drill a hole slightly smaller in diameter than the nail.

APPLYING BASE MOLDINGS

Molded or *S4S* stock may be used for base. If S4S base is used, it should be backed out. A base cap should be applied to its top edge. The base cap conforms easily to the wall surface, resulting in a tight fit against the wall. The base trim should be thinner than the door casings against which it butts. This makes a more attractive appearance.

The base is applied in a manner similar to wall and ceiling molding. However, copes are laid out for joints in interior corners. Instead of back-mitering to outline the cope, it is usually more accurate to lay out the cope by **scribing (Figure 79–7)**. When placed against the wall, the face of the base may not

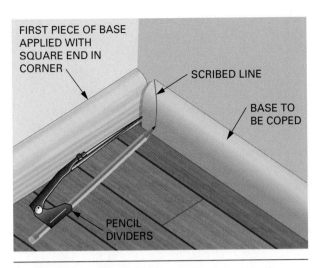

FIGURE 79–7   Laying out a coped joint on base molding by scribing.

always be square with the floor. Therefore, if the base is tilted slightly, back-mitering to obtain the outline of the cope will result in a poor fit against it.

Apply the base to the first wall with square ends in each corner. Drive and set two finishing nails, of sufficient length, at each stud location. Nailing blocks previously installed during framing provide solid wood for fastening the ends of the base in interior corners.

Cut the base to go on the next wall about an inch longer than required. Lay the base against the wall by bending it so the end to be scribed lies flat against the wall and against the first base. Set the dividers to

scribe about ½ inch. Lay out the cope by riding the dividers along the face of the base on the first wall.

Hold the dividers while scribing so that a line between the two points is parallel to the floor. Twisting the dividers while making the scribe results in an inaccurate layout. Cut the end to the scribed line with a slight undercut. Bend the base back in position and try the fit. If scribed and cut accurately, no adjustments should be necessary. Its overall length must now be determined.

Cutting the Base to Length

The length of a baseboard that fits between two walls may be determined by measuring from corner to corner. Then, transfer the measurement to the baseboard. Another method of determining its length eliminates using the rule. It may be faster and more accurate.

With the base in the last position described above, place marks, near the center, on the top edge of the base and the wall so they line up with each other. Place the other end in the opposite corner. Press the base against the wall at the mark. The difference between the mark on the wall and the mark on the base is the amount to scribe off the end in the corner. Set the dividers to this distance. Scribe the end. Cut to the scribed line **(Procedure 79–C).** If a tighter fit is desired, set the dividers slightly less than the distance between the marks. This method of fitting long lengths between walls may be applied to other kinds of trim. However, this works especially well with the base.

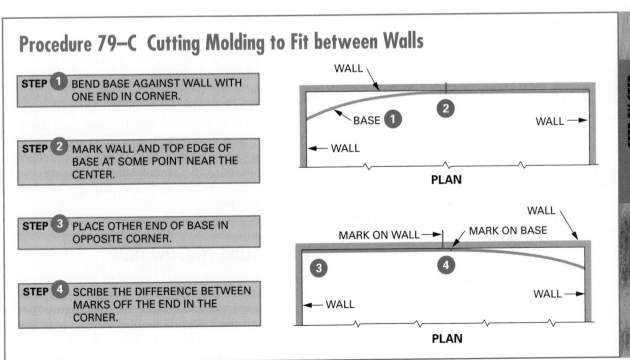

Procedure 79–C Cutting Molding to Fit between Walls

STEP ❶ BEND BASE AGAINST WALL WITH ONE END IN CORNER.

STEP ❷ MARK WALL AND TOP EDGE OF BASE AT SOME POINT NEAR THE CENTER.

STEP ❸ PLACE OTHER END OF BASE IN OPPOSITE CORNER.

STEP ❹ SCRIBE THE DIFFERENCE BETWEEN MARKS OFF THE END IN THE CORNER.

WALL

BASE ❶   ❷   WALL →

← WALL

PLAN

WALL

MARK ON WALL →   MARK ON BASE

❸   ❹

← WALL   WALL →

PLAN

STEP BY STEP PROCEDURES

Procedure 79–D Cutting Outside Miters

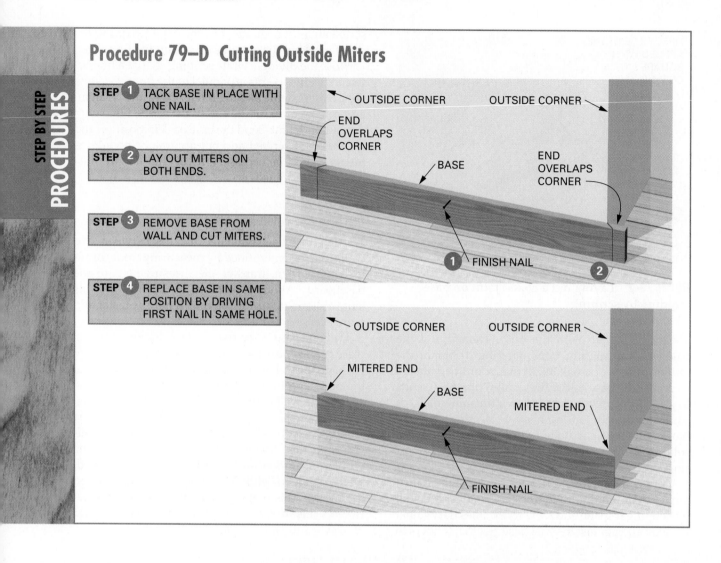

STEP 1 TACK BASE IN PLACE WITH ONE NAIL.

STEP 2 LAY OUT MITERS ON BOTH ENDS.

STEP 3 REMOVE BASE FROM WALL AND CUT MITERS.

STEP 4 REPLACE BASE IN SAME POSITION BY DRIVING FIRST NAIL IN SAME HOLE.

OUTSIDE CORNER   OUTSIDE CORNER

END OVERLAPS CORNER

BASE

END OVERLAPS CORNER

1 FINISH NAIL

2

OUTSIDE CORNER   OUTSIDE CORNER

MITERED END

BASE

MITERED END

FINISH NAIL

Place one end in the corner, and bow out the center. Place the other end in the opposite corner, and press the center against the wall. Fasten in place. Continue in this manner around the room in a previously planned order. Make regular miter joints on outside corners.

If both ends of a single piece are to have regular miters for outside corners, it is imperative that it be fastened in the same position as it was marked. Tack the rough length in position with one finish nail in the center. Mark both ends. Remove, and cut the miters. Installing the piece by first fastening into the original nail hole ensures that the piece is fastened in the same position as marked **(Procedure 79–D)**.

Applying the Base Cap and Base Shoe

The base cap is applied in the same manner as most wall or ceiling molding. Cope interior corners and miter exterior corners. The base shoe is also applied in a similar manner as other molding. However, it is ordinarily nailed into the floor and not into the base-

board. This prevents the joint under the shoe from opening should the shrinkage take place in the baseboard **(Figure 79–8)**.

Because the base shoe is a small-size molding and has solid backing, both interior and exterior corners are mitered. When the base shoe must be stopped at a door opening or other location, with nothing to butt against, its exposed end is generally back-mitered and sanded smooth **(Figure 79–9)**. No base shoe is required if carpeting is to be used as a finish floor.

INSTALLING WINDOW TRIM

Interior window trim, in order of installation, consists of the **stool,** also called *stool cap, apron, jamb extensions, casings,* and *stops* or *stop bead* **(Figure 79–10)**. Although the kind and amount of trim may differ, depending on the style of window, the application is basically the same. The procedure described in this chapter applies to most double-hung windows.

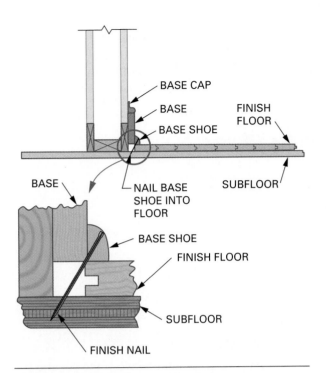

FIGURE 79-8 The base shoe is fastened into the floor, not into the baseboard. This prevents the joint at the floor from opening due to any movement of the baseboard.

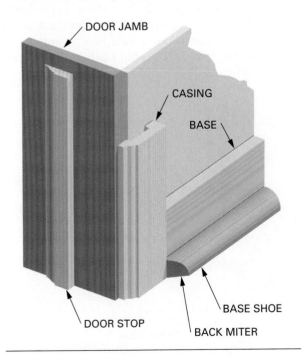

FIGURE 79-9 The exposed ends of base shoe molding are usually back-mitered and sanded smooth.

Installing the Stool

The bottom side of the *stool* is rabbeted at an angle to fit on the sill of the window frame so its top side will

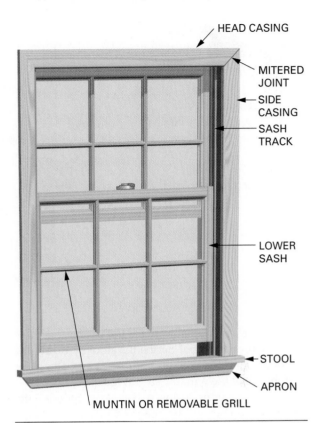

FIGURE 79-10 Component parts of window trim.

be level. Its final position has the outside edge against the sash. Both ends are notched around the side jambs of the window frame. Each end projects beyond the casings by an amount equal to the casing thickness.

The stool length is equal to the distance between the outside edges of the vertical casing plus twice the casing thickness. On both sides of the window, just above the sill, hold a scrap piece of casing stock on the wall. Its inside edge should be flush with the inside face of the side jamb of the window frame. Draw a light line on the wall along the outside edge of the casing stock. Lay out a distance outward from each line equal to the thickness of the window casing. Cut a piece of stool stock to length equal to the distance between the outermost marks.

Raise the lower sash slightly. Place a short, thin strip of wood under it, on each side, which projects inward to support the stool while it is being laid out **(Figure 79–11).** Place the stool on the strips. Raise or lower the sash slightly so the top of the stool is level. Position the stool with its outside edge against the wall. Its ends should be in line with the marks previously made on the wall.

Square lines, across the face of the stool, even with the inside face of each side jamb of the window frame. Set the pencil dividers or scribers so that, on both

ON THE JOB

Wood shingles or trim wood strips placed under the window sash support the stool in a level position during layout.

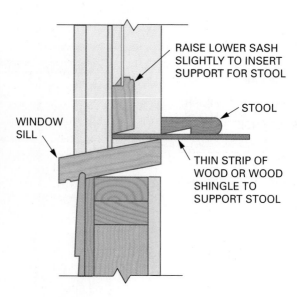

FIGURE 79–11
Technique to hold a stool for easy layout.

RAISE LOWER SASH SLIGHTLY TO INSERT SUPPORT FOR STOOL

STOOL

WINDOW SILL

THIN STRIP OF WOOD OR WOOD SHINGLE TO SUPPORT STOOL

sides, an amount equal to twice the casing thickness will be left on the stool. Scribe the stool by riding the dividers along the wall on both sides and along the bottom rail of the window sash **(Figure 79–12).**

Cut to the lines, using a handsaw. Smooth the sawed edge that will be nearest to the sash. Shape and smooth both ends of the stool the same as the inside edge. Apply a small amount of caulking compound to the bottom of the stool along its outside edge. Fasten the stool in position by driving finish nails along its outside edge into the sill. Set the nails.

Applying the Apron

The apron covers the joint between the sill and the wall. It is applied with its ends in line with the outside edges of the window casing. Cut a length of apron stock equal to the distance between the outer edges of the window casings.

Each end of the apron is then *returned upon itself.* This means that the ends are shaped the same as its face. To return an end upon itself, hold a scrap piece on the apron. Draw its profile flush with the end.

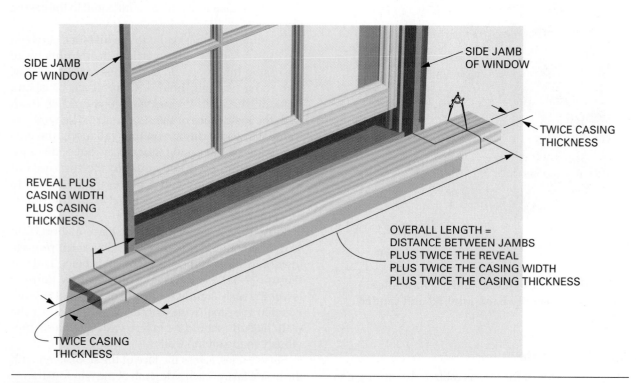

SIDE JAMB OF WINDOW

SIDE JAMB OF WINDOW

TWICE CASING THICKNESS

REVEAL PLUS CASING WIDTH PLUS CASING THICKNESS

OVERALL LENGTH = DISTANCE BETWEEN JAMBS PLUS TWICE THE REVEAL PLUS TWICE THE CASING WIDTH PLUS TWICE THE CASING THICKNESS

TWICE CASING THICKNESS

FIGURE 79–12   Method of laying out a stool.

Cut to the line with a coping saw. Sand the cut end smooth **(Figure 79–13).** Return the other end upon itself in the same manner.

FIGURE 79–13 Returning the end of an apron upon itself.

Place the apron in position with its upper edge against the bottom of the stool. Be careful not to force the stool upward. Keep the top side of the stool level by holding a square between it and the edge of the side jamb. Fasten the apron along its bottom edge into the wall. Then drive nails through the stool into the top edge of the apron. When nailing through the stool, wedge a short length of 1 × 4 stock between the apron and the floor at each nail location. This supports the apron while nails are being driven. Failure to support the apron results in an open joint between it and the stool. Take care not to damage the bottom edge of the apron with the supporting piece **(Figure 79–14).**

Installing Jamb Extensions

Windows are often installed with jambs that are narrower than the wall thickness. Strips must be fastened to these narrow jambs to bring the inside edges flush with inside wall surface. These strips are called **jamb extensions.**

Some manufacturers provide jamb extensions with the window unit. However, they are not always applied when the window is installed, but when the window is trimmed. Therefore when windows are set, these pieces should be carefully stored and then retrieved when it is time to apply the trim. They are usually precut to length and need only to be cut to width.

Measure the distance from the inside edge of the jamb to the finished wall. Rip the jamb extensions to this width with a slight back-bevel on the inside edge. Cut the pieces to length, if necessary, and apply them to the header and side jambs. Drive finish nails through the edges into the edge of the jambs **(Figure 79–15).**

ON THE JOB

Support the apron when fastening the stool to it.

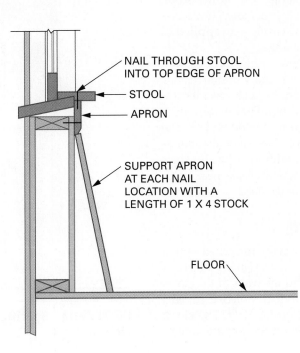

NAIL THROUGH STOOL INTO TOP EDGE OF APRON

STOOL

APRON

SUPPORT APRON AT EACH NAIL LOCATION WITH A LENGTH OF 1 X 4 STOCK

FLOOR

FIGURE 79–14
Technique for holding an apron in place for nailing.

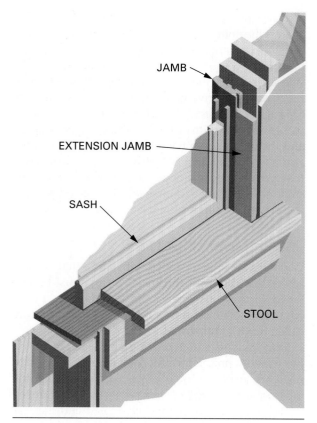

JAMB

EXTENSION JAMB

SASH

STOOL

FIGURE 79–15 Jamb extensions are used to widen the window jamb.

Applying the Casings

Window casings usually are installed with a reveal similar to that of door casings. They also may be installed flush with the inside face. In either case, the bottom ends of the side casings rest on the stool. The window casing pattern is usually the same as the door casings. Window casings are applied in the same manner as door casings.

Cut the number of window casings needed to a rough length with a miter on one end. Cut side casings with left- and right-hand miters. Install the header casing first and then the side casings. Find the length of side casings by turning them upside down with the point of the miter on the stool in the same manner as door casings. Fasten casings with their inside edges flush with the inside face of the jamb. Make neat, tight-fitting joints at the stool and at the head.

INSTALLING CLOSET TRIM

A simple clothes closet is normally furnished with a shelf and a rod for hanging clothes. Usually a piece of 1 × 5 stock is installed around the walls of the closet to support the shelf and the rod. This piece is called a **cleat.** The shelf is installed on top of it. The closet pole is installed in the center of it. Shelves are not fas-

tened to the cleat. Rods are installed for easy removal in case the closet walls need refinishing.

Shelves are usually 1 × 12 boards. Rods may be ¾-inch steel pipe, ¹⁵⁄₁₆-inch full round wood poles, or chrome plated rods manufactured for this purpose. On long spans, the rod may be supported in its center by special metal closet pole supports. On each end, the closet pole is supported by plastic or metal closet pole *sockets*. In place of sockets, holes and notches are made in the cleat to support the ends of the closet pole.

For ordinary clothes closets, the height from the floor to the top edge of the cleat is 66 inches **(Figure 79–16)**. Measure up from the floor this distance. Draw a level line on the back wall and two end walls of the closet. Ease the bottom outside corner and install the cleat so its top edges are to the line. The cleat is installed in the same manner as baseboard. Butt the interior corners. Fasten with two finish nails at each stud.

Install the closet pole sockets on the end cleats. The center of the socket should be at least 12 inches from the back wall and centered on the width of the cleat. Fasten the sockets through the predrilled holes with the screws provided.

Installing Closet Shelves

For a professional job, fit the ends and back edge of the shelf to the wall. Cut the shelf about ½ inch longer than the distance between end walls.

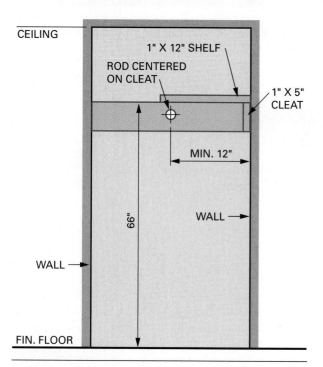

CEILING

1" X 12" SHELF

ROD CENTERED ON CLEAT

1" X 5" CLEAT

MIN. 12"

66"

WALL

WALL

FIN. FLOOR

FIGURE 79–16 Specifications for an ordinary clothes closet.

Procedure 79–E Fitting a Closet Shelf

STEP 1 CUT SHELF ABOUT ½" LONGER THAN WIDTH OF CLOSET.

STEP 2 TILT SHELF IN POSITION WITH ONE END ON CLEAT. SCRIBE ABOUT ¼" ON THIS END. REMOVE SHELF AND CUT TO SCRIBED LINE.

STEP 3 FROM SCRIBED END, LAY OUT LENGTH OF SHELF ON BACK EDGE.

STEP 4 REPLACE SHELF IN TILTED POSITION WITH OPPOSITE END ON CLEAT.

STEP 5 SET DIVIDERS FOR DISTANCE FROM WALL TO MARK INDICATING SHELF LENGTH AT BACK EDGE. SCRIBE ALONG END OF SHELF.

STEP 6 REMOVE SHELF AND CUT END TO SCRIBED LINE. REPLACE SHELF WITH BOTH ENDS ON CLEAT.

STEP 7 SCRIBE BACK EDGE TO BACK WALL. SCRIBE ONLY ENOUGH TO FIT SHELF. REMOVE SHELF AND CUT TO SCRIBED LINE. EASE CORNERS ON FRONT EDGE. REPLACE SHELF.

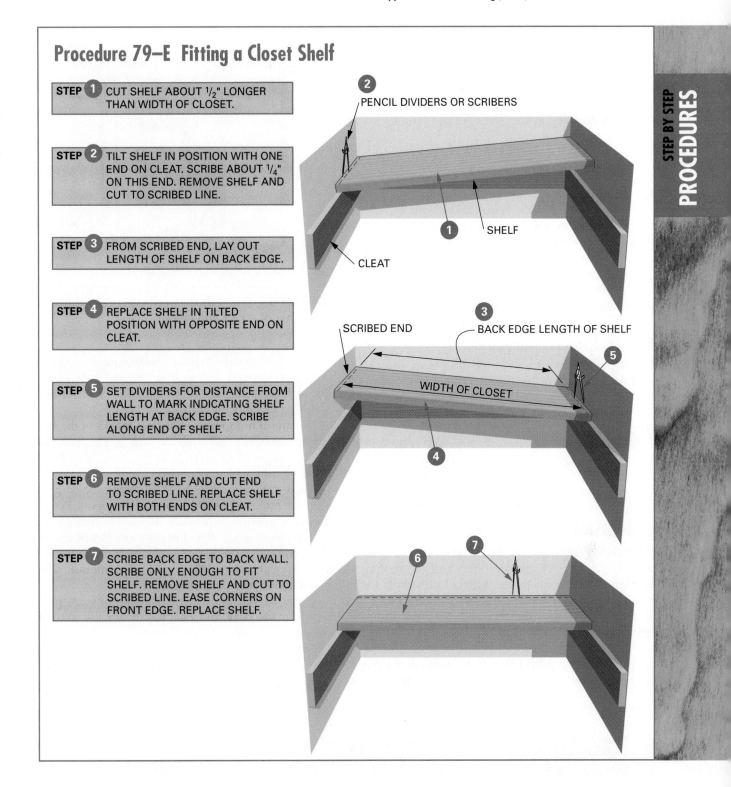

PENCIL DIVIDERS OR SCRIBERS

SHELF

CLEAT

SCRIBED END

BACK EDGE LENGTH OF SHELF

WIDTH OF CLOSET

Place the shelf in position by laying one end on the cleat and tilting the other end up and resting against the wall. Scribe about ¼ inch off the end resting on the cleat. Remove the shelf. Cut to the scribed line. Measure the distance between corners along the back wall. Transfer this measurement to the shelf, measuring from the scribed cut along the back edge of the shelf.

Place the shelf in position, tilted in the opposite direction. Set the dividers to scribe the distance from the wall to the mark on the shelf. Scribe and cut the other end of the shelf. Place the shelf into position, resting it on the cleats. Scribe the back edge to the wall to take off as little as possible. Cut to the scribed line. Ease the corners on the front edge of the shelf with a hand plane. Sand and place the shelf in position **(Procedure 79–E).**

Installing the Closet Pole

Measure the distance between pole sockets. Cut the pole to length. Install the pole on the sockets. One socket is closed. The opposite socket has an open top. Place one end of the pole in the closed socket. Then rest the other end on the opposite socket.

Linen Closets

Linen closets usually consist of a series of shelves spaced 12 to 16 inches apart. Cleats used to support shelves are ¾ × 1 stock, chamfered on the bottom outside corner. A *chamfer* is a bevel on the edge of a board that extends only partway through the thickness of the stock.

Lay out level lines for the top edges of each set of cleats. Install the cleats and shelves in the same manner as described for clothes closets.

MANTELS

Mantels are used to decorate fireplaces and to cover the joint between the fireplace and the wall. Most mantels come preassembled from the factory. They are available in a number of sizes and styles **(Figure 79–17)**.

Study the manufacturer's directions carefully. Place the mantel against the wall. Center it on the fireplace. Scribe it to the floor or wall as necessary. Carefully fasten the mantel in place and set all nails.

CONCLUSION

All pieces of interior trim should be sanded smooth after they have been cut and fitted, and before they are fastened. The sanding of interior finish provides a smooth base for the application of stains, paints, and clear coatings. Always sand with the grain, never across the grain.

All sharp, exposed corners of trim should be rounded over slightly. Use a block plane to make a slight chamfer. Then round over with sandpaper.

FIGURE 79–17   Mantels may come preassembled in a number of styles and sizes.

If the trim is to be stained, make sure every trace of glue is removed. Excess glue, allowed to dry, seals the surface. It does not allow the stain to penetrate, resulting in a blotchy finish.

Be careful not to make hammer marks in the finish. Occasionally rubbing the face of the hammer with sandpaper to clean it helps prevent it from glancing off the head of a nail.

Make sure any pencil lines left along the edge of a cut are removed before fastening the pieces. Pencil marks in interior corners are difficult to remove after the pieces are fastened in position. Pencil marks show through a stained or clear finish and make the joint appear open. When marking interior trim make light, fine pencil marks.

Note: Layout lines in the illustrations are purposely made dark and heavy only for the sake of clarity.

Make sure all joints are tight fitting. Measure, mark, and cut carefully. Do not leave a poor fit. Do it over, if necessary!

Key Terms

aprons	base shoe	cleat	door casings
back bands	bed	coped	finger-jointed
back miter	caps	corner guards	full rounds
base	casings	cove	half rounds
base cap	chair rail	crown	jamb extensions

miter boxes	quarter rounds	standard moldings	stools
mitered	reveal	stops	surfaced four sides
plinth blocks	scribing		

Review Questions

Select the most appropriate answer.

1. Bed, crown, and cove moldings are used frequently as
   a. window trim.
   b. ceiling molding.
   c. part of the base.
   d. door casings.

2. Back bands are applied to
   a. wainscoting.
   b. exterior corners.
   c. casings.
   d. interior corners.

3. A stool is part of the
   a. soffit.
   b. door trim.
   c. base.
   d. window trim.

4. The joint between moldings in interior corners is usually
   a. coped.
   b. mitered.
   c. butted.
   d. bisected.

5. The setback of door casings from the face of the jamb is often referred to as a
   a. gain.
   b. backset.
   c. reveal.
   d. quirk.

6. Find the length of door side casings by
   a. measuring the distance from floor to the header casing.
   b. marking the length on a scrap strip and transferring it to the side casing.
   c. turning the side casing upside down with the point of the miter against the floor.
   d. holding the side casing with the right end up and marking the miter.

7. If the joint between a head and side casing is not a tight fit, it is best to
   a. plane the mitered surfaces
   b. fill the gap with glue
   c. sand the casing face
   d. nail casing tighter

8. The cope on baseboard is laid out more accurately by
   a. back mitering.
   b. returning it.
   c. a combination square.
   d. scribing.

9. The base shoe is fastened
   a. to the baseboard only.
   b. to both the base and the floor.
   c. to the floor only.
   d. directly to the wall.

10. When the end of a molding has no material to butt against, its end is
    a. back-mitered.
    b. mitered.
    c. returned upon itself.
    d. coped.

UNIT 28
Stair Finish

T he staircase is usually the most outstanding feature of a building's interior. It is a showplace for architectural appeal and carpentry skills. All stair finish work must be done in a first-class manner. Joints between stair finish members must be accurate and tight-fitting. Balustrades are installed with perfectly fitting joints.

SAFETY REMINDER

S taircases are used by many different sized people. They must be built to safety standards. They are also assembled from many small pieces. Each piece should be examined for defects, looking for strength as well as appearance characteristics. Each must be installed securely enough to provide support and prevent accidents.

OBJECTIVES

After completing this unit, the student should be able to:

- name various stair finish parts and describe their location and function.
- lay out, dado, and assemble a housed-stringer staircase.
- apply finish to open and closed staircases.
- lay out treads for winding steps.
- install a post-to-post balustrade, without fittings, from floor to balcony on the open end of a staircase.
- install an over-the-post balustrade, with fittings, on an open staircase that runs from a starting step to an intermediate landing and, then, to a balcony.

80   Description of Stair Finish

Many kinds of stair finish parts are manufactured in a wide variety of wood species, such as oak, beech, cherry, poplar, pine, and hemlock. It is important to identify each of the parts, know their location, and understand their function when learning to apply stair finish.

TYPES OF STAIRCASES

The stair finish may be separated in two parts: the **stair body** and the **balustrade.** Important components of the stair body finish are *treads, risers,* and *finish stringers.* The stair body may be constructed as an **open** or **closed** staircase. In an open staircase, the ends of the treads are exposed to view. In a closed staircase, they butt against the wall. Staircases may be open or closed on one or both sides.

Major parts of the balustrade include *handrails, newel posts,* and *balusters.* Balustrades are constructed in either a **post-to-post** or **over-the-post** method. In the post-to-post method, the handrail is fitted between the newel posts **(Figure 80–1).** In the over-the-post method, the handrail runs continuously from top to bottom. It requires special curved sections, called *fittings,* where the handrail changes height or direction **(Figure 80–2).**

STAIR BODY PARTS

Many kinds of stair parts are required to finish the stair body.

Risers

Risers are vertical members that enclose the space between treads. They are manufactured in a thickness of ¾ inch and in widths of 7½ and 8 inches.

Treads

Treads are horizontal finish members on which the feet are placed when ascending or descending stairs. High-quality treads are made from oak. Others are made from poplar or hard pine. They normally come in ¾- and 1½2-inch thicknesses, in

FIGURE 80–1   **A closed staircase with a post-to-post balustrade on a kneewall.** *(Courtesy of L. J. Smith)*

FIGURE 80–2   **An open-one-side staircase with an over-the-post balustrade.** *(Courtesy of L. J. Smith)*

10½- and 11½-inch widths, and in lengths from 36 up to 72 inches.

Nosings. The outside edge of the tread beyond the riser has a half round shape. It is called the **nosing.**

A *return nosing* is a separate piece mitered to the open end of a tread to conceal its end grain. Return nosings are available in the same thickness as treads and in 1¼-inch widths. Treads are available with the return nosing already applied to one end **(Figure 80–3).**

Landing Treads. **Landing treads** are used at the edge of landings and balconies. They match the tread thickness at the nosing. However, they are rabbeted to match the finish floor thickness on the landing. They come in 3½- and 5½-inch widths. The wider landing tread is used when newel posts are more than 3½ inches wide **(Figure 80–4).**

Tread Molding. The **tread molding** is a small cove molding used to finish the joint under stair and landing treads. The molding should be the same kind of wood as the treads. Its usual size is ⅝ × ¹³⁄₁₆.

Finish Stringers

Finish stringers are sometimes called *skirt boards.* They are members of the stair body used to trim the intersection of risers and treads with the wall. They are called *closed finish stringers* when they are located above treads that butt the wall. They are termed *open finish stringers* when they are placed on the open side of a stairway below the treads **(Figure 80–5).** Finish

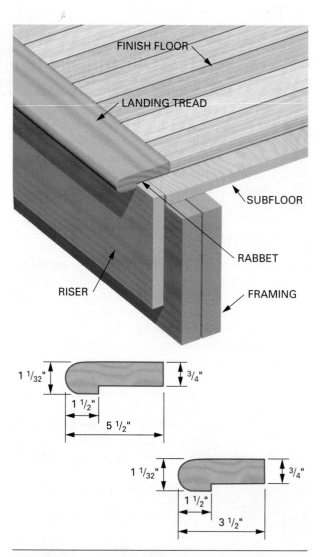

FIGURE 80–4   Landing treads are rabbeted to match the thickness of the finish floor. (*Courtesy of L. J. Smith***)**

stringer lineal stock is available in a ¾-inch thickness and in widths of 9¼ and 11¼ inches.

Starting Steps

A **starting step** or bull-nose step is the first tread-and-riser unit sometimes used at the bottom of a stairway. The starting step is used when the staircase is open on one or both sides and the handrail curves outward at the bottom. They are available in a number of styles with *bull-nosed* ends, preassembled, and ready for installation **(Figure 80–6).**

BALUSTRADE MEMBERS

Finish members of the balustrade are available in many designs that are combined to complement

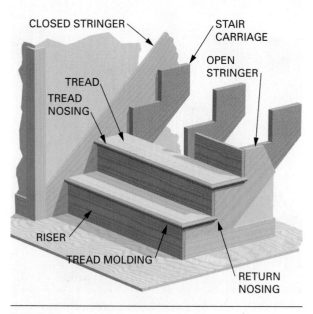

FIGURE 80–3   Treads and risers may be fastened to supporting stair carriages.

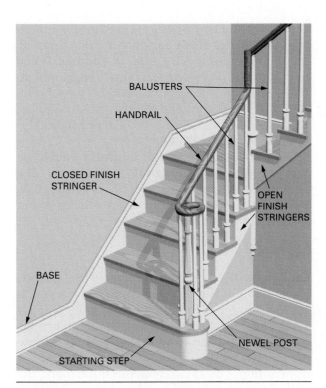

FIGURE 80–5   Open and closed finish stringers are finished trim pieces used to cover the intersections of the treads and risers with the wall and stair body.

each other. Various types of fittings are sometimes joined to straight lengths of handrail when turns in direction are required.

Newel Posts

Newel posts are anchored securely to the staircase to support the handrail. In post-to-post balustrades, the newel posts have flat, square surfaces near the top, against which the handrails are fitted, and also at the bottom for fitting and securing the post to the staircase. In between the flat surfaces, the posts may be *turned* in a variety of designs **(Figure 80–7).**

In over-the-post systems, a round pin at the top of each newel posts fits into the underside of handrail fittings. The post are tapered toward the top end in a number of turned designs **(Figure 80–8).**

Three types of newel posts are used in a post-to-post balustrade. *Starting newels* are used at the bottom of a staircase. They are fitted against the first or second riser. If fitted against the second riser, the flat, square surface at the bottom must be longer. At the top of the staircase, *second-floor newels* are used. *Intermediate landing newels* are also available. Because part of the bottom end of these newels are exposed, turned *buttons* are available to finish the end.

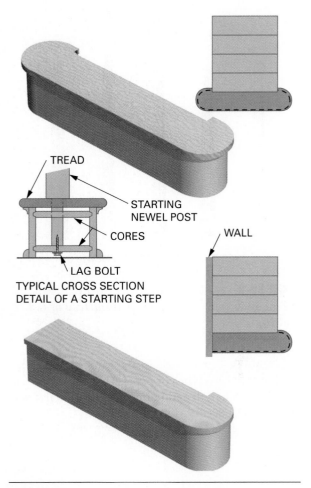

FIGURE 80–6   Starting steps are available in a number of styles and sizes. *(Courtesy of L. J. Smith)*

POST-TO-POST STARTING NEWELS

FIGURE 80–7   Newels in post-to-post balustrades must have flat surfaces where the handrails attach. *(Courtesy of L. J. Smith)*

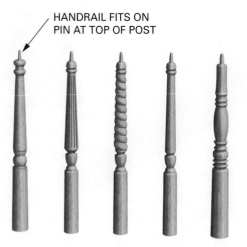

HANDRAIL FITS ON
PIN AT TOP OF POST

OVER-THE-POST STARTING NEWELS

FIGURE 80–8 **Newels in over-the-post balustrades are made with a pin at the top.** (*Courtesy of L. J. Smith*)

The same design is used in the same staircase for each of the three types of posts. They differ only in their overall length and in the length of the flat surfaces **(Figure 80–9).**

Four types of newel posts are used in an over-the-post balustrade. There are three types of *starting newels* depending on the type of handrail fitting used. If a *volute* or *turnout* is used, a newel post with a dowel at the bottom is installed on top of a required starting step. The fourth type is a longer newel for landings, where a gooseneck handrail fitting is used **(Figure 80–10).**

When the balustrade ends against a wall, a *half newel* is sometimes fastened to the wall. The handrail is then butted to it. In place of a half newel, the handrail may butt against an oval or round *rosette* **(Figure 80–11).**

Handrails

The **handrail** is the sloping finish member grasped by the hand of the person ascending or descending the stairs. It is installed horizontally when it runs along the edge of a balcony. Handrail heights are 30 to 38 inches vertically above the nosing edge of the tread. There should be a continuous 1½-inch finger clearance between the rail and the wall. Several styles of handrails come in lineal lengths that are cut to fit on the job. Some handrails are *plowed* with a wide groove on the

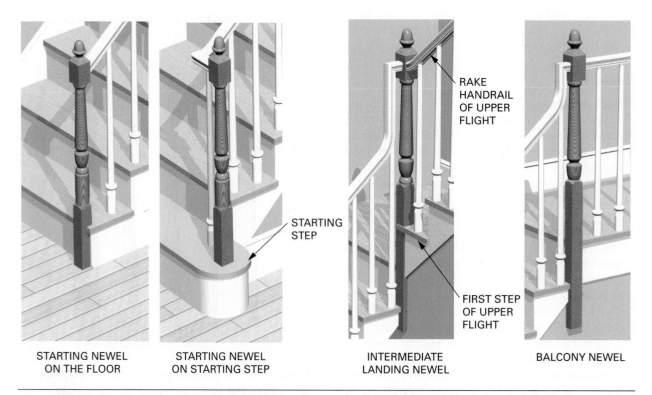

STARTING NEWEL
ON THE FLOOR

STARTING NEWEL
ON STARTING STEP

STARTING
STEP

RAKE
HANDRAIL
OF UPPER
FLIGHT

FIRST STEP
OF UPPER
FLIGHT

INTERMEDIATE
LANDING NEWEL

BALCONY NEWEL

FIGURE 80–9 **Three types of newel posts are used in a post-to-post balustrade.** (*Courtesy of L. J. Smith*)

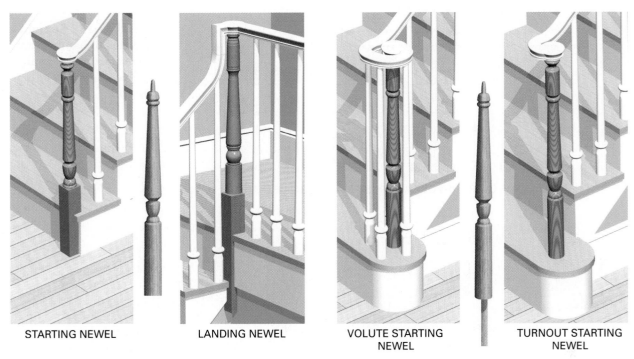

STARTING NEWEL LANDING NEWEL VOLUTE STARTING NEWEL TURNOUT STARTING NEWEL

SQUARE BOTTOM NEWELS USED AT START, LANDINGS, AND BALCONIES

NEWELS WITH PIN BOTTOMS USED ON STARTING STEPS

FIGURE 80–10 Newels for over-the-post balustrades either have pinned or square bottoms. *(Courtesy of L. J. Smith)*

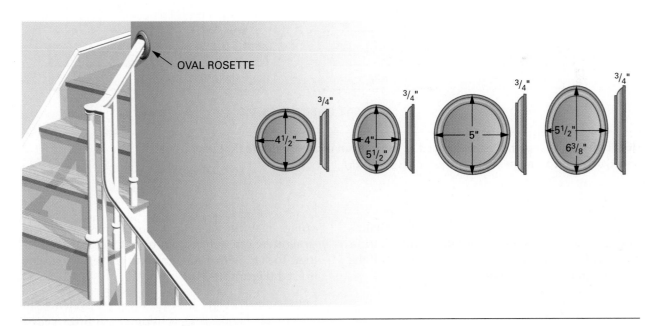

OVAL ROSETTE

FIGURE 80–11 Rosettes are fastened to the wall to provide a decorative attaching surface for the handrail. *(Courtesy of L. J. Smith)*

bottom side to hold square top balusters in place **(Figure 80–12).**

On closed staircases, a balustrade may be installed on top of a **kneewall** or buttress. In relation to stairs, a kneewall is a short wall that projects a short distance above and on the same rake as the stair body. A **shoe rail** or buttress cap, which is plowed on the top side, is usually applied to the top

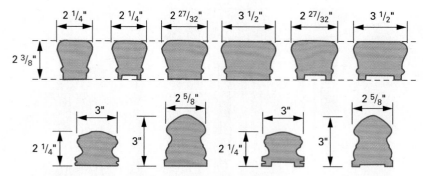

FIGURE 80–12 Straight lengths of handrail are manufactured in many styles. *(Courtesy of L. J. Smith)*

HANDRAILS FOR OVER-THE-POST AND POST-TO-POST BALUSTRADES

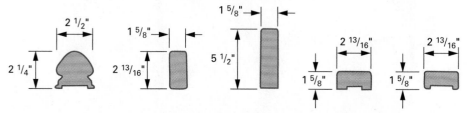

HANDRAILS FOR POST-TO-POST BALUSTRADES ONLY

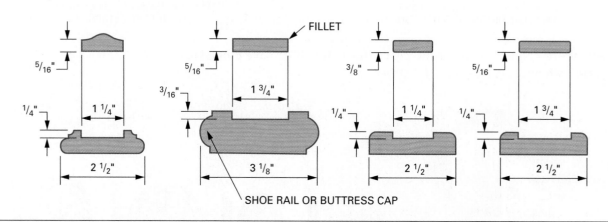

FILLET

SHOE RAIL OR BUTTRESS CAP

FIGURE 80–13 A shoe rail is often used at the bottom of a balustrade that is constructed on a kneewall. *(Courtesy of L. J. Smith)*

of the kneewall on which the bottom end of balusters are fastened **(Figure 80–13)**. Narrow strips, called **fillets,** are used between balusters to fill the plowed groove on handrails and shoe rails.

Handrail Fittings

Short sections of specially curved handrail are called **fittings.** They are used at various locations, joined to straight sections, to change the direction of the handrail. They are classified as *starting, gooseneck,* and *miscellaneous* fittings.

Starting Fittings. To start an over-the-post handrail, starting fittings called **volutes, turnouts,** or *starting*

easings may be used. In a post-to-post system, a straight length of handrail may be used at the bottom. To start with a soft, graceful curve, an *upeasing* is used **(Figure 80–14)**.

Gooseneck Fittings. In over-the-post systems, in which the handrail is continuous, fittings called **goosenecks** are required at intermediate landings and at the top. This is because of changes in the handrail height or direction. In post-to-post systems, goosenecks are not required **(Figure 80–15)**.

Goosenecks are available for handrails that continue level or sloping or that turn 90 or 180 degrees right or left. They are made with or without caps for both types of handrail systems.

VOLUTE   TURNOUT   STARTING EASING

THREE WAYS TO START AN OVER-THE-POST HANDRAIL

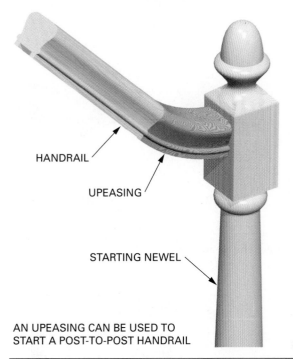

HANDRAIL

UPEASING

STARTING NEWEL

AN UPEASING CAN BE USED TO START A POST-TO-POST HANDRAIL

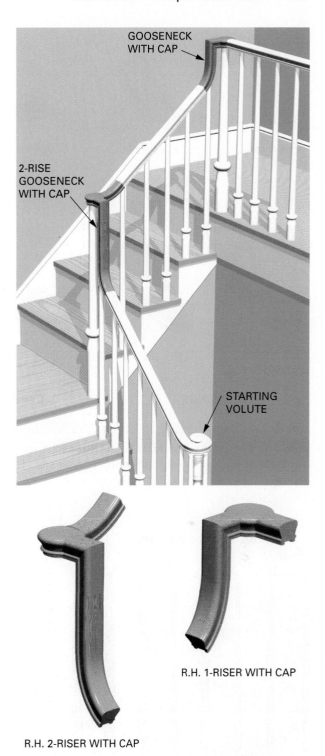

GOOSENECK WITH CAP

2-RISE GOOSENECK WITH CAP

STARTING VOLUTE

R.H. 1-RISER WITH CAP

R.H. 2-RISER WITH CAP

FIGURE 80–14 A handrail fitting is used to start an over-the-post handrail. An upeasing is sometimes used to start a post-to-post handrail. *(Courtesy of L. J. Smith)*

FIGURE 80–15 Gooseneck fittings are used at landings when handrails change direction or height. *(Courtesy of L. J. Smith)*

Miscellaneous Fittings. Among the miscellaneous handrail fittings are *easings* of various kinds, *coped* and *returned ends*, *quarterturns*, and *caps* **(Figure 80–16).** They are used where necessary to meet the specifications for the staircase. All handrail fittings are ordered to match the straight lengths of handrail being used. When the handrail being used is plowed, a matching plow is specified for the handrail fittings. A special fillet, shaped to fit the plow, comes with the handrail fitting.

Balusters

Balusters are vertical, usually decorative pieces between newel posts. They are spaced close together and support the handrail. On a kneewall, they run from the handrail to the shoe rail. On an open stair-

case, they run from the handrail to the treads (see Figures 80–1 and 80–2).

Balusters are manufactured in many styles. They should be selected to complement the newel posts

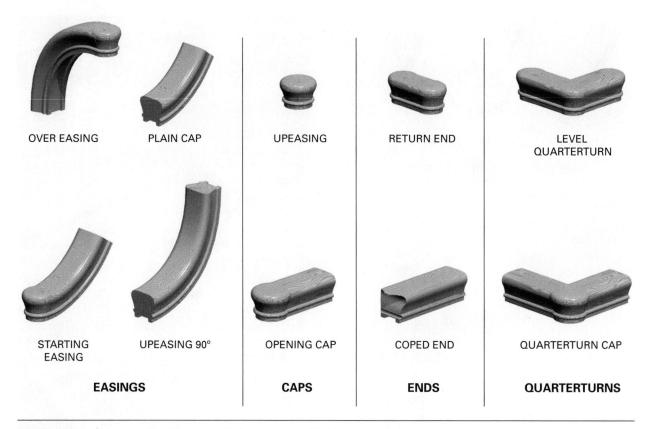

FIGURE 80–16 Many miscellaneous fittings are available to meet the needs of any type of handrail installation. *(Courtesy of L. J. Smith)*

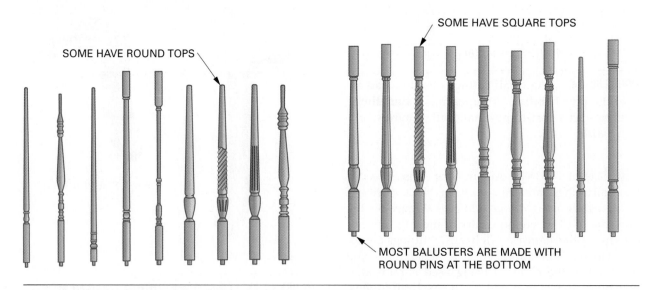

FIGURE 80–17 Balusters are made in designs that match newel post design. *(Courtesy of L. J. Smith)*

being used **(Figure 80–17)**. For example, in a post-to-post staircase, balusters with square tops are usually used. In an over-the-post staircase, balusters that are tapered at the top complement the newel posts.

Most balusters are made in lengths of 31, 34, 36, 39, and 42 inches for use in any part of the stairway. Several lengths of the same style baluster are needed for each tread of the staircase because of the rake of the handrail.

81 Finishing Open and Closed Staircases

Several methods are used to finish a set of stairs. Treads and risers may be inserted into housed stringers, fastened onto a stair carriage, or a combination of both. Both methods require the riser height (unit rise) and tread width (unit run) to be determined. Example calculations of the unit rise and unit run of the stair may be found in Chapter 47. Review Unit 16 for the description, location, and function of the finish members.

National building codes maintain requirements for variations in dimensions. Adjacent riser heights must be within ³⁄₁₆ inch of each other. The maximum variation from largest to smallest riser must be ³⁄₈ inch. Check local codes. With care, all riser dimensions of a set of stairs can be built with only ¹⁄₁₆-inch variations.

MAKING A HOUSED STRINGER

Housed stringers can be laid out using a stair router template or a pitch board and job-made router template. In either case the layout begins on the face side of the stringer stock. Draw a setback line parallel to and about 2 inches down from the top edge. The intersection of the tread and riser faces will land on this line **(Figure 81–1).**

The 2-inch distance may vary, depending on the width of the stringer stock and the desired height of the top edge of the stringer above the stair treads.

Using a Stair Router Template

Stair routing templates are manufactured to guide a router in making dadoes in stringers for treads and risers. The router must be equipped with a straight bit and a template guide of the correct size. Stair templates are adjustable for different rises and runs. They are easily clamped to the stock for routing the dadoes and then moved.

The template is shaped so the dadoes will be the exact tread width at the nosing and wider toward the backside of the stair. The template has nonparallel sides so the finished dadoes will be tapered. The treads and risers are then wedged tightly against the face side shoulders of the dadoes **(Figure 81–2).**

Full layout of all treads and risers is not necessary while using the template. An alignment gauge is designed to position the template along the setback line according to the unit length of the stair **(Figure 81–3).** Using the Pythagorean theorem, calculate the unit length. Lightly mark squared lines on the setback line spaced out the unit length distance.

Using a framing square, mark the unit rise and unit run on the board. Mark them such that a rise–run intersection lands on the setback line. Then, using these lines, mark the thickness of a tread and riser pair. Loosen template shoulder clamp bolts and position the square edges of the template to fit parallel to the tread-riser layout. Retighten shoulder clamps.

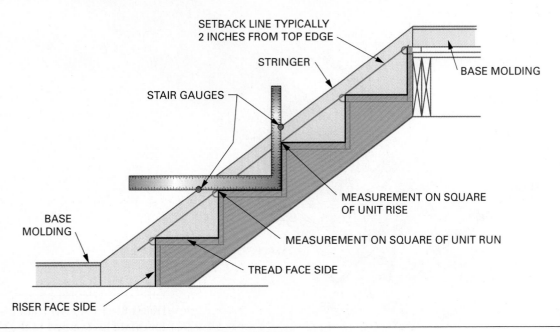

FIGURE 81–1 **Layout considerations for a housed stringer.**

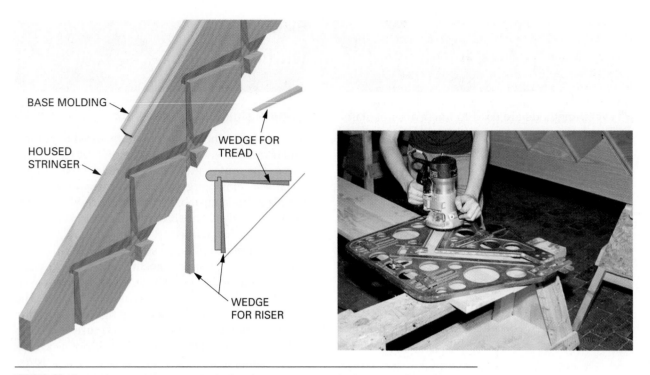

FIGURE 81–2 A housed stringer is dadoed to accept tread, risers, and wedges.

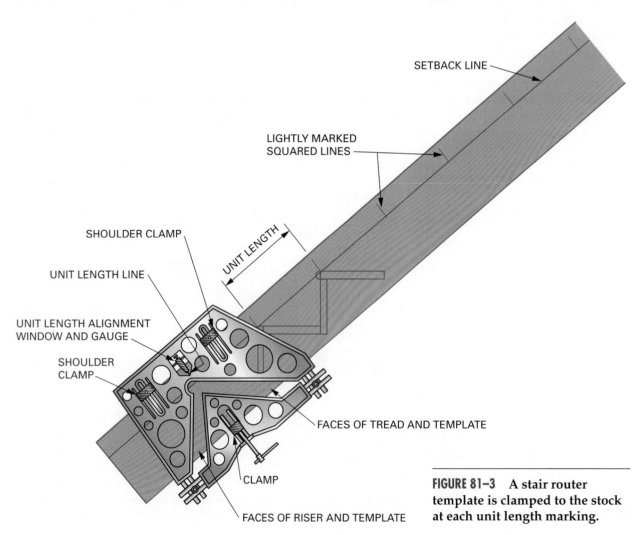

FIGURE 81–3 A stair router template is clamped to the stock at each unit length marking.

Move and clamp the template to the stringer with the alignment gauge on a unit length line. Rout the stringer, about ¼ inch deep. Place the router on the template where it does not touch the stock material. Start router and ease it into the stringer. Press the router guide firmly against the template on all four sides. This will ensure the dado is completely removed.

> **⚠ CAUTION CAUTION CAUTION CAUTION**
>
> Take great care when removing the router from the template. The bit will damage the template if it is touched and the operator is at risk from flying metal fragments. ■

Let the router come to a complete stop before removing it from the template. This will reduce the danger of personal injury and damage to the template and bit. Loosen the template clamp and move the template to the next unit length line. Make sure template is resting entirely on the face of the stock. Rout the dado and repeat for the remaining treads and risers.

Using a Pitch Board

A **pitch board** may be used to lay out a housed stringer. This process requires that each tread and riser pair be laid out. A pitch board is a piece of stock, usually ¾ inch thick. It is cut to the rise and run of the stairs. A strip of wood is fastened to the rake edge of the pitch board. This is used to hold the pitch board against the edge while laying out the stringer **(Figure 81–4).** Care should be taken

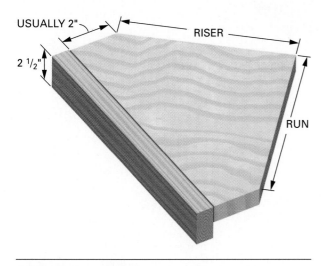

FIGURE 81–4 A pitch board can be used for laying out a housed stringer.

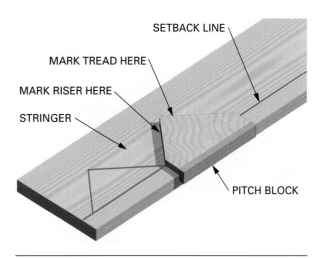

FIGURE 81–5 Use the pitch board to step off each tread and riser on a housed stringer.

when making the pitch board because many layouts will be made from it.

Using the pitch board, lay out the risers and treads for each step of the staircase. These lines show the location of the face side of each riser and tread and are the outside edges of the housing **(Figure 81–5).**

After the stringer has been laid out, make a template to guide the router by cutting out the shape of the dadoes from a piece of thin plywood or hardboard. The cut is made slightly larger than the dadoes to allow for the router guide to follow the template. Take care to make cuts smooth and clean as the router guide will transfer onto the stringer every deviation in the template.

Completing the Stringer

Cut and fit the top and bottom ends of the stringer to the floor and the top end to the landing. Equalize the bottom riser to account for the finished floor thickness. Make end cuts that will properly join with the baseboard. This joint should be made in a professional manner to provide a continuous line of finish from one floor to the next. Since the stringer is usually S4S stock and the base often is not, the stringer should be cut to allow the base to end against it.

LAYING OUT AN OPEN STRINGER

The layout of an open (or *mitered*) stringer is similar to that of a housed stringer. However, riser and tread layout lines intersect at the top edge of the stringer, instead of against a line from the edge. The riser layout line is the outside face of the riser. This layout line is mitered to fit the mitered end of the riser. The tread layout is to the face side of the

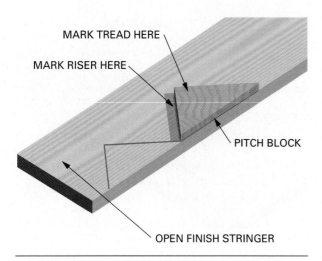

MARK TREAD HERE

MARK RISER HERE

PITCH BLOCK

OPEN FINISH STRINGER

FIGURE 81–6 **Laying out the open finish stringer of a housed staircase.**

tread. The risers and treads are marked lightly with a sharp pencil **(Figure 81–6).**

To lay out the *miter cut* for the risers, measure in at right angles from the riser layout line a distance equal to the thickness of the riser stock. Draw another plumb line at this point. Square both lines across the top edge of the stringer stock. Draw a diagonal line on the top edge to mark the miter angle **(Figure 81–7).**

To mark the tread cut on the stringer, measure down from the tread layout line a distance equal to the thickness of the tread stock. Draw a level line at this point for the tread cut. The tread cut is square through the thickness of the stringer. Fit the bottom end to the floor. Fit the top end against the landing. Make the mitered plumb cuts for the risers and the square level cuts for the treads.

Installing Risers and Treads

Cut the required number of risers to a rough length. Determine the face side of each piece. Sometimes a rabbet and a groove are made to increase the strength of the tread and riser joint forming the inside corner. This is done by cutting a ⅜- × ⅜-inch groove near the bottom edge on the face side of all but the starting riser. The groove is located so its top is a distance from the bottom edge equal to the tread thickness. The groove is made for the rabbeted inner edge of the tread to fit into it **(Figure 81–8).** Rip the risers to width by cutting the edge opposite the groove. Rip the treads to width. Rabbet their back edges to fit in the riser grooves. Cut the risers and tread to exact lengths.

On a closed staircase, the risers are installed with wedges, glue, and screws between housed stringers. On the open side of a staircase, where the riser and open stringer meet, a miter joint is made so no end

FIGURE 81–7 **Laying out the miter angle on an open finish stringer.**

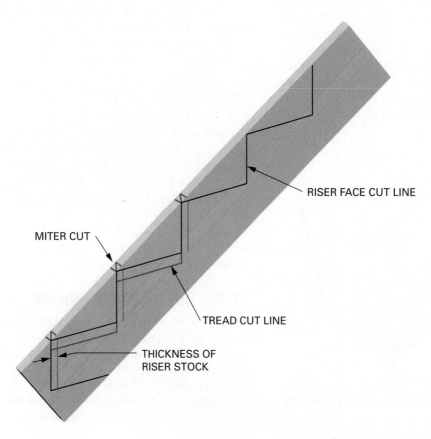

RISER FACE CUT LINE

MITER CUT

TREAD CUT LINE

THICKNESS OF RISER STOCK

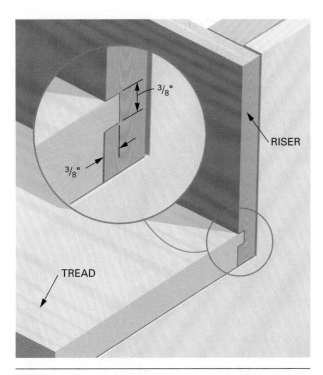

FIGURE 81–8   A groove can be made in the riser to receive the rabbeted inner edge of the tread.

grain is exposed **(Figure 81–9).** The treads are then installed with wedges, glue, and screws on the closed side and with screws through screw blocks on the open side. Screw blocks reinforce interior corners on the underside at appropriate locations and intervals.

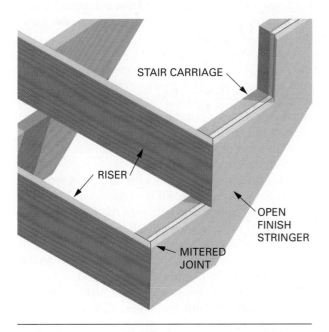

FIGURE 81–9   A mitered joint is made between the risers and open finish stringer so no end grain is exposed.

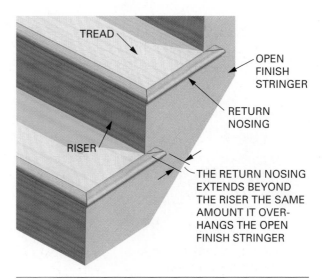

FIGURE 81–10   A return nosing is mitered to the open end of a tread.

Applying Return Nosings and Tread Molding

If the staircase is open, a return nosing is mitered to the end of the tread. The back end of the return nosing projects past the riser the same amount as the tread overhangs the riser. The end is returned on itself.

The tread molding is then applied under the overhang of the tread. If the staircase is closed on both sides, the molding is cut to fit between finish stringers. On the open end of a staircase, the molding is mitered around and under the return nosing. It is stopped and returned on itself at a point so the end assembly appears the same as at the edge **(Figure 81–10).**

After the housed-stringer staircase is assembled, it is installed in position. The balustrade is then constructed in a manner similar to that described later in this chapter for framed staircases.

FINISHING THE BODY OF A CLOSED STAIRCASE SUPPORTED BY STAIR FRAMING

The following section describes the installation of finish to a closed staircase in which the supporting stair carriages have already been installed between two walls that extend from floor to ceiling.

Applying Risers

The first trim members applied to the stair carriages are the risers. Rip the riser stock to the proper width. Cut grooves as described previously for housed-stringer construction. Cut the risers to a length, with square ends, about ¼ inch less than the distance between walls.

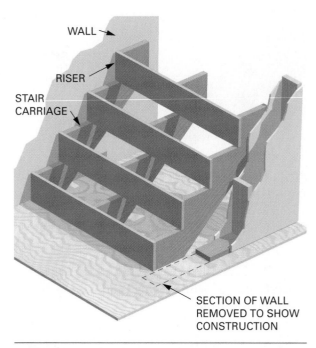

FIGURE 81-11   **Risers are typically the first finish members applied to the stair carriage in a closed staircase.**

Fasten the risers in position with three 2½-inch finish nails into each stair carriage. Start at the top and work down. Remove the temporary treads installed previously as work progresses downward **(Figure 81–11)**.

⚠ **CAUTION   CAUTION   CAUTION   CAUTION**

Put up positive barriers at the top and bottom of the stairs so that the stairs cannot be used while the finish is applied. A serious accident can happen if a person, who does not realize that the temporary treads have been removed, uses the stairs. ■

Laying Out and Installing the Closed Finish Stringer

After the risers have been installed, the *closed finish stringer* is cut around the previously installed risers. Usually 1 × 10 lumber is used. When installed, its top edge will be about 3 inches above the tread nosing. A 1 × 12 may be used if a wider finish stringer is desired.

Tack a length of stringer stock to the wall. Its bottom edge should rest on the top edges of the previously installed risers. Its bottom end should rest on the floor. The top end should extend about 6 inches beyond the landing.

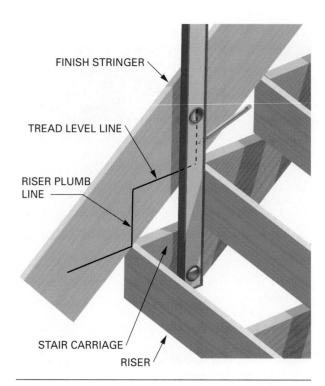

FIGURE 81-12   **The closed finish stringer is laid out using a level to extend plumb and level lines from the stair carriage.**

Lay out plumb lines, from the face of each riser, across the face of the finish stringer. If the riser itself is out of plumb, then plumb upward from that part of the riser that projects farthest outward. Then, lay out level lines on the stringer, from each tread cut of the stair carriage and also from the floor of the landing above **(Figure 81–12)**.

Remove the stringer from the wall. Cut to the layout lines. Follow the plumb lines carefully. Plumb cuts will butt against the face of the risers, so a careful cut needs to be made. Not as much care needs to be taken with level cuts because treads will later butt against and cover them. A circular saw may be used to make most of the cut and then a handsaw is used to finish the cut.

After the cutouts are made in the finish stringer, tack it back in position. Fit it to the floor. Then, lay out top and bottom ends to join the base that will later be installed on the walls. Remove the stringer. Make the end cuts. Sand the board, and place it back in position. Fasten the stringer securely to the wall with finishing nails. Do not nail too low to avoid splitting the lower end of the stringer. Install the finish stringer on the other wall in the same manner.

Drive shims, at each step, between the back side of the risers and the stair carriage. The shims force the risers tightly against the plumb cut of the finish stringer. Shim at intermediate stair carriages to straighten the risers, from end to end, between walls.

Installing Treads

Treads are cut on both ends to fit snugly between the finish stringers. The nosed edge of the tread projects beyond the face of the riser by 1⅛ inches **(Figure 81–13)**.

Along the top edge of the riser, measure carefully the distance between finish stringers. Transfer the measurement and square lines across the tread. Cut in from the nosed edge. Square through the thickness for a short distance. Then undercut slightly. Smooth the cut ends with a block plane. Rub one end with wax. Place the other end in position. Press on the waxed end until the tread lays flat on the stair carriages.

Place a short block on the nosed edge. Tap it until the inner rabbeted edge is firmly seated in the groove of the riser.

If it is possible to work from the underside, the tread may be fastened by the use of screw blocks at each stair carriage and at intermediate locations.

If it is not possible to work from the underside, the treads must be face nailed. Fasten each tread in place with three 8d finish nails into each stair carriage. It may be necessary to drill holes in hardwood treads to prevent splitting the tread or bending the nail. A little wax applied to the nail makes driving easier and helps keep the nail from bending.

Start from the bottom and work up, installing the treads in a similar manner. At the top of the stairs, install a landing tread. If 1½-inch-thick treads are used

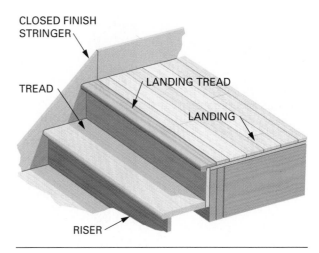

FIGURE 81–14 is illustrated with: CLOSED FINISH STRINGER, TREAD, LANDING TREAD, LANDING, RISER.

FIGURE 81–14 **A rabbeted landing tread is used at the top of the stairway.**

on the staircase, use a landing tread that is rabbeted to match the thickness of the finish floor **(Figure 81–14)**.

Tread Molding

The tread molding is installed under the overhang of the tread and against the riser. Cut the molding to the same length as the treads, using a miter box. Predrill holes. Fasten the molding in place with 4d finish nails spaced about 12 inches apart. Nails are driven at about a 45-degree angle through the center of the molding.

FINISHING THE BODY OF AN OPEN STAIRCASE SUPPORTED BY STAIR FRAMING

The following section describes the installation of finish to the stair body of a staircase, supported by stair carriages, which is closed on one side and open on the other side.

Installing the Finish Stringers

The *open finish stringer* must be installed before the *risers* and the *closed finish stringer*. To lay out the open finish stringer, cut a length of finish stringer stock. Fit it to the floor and against the landing. Its top edge should be flush with the top edge of the stair carriage. Tack it in this position to keep it from moving while it is being laid out.

First, lay out level lines on the face of the stringer in line with the tread cut on the stair carriage. Next, plumb lines must be laid out on the face of the finish stringer for making miter joints with risers.

Using a Preacher to Lay Out Plumb Lines. Use a **preacher** to lay out the plumb lines on the open finish stringer. A preacher is made from a piece of

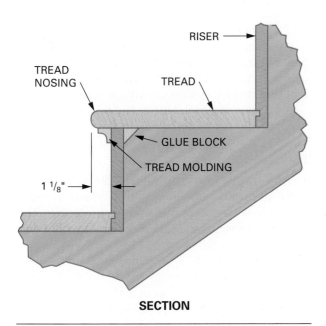

FIGURE 81–13 is labeled with: RISER, TREAD NOSING, TREAD, GLUE BLOCK, TREAD MOLDING, 1⅛", SECTION.

FIGURE 81–13 **Tread and riser details.**

Use a preacher to lay out plumb cuts on an open finish stringer.

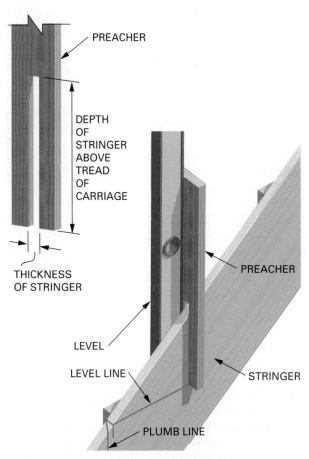

PREACHER

DEPTH
OF
STRINGER
ABOVE
TREAD
OF
CARRIAGE

THICKNESS
OF STRINGER

PREACHER

LEVEL

LEVEL LINE

STRINGER

PLUMB LINE

LAYING OUT OF THE PLUMB CUT

FIGURE 81–15
Technique for easily marking a stringer on both faces.

nominal 4-inch stock about 12 inches long. Its thickness must be the same as the riser stock. The preacher is notched in the center. It should be wide enough to fit over the finish stringer. It should be long enough to allow the preacher to rest on the tread cut of the stair carriage when held against the rise cut.

Place the preacher over the stringer and against the rise cut of the stair carriage. Plumb the preacher with a hand level. Lay out the plumb cut on the stringer by marking along the side of the preacher that faces the bottom of the staircase **(Figure 81–15).**

Mark the top edge of the stringer along the side of the preacher that faces the top of the staircase. Draw a diagonal line across the top edge of the stringer for the miter cut. Lay out all plumb lines on the stringer in this manner.

Remove the stringer. Cut to the layout lines. Make miter cuts along the plumb lines. Cut square through the thickness along the level lines. Sand the piece. Fasten it in position. To ensure the piece will be in the same position as it was when laid out, fasten it first in the same holes where the piece was originally tacked.

Installing Risers

Cut risers to length by making a square cut on the end that goes against the wall. Make miters on the other end to fit the mitered plumb cuts of the open finish stringer. Sand all pieces before installation. Apply a small amount of glue to the miters. Fasten them in position to each stair carriage. Drive finish nails both ways through the miter to hold the joint tight **(Figure 81–16).** Wipe off any excess glue. Set all nails. Lay out and install the closed finish stringer in the same manner as described previously.

Installing the Treads

Rip the treads to width. Rabbet the back edges. Make allowance for the rabbet when ripping treads to width. Cut one end to fit against the closed finish stringer. Make a cut on the other end to receive the return nosing. This is a combination square and miter cut. The square cut is made flush with the outside face of the open finish stringer. The miter starts a distance equal to the width of the return nosing beyond the square cut as shown in Figure 82–9.

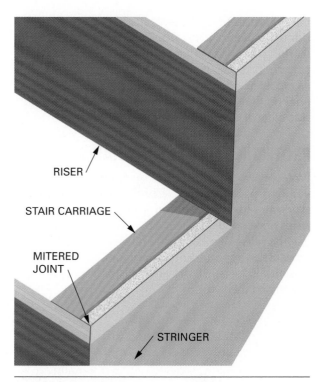

FIGURE 81–16 Open finish stringers are mitered to receive mitered risers.

Applying the Return Nosings

The return nosings are applied to the open ends of the treads. Miter one end of the return nosing to fit against the miter on the tread. Cut the back end square. Return the end on itself. The end of the return nosing extends beyond the face of the riser, the same amount as its width.

Predrill pilot holes in the return nosing for nails. Locate the holes so they are not in line with any balusters that will later be installed on the treads. Holes must be bored in the treads to receive the balusters. Any nails in line with the holes will damage the boring tool **(Figure 81–17).**

Apply glue to the joint. Fasten the return nosing to the end of the tread with three 8d finishing nails. Set all nails. Sand the joint flush. Apply all other return nosings in the same manner. Treads may be purchased with the return nosing applied in the factory. If used, the closed end of the tread is cut so the nosed end overhangs the finish stringer by the proper amount.

Applying the Tread Molding

The tread molding is applied in the same manner as for closed staircases. However, it is mitered on the open end and returned back onto the open stringer. The back end of the return molding is cut and returned on itself at a point so the end assembly shows the same as at the edge **(Figure 81–18).** Predrill pilot holes in the molding. Fasten it in place. Molding on starting and landing treads should only be tacked in case it needs to be removed for fitting after newel posts have been installed.

INSTALLING TREADS ON WINDERS

Treads on winding steps are especially difficult to fit because of the angles on both ends. A method used by many carpenters involves the use of a pattern and a scribing block. Cut a thin piece of plywood so it fits in the tread space within ½ inch on the ends and back edge. The outside edge should be straight and in line with the nosed edge of the tread when installed.

Tack the plywood pattern in position. Use a ¾-inch block, rabbeted on one side by the thickness of

ON THE JOB

Make sure nails used to fasten return nosings do not line up with balusters.

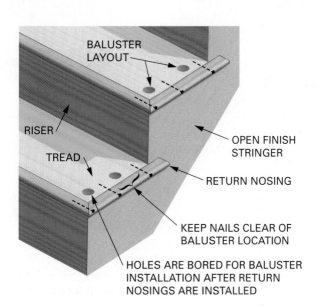

FIGURE 81–17
Alignment of trim nails should be positioned to avoid future baluster holes.

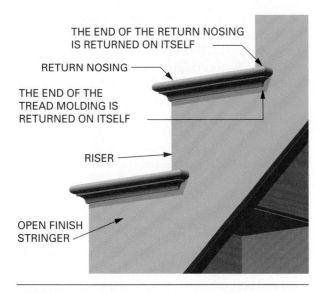

FIGURE 81–18 The back ends of the return nosing and molding are returned on themselves.

the pattern, to scribe the ends and back edge. Scribe by riding the block against stringers, riser, and post while marking its inside edge on the pattern. Remove the pattern. Tack it on the tread stock. Place the block with its rabbeted side down and inside edge to the scribed lines on the pattern. Mark the tread stock on the outside edge of the block **(Figure 81–19).**

PROTECTING THE FINISHED STAIRS

Protect the risers and treads by applying a width of building paper to them. Unroll a length down the stairway. Hold the paper in position by tacking thin strips of wood to the risers. This assumes the tack hole may be filled later before the finish is applied. Sometimes when appearance is paramount, the finished set of stairs is installed last after all tradespeople have completed their work.

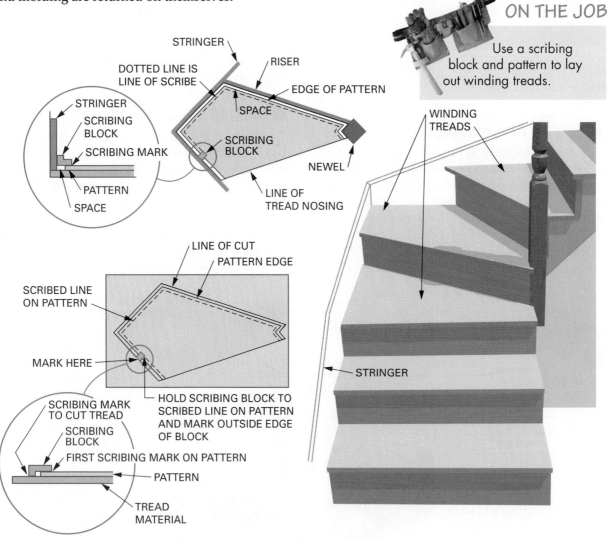

FIGURE 81–19 Technique for scribing winding treads to the stringers, risers, and posts.

82 Balustrade Installation

Balustrades are the most visible and complex component of a staircase (see Figure I–9). Installing them is one of the most intricate kinds of interior finish work. This chapter describes installation of post-to-post and over-the-post balustrades. Mastering the techniques described in this chapter will enable the student to install balustrades for practically any situation.

LAYING OUT BALUSTRADE CENTERLINES

For the installation of any balustrade, its centerline is first laid out. On an open staircase, the centerline should be located a distance inward, from the face of the finish stringer, that is equal to half the baluster width. It is laid out on top of the treads. If the balustrade is constructed on a kneewall, it is centered and laid out on the top of the wall **(Figure 82–1).**

Laying Out Baluster Centers

The next step is to lay out the baluster centers. Code requirements for maximum baluster spacing may vary. Check the local building code for allowable spacing. Most codes require that balusters be spaced so that no object 4 inches in diameter or greater can pass through.

On open treads, the center of the front baluster is located a distance equal to half its thickness back from the face of the riser. If two balusters can be used on each tread, the spacing is half the run. If codes require three balusters per tread, the spacing is one-third the run **(Figure 82–2).**

INSTALLING A POST-TO-POST BALUSTRADE

The following procedure applies to a post-to-post balustrade running, without interruption, from floor to floor.

Laying Out the Handrail

Clamp the handrail to the tread nosings. Use a short bar clamp from the bottom of the finish stringer to the top of the handrail. Clamp opposite a nosing to avoid bowing the handrail. Use only enough pressure to keep the handrail from moving. Protect the edges of the handrail and finish stringer with blocks to avoid marring the pieces. Use a framing square to mark the handrail where it will fit between starting and balcony newel posts **(Figure 82–3).**

While the handrail is clamped in this position, use a framing square at the landing nosing to measure the vertical thickness of the rake handrail. Also,

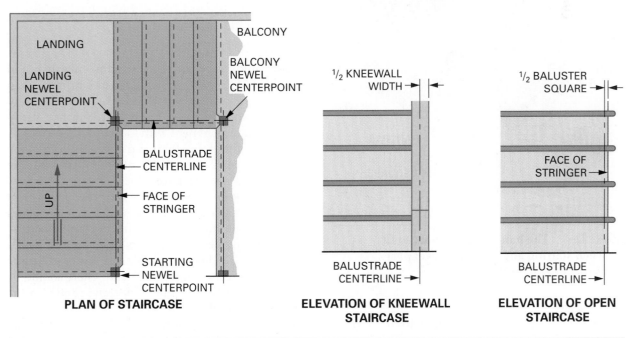

PLAN OF STAIRCASE

ELEVATION OF KNEEWALL STAIRCASE

ELEVATION OF OPEN STAIRCASE

FIGURE 82–1 The centerline of the balustrade is laid out on a kneewall or open treads. *(Courtesy of L. J. Smith)*

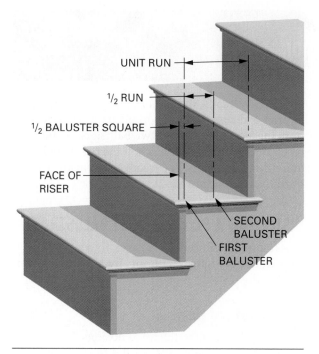

FIGURE 82–2 **Layout of baluster centers on open treads.** *(Courtesy of L. J. Smith)*

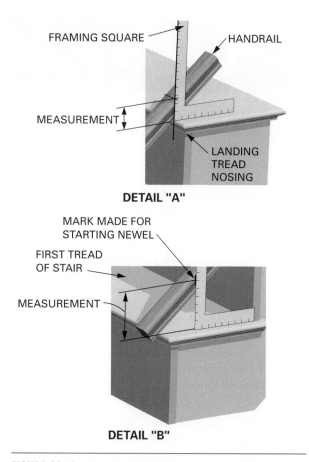

FIGURE 82–4 **Determine the two measurements shown and record for future use.** *(Courtesy of L. J. Smith)*

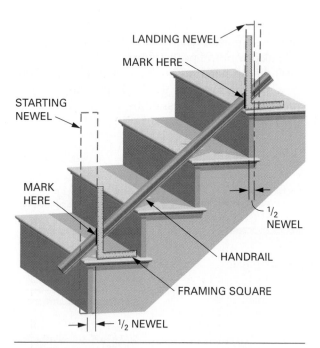

FIGURE 82–3 **The handrail is laid out to fit between starting and balcony newel posts.** *(Courtesy of L. J. Smith)*

at the bottom, measure the height from the first tread to the top of the handrail where it butts the newel post. Record and save the measurements for later use **(Figure 82–4).**

Determining the Height of the Starting Newel

Most building codes state that the rake handrail height shall be no less than 30 inches or more than 34 inches. However, some codes require heights of no less than 34 inches or more than 38 inches. Verify the height requirement with local building codes.

The height of the stair handrail is taken from the top of the tread along a plumb line flush with the face of the riser **(Figure 82–5).** Handrails are required only on one side in stairways of less than 44 inches in width. Stairways wider than 44 inches require a handrail on both sides. A center handrail must be provided in stairways more than 88 inches wide. Handrail heights are 30 to 38 inches vertically above the nosing edge of the tread. There should be a continuous 1½-inch finger clearance between the rail and the wall.

If a turned starting newel post is used, add the difference between the two previously recorded measurements above to the required rake handrail height. Add 1 inch for a block reveal. The block re-

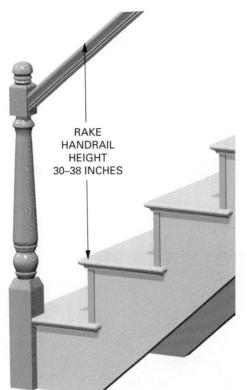

RAKE
HANDRAIL
HEIGHT
30–38 INCHES

FIGURE 82–5 The rake handrail height is the vertical distance from the tread nosing to the top of the handrail. *(Courtesy of L. J. Smith)*

veal is the distance from the top of the handrail to the top of the square section of the post. This sum is the distance from the top of the first tread to the top of the upper block. To this measurement add the height of the turned top and the distance the newel extends below the top of the first tread to the floor **(Figure 82–6).** Cut the starting newel to its total length.

Installing the Starting Newel

The starting newel is notched over the outside corner of the first step. One-half of its bottom thickness is left on from the front face of the post to the face of the riser. In the other direction, it is notched so its centerline will be aligned with the balustrade centerline **(Figure 82–7).** The post is then fastened to the first step with lag screws. The lag screws are counterbored and later concealed with wood plugs.

Newel posts must be set plumb. They must be strong enough to resist lateral force applied by persons using the staircase. The post may be slightly out of plumb after it is fastened. If so, loosen the lag screws slightly. Install thin shims, between the post and riser or finish stringer, to plumb the post. On one or both sides, install the shims, near the bottom or top of the notch as necessary, to plumb the post. When plumb, retighten the lag screws.

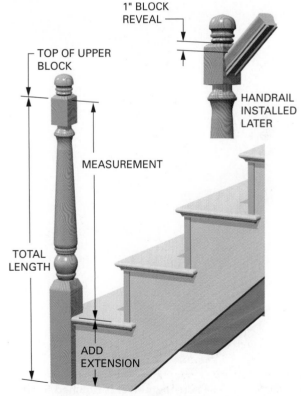

1" BLOCK REVEAL

TOP OF UPPER BLOCK

MEASUREMENT

TOTAL LENGTH

ADD EXTENSION

HANDRAIL INSTALLED LATER

TO DETERMINE MEASUREMENT, ADD THE DIFFERENCE BETWEEN DISTANCES SHOWN IN DETAILS "A" AND "B" IN FIGURE 82–4 TO THE HANDRAIL HEIGHT. ADD 1" MORE FOR BLOCK REVEAL.

FIGURE 82–6 Determining the height of the starting newel. *(Courtesy of L. J. Smith)*

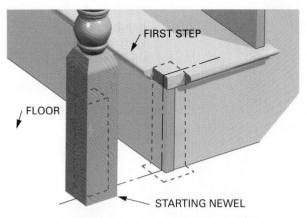

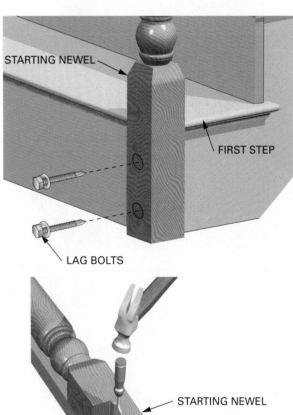

FIGURE 82–7 **The starting newel is notched to fit over the first step.** *(Courtesy of L. J. Smith)*

Installing the Balcony Newel

Generally, codes require that balcony rails for homes be no less than 36 inches. For commercial or public structures, the rails are required to be no less than 42 inches. Check local codes for requirements.

The height of the balcony newel is determined by finding the sum of the required balcony handrail height, a block reveal of 1 inch, the height of the turned top, and the distance the newel extends below the balcony floor.

Trim the balcony newel to the calculated height. Notch and fit it over the top riser with its centerlines aligned with both the rake and balcony handrail centerlines. Plumb it in both directions. Fasten it in place with counterbored lag bolts **(Figure 82–8)**.

Preparing Treads and Handrail for Baluster Installation

Bore holes in the treads at the center of each baluster. The diameter of the hole should be equal to the diameter of the pin at the bottom end of the baluster. The depth of the hole should be slightly more than the length of the pin **(Figure 82–9)**.

Cut the handrail to fit between starting and balcony newels. Lay it back on the tread nosings. The handrail can be cut with a handsaw, a radial arm saw, or a compound miter saw with the blade tilted. If using a power saw, make a practice cut to be sure the setup is correct, before cutting the handrail. Transfer the baluster centerlines from the treads to the handrail **(Figure 82–10)**.

Turn the handrail upside down and end for end. Set it back on the tread nosings with the starting newel end facing up the stairs. Bore holes at baluster centers at least ¾ inch deep, if balusters with round tops are to be used **(Figure 82–11)**.

Installing Handrail and Balusters

Prepare the posts for fastening the handrail by counterboring and drilling shank holes for lag bolts through the posts. Place the handrail at the correct height between newel posts. Drill pilot holes. Temporarily fasten the handrail to the posts **(Figure 82–12)**.

Cut the balusters to length. Allow ¾ inch for insertion in the hole in the bottom of the handrail. The handrail may have to be removed for baluster installation and then fastened permanently. The bottom pin is inserted in the holes in the treads. The top of the baluster is inserted in the holes in the handrail bottom.

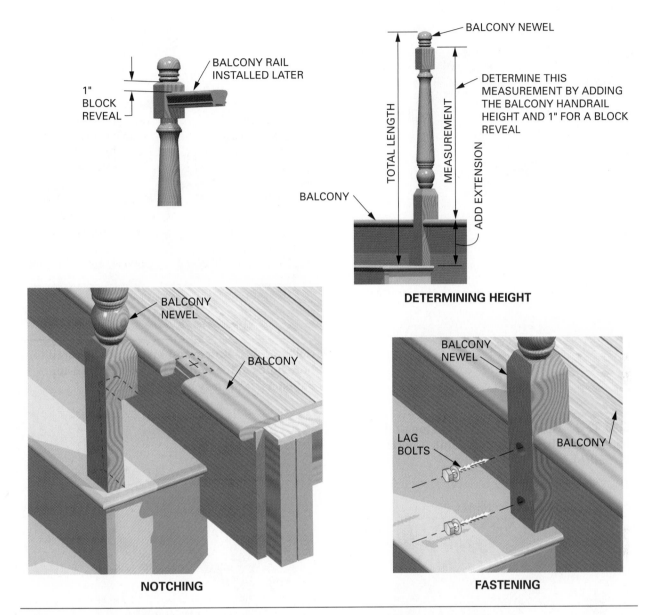

BALCONY RAIL INSTALLED LATER

1" BLOCK REVEAL

BALCONY NEWEL

DETERMINE THIS MEASUREMENT BY ADDING THE BALCONY HANDRAIL HEIGHT AND 1" FOR A BLOCK REVEAL

TOTAL LENGTH

MEASUREMENT

ADD EXTENSION

BALCONY

DETERMINING HEIGHT

BALCONY NEWEL

BALCONY

NOTCHING

BALCONY NEWEL

LAG BOLTS

BALCONY

FASTENING

FIGURE 82–8 The height of the balcony newel post is calculated, notched at the bottom, and fastened in place. *(Courtesy of L. J. Smith)*

If *square-top balusters* are used, they are trimmed to length at the rake angle. They are inserted into a *plowed handrail*. The balusters are then fastened to the handrail with finish nails and glue **(Figure 82–13)**. Care must be taken to keep the handrail in a straight line from top to bottom when fastening square-top balusters. Care must also be taken to keep each baluster in a plumb line. Install fillets in the plow of the handrail, between the balusters.

Installing the Balcony Balustrade

Cut a half newel to the same height as the balcony newel. Temporarily place it against the wall. Mark the length of the balcony handrail **(Figure 82–14)**. Cut the handrail to length. Fasten the half newel to one end of it. Temporarily fasten the half newel to the wall and the other end of the handrail to the landing newel, if they must be removed to install the balcony balusters.

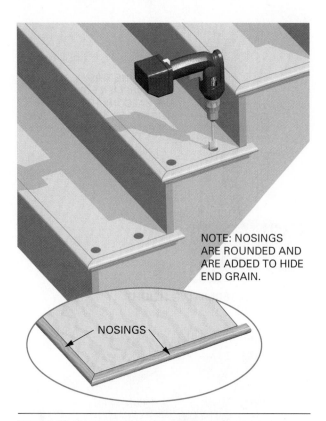

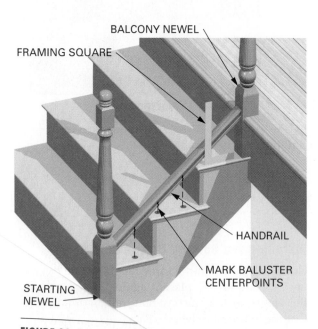

FIGURE 82-9 Holes are bored in the top of the treads at each baluster center point. *(Courtesy of L. J. Smith)*

FIGURE 82-10 The handrail is fitted between newel posts and then the baluster centers are transferred to it. *(Courtesy of L. J. Smith)*

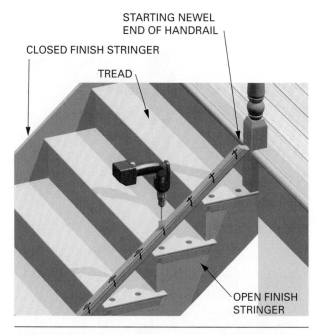

FIGURE 82-11 Rotate and invert rake handrail to bore holes in the bottom to receive round-top balusters. *(Courtesy of L. J. Smith)*

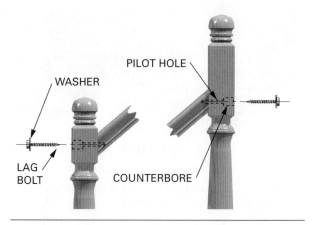

FIGURE 82-12 The handrail is fastened to newel posts with lag bolts. The use of nails when constructing balustrades is discouraged. *(Courtesy of L. J. Smith)*

If the balcony handrail ends at the wall against a rosette, first fasten the rosette to the end of the handrail. Hold the rosette against the wall. Mark the length of the handrail at the landing newel. Cut the handrail to length. Temporarily fasten it in place **(Figure 82–15).**

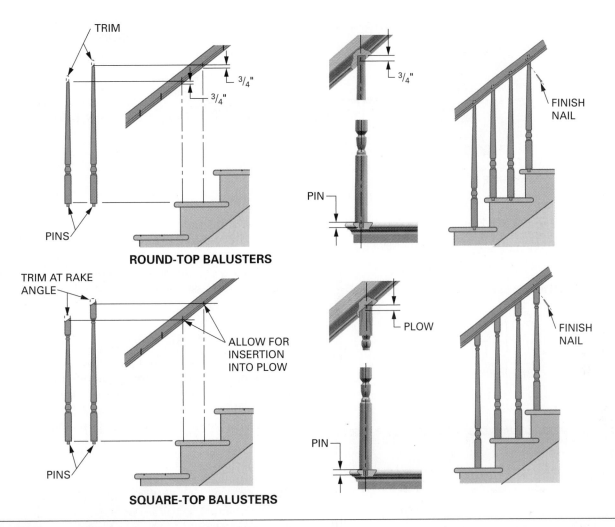

TRIM

PINS

ROUND-TOP BALUSTERS

3/4"

3/4"

FINISH NAIL

PIN

TRIM AT RAKE ANGLE

PINS

SQUARE-TOP BALUSTERS

ALLOW FOR INSERTION INTO PLOW

3/4"

PLOW

FINISH NAIL

PIN

FIGURE 82–13 Balusters are cut to length and installed between handrail and treads. (*Courtesy of L. J. Smith*)

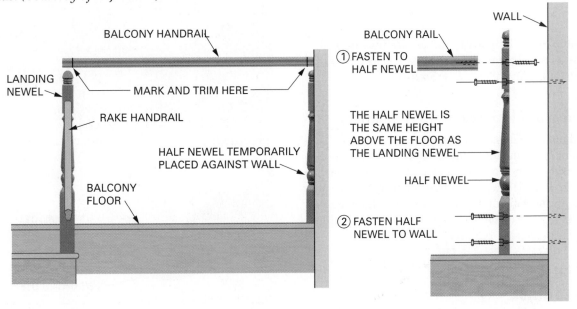

BALCONY HANDRAIL

LANDING NEWEL

MARK AND TRIM HERE

RAKE HANDRAIL

HALF NEWEL TEMPORARILY PLACED AGAINST WALL

BALCONY FLOOR

WALL

BALCONY RAIL

① FASTEN TO HALF NEWEL

THE HALF NEWEL IS THE SAME HEIGHT ABOVE THE FLOOR AS THE LANDING NEWEL

HALF NEWEL

② FASTEN HALF NEWEL TO WALL

FIGURE 82–14 The balcony rail is fitted between the landing newel and a half newel placed against the wall. (*Courtesy of L. J. Smith*)

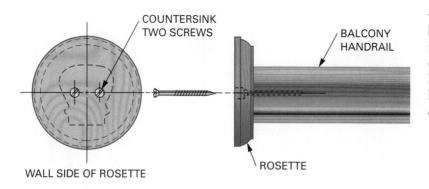

FIGURE 82–15 A rosette is sometimes used to end the balcony handrail instead of a half newel. (*Courtesy of L. J. Smith*)

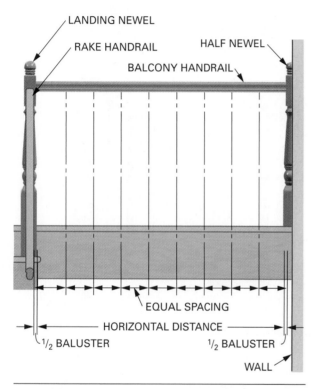

FIGURE 82–16 Balcony balusters are installed as close as possible to the same spacing as the rake balusters. (*Courtesy of L. J. Smith*)

Spacing and Installing Balcony Balusters

The balcony balusters are spaced by adding the thickness of one baluster to the distance between the balcony newel and the half newel. The overall distance is then divided into spaces that equal, as close as possible, the spacing of the rake balusters (**Figure 82–16**). The balcony balusters are then installed in a manner similar to the rake balusters.

INSTALLING AN OVER-THE-POST BALUSTRADE

The following procedures apply to an over-the-post balustrade running from floor to floor with an intermediate landing (see Figure 80–2). An over-the-post balustrade is more complicated. Handrail fittings are required to be joined to straight sections to construct a continuous handrail from start to end. The procedure consists of constructing the entire hand-rail first, then setting newel posts and installing balusters.

Constructing the Handrail

The first step is to lay out balustrade and baluster centerlines on the stair treads as described previously. If a starting step is used, lay out the baluster and starting newel centers using a template provided with the starting fitting (**Figure 82–17**).

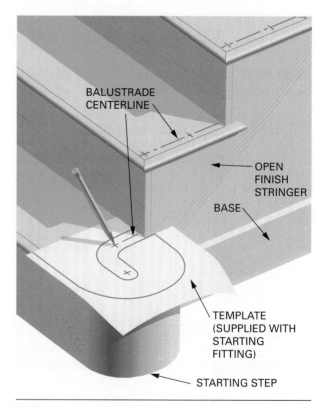

FIGURE 82–17 Baluster and newel post centers are laid out on a starting step with a template. (*Courtesy of L. J. Smith*)

Procedure 82–A Cutting the Starting Fitting *(Courtesy of L. J. Smith)*

STEP 1 MAKE A PITCH BLOCK USING UNIT RISE AND UNIT RUN OF STAIRS.

STEP 2 MARK TANGENT POINT WHERE PITCH BLOCK TOUCHES RAIL.

STEP 3 ROTATE PITCH BLOCK AND MARK ANGLE ALONG RAKE EDGE OF PITCH BLOCK.

STEP 4 TRIM CONNECTING EASEMENT USING PITCH BLOCK AS A GUIDE. BE SURE TO SECURE RAIL BEFORE CUTTING.

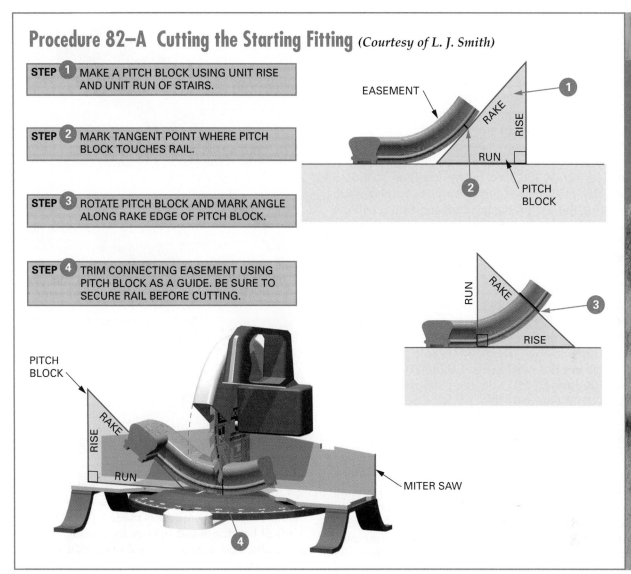

Laying Out the Starting Fitting

Make a pitch block by cutting a piece of wood in the shape of a right triangle whose sides are equal in length to the rise and tread run of the stairs. Hold the cap of the starting fitting and the run side of the pitch block on a flat surface. The rake edge of the pitch block should be against the bottom side of the fitting. Mark the fitting at the tangent point, where its curved surface touches the pitch block. A straight line, tangent to a curved line, touches the curved line at one point only.

Then turn the pitch block so its rise side is on the flat surface. Mark the fitting along the rake edge of the pitch block. Place the fitting in a power miter box, supported by the pitch block. Cut it to the layout line **(Procedure 82–A).**

> ⚠ **CAUTION CAUTION CAUTION CAUTION**
>
> When cutting handrail fittings supported by a pitch block in a power miter box, clamp the pitch block and fitting securely to make sure they will not move when cut. If either one moves during cutting, a serious injury could result. Even if no injury occurs, the fitting would most likely be ruined. ∎

Joining the Starting Fitting to the Handrail

The starting fitting is joined to a straight section of handrail by means of a special **handrail bolt.** The bolt has threads on one end designed for fastening into wood. The other end is threaded for a nut. Holes must be drilled in the end of both the fitting and the handrail in a manner that ensures their alignment when joined.

To mark the hole locations, make a template by cutting about a ⅛-inch piece of handrail. Drill a ⅟₁₆-inch hole centered on the template width and ⅝₁₆ inch from the bottom side. Mark one side rail and the other side fitting. This will ensure that the template is facing in the right direction when making the layout. If the hole in the template is drilled slightly off center and the template turned when making the layout, the handrail and fitting will not be in alignment when joined. Use the template to mark the location of the rail bolt on the end of the fitting and the handrail.

Drill all holes to the depth and diameter shown in **Figure 82–18.** Double nut the rail bolt and turn it into the fitting. Remove the nuts, place the handrail on the bolt. Install a washer and nut, and tighten to join the sections. Clamp the assembly to the tread nosings. The newel center points on the fitting and the starting tread should be in a plumb line **(Figure 82–19).** Note the measurement in Detail "C" of Figure 82–19.

Installing the Landing Fitting

The second flight of stairs turns at the landing, so a right hand, two-riser gooseneck fitting is used. It is laid out and joined to the bottom end of the handrail of the second flight in a similar manner as the starting fitting **(Procedure 82–B).** The assembled fitting and rail are then clamped to the nosings of the treads in the second flight. Position the rail so that the *newel cap* of the fitting is in a plumb line with the baluster centerlines of both flights **(Figure 82–20).**

Joining First- and Second-Flight Handrails

In preparation for laying out the **easement** used to join the first- and second-flight handrails, tack a piece of plywood about 5 inches wide to the bottom side of the gooseneck fitting and the handrail of the first flight. These pieces are used to rest the connecting easement against when laying out the joint.

Clamp the handrails back on the treads. Make sure the newel cap centers are plumb with starting and landing newel post centers. Rest the connecting easement on the plywood blocks. Level its upper end. Mark the gooseneck in line with the upper end. Mark the lower end of the easement at a point tangent with the block under the handrail **(Procedure 82–C).** Note the measurement in Detail "D" of Procedure 82–C.

Make a square cut at the mark laid out on the lower end of the gooseneck. Join the gooseneck and

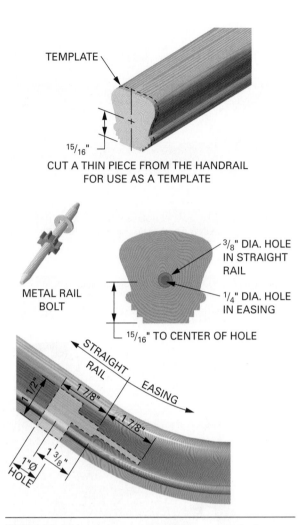

CUT A THIN PIECE FROM THE HANDRAIL FOR USE AS A TEMPLATE

TEMPLATE

15/16"

METAL RAIL BOLT

3/8" DIA. HOLE IN STRAIGHT RAIL

1/4" DIA. HOLE IN EASING

15/16" TO CENTER OF HOLE

STRAIGHT RAIL   EASING

1 1/2"

1 7/8"

1 7/8"

1 3/8"

1" Ø HOLE

FIGURE 82–18 **A template is made to mark the ends of handrails and fittings for joining with rail bolts. The ends are drilled to specific depths and diameters.**

easement with a rail bolt. Lay out the lower end of the easement with the pitch block. Cut it using the pitch block for support **(Procedure 82–D).**

Place the assembled landing fitting and handrail back on the stair nosings. Mark the handrail of the lower flight where it meets the end of the fitting. Cut the handrail square at the mark. Join it to the landing fitting with a rail bolt. Clamp the entire handrail assembly to the tread nosings with newel cap centers plumb with newel post centerlines **(Figure 82–21).**

Installing the Balcony Gooseneck Fitting

A one-riser balcony gooseneck fitting is used when balcony rails are 36 inches high. Two-riser fittings are used for rails 42 inches high. In this case, a one-riser fitting is used at the balcony. Laying out and fitting a two-riser fitting at the landing has been previously described.

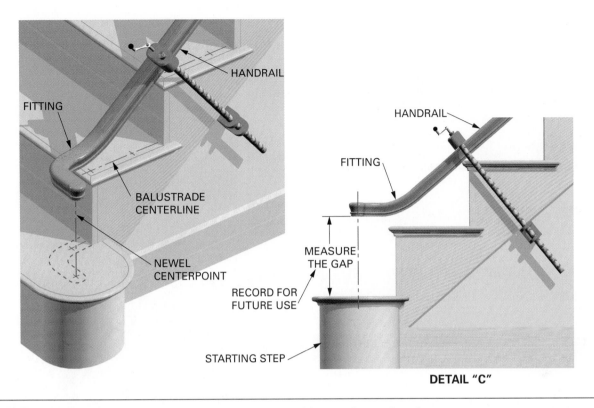

FIGURE 82–19   The starting fitting and handrail assembly are clamped to the tread nosings of the first flight of stairs. (*Courtesy of L. J. Smith*)

Procedure 82–B Cutting the Landing Fitting to the Second-Flight Handrail

(*Courtesy of L. J. Smith*)

STEP ① USING PITCH BOARD, MARK TANGENT POINT.

STEP ② ROTATE PITCH BOARD TO MARK ANGLE ALONG RAKE EDGE.

STEP ③ TRIM CONNECTING EASEMENT. MAKE SURE HANDRAIL IS SECURE BEFORE CUTTING.

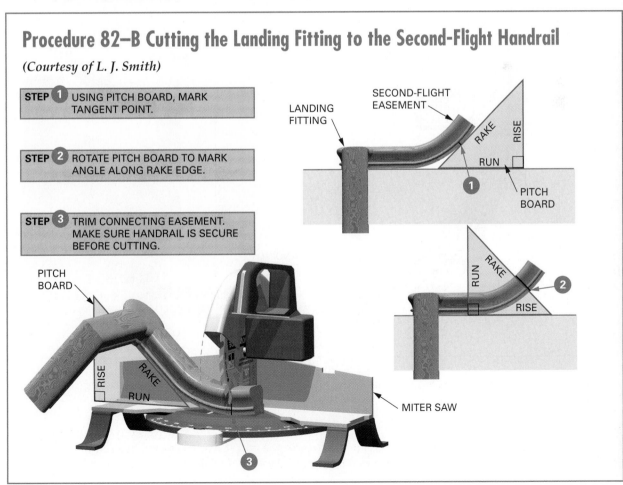

STEP BY STEP PROCEDURES

Procedure 82–C Marking the Landing Fitting *(Courtesy of L. J. Smith)*

**STEP ① ** PLACE EASEMENT BLOCKS UNDER RAILING AND AGAINST STRINGER AND ON TOP OF TREAD NOSINGS.

**STEP ② ** ADJUST EASEMENT SO TOP END IS LEVEL.

**STEP ③ ** MARK THE LANDING FITTING AT THE TOP OF THE EASEMENT.

**STEP ④ ** MARK THE LOWER TANGENT POINT OF EASEMENT AND THE BLOCK.

**STEP ⑤ ** MEASURE AND RECORD THE HEIGHT OF THE LANDING FITTING.

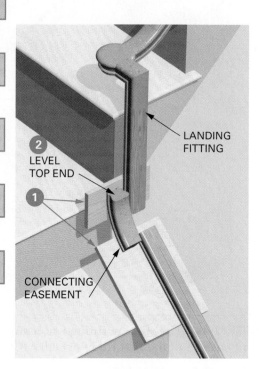

2 LEVEL TOP END

1

LANDING FITTING

CONNECTING EASEMENT

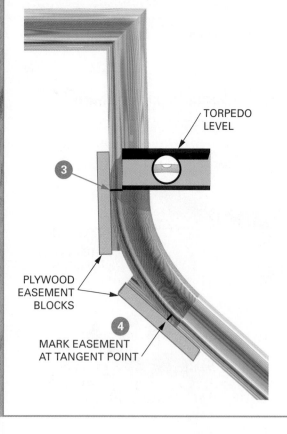

TORPEDO LEVEL

3

PLYWOOD EASEMENT BLOCKS

4

MARK EASEMENT AT TANGENT POINT

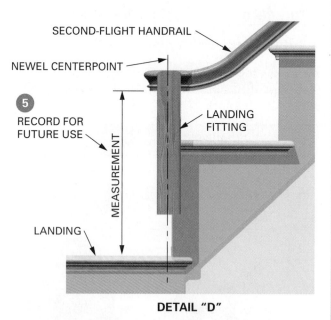

SECOND-FLIGHT HANDRAIL

NEWEL CENTERPOINT

5 RECORD FOR FUTURE USE

LANDING FITTING

MEASUREMENT

LANDING

DETAIL "D"

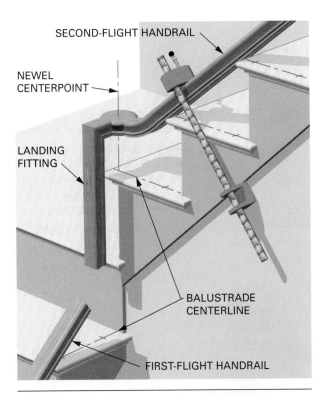

FIGURE 82–20 The second-flight handrail is clamped to the tread nosings after the landing fitting is joined to it. *(Courtesy of L. J. Smith)*

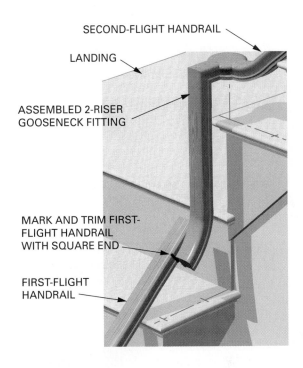

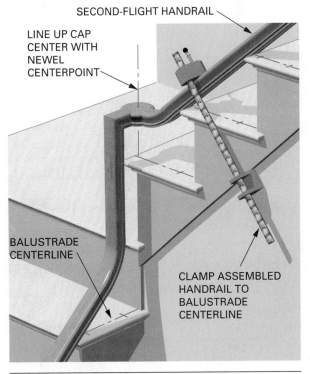

FIGURE 82–21 Lower and upper handrail assemblies are joined. They are then clamped to the tread nosings. *(Courtesy of L. J. Smith)*

Hold the fitting so the center of its cap is directly above the balcony newel post centerline. Hold it against the handrail of the upper flight. Mark it and the handrail at the point of tangent **(Figure 82–22).** Lay out and make the cut on the gooseneck with the use of a pitch block **(Procedure 82–E).** Cut the handrail square at the mark. Join the gooseneck fitting and the handrail with a rail bolt.

Clamp the entire rail assembly back on the nosing of the treads in line with the balustrade centerlines and the three newel post centers. Use a framing square to transfer the baluster centers from the treads to the side of the handrails. Remove the handrail assembly out of the way of newel post installation.

INSTALLING NEWEL POSTS

In this staircase, the starting newel is installed on a starting step. The height of the rake handrail is calculated, from the height of the starting newel, to make sure the handrail will conform to the height required by the building code.

From the height of the starting newel to be used, subtract the previously recorded distance between the starting fitting and the starting tread as shown in Detail "C" of Figure 82–19. Then, add the vertical thickness of the rake handrail shown in Figure 82–4, Detail "A". The result is the rake handrail height **(Procedure 82–F).** If the height does not conform to the building code, the starting newel post height must be changed.

Procedure 82–D Connecting the Easement and Landing Fitting

(Courtesy of L. J. Smith)

STEP 1 SQUARE CUT THE LANDING FITTING AND ATTACH THE EASEMENT FITTING.

STEP 2 MARK THE ANGLE ALONG THE RAKE AT THE TANGENT POINT.

STEP 3 TRIM CONNECTING EASEMENT. MAKE SURE THE RAIL IS SECURE BEFORE CUTTING.

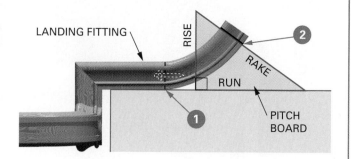

LANDING FITTING

RISE

RAKE

RUN

PITCH BOARD

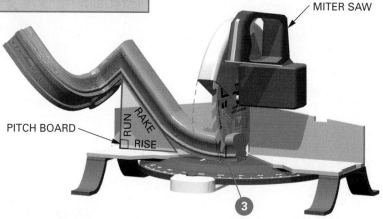

MITER SAW

PITCH BOARD

RUN

RAKE

RISE

Procedure 82–E Cutting the Balcony Gooseneck Fitting *(Courtesy of L. J. Smith)*

STEP 1 USING A PITCH BOARD, MARK THE ANGLE AT THE TANGENT POINT.

STEP 2 TRIM CONNECTING EASEMENT. BE SURE TO SECURE HANDRAIL BEFORE CUTTING.

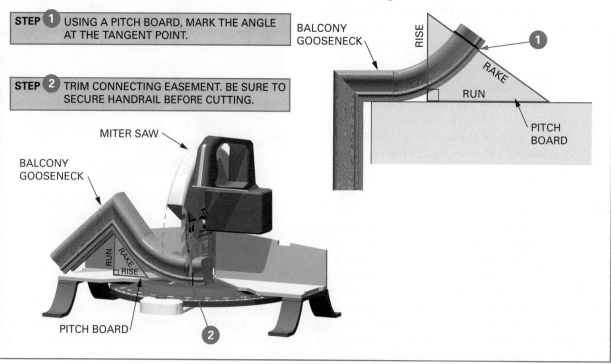

BALCONY GOOSENECK

RISE

RAKE

RUN

PITCH BOARD

MITER SAW

BALCONY GOOSENECK

RUN

RAKE

RISE

PITCH BOARD

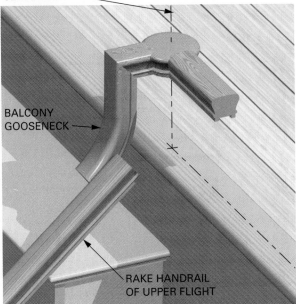

SHIFT GOOSENECK UP OR DOWN
TO LINE UP WITH NEWEL
CENTERLINE

BALCONY
GOOSENECK

RAKE HANDRAIL
OF UPPER FLIGHT

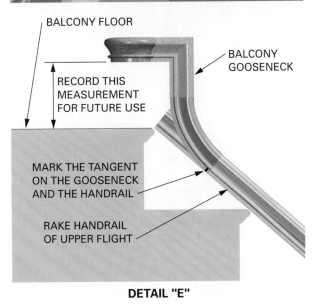

BALCONY FLOOR

BALCONY
GOOSENECK

RECORD THIS
MEASUREMENT
FOR FUTURE USE

MARK THE TANGENT
ON THE GOOSENECK
AND THE HANDRAIL

RAKE HANDRAIL
OF UPPER FLIGHT

DETAIL "E"

FIGURE 82–22 **Method used to mark the balcony gooseneck fitting and handrail of the upper flight.** *(Courtesy of L. J. Smith)*

Installing the Starting Newel Post

Before installing the starting step, measure the diameter of the dowel at the bottom of the starting newel post. At the centerpoint of the newel post, bore a hole for the dowel through the tread and floor. Install the post. Wedge it under the floor. Wedges are driven in a through mortise cut in the doweled end of the post.

An alternate method is used when there is no access under the floor. Bore holes only through the tread and upper riser block of the starting step. Cut

the dowel to fit against the lower riser block so the newel post rests snugly on the tread. Fasten the end of the dowel with a lag screw through the lower riser block **(Figure 82–23).** Fasten the assembled starting newel and starting step in position.

Installing the Landing Newel Post

The height of the landing newel above the landing is found by subtracting the previously recorded distance between the starting fitting and the starting tread (Figure 82–19, Detail "C") from the height of the starting newel. Then add the distance between the landing fitting and the landing as previously recorded and shown in Procedure 82–C, Detail "D". To this length, add the distance that the landing newel extends below the landing.

Notch the landing newel to fit over the landing and the first step of the upper flight. Fasten the post in position with lag bolts in counterbored holes **(Procedure 82–G).**

Installing the Balcony Newel Post

The height of the balcony handrail must be calculated before the height of the balcony newel can be determined. The height of the balcony handrail is found by subtracting the previously recorded distance between the starting fitting and the starting tread (Figure 82–19, Detail "C") from the height of the starting newel. Then add the previously recorded distance between the balcony gooseneck fitting and the landing (Figure 82–22, Detail "E"). Then, add the thickness of the balcony handrail. The balcony handrail height must conform to the building code. If not, substitute a two-riser gooseneck fitting instead of a one-riser fitting.

The height of the balcony newel above the balcony floor is found by subtracting the handrail thickness from the handrail height. To this length, add the distance the post extends below the floor. Notch and install the post over the balcony riser in line with balustrade centerlines **(Figure 82–24).**

Installing Balusters

Bore holes for balusters in the tread and bottom edge of the handrail. No holes are bored in the handrail if square-top balusters are used. Install the handrail on the posts. Cut the balusters to length. Install them in the manner described previously for post-to-post balustrades.

Installing the Balcony Balustrade

Cut a half newel to the same height as the balcony newel extends above the floor. Install it against the wall on the balustrade centerline. Cut an opening cap so it fits on top of the half newel. Join the cap to

STEP BY STEP PROCEDURES

Procedure 82–F Determining the Height of the Starting Newel Post

(Courtesy of L. J. Smith)

STEP 1 FROM DIMENSION T, SUBTRACT PREVIOUSLY RECORDED DIMENSION C. (SEE FIGURE 82–19, DETAIL "C".)

STEP 2 ADD VERTICAL THICKNESS OF HANDRAIL. (SEE FIGURE 82–9, DETAIL "A".)

STEP 3 RESULT EQUALS HEIGHT OF HANDRAIL FOR STARTING NEWEL SELECTED.

PIN

1 C

T

1

RAKE HANDRAIL

2 A

TREAD NOSING

DOWEL

3

RAKE HANDRAIL HEIGHT

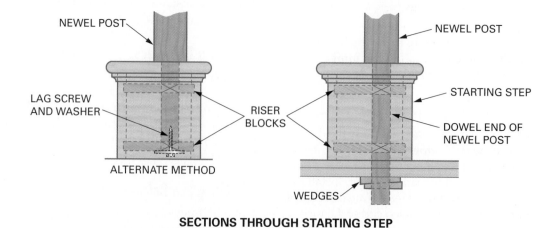

NEWEL POST

LAG SCREW AND WASHER

RISER BLOCKS

ALTERNATE METHOD

NEWEL POST

STARTING STEP

DOWEL END OF NEWEL POST

WEDGES

SECTIONS THROUGH STARTING STEP

FIGURE 82–23   Methods of installing the starting newel on a starting step.

STEP BY STEP PROCEDURES

Procedure 82–G Determining the Height of the Landing Newel

(Courtesy of L. J. Smith)

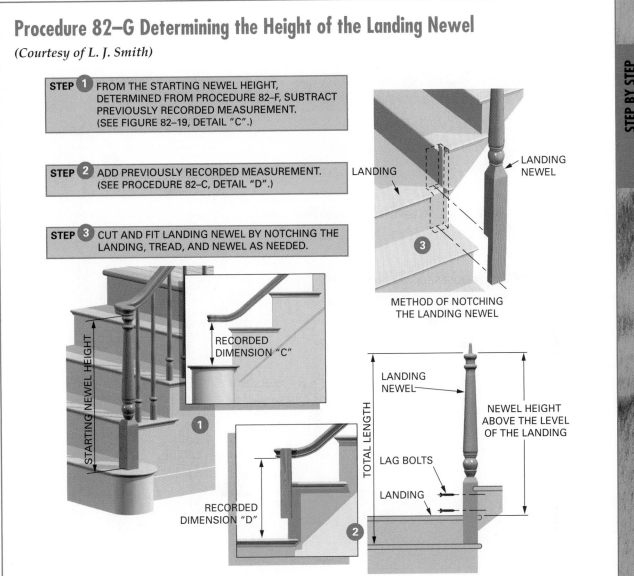

STEP 1 FROM THE STARTING NEWEL HEIGHT, DETERMINED FROM PROCEDURE 82–F, SUBTRACT PREVIOUSLY RECORDED MEASUREMENT. (SEE FIGURE 82–19, DETAIL "C".)

STEP 2 ADD PREVIOUSLY RECORDED MEASUREMENT. (SEE PROCEDURE 82–C, DETAIL "D".)

STEP 3 CUT AND FIT LANDING NEWEL BY NOTCHING THE LANDING, TREAD, AND NEWEL AS NEEDED.

LANDING

LANDING NEWEL

METHOD OF NOTCHING THE LANDING NEWEL

STARTING NEWEL HEIGHT

RECORDED DIMENSION "C"

RECORDED DIMENSION "D"

TOTAL LENGTH

LANDING NEWEL

NEWEL HEIGHT ABOVE THE LEVEL OF THE LANDING

LAG BOLTS

LANDING

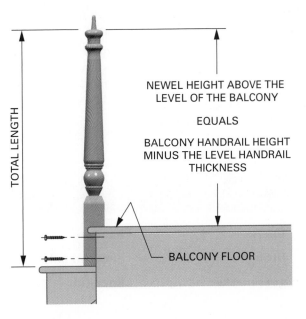

FIGURE 82–24 The height of the balcony newel is determined. The post is then fastened in place. *(Courtesy of L. J. Smith)*

TOTAL LENGTH

NEWEL HEIGHT ABOVE THE LEVEL OF THE BALCONY

EQUALS

BALCONY HANDRAIL HEIGHT MINUS THE LEVEL HANDRAIL THICKNESS

BALCONY FLOOR

Procedure 82–H Installing a Half Newel and the Balcony Handrail

(Courtesy of L. J. Smith)

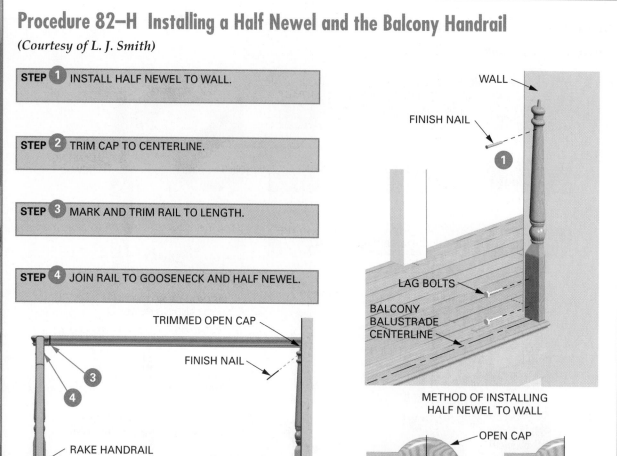

STEP 1 INSTALL HALF NEWEL TO WALL.

STEP 2 TRIM CAP TO CENTERLINE.

STEP 3 MARK AND TRIM RAIL TO LENGTH.

STEP 4 JOIN RAIL TO GOOSENECK AND HALF NEWEL.

WALL

FINISH NAIL

1

LAG BOLTS

BALCONY BALUSTRADE CENTERLINE

METHOD OF INSTALLING HALF NEWEL TO WALL

TRIMMED OPEN CAP

FINISH NAIL

3

4

RAKE HANDRAIL

FASCIA

OPEN CAP

PLAN

2

the end of the balcony handrail with a rail bolt. Place the cap on the half newel. Mark the length of the handrail at the balcony newel. Cut the handrail and join it, on one end, to the balcony newel post and, on the other end to the half newel **(Procedure 82–H).**

A rosette may be used against the wall instead of the half newel and opening cap. The procedure for installing a rosette has been previously described (see Figure 82–15).

Installing Balcony Balusters

Balcony balusters are laid out and installed in the same manner as described previously for post-to-post balustrades. Balconies with a span of 10 feet or more should have intermediate balcony newels installed every 5 or 6 feet.

Key Terms

balustrade	fillets	handrail	newel posts
closed	fittings	handrail bolt	open
easement	goosenecks	kneewall	over-the-post

pitch board	risers	stair body	turnouts
post-to-post	shoe rail	treads	volutes
preacher			

Review Questions

Select the most appropriate answer.

1. The rounded outside edge of a tread that extends beyond the riser is called a
   a. housing.
   b. turnout.
   c. coving.
   d. nosing.

2. Finish boards between the stairway and the wall are called
   a. returns.
   b. balusters.
   c. stringers.
   d. casings.

3. Treads are rabbeted on their back edge to fit into
   a. risers.
   b. housed stringers.
   c. return nosings.
   d. newel posts.

4. An open stringer is
   a. housed to receive risers.
   b. mitered to receive risers.
   c. housed to receive treads.
   d. mitered to receive treads.

5. A volute is part of a
   a. tread.
   b. baluster.
   c. newel post.
   d. handrail.

6. The entire rail assembly on the open side of a stairway is called a
   a. baluster.
   b. balustrade.
   c. guardrail.
   d. finish stringer assembly.

7. In a framed staircase, the treads and risers are supported by
   a. stair carriages.
   b. housed stringers.
   c. each other.
   d. blocking.

8. One of the first things to do when trimming a staircase is
   a. check the rough framing for rise and run.
   b. block the staircase so no one can use it.
   c. straighten the stair carriages.
   d. install all the risers.

9. Return nosings usually project beyond the face of the stringer
   a. ¾ inch.
   b. 1⅛ inches.
   c. 1¼ inches.
   d. 1⅜ inches.

10. Newel posts are notched around the stairs so that their centerline lines up with the
    a. centerline of the stair carriage.
    b. centerline of the balustrade.
    c. outside face of the open stringer.
    d. outside face of the stair carriage.

UNIT 29

Finish Floors

Many times a layer of underlayment is installed over the subfloor as part of the finished floor. A number of materials are then used for finish floors. Each may be applied by a flooring specialist, such as carpet by carpet installers. But in some areas carpenters are asked to install various flooring materials.

Resilient flooring sheets and tiles are often used in kitchens and bathrooms. Wood flooring comes as strips of solid wood or a wood-plastic combination. Solid wood flooring is a long-time favorite because of its durability, beauty, and warmth.

SAFETY REMINDER

Installing flooring requires the installer to bend or work on knees for long periods of time. Take care to protect your body from injury due to repetitive movements and contact with hard surfaces.

OBJECTIVES

After completing this unit, the student should be able to:

- describe the kinds, sizes, and grades of hardwood finish flooring.
- apply strip, plank, and parquet finish flooring.
- estimate quantities of wood finish flooring required for various installations.
- apply underlayment and resilient tile flooring.
- estimate required amounts of underlayment and resilient tile for various installations.

83 Description of Wood Finish Floors

Most hardwood finish flooring is made from white or red oak. Beech, birch, hard maple, and pecan finish flooring are also manufactured. For less expensive finish floors, some softwoods such as Douglas fir, hemlock, and southern yellow pine are used.

KINDS OF HARDWOOD FLOORING

The four basic types of solid wood finish flooring are **strip, plank, parquet strip,** and **parquet block.** **Laminated strip** wood flooring is a relatively new type that is gaining in popularity. *Laminated parquet blocks* are also manufactured.

Solid Wood Strip Flooring

Solid wood **strip** flooring is probably the most widely used type. Most strips are tongue-and-grooved on edges and ends to fit precisely together.

Unfinished strip flooring is milled with square, sharp corners at the intersections of the face and edges. After the floor is laid, any unevenness in the faces of adjoining pieces is removed by sanding the surface so strips are flush with each other.

Prefinished strips are sanded, finished, and waxed at the factory. They cannot be sanded after installation. A **chamfer** is machined between the face side and edges of the flooring prior to prefinishing. When installed, these chamfered edges form small V-grooves between adjoining pieces. This obscures any unevenness in the surface.

The most popular size of hardwood strip flooring is ¾ inch thick with a face width of 2¼ inches. The face width is the width of the exposed surface between adjoining strips. It does not include the tongue. Other thicknesses and widths are manufactured **(Figure 83–1).**

Laminated Wood Strip Flooring

Laminated strip flooring is a five-ply prefinished wood assembly. Each board is ⁹⁄₁₆ inch thick, 7½ inches wide, and 7 feet, 11½ inches long. The board consists of a bottom veneer, a three-ply cross-laminated core, and a face layer. The face layer consists of three rows of hardwood strips joined snugly edge to edge **(Figure 83–2).**

The uniqueness of this flooring is in the exact milling of edge and end tongue and grooves. This precision allows the boards to be joined with no noticeable unevenness of the prefinished surface. This eliminates the need to chamfer the edges. Without the chamfers, a smooth continuous surface without V-grooves results when the floor is laid **(Figure 83–3).**

Plank Flooring

Solid wood **plank** flooring is similar to strip flooring. However, it comes in various mixed combinations ranging from 3 to 8 inches in width. For instance, plank flooring may be laid with alternating widths of 3 and 4 inches, 3, 4, and 6 inches, 3, 5, and 7 inches, or any random-width combination.

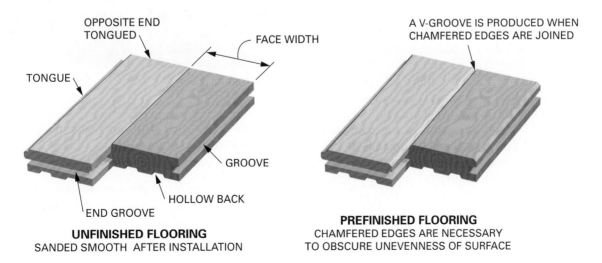

FIGURE 83–1   Hardwood strip flooring is edge and end matched. The edges of prefinished flooring are chamfered.

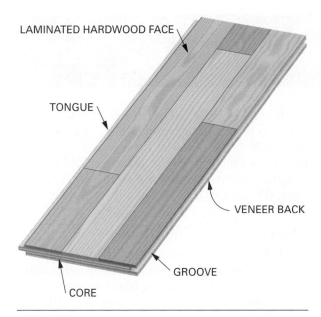

FIGURE 83–2 Cross-section of a board of laminated strip flooring.

FIGURE 83–3 Completed installation of a laminated strip floor.

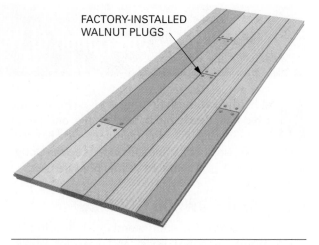

FIGURE 83–4 Plank flooring usually is applied in rows of alternating widths. Plugs of contrasting color may be added to simulate screw fastening.

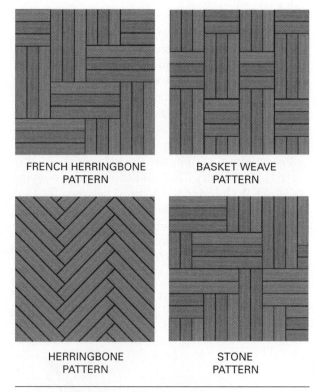

FRENCH HERRINGBONE PATTERN

BASKET WEAVE PATTERN

HERRINGBONE PATTERN

STONE PATTERN

FIGURE 83–5 Parquet strips are made in lengths that are multiples of its width.

Like strips, planks are available unfinished or prefinished. The edges of some prefinished planks have deeper chamfers to accentuate the plank widths. The surface of some prefinished plank flooring may have plugs of contrasting color already installed to simulate screw fastening. One or more plugs, depending on the width of the plank, are used across the face at each end **(Figure 83–4).**

Unfinished plank flooring comes with either square or chamfered edges and with or without plugs. The planks may be bored for plugs on the job, if desired.

Parquet Strips

Parquet strip flooring has short strips that are laid to form various mosaic designs. The original parquet floors were laid by using short strips. Some, at the

present time, are laid in the same manner. This type is manufactured in precise, short lengths, which are multiples of its width. For instance, 2¼-inch parquet strips come in lengths of 9, 11¼, 13½, and 15¾ inches. Each piece is tongue-and-grooved on the edges and ends. Herringbone, basket weave, and other interesting patterns can be made using parquet strips **(Figure 83–5).**

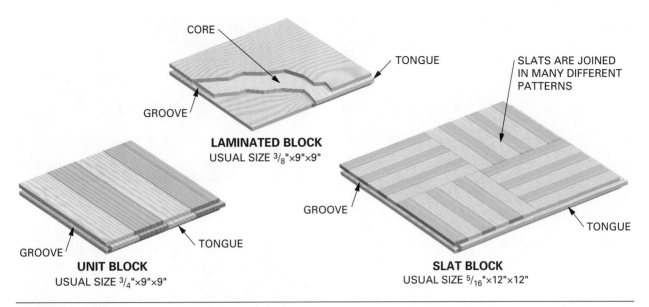

CORE

TONGUE

SLATS ARE JOINED IN MANY DIFFERENT PATTERNS

GROOVE

LAMINATED BLOCK
USUAL SIZE ³/₈"×9"×9"

GROOVE

TONGUE

GROOVE

TONGUE

UNIT BLOCK
USUAL SIZE ³/₄"×9"×9"

SLAT BLOCK
USUAL SIZE ⁵/₁₆"×12"×12"

FIGURE 83–6 Three basic types of parquet block.

Parquet Blocks

Parquet block flooring consists of square or rectangular blocks, sometimes installed in combination with strips, to form mosaic designs. The three basic types are the *unit, laminated,* and *slat* block **(Figure 83–6).**

Unit Blocks. The highest-quality parquet block is made with ¾-inch-thick, tongue-and-groove solid hardwood, usually oak. The widely used 9 × 9 **unit block** is made with six strips, 1½ inches wide, or with four strips 2¼ inches wide. Unit blocks are laid with the direction of the strips at right angles to adjacent blocks **(Figure 83–7).** Unit blocks are made in other sizes and used in combination with parquet strips. Several patterns have gained popularity.

Monticello is the name of a parquet originally designed by Thomas Jefferson, the third president of the United States. The pattern consists of a 6 × 6 center unit block surrounded by 2¼-inch-wide pointed *pickets.* Each center unit block is made of four 1½-inch-wide strips **(Figure 83–8).** Each block comes with three pickets joined to it at the factory.

Another popular parquet, called the *Marie Antoinette,* is copied from part of the Versailles Palace in France. Square center unit blocks are enclosed by strips applied in a basket weave design **(Figure 83–9).**

Rectangular parquet blocks are often used in a *herringbone* pattern. One commonly used block is 4½ × 9 and made of three strips each 1½ inches wide and 9 inches long **(Figure 83–10).**

FIGURE 83–7 Unit blocks are widely used in an alternating pattern.

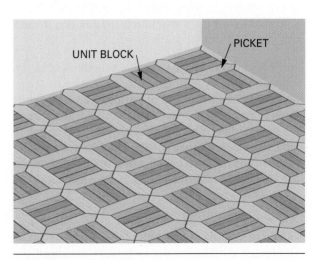

UNIT BLOCK

PICKET

FIGURE 83–8 The Monticello pattern is a famous parquet that uses square center blocks and picket-shaped strips.

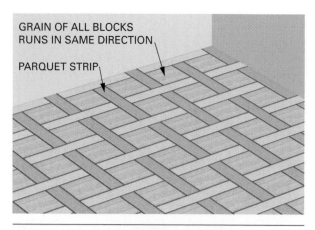

FIGURE 83–9   A popular parquet called the Marie Antoinette uses center blocks and parquet strips.

FIGURE 83–10   Rectangular parquet blocks may be used to make herringbone and similar patterns.

Laminated Blocks. **Laminated blocks** are generally made of three-ply laminated oak in a ⅜-inch thickness. Most blocks come in 8 × 8 or 9 × 9 sizes. A 2 × 12 laminated strip is manufactured for use in herringbone and similar patterns.

Slat Blocks. **Slat blocks** are also called *finger blocks*. They are made by joining many short, narrow strips together in small squares of various patterns. Some strips may be as narrow as ⅝ inch and as short as 2 inches or less. Several squares are assembled to make the block **(Figure 83–11)**. The squares are held together with a mesh backing or with a paper on the face side that is removed after the block is laid.

Presanded Flooring

Some strip, plank, and parquet finish flooring can be obtained presanded at the factory, but without the finish. The presanded surface eliminates the necessity for all but a touch-up sanding after installation to remove surface marks before the finish is applied.

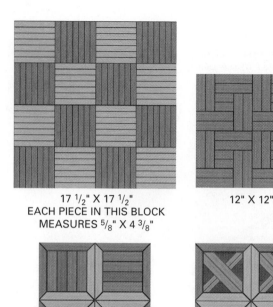

FIGURE 83–11   Slat blocks are made of short, narrow strips joined in various patterns. *(Courtesy of Bruce Hardwood Floors)*

GRADES OF HARDWOOD FLOORING

Uniform grading rules have been established for strip and plank solid hardwood flooring by the National Oak Flooring Manufacturers Association. The association's trademark on flooring assures consumers that the flooring is manufactured and graded in compliance with established quality standards. Other types of wood finish flooring, such as parquet and laminated flooring, have no official grade rules.

Unfinished Flooring

Oak flooring is available quarter-sawed and plainsawed. The grades for unfinished oak flooring, in declining order, are clear, select, no. 1 common, no. 2 common, and 1¼-foot shorts. Quarter-sawed flooring is available in clear and select grades only.

Birch, beech, and hard maple flooring are graded in declining order as first grade, second grade, third grade, and special grade. Grades of pecan flooring are first grade, first grade red, first grade white, second grade, second grade red, and third grade. Red grades contain all heartwood. White grades are all bright sapwood.

In addition to appearance, grades are based on length. For instance, bundles of 1¼-foot shorts contain pieces from 9 to 18 inches long. The average length of clear bundles is 3¾ feet. The flooring comes in bundles

Unfinished Oak Flooring (Red & White Separated)	Unfinished Hard Maple (Beech & Birch)	Unfinished Pecan Flooring	Prefinished Oak Flooring (Red & White separated, graded after finishing)
Clear Plain or CLear Quartered* Best appearance. Best grade, most uniform color, limited small character marks. Bundles 1¼ ft. and up. Average length 3¾ ft.	**First Grade** Best appearance. Natural color variation, limited character marks, unlimited sap. Bundles 1¼ ft. and up. 1¼ ft., 2 ft., & 3 ft. bundles up to 45% footage. 2 ft. bundles up to 25% footage. 1¼ ft. bundles up to 5% footage.	**First Grade** Excellent appearance. Natural color variation, limited character marks, unlimited sap. Bundles 2 ft. & up. 2 & 3 ft. bundles up to 25% footage. **First Grade Red** (Special Order) **First Grade White** (Special Order)	**Prime Grade (Special Order)** Excellent appearance. Natural color variation, limited character marks, unlimited sap. Bundles 1¼ ft. & up. Average length 3½ ft.
Select Plain or Select Quartered Excellent appearance. Limited character marks, unlimited sound sap. Bundles 1¼ ft. and up. Average length 3¼ ft. **Select & Better** A combination of Clear and Select grades.	**Second Grade** Variegated appearance. Varying sound wood characteristics of species. Bundles 1¼ ft. and up. 1¼ ft. 2 ft., & 3 ft. bundles up to 55% footage. 2 ft. bundles up to 27% footage. 1¼ ft. bundles up to 10% footage.		
No. 1 Common Variegated appearance. Light and dark colors; knots, flags, worm holes and other character marks allowed to provide a variegated appearance after imperfections are filled and finished. Bundles 1¼ ft. and up. Average length 2¾ ft.	**Second & Better Grade** A combination of First & Second Grades. Lengths equivalent to Second Grade. **Third Grade** Rustic appearance. All wood characteristics of species. Serviceable economical floor after filling. Bundles 1¼ ft. and up. 1¼ ft. to 3 ft. bundles as produced up to 75% footage. 1¼ ft. bundles up to 45% footage.	**Second Grade** Variegated appearance. Varying sound wood characteristics of species. Bundles 1¼ ft. and up. 1¼ ft. to 3 ft. bundles as produced up to 40% footage. **Second & Better Grade** A combination of FIRST and SECOND GRADES.	**Standard Grade** Variegated appearance. Varying sound wood characteristics of species. A sound floor. Bundles 1¼ ft. & up. Average length 2¾ ft. **Standard & Better Grade** Combination of STANDARD and PRIME. Bundles 1¼ ft. & up. Average length 3 ft.
No. 2 Common Rustic appearance. All wood characteristics of species. A serviceable economical floor after knot holes, worm holes, checks and other imperfections are filled and finished. Bundles 1¼ ft. and up. Average length 2¼ ft. Red and White may be mixed.	**Third & Better Grade** A combination of First, Second & Third Grades. Bundles 1¼ ft. and up. 1¼ ft. to 3 ft. bundles as produced up to 60% of footage.	**Third Grade** Rustic appearance. All wood characteristics of species. A serviceable, economical floor after filling. Bundles 1¼ ft. and up. 1¼ ft. to 3 ft. bundles as produced up to 60% footage. **Third & Better Grade** A combination of FIRST, SECOND and THIRD GRADES.	**Tavern Grade** Rustic appearance. All wood characteristics of species. A serviceable, economical floor. Bundles 1¼ ft. & up. Average length 2¼ ft. **Tavern & Better Grade** (Special Order) Combination of PRIME, STANDARD and TAVERN. All wood characteristics of species. Bundles 1¼ ft. & up. Average length 3 ft.
1¼ Shorts Pieces 9 to 18 inches. Bundles average nominal 1¼ ft. **No. 1 Common & Better Shorts** A combination grade of CLEAR, SELECT, & NO. 1 COMMON **No. 2 Common Shorts** Same as No. 2 Common.			**Prefinished Beech & Pecan Flooring Tavern & Better Grade** (Special Order) Combination of PRIME, STANDARD and TAVERN. All wood characteristics of species. Bundles 1¼ ft. & up. Average length 3 ft.

Vertical labels between columns: SELECT AND BETTER | SECOND AND BETTER | THIRD AND BETTER | STANDARD AND BETTER | TAVERN AND BETTER

FIGURE 83–12 Guide to hardwood flooring grades. *(Courtesy of National Oak Flooring Manufacturers Association)*

in lengths of 1¼ feet and up. Pieces in each bundle are not of equal lengths. A bundle may include pieces from 6 inches under to 6 inches over the nominal length of the bundle. No pieces shorter than 9 inches are allowed.

Prefinished Flooring

Grades of prefinished flooring are determined after it has been sanded and finished. In declining order, they are prime, standard and better, standard, and tavern. Prefinished beech and pecan are furnished only in a combination grade called tavern or better. A guide to hardwood flooring grades is shown in **Figure 83–12.**

ESTIMATING HARDWOOD FLOORING

To estimate the amount of hardwood flooring material needed, first determine the area to be covered.

Add to this a percentage of the area depending on the width of the flooring to be used. The percentages include an additional 5 percent for end matching and normal waste:

    55 percent for flooring 1½ inches wide
    42.5 percent for flooring 2 inches wide
    38.33 percent for flooring 2¼ inches wide.

For example, the area of a room 16 feet by 24 feet is 384 square feet. If 2¼-inch flooring is to be used, multiply 384 by 0.3833 to get 147.18. Round this off to 147 square feet. Add 147 to 384 to get 531, which is the number of board feet of flooring required.

84 Laying Wood Finish Floor

Lumber used in the manufacture of hardwood flooring has been air dried, kiln dried, cooled, and then accurately machined to exacting standards. It is a fine product that should receive proper care during handling and installation.

HANDLING AND STORAGE OF WOOD FINISH FLOOR

Maintain moisture content of the flooring by observing recommended procedures. Flooring should not be delivered to the jobsite until the building has been closed in. Outside windows and doors should be in place. Cement work, plastering, and other materials must be thoroughly dry. In warm seasons, the building should be well ventilated. During cold months, the building should be heated, not exceeding 72°F, for at least five days before delivery and until flooring is installed and finished.

Do not unload flooring in the rain. Stack the bundles in small lots in the rooms where the flooring is to be laid. Leave adequate space around the bundles for good air circulation. Let the flooring become acclimated to the atmosphere of the building for four or five days or more before installation.

Concrete Slab Preparation

Wood finish floors can be installed on an on-grade or above-grade concrete slab. Floors should not be installed on below-grade slabs. New slabs should be at least ninety days old. Flooring should not be installed when tests indicate excessive moisture in the slab.

Testing for Moisture

A test can be made by laying a smooth rubber mat on the slab. Put weight on it to prevent moisture from escaping. Allow the mat to remain in place for twenty-four hours. If moisture shows when the mat is removed, the slab is too wet. Another method of testing is by taping and sealing the edges of about a one-foot square of 6-mil clear polyethylene film to the slab. If moisture is present, it can be easily seen on the inside of the film after twenty-four hours.

If no moisture is present, prepare the slab. Grind off any high spots. Fill low spots with leveling compound. The slab must be free of grease, oil, or dust.

Applying a Moisture Barrier

A moisture barrier must be installed over all concrete slabs. This ensures a trouble-free finish floor installation. Spread a skim coat of mastic with a straight trowel over the entire slab. Allow to dry at least two to three hours. Then, cover the slab with polyethylene film. Lap the edges of the film 4 to 6 inches and extending up all walls enough to be covered by the baseboard, when installed.

When the film is in place, walk it in. Step on the film, over every square inch of the floor, to make sure it is completely adhered to the cement. Small bubbles of trapped air may appear. The film may be punctured, without concern, to let the air escape.

Applying Plywood Subfloor

A plywood subfloor may be installed over the moisture barrier on which to fasten the finish floor. Exterior grade sheathing plywood of at least ¾-inch thickness is used. The plywood is laid with staggered joints. Leave a ¾-inch space at walls and ¼- to ½-inch space between panel edges and ends. Fasten the plywood to the concrete with at least nine nails per panel **(Figure 84–1)**.

Instead of driving fasteners, the plywood may be cemented to the moisture barrier. Cut the plywood in 4 × 4 squares. Use a portable electric circular saw. Make scores ⅜ inch deep on the back of each panel to

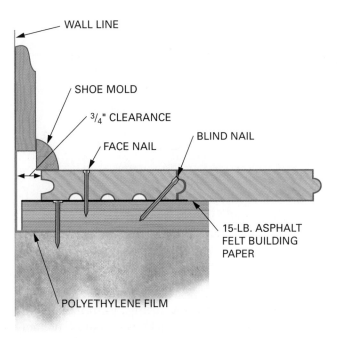

FIGURE 84–1 Installation details of a plywood subfloor over a concrete slab.

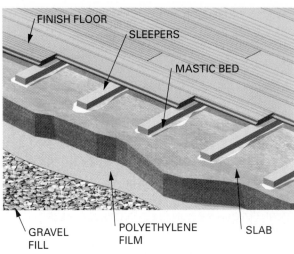

FIGURE 84–2 Sleepers are cemented to a concrete slab to provide fastening for strip or plank finish flooring.

form a 12 × 12 grid. Lay the panel in asphalt cement spread with a ¼-inch notched trowel.

Applying Sleepers

Finish flooring may also be fastened to **sleepers** installed on the slab. Sleepers are short lengths of lumber cemented to the slab. They must be pressure-treated and dried to a suitable moisture content. Usually, 2 × 4 lumber, from 18 to 48 inches long, is used.

Sleepers are laid on their side and cemented to the slab with mastic. They are staggered, with end laps of at least 4 inches, in rows 12 inches on center and at right angles to the direction of the finish floor. If the slab was not installed with a vapor retarder, a polyethylene vapor barrier is then placed over the sleepers. The edges are lapped over the rows **(Figure 84–2).** With end-matched flooring, end joints need not meet over the sleepers.

SUBFLOORS ON JOISTS

Exterior plywood or boards are recommended for use as subfloors on joists when wood finish floors are installed on them. If plywood is used, a full ½-inch thickness is required. Thicknesses of ⅝ inch or more are preferred for ¾-inch strip finish flooring. Use ¾-inch-thick subfloor for ½-inch strip flooring. The National Oak Flooring Manufacturers Association does not recommend fastening finish flooring to subfloors of nonveneered panels.

LAYING WOOD FINISH FLOOR

In new construction, the base or door casings are not usually applied yet for easier application of the finish floor. In remodeling, the base and base shoe must be removed. Use a scrap piece of finish flooring as a guide on which to lay a handsaw. Cut the ends of any door casings that are extending below the finish floor surface.

Before laying any type of wood finish flooring, nail any loose areas. Sweep the subfloor clean. Scraping may be necessary to remove all plaster, taping compound, or other materials.

Laying Strip Flooring

Strip flooring laid in the direction of the longest dimension of the room gives the best appearance. The flooring may be laid in either direction on a plywood subfloor **(Figure 84–3).**

When the subfloor is clean, cover it with building paper. Lap it 4 inches at the seams, and at right angles to the direction of the finish floor. The paper helps keep out dust, prevents squeaks in dry seasons, and retards moisture from below that could cause warping of the floor.

Snap chalk lines on the paper showing the centerline of floor joists so flooring can be nailed into them. For better holding power, fasten flooring through the subfloor and into the floor joists whenever possible. On ½-inch plywood subfloors, flooring fasteners must penetrate into the joists.

Starting Strip. The location and straight alignment of the first course is important. Place a strip of flooring

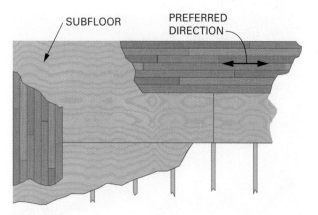

FINISH FLOOR MAY RUN IN EITHER DIRECTION, BUT
PERPENDICULAR TO JOISTS IS PREFERRED

PANEL SUBFLOOR

FIGURE 84–3   Several factors determine the
direction in which strip flooring is laid.

on each end of the room, ¾ inch from the starter wall
with the groove side toward the wall. Mark along the
edge of the flooring tongue. Snap a chalk line between
the two points. The gap between the flooring and the
wall is needed for expansion. It will eventually be covered by the base.

Hold the strip with its tongue edge to the chalk
line. Face nail it with 8d finish nails, alternating from
one edge to the other, not less than 8 inches apart.
Work from left to right with the grooved end of the
first piece toward the wall. Left is determined by having the back of the person laying the floor to the wall
where the starting strip is laid. Make sure end joints
between strips are driven up tight (**Figure 84–4**).

When necessary to cut a strip to fit to the right
wall, use a strip long enough so that the cut-off piece
is 8 inches or longer. Start the next course on the left
wall with this piece.

Blind Nailing

Flooring is **blind nailed** by driving nails at about a
45-degree angle through the flooring. Start the nail
in the corner at the top edge of the tongue. Usually
2¼-inch hardened cut or spiral screw nails are used.
Recommendations for fastening are shown in **Figure 84–5**.

For the first two or three courses of flooring, a
hammer must be used to drive the fasteners. For
floor laying, a heavier than usual hammer, from
20 to 28 ounces, is generally used for extra driving
power. Care must be taken not to let the hammer
glance off the nail. This may damage the edge of the
flooring. Care also must be taken that, on the final
blows, the hammer head does not hit the top corner

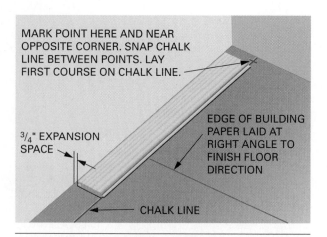

FIGURE 84–4   A chalk line is snapped on the floor
for alignment of the starting row of strip flooring.

of the flooring. To prevent this, raise the hammer
handle slightly on the final blow so that the hammer
head hits the nail head and the tongue, but not the
corner of the flooring (**Figure 84–6**).

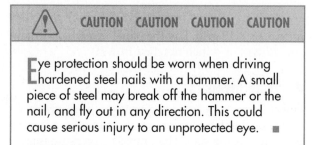

⚠ **CAUTION   CAUTION   CAUTION   CAUTION**

Eye protection should be worn when driving
hardened steel nails with a hammer. A small
piece of steel may break off the hammer or the
nail, and fly out in any direction. This could
cause serious injury to an unprotected eye. ■

After the nail is driven home, its head must be set
slightly. This allows adjoining strips to come up
tightly against each other. Floor layers use the head
of the next nail to be driven to set the nail just driven.

When fastening flooring, the floor layer holds a
hammer in one hand and a number of nails in the
other. While driving one nail, the floor layer fingers the
next nail to be driven into position to be used as a set.
When the nail being driven is most of the way, the fin-
gered nail is laid on edge with its head on the nail to be
set. With one sharp blow, the nail is set (**Figure 84–7**).
The setting nail is then the next nail to be driven. In this
manner, the floor layer maintains a smooth, continuous motion when fastening flooring.

Note: A nail set should not be used to set hardened flooring nails. If used, the tip of the nail set will
be flattened, thus rendering the nail set useless. Do
not lay the nail set flat along the tongue, on top of the
nail head, and then set the nail by hitting the side of
the nail set with a hammer. Not only is this method
slower, but it invariably breaks the nail set, possibly
causing an injury.

Strip T & G Size Flooring	Size Nail to Be Used	Blind Nail Shading Along the Length of Strips. Minimum 2 Nails Per Piece Near the Ends. (1″–3″)
¾ × 1½″, 2¼″, & 3¼″	2″ Serrated edge barbed fastener, 2¼″ or 2½″ screw or cut nail, 2″ 15 gauge staples with ½″ crown. On slab with ¾″ plywood subfloor use 1½″ barbed fastener, ½″ plywood subfloor with joists a maximum 16″ OC, fasten into each joist with additional fastening between, or 8″ apart.	In addition-10–12″ apart-8–10″ preferred
½ × 1½″ & 2″	1½″ serrated edge barbed fastener, 1½″ screw, cut steel, or wire casing nail.	10″ apart ½″ flooring must be installed over a minimum ⅝″ thick plywood subfloor.
⅜ × 1¼″ & 2″	1¼″ Serrated edge barbed fastener, 1½″ bright wire casing nail.	8″ apart
Square-Edge Flooring		
5/16 × 1½″ & 2″	1″ 15 gauge fully barbed flooring brad.	2 nails every 7″
5/16 × 1⅓″ flooring brad.	1″ 15 gauge fully barbed alternate sides of strip	1 nail every 5″ on
Plank ¾ × 3″ to 8″	2″ serrated edge barbed fastener, 2¼″ or 2½″ screw, or cut nail, use 1½″ length with ¾″ plywood subfloor on slab.	8″ apart
	FOLLOW MANUFACTURER'S INSTRUCTIONS FOR INSTALLING PLANK FLOORING	

FIGURE 84–5 **Nailing guide for strip and plank finish flooring.** *(Courtesy of National Oak Flooring Manufacturers Association)*

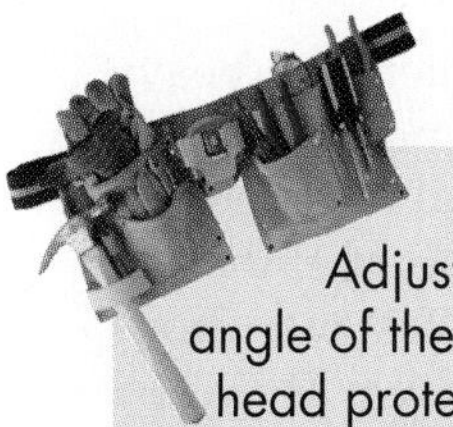

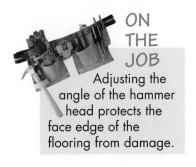

ON THE JOB

Adjusting the angle of the hammer head protects the face edge of the flooring from damage.

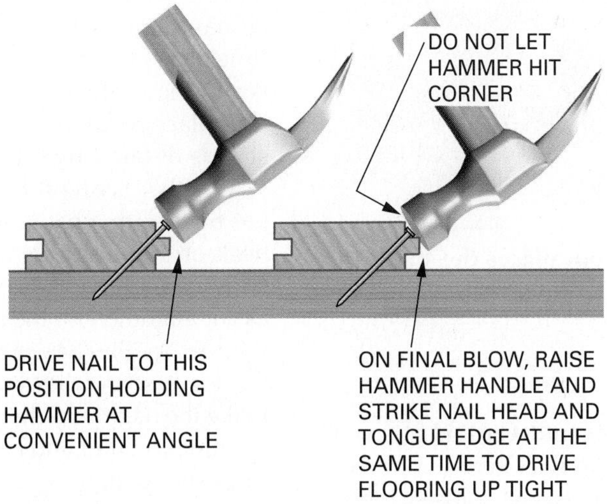
DO NOT LET HAMMER HIT CORNER
DRIVE NAIL TO THIS POSITION HOLDING HAMMER AT CONVENIENT ANGLE
ON FINAL BLOW, RAISE HAMMER HANDLE AND STRIKE NAIL HEAD AND TONGUE EDGE AT THE SAME TIME TO DRIVE FLOORING UP TIGHT

FIGURE 84–6 **Technique for driving a blind nail.**

Racking the Floor

After the second course of flooring is fastened, lay out seven or eight loose rows of flooring, end to end. Lay out in a staggered pattern. End joints should be at least 6 inches apart. Find or cut pieces to fit within ½ inch of the end wall. Distribute long and short pieces evenly for the best appearance. Avoid clusters of short strips. Laying out loose flooring in this manner is called **racking the floor.** Racking is done to save time and material **(Figure 84–8).**

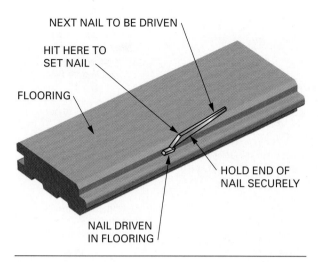

FIGURE 84–7   Method to set nails driven by hand.

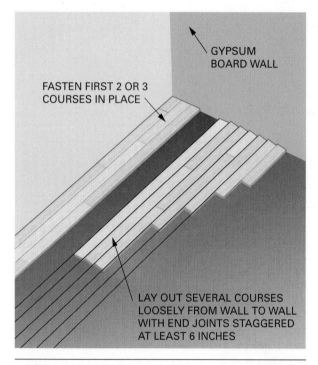

FIGURE 84–8   Racking the floor places the strips in position for efficient installation.

Using the Power Nailer

At least two courses of flooring must be laid by hand to provide clearance from the wall before a power nailer can be used. The power nailer holds strips of special barbed fasteners. The fasteners are driven and set through the tongue of the flooring at the proper angle. Although it is called a power nailer, a heavy hammer is swung by the operator against a plunger to drive the fastener **(Figure 84–9).**

The hammer is double ended. One end is rubber and the other end is steel. The flooring strip is placed

FIGURE 84–9   A power nailer is widely used to fasten strip flooring.

in position. The rubber end of the hammer is used to drive the edges and ends of the strips up tight. The steel end is used against the plunger of the power nailer to drive the fasteners. Slide the power nailer along the tongue edge. Drive fasteners about 8 to 10 inches apart or as needed to bring the strip up tight against previously laid strips.

Note: When using the power nailer, one heavy blow is used to drive the fastener. Some nailer models have a ratcheting drive shaft that allows for multiple blows to set the nail. In either case, after the nail is set, the shaft returns and another fastener drops into place ready to be driven. Make sure the wood strip is fit fairly tight before nailing.

Whether laying floor with a power nailer or driving nails with a hammer, the floor layer stands with heels on strips already fastened, and toes on the loose strip to be fastened. With weight applied to the joint, easier alignment of the tongue and groove is possible **(Figure 84–10).** The weight of the worker also prevents the loose strip from bouncing when it is driven to make the edge joint tight. Avoid using a power nailer, pneumatic nailer, and hammer-driven fasteners on the same strip of flooring. Each method of fastening places the strips together with varying degrees of tightness. This variation, compounded over multiple strips, will cause waves in the straightness of the flooring.

Ending the Flooring

Continue across the room. Rack seven or eight courses as work progresses. The last three or four courses from

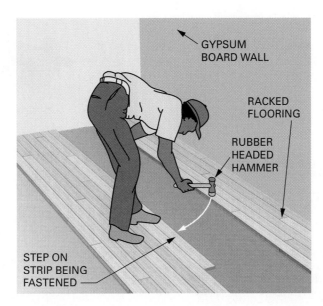

FIGURE 84–10
Technique of aligning boards together tightly before nailing.

GYPSUM BOARD WALL

RACKED FLOORING

RUBBER HEADED HAMMER

STEP ON STRIP BEING FASTENED

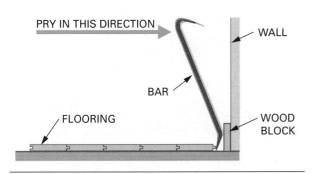

PRY IN THIS DIRECTION

WALL

BAR

FLOORING

WOOD BLOCK

FIGURE 84–11 The last two courses of strip flooring may be brought tight with a pry bar.

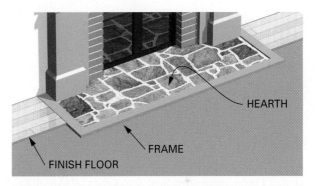

HEARTH

FRAME

FINISH FLOOR

FIGURE 84–12 Frame around floor obstructions, such as hearths, with strips that are mitered at the corners.

the opposite wall must be nailed by hand. This is because of limited room to place the power nailer and swing the hammer. The next-to-the-last row can be blind nailed if care is taken. However, the flooring must be brought up tightly by prying between the flooring and the wall. Use a bar to pry the pieces tight at each nail location **(Figure 84–11)**.

The last course is installed in a similar manner. However, it must be face nailed. It may need to be ripped to the proper width. If it appears that the installation will end with an undesirable, difficult to apply, narrow strip, lay wider strips in the last row **(Procedure 84–A)**.

Framing around Obstructions

A much more professional and finished look is given to a strip flooring installation if **hearths** and other floor obstructions are framed. Use flooring, with mitered joints at the corners, as framework around the obstructions **(Figure 84–12)**.

Changing Direction of Flooring

Sometimes it is necessary to change direction of flooring when it extends from a room into another room, hallway, or closet. To do this, face nail the extended piece to a chalk line. Change directions by joining groove edge to groove edge and inserting a *spline*, ordinarily supplied with the flooring **(Figure 84–13)**. For best appearance, avoid bunching short or long

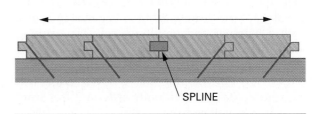

SPLINE

FIGURE 84–13 The direction of strip flooring can be changed by the use of a spline.

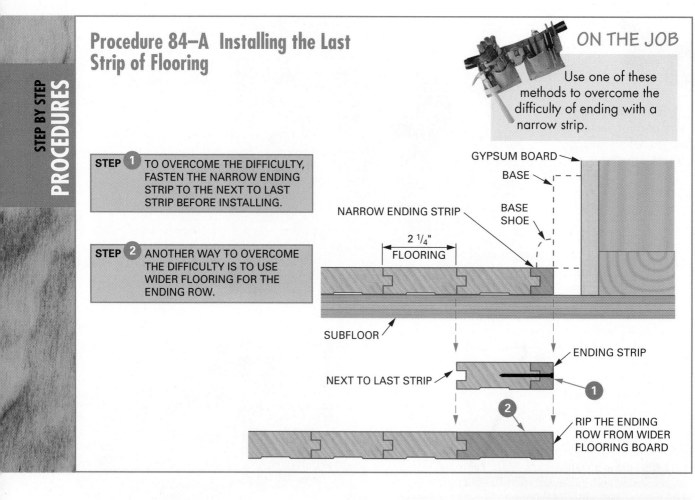

Procedure 84–A Installing the Last Strip of Flooring

ON THE JOB

Use one of these methods to overcome the difficulty of ending with a narrow strip.

STEP 1 TO OVERCOME THE DIFFICULTY, FASTEN THE NARROW ENDING STRIP TO THE NEXT TO LAST STRIP BEFORE INSTALLING.

STEP 2 ANOTHER WAY TO OVERCOME THE DIFFICULTY IS TO USE WIDER FLOORING FOR THE ENDING ROW.

GYPSUM BOARD
BASE
NARROW ENDING STRIP
BASE SHOE
2 1/4"
FLOORING
SUBFLOOR
ENDING STRIP
NEXT TO LAST STRIP
1
2
RIP THE ENDING ROW FROM WIDER FLOORING BOARD

strips. Open extra bundles, if necessary, to get the right selection of lengths.

Laying Laminated Strip Flooring

Before laying laminated strip flooring, a ⅛-inch foam underlayment, supplied or approved by the manufacturer, is applied to the subfloor or slab. The flooring is not fastened or cemented to the floor. However, the boards must be glued to each other along the edges and ends. Apply glue on edges in 8-inch-long beads with 12-inch spaces between them. Apply a full bead across the ends.

The first row is laid in a straight line with end joints glued. The groove edge is placed toward the wall, leaving about a ½-inch expansion space. Subsequent courses are installed. Edges and ends are glued. Each piece is brought tight against the other with a hammer and tapping block. The tapping block should be used only against the tongue. It should never be used against the grooved edge **(Figure 84–14)**. Tapping the grooved edge will damage it. Stagger end joints at least 2 feet from those in adjacent courses.

Usually the last course must be cut to width. To lay out the last course to fit, lay a complete row of boards,

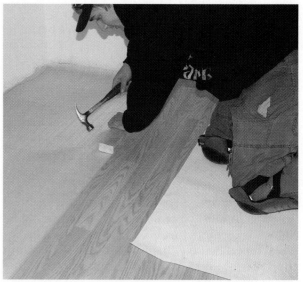

FIGURE 84–14   A tapping block is used to drive laminated strip flooring boards tight against those already installed.

unglued, tongue toward wall, directly on top of the already installed next to last course. Cut a short piece of flooring for use as a scribing block. Hold the tongue

FIGURE 84–15 **Technique for scribing the last strip on flooring using a scrap piece of flooring.**

edge against the wall. Move the block along the wall while holding a pencil against the other edge to lay out the flooring. The width of the tongue of the scribing block provides the necessary expansion space **(Figure 84–15)**. The last row, when cut, can be glued and wedged tightly in place with a pry bar.

Installing Plank Flooring

Plank flooring is installed like strip flooring. Alternate the courses by widths. Start with the narrowest pieces. Then use increasingly wider courses, and repeat the pattern. Stagger the joints in adjacent courses. Use lengths so they present the best appearance.

Manufacturers' instructions for fastening the flooring vary and should be followed. Generally, the flooring is blind nailed through the tongue of the plank and at intervals along the plank in a manner similar to strip flooring.

INSTALLING PARQUET FLOORING

Procedures for the application of parquet flooring vary with the style and the manufacturer. Detailed installation directions are usually provided with the flooring. Generally, both parquet blocks and strips are laid in *mastic*. Use the recommended type. Apply with a notched trowel. The depth and spacing of the notches are important to leave the correct amount of mastic on the floor. Parquet may be installed either square with the walls or diagonally.

Square Pattern

When laying unit blocks in a square pattern, two layout lines are snapped, at right angles to each other, and parallel to the walls. Blocks are laid with their edges to the lines. Lines are usually laid out so

that rows of blocks are either centered on the floor or half the width of a block off center. This depends on which layout produces border blocks of equal and maximum widths against opposite walls.

To determine the location of the layout lines, measure the distance to the center of the room's width. Divide this distance by the width of a block. If the remainder is half or more, snap the layout line in the center. If the remainder is less than half, snap the layout line off center by half the width of the block. Find the location of the other layout line in the same way. It is possible that one of the layout lines will be centered, while the other must be snapped off center.

Other factors may determine the location of layout lines, such as ending with full blocks under a door, or where they meet another type of floor. Regardless of the location, two lines, at right angles to each other, must be snapped.

Place one unit at the intersection of the lines. Position the grooved edges exactly on the lines. Lay the next units ahead and to one side of the first one and along the lines. Install blocks in a pyramid. Work from the center outward toward the walls in all directions. Make adjustments as installation progresses to prevent misalignment **(Figure 84–16)**.

Diagonal Pattern

To lay unit blocks in a diagonal pattern, measure an equal distance from one corner of the room along both walls. Snap a starting line between the two marks. At the center of the starting line, snap another line at right angles to it. The location of both lines may need to be changed in order to end with border blocks of equal and largest possible size against opposite walls. The diagonal pattern is then laid in a manner similar to the square pattern **(Procedure 84–B)**.

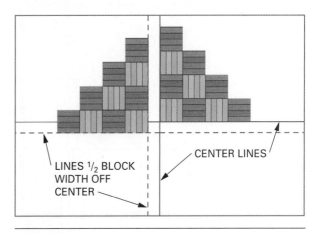

FIGURE 84–16 **Parquet blocks are laid to the center line or off the center lines. This depends on which produces the best size border blocks.**

STEP BY STEP PROCEDURES

Procedure 84–B Laying Out a Diagonal Pattern

STEP 1 LAY OUT EQUAL DISTANCES FROM A SQUARE CORNER.

STEP 2 SNAP A LAYOUT LINE BETWEEN THE TWO POINTS.

STEP 3 SNAP A LAYOUT LINE AT RIGHT ANGLES TO THE FIRST LAYOUT LINE.

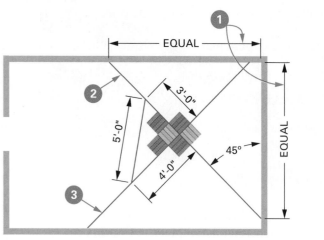

Special Patterns

Many parquet patterns can be laid out with square and diagonal layout lines. The *herringbone* pattern requires three layout lines. One will be the 90-degree line used for a square pattern. The other line crosses the intersection at a 45-degree angle. Align the first block or strip with its edge on the diagonal line. The corner of the piece should be lined up at the intersection. Continue the pattern in rows of three units wide, aligning the units with the layout lines **(Figure 84–17).**

The *Monticello* pattern can be laid square or diagonally in the same way as unit blocks. For best results, lay the parquet in a pyramid pattern. Alternate the grain of the center blocks. Keep *picket* points in precise alignment **(Figure 84–18).**

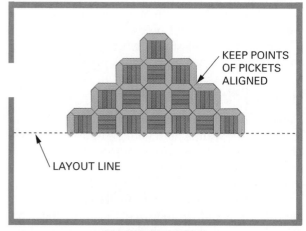

SQUARE PATTERN

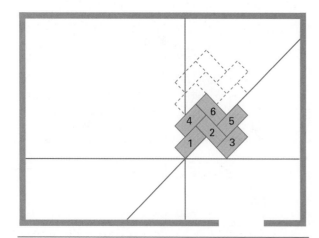

FIGURE 84–17 The herringbone pattern requires 90- and 45-degree layout lines. (*Courtesy of Chickasaw Hardwood Floors*)

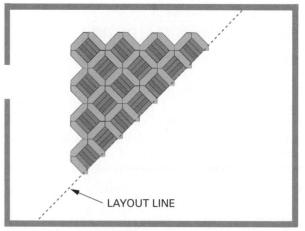

DIAGONAL PATTERN

FIGURE 84–18 Layout of the Monticello parquet pattern. (*Courtesy of Chicasaw Hardwood Floors*)

The *Marie Antoinette* pattern may also be laid square or diagonally. It is started by laying a *band* with its grooved edge and end aligned with the layout lines with the tongued edge to the right. Continue laying the pattern by placing center blocks and bands in a sequence so that bands appear woven **(Figure 84–19)**. Tongues of all members face toward the right or ahead. In this pattern, the grain of center blocks runs in the same direction.

Many other parquet patterns are manufactured. With the use of parquet blocks and strips, the design possibilities are almost endless. Competence in laying the popular patterns described in this chapter will enable the student to apply the principles for professional installations of many different parquet floors.

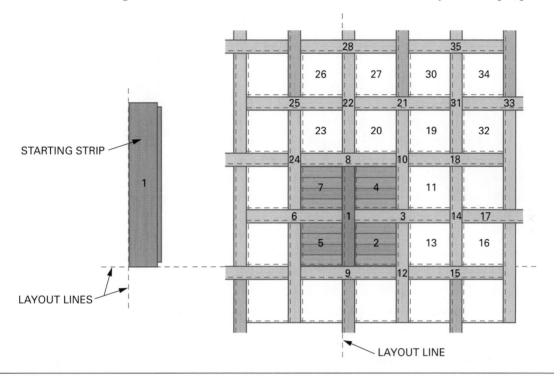

FIGURE 84–19 Layout of the Marie Antoinette parquet pattern. *(Courtesy of Chickasaw Hardwood Floors)*

85 Underlayment and Resilient Tile

Resilient flooring is widely used in residential and commercial buildings. It is a thin, flexible material that comes in sheet or tile form. It is applied on a smooth concrete slab or an underlayment.

UNDERLAYMENT

Plywood, strandboard, hardboard, or particleboard may be used for **underlayment.** It is installed on top of the subfloor. This provides a base for the application of resilient sheet or tile flooring. The number of underlayment panels required is found by dividing the area to be covered by the area of the panel.

Underlayment thickness may range from ¼ to ¾ inch, depending on the material and job requirements. In many cases, where a finish wood floor meets a tile floor, the underlayment thickness is determined by the difference in the thickness of the two types of finish floor. Both floor surfaces should come flush with each other.

Installing Underlayment

Sweep the subfloor as clean as possible. Cover with asphalt felt lapped 4 inches at the edges. Stagger joints between the subfloor and underlayment. If installation is started in the same corner as

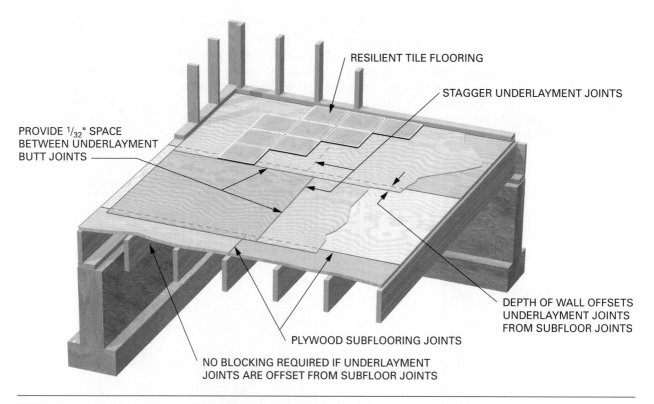

FIGURE 85–1 The joints between underlayment and subfloor are offset.

the subfloor, the thickness of the walls will be enough to offset the joints. Leave about ¹⁄₃₂ inch between underlayment panels to allow for expansion.

If the underlayment is to go over a board subfloor, install the underlayment with its face grain across the boards. In all other cases, a stiffer and stronger floor is obtained if the face grain is across the floor joists **(Figure 85–1)**.

Fasten the first row in place with staples or nails. Underlayment requires more nails than subfloor to provide a squeak-free and stiff floor. Nail spacing and size are shown in **Figure 85–2** for various thick-

nesses of plywood underlayment. Similar nail spacing and size are appropriate for other types of underlayment.

Install remaining rows of underlayment with end joints staggered from the previous courses. The last row of panels is ripped to width to fit the remaining space.

If APA Sturd-I-Floor is used, it can double as a subfloor and underlayment. If square-edged panels are used, blocking must be provided under the joints to support the edges. If tongue-and-groove panels are used, blocking is not required **(Figure 85–3)**.

RESILIENT TILE FLOORING

Most resilient floor tiles are made of vinyl. Many different colors, textures, and patterns are available. Tiles come in 12 × 12 squares. They are applied to the floor in a manner similar to that used for applying parquet blocks. The most common tile thicknesses are ¹⁄₁₆ and ⅛ inch. Thicker tiles are used in commercial applications subjected to considerable use.

Long strips of the same material are called feature strips. They are available from the manufacturer. The strips vary in width from ¼ inch to 2 inches. They are used between tiles to create unique floor patterns.

Plywood Thickness (inch)	Fastener Size (approx.) and Type	Fastener Spacing (inches)	
		Panel Edges	Panel Interior
¼	1¼″ ring-shank nails	3	6 each way
	18 gauge staples	3	6 each way
⅜ or ½	1¼″ ring-shank nails	6	8 each way
	16 gauge staples	3	6 each way
⅝ or ¾	1½″ ring-shank nails	6	8 each way
	16 gauge staples	3	6 each way
¼	1¼″ ring-shank nails	3	6 each way
	18 gauge staples	3	6 each way

FIGURE 85–2 Nailing specifications for plywood underlayment.

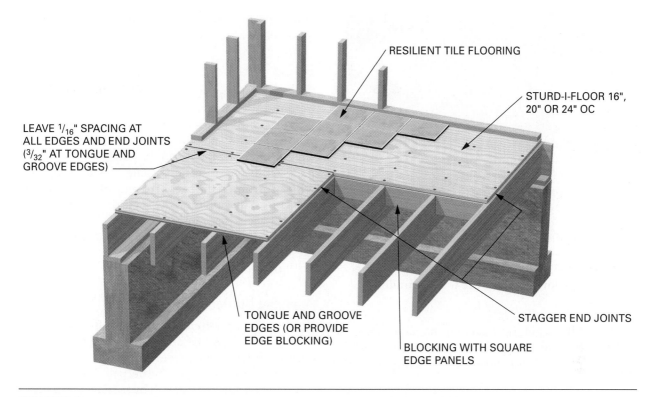

LEAVE $^1/_{16}$" SPACING AT ALL EDGES AND END JOINTS ($^3/_{32}$" AT TONGUE AND GROOVE EDGES)

RESILIENT TILE FLOORING

STURD-I-FLOOR 16", 20" OR 24" OC

TONGUE AND GROOVE EDGES (OR PROVIDE EDGE BLOCKING)

BLOCKING WITH SQUARE EDGE PANELS

STAGGER END JOINTS

FIGURE 85–3 APA Sturd-I-Floor requires no underlayment for the installation of resilient floors.

INSTALLING RESILIENT TILE

⚠ CAUTION   CAUTION   CAUTION   CAUTION

If the application involves the removal of existing resilient floor covering, be aware that it, and the adhesive used, may contain asbestos. The presence of asbestos in the material is not easily determined. If there is any doubt, always assume that the existing flooring and adhesive do contain asbestos. Practices for removal of existing flooring or any other building material containing asbestos should comply with standards set by the U.S. Occupational Safety and Health Administration (OSHA) or corresponding authorities in other jurisdictions.

If the application is over an existing resilient floor covering, do not sand the existing surface unless absolutely sure it does not contain asbestos. Inhalation of asbestos dust can cause serious harm. ■

Before installing resilient tile, make sure underlayment fasteners are not projecting above the surface. Fill any open areas, such as splits, with a floor leveling compound **(Figure 85–4).** On underlayment, use

FIGURE 85–4 Uneven floor surfaces are patched with a patching compound before tile is installed.

a portable disc sander to bring the joints flush. Skim the entire surface with the sander to make sure the surface is smooth. Sweep the floor clean.

Check to see if the spaces between the door stops and casing are sufficient to allow tile to slide under. They may be trimmed using a handsaw on a scrap piece of tile as a gauge **(Figure 85–5).**

Layout

Snap layout lines across the floor at right angles to each other. The lines may be centered if border tiles are a half width or more. If border tiles are less than

FIGURE 85–5 The bottom end of molding may be trimmed to allow the floor tile to slide under.

a half width, layout lines may be snapped half the width of a tile off center. The important thing is to have two perpendicular lines that serve to guide the joints in the tile.

Applying Adhesive

Adhesive is applied to the entire floor area **(Figure 85–6).** When the adhesive becomes transparent and appears to be dry, it is ready for tile to be installed. It will remain tacky and ready for application for up to 72 hours.

A notched trowel is used to spread the adhesive. It must be properly sized according to the manufacturer's recommendation. If the notches are too deep, more adhesive than necessary will be applied. This will result in the adhesive squeezing up through the

joints onto the face of the tile and will probably require removal of the application.

Laying Tiles

Start by applying a tile to the intersection of the layout lines with the two adjacent edges on the lines. Lay tiles with edges tight. Work toward the walls **(Figure 85–7).** Watch the grain pattern. It may be desired to alternate the run of the patterns or to lay the patterns in one direction. Some tiles are stamped with an arrow on the back. They are placed so the arrows on all tiles point in the same direction.

Lay tiles in place instead of sliding them into position. Sliding the tile pushes the adhesive through the joint. With most types of adhesives, it may be difficult or impossible to slide the tile.

Applying Border Tiles

Border tiles are often cut using a tile cutter **(Figure 85–8).** It cuts fast and the cut edges are clean and straight. Each piece is measured, marked, and cut by rotating the handle.

Border tiles may also be cut by scoring with a sharp utility knife and bending. To lay out and score a border tile to fit snugly, first place the tile to be cut directly on top of the last tile installed. Make sure all edges are in line. Place a full tile with its edge against the wall and on top of the one to be fitted. Score the border tile along the outside edge of the top tile **(Figure 85–9).** Bend and break the tile along the scored line.

If the base has not been installed yet, the border tiles are fit roughly into place. Then when the base is installed later, it covers the cut edge. If the base has already been installed, the tile must be fit closely.

FIGURE 85–6 Floor adhesive is often troweled onto the entire floor area before tile application begins.

FIGURE 85–7 Tile is applied from the layout lines towards the walls.

FIGURE 85–8 A tile cutter is often used to install border tile.

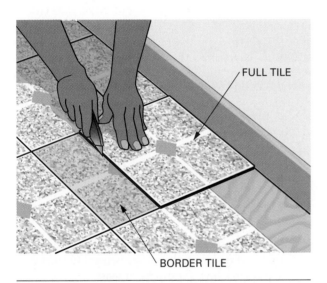

FIGURE 85–9 Fitting a border tile by placing it under a guide piece of tile while scoring.

Scored cuts may need to be smoothed with a file or sandpaper to improve their look.

For tiles that require curved cuts, a propane torch can be used on the back side of the tile to warm it. This makes the knife cut easily and the resulting cut is smooth **(Figure 85–10).** Use care not to overheat and burn the tile. This is also done when fitting a tile around a corner where two or more cuts are required **(Figure 85–11).**

Applying a Vinyl Cove Base

Many times a vinyl cove base is used to trim a tile floor. A special vinyl base cement is applied to its back. The base is pressed into place **(Figure 85–12).**

FIGURE 85–10 A propane torch is used to soften the tile for cutting. Use care not to overheat and burn the tile.

FIGURE 85–11 Fitting a tile around a corner involves two cuts.

FIGURE 85–12 Vinyl base may be bent to fit around a corner.

ESTIMATING RESILIENT TILE

To estimate the amount of tile flooring needed, find the area of the room in square feet. To do this, measure the length and width of the room to the next whole foot. Multiply these figures to find the area. For 12 × 12 tiles, the result is the number of pieces needed. Divide the number of pieces by 45, or whatever the number of pieces in the box is, to determine the number of boxes. Rounding this number up to the next whole box usually takes care of any waste factor.

Key Terms

blind nailed	parquet block	racking the floor	strip flooring
chamfer	parquet strip	resilient	underlayment
hearths	plank	slat block	unfinished strip
laminated block	plank flooring	sleepers	unit block
laminated strip flooring	prefinished strip	strip	

Review Questions

Select the most appropriate answer.

1. If hardwood flooring is stored in a heated building, the temperature should not exceed
   a. 72°F.   c. 85°F.
   b. 78°F.   d. 90°F.

2. Most hardwood finish flooring is made from
   a. Douglas fir.   c. southern pine.
   b. hemlock.   d. oak.

3. Bundles of strip flooring may contain pieces over and under the nominal length of the bundle by
   a. 4 inches.   c. 8 inches.
   b. 6 inches.   d. 9 inches.

4. No pieces are allowed in bundles of hardwood strip flooring shorter than
   a. 4 inches.   c. 8 inches.
   b. 6 inches.   d. 9 inches.

5. The edges of prefinished strip flooring are chamfered to
   a. prevent splitting.
   b. apply the finish.
   c. simulate cracks between adjoining pieces.
   d. obscure any unevenness in the floor surface.

6. The best grade of unfinished oak strip flooring is
   a. prime.   c. select.
   b. clear.   d. quarter-sawed.

7. To estimate the amount of 2¼-inch face hardwood flooring, add to the area to be covered a percentage of the area of
   a. 42.5.   c. 29.
   b. 38.33.   d. 33.33.

8. When it is necessary to cut the last strip in a course of flooring, the waste is used to start the next course and should be at least
   a. 8 inches long.   c. 12 inches long.
   b. 10 inches long.   d. 16 inches long.

9. For floor laying, the hammer weight is generally
   a. 13 to 16 ounces.   c. 20 to 28 ounces.
   b. 16 to 20 ounces.   d. 25 to 30 ounces.

10. To change direction of strip flooring,
    a. face nail both strips.
    b. turn the extended strip around.
    c. blind nail both strips.
    d. use a spline.

UNIT 30

Cabinets and Countertops

CHAPTER 86 Description and Installation of Manufactured Cabinets

CHAPTER 87 Countertops and Cabinet Components

Cabinets and countertops usually are purchased in preassembled units and may be installed by a carpenter. Manufactured cabinets are often used because of the great variety and shorter installation time than job-built cabinets. Cabinets can be custom-made to meet the specifications of most any job, but they are usually made in a cabinet shop. Countertops, cabinet doors, and drawers may be customized in a wide variety of styles and sizes.

SAFETY REMINDER

Kitchen cabinets are installed in large units that are cumbersome an often heavy to move. Watch that other wall and floor finishes are not marred when positioning the cabinets. Also, remember to lift with your legs and not your back.

OBJECTIVES

After completing this unit, the student should be able to:

- state the sizes and describe the construction of typical base and wall kitchen cabinet units.
- plan, order, and install manufactured kitchen cabinets.
- construct, laminate, and install a countertop.
- identify cabinet doors and drawers according to the type of construction and method of installation.
- identify overlay, lipped, and flush cabinet doors and proper drawer construction.
- apply cabinet hinges, pulls, and door catches.

86 Description and Installation of Manufactured Cabinets

anufactured kitchen and bath cabinets come in a wide variety of styles, materials, and finishes **(Figure 86–1).** The carpenter must be familiar with the various kinds, sizes, uses, and construction of the cabinets to know how to plan, order, and install them.

DESCRIPTION OF MANUFACTURED CABINETS

For commercial buildings, many kinds of specialty cabinets are manufactured. They are designed for specific uses in offices, hospitals, laboratories, schools, libraries, and other buildings. Most cabinets used in residential construction are manufactured for the kitchen or bathroom. All cabinets, whether for commercial or residential use, consist of a case which is fitted with shelves, doors, and/or drawers. Cabinets are manufactured and installed in essentially the same way. Designs vary considerably with the manufacturer, but sizes are close to the same.

Kinds and Sizes

One method of cabinet construction utilizes a **face frame.** This frame provides openings for doors and drawers. Another method, called *European* or **frameless,** eliminates the face frame **(Figure 86–2).** Faceframed cabinets usually give a traditional look. Frameless cabinets are used when a contemporary appearance is desired.

The two basic kinds of kitchen cabinets are the **wall unit** and the **base unit.** The surface of the countertop is usually about 36 inches from the floor. Wall units are installed about 18 inches above the countertop. This distance is enough to accommodate such articles as coffeemakers, toasters, blenders, and mixers. Yet it keeps the top shelf within reach, not over 6 feet from the floor. The usual overall height of a kitchen cabinet installation is 7'–0" **(Figure 86–3).**

Wall Cabinets. Standard wall cabinets are 12 inches deep. They normally come in heights of 42, 30, 24, 18, 15, and 12 inches. The standard height is 30 inches. Shorter cabinets are used above sinks, refrigerators, and ranges. The 42-inch cabinets are for use in kitchens without soffits where more storage space is desired. A standard height wall unit usually contains two adjustable shelves.

Usual wall cabinet widths range from 9 to 48 inches in 3-inch increments. They come with single or double

FIGURE 86–1   Kitchen cabinets are available in a wide variety of styles and sizes.

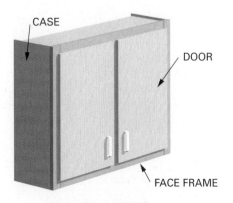

FRAMED CABINET

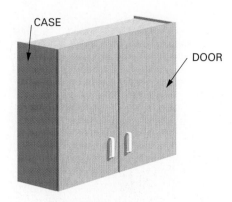

FRAMELESS CABINET

FIGURE 86–2   Two basic methods of cabinet construction.

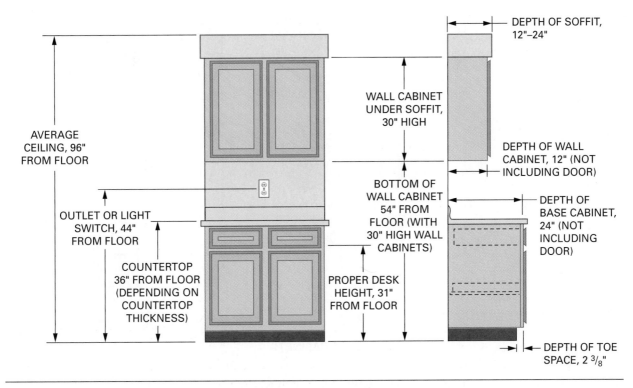

FIGURE 86–3 Common kitchen cabinet heights and dimensions. (*Courtesy of Merillat Industries*)

doors depending on their width. Single-door cabinets can be hung so doors can swing in either direction.

Wall corner cabinets make access into corners easier. **Double-faced cabinets** have doors on both sides for use above island and peninsular bases. Some wall cabinets are made 24 inches deep for installation above refrigerators. A microwave oven case, with a 30-inch wide shelf, is available **(Figure 86–4).**

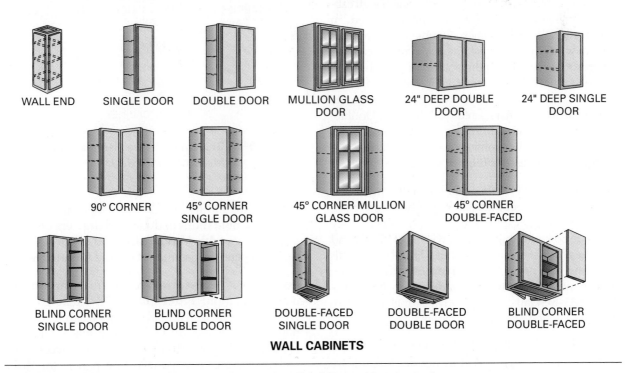

FIGURE 86–4 Kinds and sizes of manufactured wall cabinets. (*Courtesy of Merillat Industries*)

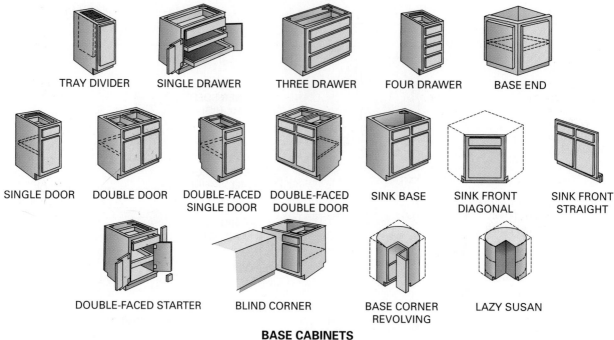

TRAY DIVIDER   SINGLE DRAWER   THREE DRAWER   FOUR DRAWER   BASE END

SINGLE DOOR   DOUBLE DOOR   DOUBLE-FACED SINGLE DOOR   DOUBLE-FACED DOUBLE DOOR   SINK BASE   SINK FRONT DIAGONAL   SINK FRONT STRAIGHT

DOUBLE-FACED STARTER   BLIND CORNER   BASE CORNER REVOLVING   LAZY SUSAN

BASE CABINETS

FIGURE 86–5 **Most base cabinets are manufactured to match wall units.**
(Courtesy of Merillat Industries)

Base Cabinets. Most base cabinets are manufactured 34½ inches high and 24 inches deep. By adding the usual countertop thickness of 1½ inches, the work surface is at the standard height of 36 inches from the floor. Base cabinets come in widths to match wall cabinets. Single-door cabinets are manufactured in widths from 9 to 24 inches. Double-door cabinets come in widths from 27 to 48 inches. A recess called a **toe space** is provided at the bottom of the cabinet.

The standard base cabinet contains one drawer, one door, and an adjustable shelf. Some base units have no drawers; others contain all drawers. Double-faced cabinets provide access from both sides. Corner units, with round revolving shelves, make corner storage easily accessible **(Figure 86–5)**.

Tall Cabinets. Tall cabinets are usually manufactured 24 inches deep, the same depth as base cabinets. Some utility cabinets are 12 inches deep. They are made 66 inches high and in widths of 27, 30, and 33 inches for use as oven cabinets. Single-door utility cabinets are made 18 and 24 inches wide. Double-door pantry cabinets are made 36 inches wide **(Figure 86–6)**. Wall cabinets with a 24-inch depth are usually installed above tall cabinets.

Vanity Cabinets. Most vanity base cabinets are made 31½ inches high and 21 inches deep. Some are made in depths of 16 and 18 inches. Usual widths range from 24 to 36 inches in increments of 3 inches,

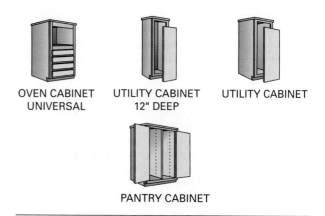

OVEN CABINET UNIVERSAL   UTILITY CABINET 12" DEEP   UTILITY CABINET

PANTRY CABINET

FIGURE 86–6 **Tall cabinets are also manufactured.**
(Courtesy of Merillat Industries)

then 42, 48, and 60 inches. They are available with several combinations of doors and drawers depending on their width. Various sizes and styles of vanity wall cabinets are also manufactured **(Figure 86–7)**.

Accessories. **Accessories** are used to enhance a cabinet installation. **Filler** pieces fill small gaps in width between wall and base units when no combination of sizes can fill the existing space. They are cut to necessary widths on the job. Other accessories include cabinet end panels, face panels for dishwashers and refrigerators, open shelves for cabinet ends, and spice racks.

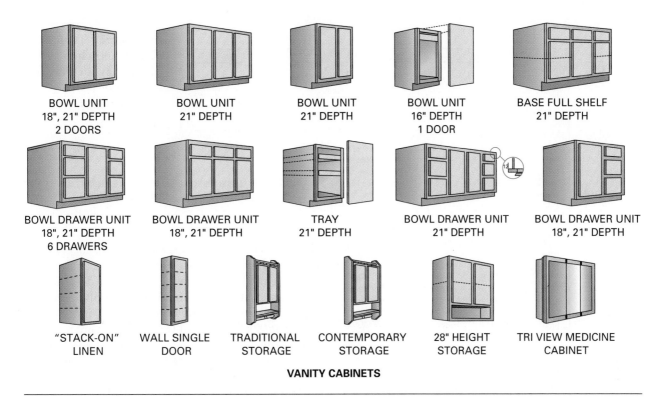

BOWL UNIT
18", 21" DEPTH
2 DOORS

BOWL UNIT
21" DEPTH

BOWL UNIT
21" DEPTH

BOWL UNIT
16" DEPTH
1 DOOR

BASE FULL SHELF
21" DEPTH

BOWL DRAWER UNIT
18", 21" DEPTH
6 DRAWERS

BOWL DRAWER UNIT
18", 21" DEPTH

TRAY
21" DEPTH

BOWL DRAWER UNIT
21" DEPTH

BOWL DRAWER UNIT
18", 21" DEPTH

"STACK-ON"
LINEN

WALL SINGLE
DOOR

TRADITIONAL
STORAGE

CONTEMPORARY
STORAGE

28" HEIGHT
STORAGE

TRI VIEW MEDICINE
CABINET

VANITY CABINETS

FIGURE 86–7 Vanity cabinets are made similar to kitchen cabinets, but differ in size. *(Courtesy of Merillat Industries)*

LAYING OUT MANUFACTURED KITCHEN CABINETS

The blueprints for a building contain plans, elevations, and details that show the cabinet layout. Architects may draw the layout. But they may not specify the size or the manufacturer's identification for each individual unit of the installation. In residential construction, particularly in remodeling, no plans are usually available to show the cabinet arrangement. In addition to installation, it becomes the responsibility of the carpentry contractor to plan, lay out, and order the cabinets, in accordance with the customer's specifications.

The first step is to measure carefully and accurately the length of the walls on which the cabinets are to be installed. A plan is then drawn to scale. It must show the location of all appliances, sinks, windows, and other necessary items **(Figure 86–8).**

Next, draw elevations of the base cabinets, referring to the manufacturer's catalog for sizes. Always use the largest size cabinets available instead of two or three smaller ones. This reduces the cost and makes installation easier.

Match up the wall cabinets with the base cabinets, where feasible. If filler strips are necessary, place them between a wall and a cabinet or between cabinets in the corner. Identify each unit on the elevations with the manufacturer's identification **(Figure 86–9).** Make a list of the units in the layout. Order from the distributor.

Computer Layouts

Computer programs are available to help in laying out manufactured kitchen cabinets. When the required information is fed into the computer, a number of different layouts can be quickly made. When a acceptable layout is made, it can be printed with each of the cabinets in the layout identified and priced. Most large kitchen cabinet distributors will supply computerized layouts on request.

INSTALLING MANUFACTURED CABINETS

Cabinets must be installed level and plumb even though floors are not always level and walls not always plumb. Level lines are first drawn on the wall for base and wall cabinets. To level base cabinets that set on an unlevel floor, either shim the cabinets from the high point of the floor or scribe and fit the cabinets to the floor from the lowest point on the floor. Shimming the base cabinets leaves a space that must be later covered by a molding. Scribing and fitting the cabinets to the floor eliminate the need for a molding. The method used depends on various conditions of the job. If shimming base cabinets, lay out the level lines on the wall from the highest point on the floor where

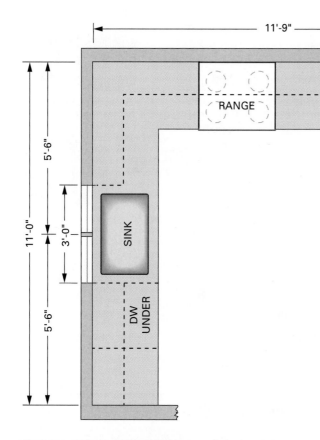

FIGURE 86–8 Typical plan of a kitchen cabinet layout showing location of walls, windows, and appliances

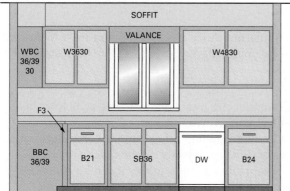

SINK WALL ELEVATION

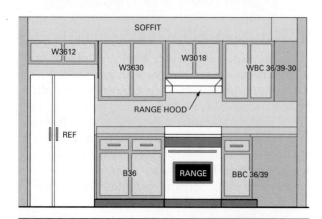

FIGURE 86–9 Elevations of the installation are sometimes drawn and the cabinets identified.

cabinets are to be installed. If fitting cabinets to the floor, measure up from the low point.

Laying Out the Wall

Measure 34½ inches up the wall. Draw a level line to indicate the tops of the base cabinets. Use the most accurate method of leveling available (described in Chapter 25, Leveling and Layout Tools). Another level line must be made on the wall 54 inches from the floor. The bottoms of the wall units are installed to this line. It is more accurate to measure 19½ inches up from the first level line and snap lines parallel to it than to level another line.

The next step is to mark the stud locations in a framed wall. (Cabinet mounting screws will be driven into the studs.) An electronic stud finder works well to locate framing. The other method is to lightly tap on and across a short distance of the wall with a hammer. Drive a finish nail in at the point where a solid sound is heard. Drive the nail where holes are later covered by a cabinet. If a stud is found, mark the location with a pencil. If no stud is found, try a little over to one side or the other.

Measure at 16-inch intervals in both directions from the first stud to locate other studs. Drive a finish nail to test for solid wood. Mark each stud location. If studs are not found at 16-inch centers, try 24-inch centers. At each stud location, draw plumb lines on the wall. Mark the outlines of all cabinets on the wall to visualize and check the cabinet locations against the layout **(Figure 86–10).**

Installing Wall Units

Many installers prefer to install the soffit and wall cabinets first so the work does not have to be done leaning over the base units. The soffit may be framed using any of several methods **(Figure 86–11).** One uses drywall to cover 2 × 2 framing. Another method uses paneling or wood strips to cover a 2 × 2 frame. In either case, the ceiling and wall drywall should be installed completely to the corner beforehand. This makes the house more airtight.

A **cabinet lift** may be used to hold the cabinets in position for fastening to the wall. If a lift is not available, the doors and shelves may be removed to make the cabinet lighter and easier to clamp together. If

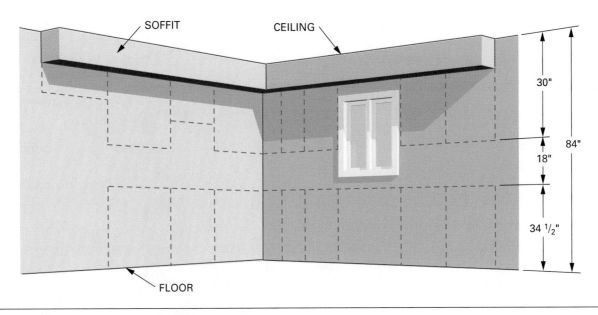

FIGURE 86-10 The wall is laid out with outlines for the cabinet locations.

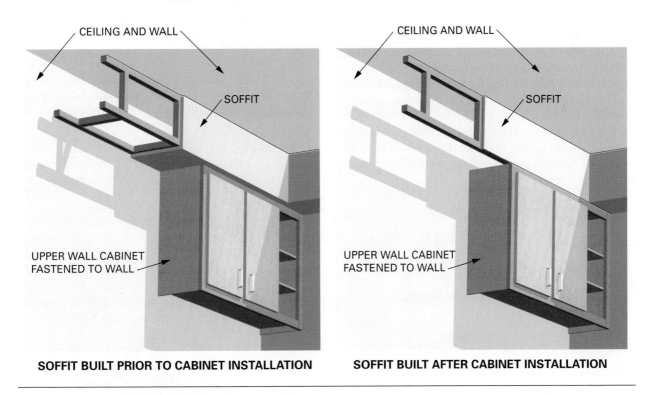

SOFFIT BUILT PRIOR TO CABINET INSTALLATION

SOFFIT BUILT AFTER CABINET INSTALLATION

FIGURE 86-11 Two methods of finishing a soffit.

possible, screw a 1 × 3 strip of lumber so its top edge is on the level line for the bottom of the wall cabinets. This is used to support the wall units while they are being fastened. If it is not possible to screw to the wall, build a stand on which to support the unit near the line of installation.

Start the installation of wall cabinets in a corner. On the wall, measure from the line representing the outside of the cabinet to the stud centers. Transfer the measurements to the cabinets. Drill shank holes for mounting screws through mounting rails usually installed at the top and bottom of the cabinet. Place the cabinet on the supporting strip or stand so its bottom is on the level layout line. Wood and steel framing typically do not need predrilled screw holes. Concrete screws used in masonry walls do need to have predrilled holes. This is normally done with the cabinet held in place. Make sure holes are sufficiently

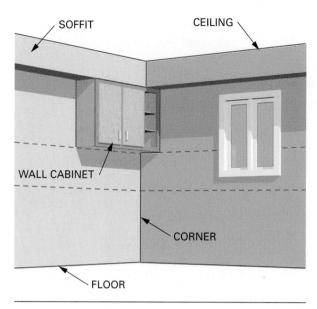

FIGURE 86–12 **Installation of wall cabinets is started in the corner.**

deep to prevent the screw from bottoming out in the hole. Fasten the cabinet in place with mounting screws of sufficient length to hold the cabinet securely. Do not fully tighten the screws **(Figure 86–12).**

The next cabinet is installed in the same manner. Align the adjoining stiles so their faces are flush with each other. Clamp them together with C-clamps. Screw the stiles tightly together **(Figure 86–13).** Continue this procedure around the room. Tighten all mounting screws.

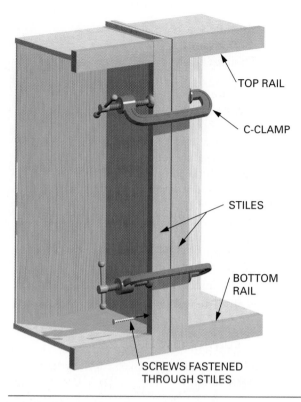

FIGURE 86–13 **The stiles of adjoining cabinets are joined together with screws.**

If a filler needs to be used, it is better to add it next to a blind corner cabinet or at the end of a run. It may be necessary to scribe the filler to the wall **(Procedure 86–A).**

Procedure 86–A Scribing a Filler Piece Using a Block

STEP ➊ SELECT A SCRIBING BLOCK THE SAME WIDTH AS STILE.

STEP ➋ CLAMP FILLER TO STILE WITH INSIDE EDGES FLUSH WITH EACH OTHER.

STEP ➌ RIDE SCRIBING BLOCK AGAINST WALL TO MARK FILLER.

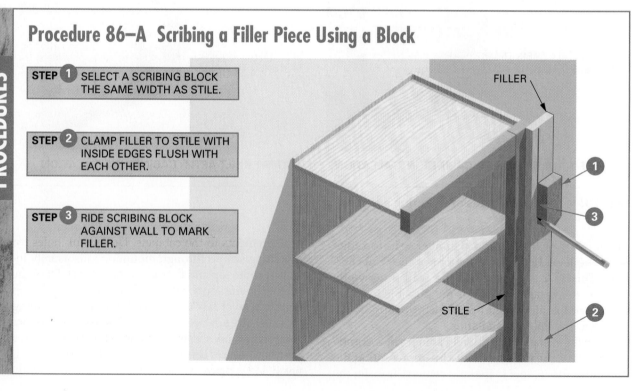

STEP BY STEP PROCEDURES

Installing Base Cabinets

Start the installation of base cabinets in a corner. Shim the bottom until the cabinet top is on the layout line. Then level and shim the cabinet from back to front.

If cabinets are to be fitted to the floor, shim until their tops are level across width and depth. This will bring the tops above the layout line that was measured from the low point of the floor. Adjust the pencil dividers so the distance between the points is equal to the amount the top of the unit is above the layout line. Scribe this amount on the bottom end of the cabinets by running the dividers along the floor **(Figure 86–14).**

Cut both ends and toeboard to the scribed lines. There is no need to cut the cabinet backs because they do not, ordinarily, extend to the floor.

Place the cabinet in position. The top ends should be on the layout line. Fasten it loosely to the wall.

The remaining base cabinets are installed in the same manner. Align and clamp the stiles of adjoining cabinets. Fasten them together. Finally, fasten all units securely to the wall **(Figure 86–15).**

FIGURE 86–15 Base cabinets are secured with screws to wall studs.

INSTALLING MANUFACTURED COUNTERTOPS

Countertops are manufactured in various standard lengths. They can be cut to fit any installation against walls. They are also available with one end precut at a 45-degree angle for joining with a similar one at corners. Special hardware is used to join the sections. The countertops are covered with a thin, tough *high-pressure plastic laminate.* This is generally known as **mica.** It is available in many colors and patterns. The countertops are called **postformed** countertops. This term comes from the method of forming the mica to the rounded edges and corners of the countertop **(Figure 86–16).** Postforming is bending the mica with heat to a radius of ¾ inch or less. This can only be done with special equipment.

After the base units are fastened in position, the countertop is cut to length. It is fastened on top of the

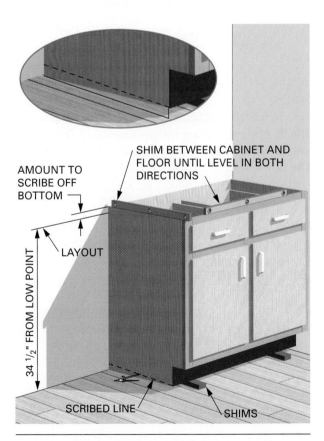

FIGURE 86–14 Method of scribing base cabinets to the floor.

FIGURE 86–16 A section of a manufactured postformed countertop. The edges and interior corner are rounded.

base units and against the wall. The backsplash can be scribed, limited by the thickness of its scribing strip, to an irregular wall surface. Use pencil dividers to scribe a line on the top edge of the backsplash. Then plane or belt sand to the scribed line.

Fasten the countertop to the base cabinets with screws up through triangular blocks usually installed in the top corners of base units. Use a stop on the drill bit. This prevents drilling through the countertop. Use screws of sufficient length, but not so long that they penetrate the countertop.

Exposed cut ends of postformed countertops are covered by specially shaped pieces of plastic laminate.

Sink cutouts are made by carefully outlining the cutout and cutting with a saber saw. The cutout pattern usually comes with the sink. Use a fine-tooth blade to prevent chipping out the face of the mica beyond the sink. Some duct tape applied to the base of the saber saw will prevent scratching of the countertop when making the cutout.

87   Countertops and Cabinet Components

Custom cabinets usually are made at cabinet shops. Sometimes, though, countertops, doors, and drawers are produced or modified on the job.

MAKING THE COUNTERTOP

Use the pieces of ¾- or ⅝-inch panel material left from making the cabinet bottoms to make the *countertop*. If more than one length is required, join them with glue and screws to a short piece of backing plywood. The width of the pieces should be about 24½ inches.

Fitting the Countertop

Place the countertop on the base cabinets, against the wall. Its outside edge should overhang the face frame the same amount along the entire length. Open the pencil dividers or scribers to the amount of overhang. Scribe the back edge of the countertop to the wall. Cut the countertop to the scribed line. Place it back on top of the base cabinets. The ends should be flush with the ends of the base cabinets. The front edge should be flush with the face of the face frame **(Figure 87–1).** Install a 1 × 2 on the front edge and at the ends if an end overhang is desired. Keep the top edge flush with the top side of the countertop.

Applying the Backsplash

If a backsplash is used, rip a 4-inch-wide length of ¾-inch stock the same length as the countertop. Use lumber for the backsplash, if lengths over 8 feet are required, to eliminate joints. Fasten the backsplash on top of and flush with the back edge of the countertop by driving screws up through the countertop and into the bottom edge of the backsplash **(Figure 87–2).** In

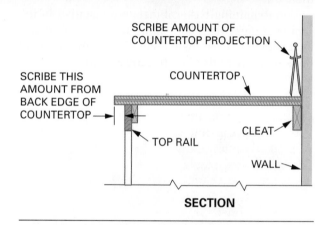

FIGURE 87–1   **Scribing the countertop to fit the wall with its outside edge flush with the face of the cabinet.**

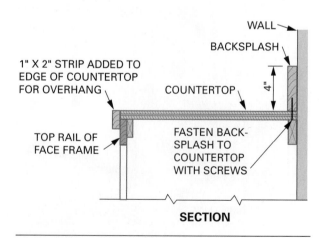

FIGURE 87–2   **Technique for fastening the backsplash to the countertop.**

corners, fasten the ends of the backsplash together with screws.

LAMINATING A COUNTERTOP

Countertops may be covered with plastic laminate. Before laminating a countertop, make sure all surfaces are flush. Check for protruding nailheads. Fill in all holes and open joints. Lightly hand or power sand the entire surface, making sure joints are sanded flush.

Laminate Trimming Tools and Methods

Pieces of laminate are first cut to a **rough size,** about ¼ to ½ inch wider and longer than the surface to be covered. A strip is then cemented to the edge of the countertop. Its edges are flush trimmed even with the top and bottom surfaces. Laminate is then cemented to the top surface, overhanging the edge strip. The overhang is then bevel trimmed even with the laminated edge. A **laminate trimmer** or a small router fitted with laminate trimming bits is used for rough cutting and flush and bevel trimming of the laminate **(Figure 87–3).**

Cutting Laminate to Rough Sizes

Sheets of laminate are large, thin, and flexible. This makes them difficult to cut on a table saw. One method of cutting laminates to rough sizes is by clamping a straightedge to the sheet. Cut it by guiding a laminate trimmer with a flush-trimming bit along the straightedge **(Figure 87–4).** It is easier to run the trimmer across the sheet than to run the sheet across the table saw. Also, the router bit leaves a smooth, clean cut edge. Use a solid carbide trimming bit, which is smaller in diameter than one with ball bearings. It makes a narrower cut. It is easier to control and creates less waste. With this method, cut all the pieces of laminate needed to a rough width

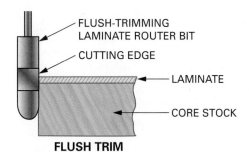
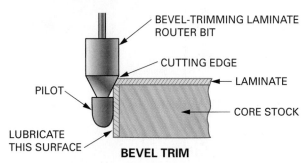

FLUSH-TRIMMING LAMINATE ROUTER BIT

CUTTING EDGE

LAMINATE

CORE STOCK

FLUSH TRIM

BEVEL-TRIMMING LAMINATE ROUTER BIT

CUTTING EDGE

PILOT

LAMINATE

CORE STOCK

LUBRICATE THIS SURFACE

BEVEL TRIM

ADJUST BEVEL-TRIMMING BIT TO CUT FLUSH WITH, BUT NOT INTO, EDGE LAMINATE. THE BEVEL KEEPS THE CUTTING EDGE FROM GRAZING THE FIRST LAYER OF LAMINATE.

NOTE: THE TOP LAMINATE OVERLAPS SIDE TO HELP PREVENT MOISTURE FROM GETTING UNDER LAMINATE.

FIGURE 87–3 The laminate trimmer is used with flush and bevel bits to trim overhanging edges of laminate.

and length. Cut the narrow edge strips from the sheet first.

Another method for cutting laminate to rough size is to use a carbide-tipped hand cutter **(Figure 87–5).** It scores the laminate sufficiently in three passes to be able to break the piece away. Make the cut on the face side and then bend the piece up to create the cleanest break. When used with a straightedge, the cutter is fast and effective.

ON THE JOB
Clamp the laminate to a straightedge. Cut rough sizes with a laminate trimmer.

FIGURE 87–4 Technique for using a laminate trimmer to cut laminate to size.

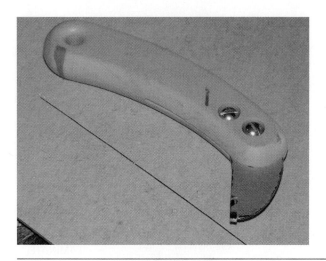

FIGURE 87–5   A hand cutter is often used to cut laminate to a rough size.

Using Contact Cement

Contact cement is used for bonding plastic laminates and other thin, flexible material to surfaces. A coat of cement is applied to the back side of the laminate and to the countertop surface. The cement must be dry before the laminate is bonded to the core. The bond is made on contact without the need to use clamps. A contact cement bond may fail for several reasons:

- Not enough cement is applied. If the material is porous, like the edge of particleboard or plywood, a second coat is required after the first coat dries. When enough cement has been applied, a glossy film appears over the entire surface when dry.

- Too little time is allowed for the cement to dry. Both surfaces must be dry before contact is made. To test for dryness, lightly press your finger on the surface. Although it may feel sticky, the cement is dry if no cement remains on the finger.

- The cement is allowed to dry too long. If contact cement dries too long (more than about 2 hours, depending on the humidity), it will not bond properly. To correct this condition, merely apply another coat of cement and let it dry.

- The surface is not rolled out or tapped after the bond is made. Pressure must be applied to the entire surface using a 3-inch **J-roller** or by tapping with a hammer on a small block of wood.

⚠️ CAUTION   CAUTION   CAUTION   CAUTION

Some contact cements are flammable. Apply only in a well-ventilated area around no open flame. Avoid inhaling the fumes. ∎

Laminating the Countertop Edges

Remove the backsplash from the countertop. Apply coats of cement to the countertop edges and the back of the edge laminate with a narrow brush or small paint roller. After the cement is dry, apply the laminate to the front edge of the countertop **(Figure 87–6)**. Position it so the bottom edge, top edge, and ends overhang. A permanent bond is made when the two surfaces make contact. A mistake in positioning means removing the bonded piece—a time-consuming, frustrating, and difficult job. Roll out or tap the surface.

Apply the laminate to the ends in the same manner as to the front edge piece. Make sure that the square ends butt up firmly against the back side of the overhanging ends of the front edge piece to make a tight joint.

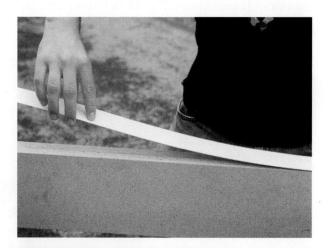

FIGURE 87–6   Applying laminate to the edge of the countertop.

Trimming Laminated Edges

The overhanging ends of the edge laminate must be trimmed before the top and bottom edges. If the laminate has been applied to the ends, a bevel trimming bit must be used to trim the overhanging ends.

Using a Bevel-Trimming Bit. When using a bevel-trimming bit, the router base is gradually adjusted to expose the bit so that the laminate is trimmed flush with the first piece but not cutting into it. The bevel of the cutting edge allows the laminate to be trimmed without cutting into the adjacent piece (see Figure 87–3). A flush-trimming bit cannot be used when the pilot rides against another piece of laminate because the cutting edge may damage it.

Ball bearing trimming bits have *live pilots.* Solid carbide bits have *dead pilots* that turn with the bit. When using a trimming bit with a dead pilot, the laminate must be lubricated where the **pilot** will ride. Rub a short piece of white candle or some solid shortening on the laminate to prevent marring the laminate with the bit.

Using the bevel-trimming bit, trim the overhanging ends of the edge laminate. Then, using the flush-trimming bit, trim off the bottom and top edges of both front and end edge pieces **(Figure 87–7).** To save the time required to change and adjust trimming bits, some installers use two laminate trimmers, one with a flush bit and the other with a bevel bit.

Use a belt sander or a file to smooth the top edge flush with the surface. Sand or file *flat* on the countertop core so a sharp square edge is made. This ensures a tight joint with the countertop laminate. Sand or file *toward* the core to prevent chipping the laminate. Smooth the bottom edge. Ease the sharp outside corner with a sanding block.

FIGURE 87–8 Position the laminate on the countertop using strips before allowing cemented surfaces to contact each other.

Laminating the Countertop Surface

Apply contact bond cement to the countertop and the back side of the laminate. Let dry. To position large pieces of countertop laminate, first place thin strips of wood or metal venetian blind slats about a foot apart on the surface. Lay the laminate to be bonded on the strips or slats. Then position the laminate correctly **(Figure 87–8).**

Make contact on one end. Gradually remove the slats one by one until all are removed. The laminate should then be positioned correctly with no costly errors. Roll out the laminate **(Figure 87–9).** Trim the overhanging back edge with a flush-trimming bit. Trim the ends and front edge with a bevel-trimming bit **(Figure 87–10).** Use a flat file to smooth the trimmed edge. Slightly ease the sharp corner.

FIGURE 87–7 Flush trimming the countertop edge laminate.

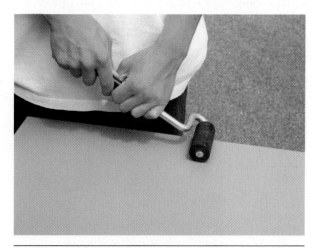

FIGURE 87–9 Rolling out the laminate with a J-roller is required to ensure a proper bond.

FIGURE 87–10   **The outside edge of the countertop laminate is bevel trimmed flush to the edge laminate.**

LAMINATING A COUNTERTOP WITH TWO OR MORE PIECES

When the countertop is laminated with two or more lengths, tight joints must be made between them. Tight joints can be made by clamping the two pieces of laminate in a straight line on some strips of ¾-inch stock. Butt the ends together or leave a space less than ¼ inch between them.

Using one of the strips as a guide, run the laminate trimmer, with a flush-trimming bit installed, through the joint. Keep the pilot of the bit against the straightedge. Cut the ends of both pieces at the same time to ensure making a tight joint **(Figure 87–11).** Bond the sheets as previously described. Apply **seam-filling compound,** especially made for laminates, to make a practically invisible joint. Wipe off excess compound with the recommended solvent.

Laminating Backsplashes

Backsplashes are laminated in the same manner as countertops. Laminate the backsplash. Then reattach it to the countertop with the same screws. Use a little caulking compound between the backsplash and countertop. This prevents any water from seeping through the joint **(Figure 87–12).**

Laminating Rounded Corners

If the edge of a countertop has a rounded corner, the laminate can be bent. Strips of laminate can be cold bent to a minimum radius of about 6 inches. Heating the laminate to 325°F uniformly over the entire bend will facilitate bending to a minimum radius of about 2½ inches. Heat the laminate carefully with a *heat gun.* Bend it until the desired radius is obtained **(Figure 87–13).** Experimentation may be necessary until success in bending is achieved.

FIGURE 87–12   **Apply the laminate to the backsplash and then fasten it to the laminated countertop.**

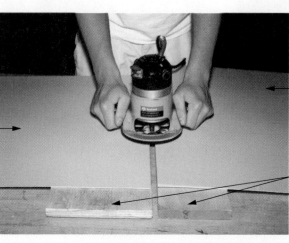

A GUIDE STRIP FOR ROUTER MUST BE INSTALLED ON OPERATOR'S RIGHT WHEN PULLING ROUTER THROUGH CUT.

BOTH PIECES ARE HELD SECURELY AND CUT AT THE SAME TIME.

SUPPORTING STRIPS

FIGURE 87–11   **Making a tight laminate butt seam.**

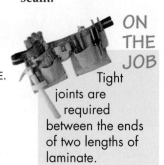

 ON THE JOB   Tight joints are required between the ends of two lengths of laminate.

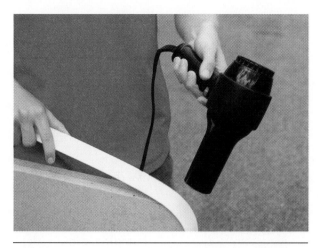

FIGURE 87–13 A heat gun makes laminate bend easily.

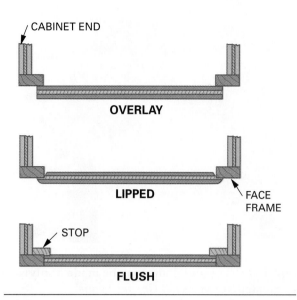

FIGURE 87–14 Plan views of types of cabinet doors.

⚠ CAUTION CAUTION CAUTION CAUTION

Keep fingers away from the heated area of the laminate. Remember that the laminate retains heat for some time. ∎

KINDS OF DOORS

Cabinet doors are classified by their construction and also by the method of installation. Sliding doors are occasionally installed, but most cabinets are fitted with hinged doors that swing.

Hinged cabinet doors are classified as overlay, lipped, and flush, based on the method of installation **(Figure 87–14)**. The overlay method of hanging cabinet doors is the most widely used.

Overlay Doors

The **overlay door** laps over the opening, usually ⅜ inch on all sides. However, it may overlay any amount. In many cases, it may cover the entire face frame. The overlay door is easy to install. it does not require fitting in the opening, and the face frame of the cabinet acts as a stop for the door. *European-style* cabinets omit the face frame. Doors completely overlay the front edges of the cabinet **(Figure 87–15).**

Lipped Doors

The **lipped door** has rabbeted edges that overlap the opening by about ⅜ inch on all sides. Usually the ends and edges are rounded over to give a more pleasing appearance. Lipped doors and drawers are easy to install. No fitting is required and the rabbeted edges stop against the face frame of the cabi-

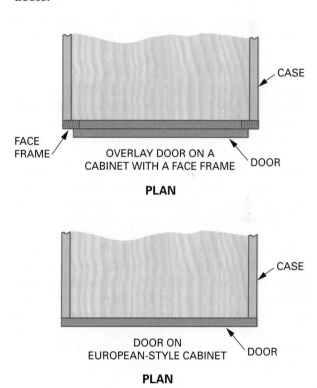

FIGURE 87–15 Overlay doors lap the face frame by varying amounts.

net. However, a little more time is required to shape the rabbeted edges.

Flush Type

The **flush type door** fits into and flush with the face of the opening. They are a little more difficult to hang because they must be fitted in the opening. A fine

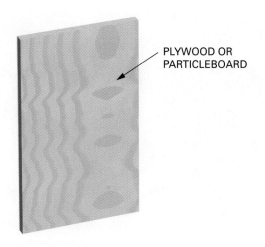

PLYWOOD OR
PARTICLEBOARD

FIGURE 87–16 **Solid doors may be made of plywood or particleboard and then laminated.** *(Courtesy of Gaspar Lewis)*

joint, about the thickness of a dime, must be made between the opening and the door. Stops must be provided in the cabinet against which to close the door.

Door Construction

Doors are also classified, by their construction, as **solid** or **paneled.** Solid doors are made of plywood, particleboard, or solid lumber. Particleboard doors are ordinarily covered with plastic laminate. Matched boards with V-grooves and other designs, such as those used for wall paneling, are often used to make solid doors **(Figure 87–16).** Designs may be grooved into the face of the door with a router. Small moldings may be applied for a more attractive appearance.

Paneled doors have an exterior framework of solid wood with panels of solid wood, plywood, hardboard, plastic, glass, or other panel material. Many complicated designs are manufactured by millworkers with specialized equipment. With the equipment available, carpenters can make paneled doors of simple design only **(Figure 87–17).** Both solid doors and paneled doors may be hinged in overlay, lipped, or flush fashion.

TYPES OF HINGES

Several types of cabinet hinges are *surface, offset, overlay, pivot,* and *butt.* For each type there are many styles and finishes **(Figure 87–18).** Some types are *self-closing* hinges that hold the door closed and eliminate the need for door catches.

Surface Hinges

Surface hinges are applied to the exterior surface of the door and frame. The back side of the hinge leaves may lie in a straight line for flush doors. One

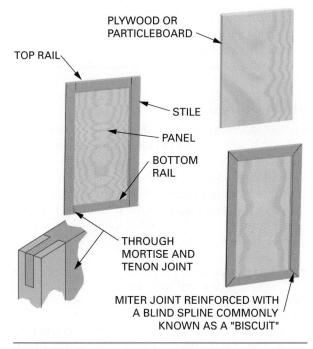

PLYWOOD OR
PARTICLEBOARD

TOP RAIL

STILE

PANEL

BOTTOM
RAIL

THROUGH
MORTISE AND
TENON JOINT

MITER JOINT REINFORCED WITH
A BLIND SPLINE COMMONLY
KNOWN AS A "BISCUIT"

FIGURE 87–17 **Panel doors of simple design can be made on the job.**

leaf may be offset for lipped doors **(Figure 87–19).** The surface type is used when it is desired to expose the hardware, as in the case of wrought iron and other decorative hinges.

Offset Hinges

Offset hinges are used on lipped doors. They are called *offset surface* hinges when both leaves are

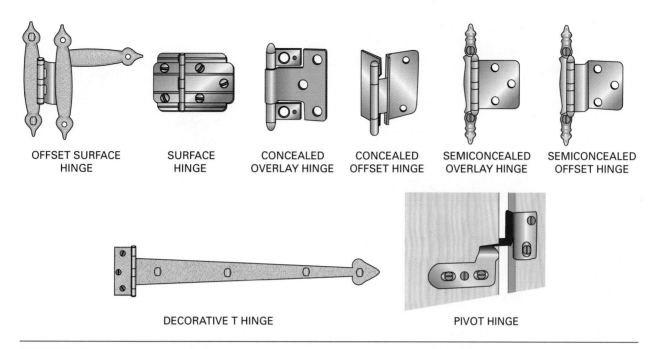

OFFSET SURFACE HINGE   SURFACE HINGE   CONCEALED OVERLAY HINGE   CONCEALED OFFSET HINGE   SEMICONCEALED OVERLAY HINGE   SEMICONCEALED OFFSET HINGE

DECORATIVE T HINGE   PIVOT HINGE

FIGURE 87–18   Cabinet door hinges come in many styles and finishes. *(Courtesy of Amerock Corporation)*

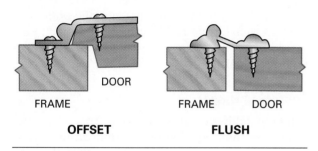

OFFSET   **FLUSH**

FIGURE 87–19   Surface hinges.

fastened to outside surfaces. The *semiconcealed off-set* hinge has one leaf bent to a ⅜-inch offset that is screwed to the back of the door. The other leaf screws to the exterior surface of the face frame. A *concealed offset* type is designed in which only the pin is exposed when the door is closed **(Figure 87–20)**.

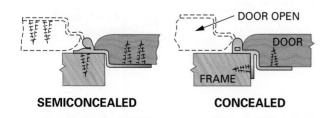

SEMICONCEALED   **CONCEALED**

FIGURE 87–20   Offset hinges.

Overlay Hinges

Overlay hinges are available in *semiconcealed* and *concealed* types. With semiconcealed types, the amount of overlay is variable. Certain concealed overlay hinges are made for a specific amount of overlay, such as ¼, ⁵⁄₁₆, ⅜, and ½ inch. European-style hinges are completely concealed. They are not usually installed by the carpenter because of the equipment needed to bore the holes to receive the hinge. Some overlay hinges, with one leaf bent at a 30-degree angle, are used on doors with reverse beveled edges **(Figure 87–21)**.

Pivot Hinges

Pivot hinges are usually used on overlay doors. They are fastened to the top and bottom of the door and to the inside of the case. They are frequently used when there is no face frame and the door completely covers the face of the case **(Figure 87–22)**.

Butt Hinges

Butt hinges are used on flush doors. Butt hinges for cabinet doors are a smaller version of those used on entrance doors. The leaves of the hinge are set into **gains** in the edges of the frame and the door, in the same manner as for entrance doors. Butt hinges are used on flush doors when it is desired to conceal most of the hardware. They are not often used on cabinets because they take more time to install than other types **(Figure 87–23)**.

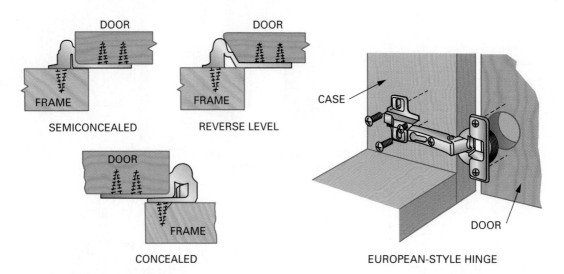

FIGURE 87–21 Overlay hinges.

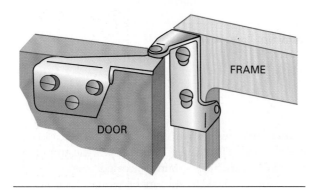

FIGURE 87–22 Pivot hinges for an overlay door.

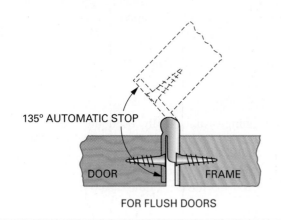

FIGURE 87–23 Butt hinges.

HANGING CABINET DOORS

Surface Hinges

To hang cabinet doors with surface hinges, first apply the hinges to the door. Then shim the door in the opening so an even joint is obtained all around. Screw the hinges to the face frame.

Semiconcealed Hinges

For semiconcealed hinges, screw the hinges to the back of the door. Then center the door in the opening. Fasten the hinges to the face frame. When more than one door is to be installed side by side, clamp a straightedge to the face frame along the bottom of the openings for the full length of the cabinet. Rest the doors on the straightedge to keep them in line **(Figure 87–24).**

Concealed Hinges

When installing concealed hinges, first screw the hinges on the door. Center the door in the opening. Press or tap on the hinge opposite the face frame. Small projections on the hinge make indentations to mark its location on the face frame.

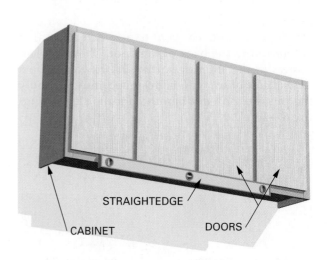

FIGURE 87–24 When installing doors, use a straightedge to keep them in line.

FIGURE 87-25 The VIX bit is a self-centering drill stop used for drilling holes for cabinet hinges.

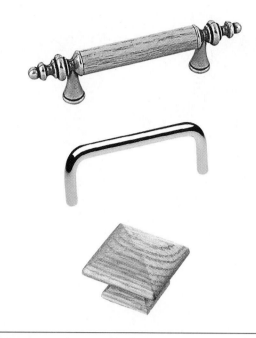

FIGURE 87-26 A few of the many styles of pulls and knobs used on cabinet doors and drawers. *(Courtesy of Amerock Corporation)*

Open the door. Place the projections of the hinges into the indentations. Screw the hinges to the face frame.

Butt Hinges

Hanging flush cabinet doors with butt hinges is done in the same manner as hanging entrance doors. Drill pilot holes for all screws so they are centered on the holes in the hinge leaf. Drilling the holes off center throws the hinge to one side when the screws are driven. This usually causes the door to be out of alignment when hung. Many carpenters use a self-centering tool, called a *VIX bit,* when drilling pilot holes for screw fastening of cabinet door hinges of all types **(Figure 87-25).** The tool centers a twist drill on the hinge leaf screw hole. It also stops at a set depth to prevent drilling through the door or face frame.

INSTALLING PULLS AND KNOBS

Cabinet pulls or knobs are used on cabinet doors and drawers. They come in many styles and designs. They are made of metal, plastic, wood, porcelain, or other material **(Figure 87-26).**

Pulls and knobs are installed by drilling holes through the door. Then fasten them with machine screws from the inside. When two screws are used to fasten a pull, the holes are drilled slightly oversize in case they are a little off center. This allows the pulls to be fastened easily without cross-threading the screws. Usually $\frac{3}{16}$-inch-diameter holes are drilled for $\frac{1}{8}$-inch machine screws. To drill holes quickly and accurately, make a template from scrap wood that fits over the door. The template can be made so that

holes can be drilled for doors that swing in either direction **(Figure 87-27).**

DOOR CATCHES

Doors without self-closing hinges need **catches** to hold them closed. Many kinds of catches are available **(Figure 87-28).** Catches should be placed where they are not in the way, such as on the bottom of shelves, instead of the top.

Magnetic catches are widely used. They are available with single or double magnets of varying holding power. An adjustable magnet is attached to the inside of the case. A metal plate is attached to the door. First attach the magnet. Then place the plate on the magnet. Close the door and tap it opposite the plate. Projections on the plate mark its location on the door. Attach the plate to the door where marked. Try the door. Adjust the magnet, if necessary.

Friction catches are installed in a similar manner to that used for magnetic catches. Fasten the adjustable section to the case and the other section to the door.

Elbow catches are used to hold one door of a double set. They are released by reaching to the back side of the door. These catches are usually used when one of the doors is locked against the other.

Bullet catches are spring loaded. They fit into the edge of the door. When the door is closed, the catch fits into a recessed plate mounted on the frame.

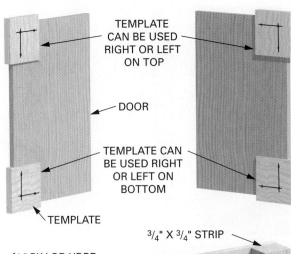

TEMPLATE CAN BE USED RIGHT OR LEFT ON TOP

DOOR

TEMPLATE CAN BE USED RIGHT OR LEFT ON BOTTOM

TEMPLATE

FIGURE 87–27
Technique for making a jig (template) to speed installation of door pulls.

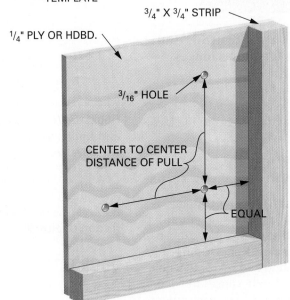

³/₄" X ³/₄" STRIP

¹/₄" PLY OR HDBD.

³/₁₆" HOLE

CENTER TO CENTER DISTANCE OF PULL

EQUAL

DOOR SIDE OF TEMPLATE

FIGURE 87–28   Several types of catches are available for use on cabinet doors. (*Courtesy of Amerock Corporation*)

DRAWER CONSTRUCTION

Drawers are classified as overlay, lipped, and flush in the same way as doors. In a cabinet installation, the drawer type should match the door type.

Drawer fronts are generally made from the same material as the cabinet doors. Drawer sides and backs are generally ½ inch thick. They may be made of solid lumber, plywood, or particleboard.

Medium-density fiberboard with a printed wood grain is also manufactured for use as drawer sides and backs. The drawer bottom is usually made of ¼-inch plywood or hardboard. Small drawers may have ⅛-inch hardboard bottoms.

Drawer Joints

Typical **joints** between the front and sides of drawers are the *dovetail*, *lock*, and *rabbet* joints. The dovetail joint is used in higher-quality drawer construction. It

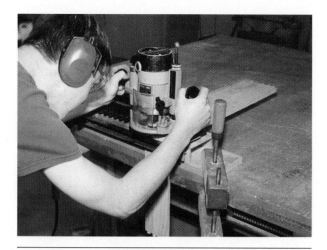

FIGURE 87–29 Dovetail joints can be made with a router and a dovetail template.

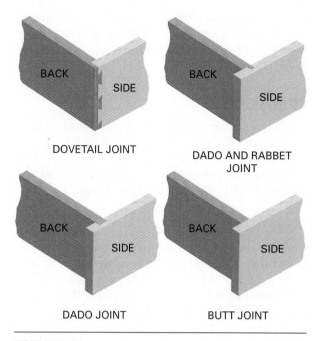

FIGURE 87–31 Various joints between drawer back and side.

takes a longer time to make, but is the strongest. Dovetail drawer joints may be made using a router and a dovetail template **(Figure 87–29).** The lock joint is simpler. It can be easily made using a table saw. The rabbet joint is the easiest to make. However, it must be strengthened with fasteners in addition to glue **(Figure 87–30).**

Joints normally used between the sides and back are the *dovetail, dado and rabbet, dado,* and *butt* joints. With the exception of the dovetail joint, the drawer back is usually set in at least ½ inch from the back ends of the sides to provide added strength. This

helps prevent the drawer back from being pulled off if the contents get stuck while opening the drawer **(Figure 87–31).**

Drawer Bottom Joints

The drawer bottom is fitted into a groove on all four sides of the drawer **(Figure 87–32).** In some cases,

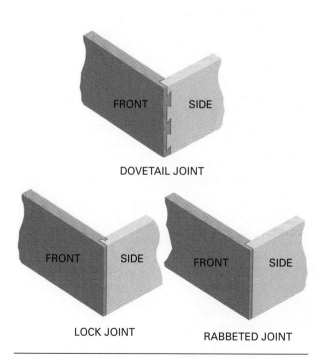

FIGURE 87–30 Various joints between drawer front and side.

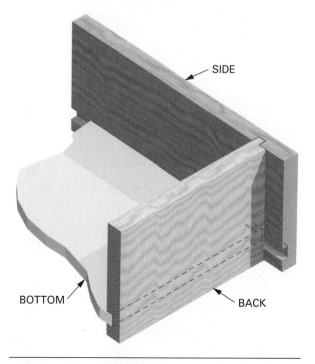

FIGURE 87–32 A drawer bottom may be fitted into a groove at the drawer back.

the drawer back is made narrower, the four sides assembled, the bottom slipped in the groove, and its back edge fastened to the bottom edge of the drawer back **(Figure 87–33)**.

DRAWER GUIDES

There are many ways of guiding drawers **(Figure 87–34)**. The type of drawer guide selected affects the size of the drawer. The drawer must be supported level and guided sideways. It must also be kept from tilting down when opened.

Wood Guides

Probably the simplest wood guide is the center strip. It is installed in the bottom center of the opening from front to back **(Figure 87–35)**. The strip projects above the bottom of the opening about ¼ inch. The bottom edge of the drawer back is notched to ride in the guide. A **kicker** is installed. It is centered above the drawer to keep it from tilting downward when opened.

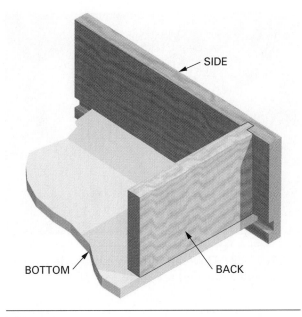

FIGURE 87–33 **A drawer bottom may be fastened to bottom edge of drawer back.**

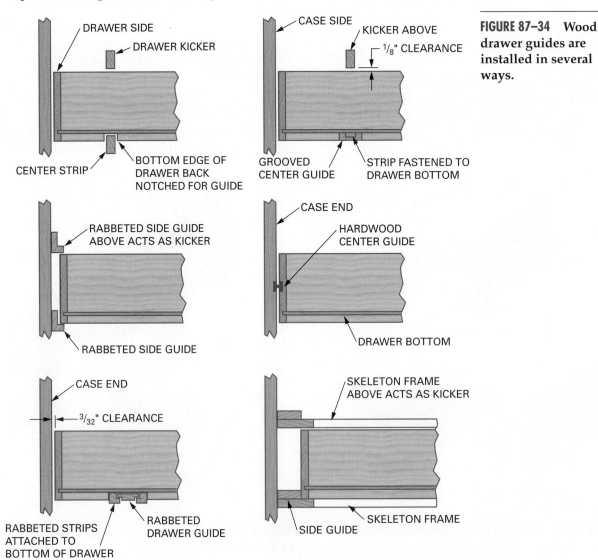

FIGURE 87–34 **Wood drawer guides are installed in several ways.**

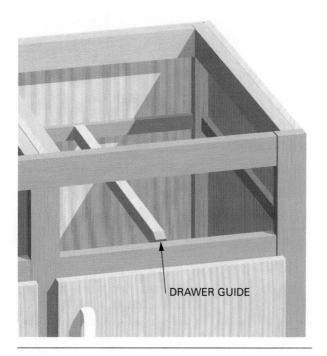

FIGURE 87–35 **Simple center wood drawer guide. The back of the drawer is notched to run on the glide.**

Another type of wood guide is the grooved center strip **(Figure 87–36).** The strip is placed in the center of the opening from front to back. A matching strip is fastened to the drawer bottom. In addition to guiding the drawer, this system keeps it from tilting when opened, eliminating the need for drawer kickers.

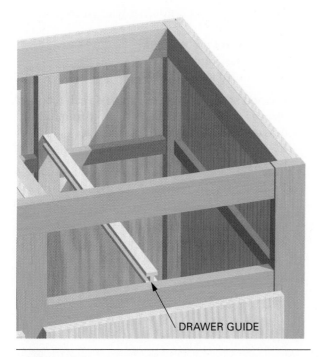

FIGURE 87–36 **The grooved center wood drawer guide eliminates the need for a kicker.**

Another type of wood guide is a rabbeted strip. Strips are used on each side of the drawer opening **(Figure 87–37).** The drawer sides fit into and slide along the rabbeted pieces. Sometimes these guides are made up of two pieces instead of rabbeting one piece. A kicker above the drawer is necessary with this type of guide.

Metal Drawer Guides

Many different types of metal drawer guides are available. Some have a single track mounted on the bottom center of the opening. Others may be centered above or on each side of the drawer. Nylon rollers mounted on the drawer ride in the track of the guide **(Figure 87–38).**

Instructions for installation differ with each type and manufacturer. When using commercially made drawer guides, read the instructions first, before making the drawer, so proper allowances for the drawer guide can be made.

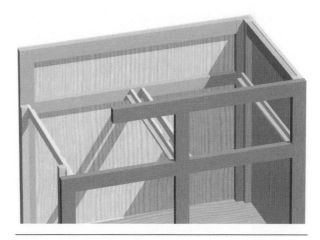

FIGURE 87–37 **Rabbeted wood guides are installed on each side of the drawer.**

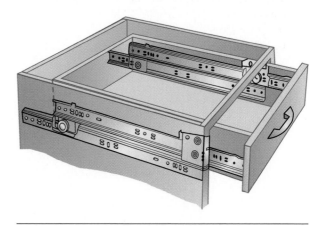

FIGURE 87–38 **Metal drawer guides. (*Courtesy of Knape and Vogt Mfg. Co.*)**

Key Terms

accessories	face frame	lipped door	postformed
base unit	frameless	mica	rough size
cabinet lift	filler	offset hinges	seam-filling
catches	flush type door	overlay door	solid
compound	J-roller	overlay hinges	surface hinges
contact cement	joints	paneled	toe space
double-faced cabinets	kicker	pilot	wall unit
	laminate trimmer	pivot	

Review Questions

Select the most appropriate answer.

1. The vertical distance between the base unit and a wall unit is usually
   a. 12 inches.   c. 18 inches.
   b. 15 inches.   d. 24 inches.

2. The distance from the floor to the surface of the countertop is usually
   a. 30 inches.   c. 36 inches.
   b. 32 inches.   d. 42 inches.

3. To accommodate sinks and provide adequate working space, the width of the countertop is
   a. 25 inches.   c. 30 inches.
   b. 28 inches.   d. 32 inches.

4. Standard wall cabinet height is
   a. 24 inches.   c. 32 inches.
   b. 30 inches.   d. 36 inches.

5. The height of most manufactured base kitchen cabinets is
   a. 30¾ inches.   c. 34½ inches.
   b. 32½ inches.   d. 35¼ inches.

6. A drawer front or door with its edges and ends rabbeted to fit over the opening is called
   a. an overlay type.   c. a lipped type.
   b. a flush type.   d. a rabbeted type.

7. The offset hinge is used on
   a. paneled doors.   c. lipped doors.
   b. flush doors.   d. overlay doors.

8. Butt hinges are used on
   a. flush doors.   c. overlay doors.
   b. lipped doors.   d. solid doors.

9. The joint used on high-quality drawers is the
   a. dado joint.
   b. dado and rabbet joint.
   c. dovetail joint.
   d. rabbeted joint.

10. A wood strip installed in cabinets with wood drawer guides to prevent a drawer from tilting downward when opened is called a
   a. top guide.   c. sleeper.
   b. kicker.   d. tilt strip.

Glossary

Accessories   in cabinetry, items used to enhance cabinet installation such as end panels, face panels for dishwashers and refrigerators, open shelves for cabinet ends, and spice racks

Acoustical board   material used to control or deaden sound

Acoustical tile   a fiberboard ceiling tile whose surface consists of small holes and/or fissures that act as sound traps

Actual size   the size of lumber after it has been surfaced

Adjustable wrench   wrench with jaws that move to various widths by turning a screw

Admixture   material used in concrete or mortar to produce special qualities

Aggregate   materials, such as sand, rock, and gravel, used to make concrete

Air entrainment   process of adding tiny air bubbles to concrete to increase its resistance to freeze and thaw cycles

Air-dried lumber   lumber that has been seasoned by drying in the air

Anchor   a device used to fasten structural members in place

Annular ring   the rings seen when viewing a cross-section of a tree trunk; each ring constitutes one year of tree growth

Apprentice   a beginner who serves for a stated period of time to learn a trade

Apron   a piece of the window trim used under the stool

Apron flashing   shingle courses that are applied under the flashing on the lower side of a chimney

Arbor   a shaft on which circular saw blades are inserted

Arch windows   windows that have a curved top or head that make them well suited to be joined in combination with a number of other types of windows or doors

Architect's scale   ruler used to draw and read measurements in various proportions or scales

Areaway   below-grade, walled area around basement windows

Argon gas   a gas used to fill the space between layers of insulated glass to increase the R-value

Asphalt felt   a building paper saturated with asphalt for waterproofing

Asphalt shingles   a type of shingle surfaced with selected mineral granules and coated with asphalt to provide weatherproofing qualities

Astragal   a semicircular molding often used to cover a joint between doors

Auger bits   wood-boring bit with a piloting screw tip

Aviation snips   metal shears with lever-action handles to increase cutting power.

Awning window   a type of window in which the sash is hinged at the top and swing outward

Back band   molding applied around the sides and tops of windows and doors

Back bevel   a bevel on the edge or end of stock toward the back side

Backing   strips or blocks installed in walls or ceilings for the purpose of fastening or supporting trim or fixtures

Backing board   a type of gypsum board designed to be used as a base layer in multilayer systems

Back miter   an angle cut starting from the end and coming back on the face of the stock

Backing the hip   beveling the top edge of a hip rafter to line it up with adjacent roof surfaces

Backset   the distance an object is set back from an edge, side, or end of stock, such as the distance a hinge is set back from the edge of a door

Backsplash   a raised portion on the back edge of a countertop to protect the wall

Balloon frame   a type of frame in which the studs are continuous from foundation to roof

Baluster   vertical members of a stair rail, usually decorative and spaced closely together

Balustrade   the entire stair rail assembly, including handrail, balusters, and posts

Barge rafter   see *rake rafter*

Base cap   a molding applied to the top edge of the baseboard

Base   see *base shoe*

Base shoe   a molding applied between the baseboard and floor

Baseboard   finish board used to cover the joint at the intersection of wall and floor

Base unit   one of two basic kinds of kitchen cabinets; the base unit is on the bottom

Batten   a thin, narrow strip usually used to cover joints between vertical boards

Batter board   a temporary framework erected to hold the stretched lines of a building layout

Bay window   a window, usually three-sided, which projects out from the wall line

Beam pocket   an indentation in a foundation wall where a girder rests

Bearer   horizontal members of a wood scaffold that support scaffold plank

Bearing   the surface area of a structural member where weight or load is transferred

Bearing partition   an interior wall that supports the floor above

Bed   a standard molding often used at the intersections of walls and ceilings

Bedding   filling of mortar, putty, or similar substance used to secure a firm bearing, as in a bed of putty for glass panes

Belt sander   a two-handed electric sanding tool with a looped strip of sandpaper

Benchmark   a reference point for determining elevations during the construction of a building

Bench plane iron   cutting portion of a bench plane

Bench planes   hand tool used to shave thin layers of wood

Bevel   the sloping edge or side of a piece for which the angle formed by the slope is not a right angle

Bevel ripping   term used to describe cutting wood with the grain and at an angle

Bifold door   doors that are hinged to each other in pairs as well as one being hinged to the jamb

Bird's mouth   a notch cut in the underside of a rafter to fit on top of the wall plate

Bit brace   two-handed drilling tool used to spin auger bits

Blind joint   a type of joint in which the cuts do not go all the way through

Blind nail   a method of fastening that conceals the nails

Blind stop   part of a window finish applied just inside the exterior casing

Blocking   short pieces of lumber installed in floor, ceiling, or wall construction to provide weather tightness, firestopping, or support for parts of the structure

Block planes   a small hand tool with a low-blade angle used to shave thin layers of wood.

Blueprinting   older process of creating copies of construction drawings where the result is blue with white lines and letters

Board   lumber usually 8 inches or more in width and less than 2 inches thick

Board foot   a volume of wood that measures 1 foot square and 1 inch thick or any equivalent lumber volume

Board paneling   paneling used on interior walls when the warmth and beauty of solid wood is desired

Boring   (what teachers are to students) term used to describe larger drilled holes

Boring jig   a tool frequently used to guide bits when boring holes for locksets

Bow   a type of warp in which the side of lumber is curved from end to end

Bow window   window units that project out from the building, often creating more floor space

Box header   in platform construction, framing members that cap the ends of the floor joists; also called a *band joist*, *rim joist*, or *joist header*

Box nail   a thin nail with a head, usually coated with a material to increase its holding power

Braces   diagonal members of a scaffold that prevent the poles from moving or buckling

Brad   a thin, short, finishing nail

Break joints   to stagger joints in adjacent rows of sheathing, siding, roofing, flooring, and similar materials

Bridging   diagonal braces or solid wood blocks between floor joists used to distribute the load imposed on the floor

Bright   term used to describe uncoated nails

Buck   a rough frame used to form openings in concrete walls

Bucks   in masonry construction, wood pieces that are fastened to the sides of a window opening against which the window may be installed

Builder's level   a telescope to which a spirit level is mounted

Building codes   see national building codes

Building permits   proof of permission to build a structure granted by the local authorities

Bullnose   a starting step that has one or both ends rounded

Butt   the joint formed when one square cut piece is placed against another; also, a type of hinge

Butt hinge   a hinge composed of two plates attached to abutting surfaces of a door and door jamb, joined by a pin

Butt hinge template   a metal template positioned and tacked into place with pins, used as a guide to rout hinge gains on doors and frames

Butt markers   tool used to outline a mortise for butt hinges

Bypass doors doors, usually two for a single opening, that are mounted on rollers and tracks so that they slide by each other

Cabinet lift a lift used to hold cabinets in position for fastening to the wall

Café doors short panel or louver doors hung in pairs that swing in both directions; used to partially screen an area yet allow easy and safe passage through the opening

Cambium layer a layer just inside the bark of a tree where new cells are formed

Cant strip a thin strip of wood placed under a piece to tilt the piece at a slant, or triangular-shaped ripping used to blunt an inside corner

Caps molding used to finish the top edge of a molding application

Carbide blade see Tungsten carbide-tipped blade

Carbide-tipped refers to cutting tools that have small, extremely hard pieces of carbide steel welded to the tips

Carpenter's level two-foot-long tool with spirit vials used to determine level and plumb

Carriage Bolt a machine-threaded bolt having a rounded head and no screw slot

Casement window a type of window in which the sash is hinged at the edge, and usually swings outward

Casing molding used to trim around doors, windows, and other openings

Catches hardware used to keep cabinet doors shut when they do not have self-closing hinges

Caulking putty-like mastic used to seal cracks and crevices

Cement board a panel product with a core of port-land cement reinforced with a glass fiber mesh embedded in both sides

Cements adhesives that bond asphalt roofing products and flashings

Centering punch a tool used to make an indentation at the centerline of holes

Ceramic tile hard, brittle, heat-resistant, and corrosion-resistant tiles made by firing a nonmetallic mineral, such as clay; ceramic tiles come in a variety of geometric shapes, sizes, patterns, colors, and designs

Chair rail molding applied horizontally along the wall to prevent chair backs from marring the wall

Chalk line string in a box with colored powder used to establish straight lines

Chamfer a beveled edge, such as that machined between the face side and edges of strip flooring prior to finishing to obscure any unevenness after installation

Chase a channel formed in buildings to run electrical, plumbing, or mechanical lines

Chase wall two closely-spaced, parallel walls constructed for the running of plumbing, heating and cooling ducts, and similar items

Check lengthwise split in the end or surface of lumber, usually resulting from more rapid drying of the end than the rest of the piece

Cheek cut a compound miter cut on the end of certain roof rafters

Clamps tools used to temporarily hold material together

Claw hammer carpenter's tool used to drive and pull nails

Cleat a piece of 1" x 5" stock that is installed around the walls of a closet to support the shelf and the rod; also, blocks fastened to the stair carriage to support the treads

Closed a type of staircase where the treads butt against the wall and are not exposed to view

Closed stairway a type of stair installed between walls where the ends of treads and risers are hidden from view

Closed valley a roof valley in which the roof covering meets in the center of the valley, completely covering the valley

Close grained wood in which the pores are small and closely spaced

Coatings a product used to resurface old roofing or metal that has become weathered

Coil stock aluminum stock sold in 50-foot-long rolls of various widths ranging from 12 to 24 inches

Cold-rolled channels (CRCs) channels formed from 54-mil steel that are often used in suspended ceilings and as bridging for lateral support in walls

Collar tie a horizontal member placed close to the ridge at right angles to the plate

Combination blade circular saw blade designed to cut across and with wood grain

Combination pliers pliers with adjustable jaws

Combination square a squaring tool with a sliding blade

Common nails fasteners used for rough work such as framing

Common rafter extends from the wall plate to the ridge board, where its run is perpendicular to the plate

Common screw multi-purpose slot-headed fastener

Compass saw short hand saw with a tapered blade used to cut curves or irregular shapes

Competent person an individual who supervises and directs scaffold erection, and who has the authority to take corrective measures to ensure that scaffolding is safe to use

Compound miter a bevel cut across the width and also through the thickness of a piece

Compression glazing a term used to describe various means of sealing monolithic and insulating glass in the supporting framing system with synthetic rubber and other elastomeric gasket materials; this method virtually eliminates the possibility of any water seeping through the joints

Concrete a building material made from portland cement, aggregates, and water

Condensation when water, in a vapor form, changes to a liquid due to cooling of the air; the resulting drops of water that accumulate on the cool surface

Conductor a vertical member used to carry water from the gutter downward to the ground; also called *downspout* or *leader*

Conical screw hollow wall fastener with deep threads used as a holder for screws

Coniferous cone-bearing tree; also known as *evergreen* tree

Connectors term used to describe a large metal fastener used to join framing members

Contact adhesives adhesives used for laminating gypsum boards to each other; the board cannot be moved after contact has been made

Contact cement an adhesive used to bond plastic laminates or other thin material; so called because the bond is made on contact, eliminating the need for clamps

Contour interval scale used between lines of a contour map

Contour line lines on a drawing representing a certain elevation of the land

Control joints metal strips with flanges on both sides of a 1/4-inch, V-shaped slot placed in large drywall areas to relieve stresses induced by expansion or contraction

Convection fluid (air) movement caused by differences in warm and cool fluid (air) densities; denser cool air is pulled downward by gravity displacing the less dense warm air

Coped joint a type of joint between moldings in which the end of one piece is cut to fit the molded surface of the other

Coping saw hand saw with short thin blade held by a bow-shaped frame used to cut irregular shapes in trim

Cordless nailer tool which drives fasteners by use of a fuel cell instead of air pressure

Corner bead metal trim used on exterior corners of walls to trim and reinforce them

Corner boards boards used to trim corners on the exterior walls of a building

Corner brace diagonal member of the wall frame used at the corners to stiffen and strengthen the wall

Corner guards molding applied to wall corners

Cornerite metal lath, cut into strips and bent at right angles, used in interior corners of walls and ceilings, on top of lath to prevent cracks in plaster

Corner post built-up stud used in the corner of a wall frame

Cornice the entire finished assembly where the walls of a structure meet the roof; sometimes called the eaves

Counterbore boring a larger hole partway through the stock so that the head of a fastener can be recessed

Countersink making a flared depression around the top of a hole to receive the head of a flathead screw; also, the tool used to make the depression

Course a continuous row of building material, such as brick, siding, roofing, or flooring

Coverage the number of overlapping layers of roofing and the degree of weather protection offered by roofing material

Cove a concave-shaped molding

Crawl space foundation type creating a space under the first floor which is not tall enough to allow a full basement

Cricket a small, false roof built behind a chimney or other roof obstacle for the purpose of shedding water; also called *saddle*

Cripple jack rafter a common rafter cut shorter that does not contact either a top plate or a ridge

Crook a type of warp in which the edge of lumber is not straight

Crosscut a cut made across the grain of lumber

Crosscut circular saw blade saw blade designed to cut best when cutting across wood grain

Crosscut saws cutting tools designed to cut across wood grain

Cross tees metal pieces that come in 2- and 4-foot lengths with connecting tabs on each end; the tabs are inserted and locked into main runners and other cross tees in a suspended ceiling grid

Crown see *bed*

Crown usually referred to as the high point of the crooked edge of joists, rafters, and other framing members

Cup a type of warp in which the side of a board is curved from edge to edge

Cylindrical lockset hardware for shutting or locking a door that is often called a *key-in-the-knob lockset*

Dado a cut, partway through, and across the grain of lumber

Deadbolt door-locking bolt operated by a key from the outside and by a handle or key from the inside

Deadman a T-shaped wood device used to support ceiling drywall panels and other objects

Deciduous trees that shed leaves each year

Deck the wood roof surface to which roofing materials are applied

Deck board lumber that covers the joist and girder platform to provide the surface and walking area of the deck

Detail a drawing showing a close-up or zoomed-in view of part of another drawing

Dew point temperature at which moisture in the air condenses into drops

Diagonal   at an angle, usually from corner to corner in a straight line

Dimension lumber   wood used for framing having a nominal thickness of two inches

Door casings   the molding used to trim the sides and top of a door

Door jack   a manufactured or job-made jack that holds a door steady during installation

Door schedules   informational chart found on a set of prints providing pertinent information on doors of the building

Door stop   molding fastened to door jambs for the door to stop against

Dormer   a structure that projects out from a sloping roof to form another roofed area to provide a surface for the installation of windows

Double coverage   a concealed-nail type of roll roofing

Double pole   a type of wood scaffold used when the scaffolding must be kept clear of the wall for the application of materials or for other reasons

Double-strength glass   glass about 1/8 inch thick and used for larger lights of glass

Double-acting doors   doors that swing in both directions or the hinges used on these doors

Doubled-faced cabinets   cabinets that have doors on both sides for use above island and peninsular bases

Double-hung window   a window in which two sashes slide vertically by each other

Dovetail   a type of interlocking joint resembling the shape of a dove's tail

Dowel   hardwood rods of various diameters

Downspout   see *conductor*

Draft   see *firestop*

Draftstop blocking   see *firestop blocking*

Drilling   term uses to describe cutting holes with turning bits

Drip   that part of a cornice or a course of horizontal siding that projects below another part; also, a channel cut in the underside of a windowsill that causes water to drop off instead of running back and down the wall

Drip cap   a molding placed on the top of exterior door and window casings for the purpose of shedding water away from the units

Drip edge   metal edging strips placed on roof edges to provide a support for the overhang of the roofing material

Drive anchor   a solid wall fastener that is set with a hammer

Drop-in anchor   a solid wall fastener that requires a setting tool used to support bolt threads

Dropping the hip   increasing the depth of the hip rafter seat cut so that the centerline of its top edge will lie in the plane of adjacent roof surfaces

Dry rot   dry, powdery residue of wood left after fungus destruction of wood due to excessive moisture

Dry well   gravel or stone-filled excavation for catching water so it can be absorbed into the earth

Drying potential   a building constructed so that building materials can dry easily after exposure to severe weather

Drywall   a type of construction usually referred to as the installation of gypsum board

Drywall cutout tool   a portable electric tool used to make cutouts in wall panels for electrical boxes, ducts, and similar objects

Drywall foot lifter   a tool used to lift gypsum panels in the bottom course up against the top panels

Duplex nail   a double-headed nail used for temporary fastening such as in the construction of wood scaffolds

Dutch hip   a roof consisting of a partial hip and partial gable roof

Dutchman   an odd-shaped piece usually used to fill or cover an opening

Eased edge   an edge of lumber whose sharp corners have been rounded

Easement   the curved piece that joins the first- and second-flight handrails in a balustrade

Eaves   that part of a roof that extends beyond the sidewall

Edge   the narrow surface of lumber running with the grain

Edge grain   boards in which the annular rings are at or near perpendicular to the face; sometimes called vertical grain

Electrolysis   the decomposition of the softer of two unlike metals in contact with each other in the presence of water

Elevation   a drawing in which the height of the structure or object is shown; also, the height of a specific point in relation to another point

Emissivity   a measure of a material's ability to give off heat

End   the extremities of a piece of lumber

End lap   the horizontal distance that the ends of roofing in the same course overlap each other

Engineered Lumber Products (ELP)   manufactured lumber substitutes, such as wood I-beams, glue-laminated beams, laminated veneer lumber, parallel strand lumber, and laminated strand lumber

Equilibrium   a state of balance; heat energy is thought to move from warmer materials into cooler ones in an attempt to reach this state

Equilibrium moisture content   the point at which the moisture content of wood is equal to the moisture content of the surrounding air

Escutcheon   protective plate covering the knob or key hole in doors

Expanded polystyrene   a type of rigid foam plastic insulation that will absorb moisture if it comes in contact with water; also called white bead board

Expansion anchors hollow wall fastener that spreads out behind the drywall to secure screws

Expansive bit wood-drilling tool that adjusts to various diameters

Exposure the amount that courses of siding or roofing are exposed to the weather

Extension jambs extensions applied to the inside edge of window jambs to accommodate various wall thicknesses

Extension ladder a ladder whose length can be extended by pulling on the rope and raising it to the desired height

Extruded polystyrene a type of rigid foam plastic insulation with a closed-cell structure that will not absorb water

Face the best-appearing side of a piece of wood or the side that is exposed when installed, such as finish flooring

Face frame a framework of narrow pieces on the face of a cabinet containing door and drawer openings

Face nailing driving a fastener nearly perpendicular to the material surface

Faceplate marker a tool used to lay out the mortise for the latch faceplate

Fall protection an OSHA requirement when using scaffolding where workers are at heights above 10 feet; consists of either guardrails or a personal fall protection system

Fascia a vertical member of the cornice finish installed on the tail end of rafters

Feather boards guide tool used to secure material against a fence while it is being fed into stationary power tools

Feather edge the edge of material brought down in a long taper to a very thin edge, such as a wood shingle tip

Fence a guide for ripping lumber on a table saw

Fiberboard building material made from fine wood chips pressed into sheets

Fiber-saturation point the moisture content of wood when the cell cavities are empty but the cell walls are still saturated

Fill coat another name for the second coat of drywall compound that is applied to cover wall joints, metal trims, and fasteners

Filler in cabinetry, pieces that are used to fill small gaps in width between wall and base units when no combination of sizes can fill the existing space

Fillet small strips used to fill a space, such as between balusters in a plowed handrail

Fillets narrow strips used between balusters to fill the plowed groove on handrails and shoe rails

Finish carpentry that part of the carpentry trade involved with the application of exterior and interior finish

Finger joint joints made in a mill used to join short lengths together to make long lengths

Finish nail a thin nail with a small head designed for setting below the surface of finish material

Finish schedules informational chart found on a set of prints providing pertinent information on interior design of a building

Finish stringer the finish board running with the slope of the stairs and covering the joint between the stairs and the wall; also called a *skirt board*

Finishing coat the third and final coat of drywall compound applied to cover wall joints, metal trims, and fasteners

Firecut an angle cut made on the ends of floor joists bearing in a masonry wall designed to prevent the masonry wall from toppling in case the joists are burned through and collapse

Firestop blocking material installed to slow the movement of fire and smoke within smaller cavities of a building frame during a fire; also called *draftstop blocking*

Firestop material used to fill air passages in a frame to prevent the spread of fire; and hot gases; also called *draft stop*

Firsts and Seconds the best grade of hardwood lumber

Fissured irregularly shaped grooves made in material, such as ceiling tile, for acoustical purposes

Fittings short sections of specially curved handrail

Fixed windows windows that consist of a frame in which the sash is fitted in a fixed position

Flange size the steel term that is similar to the thickness of a piece of wood

Flashing material used at intersections such as roof valleys, dormers, and above windows and doors to prevent the entrance of water

Flat bar general purpose prying tool with flatten claws

Flat grain grain in which the annular rings of lumber lie close to parallel to the sides; opposite of edge grain

Flexible insulation a type of insulation manufactured in blanket or batt form

Floor guide hardware installed at the bottom of bypass doors to keep the doors in alignment

Floor joists horizontal members of a frame that rest on and transfer the load to sills and girders

Floor plans pages of a set of construction drawings showing the walls as viewed from above

Flush term used to describe when the surface joint between two materials is perfectly aligned

Flush door a door with a smooth, flat surface of wood veneer or metal

Flush type door a type of cabinet door that fits into and flush with the face of the opening

Flute concave groove in lumber; usually a number are closely spaced for decorative purposes as in a column, post, or pilaster

Fly rafter see *rake rafter*

Foamed-in-place a urethane foam insulation produced by mixing two chemicals that are injected into place to expand on contact with surfaces

Footing a foundation for a column, wall, or chimney made wider than the object it supports, to distribute the weight over a greater area

Foundation that part of a wall on which the major portion of the structure is erected

Foundation wall the supporting portion of a structure below the first floor construction, or below grade, including footings

Frameless a type of cabinet construction that does not use a face frame; also called *European method*

Framing hammer heavier (20-32 ounce) hammers used in framing

Framing square L-shaped steel tool, 24 inches long, used to lay out rafters and stairs

French door a door, usually one of a pair, of light construction with glass panes extending for most of its length

Frieze a part of the exterior finish applied at the intersection of an overhanging cornice and the wall

Frost line the depth to which frost penetrates into the ground in a particular area; footings must be placed below this depth

Full rounds a round piece of molding often used for such things as closet poles

Full sections pages of a set of construction drawings showing the cross section of the building

Furring channels hat-shaped pieces made of 18- and 33-mil steel that are applied to walls and ceilings for the screw attachment of gypsum panels

Furring strips strips that are usually attached directly to exposed joists to provide a surface for fastening ceiling tiles

Gable end the triangular-shaped section on the end of a building formed by the rafters in a common or gable roof and the top plate line

Gable end studs studs that form the wall closing in the triangular area under a gable roof

Gable roof a type of roof that pitches in two directions

Gables the triangular areas formed by the rake rafters and the wall plate at the ends of the building

Gable studs a stud whose bottom end is cut square and fit against the top of the wall plate, and whose top end fits snugly against the bottom edge and inside face of the end rafter to frame the gable

Gain a cutout made in a piece to receive another piece, such as a cutout for a butt hinge

Galvanized protected from rusting by a coating of zinc

Galvanized steel steel that is coated with zinc to help protect against corrosion

Gambrel roof a type of roof that has two slopes of different pitches on each side of center

Girder a heavy timber or beam used to support vertical loads

Glass bead small molding used to hold lights of glass in place

Glazier a person who installs glass in a frame

Glazing the act of installing glass in a frame

Glazing compound a soft, plastic-type material, similar to putty, used for sealing lights of glass in a frame

Glazing points small, triangular, or diamond-shaped pieces of metal used to secure and hold lights of glass in a frame

Glue-laminated lumber (glulam) large beams or columns made by gluing smaller-dimension lumber together side to side

Gooseneck a curved section of handrail used when approaching a landing; also, an outlet in a roof gutter

Grade the level of the ground; also identifies the quality of lumber

Grade rod the height of the instrument minus the control point elevation

Grain in wood, the design on the surfaces caused by the contrast, spacing, and direction of the annular rings

Graphite a mineral used as pencil lead and also as a lubricant for the working parts of locks and certain tools

Green a term applied to concrete that has not fully cured

Green concrete a term used to describe concrete that has not yet cured to higher strength

Green lumber lumber that has not been dried to a suitable moisture content

Green space areas of a building site devoted to natural vegetation

Grilles on a window, false muntins applied as an overlay to simulate small lights of glass

Grit material of sandpaper that actually provides the sanding action

Groove a cut, partway through, and running with the grain of lumber

Ground strips of wood placed at the base of walls and around openings and used as a guide for the application of an even thickness of plaster; also, a system used for electrical safety

Ground Fault Circuit Interrupter (GFCI) device used in electrical circuits for protection against electrical shock; it detects a short circuit instantly and shuts off the power automatically

Grout a mixture of cement, fine aggregate, and water used to fill joints in masonry and tile

Guardrails rails installed on all open sides and ends of scaffolds that are more than 10 feet in height

Gusset a pad of wood or metal used over a joint to stiffen and strengthen it

Gutter a wood or metal trough used at the roof edge to carry off rain water and water from melting snow

Gypsum   a chalky type rock that is the basic ingredient of plaster

Gypsum board   a sheet product made by encasing gypsum in a heavy paper wrapping

Gypsum lath   a type of gypsum board used as a base to receive conventional plaster

Hacksaws   saws designed to cut metal

Half round   a molding with its end section in the shape of a semicircle

Hammer-drills   electric drilling tools that also provide an impact action to speed drilling holes in masonry

Handrail   a railing on a stairway intended to be grasped by the hand to serve as a support and guard

Handrail bolt   a bolt with threads on one end designed for fastening into wood and threads for a nut on the other end; used to join the starting fitting to a straight section of handrail on the balustrade

Hand scraper   tool designed to remove thin layers of material from various surfaces

Hardboard   a building product made by compressing wood fibers into sheet form

Hardwoods   the wood of broad-leaved dicotyledonous trees (as distinguished from the wood of conifers)

Head casing   the top member of a window unit's exterior casing

Head flashing   the flashing on the upper side of a chimney

Head jamb   the top horizontal member of a window frame

Head lap   the distance from the bottom edge of an overlapping shingle to the top of a shingle two courses under, measured up the slope

Header   pieces placed at right angles to joists, studs, and rafters to form and support openings in a wood frame

Hearth   an area near a fireplace, usually paved and extending out into a room, around which a wood floor installation must be framed

Heartwood   the wood in the inner part of a tree, usually darker and containing inactive cells

Heel   the back end of objects, such as a handsaw or hand plane

Height of the instrument   measurement found by placing the leveling rod on the benchmark, then adding the rod reading to the elevation of the benchmark

Herringbone   a pattern used in parquet floors

Hexagon   a plane figure having six sides

High-speed steel blade   saw blades with no carbide cutting tips

Hinge gain   the recessed area for the hinge on a door

Hinge mortise   see *hinge gain*

Hip jack rafter   a shortened common rafter that spans from the wall plate to a hip rafter

Hip rafter   extends diagonally from the corner of the plate to the ridge at the intersection of two surfaces of a hip roof

Hip roof   a roof that slopes upward toward the ridge from four directions

Hip-valley cripple jack rafter   a short rafter running parallel to common rafters, cut between hip and valley rafters

Hole saws   drills that cut with saw teeth along its perimeter

Hopper window   a type of window in which the sash is hinged at the bottom and swings inward

Horizontal circle scale   a scale that is divided into quadrants of 90 degrees each and remains stationary as the telescope is turned

Horizontal vernier   a scale that rotates with the telescope and is used to read minutes of a degree

Horn   an extension of the stiles of doors or the side jambs of window and door frames

Hose bibbs   external water faucets of a building

Housed stringer   a finished stringer that is dadoed to receive treads and risers of a stairway

Housewrap   type of building paper with which the entire sidewalls of a building are covered

Hydration   chemical reaction of cement and water causing concrete or mortar to harden

Ice and water shield flashing   flashing applied whenever there is a possibility of ice dams forming along the eaves and causing a backup of water

Ice dam   ice that forms on an overhang, causing water buildup behind it to back up under roofing material

Impact noise   noise caused by the vibration of an impact, such as dropped objects or footsteps on a floor

Impact noise rating (INR)   a rating that shows the resistance of various types of floor-ceiling construction to impact noises

Independent slab   a concrete slab that is separate from the foundation

Insulated glass   double- or triple-glazed windows; improved insulated glass design uses argon gas between two panes of glass

Insulation   material used to restrict the passage of heat or sound

Intersecting roof   the roof of irregularly shaped buildings; valleys are formed at the intersection of the roofs

Isometric   a drawing in which three surfaces of an object are seen in one view, with the base of each surface drawn at a 30° angle

Jack rafter   part of a common rafter, shortened for framing a hip rafter, a valley rafter, or both

Jack studs   shortened studs that support the headers

Jamb   the sides and top of window and door frames

Jamb extension narrow strips of wood fastened to the edge of window jambs to increase their width

Jig any type of fixture designed to hold pieces or guide tools while work is being performed

Jigsaw a small hand-held electric saw that cuts with a stroking action

Joint compound a substance similar to plaster used to cover joints or the heads of screws or nails in plasterboard

Jointer plane largest of the bench planes used to finish wood by removing thin strips of wood

Joint reinforcing tape material used to cover, strengthen, and provide crack resistance to drywall joints

Joint as a verb, denotes straightening the edge of lumber; as a noun, means the place where parts meet and unite

Joist horizontal framing members used in a spaced pattern that provide support for the floor or ceiling system

Joist hanger metal stirrups used to support the ends of joists

Joist header see *box header*

Journeyman a tradesman who has completed an apprenticeship or who has gained enough experience to perform work without instruction

J-roller a 3-inch-wide rubber roller used to apply pressure over the surface of contact cement-bonded plastic laminates

Juvenile wood the portion of wood that contains the first seven to fifteen growth rings of a log. They are located in the pith.

Kerf the width of a cut made with a saw

Keyhole saw thinner version of a compass saw

Keyway a groove made in concrete footings for tying in the concrete foundation wall

Kicker in drawer construction, a piece centered above the drawer to keep it from tilting downward when opened

Kiln-dried lumber dried by placing it in huge ovens called kilns

Kneewall in relation to stairs, a short wall that projects a short distance above and on the same rake as the stair body

Knot a defect in lumber caused by cutting through a branch or limb embedded in the log

Ladder jacks metal brackets installed on ladders to hold scaffold planks

Lag screws large threaded screw with a hex head

Lag shield anchor used to secure lags in solid walls.

Laminate a thin layer of plastic often used as a finished surface for countertops

Laminate trimmer small router used to cut and fit plastic laminate in cabinet making

Laminated block a type of parquet block generally made of three-ply laminated oak in a 3/8-inch thickness

Laminated glass a type of safety glass that has two or more layers of glass with inner layers of transparent plastic bonded together

Laminated Strand Lumber (LSL) lumber manufactured by bonding thin strands of wood, up to 12 inches long, with adhesive and pressure

Laminated strip flooring a five-ply prefinished wood assembly

Laminated Veneer Lumber (LVL) lumber manufactured by laminating many veneers of plywood with the grain of all running in the same direction

Landing an intermediate-level platform between flights

Laser level a level that emits a red beam and can rotate through a full 360 degrees, creating a level plane of light

Laser a concentrated, narrow beam of light; laser-equipped devices are used in building construction

Lateral bracing temporary or permanent bracing that runs perpendicular to the braced members

Lath a base for plaster; usually gypsum board or expanded metal sheets

Lattice thin strips of wood, spaced apart and applied in two layers at angles to each layer resulting in a kind of grillwork

Lazy Susan a set of revolving circular shelves; used in kitchen cabinets and other places

Ledger a horizontal member of a wood scaffold that ties the scaffold posts together and supports the bearers; a temporary or permanent supporting member for joists or other members running at right angles

Level line any line on the rafter that is horizontal when the rafter is in position

Level horizontal or perpendicular to the force of gravity

Light a pane of glass or an opening for a pane of glass

Lignin the natural glue in wood that holds together the wood cells and fibers

Line length the length of a rafter along a measuring line without consideration to the width or thickness of the rafter

Line level device suspended from a string to determine level

Linear feet a one-dimensional measurement of length in foot increments

Linear measure a measurement of length

Lintel horizontal load-bearing member over an opening; also called a *header*

Lipped door a type of cabinet door that has rabbeted edges that overlap the opening by about 3/8 inch on all sides

Load-bearing partitions (LBP) partitions that support the inner ends of ceiling or floor joists

Lock rail a rail situated at lockset height, usually 38 inches from the finish floor to its center

Lockset a set of hardware for shutting or locking a door

Lookout horizontal framing pieces in a cornice, installed to provide fastening for the soffit

Loose-fill insulation a type of insulation usually composed of materials in bulk form and is supplied in bags or bales

Louver door a door made with spaced horizontal slats called louvers used in place of panels; the louvers are installed at an angle to obstruct vision but permit the flow of air through the door

Louver an opening for ventilation consisting of horizontal slats installed at an angle to exclude rain, light, and vision, but to allow the passage of air

Low-emissivity coating a coating on the inner glass surface of a window that allows ultraviolet rays from the sun to pass but reflects the inside infrared (radiant) rays back into the building

Low-emissivity glass (LoE) a coating on double-glazed windows designed to raise the insulating value by reflecting heat

Lumber wood that is cut from the log to form boards, planks, and timbers

Lumber grades numbers and letters used to rank wood according to quality

Machine bolt hex-headed threaded fastener typically used with a nut and washer

Magazine a container in power nailers and staplers in which the fasteners to be ejected are placed

Main runners metal pieces shaped in the form of an upside-down T that are the primary support of a suspended ceiling's weight; main runners extend from wall to wall

Major span width of the larger portion of a building with six or eight corners

Mansard roof a type of roof that has two different pitches on all sides of the building, with the lower slopes steeper than the upper

Mantel the ornamental finish around a fireplace, including the shelf above the opening

Masonry any construction of stone, brick, tile, concrete plaster, and similar materials

Masonry drill bits drill bits with a carbide tip designed primarily to drill masonry products

Masonry nails hardened steel fasteners used in cementitious materials

Mastic a thick adhesive

Matched boards boards that have been finished with tongue-and-grooved edges

Medullary ray bands of cells radiating from the cambium layer to the pith of a tree to transport nourishment toward the center

Metal corner tape material used to protect corners on arches, windows with no trim, and other locations; applied by embedding in joint compound

Metes and bounds boundaries established by distances and compass directions

Millwork any wood products that have been manufactured, such as moldings, doors, windows, and stairs for use in building construction; sometimes called joinery

Minor span width of the smaller portion of a building with six or eight corners

Miter the cutting of the end of a piece at any angle other than a right angle

Miter boxes a fixed or adjustable tool for guiding handsaws in cutting miter joints or in making crosscuts

Miter gauge a guide used on the table saw for making miters and square ends

Miter joint the joining of two pieces by cutting the end of each piece by bisecting the angle at which they are joined

Mobile scaffold scaffold whose components include casters and horizontal diagonal bracing; also called a *rolling tower*

Modular construction a method of construction in which parts are preassembled in convenient-sized units

Modular measurement process of designing structures to best fit standard material sizes

Moisture content the amount of moisture in wood expressed as a percentage of the dry weight

Moisture meter a device used to determine the moisture content of wood

Molding decorative strips of wood used for finishing purposes

Monolithic slab a combined slab and foundation; also referred to as a thickened edge slab

Mortar a mixture of portland cement, lime, sand, and water used to bond masonry units together

Mortise a rectangular cavity cut in a piece of wood to receive a tongue or tenon projecting from another piece

Mud sill typically a 2 x 10 board approximately 18 to 24 inches long upon which scaffold baseplates rest

Mullion a vertical division between windows or panels in a door

Multispur bit a power-driven bit, guided by a boring jig, that is used to make the hole in a door for the lockset

Muntin slender strips of wood between lights of glass in windows or doors

Nail claw prying tool used to remove nails that have been driven all the way

Nailers general term use to describe power tools used to drive nails

Nail set tool used to drive nail heads below the surface

National building codes rules and regulations guiding the construction industry as set by national agencies

Newel posts a post to which the end of a stair railing or balustrade is fastened; also, any post to which a railing or balustrade is fastened

No. 1 common a lower grade of hardwood lumber

Nominal size the stated size of the thickness and width of lumber even though it differs from its actual size; the approximate size of rough lumber before it is surfaced

Nonconforming term used to describe buildings that do not fit the local zoning laws

Non-load-bearing partitions (NLBP) partitions built to divide a space into rooms of varying size carrying only the load of the partition material itself and no structural load from the rest of the building

Nosing the rounded edge of a stair tread projecting over the riser

OC ("on center") the distance from the center of one structural member to the center of the next one

Octagon a plane figure with eight sides

Offset hinges a type of cabinet hinge used on lipped doors

Ogee a molding with an S-shaped curve

Open a type of staircase where the ends of the treads are exposed to view

Open-grained a texture quality of wood where wood cells or pores are open to the surface

Open stairway type of stairway with the ends of the treads and risers visible

Open valley a roof valley in which the roof covering is kept back from the centerline of the valley

Optical levels commonly used instruments for leveling, plumbing, and angle layout; includes the builder's level and transit-level

Orthographic multiview drawings

Overlay door a type of cabinet door that laps over the opening, usually 3/8 inch on all sides

Overlay hinges a type of cabinet hinge that is available in either semiconcealed or concealed

Over-the-post a balustrade system which utilizes fittings to go over newels for an unbroken, continuous handrail

Panel a large sheet of building material

Panel door a door that consists of a frame with usually one to eight wood panels in various designs

Paneled a type of cabinet door construction that has an exterior framework of solid wood with panels of solid wood, plywood, hardboard, plastic, glass, or other panel material

Parallel Strand Lumber (PSL) lumber manufactured by bonding strands of structural lumber, up to 8 feet long, with adhesive, heat, and pressure

Parquet a floor made with strips or blocks to form intricate designs

Parquet block a type of flooring consisting of square or rectangular blocks; sometimes installed in combination with strips, to form mosaic designs

Parquet strip a type of flooring with short strips that are laid to form various mosaic designs

Partial section elevations orthographic drawing showing only one side of the outside of the building at a distance of about 100 feet

Particleboard a building product made by compressing wood chips and sawdust with adhesives to form sheets

Parting strip a small strip of wood separating the upper and lower sash of a double-hung window

Partition an interior wall separating one portion of a building from another

Passage lockset a lockset with knobs on both sides that are turned to unlatch the door; used when it is not desirable to lock the door

Patio door a door unit normally consisting of two or three sliding or swinging glass doors completely assembled in the frame

Penny (d) a term used in designating nail sizes

Perforated material that has closely spaced holes in a regular or irregular pattern

Perlite loose-fill insulation sometimes mixed into concrete to make it lighter in weight and a better insulator

Perm a measure of water vapor movement through a material

Phillips head a type of screw head with a cross-slot

Pier a column of masonry, usually rectangular in horizontal cross-section, used to support other structural members

Pilaster column built within and usually projecting from a wall to reinforce the wall

Pile concrete, metal, or wood pillar forced into the earth or cast in place as a foundation support

Pilot a guide on the end of edge-forming router bits used to control the amount of cut

Pilot hole a small hole drilled to receive the threaded portion of a wood screw

Pitch the amount of slope to a roof expressed as a ratio of the total rise to the span

Pitch block a piece of wood cut in the shape of a right triangle; used as a pattern for laying out stair stringers, rafters, and handrail fittings

Pitch board a block of wood that is cut to form a right triangle using the rise and run dimensions of the stair

Pitch pocket an opening in lumber between annular rings containing pitch in either liquid or solid form

Pith the small, soft core at the center of a tree

Pivot revolving around a point

Plain-sawed a method of sawing lumber that produces flat-grain

Plan in an architectural drawing, a object drawn as viewed from above

Plan view blueprint that shows the building from above looking down

Plancier the finish member on the underside of a box cornice; also called *soffit*

Plank   lumber that is 6 or more inches in width and from 1½ to 6 inches in thickness

Plank flooring   a type of solid wood flooring that may be laid using alternating widths

Plaster   a mixture of portland cement, sand, and water used for covering walls and ceilings of a building

Plastic laminate   a very tough, thin material in sheet form used to cover countertops; available in a wide variety of colors and designs

Plastic toggle   hollow wall anchor use to secure screws

Plate   top or bottom horizontal member of a wall frame

Platform frame   method of wood frame construction in which walls are erected on a previously constructed floor deck or platform

Plinth block   a small, decorative block, thicker and wider than a door casing, used as part of the door trim at the base and at the head

Plot plan   a drawing showing a bird's-eye view of the lot, the position of the building, and other pertinent information; also called *site plan*

Plow   a wide groove cut with the grain of lumber

Plumb   vertical; at right angles to level

Plumb bob   a pointed weight attached to a line for testing plumb

Plumb line   any line on the rafter that is vertical when the rafter is in position

Plunge cut   an interior cut made with a portable saw by a method that does not require holes to be bored before making the cut

Ply   one thickness of several layers of built-up material, such as one of the layers of plywood

Plywood   a building material in which thin sheets of wood are glued together with the grain of adjacent layers at right angles to each other

Pneumatic   powered by compressed air

Pocket door   a door that slides sideways into the interior of a partition; when opened, only the lockedge of the door is visible

Pocket tapes   measuring device made of a steel self-retracting strip

Pocket   a recess in a wall to receive a piece, such as a recess in a concrete foundation wall to receive the end of a girder

Point of beginning   mark on plot plan indicating the start point for laying out the lot. Usually a large object that is unlikely to move during construction such as a large rock or tree is used

Poles   the vertical members of a scaffold

Polyethylene film   a thin plastic sheet used as a vapor barrier

Polyurethane   a type of rigid foam insulation made with a facing of foil or building paper

Portland cement   an improved type of cement that is a fine gray powder

Postforming   method used to bend plastic laminate to small radii

Post-to-post   a balustrade system where the handrail is cut and attached between newels

Powder-actuated drivers   tool that uses gunpowder to drive fasteners into metal and masonry

Power miter saw   tool used to cut trim material at various angles

Preacher   a small piece of wood of the same thickness as stair risers; it is notched in the center to fit over the finish stringer and rests on the tread cut of the stair carriage; it is used to lay out open finished stringers in a staircase

Prefinished strip   solid wood flooring that is sanded, finished, and waxed at the factory; it cannot be sanded after installation

Prehung door   a door that comes already fitted and hinged in the door frame with the lockset and outside casings installed

Preservative   a substance applied to wood to prevent decay

Pressure-treated   treatment given to lumber that applies preservative under pressure to penetrate the total piece

Primer   the first coat of paint applied to the surface or the paint used to prime a surface

Privacy lock   a lock often used on bathroom and bedroom doors that functions by pushing or turning a button on the room side of the doorknob

Pullsaws   hand saws that cut while being drawn towards the user

Pump jack   a type of scaffold that consists of 4 x 4 poles, a pump jack mechanism, and metal braces for each pole

Purlin   horizontal roof member used to support rafters between the plate and ridge

Purlin knee wall   supporting timbers at which rafters may intersect each other in a gambrel roof

Putty   glazing compound used to cover and seal around the edges of glass panes

Pythagorean theorem   a theorem that can be used to determine the diagonal of any right triangle: $a^2 + b^2 = c^2$

Quarter round   a type of molding, an end section of which is in the form of a quarter-circle

Quarter-sawed   a method of sawing lumber parallel to the medullary rays to produce edge-grain lumber; see *edge grain*

Rabbet   a cutout along the edge or end of lumber

Rabbeted jamb   a type of side door jamb in which the rabbet acts as the stop; on a double-rabbeted jamb, one rabbet is used as a stop for the main door and the other as a stop for storm and screen doors

Racking   term used to describe the lateral movement of a wall where the 90-degree angle between the studs and plates is affected

Racking the floor   the practice of laying out seven or eight loose rows of flooring, end to end, after the second course of flooring has been fastened; done to save time and material

Radial arm saw   older style stationary saw used for crosscuting and ripping

Radiation   a general word referring to electromagnetic radiation, which includes microwaves, radio waves, infrared, visible light, ultraviolet light, x rays, and cosmic rays

Radius edge decking   specially shaped lumber that is usually used to provide the surface and walking area of the deck

Rafter   a sloping structural member of a roof frame that supports the roof sheathing and covering

Rafter tables   information found printed on the body of a framing square; used to calculate the lengths of various components of a roof system

Rail   the horizontal member of a frame

Rake   the sloping portion of the gable ends of a building

Rake rafter   the first and last rafter of a gable roof, usually having a finish or trim applied to it; also called *barge* or *fly rafter*

Rebar   steel reinforcing rod used in concrete

Reciprocating   a back-and-forth action, as in certain power tools

Reciprocating saw   general term used to describe saws which cut in a back and forth or stroking motion

Reinforced concrete   concrete that has been reinforced with steel bars

Relative humidity   percentage of moisture suspended in air compared to the maximum amount it could hold at the same temperature

Resilient   a type of thin, flexible flooring material widely used in residential and commercial buildings that comes in sheet or tile form

Return   a turn and continuation for a short distance of a molding, cornice, or other kind of finish

Return nosing   a separate piece mitered to the open end of a stair tread for the purpose of returning the tread nosing

Reveal   the amount of setback of the casing from the face side of window and door jambs or similar pieces

Ribbon   a narrow board let into studs of a balloon frame to support floor joists

Ridge   the highest point of a roof that has sloping sides

Ridgeboard   a horizontal member of a roof frame that is placed on edge at the ridge and into which the upper ends of rafters are fastened

Rim Joist   floor member that is nailed perpendicular to the ends of joists

Rip   sawing lumber in the direction of the grain

Rip fence   guide tool used to support and maintain a straight cut in material being cut on by stationary power tools

Ripsaw blade   saw designed to cut best when cutting with the grain of the wood

Ripsaws   saws designed to cut in the direction that is parallel to wood grain

Riser   the finish member in a stairway covering the space between treads

Rise   in stairs, the vertical distance of the step; in roofs, the vertical distance from plate to ridge; may also be the vertical distance through which anything rises

Risers   vertical members that enclose the space between treads

Roll roofing   a type of roofing made of the same material as asphalt shingles that comes in rolls that are 36 inches wide

Roofing brackets   a metal bracket used when the pitch of the roof is too steep for carpenters to work on without slipping

Roofing nails   large headed galvanized fastener used to attach roofing materials

Roof window   a window for the roof that contains an operating sash to provide light and ventilation

Rosette   a round or oval decorative wood piece used to end a handrail against a wall

Rough carpentry   that part of the trade involved with construction of the building frame or other work that will be dismantled or covered by the finish

Rough opening   opening in a wall in which windows or doors are placed

Rough sills   members that form the bottom of a window opening, at right angles to the studs

Rough size   approximate size; for example, cutting countertop laminate about ¼ to ½ inch wider and longer than the surface to be covered

Rough stringer   cutout supports for the treads and risers in a staircase; also called *stair carriage* and *stair horse*

Router   hand-held electric tool used to shape material with high spinning cutters

Run   the horizontal distance over which rafters, stairs, and other like members travel

R-value   a number given to a material to indicate its resistance to the passage of heat

S2S   surfaced two sides

S4S   surfaced four sides

Saddle   see *cricket*

Safety glass   glass constructed, treated, or combined with other materials to minimize the possibility of injuries resulting from contact with it

Sapwood   the outer part of a tree just beneath the bark containing active cells

Sash that part of a window into which the glass is set

Sash balance a device, usually operated by a spring or tensioned weatherstripping, designed to counterbalance double-hung window sashes

Sawhorses frameworks for holding material that is being laid out or cut to size

Sawyer a person whose job is to cut logs into lumber

S-beam an I-shaped steel beam

Scab a length of lumber applied over a joint to stiffen and strengthen it

Scaffold an elevated, temporary working platform

Scratch coat the first coat of plaster applied to metal lath

Screed strips of wood, metal, or pipe secured in position and used as guides to level the top surface of concrete

Screwdriver bits interchangeable tips or various driving style heads

Screwdrivers tools used to turn fasteners into and out of materials

Screwguns electric drills with a special nose piece to drive wallboard screws to an exact depth

Scribe laying out woodwork to fit against an irregular surface

Scribing fitting woodwork to an irregular surface; with moldings, cutting the end of one piece to fit the molded face of the other at an interior angle to replace a miter joint

Scuttle attic access or drain through a parapet wall

Seam-filling compound a compound made especially for laminates used to make a practically invisible joint

Seasoned lumber lumber that has been dried to a suitable moisture content

Seat cut a cut on the rafter that is a combination of a level cut and a plumb cut; also called the *bird's mouth*

Section view building plan that shows a cross-section of the building as if it were sliced to reveal its skeleton

Section drawing showing a vertical cut through an object or part of an object

Self-tapping screw screws designed to drill a pilot hole as it is turned

Selvage the ungranulated or unexposed part of roll roofing covered by the course above

Setbacks distance buildings must be kept from property lines

Set alternate bending of saw teeth to provide clearance in the saw cut

Shake a defect in lumber caused by a separation of the annular rings; also, a type of wood shingle

Shank hole a hole drilled for the thicker portion of a wood screw

Shark tooth handsaw blade style with aggressive teeth that cut wood fast in both directions

Shear walls wall bracing, framed of 2 x 6s anchored to the slab, with OSB heavily nailed to them; used in regions of severe seismic activity

Sheathing boards or sheet materials that are fastened to roofs and exterior walls and on which the roof covering and siding are applied

Shed roof a type of roof that slopes in one direction only

Sheet paneling paneling, such as prefinished plywood and hardboard, that is usually applied to walls with long edges vertical

Shim a thin, wedge-shaped piece of material used behind pieces for the purpose of straightening them, or for bringing their surfaces flush at a joint

Shingle butt the bottom exposed edge of a shingle

Shingle tip the thin end of a wood shingle

Shingling hatchet a lightweight hatchet with both a sharp blade and a heel that is used to apply wood shakes and shingles

Shoe rail a plowed, lineal molding designed to receive the bottom square of a baluster

Shortened valley rafter a valley rafter that runs from the plate to the supporting valley rafter

Side the wide surfaces of a board, plank, or sheet

Side casings the side members of a window unit's exterior casing

Side flashing flashing tucked in between shingles along the sides of a chimney

Side jambs the vertical sides of a window frame

Siding exterior sidewall finish covering

Sidelight a framework containing small lights of glass placed on one or both sides of the entrance door

Sill horizontal timbers resting on the foundation supporting the framework of a building; also, the lowest horizontal member in a window or door frame

Single pole a type of wood scaffold used when it can be attached to the wall and does not interfere with the work

Single-hung window a window in which the lower sash slides vertically, but the upper sash is fixed

Single-strength glass glass about 3/32 inch thick used for small lights of glass

Skilsaw general term use to describe portable electric circular saws

Skylight a fixed-sash window for the roof that provides light only, no ventilation

Slab-on-grade foundation a solid concrete building base used instead of a foundation because it saves on material and labor

Slat block a type of parquet block made by joining many short, narrow strips together in small squares of various patterns

Sleeper strips of wood laid over a concrete floor to which finish flooring is fastened

Sleeve anchor solid wall fastener used to secure bolts

Sliding a type of exterior door. Doors are opened by sliding the panels along a track horizontally

Sliding gauge a gauge used for fast and accurate checking of shingle exposure that permits several shingle courses to be laid at a time without snapping a chalk line

Sliding windows windows that have sashes that slide horizontally in separate tracks located on the header jamb and sill

Slope term used to indicate the steepness of a roof; stated as unit rise on unit run, e.g., 6 on 12

Slump test a test given to concrete to determine its consistency

Snaplock punch a special tool used to produce raised lugs on a piece of siding so that it will snap snugly into the trim when fitting around windows

Snap tie a metal device to hold concrete wall forms the desired distance apart

Soffit the underside trim member of a cornice or any such overhanging assembly

Softboard a low-density fiberboard

Softwood wood from coniferous (cone-bearing) trees

Soil stack part of the plumbing; a vertical pipe extending up through the roof to vent the system

Sole plate the bottom horizontal member of a wall frame

Solid a type of cabinet door construction with doors are made of plywood, particle board, or solid lumber

Solid sheathing usually APA-rated panels over which shingles or shakes may be applied

Solid wall anchors metal fasteners that tighten inside a hole as a bolt or screw is turned

Sound-deadening insulating board a sound resistance system using a gypsum board outer covering and resilient steel channels placed at right angles to the studs to isolate the gypsum board from the stud

Sound transmission class (STC) a rating given to a building section, such as a wall, that measures the resistance to the passage of sound

Spaced sheathing usually 1 x 4 or 1 x 6 boards over which shingles or shakes may be applied

Specifications sometimes referred to as specs, written or printed directions of construction details for a building

Specifications writer a person who writes supplemental information for construction projects to include any information that cannot be communicated in drawings or schedules

Speed square tool used for quick layout of angles, particularly for rafters

Spike a large nail 4 inches or longer

Spline a thin, flat strip of wood inserted into the grooved edges of adjoining pieces

Split fast anchor fastener that tightens in a masonry hole by the opposing pressure of shaft legs

Spray foams a type of foam insulation in an aerosol can that is sprayed into gaps to expand and create an airtight seal

Spreader a narrow strip of wood tacked to the side jambs of a door frame a few inches up from the bottom so that the frame width is the same at the bottom as it is at the top

Spud vibrator an immersion type vibrator with a metal tube on its end that is commonly used to vibrate and consolidate concrete

Square the amount of roof covering that will cover 100 square feet of roof area

Staging see *scaffold*

Staging planks planks that rest on scaffold bearers with edges close together to form a tight platform

Stair body the part of the stair consisting mainly of the treads, risers, and finish stringers

Stair carriage see *rough stringer*

Staircase term used to refer to an entire set of stairs, which include treads, risers, stringers, and balustrade

Stair horse see *rough stringer*

Stairwell an opening in the floor for climbing or descending stairs or the space of a structure where the stairs are located

Standard moldings molding manufactured in standard patterns with each type available in several sizes; designated as bed, crown, cove, full round, half round, quarter round, base, base shoe, cap, casing, chair rail, back band, apron, stool, stop, and others.

Standing cut a cut made through the thickness of stock at more than a 90° angle between the side and edge or end

Staplers tools that drive U-shaped fasteners by either squeezing a spring-loaded lever or swinging like a hammer

Stapling guns tools that drive U-shaped fasteners by use of power provided by electricity, air pressure, or a fuel cell

Starter course usually used in reference to the first row of shingles applied to a roof or wall

Starting step the first step in a flight of stairs

Steel buck a metal door frame using all steel studs to frame around door openings

Steel tapes term used to describe long measuring tools that are rewound by hand

Step ladder a folding portable ladder hinged at the top

Stickering machine that makes moldings or a thin strip placed between layers of lumber to create an air space for drying

Sticking the molded inside edge of the frame of a panel door

Stile the outside vertical members of a frame, such as in a paneled door

Stockade fence a higher fence with picket edges applied tightly together to provide privacy

Stool cap a horizontal finish piece covering the stool or sill of a window frame on the interior; also called *stool*

Stool the bottom member of a door or window frame; also called a *sill*

Stop a strip of wood applied vertically to the door jamb upon which the door rests when closed

Stop bead a vertical member of the interior finish of a window against which the sash butts or sides

Storm sash an additional sash placed on the outside of a window to create a dead air space to prevent the loss of heat from the interior in cold weather

Storm window a secondary window attached over the usual window to protect against the wind and cold

Story the distance between the upper surface of any floor and the upper surface of the floor above

Story pole a narrow strip of wood used to lay out the heights of members of a wall frame or courses of siding

Stove bolt bolt with a machine thread and a screw-slot head

Straight tin snips shearlike tool used for all-purpose cutting of sheet material

Straightedge a length of wood or metal having at least one straight edge to be used for testing straight surfaces

Strapping application of furring strips at specified spacings for the purpose of attaching wall or ceiling finish; called stripping in some locations

Striated finish material with random and finely spaced grooves running with the grain

Striker plate a plate installed on the door jamb against which the latch on the door engages when the door is closed

Stringer the finish material applied to cover the stair carriage

Strip see *strip flooring*

Strip flooring a type of widely-used solid wood flooring most often tongue-and-grooved on edges and ends to fit precisely together

Strongback a member placed on edge and fastened to others to help support them

Stub joists joists that are shorter and run perpendicular to the normal joists

Stud vertical framing member in a wall running between plates

Supporting valley rafter a rafter that runs from the plate to the ridge of the main roof

Surfaced four sides smooth, square-edge lumber used for door casings; abbreviated as *S4S*

Surface hinges a type of cabinet hinge that is applied to the exterior surface of the door and frame

Swing direction an installed door will open

Swinging doors doors that are hinged on one end and swing out of an opening; when closed, they cover the total opening

Table saw stationary circular power saw with a large bed or table

Tack to fasten temporarily in place; also, a short nail

Tail cut a cut on the extreme lower end of a rafter

Tail joist short joist running from an opening to a bearing

Tangent straight line touching the circumference of a circle at one point

Taper becoming thinner from one end to the other

Taper ripping technique of ripping material with grain where the width varies from end-to-end

Tempered treated in a special way to be harder and stronger

Tempered glass a type of safety glass treated with heat or chemicals which cause the entire piece of glass to immediately disintegrate into a multitude of small granular pieces when broken

Template a pattern or a guide for cutting or drilling

Tenon a tongue cut on the end of a piece usually to fit into a mortise

Tensile strength the greatest longitudinal stress a substance can bear without tearing apart

Termites insects that live in colonies and feed on wood

Termite shield metal flashing plate over the foundation to protect wood members from termites

Thermal envelope the part of a building that creates the boundary between conditioned and unconditioned air

Thin set a mortar-type adhesive often used to set ceramic tile; used when tile is likely to get soaked with water

Threshold a piece with chamfered edges placed on the floor under a door; also called a *sill*

Tile square or rectangular blocks placed side by side to cover an area

Tile grout a thin mortar, mixed from a powder and water, to form a paste that is worked into tile joints with a trowel

Timber large pieces of lumber over 5 inches in thickness and width

Toe the forward end of tools, such as a hand saw and hand plane

Toeboard a strip of material located at the back of the toe space under a base cabinet; also the bottom horizontal member of a scaffold guardrail

Toenail nail driven diagonally to fasten the end of framing

Toe space a recess provided at the bottom of a cabinet

To the weather a term used to indicate the exposure of roofing or siding

Toggle bolts   hollow wall anchor with spring-loaded wings to secure a machine-treaded bolt

Top lap   the height of the shingle or other roofing minus the exposure

Topography   a detailed description of the land surface

Total rise   the vertical distance that the roof rises from plate to ridge

Total span   the horizontal distance covered by the roof

Trammel points   tool with sharp points that is clamped to a strip of wood to lay out arcs

Transit-level   similar to a builder's level, but with a telescope that can be moved up and down 45 degrees in each direction

Transom   small sash above a door

Tread   horizontal finish members in a staircase on which the feet of a person ascending or descending the stairs are placed

Trestle   similar to a sawhorse used to support scaffold plank

Trimmer joist   full-length joists that run along the inside of an opening

Trimmers   members of a frame placed at the sides of an opening running parallel to the main frame members

Truss   an engineered assembly of wood or wood and metal members used to support roofs or floors

Tungsten carbide   material used to make cutting edges stay sharper longer

Turnbuckle   piece of hardware that allows two threaded rods to be tightened or loosened by rotating a threaded sleeve

Turnout   a type of handrail fitting

Twist   lumber defect in wood

Twist drills   term used for drilling bits typically used for making holes in metal

Type   a type of gypsum board with greater resistance to fire because of special additives in the core; typically known as *firecode board*

Undercourse   the bottom layer of less expensive wood shingles applied when double coursing sidewalls

Undercut   a cut made through the thickness of finished material at slightly less than 90 degrees, such that butt joints in the material will fit tightly on the face sides

Underlayment   material placed on the subfloor to provide a smooth, even surface for the application of resilient finish floors

Unfinished strip   solid wood flooring that is milled with square, sharp corners at the intersections of the face and edges; once laid, any unevenness in the faces of adjoining pieces is removed by sanding the surface

Unit block   a parquet block made up of smaller strips of wood

Unit length   the length of a stair stringer or rafter per unit of run

Unit rise   the amount a stair or rafter rises per unit of run

Unit run   a horizontal distance of a stair tread or horizontal segment of the total run of a rafter

Urethane glue   a high performance glue that is strong, flexible, and durable, and is used for exterior applications

Utility knife   all-purpose cutting tool, typically with a retractable blade

Valley   the intersection of two roof slopes at interior corners

Valley cripple jack rafter   a rafter running between two valley rafters

Valley jack rafter   a rafter running between a valley rafter and the ridge

Valley rafter   the rafter placed at the intersection of two roof slopes in interior corners

Vapor   normally invisible, a cool gaseous state of water

Vapor barrier   plastic sheet used to prevent moisture from penetrating the building surface

Vapor retarder   a material used to prevent the passage of vapor; also called *vapor barrier*

Variance   a notion granted by the Zoning Board of Appeals in a community to change the zoning code due to hardships imposed by zoning regulations

Veneer   a very thin sheet or layer of wood

Vermiculite   a mineral closely related to mica with the ability to expand on heating to form a lightweight material with insulating qualities

Vertical arc   a scale, attached to a telescope, that measures vertical angles to 45 degrees above and below the horizontal

Vinyl-clad   windows whose exposed wood parts are encased in vinyl

Volute   a spiral fitting at the beginning of a handrail

Wainscoting   a wall finish applied partway up the wall from the floor

Waler   horizontal or vertical members of a concrete form used to brace and stiffen the form and to which ties are fastened

Wall angle   metal L-shaped pieces that are fastened to the wall to support the ends of main runners and cross tees in a suspended ceiling grid

Wallboard saw   hand saw designed to cut gypsum board

Wall sheathing   exterior wall covering that may consist of boards, rated panels, fiberboard, gypsum board, or rigid foam board

Wall unit   one of two basic kinds of kitchen cabinets; the wall unit is installed about 18 inches above the countertop

Wane   bark, or lack of wood, on the edge of lumber

Warp   any deviation from straightness in a piece of lumber

Waste factor an amount added to a material order beyond the exact calculation to ensure sufficient material supply for job completion

Water table finish work applied just above the foundation that projects beyond it and sheds water away from it

W-beam a wide-flanged, I-shaped steel beam

Weatherstripping narrow strips of thin metal or other material applied to windows and doors to prevent the infiltration of air and moisture

Web wood or metal members connecting top and bottom chords in trusses; also, the center section of a wood or steel I-beam

Web stiffener a wood block that is used to reinforce the web of an I-joist, often at locations where the I-joist is supported in a hanger and the sides of the hanger do not extend up to the top flange

Wedge anchor fastener used to secure bolts in solid walls

Whet the sharpening of a tool on a sharpening stone by rubbing the tool on the stone

Wind a defect in lumber caused by a twist in the stock from one end to the other

Winder a tread in a stairway, wider on one end than the other, that changes the direction of travel

Window frame the stationary part of a window unit; the window sash fits into the window frame

Window schedules drawings that give information about the location, size, and kind of windows to be installed in the building

Wing dividers compasslike tool use to lay out circles and perform incremental step-offs in various layouts

Wire nails general term for most nails which are made from coils of wire

Wood chisel metal tool designed to be driven by a hammer, and which is used to make mortises and other rectangular holes in wood

Wood shingles a common roof covering most often produced from western red cedar

Woven valley valleys flashed by applying asphalt shingles on both sides of the valley and alternately weaving each course over and across the valley

Wrecking bar large pry bar used in demolition; often called a *crow* bar

Yoking applying vertical members (2 x 4s) that hold outside form corners together

Zones areas communities are divided into to separate the types of buildings that can be built in that area

Zoning regulations keeps buildings of similar size and purpose in areas for which they have been planned

Index